Plant-Microbe Interaction: An Approach to Sustainable Agriculture

Devendra K. Choudhary
Ajit Varma • Narendra Tuteja
Editors

Plant-Microbe Interaction: An Approach to Sustainable Agriculture

Springer

Editors
Devendra K. Choudhary
Amity Institute of Microbial
Technology (AIMT)
Amity University Uttar Pradesh
Noida, UP, India

Ajit Varma
Amity Institute of Microbial
Technology (AIMT)
Amity University Uttar Pradesh
Noida, UP, India

Narendra Tuteja
Amity Institute of Microbial
Technology (AIMT)
Amity University Uttar Pradesh
Noida, UP, India

ISBN 978-981-10-2853-3 ISBN 978-981-10-2854-0 (eBook)
DOI 10.1007/978-981-10-2854-0

Library of Congress Control Number: 2016963687

Printed on acid-free paper

This Springer imprint is published by Springer Nature
The registered company is Springer Nature Singapore Pte Ltd.
The registered company address is: 152 Beach Road, #21-01/04 Gateway East, Singapore 189721, Singapore

Preface

Sustainable agriculture involves designing a farm system employing nature as a model. In most natural ecosystems, the greater the diversity, the more resistant an ecosystem is to change and better able to recover from disturbances. In an agricultural ecosystem or the so-called agroecosystems (AESs), disturbance is much more frequent, regular, and intense. The ecological concepts of disturbance and their recovery through succession play an important role in AES management. AESs are undergoing disturbances in the form of cultivation, soil preparation, sowing, planting, irrigation, fertilizer application, pest management, pruning, harvesting, and burning. The diversity and intensity of AESs in developing and developed countries have been changing over time in response to a number of interacting biophysical and social factors at the local, regional, and global levels. The impact of increased spatiotemporal climate variability on AESs is likely to be intensified by climate change, which will disrupt many ecosystem functions, altering their capacity to provide goods and services and rendering them more susceptible to degradation. In addition, the security of food supply to an increasing world population has turned into a pressing issue worldwide. Sustainable food production can be achieved by avoiding excessive disturbance and allowing successional processes to generate greater AES stability. One can enhance the ability of AESs to maintain both fertility and productivity through appropriate management of disturbance and recovery.

Plant productivity is often limited by soil nutrient availability and the interface between living roots and soils, i.e., rhizosphere, which is a central commodity of exchange where organic C flux from root fuels and microbial decomposers can provide nutrients available to roots. It is virtually impossible to investigate the intricacies of potential rhizosphere interaction in every environmental condition by virtue of tremendous diversity of soil microbes, soil fauna, and plants. In addition, the physicochemical and structural properties of soils including development have been strongly affected by the action of rhizosphere over consecutive evolutionary time frame, and the evolution of true plant roots along with their extension deep into substrate is considerably hypothesized to have led to a revolution in planetary C and water cycling that reflects on the biogeochemical functions of the rhizosphere on Earth today. Understanding the complex microbial community in the rhizosphere environment has proven to be a challenging task because of the vast diversity and the enormity of the population inhabiting this unique habitat. Extensive studies have investigated perturbation of microbial community equilibrium population by changes in environmental conditions and soil management practices. It has long been recognized that the activity of soil microorganisms plays an intrinsic role in residue decomposition, nutrient cycling, and crop production. Any shift in microbial community structure can be reflected in the implementation of various land use and management systems that lead to development of best management practices for an AES.

In subsistence AESs, crop yields are directly dependent on the inherent soil fertility and on microbial processes that govern the mineralization and mobilization of nutrients required for plant growth. In addition, the impact of different crop species that are used in various combinations is likely to be an important factor in determining the structure of plant-beneficial microbial communities that function in nutrient cycling, the production of plant growth hormones, and the suppression of root diseases. Microorganisms represent a substantial portion of the standing biomass in terrestrial ecosystem and contribute in regulation of C sequestration, N availability and losses, and P dynamics. The size and physiological state of the standing microbial biomass are influenced by management practices including rotational diversity, tillage, and the quality and quantity of C inputs to the soils. In AES, sustainability is dependent on the biological balance in the soils that is governed by the activity of microbial communities. Soil microbial populations are immersed in a framework of interactions known to affect plant fitness and soil quality; thereby, the stability and productivity of both AES and natural ecosystem are enhanced. The global necessity to increase agricultural productivity from steadily decreasing and degrading land resource base has placed significant strain on the fragile agroecosystems. Therefore, it is necessary to adopt strategies to maintain and improve agricultural productivity employing high-input practices. Improvement in agricultural sustainability requires the optimal use and management of soil fertility and soil physical properties and relies on soil biological processes and soil biodiversity. It is necessary to understand the perspectives of microbial diversity in agricultural context that is important and useful to arrive at measures that can act as indicators of soil quality and plant productivity.

Sustainable agriculture has currently to cope with serious threats that compromise the food security for a human population under continuous growth, all these exacerbated by climate change. Some of these include the loss of usable land through overuse, deforestation, and poor irrigation practices, which have led to desertification and salinization of soils, especially in dry lands. Approaches currently being taken to face this situation come from the development of stress-tolerant crops, e.g., by genetic modification or breeding traits from wild plants. Genetic engineering has been proposed as the solution to these problems through a rapid improvement of crops. Crop genetic modification has generated a great public concern regarding their potential threats to the environmental and public health. As a consequence, legislation of several countries has restricted their use in agriculture. On the other hand, exotic libraries from wild plants for clever plant breeding could overcome the problem of narrowed genetic variability of today's high-yield crops. Plant breeding driven by selection marker has also been a major breakthrough. However, these approaches have met limited success, probably because stress tolerance involves genetically complex processes and the ecological and evolutionary mechanisms responsible for stress tolerance in plants are poorly defined.

Heavy metal contamination in soils is one of the world's major environmental problems, posing significant risks to public health and ecosystems. Therefore, the development of a remediation strategy for metal-contaminated soils is urgent for environmental conservation and human health. Phytoremediation offers significantly more benefits than conventional technology to accumulate heavy metals from the soil due to it being less expensive and safer for humans and the environment. But slow growth and low biomass of plants in heavy metal-contaminated soil may limit the efficiency of phytoremediation. This has prompted us to explore the possibilities of enhancing the biomass of metal accumulators using bacteria as plant growth-promoting bioinoculants. Bacteria that can produce IAA, siderophores, and ACC-deaminase are capable of stimulating plant growth; lowering the level of ethylene by consuming ACC, the immediate precursor of ethylene in plants growing in the presence of heavy metals; and helping plants acquire sufficient iron for optimal growth. Most of the heavy metals have low mobility in soil and are not easily absorbed by plant roots. Plant roots and soil microbes and their interaction can improve metal bioavailability in rhizosphere and lead to host adaptation to a changing environment.

Pathogen suppression by antagonistic microorganisms can result from one or more mechanisms depending on the antagonist involved. Direct effects on the pathogen include competition for colonization or infection sites, competition for carbon and nitrogen sources as nutrients and signals, competition for iron through the production of iron-chelating compounds or siderophores, inhibition of the pathogen by antimicrobial compounds such as antibiotics and HCN, degradation of pathogen germination factors or pathogenicity factors, and parasitism. These effects can be accompanied by indirect mechanisms, including improvement of plant nutrition and damage compensation, changes in root system anatomy, microbial changes in the rhizosphere, and activation of plant defense mechanisms, leading to enhanced plant resistance. Nowadays, it is well known that some soils are naturally suppressive to some soilborne plant pathogens including *Fusarium*, *Gaeumannomyces*, *Rhizoctonia*, *Pythium*, and *Phytophthora*. Although this suppression relates to both physicochemical and microbiological features of the soil, in most systems, the biological elements are the primary factors in disease suppression, and the topic of "biological control of plant pathogens" gained feasibility in the context of sustainable issues. The groups of microorganisms with antagonistic properties toward plant pathogens are diverse, including plant-associated prokaryotes and eukaryotes. Among the prokaryotes, a wide range of bacteria such as *Agrobacterium*, *Bacillus* spp. (e.g., *B. cereus*, *B. pumilus*, and *B. subtilis*), *Streptomyces*, and *Burkholderia* have been shown to be effective antagonists of soilborne pathogens. The most widely studied bacteria by far in relation to biocontrol are *Bacillus* spp. and *Pseudomonas* spp., viz., *P. aeruginosa* and *P. fluorescens*, which are probably among the most effective root-colonizing bacteria.

Sustainable agriculture has a long history of research targeted at understanding how to improve the effectiveness of root symbionts, viz., rhizobia and mycorrhiza.

A promising approach has been employed to understand how natural selection regulates changes in mutualistic interactions. A descriptive knowledge of basic evolutionary processes can be employed to develop agricultural management practices that favor the most effective symbionts. Mutually beneficial interactions between plant and associated rhizospheric microorganisms are ubiquitous which is important for ecosystem functioning. Symbiotic nitrogen fixation by bacteria, e.g., *Rhizobium*, *Bradyrhizobium*, *Mesorhizobium*, *Sinorhizobium*, and *Azorhizobium* spp., that are collectively known as rhizobia or by *Frankia* spp. is the major N input to many natural and agricultural ecosystems in the root nodules of legumes or actinorhizal plants, respectively. In addition, mycorrhizal fungi supply their host plants with mineral nutrients, viz., P, and other benefits. Several rhizospheric microorganisms cause severe infection to roots, and these so-called root pathogens can be suppressed by *Pseudomonas fluorescens* after colonization of the roots thereby improving plant health.

The exploitation of plant–fungal symbiosis appears as a smart alternative for plant adaptation due to their great quantity, ubiquity, diversity, and broad range of ecological functions they play in the natural ecosystem. Recent studies have shown that symbiotic microbes are of crucial importance in the distribution of plant communities worldwide and are responsible of their adaptation to environments under highly selective pressure. These indicate that some microbes confer tolerance to specific stresses and are responsible of the survival of plants to environments submitted to these particular conditions. The stress tolerance conferred by the symbiosis is a habitat-specific phenomenon, which has been defined as habitat-adapted symbiosis that confers tolerance to heat but not salt and coastal symbiotic microbes conferring tolerance to salt but not to heat. The same fungal species isolated from plants in habitats devoid of salt or heat stress did not appear to confer tolerance to these stresses. It is currently thought that each plant in natural ecosystems comprises a community of organisms, including mycorrhizae and bacteria. The ability of the symbiotic fungi to confer tolerance to stress may provide a new strategy to mitigate the impacts of global climate change on agriculture and natural plant communities. Such symbiotic lifestyles suppose a potential source for the improvement of food crops through adapting them to situations of increasing desertification and drought on global crop lands. It appears therefore as a sustainable alternative to the use of genetically modified organisms, which on the other hand did not yield the expected results.

Finally, plant-associated microorganisms can play an important role in conferring resistance to abiotic stresses. These organisms could include rhizoplane and symbiotic bacteria and fungi that operate through a variety of mechanisms like triggering osmotic response and induction of novel genes in plants. The development of stress-tolerant crop varieties through genetic engineering and plant breeding is an essential but a long-drawn process, whereas microbial inoculation to alleviate stresses in plants could be a more cost-effective environmental friendly option which could be available in a shorter time frame. Taking the current leads available, concerted future research is needed in this area, particularly on field evaluation and application of potential organisms. It is our contention that native plants survive and flourish in stressed ecosystems because of endosymbiotic organisms that have coevolved and were essential for their adaptation to changing environments. Plant growth and development cannot be adequately described without acknowledging microbial interactions. We need to determine the extent of microbial associations in the plant kingdom. This question will only be answered as technology is developed to detect their presence in plant tissues. What we have learned is that there is a need to understand how plant and microbes communicate in these endosymbiotic relationships and how they regulate basic genetic and physiological functions.

Hence, in the present book, editors compiled researches carried out by researchers in three sections with elaborate description related to "plant–microbe interaction for sustainable agriculture."

Part I: An Introduction to Plant-Microbe Interaction

Chapter 1 summarizes an exposition of plant–microbe and microbe–plant interactions describing the interplay of chemicals and signals that participate in the complex domain of the rhizosphere. The information derived from the current studies and the utilization of current technological platforms will enable researchers to explore and garner more information at the plant–microbe and plant–microbiome levels.

Chapter 2 briefly describes the various physical and chemical processes occurring in the rhizosphere, how the change in environment hampers these factors, and how that affects the rhizospheric diversity in modifying the microbial ecology and root architecture.

Chapter 3 emphasizes the insight of the rhizosphere and plant growth-promoting rhizobacteria under the current viewpoints. Conclusively, the applicability of these favorable rhizobacteria in different agroecosystems has been offered systematically under both normal and stress circumstances to focus the recent trends with the objective to improve upcoming visions.

Chapter 4 describes a holistic perception of rhizosphere functioning with a highlight on the ecological drivers that promote colonization of coherent functional groups of microorganisms influencing plant life through several direct and indirect mechanisms.

Chapter 5 describes the concept of rhizosphere, hyphosphere, and mycorrhizosphere and the various activities involved in understanding the functional diversity of microorganisms inhabiting the mycorrhizosphere necessary to optimize soil microbial technology for the benefit of plant growth and health.

Chapter 6 highlights the importance of mycorrhizae with beneficial microbes in plant growth promotion, nutrient uptake, and stress tolerance either biotic or abiotic. The presence of bacteria in the rhizosphere synchronizes with mycorrhizae termed as "mycorrhizae helper bacteria" that increase plant growth by focusing on N and P in particular while micronutrients in general.

Part II: Plant-Microbe Interaction Under Abiotic and Biotic Stress

Chapter 7 describes deployment of microbe–plant interactions that results in the promotion of plant health in arid and semiarid regions with reference to India under abiotic stress.

Chapter 8 briefly describes an attempt to explore current knowledge of bacterial ACC-deaminase-mediated physiological and molecular changes in the plants under diverse environmental conditions (drought and high salinity), mode of ACC-deaminase enzyme action, and drastic effects of salinity and drought on plant growth with a special reference to ethylene evolution.

Chapter 9 highlights the success and efficiency of phytoremediation with association of heavy metal-resistant plant growth-promoting rhizobacteria.

Chapter 10 briefly describes the importance of microbe–plant interaction under salt stress. It describes strategies that plants adapt to deal with salinity, and current biotechnological efforts toward producing salt-tolerant crops are summarized.

Chapter 11 summarizes the comprehensive understanding that required learning the mechanisms and critical factors influencing the plant–microbe–toxicant interaction in soils for success of phytoremediation.

Chapter 12 elaborately describes priming of benign microbes especially bacteria for plant growth promotion under biotic stresses to unravel the mystification of mechanisms involved in plant defense including ISR and SAR using sustainable development of plants.

Chapter 13 discusses on the susceptibility of most important bacterial and fungal plant pathogens toward different essential oils and their constituents responsible for biological activities such as antibacterial and antifungal. In addition, the potential effectiveness of herb essential oils against different plant pathogenic fungi and bacteria has been verified.

Chapter 14 elaborately describes the use of halophilic bacteria in agriculture system toward producing salt stress-tolerant crops and understanding the mechanisms of plant and halophilic bacterial interaction.

Chapter 15 describes that PGPR has the ability to mitigate most effectively the impact of abiotic stresses on plants through degradation of the ethylene precursor ACC by bacterial ACC-deaminase and through biofilm and exopolysaccharide production.

Part III: Plant-Microbe Interaction and Plant Productivity

Chapter 16 presents an overview of the importance of the microbiome to the plant growth promotion, focusing on the diversity, functional and taxonomic, of the microbiota associated to maize, and the desirable characteristics of microorganism's candidates to the use in PGP formulations.

Chapter 17 describes the role of biofertilizers that not only exhibit plant growth promotion but also are effective in bioremediation by detoxifying detrimental pollutants such as pesticides and heavy metal pollutants. Besides, PGPR-mediated organic farming would pave the way to prosperous, healthy, and sustainable nation. Thus, this trend of least possible input of chemicals in sustainable agricultural systems may help to achieve the goal of holistic well-being of planet Earth.

Chapter 18 describes the mechanisms underlying beneficial impacts of fungi on growth promotion of the host plant. It involves benign fungi that are potentially useful in agriculture sector to avail several services to crop plants such as water status, nutrient enrichment, stress tolerance, protection, weed control, and biocontrol.

Chapter 19 summarizes various microbial interactions between mycorrhizal fungi and other soil microbial communities. This chapter discusses on the following: (1) microbial communities in the soil, (2) arbuscular mycorrhiza fungi interaction with plants, (3) interaction with rhizosphere microorganisms, (4) interaction with soilborne pathogens, (5) potential benefits of arbuscular mycorrhizal fungi in plant growth and disease control, and (6) effect of soil microorganisms on mycorrhizal symbiosis. The main conclusion is that the microbial population interactions with arbuscular mycorrhizal fungi in the rhizosphere majorly influence plant health, crop productivity, and soil fertility.

Chapter 20 addresses general characterization of phyllospheric environment, microbial association process, microbial population structure, quorum sensing, and cross talk between plant and microbes. This chapter provides information on the microbial diversity of the phyllosphere bioenergy crop *Jatropha curcas*. Major bacterial groups prevalent on the *J. curcas* phyllosphere and plant growth-promoting activities are addressed.

Chapter 21 ascribes more complex physiological and ecological role to tree roots with soil profile. This is more particularly so in many tree species where roots have a characteristic dimorphic spread having (i) the surface roots that have a subterranean horizontal spread a few meters around the trunk and (ii) sinker roots that go vertically downward to 10 m and beyond. The sinker root system, therefore, enables hydraulic redistribution sustaining dry-season transpiration and photosynthetic rates of the parent tree and surrounding shallow-rooted vegetation, prolonging the life span of fine roots and maintaining root–soil contact in dry soils, and storing rainwater down into deeper soil layers for dry-season utilization.

Chapter 22 summarizes the information related to biosynthesis, metabolism, regulation, physiological role, and agronomical impact of phytohormone produced by the plant growth-promoting rhizobacteria.

Chapter 23 describes an overview of diversity of endophytes associated with sugarcane with special reference to *Gluconacetobacter diazotrophicus*. It also describes the role of different bacterial traits that are necessary for successful colonization of the plant interior part.

Finally, we'd like to express our gratitude to the contributors upon their consent to be a part of this book.

Noida, UP, India

Devendra K. Choudhary
Ajit Varma
Narendra Tuteja

Contents

Part I An Introduction to Plant-Microbe Interaction

1 **Rhizosphere Interactions: Life Below Ground** 3
Kalaivani K. Nadarajah

2 **Shaping the Other Sides: Exploring the Physical Architecture of Rhizosphere** 25
Madhurima Chatterjee, Raktim Bhattacharya, and Rabindranath Bhattacharyya

3 **Applications and Mechanisms of Plant Growth-Stimulating Rhizobacteria** 37
Prem Chandra and Enespa Singh

4 **Microbial Ecology at Rhizosphere: Bioengineering and Future Prospective** 63
Shyamalina Haldar and Sanghamitra Sengupta

5 **Mycorrhizosphere: The Extended Rhizosphere and Its Significance** 97
P. Priyadharsini, K. Rojamala, R. Koshila Ravi, R. Muthuraja, K. Nagaraj, and T. Muthukumar

6 **Arbuscular Mycorrhizae: Effect of Rhizosphere and Relation with Carbon Nutrition** 125
Ibrahim Ortaş, Somayyeh Razzaghi, and Mazhar Rafique

Part II Plant-Microbe Interaction Under Abiotic and Biotic Stress

7 **Microbial-Mediated Amelioration of Plants Under Abiotic Stress: An Emphasis on Arid and Semiarid Climate** 155
Amrita Kasotia, Ajit Varma, Narendra Tuteja, and Devendra Kumar Choudhary

8 **Bacterial ACC-deaminase: An Eco-friendly Strategy to Cope Abiotic Stresses for Sustainable Agriculture** 165
Sarita Kumari, Ajit Varma, Narendra Tuteja, and Devendra Kumar Choudhary

9 Increasing Phytoremediation Efficiency of Heavy Metal-Contaminated Soil Using PGPR for Sustainable Agriculture 187
Payman Abbaszadeh-Dahaji, Mahtab Omidvari, and Mansour Ghorbanpour

10 PGPR-Mediated Amelioration of Crops Under Salt Stress 205
Anukool Vaishnav, Ajit Varma, Narendra Tuteja, and Devendra Kumar Choudhary

11 Plant–Microbe Interaction for the Removal of Heavy Metal from Contaminated Site 227
Asit Mandal, J.K. Thakur, Asha Sahu, Sudeshna Bhattacharjya, M.C. Manna, and Ashok K. Patra

12 Bacteria-Mediated Elicitation of Induced Resistance in Plants upon Fungal Phytopathogen 249
Shekhar Jain, Ajit Varma, Narendra Tuteja, and Devendra Kumar Choudhary

13 Essential Oils as Antimicrobial Agents Against Some Important Plant Pathogenic Bacteria and Fungi 271
Bachir Raho Ghalem

14 Halophilic Bacteria: Potential Bioinoculants for Sustainable Agriculture and Environment Management Under Salt Stress 297
Anjney Sharma, Anukool Vaishnav, Hena Jamali, Anchal Kumar Srivastava, Anil Kumar Saxena, and Alok Kumar Srivastava

15 Abiotic Stress Mitigation Through Plant-Growth-Promoting Rhizobacteria 327
Palika Sharma, Veena Khanna, and Suman Kumari

Part III Plant-Microbe Interaction and Plant Productivity

16 Growth Promotion Features of the Maize Microbiome: From an Agriculture Perspective 345
Ubiana de Cássia Silva, Christiane Abreu de Oliveira, Ubiraci Gomes de Paula Lana, Eliane Aparecida Gomes and Vera Lúcia dos Santos

17 Biofertilizers: A Timely Approach for Sustainable Agriculture 375
Supriya Tomer, Deep Chandra Suyal, and Reeta Goel

18 **Role of Beneficial Fungi in Sustainable Agricultural Systems** 397
Mehrnaz Hatami and Fereshteh Ahangarani

19 **Significance of Arbuscular Mycorrhizal Fungi and Rhizosphere Microflora in Plant Growth and Nutrition** 417
Hindumathi Amballa and Narasimha Reddy Bhumi

20 **Prospect of Phyllosphere Microbiota: A Case Study on Bioenergy Crop *Jatropha curcas*** 453
Santosh Ranjan Mohanty, Garima Dubey, Usha Ahirwar, Ashok Kumar Patra, and Bharati Kollah

21 **Sinker Root System in Trees with Emphasis on Soil Profile** 463
S. Devi, R. Angrish, S. Madaan, O.P. Toky, and S.S. Arya

22 **Plant Growth-Promoting Rhizobacteria Play a Role as Phytostimulator for Sustainable Agriculture** 475
Sapna Gupta, Ruchi Seth, and Anima Sharma

23 **Diversity, Quorum Sensing, and Plant Growth Promotion by Endophytic Diazotrophs Associated with Sugarcane with Special Reference to *Gluconacetobacter diazotrophicus*** 495
Iqbal Ahmad, Mohd. Musheer Altaf, Jyoti Sharma, and Abdullah Safar Al-thubiani

About the Editors

Devendra K. Choudhary Dr. Choudhary has 15 years of experience in microbial ecology and is currently working as an assistant professor in Grade III at Amity University, Noida. Before joining Amity University, Dr. Choudhary spent several years at Mody University, Lakshmangarh, as an assistant professor, preceded by work at People's and Barkatullah University, Bhopal, as a lecturer cum scientist. Dr. Choudhary received his Ph.D. in microbiology in the year 2005 from GB Pant University of Agriculture and Technology, Pantnagar, after having received his M.Sc. in microbiology from MDS University, Ajmer, and qualifying CSIR-UGC-NET in 2002.

Dr. Choudhary has worked on GOI-sponsored major projects as principal investigator (PI) wherein recently he worked on two major projects sponsored by DBT and SERB, New Delhi, at Amity University until the year 2015 preceded by Mody University and DST Fast Track project at the Department of Biotechnology, Barkatullah University, Bhopal.

As an active researcher, Dr. Choudhary has published research and review articles along with several book chapters for reputed journals and edited books. In addition, he has served as Ph.D. supervisor/co-supervisor for several research scholars. Under the supervision of Dr. Choudhary, his scientific team has assigned three accession numbers for bacterial cultures wherein two from MTCC (12057 and 12058), IMTECH, and one with MCC no. 2607. Most recently, his team has filed three patents with the India Patent Office, New Delhi. Recently, in association with senior colleagues, Dr. Choudhary has edited the book *Microbial-Mediated Induced Systemic Resistance in Plants* with emphasis on global food security.

Dr. Choudhary has been recognized as a member of the National Academy of Sciences (MNASc) in 2016. Besides, he has been selected for Indian National Science Academy (INSA) visiting and summer research fellowship in 2014. Further, he received the Dr. RS Rana Memorial best research award in the year 2013, sponsored by the Association of Microbiologists of India. Besides, several other achievements have been credited in his account.

Ajit Varma Dr. Varma completed his M.Sc. (1959) and Ph.D. (1964) degrees at Allahabad University, Allahabad, India. In the course of his professional career, he has also served as a microbiologist (assistant professor) of IARI, New Delhi

(1963–1971); senior microbiologist (associate professor) of IARI, New Delhi (1971–1974); associate professor of JNU, New Delhi (1975–1984); and professor of JNU, New Delhi (1985–2004). He has been a visiting professor and visiting research scientist at the Graz University of Technology, Graz (Austria); University of Tübingen, Tübingen (Germany); Friedrich Schiller University, Jena (Germany); Philipps-Universität Marburg, Marburg (Germany); Technical University of Munich, Munich (Germany); Kingston (Jamaica); Max Planck Visiting Professorship (Germany); Helmholtz Zentrum, München (Germany); Gutenberg University, Mainz (Germany); CSIC, Madrid (Spain); University of Dundee (Scotland); University of Ljubljana (Slovenia); and ICGEB (Italy). His international awards/fellowships include the Commonwealth fellowship (Australia), National Research Council (Canada), Alexander von Humboldt Foundation (Germany), National Science Foundation (USA), Indo-Czechoslovakia Exchange Programme (Prague), DAAD fellowship (Germany), and the Deutsches BMFT Programme, Georg-August-Universität Göttingen, Gottingen (Germany), RAISA. He was awarded a fellowship for Innovative Research in Biotechnology (Italy), Swiss federal research fellowship (Switzerland), the BP Koirala award (Nepal), and DFG-INSA fellowship (Indo-Germany), as well as the FAMI Award-Association of Microbiologists of India and honorary diploma of the UMF, Cluj-Napoca, Romania. Dr. Varma has successfully completed major projects as PI sponsored by DBT, DST, DRDO, and ICAR. Besides, he has supervised more than 60 Ph.D. students and published over 300 research articles for national and international journals of repute, as well as several major review articles and chapters in books. He has published 50 books in the area of microbial technology. Dr. Varma has been the series editor and has edited 50 volumes on *Soil Biology*. He was also nominated as editor in chief by IK International to make a series of books on microbial and biotechnological research. Dr. Varma has been a member of the National Academy of Agricultural Sciences (FNAAS); International Society of Symbiosis, Boston, USA; Indian Science Congress Association; Executive Council, Amity University Uttar Pradesh; University Research Council, Amity University Uttar Pradesh; Academic Council, Amity University Rajasthan; ASSOCHAM Knowledge Millennium Council; ASSOCHAM Expert Committee on Agriculture and Food Processing; and ASSOCHAM Expert Committee on S&T and Innovation. He has vast experience in organizing national and international training workshops/symposia and congresses.

Narendra Tuteja Dr. Narendra Tuteja is presently acting as director of Amity Institute of Microbial Technology (AIMT), Amity University Uttar Pradesh (AUUP), Noida. Before joining AIMT, he spent years as group leader and senior scientist in the International Centre for Genetic Engineering and Biotechnology (ICGEB), New Delhi. Academically, Dr. Tuteja completed his M.Sc. (1977), Ph.D. (1982), and D.Sc. (2008) in the subject biochemistry from the University of Lucknow. Professionally, he served as group leader and senior scientist (2012–2015) of the Plant Molecular Biology (PMB) group, International Centre for Genetic Engineering and Biotechnology (ICGEB), New Delhi; senior scientist (2008–2011) of the PMB group, ICGEB, New Delhi; associate scientist (1997–2007) of the PMB group, ICGEB, New Delhi; group leader and associate scientist (1995–1997) of the PMB group, ICGEB, New Delhi; assistant research scientist (1988–1995) of the Molecular Biology group, ICGEB, Trieste, Italy; postdoctoral fellow (1986–1988) of the UCLA, Los Angeles, California, USA; visiting scientist (1983–1986) of the NIH (NCI and NICHHD), Bethesda, Maryland, USA; and research assistant (1977–1983) of K.G. Medical College, Lucknow, UP, India. Based on his contribution to science and society, he has been elected as fellow for various science academies in India/abroad such as fellow of "Third World Academy of Sciences" (FTWAS), Trieste, Italy; fellow of "National Academy of Agricultural Sciences" (FNAAS), New Delhi; fellow of "National Environmental Science Academy" (FNESA), New Delhi; fellow of "Indian Academy of Sciences" (FASc.), Bangalore; fellow of "Indian National Science Academy"; and fellow of "National Academy of Sciences" (FNASc.), Allahabad. Besides, he has served as an expert member for various advisory committees and task force constituted by GOI, advisory committee of "Bejo Sheetal Bioscience Foundation," Jalna, MR, India; task force expert member of the Plant Sciences Research Committee of the Council of Scientific and Industrial Research (CSIR), New Delhi; expert member of the academic committee meeting of the Central Institute of Medicinal and Aromatic Plants (CIMAP), CSIR, Lucknow; task force expert member of the Department of Information Technology (DIT), Ministry of Communication and Information Technology, New Delhi; and expert member of the Advisory Board, Agricultural Scientists Recruitment Board, New Delhi. Dr. Tuteja has become associate editor for *Plant Signaling and Behavior* (PSB) and editor in chief for *Plant Signaling and Behavior* (PSB) Vol. 6, Issue 2, Feb. 2011, special issue on *Plant Abiotic Stress*. In addition, his name was included in the list of members of the Editorial Advisory Board for *The Open Plant Science Journal*, *Archives of Biochemistry and Biophysics* (ABB), and *Plant Physiology and Biochemistry* (PPB).

Part I

An Introduction to Plant-Microbe Interaction

Rhizosphere Interactions: Life Below Ground

1

Kalaivani K. Nadarajah

Abstract

The interface between roots and soil is a region with high interaction among a myriad of organisms that affect biogeochemical cycles, plant growth, and stress tolerance. Similarly chemical compounds secreted within the rhizosphere act as attractants to microorganisms. Due to its dynamic nature and complexity, understanding rhizospheric biology and activity is essential in ensuring improved plant function and productivity within an ecosystem. Sustainable agricultural practices are dependent on studies conducted with regards to plant–microbe interactions in the rhizosphere. This chapter is an exposition of rhizospheric interactions spanning the chemistry of exudates and signals that contribute towards the complexity of the rhizosphere. The information derived from recent studies and the utilization of current technological platforms will enable us to explore and gather more information at the plant and microbiome level.

1.1 Introduction

The rhizosphere was described by Lorenz Hiltner a century ago as a microbial hotspot that is dependent on plant roots (Hartmann et al. 2008). The interactions and activities within have been researched extensively due to the dynamic nature of this region (Bakker et al. 2013). Studies have shown that the microbial communities within the rhizosphere can affect the well-being of plants (Mendes et al. 2011) by either directly or indirectly affecting the biomass and composition within the plant's natural ecosystem (Schnitzer et al. 2011). The microbiota contributing towards

K.K. Nadarajah (✉)
School of Environmental and Natural Resource Sciences, Faculty of Science and Technology, Universiti Kebangsaan Malaysia, Bangi, Malaysia
e-mail: vani@ukm.edu.my

D.K. Choudhary et al. (eds.), *Plant-Microbe Interaction: An Approach to Sustainable Agriculture*, DOI 10.1007/978-981-10-2854-0_1

these processes in the rhizosphere involve antagonists, mutualists, symbionts, and the rich plant root system (Kardol et al. 2007). The microbial activity is essential as they contribute towards physiological processes such as nutrient uptake and plant defense responses (Berendsen et al. 2012). Although much has been done to unravel the mysteries of these underground plant–microbe interactions, the complexity of these interactions leaves gaps in knowledge that requires further investigation (Urich et al. 2008; Jansson et al. 2011).

The variety of low molecular weight (LMW) exudates secreted into the plant's surrounding soil environment influences the complex interaction between the root and plant. These exudates when secreted into the environment contributed towards the highly interactive nature of this region. Though enormous strides have been made in understanding the interactions down under, much still remains elusive in our understanding with regard to the root–microbe–insect–nematode interactions within the rhizosphere (Weir et al. 2004; Walker et al. 2003). As plant roots remain hidden below ground, most of these interactions remain unnoticed especially the chemical components facilitating these belowground interactions (Bais et al. 2006).

Root chemicals result in varying consequences and responses in different plant systems. However to date, the mechanism underlying the chemical signal perception and response between the soil, roots, and invertebrates remains largely obscure. Ultimately the positive or negative way in which these chemicals are perceived will determine the plant and soil community's dynamics. We anticipate that deciphering the processes that direct the variety of activities within the rhizospheric microbiome will provide new avenues of crop manipulation for plant fitness and yield. Initial reports into these insights have been obtained through studies of *Arabidopsis thaliana* and *Medicago truncatula* plant systems. These studies have shown us how microbial ecosystems in the rhizosphere influence allocation, diversity, and belowground interactions (Berendsen et al. 2012; Bakker et al. 2012).

Here we have outlined current advances in deciphering the rhizospheric interactions, paying special emphasis on how these exudates mediate the various interactions below ground. In addition this chapter addresses how these beneficial interactions will influence plant growth, yield, and therefore contribute towards sustainable agriculture.

1.2 Rhizosphere and Root Exudates

The adaptability and survival of plants in any given environment is dependent on acquisition of resources from the soil environment (Badri et al. 2009b, 2013a; Chaparro et al. 2013a; Nihorimbere et al. 2011). The variation in amount of root exudates within the soil will determine the nutrient dynamics and hence affect the microbial population and diversity (Paterson et al. 2006). It has been reported that plants exude their photosynthetic components (5–21 %) such as sugars, proteins, and secondary metabolite into the root environment (Badri et al. 2013b; Badri and Vivanco 2009; Chaparro et al. 2013b). There are two groups of root exudates: (i) LMW exudates, e.g., amino acids, sugars, phenolics, secondary metabolites, and organic acids, and (ii) the HMW exudates, e.g., proteins and complex carbohydrates (Bais et al. 2006; Narasimhan et al. 2003). LMW and HMW compounds that are exuded into the soil environment are largely dependent on the plant cultivar and species, the developmental stages of the plant, soil chemistry, and microbial diversity (Badri and Vivanco 2009; Huang et al. 2015; Uren 2000). Recent reports have implicated root cells in the cap and root hairs as secretors of compounds from roots into the soil (Czarnota et al. 2003; Pineros et al. 2002; Nguyen 2003). In addition to secretion, root hairs are involved in anchoring and nutrient-water intake (Fan et al. 2001). The relationship between root exudates and microorganisms are chemotactically disposed, i.e., where plant roots secrete glucose, sugars, organic, and amino acids into the soil; microbes migrate chemotactically toward these exudates (Kumar et al. 2007).

1.3 System of Root Emission

Despite the huge strides made by scientists in investigating exudates within the rhizospheric domain, the mechanisms involved in root secretions are poorly understood. The synthesis and release of root-derived components are generally constitutive, while the secretion mechanisms of these exudates are thought to be passive involving three separate pathways such as dissemination, vesicle transport, and particle channels (Dennis et al. 2010).

1.3.1 Diffusion

Membrane permeability and the cytosolic pH largely influences the passive diffusion of small polar and uncharged molecules produced by plants across the cell's lipid membranes (Marschner 1995; Sanders and Bethke 2000). This is the simplest form of mobilizing molecules across the membrane.

1.3.2 Vesicular Transport

High molecular weight root exudates are secreted through different mechanisms such as vesicular transport (Battey and Blackbourn 1993). Field et al. (2006) reviewed vesicle-mediated trafficking of proteins, but this review however did not involve the mechanism of transport for phytochemicals (Grotewold 2004). While there are extensive reports on the phytochemical exudates in leaf tissue, little has been reported with regards to phytochemical exudates from roots. Vesicular secretion has been implicated in the transportation of antimicrobial products at the location of bacterial or fungal infections. One such example is the pigmented 3-deoxyanthocyanidins, an antimicrobial flavonoid observed in fungal infection sites of sorghum leaves (Snyder et al. 1991). Roots of knapweed plants have been reported to secrete cytotoxic and antimicrobial catechin flavonoids (Bais et al. 2002).

Although certain researchers have implicated the cytoplasmic surface of the endoplasmic reticulum (ER) as the site of synthesis for certain root exudates from the phenylpropanoids and flavonoids families (Winkel-Shirley 2001), the mechanism of transport from the ER to the membrane is unknown. However there is a possibility that these compounds are transported through ER-originating vesicles that secrete their contents once bound to the cell's membrane.

1.3.3 Transporter Proteins

Transporter proteins are responsible for the transportation or passage of amino acids, sugars, and carboxylate anions from root cell cytoplasm to soil (Colangelo and Guerinot 2006; Hirner et al. 2006; Lee et al. 2007; Svennerstam et al. 2007). ABC transporter proteins are implicated in various cellular processes, spanning the discharge of harmful compounds, translocation of lipids, disease resistance, salt stress, nutrient transport, and substantial metal resilience (Stein et al. 2006; Kobae et al. 2006). The utilization of *Arabidopsis* ABC transporter knockout mutants proved that these transporters were involved in root secretions. What's more, the ABC transporters are confined to the plasma membrane (Sidler et al. 1998) and are involved in auxin pumping and secretion of resistance metabolites (Badri et al. 2009a).

Another transporting system, MATE, is involved in the discharge of phytochemicals. MATEs, through electrochemical gradient of other ions, are effectively able to transport substrates across cell membranes. Numerous MATE genes involved in transporting compounds such as toxic materials, plant-inferred alkaloids, antimicrobials, phenolics, and anions have been identified and characterized in the root cells of sorghum, *Arabidopsis*, rice, and grain (Furukawa et al. 2007; Ishimaru et al. 2011; Liu et al. 2009; Magalhaes et al. 2007; Weston et al. 2012).

Further, MFS transporter proteins assist with the release of secondary metabolites such as phytosiderophores from root cells (Kim and Guerinot 2007). These proteins can work as uniporters, co-transporters, or antiporters. In rice for instance, deoxymugineic and avenic acids are aided by *TOM1* (transporter of mugineic corrosive family phytosiderophores1) (Nozoye et al. 2011) in translocation of proteins. Through transgenic studies it was proven that the expression of *TOM1* is induced in the state of limited iron supply where overexpressing *TOM1* showed improved deoxymugineic acid release and enhanced resilience to a limited iron supply. ALMT transporter proteins belongs to the *ALMT* gene family that enables malate efflux from plants. *ALMT* genes encode the pore-forming anion channels within the membranes that facilitate the passive transport of substances across the membranes (down their electrochemical slopes) (Ryan et al. 2011; Weston et al. 2012). Other than the above transporters, monosaccharide transporters have been associated with hexose, pentose, ribose, and polyols transport (Klepek et al. 2005; Buttner 2007), while silicon efflux transporters have been associated with the excretion of silicon from rice root cells to soil (Ma and Yamaji 2008).

1.4 Rhizospheric Plant–Microbe Interactions

Root-secreted phytochemicals can result in beneficial, deleterious, or neutral interactions (Raaijmakers et al. 2009; Mercado-Blanco and Bakker 2007). Likewise, microbes are also able to transition from pathogenic to symbiotic in response to differing environments (Newton et al. 2010). Hence we can anticipate that the chemical diversity exhibited by root exudates will be an excellent source to look for novel, biologically active compounds, including antimicrobials (Huang et al. 2014). Previous reports have highlighted that the association of plants and the microbial community in the soil is important for plant health. These communities are dependent on the root exudates that positively or negatively affect the microorganisms within the soil. In the following sections, the integral role played by the exudates in plant–microbe and microbe–plant interactions will be expounded. Figure 1.1 presents the various underground processes that occur within the rhizosphere (Huang et al. 2014; Zhuang et al. 2013).

1.4.1 Positive Plant–Microbe Interactions

(a) *Nitrogen fixation*

The nitrogen levels within the rhizosphere will determine the diversity of nitrogen (N)-fixing bacteria within the soil (Zahran 1999). In nitrogen-limiting conditions, the nodule containing nitrogen-fixing bacteria produces flavonols and flavones that attract and initiate legume–rhizobia symbiosis (Zhang et al. 2009; Coronado et al. 1995). The flavones and flavonols induced bacterial nod gene expression, which lead to the initiation of root nodulation. The aerobic N_2-fixing bacterium also exhibited N_2ase activity when inoculated into the rhizosphere of rice, wheat, and oat seedlings. Further, microscopic observations of this N_2-fixing bacterium in barley roots suggest that this organism is an endophyte that associates with root tissue to form vesicle-like structures (Santi et al. 2013). The aggregation of rhizobia to legume root tissues is dependent on the association to specific sugar-binding sites. During nitrogen fixation, lectins (functions as binding protein) bind polysaccharides to stimulate aggregation. Lectins sustain increased nod factor concentrations and mitotic activity necessary for nodulation (Mathesius and Watt 2010). In general mixed cultures have been shown to increase nitrogen-fixing capacity as observed in the association between *Staphylococcus* sp. and diazotrophic bacteria that increased the nitrogen-fixing capacity of *L. anguillarum* by 17 %. Another example is the production of nodulating compounds such as exopolysaccharides (EPS and EPS II) by a mixed culture of *Rhizobium* sp. and *Sinorhizobium* sp. Exopolysaccharide-deficient mutants were incapable of invading legumes and establishing symbiotic relations (Jones et al. 2008). Legume-secreted isoflavonoids such as daidzein and genistein have been reported to effectively induce *Bradyrhizobium japonicum* nod genes, while nod genes in *Sinorhizobium meliloti* were induced by luteolin (Juan et al. 2007). The level of specificity exhibited enables the rhizobial community to identify their specific host accurately (Bais et al. 2006) (Table 1.1).

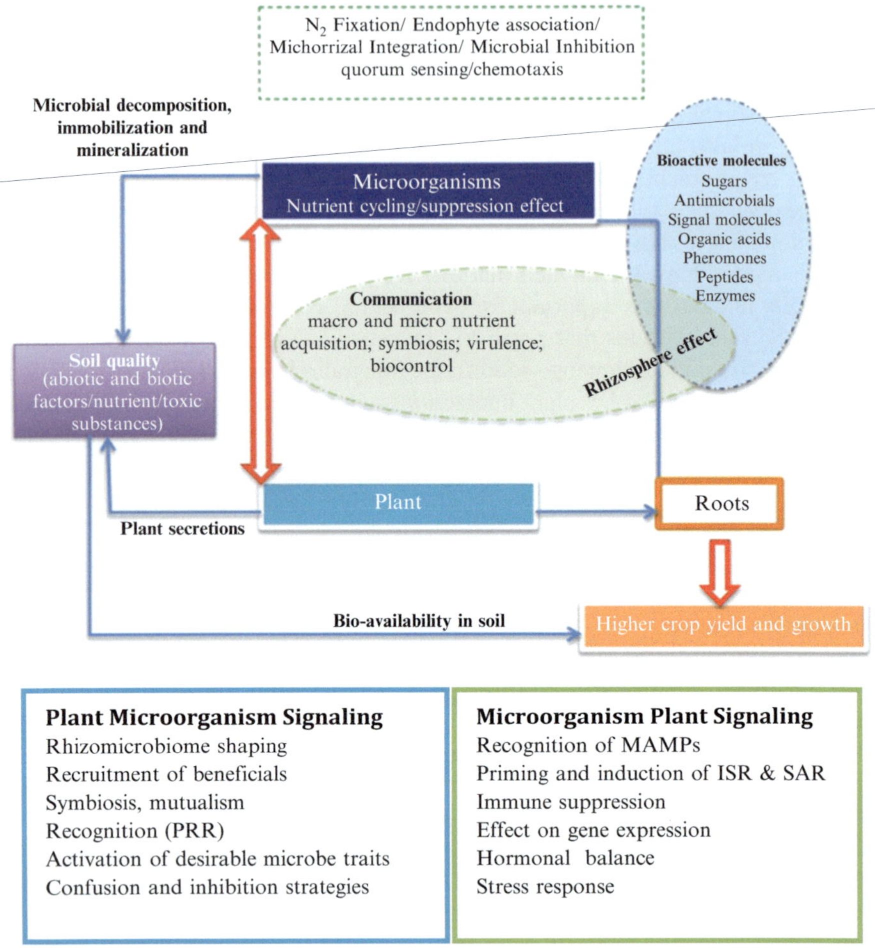

Fig. 1.1 Plant-microbe interactions and their role in belowground ecosystem and sustainable agriculture (Modified from Zhuang et al. 2013)

A mixed inoculation of *Rhizobium* sp. and *Azotobacter* sp. resulted in *Azotobacter* sp. significantly increasing *Rhizobium* nodulation. Both microorganisms enhanced growth and yield in various soil and mineral compositions. These findings suggest that there exist a mutualistic relationship between *Azotobacter*, *Azospirillum*, and *Rhizobium*, which results in improved yields in crops (Parmar 1995; Parmar and Dadarwal 1997). Researchers have reported that *Azotobacter* and *Azospirillum* contribute towards a plethora of positive responses in plants that include good root development, increase in nutrient and water uptake, inhibition of pathogenic and non-beneficial interactions, and a small contribution towards nitrogen fixation (Okon and Itzigsohn 1995; Steenhoudt and Vanderleyden 2000).

Table 1.1 Biomolecules involved in direct and indirect microbe and root-based activity

Activity	Biomolecules	Function
Direct microorganism-based activity		
Nitrogen fixation	EPS, EPS II, lipochitooligosaccharides, flavanols, flavanones, nodulating factors	Division of root cortical cells and nodule morphogenesis
Mycorrhizal association	Sesquiterpene, Myc factor	Fungal factors that trigger mycorrhization
Metal uptake	Glutathione, metallothioneins, and acid such as ferulic, chorismic, mugineic, caffeic, p-coumaric, oxalic	Metallic bioavailability
Virulence factors	Extracellular polysaccharide, phytotoxins, effector proteins	Crucial for virulence and suppression of resistance reactions
PGPR	LPS, EPS, antimicrobials, siderophores, lipopeptides, cell wall-degrading enzyme (CWDE)	Protection of plants against pathogens Improved nutrient uptake and growth
Direct root-based activity		
Bacterial and fungal symbionts	Flavonoids (glyceollin, coumestrol, daidzein, glyceollin, coumestrol, genistein), strigolactones, jasmonates, auxins, abscisic acid, ethylene, gibberellin	Stimulating pre-symbiotic processes and enhanced Arbuscular mycorrhiza fungi (AMF) colonization of roots
Carbon uptake	Arabinose, fructose, ribose, hexose	Carbon utilization and metabolism
Pathogenicity factors and defense response	Phytoalexins, naphthoquinones, indole, saponins, benzoxazinone, flavonoid, terpenoid, rosmarinic acid, glucosinolates	Protection against pathogenic microorganisms
Indirect microorganism-based activity		
Quorum sensing	Peptide molecules, N-acyl homoserine lactones (AHLs), quinolone, p-coumarate	Cellular communication, swarming, biofilm, and antibiotic production
Indirect root-based activity		
Defense	Phospholipases, phosphatases, MAP kinases: Lipoxygenase, linolenic acid, jasmonate, methyl jasmonate	Activation of other defense reactions

(b) *Mycorrhizal interactions*

The "fair-trade" between plant and mycorrhiza involves the provision of N by mycorrhiza and carbon by the plant (Fellbaum et al. 2012). A quantitative and qualitative change in the chemical content of soil and plant observed during AMF establishment includes the transient rise in phytoalexin levels during colonization (Leyval and Berthelin 1993). The beneficial fungal isolates or plant cultivars involved in AMF symbiosis can influence the concentration and types of flavonoids produced. The type of flavonoids produced influences the mycorrhizal spore germination, hyphal growth and root colonization. For example, strigolactone, a sesquiterpene lactone, is essential in inducing AMF hyphal branching (Akiyama et al. 2005; Siegrid et al. 2007). Morandi et al. (1984) reported that flavonoids such as glyceollin, coumestrol, and daidzein stimulates AMF colonization in soybean and thus implicates flavonoids as signaling compounds involved in AMF root colonization. In contrast there are chemicals that inhibit hyphal growth of mycorrhiza such as observed within a non-nodule-forming legumes (Oba et al. 2002). Further, it has also been reported that sugars, carbohydrates, and strigolactone 5-deoxygol facilitate the symbiotic associations between the mycorrhiza and non-legume crops (Yoneyama et al. 2008; Fang and St. Leger 2010; Kiers et al. 2011).

Vesicular-arbuscular mycorrhizae (VAM) on the other hand are a group of fungi that are involved in the mobilization of phosphorus from soil with low levels of available phosphorous. The associative relationship of these fungi with legumes influences the root and shoot development as well as the phosphorous uptake that eventually results in enhanced nodulation and nitrogen fixation. Combinatorial inoculation of soil systems with *Rhizobium* and VAM has unequivocally contributed towards plant growth enhancement, nodulation, and N_2 fixation. The effectiveness of *Rhizobacterium* sp. as nodulating and N_2-fixing fungi in the mycotrophic legume, *Anthyllis cytisoides*, further substantiates AM's function in supplying P to root nodules (Requena et al. 2001). Research shows that other root–microbe symbionts share the same symbiotic genetic pathway as the N_2-fixing rhizobia. "Myc" triggers gene activation in roots through a diffusible signaling factor that is required for mycorrhization. This signaling factor results in elevated calcium levels which inevitably caused calcium fluctuations required for epidermal root cell priming for fungal colonization (Meier et al. 2013; Zhuang et al. 2013). These specific interactions have provided insights into functional compatibility between AMF and PGPR as plant growth promoters.

(c) *Endophytic associations*

Plants have supported endophytes that are either nonpathogenic bacterial or fungal species with no detrimental effects to the host. Although this is a long-standing interaction, but it has not been well studied and documented. Hosts that harbor these endophytes have shown increased resistance to plant stresses. The presence of these endophytes can result in the alteration of root exudates causing a change in the secretion of phenolics and hence altering the pH within the rhizosphere and elevating tolerance toward mineral deficiencies. While endophytic relationships are largely beneficial, there are however some opportunistic associations. The altered exudates from endophytic plants may affect the microbial community within the soil and influence the biology and ecology of the system (Malinowski and Belesky 2000). Plants involved in symbiotic relations with endophytes have also been reported to enhance AMF interactions through root exudates (Novas et al. 2011).

(d) *PGPR*

PGPRs have been characterized as organisms that colonize and suppress plant pathogens. This group of organisms has been exploited extensively for economic gains due to its inhibitory potential (Parmar 1995). Through a plethora of direct and indirect mechanisms, the PGPR is found to positively influence plants. It is believed that soil microorganisms involved in this interaction are recruited by cues exuded by the host roots hence establishing the PGPRs population and activities. PGPRs on the other hand are reported to produce chemicals that affect plant growth and resistance indicating a two-way relationship between plants and PGPR for improved plant health (Ryu et al. 2004). The involvement of rhizospheric PGPRs in triggering the host immune response through various pathways such as jasmonate and salicylic acid has been previously reported and associated with plant fitness (Compant et al. 2010; Saharan and Nehra 2011). Chemical agents such as amino acids and carbohydrates were reported to be the signals involved in the mobilization of PGPRs to specific roots (de Weert et al. 2002).

Plant growth has been enhanced by bacterial communities that include *Azotobacter*, *Bacillus*, *Azospirillum*, *Enterobacter*, *Serratia*, *Klebsiella*, and *Pseudomonas*. Compared to single inoculums, dual inoculations significantly improved plant weight, dry mass, protein content, and grain yield. Yadegari et al. (2008) reported that combined inoculation of PGPRs increased growth, development, nodulation, and nitrogenase activity. The cumulative effects of growth-promoting substances exuded by organisms such as *Pseudomonas* sp. CRP55b, *Rhizobium* Ca181, *Pseudomonas* sp. CRP55b, *Azospirillum* spp., and *Pseudomonas fluorescens* P21 resulted in an increase of apical and root growth, plant biomass, and crop yield (Rokhzadi et al. 2008). The mechanisms contributing toward the increase in yield and growth are multitudinous, where substances or processes such as phytohormones, plant growth-regulating substances (PGRs), mineralization, cyanogens, siderophores, and phytoalexins/flavonoids collectively resulted in enhanced agricultural output (Mukerji et al. 2006; Nadarajah 2016).

Rhizobacteria produce phytostimulators in the absence of pathogens. These compounds include hormone analogues such as gibberellic acid, indole acetic acid (IAA), ethylene, and cytokinins (Lambrecht et al. 2000). The production of IAA is a plant growth-promoting trait among PGPRs. Tryptophan-dependent and tryptophan-independent pathways have been identified as contributing toward IAA biosynthesis in rhizobacteria (Steenhoudt and Vanderleyden 2000). Shoot development is stimulated in response to the action by cytokinins and gibberellins. Additionally cytokinins are also involved in cell division, primary root development, nodulation, and branching (Murray et al. 2007; Tirichine et al. 2007; Ortiz-Castro et al. 2009). N-Acyl-L-homoserine lactones, another class of phytostimulants, are associated with cellular communication and modulation of gene expression in plants (Choi et al. 2008; Ortiz-Castro et al. 2009).

A multitude of plant responses including stress is regulated by ethylene. Various factors such as temperature, nutrition, gravity, and plant hormone levels influence ethylene production (Glick 2005). In incidences of high ethylene levels, the plant undergoes stress and exhibits impaired root growth (Argueso et al. 2007). However the modulation of ethylene via ACC-deaminase is crucial in the degradation of 1-aminocyclopropane-1-carboxylic acid (ACC ethylene precursor). Various microbes have been reported to cleave ACC to ketobutyrate and ammonia as a means of improving plant stress response to ethylene production (Glick 2005; Stearns et al. 2012). Further, ACC-deaminase activity in *Achromobacter piechaudii* ARV8 improved seedling biomass in tomato and pepper (Mayak et al. 2004). Similarly a study of ACC-deaminase in *Brassica napus* revealed a downregulation of ethylene stress response while recording upregulated gene expression of auxin production genes (Stearns et al. 2012). Arshad et al. (2008) and Mayak et al. (2004) in their reports indicated a role for ACC-deaminase in reducing ethylene levels and thus contributing toward management of drought, salinity, and generally various other abiotic stresses. This therefore clearly indicates that microorganisms with ACC-deaminase activity benefits the overall well-being of plants. Understanding the overall contribution of microbial communities in reducing and mediating ethylene stress in plants may be utilized to generate technologies for plant abiotic stress management.

(e) *Enzymes and proteins*

While it has been reported that plants secrete enzymes and proteins in addition to primary and secondary metabolites into the rhizosphere (Charmont et al. 2005), information is lacking on how these substances influence the rhizosphere (De Hoff et al. 2009; De-la-Peña et al. 2008). A proteomic analysis on *A. thaliana* root exudates indicates that there is a difference in the secreted proteins according to developmental stages. During the flowering stage, defense-related proteins such as glucanases, chitinases, and myrosinases were produced (De-la-Peña et al. 2010). Higher levels of defense-related proteins such as peroxidases, hydrolase, and chitinase have been reported as secretomes into the plant root systems of *A. thaliana* in response to an infection by pathogenic *Pseudomonas syringae* pv. tomato DC3000. However when inoculated with a nonpathogenic isolate, *S. meliloti* Rm1021, no defense response proteins were secreted into the rhizosphere.

Arabinogalactan protein (AGP) is a hydroxyproline-rich glycoprotein superfamily that is found in plant cell wall proteins. AGPs play a vital role in root and rhizospheric microbe interaction (Nguema-Ona et al. 2013). Cannesan et al. (2012) and Vicré et al. (2005) reported that root tip cells and AGP containing mucilage was observed in the rhizosphere. This glycoprotein acts as an attractant to root pathogen inhibiting microbes and is implicated in the colonization by *Rhizobium* sp. through recognition and attachment to root surfaces (Vicré et al. 2005; Cannesan et al. 2012; Xie et al. 2012). Xie et al. (2012) reported on a similar glycoprotein, which promotes surface attachment of *Rhizobium leguminosarum* to roots of legumes and non-legumes. *P. fluorescens* strain WCS365 colonization of tomato roots involves a plethora of amino acids which includes aspartic acid, glutamic acid, leucine, lysine, and isoleucine (Simons et al. 1997). In another study, the exposure of plants to *Rhizobium* sp. (Sb16) and *Cyanobacterium* sp. (Sb26) (Naher et al. 2008) resulted in higher levels of amino acid exudates in rice. This may perhaps be a consequence to secretion of microbial products that result in amino acid exudates (Chaparro et al. 2013a, b; Phillips et al. 2004). However, the influence of these enzymes and proteins in the establishment, colonization, and configuration of microbial communities remains elusive.

(f) *Sugars*

Chaparro et al. (2013a) reported that the rate of sugars exuded decreased with the plant's development. This could possibly be the consequence of pathways and cycles utilizing sugars being synergistically regulated by sugars and amino acids (Poysti et al. 2007). Considering the large number of genes (27) identified and correlated to carbohydrate metabolism in microbes, sugars are probably actively utilized by these organisms. Metabolic priming of soil microbes enhanced degradation and mineralization of soil organic matter in the presence of fructose and alanine (Hamer and Marschner 2005). The observed priming effect is due to the ability of these substrates to trigger metabolism and enzyme production (Kuzyakov 2002). The priming of enzyme activities results in increased metabolic capabilities of the soil microbiome, which improves the plant acquisition of various limiting nutrients.

1.4.2 Antagonistic Plant–Microbe Interactions

(a) *Quorum sensing (QS)*

QS involves cell-to-cell communication between microorganisms in an environment. It has been implied that the plant's root systems have developed the mechanism to exude chemical signals (mimics, blockers, and or degrading enzymes) that have the ability to affect microbial QS (Gao et al. 2003). Diffusion of these small signal molecules (autoinducers), which are present in both Gram-negative and Gram-positive bacteria, is known to mediate QS. QS is essential in the development of plant–microbe interactions regardless if it's beneficial or non-beneficial. These QS-mimicking or quenching signals are potential targets for the discovery and development of new antimicrobial molecules.

Molecules that imitate acylated homo-Ser lactones (AHLs) with specific effects on QS-controlled behavior have been reported in *Oryza sativa* L. (rice), *Pisum sativum* L. (pea), and *Glycine max* (L.) Merr. (soybean). The lasIR system of QS sensing in *P. aeruginosa* regulates virulence factors such as toxins and extracellular enzymes. A second system, rhlIR, also modulates expression of virulence factors. In PUPa3, both systems form useful associations with plants. AHL signaling in *Chromobacterium violaceum* was inhibited by an arginine analog, L-canavanine, that did not interfere with its growth in alfalfa or other legumes. L-Canavanine also regulates QS ability in *S. meliloti* and is also responsible for the control of EPS II biosynthesis in this organism (Daniels et al. 2002; Teplitski et al. 2000; 2004; Zhuang et al. 2013).

The pcoIR system in *P. fluorescens* is connected to the biosynthesis of antimicrobial compounds, e.g., pyrrolnitrin, phenazines, hydrogen cyanide, and pyoluteorin. Similarly the pcoIR system in *P. fluorescens* 2P24 indirectly regulates the production of metabolites, including siderophores, 2,4-diacetylphloroglucinol, and hydrogen cyanide. Tyrosol, farnesol, trisporic acid, and dimethoxycinnamate are some of the signal molecules produced by *Uromyces phaseoli*, *Candida albicans*, and zygomycetes in their host–microbe interactions. 3-oxo-C12-HSL from *P. aeruginosa* inhibits structural changes from yeast-like to filamentous in *C. albicans* (required for virulence). In turn, AHL synthesis in *P. aeruginosa* is strongly suppressed by farnesol. However, the pathways and specific mechanisms involved in fungal QS remain obscure (Hogan 2006; Sanchez-Contreras et al. 2007; Wu et al. 2010; Zhuang et al. 2013).

GABA is another component involved in cellular communication. GABA quenches QS and reduces the virulence of *A. tumefaciens* (Chevrot et al. 2006) while utilizing GABA as sole nutrient source in *P. putida* (Ramos-González et al. 2005). Proline however reverses GABA's ability to quench QS (Haudecoeur et al. 2009). These opposing signals require further investigation to understand the interplay involved in the complex rhizospheric interaction.

(b) *Antimicrobial*

Plant secondary metabolites are compounds that attract beneficial microbes and defend plants against negative interactions. Plants synthesize secondary metabolites such as phenols or their oxygen-substituted derivatives in a limitless manner (Badri et al. 2008; Neal et al. 2012). One such example is rosmarinic acid (RA) (Bais et al. 2002). Basil roots, for instance, have been reported to exude RA when induced or challenged by fungi. RA demonstrates powerful antimicrobial activity against a vast selection of soil microbes, including *P. aeruginosa* (Bais et al. 2002). Fungal (*Phytophthora cinnamomi* and *Pythium ultimum*) elicitation of basil roots produced naphthoquinones and RA that are strong inhibitors of pathogenic and opportunistic microorganisms in the soil including the opportunist plant pathogen *P. aeruginosa*. In addition, grafted watermelon roots with high levels of chlorogenic and caffeic acid exudates and low levels of cinnamic acid (Ling et al. 2013) were resistant towards *Fusarium oxysporum* f.sp. *niveum* infections. Cai et al. (2009) reported that the antimicrobial agent canavanine obtained from leguminous plants inhibits rhizospheric bacteria excluding rhizobia. This suggests canavanine's involvement in the selection of beneficial microbes.

Most antimicrobial products are broad spectrum, and their specificity is determined by the existence of enzymatic machinery to detoxify any of the host products. Antimicrobial compounds are induced through the activation of linked signal transduction pathways as a consequence of pathogen perception by host resistance gene-encoded receptors. However, most studies have not looked into the mechanism of accumulation of these secondary metabolites within the plants and its excretion into the soil environment. In a study conducted on root exudates from *Gladiolus* spp. L., the resistant varieties produced root exudates that had antimicrobial effects against *Fusarium oxysporum* sp. *gladioli*, while the susceptible lines showed no reduction on conidial germination (Taddei et al. 2002). The inhibition of conidial germination of *F. oxysporum gladioli* by resistant cultivars is mainly regulated by the presence of aromatic-phenolic compounds.

Fungal communities in the rhizosphere produced abundant antimicrobial substances (Hoffmeister and Keller 2007; Brakhage and Schroeckh 2011). For example, *Trichoderma* species have been reported to produce a large array of antimicrobials (Elad et al. 2008) among other bioactive compounds. Fungal and bacterial biocontrol strains produced several antimicrobial compounds with similar or varying degree of activity. Bacteriocin such as agrocin 84 (Kim et al. 2006) exhibits narrow-spectrum antimicrobial activity against closely related genera, while polyketides and peptides exhibit broad-spectrum activity (Raaijmakers et al. 2010). The effectiveness of these compounds varies from microbe to microbe. The antimicrobial compounds found within the root cells differ in composition to the antimicrobials found in root exudates (Bednarek and Osbourn 2009).

1.5 Multitrophic Interactions in the Rhizosphere

From the one-to-one interactions observed in the rhizosphere, here we look into the multipartite interactions that present the complexity within the rhizosphere. In the root soil environment of plants such as switch grass, endophytic associations of microbe–insect–plant enhanced N availability for the plant (Behie et al. 2012). The presence of raffinose and sucrose in root exudates of switch grass attracted *Metarhizium robertsii* and enabled the tripartite interaction. In addition, plant volatiles from the legume *M. truncatula* attracted *Caenorhabditis elegans*, a nematode that transported *S. meliloti* to the plant's roots to initiate symbiosis (Fang and St. Leger 2010; Horiuchi et al. 2005). Similarly the tripartite relations between PGPR–mycorrhizae and PGPR–rhizobia resulted in the efficient colonization of mycorrhizae and nodulation of rhizobia, respectively (Guiñazú et al. 2010). Due to the complexity of the multipartite interactions, very little is known of the mechanisms involved, and hence more studies are needed to elucidate these mechanisms, colonization, establishment, and benefits of the interaction.

1.6 Concluding Remarks

The above segments have dealt with the various ways in which the plant–microbe interaction in the rhizosphere affects both the plant and the soil microbial community. These interactions have been known to effect soil fertility, thus affecting plant health, overall yield, and growth. Hence, it is evident that microorganisms are key players in plant productivity and should be given due attention in the interest of advancing our knowledge in rhizosphere biology. As we transition from conventional agriculture to sustainable agriculture, it is important to understand the differences and the benefits of this transition.

Conventional agriculture practices selection of high yielding genotypes coupled with high fertilizers inputs and pesticides to reduce losses from biotic infestations while enhancing growth and yield. Rhizospheric microorganisms play a minor role

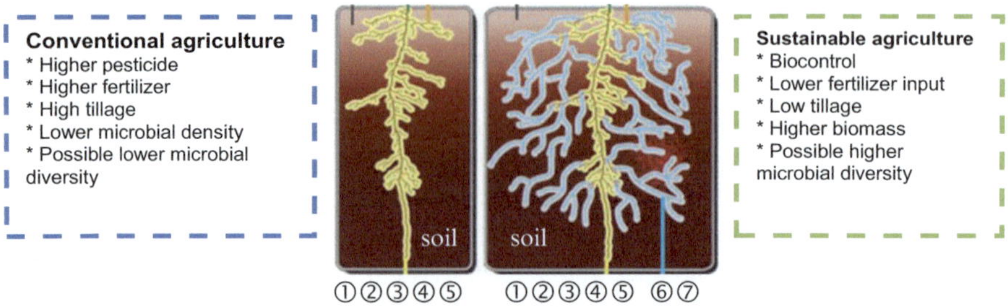

Fig. 1.2 Conventional vs. sustainable agriculture. The above diagram differentiates between conventional and sustainable agriculture, highlighting the contribution of microbes in sustainable agriculture (① rhizosphere ② N_2 fixation ③ root pathogens ④ endophytes ⑤ arbuscular mycorrhiza ⑥ hydrosphere ⑦ mycorrhizosphere)

in conventional agriculture unless they are pathogens. By excluding the microorganisms from the equation, agriculture has been dependent on plant genotypes which may not be as well adapted to adverse conditions. However, in sustainable agriculture, the microorganisms within the rhizosphere are important in crop production (Fig. 1.2). Hence through sustainable agriculture, one could select for plant genotypes that are able to mobilize nutrients from their environments directly or indirectly through interactions with rhizospheric organisms. The results from sustainable agriculture can be further enhanced through the application of good management practices, inclusive of crop rotation, mulching, and utilization of PGPRs.

In this chapter we have provided a comprehensive outline of the major interactions within the rhizosphere and how these interactions affect the plant and the microbial population. Understanding the microbial community and the potential that it carries in enhancing plant processes that leads to enhanced yield and growth would be beneficial to end users, i.e., the farmers. Enhanced yield may be attained through exploiting soil biological fertility, where lesser pesticides and fertilizers are required for improved yield and growth. Therefore through the utilization of existing knowledge and modern technologies, it is expected that valuable insight may be garnered to fill in the gaps in knowledge and information required to provide new opportunities and practices that increase crop production.

References

Akiyama K, Matsuzaki K-I, Hayashi H (2005) Plant sesquiterpenes induce hyphal branching in arbuscular mycorrhizal fungi. Nature 435(7043):824–827. doi:10.1038/nature03608, PMID: 15944706

Argueso CT, Hansen M, Kieber JJ (2007) Regulation of ethylene biosynthesis. J Plant Growth Regul 26(2):92–105. doi:10.1007/s00344-007-0013-5

Arshad M, Shaharoona B, Mahmood T (2008) Inoculation with *Pseudomonas* spp. containing ACC-deaminase partially eliminates the effects of drought stress on growth, yield, and ripening of pea (*Pisum sativum* L.). Pedosphere 18(5):611–620. doi:10.1016/S1002-0160(08)60055-7

Badri DV, Vivanco JM (2009) Regulation and function of root exudates. Plant Cell Environ 32(6):666–681. doi:10.1111/j.1365-3040.2009.01926.x, PMID: 19143988

Badri DV, Loyola-Vargas VM, Broeckling CD et al (2008) Altered profile of secondary metabolites in the root exudates of Arabidopsis ATP-binding cassette transporter mutants. Plant Physiol 146(2):762–771

Badri DV, Quintana N, El Kassis EG et al (2009a) An ABC transporter mutation alters root exudation of phytochemicals that provoke an overhaul of natural soil microbiota. Plant Physiol 151(4):2006–2017. doi:10.1104/pp.109.147462, PMID: 19854857

Badri DV, Weir TL, van der Lelie D et al (2009b) Rhizosphere chemical dialogues: plant–microbe interactions. Curr Opin Biotechnol 20(6):642–650. doi:10.1016/j.copbio.2009.09.014, PMID: 19875278

Badri DV, Chaparro JM, Zhang R et al (2013a) Application of natural blends of phytochemicals derived from the root exudates of Arabidopsis to the soil reveal that phenolic-related compounds predominantly modulate the soil microbiome. J Biol Chem 288(7):4502–4512. doi:10.1074/jbc.M112.433300, PMID: 23293028

Badri DV, Zolla G, Bakker MG et al (2013b) Potential impact of soil microbiomes on the leaf metabolome and on herbivore feeding behavior. New Phytol 198(1):264–273. doi:10.1111/nph.12124, PMID: 23347044

Bais HP, Walker TS, Schweizer HP et al (2002) Root specific elicitation and antimicrobial activity of rosmarinic acid in hairy root cultures of sweet basil (*Ocimum basilicum* L.). Plant Physiol Biochem 40:983–995

Bais HP, Weir TL, Perry LG et al (2006) The role of root exudates in rhizosphere interactions with plants and other organisms. Annu Rev Plant Biol 57(1):233–266. doi:10.1146/annurev.arplant.57.032905.105159, PMID: 16669762

Bakker M, Manter D, Sheflin A et al (2012) Harnessing the rhizosphere microbiome through plant breeding and agricultural management. Plant Soil 360(1–2):1–13. doi:10.1007/s11104-012-1361-x

Bakker PAHM, Berendsen RL, Doornbos RF et al (2013) The rhizosphere revisited: root microbiomics. Front Plant Sci 4(165). doi:10.3389/fpls.2013.00165

Battey NH, Blackbourn HD (1993) The control of exocytosis in plant cells. New Phytol 125:307–308

Bednarek P, Osbourn A (2009) Plant-microbe interactions: chemical diversity in plant defense. Science 324:746–748. doi:10.1126/science.1171661

Behie SW, Zelisko PM, Bidochka MJ (2012) Endophytic insect-parasitic fungi translocate nitrogen directly from insects to plants. Science 336(6088):1576–1577. doi:10.1126/science.1222289, PMID: 22723421

Berendsen RL, Pieterse CM, Bakker PA (2012) The rhizosphere microbiome and plant health. Trends Plant Sci 17(8):478–486. doi:10.1016/j.tplants.2012.04.001, PMID: 22564542

Brakhage AA, Schroeckh V (2011) Fungal secondary metabolites – strategies to activate silent gene clusters. Fungal Genet Biol 48:15–22. doi:10.1016/j.fgb.2010.04.004

Buttner M (2007) The monosaccharide transporter (–like) gene family in Arabidopsis. FEBS Letters 581:2318–2324

Cai T, Cai W, Zhang J et al (2009) Host legume-exuded antimetabolites optimize the symbiotic rhizosphere. Mol Microbiol 73(3):507–517. doi:10.1111/j.1365-2958.2009.06790.x, PMID: 19602148

Cannesan MA, Durand C, Burel C et al (2012) Effect of Arabinogalactan Proteins from the root caps of pea and Brassica napus on *Aphanomyces euteiches* zoospore chemotaxis and germination. Plant Physiol 159(4):1658–1670. doi:10.1104/pp.112.198507, PMID: 22645070

Chaparro JM, Badri DV, Bakker MG et al (2013a) Root exudation of phytochemicals in Arabidopsis follows specific patterns that are developmentally programmed and correlate with soil microbial functions. PloS ONE 8(2):e55731. doi:10.1371/journal.pone.0055731, PMID: 23383346

Chaparro JM, Badri DV, Vivanco JM (2013b) Rhizosphere microbiome assemblage is affected by plant development. ISME J 8(4):790–803. doi:10.1038/ismej.2013.196

Charmont S, Jamet E, Pont-Lezica R et al (2005) Proteomic analysis of secreted proteins from *Arabidopsis thaliana* seedlings: improved recovery following removal of phenolic compounds. Phytochemistry 66(4):453–461. doi:10.1016/j.phytochem.2004.12.013, PMID: 15694452

Chevrot R, Rosen R, Haudecoeur E et al (2006) GABA controls the level of quorum-sensing signal in *Agrobacterium tumefaciens*. Proc Natl Acad Sci USA 103:7460–7464. doi:10.1073/pnas.0600313103

Choi O, Kim JG, Joeng Y et al (2008) Pyrroloquinoline quinine is a plant growth promotion factor by Pseudomonas fluorescens B16. Plant Physiol 146:657–668

Colangelo EP, Guerinot ML (2006) Put the metal to the petal: metal uptake and transport throughout plants. Curr Opin Plant Biol 9:322–330

Compant S, Clément C, Sessitsch A (2010) Plant growth-promoting bacteria in the rhizo- and endosphere of plants: their role, colonization, mechanisms involved and prospects for utilization. Soil Biol Biochem 42(5):669–678. doi:10.1016/j.soilbio.2009.11.024

Coronado C, Zuanazzi J, Sallaud C et al (1995) Alfalfa root flavonoid production is nitrogen regulated. Plant Physiol 108(2):533–542. doi:10.1104/pp.108.2.533, PMID: 12228491

Czarnota MA, Paul RN, Weston LA et al (2003) Anatomy of sorgoleone-secreting root hairs of *Sorghum* species. Int J Plant Sci 164:861–866

Daniels R, De Vos DE, Desair J et al (2002) The cin quorum sensing locus of *Rhizobium etli* CNPAF512 affects growth and symbiotic nitrogen fixation. J Biol Chem 277(1):462–468. doi:10.1074/jbc.M106655200

De Hoff P, Brill L, Hirsch A (2009) Plant lectins: the ties that bind in root symbiosis and plant defense. Mol Genet Genomics 282(1):1–15. doi:10.1007/s00438-009-0460-8

de Weert S, Vermeiren H, Mulders IHM et al (2002) Flagella-driven chemotaxis towards exudate components is an important trait for tomato root colonization by *Pseudomonas fluorescens*. Mol Plant Microbe In 15(11):1173–1180. doi:10.1094/MPMI.2002.15.11.1173, PMID: 12423023

De-la-Peña C, Lei Z, Watson BS et al (2008) Root – microbe communication through protein secretion. J Biol Chem 283(37):25247–25255. doi:10.1074/jbc.M801967200, PMID: 18635546

De-la-Peña C, Badri DV, Lei Z et al (2010) Root secretion of defense-related proteins is development-dependent and correlated with flowering time. J Biol Chem 285(40):30654–30665. doi:10.1074/jbc.M110.119040, PMID: 20682788

Dennis PG, Miller AJ, Hirsch PR (2010) Are root exudates more important than other sources of rhizodeposits in determining the structure of rhizosphere bacterial communities? FEMS Microbiol Ecol 72:313–327

Elad Y, Barak R, Chet I et al (2008) Ultra structural studies of the interaction between *Trichoderma* spp. and plant pathogenic fungi. J Phytopathol 107:168–175. doi:10.1111/j.1439-0434.1983.tb00064.x

Fan TWM, Lane AN, Shenkar M et al (2001) Comprehensive chemical profiling of gramineous plant root exudates using high-resolution NMR and MS. Phytochem 57:209–221

Fang W, St. Leger RJ (2010) Mrt, a gene unique to fungi, encodes an oligosaccharide transporter and facilitates rhizosphere competency in *Metarhizium robertsii*. Plant Physiol 154(3):1549–1557. doi:10.1104/pp.110.163014, PMID: 20837701

Fellbaum CR, Gachomo EW, Beesetty Y et al (2012) Carbon availability triggers fungal nitrogen uptake and transport in arbuscular mycorrhizal symbiosis. Proc Natl Acad Sci 109(7):2666–2671. doi:10.1073/pnas.1118650109, PMID: 22308426

Field B, Jordan F, Osbourn A (2006) First encounters – deployment of defence-related natural products by plants. New Phytol 172:193–207

Furukawa J, Yamaji N, Wang H et al (2007) An aluminum-activated citrate transporter in barley. Plant Cell Physiol 48(8):1081–1091. doi:10.1093/pcp/pcm091, PMID: 17634181

Gao M, Teplitski M, Robinson JB et al (2003) Production of substances by *Medicago truncatula* that affect bacterial quorum sensing. Mol Plant Microbe In 16(9):827–834. doi:10.1094/MPMI.2003.16.9.827

Glick BR (2005) Modulation of plant ethylene levels by the bacterial enzyme ACC deaminase. FEMS Microbiol Lett 251(1):1–7. doi:10.1016/j.femsle.2005.07.030, PMID: 16099604

Grotewold E (2004) The challenges of moving chemicals within and out of cells: insights into the transport of plant natural products. Planta 219:906–909

Guiñazú L, Andrés J, Del Papa M et al (2010) Response of alfalfa (*Medicago sativa* L.) to single and mixed inoculation with phosphate-solubilizing bacteria and *Sinorhizobium meliloti*. Biol Fertil Soils 46(2):85–190. doi:10.1007/s00374-009-0408-5

Hamer U, Marschner B (2005) Priming effects in different soil types induced by fructose, alanine, oxalic acid and catechol additions. Soil Biol and Biochem 37:445. doi:10.1016/j.soilbio.2004.07.037

Hartmann A, Rothballer M, Schmid M (2008) Lorenz Hiltner, a pioneer in rhizosphere microbial ecology and soil bacteriology research. Plant Soil 312(1–2):7–14. doi:10.1007/s11104-007-9514-z

Haudecoeur E, Planamente S, Cirou A et al (2009) Proline antagonizes GABA-induced quenching of quorum-sensing in *Agrobacterium tumefaciens*. Proc Natl Acad Sci USA 106:14587–14592. doi:10.1073/pnas.0808005106

Hirner A, Ladwig F, Stransky H et al (2006) Arabidopsis LHT1 is a high-affinity transporter for cellular amino acid uptake in both root epidermis and leaf mesophyll. The Plant Cell 18:1931–1946

Hoffmeister D, Keller NP (2007) Natural products of filamentous fungi: enzymes, genes, and their regulation. Nat Prod Rep 24:393–416. doi:10.1128/EC.5.4.613-619.2006

Hogan DA (2006) Talking to themselves: autoregulation and quorum sensing in fungi. Eukaryot Cell 5(4):613–619. doi:10.1128/EC.5.4.613-619.2006

Horiuchi J-I, Prithiviraj B, Bais H et al (2005) Soil nematodes mediate positive interactions between legume plants and rhizobium bacteria. Planta 222(5):848–857. doi:10.1007/s00425-005-0025-y, PMID: 16025342

Huang XF, Chaparro JM, Reardon KF et al (2014) Rhizosphere interactions: root exudates, microbes, and microbial communities. Botany 92:267–275. http://dx.doi.org/10.1139/cjb-2013-0225

Ishimaru Y, Kakei Y, Shimo H et al (2011) A rice phenolic efflux transporter is essential for solubilizing precipitated apoplasmic iron in the plant stele. J Biol Chem 286(28):24649–24655. doi:10.1074/jbc.M111.221168, PMID:21602276

Jansson JK, Neufeld JD, Moran MA et al (2011) Omics for understanding microbial functional dynamics. Environ Microbiol 14(1):1–3. doi:10.1111/j.1462-2920.2011.02518.x

Jones KM, Sharopova N, Lohar DP et al (2008) Differential response of the plant *Medicago truncatula* to its symbiont *Sinorhizobium meliloti* or an exopolysaccharide-deficient mutant. Proc Natl Acad Sci USA 105(2):704–709. doi:10.1073/pnas.0709338105

Juan Z, Subramanian S, Zhang Y et al (2007) Flavone Synthases from *Medicago truncatula* are flavanone-2-hydroxylases and are important for nodulation. Plant Physiol 144:741–751

Kardol P, Cornips NJ, van Kempen MML et al (2007) Microbe-mediated plant–soil feedback causes historical contingency effects in plant community assembly. Ecol Monogr 77:147–162

Kiers ET, Duhamel M, Beesetty Y et al (2011) Reciprocal rewards stabilize cooperation in the mycorrhizal symbiosis. Science 333(6044):880–882. doi:10.1126/science.1208473, PMID: 21836016

Kim SA, Guerinot ML (2007) Mining iron: iron uptake and transport in plants. FEBS Letters 581(12):2273–2280. doi:10.1016/j.febslet.2007.04.043

Kim JG, Park BK, Kim SU et al (2006) Bases of biocontrol: sequence predicts synthesis and mode of action of agrocin 84, the Trojan Horse antibiotic that controls crown gall. Proc Natl Acad Sci USA 103(23):8846–8851

Klepek YS, Geiger D, Stadler R et al (2005) Arabidopsis polyol transporters, a new member of the monosaccharide transporter-like superfamily, mediates H + −symport of numerous substrates including myo-inositol, glycerol and ribose. The Plant Cell 17:204–218

Kobae Y, Sekino T, Yoshioka H et al (2006) Loss ofAtPDR8, a plasma membrane ABC transporter of *Arabidopsis thaliana*, causes hypersensitive cell death upon pathogen infection. Plant Cell Physiol 47:309–318

Kumar R, Bhatia R, Kukreja K et al (2007) Establishment of Azotobacter on plant roots: chemotactic response, development and analysis of root exudates of cotton (*Gossypium hirsutum* L.) and wheat (*Triticum aestivum* L.). J Basic Microbiol 47:436–439

Kuzyakov Y (2002) Review: factors affecting rhizosphere priming effects. J Plant Nutr Soil Sc 165(4):382–396

Lambrecht M, Okon Y, Vande BA et al (2000) Indole-3-acetic acid: a reciprocal signalling molecule in bacteria-plant interactions. Trends Microbiol 8:298–300

Lee YH, Foster J, Chen J et al (2007) AAP1 transports uncharged amino acids into roots of Arabidopsis. The Plant Journal 50:305–319

Leyval C, Berthelin J (1993) Rhizodeposition and net release of soluble organic compounds by pine and beech seedlings inoculated with rhizobacteria and ectomycorrhizal fungi. Biol Fertil Soils 15(4):259–267. doi:10.1007/bf00337210

Ling N, Zhang W, Wang D et al (2013) Root exudates from grafted-root watermelon showed a certain contribution in inhibiting *Fusarium oxysporum* f. sp. *niveum*. PloS ONE 8(5):e63383. doi:10.1371/journal.pone.0063383

Liu J, Magalhaes JV, Shaff J et al (2009) Aluminum-activated citrate and malate transporters from the MATE and ALMT families function independently to confer Arabidopsis aluminum tolerance. Plant J 57(3):389–399. doi:10.1111/j.1365-313X.2008.03696.x, PMID:1882642

Ma JF, Yamaji N (2008) Functions and transport of silicon in plants. Cell Mol Life Sci 65:3049–3057

Magalhaes JV, Liu J, Guimarães CT et al (2007) A gene in the multidrug and toxic compound extrusion (MATE) family confers aluminum tolerance in sorghum. Nat Genet 39(9):1156–1161. doi:10.1038/ng2074, PMID: 17721535

Malinowski DP, Belesky DP (2000) Adaptations of endophyte-infected cool-season grasses to environmental stresses: mechanisms of drought and mineral stress tolerance. Crop Sci 40(4):923–940. doi:10.2135/cropsci2000.404923x

Marschner H (1995) Mineral nutrition of higher plants. Academic, London

Mathesius U, Watt M (2010) Rhizosphere signals for plant-microbe interactions: implications for field-grown plants. In: Lüttge UE, Beyschlag W (eds) Progress in botany, vol 72. Springer, Berlin, pp 125–161. doi:10.1007/978-3-642-13145-5_5

Mayak S, Tirosh T, Glick BR (2004) Plant growth-promoting bacteria confer resistance in tomato plants to salt stress. Plant Physiol Biochem 42(6):565–572. doi:10.1016/j.plaphy.2004.05.009, PMID: 15246071

Meier IC, Avis PG, Phillips RP (2013) Fungal communities influence root exudation rates in pine seedlings. FEMS Microbiol Ecol 83(3):585–595. doi:10.1111/1574-6941.12016, PMID: 23013386

Mendes R, Kruijt M, de Bruijn I et al (2011) Deciphering the rhizosphere microbiome for disease suppressive bacteria. Science 332(6033):1097–1100. doi:10.1126/science.1203980, PMID: 21551032

Mercado-Blanco J, Bakker P (2007) Interactions between plants and beneficial *Pseudomonas* spp.: exploiting bacterial traits for crop protection. Antonie van Leeuwenhoek 92(4):367–389. doi:10.1007/s10482-007-9167-1

Morandi D, Bailey J, Gianinazzi-Pearson V (1984) Isoflavonoid accumulation in soybean roots infected with vesicular-arbuscular mycorrhizal fungi. Physiol Plant Pathol 24(3):357–364. doi:10.1016/0048-4059(84)90009-2

Mukerji KG, Manoharachary C, Singh J (2006) Microbial activity in the rhizospere, vol 7. Springer Science & Business Media, New York

Murray JD, Karas BJ, Sato S et al (2007) A cytokinin perception mutant colonized by Rhizobium in the absence of nodule organogenesis. Science 315:101–104

Nadarajah K (2016) Induced systemic resistance in rice. In: Choudhary KD, Varma A (eds) Microbial-mediated induced systemic resistance in plants. Springer, Singapore, pp 103–124. doi:10.1007/978-981-10-0388-2_7

Naher UA, Othman R, Mohd Saud H et al (2008) Effect of inoculation on root exudates carbon sugar and amino acids production of different rice varieties. Res J Microbiol 3(9):580–587

Narasimhan K, Basheer C, Bajic VB et al (2003) Enhancement of Plant–microbe interactions using a rhizosphere metabolomics-driven approach and its application in the removal of polychlorinated biphenyls. Plant Physiol 132(1):146–153. doi:10.1104/pp.102.016295, PMID: 12746520

Neal AL, Ahmad S, Gordon-Weeks R et al (2012) Benzoxazinoids in root exudates of maize attract *Pseudomonas putida* to the rhizosphere. PloS ONE 7(4):e35498. doi:10.1371/journal.pone.0035498, PMID: 22545111

Newton AC, Fitt BDL, Atkins SD et al (2010) Pathogenesis, parasitism and mutualism in the trophic space of microbe–plant interactions. Trends Microbiol 18(8):365–373. doi:10.1016/j.tim.2010.06.002, PMID: 20598545

Nguema-Ona E, Vicré-Gibouin M, Cannesan M-A et al (2013) Arabinogalactan proteins in root–microbe interactions. Trends Plant Sci 18(8):440–449. doi:10.1016/j.tplants.2013.03.006, PMID: 23623239

Nguyen C (2003) Rhizodeposition of organic C by plants: mechanisms and controls. Agronomoie 23:375–396

Nihorimbere V, Ongena M, Smargiassi M et al (2011) Beneficial effect of the rhizosphere microbial community for plant growth and health. Biotechnol Agron Soc 15:327–337

Novas MV, Iannone LJ, Godeas AM et al (2011) Evidence for leaf endophyte regulation of root symbionts: effect of Neotyphodium endophytes on the pre-infective state of mycorrhizal fungi. Symbiosis 55(1):19–28. doi:10.1007/s13199-011-0140-4

Nozoye T, Nagasaka S, Kobayashi T et al (2011) Phytosiderophore efflux transporters are crucial for iron acquisition in graminaceous plants. J Biol Chem 286(7):5446–5454. doi:10.1074/jbc.M110.180026

Oba H, Tawaraya K, Wagatsuma T (2002) Inhibition of pre-symbiotic hyphal growth of arbuscular mycorrhizal fungus *Gigaspora margarita* by root exudates of *Lupinus* spp. Soil Sci Plant Nutr 48(1):117–120. doi:10.1080/00380768.2002.10409180

Okon Y, Itzigsohn R (1995) The development of Azospirillum as a commercial inoculant for improving crop yields. Biotechnol Adv 13:415–424

Ortiz-Castro R, Contreras-Cornejo HA, Macias-Rodriguez L et al (2009) The role of microbial signals in plant growth and development. Plant Signal Behav 4:701–712

Parmar N (1995) Interactions of rhizosphere bacteria with Cicer-Rhizobium symbiosis. CCS Haryana Agricultural University, Hisar

Parmar N, Dadarwal KR (1997) Rhizobacteria from rhizosphere and rhizoplane of chick pea (*Cicer arietinum* L.). Indian J Microbiol 37:205–210

Paterson E, Sim A, Standing D et al (2006) Root exudation from *Hordeum vulgare* in response to localized nitrate supply. J Exp Bot 57:2413–2420

Phillips DA, Fox TC, King MD et al (2004) Microbial products trigger amino acid exudation from plant roots. Plant Physiol 136(1):2887–2894. http://dx.doi.org/10.1104/pp.104.044222

Pineros MA, Magalhaes JV, Alves VMC et al (2002) The physiology and biophysics of an aluminum tolerance regulation and function of root exudates mechanism based on root citrate exudation in maize. Plant Physiol 129:1194–1206

Poysti NJ, Loewen ED, Wang Z et al (2007) *Sinorhizobium meliloti* pSymB carries genes necessary for arabinose transport and catabolism. Microbiol 153(3):727–736. doi:10.1099/mic.0.29148-0

Raaijmakers J, Paulitz T, Steinberg C et al (2009) The rhizosphere: a playground and battlefield for soilborne pathogens and beneficial microorganisms. Plant Soil 321(1–2):341–361. doi:10.1007/s11104-008-9568-6

Raaijmakers JM, de Bruijn I, Nybroe O et al (2010) Natural functions of lipopeptides from Bacillus and Pseudomonas: more than surfactants and antibiotics. FEMS Microbiol Rev 34(6):1037–1062. doi:10.1111/j.1574-6976.2010.00221.x

Ramos-González MI, Campos MJ, Ramos JL (2005) Analysis of *Pseudomonas putida* KT2440 gene expression in the maize rhizosphere: in vivo expression technology capture and identification of root-activated promoters. J Bacteriol 187(12):4033–4041. doi:10.1128/JB.187.12.4033-4041.2005

Requena N, Perez-Solis E, Azcon-Aguilar C et al (2001) Management of indigenous plant-microbe symbioses aids restoration of desertified ecosystems. Appl Environ Microbiol 67:495–498

Rokhzadi A, Asgharzadeh A, Darvish F et al (2008) Influence of plant growth promoting rhizobacteria on dry matter accumulation of chickpea (*Cicer arietinum* L) under field conditions. JAES 3(2):253–257

Ryan PR, Tyerman SD, Sasaki T et al (2011) Identification of aluminium-resistance genes in plants provides an opportunity for enhancing the acid-soil tolerance of crop species. J Exp Bot 62:9–20

Ryu CM, Farag MA, Hu CH et al (2004) Bacterial volatiles induce systemic resistance in Arabidopsis. Plant Physiology 134:1–10

Saharan B, Nehra V (2011) Plant growth promoting rhizobacteria: a critical review. Life Sci Med Res LSMR-21:1–30

Sanchez-Contreras M, Bauer WD, Gao M et al (2007) Quorum-sensing regulation in rhizobia and its role in symbiotic interactions with legumes. Philos Trans R Soc Lond B 362(1483):1149–1163. doi:10.1098/rstb.2007.2041

Sanders D, Bethke P (2000) Membrane transport. In: Buchanan BB, Gruisham W, Jones RL (eds) Biochemistry and molecular biology of plants. ASPP, Rockville, pp 110–158

Santi C, Bogusz D, Franche C (2013) Biological nitrogen fixation in non-legume plants. Annals of Botany 111:743–767. doi:10.1093/aob/mct048

Schnitzer SA, Klironomos JN, HilleRisLambers J et al (2011) Soil microbes drive the classic plant diversity-productivity pattern. Ecology 92(2):296–303

Sidler M, Hassa P, Hasan S et al (1998) Involvement of an ABC transporter in a developmental pathway regulating hypocotyl cell elongation in the light. Plant Cell 10:1623–1636

Siegrid S, Lendzemo V, Langer I et al (2007) Flavonoids and strigolactones in root exudates as signals in symbiotic and pathogenic plant-fungus interactions. Molecule 12:1290–1306

Simons M, Permentier HP, de Weger LA et al (1997) Amino acid synthesis is necessary for tomato root colonization by *Pseudomonas fluorescens* strain WCS365. Mol Plant Microbe Interac 10(1):102–106. doi: http://dx.doi.org/10.1094/MPMI.1997.10.1.102

Snyder BA, Leite B, Hipskind J et al (1991) Accumulation of sorghum phytoalexins induced by *Colletotrichum graminicola* at the infection site. Physiol Mol Plant P 39:463–470

Stearns JC, Woody OZ, McConkey BJ et al (2012) Effects of bacterial ACC deaminase on *Brassica napus* gene expression. Mol Plant Microbe Interac 25(5):668–676. doi:10.1094/MPMI-08-11-0213, PMID: 22352713

Steenhoudt O, Vanderleyden J (2000) Azospirillum, a free living nitrogen fixing bacterium closely associated with grasses. FEMS Microbiol Lett 24:506

Stein M, Dittgen J, Sanchez-Rodriguez C et al (2006) Arabidopsis PEN3/PDR8, an ATP binding cassette transporter, contributes to nonhost resistance to inappropriate pathogens that enter by direct penetration. The Plant Cell 18:731–746

Svennerstam H, Ganeteg U, Bellini C et al (2007) Comprehensive screening of Arabidopsis mutants suggests the lysine histidine transporter 1 to be involved in plant uptake of amino acids. Plant Physiol 143:1853–1860

Taddei P, Tugnoli V, Bottura G et al (2002) Vibrational, 1H-NMR spectroscopic, and thermal characterization of gladiolus root exudates in relation to *Fusarium oxysporum f. sp. gladioli* resistance. Biopolymers 67(6):428–439. doi:10.1002/bip.10170

Teplitski M, Robinson JB, Bauer WD (2000) Plants secrete substances that mimic bacterial N-acyl homoserine lactone signal activities and affect population density-dependent behaviors in associated bacteria. Mol Plant Microbe Interact 13(6):637–648. doi: http://dx.doi.org/10.1094/MPMI.2000.13.6.637

Teplitski M, Chen H, Rajamani S et al (2004) *Chlamydomonas reinhardtii* secretes compounds that mimic bacterial signals and interfere with quorum sensing regulation in bacteria. Plant Physiol 134(1):137–146. doi:10.1104/pp.103.029918, PMID: 14671013

Tirichine L, Sandal N, Madsen LH et al (2007) A gain-of-function mutation in a cytokinin receptor triggers spontaneous root nodule organogenesis. Science 315:104–107

Uren NC (2000) Types, amounts, and possible functions of compounds released into the rhizosphere by soil-grown plants. In: Pinton R, Varanini Z, Nannipieri P (eds) The rhizosphere: biochemistry and organic substances at the soil–plant interface. Marcel Dekker, Inc., New York, pp 19–40

Urich T, Lanzén A, Qi J et al (2008) Simultaneous assessment of soil microbial community structure and function through analysis of the meta-transcriptome. PloS One 3(6):e2527. doi: http://dx.doi.org/10.1371/journal.pone.0002527

Vicré M, Santaella C, Blanchet S et al (2005) Root border-like cells of Arabidopsis. Microscopical characterization and role in the interaction with rhizobacteria. Plant Physiol 138(2):998–1008. doi:10.1104/pp.104.051813

Walker TS, Bais HP, Grotewold E et al (2003) Root exudation and rhizosphere biology. Plant Physiol 132(1):44–51. doi: http://dx.doi.org/10.1104/pp.102.019661

Weir TL, Park S-W, Vivanco JM (2004) Biochemical and physiological mechanisms mediated by allelochemicals. Curr Opin Plant Biol 7(4):472–479. doi:10.1016/j.pbi.2004.05.007

Weston LA, Ryan PR, Watt M (2012) Mechanisms for cellular transport and release of allelochemicals from plant roots into the rhizosphere. J Exp Bot 63:3445–3454. doi:10.1093/jxb/ers054, PMID: 22378954

Winkel-Shirley B (2001) Flavonoid biosynthesis: a colorful model for genetics, biochemistry, cell biology and biotechnology. Plant Physiol 126:485–493

Wu X-G, Duan H-M, Tian T et al (2010) Effect of the hfq gene on 2,4-diacetylphloroglucinol production and the PcoI/PcoR quorum-sensing system in *Pseudomonas fluorescens* 2P24. FEMS Microbiol Lett 309(1):16–24. doi:10.1111/j.1574-6968.2010.02009.x

Xie F, Williams A, Edwards A et al (2012) A plant arabinogalactan like glycoprotein promotes a novel type of polar surface attachment by *Rhizobium leguminosarum*. Mol Plant–Microbe Interact 25(2):250–258. doi:10.1094/MPMI-08-11-0211, PMID: 21995765

Yadegari M, Rahmani HA, Noormohammadi G et al (2008) Evaluation of bean (*Phaseolus vulgaris*) seeds inoculation with *Rhizobium phaseoli* and plant growth promoting rhizobacteria on yield and yield components. Pak J Biol Sci 11:1935–1939

Yoneyama K, Xie X, Sekimoto H et al (2008) Strigolactones, host recognition signals for root parasitic plants and arbuscular mycorrhizal fungi, from Fabaceae plants. New Phytol 179(2):484–494. doi:10.1111/j.1469-8137.2008.02462.x

Zahran HH (1999) Rhizobium–legume symbiosis and nitrogen fixation under severe conditions and in an arid climate. Microbiol Mol Biol Rev 63(4):968–989, PMID: 10585971

Zhang J, Subramanian S, Stacey G et al (2009) Flavones and flavonols play distinct critical roles during nodulation of *Medicago truncatula* by *Sinorhizobium meliloti*. Plant J 57(1):171–183. doi:10.1111/j.1365-313X.2008.03676.x

Zhuang X, Gao J, Ma A et al (2013) Bioactive molecules in soil ecosystems: masters of the underground. Int J Mol Sci 14(5):8841–8868. doi:10.3390/ijms14058841

Shaping the Other Sides: Exploring the Physical Architecture of Rhizosphere

2

Madhurima Chatterjee, Raktim Bhattacharya, and Rabindranath Bhattacharyya

Abstract

The root system is immediately surrounded by a narrow zone of soil called the rhizosphere. A major proportion of biodiversity of the soil resides in the rhizosphere, hence accounting for the various activities found in that area. There are various abiotic and biotic factors which help in modifying the physical structure of the rhizosphere. The main abiotic factors are light, temperature, humidity, carbon dioxide, water uptake, pH change, etc. The physical architecture determines the richness of the microbial community which in turn affects the plant growth. In this chapter, the various physical and chemical processes occurring in the rhizosphere and how the change in environment hampers these factors and how that affects the rhizospheric diversity in modifying the microbial ecology and root architecture will be discussed.

2.1 Introduction

Soil is considered the habitat of most of the organisms on earth, ranging from prokaryotes to eukaryotes. It consists of the important organisms, like various soil microflora and fungi including invertebrates (like protozoa, mites, nematodes, earthworms and insects) (Hinsinger et al. 2009). It is estimated that the number of prokaryotes inhabiting the soil ecosystem is three times more than the combination of all the other environmental counterparts of the earth's ecosystem (Curtis et al. 2002; Crawford et al. 2005; Curtis and Sloan 2005). Soil is the platform where the

M. Chatterjee • R. Bhattacharya • R. Bhattacharyya (✉)
Department of Biological Sciences, Presidency University,
86/1, College Street, Kolkata 700073, West Bengal, India
e-mail: rabindranathbpc@yahoo.co.in

D.K. Choudhary et al. (eds.), *Plant-Microbe Interaction: An Approach to Sustainable Agriculture*, DOI 10.1007/978-981-10-2854-0_2

plant grows; it is from the soil that most of the members of plant community acquired the required amount of nutrition and water in the form of sap.

The rhizosphere is defined to be the volume of soil around the living roots, influenced by certain activity of the root (Darrah 1993; Hinsinger 1998a). The rhizosphere is considered to be the hot spot of activity in soils, constituting its unique environment. It is believed that most of the diversity of soils resides in the rhizosphere (Jones and Hinsinger 2008). Different physical, chemical and biochemical processes occur in the rhizosphere as a result of root growth, water and mineral uptake, rhizodeposition and respiration. These factors distinguish it from the bulk soil. Microbial ecology and plant physiology are also affected owing to these processes.

The term rhizosphere was first coined by German plant physiologist and agronomist Lorenz Hiltner. The word rhizosphere originated from the Greek word rhiza (meaning root) (Hiltner 1904; Hartmann et al. 2008). From that time the definition and explanation of the term rhizosphere has undergone a thorough modification. And nowadays it has been considered that rhizosphere actually consists of three basic parts, and they are:

1. Endorhizosphere, the region which includes the cortex and endodermis, where microbes along with the cations can occupy the apoplastic space
2. The rhizoplane, which is the middle portion of the root situated just beside the root, consisting of the epidermis (epiblema) and mucilage
3. The ectorhizosphere, which is the outermost region, which spans from the rhizoplane to the outer bulk soil

Now it has been understandably clear that by the term rhizosphere, we cannot define a specific area, but it defines a zone of gradient spanning across the root along with the soil microflora as well as some physical and chemical factors.

The Physical Properties and Processes of the Rhizosphere The physical processes occurring in the rhizosphere are responsible for the movement of water and minerals inside and outside of the root.

The factors which alter the physical properties of the rhizosphere are primarily the root activities necessary for the root's growth and uptake of water. The bulk density, porosity and soil strength are some of the notable physical factors which are affected by the forces exerted by the growth of the root (Dexter 1987; Czarnes et al. 1999). Changes in soil structure are found due to changes in the physical properties of the rhizosphere. This is caused mainly by polysaccharides which constitute a major proportion of rhizodeposits (Czarnes et al. 2000). Rhizodeposits are basically certain organic compounds (mostly C-rich substrates) which are released by the plant roots, to be fed by heterotrophic bacteria (Lynch and Whipps 1990; Jones et al. 2004, 2009). Rhizodeposition alters the nutrient abundance in the rhizosphere. Rhizosheaths, which are certain unique structural features restricted to rhizosphere, are formed by the mucilage produced from the roots and also root hairs (McCully

1999). Rhizosheaths are mainly found in the grasses. Aggregation of soil around roots can be attributed to the mycorrhizal hyphae and also the exopolysaccharides secreted by the microorganisms residing in the rhizosphere (Amellal et al. 1998). These processes facilitate or retard the transport of water, solutes and toxic compounds and also affect plant nutrition and health.

Water uptake can also have dramatic effect on the rhizosphere architecture (Doussan et al. 2003). This alters the water potential around the roots by affecting the microbial activities happening in the rhizosphere as well as the radial movement of water particles.

Water captured in the rhizospheric region will be supplied throughout the whole plant, so change in water potential can greatly affect the rhizosphere. A very recent study on cavitation and its effect on cohesion-tension theory depicted that the small roots are more vulnerable to cavitation, so these small roots may be the weakest link in the soil-plant-atmosphere continuum (Hacke and Sauter 1996). It has also been found that, in *Acer grandidentatum*, if the negative water potential (ψ) remains constant, then the safety margin of cavitation of the roots is smaller than that of the shoots (Alder et al. 1996). As a result of this, at the time of drought, extensive cavitation in the roots inhibits the gas exchange in the shoots.

Moreover, the soil-water relationships, viscosity and surface tension properties of the soil are affected by the mucilage secreted by the roots.

The soil associated with roots confers a resistance to the external, mechanical stress in comparison to the bulk soil, thus exhibiting increased stability. The enhanced soil stability within and outside the rhizosphere can also be due to certain biological activity.

2.1.1 Chemical Processes and Properties of the Rhizosphere

Plant roots perform various functions like absorption, respiration and exudation. These functions are responsible for changes in nutrient and toxic elements' concentrations, pH, redox potential, partial pressure of oxygen and partial pressure of carbon dioxide in the rhizosphere (Hinsinger 1998a).

Due to rhizodeposition, there occurs a carbon flow in the rhizosphere, triggering the growth of bacteria. So, it seems that the rhizosphere gets enriched by C, but it lacks nitrogen. Nutrient uptake by roots often results in nutrient depletion in the rhizosphere. Nutrients such as calcium, magnesium and potassium occurring as solutes in the soil solution get transferred and are accumulated near the root surface by mass flow, when the flow of the nutrient transferred is more than the plant's demand (Lorenz et al. 1994; Barber 1995; Hinsinger 2004). Similarly, the nutrient concentration decreases in the rhizosphere when the flow is less than that required by the plant. This type of decrease is mainly found for phosphorus, nitrogen and potassium (Hendriks et al. 1981; Kuchenbuch and Jungk 1982; Gahoonia et al. 1992; Hinsinger et al. 2005). Hence, based on the two conditions, the rhizosphere may become nutrient-enriched or nutrient-depleted zone.

The pH of the rhizosphere generally increases when the plants absorb nitrogen as NO_3^-. Hence the rhizosphere becomes alkaline in this condition. But when the plants absorb nitrogen as N_2 or NH_4^+, the pH decreases (Nye 1981). The pH change can be calculated from efflux of H^+ efflux and radius of the root, initial pH, pH buffering capacity, partial pressure of CO_2 in the soil and also moisture content (Nye 1981). pH changes also affect nutrient availability. Protons in the rhizosphere compete for metal cations (e.g. Cu and Zn) (Loosemore et al. 2004; Michaud et al. 2007) on cation binding sites, thus altering nutrient composition. pH change sometimes has a dramatic effect on phosphorus acquisition.

Again, it is found that roots and rhizosphere microorganisms often exude certain organic ligands which help in increment of nutrient availability by desorption of anions in exchange of the ligands (Hinsinger 2001a; Ryan et al. 2001; Read et al. 2003; Dunbabin et al. 2006).

Usually in the rhizosphere, increased activity of certain enzymes like phosphatases, proteases and arylsulfatases, released by ectomycorrhizal fungi and microorganisms, is found, in comparison to bulk soil. These help in cycling of P, N and S.

2.1.2 Rhizosphere Architecture

Rhizosphere architecture can vary between species or between genotypes of a given species (Ge et al. 2000). It also changes in response to environmental cues; e.g. roots proliferate in nutrient-rich patches or roots form cluster in phosphorus deficiency.

It is found that root hairs by extending up to a few millimetres away from the root surface can increase the rhizosphere volume (Bhat et al. 1976). Similarly, the mycorrhizal hyphae in phosphate-deficient soil can extend several centimetres above the soil surface, thus enhancing the volume of the rhizosphere of the plants having symbiotic relationships with mycorrhizae (Li et al. 1991; Jakobsen et al. 1992; Read and Perez-Moreno 2003).

Following are the factors which affect the rhizosphere architecture:

2.1.2.1 Effects of Elevated Atmospheric CO_2 on the Microbial Structure and Rhizosphere Architecture

When concentration of atmospheric CO_2 increases, then it alters the plant C allocation. Several biochemical and physiological reactions start occurring in the fine roots, ultimately affecting the rhizosphere food webs, also the rates at which the C and N cycle take place. Mycorrhiza and the fine roots are mainly responsible for the mineral nutrition, input of soil C as well as microbial activity inside the soil. With increased CO_2 concentrations, fine root growth is enhanced (Curtis 1996; Curtis and Wang 1998; Pendall et al. 2004; Rillig et al. 1997). Increased rate of nutrient uptake and mycorrhizal activities are also found which sometimes can alter the dynamic equilibrium existing between the rhizosphere microbial community and plant roots (Hu et al. 1999; Klironomos et al. 1996). Mycorrhizal biomass increases due to limited C and nutrient availability. So, mycorrhizae are indirectly affected due to changes in C allocation from their host plants (Allen et al. 2005; Gamper et al. 2004, 2005; Parrent et al. 2006; Sanders et al. 1998; Staddon et al. 2002; Treseder and Allen 2000).

With increase in CO_2 concentration, the existing C dynamics in the rhizosphere is altered resulting in an elevated C/N ratio of rhizodeposition, despite no increase in the total plant biomass (Hu et al. 1999; Paterson et al. 1997).

It is found that the soil respiration consisting of root and microbial respiration increased when the plants were exposed to higher CO_2 concentrations.

Microbial growth and activity are usually stimulated in response to elevated CO_2 levels (Cotrufo and Gorissen 1997; Diaz et al. 1993; Paterson et al. 1997; Sadowsky and Schortemeyer 1997; Zak et al. 1993, 2000). This results in an increase in grazing which again results in a quicker nutrient recycling from the bacterial biomass, thus increasing nutrient flux to the plant.

2.1.2.2 Effect of Light on Rhizosphere Architecture

It is often found that the light intensity is directly proportional to the production of roots. It means that the plants which are subjected to low light intensity show slower growth rate of roots in comparison to the plants which are treated with high light intensity (Biswell 1935; Haig 1936). Hence plants receiving sunlight show a good response to root development than plants remaining under shade. An increment in the ratio of dry weights of roots to tops was observed in several species (Shirley 1936).

Infection by the vesicular-arbuscular mycorrhizae (VAM) is influenced by the photon flux density and photoperiod. When the plants, especially at higher altitudes, are subjected to higher light intensity, in sunlit areas and during springtime or summer, greater mycorrhizal infection was observed (Winter and Meloh 1958). Light intensity not only affects the shaping of the root, but also it helps in the rhizosphere microflora maintenance. In one study it has been found that increasing light intensity is actually helping *Glomus fasciculatum*, vesicular-arbuscular mycorrhizal (VAM) fungi, to increase colonization in the roots of Sudan grass (Ferguson and Menge 1982). All these are the evidences, which clearly indicate that root growth is actually directly proportional to the photosynthetic carbon production by the shoot and which ultimately depends on the light. So, the more a plant will get light, the more there will be photosynthesis, hence more carbon, thus more root growth. Similarly, it is found that photoperiods of longer duration cause increased infection of mycorrhiza in comparison to photoperiods of shorter duration (Hayman 1974).

2.1.2.3 Effect of Temperature on Rhizosphere Architecture

Soil temperature has a positive correlation with the root growth. The higher the temperature, the higher the root growth, and reduced growth of the root is observed during lower temperature prevailing in the soil. An experiment performed on young loblolly pine seedlings shows that with increasing temperature, initially there was a uniform rise in root growth rate at 5 till 25 °C, after which the root growth declined (Barney 1951). The length of the individual root is expressed as a function of time. This root growth can be calculated by the following formula:

$$Y = \overline{y} + b(x - \overline{x}),$$

where

Y = length of any individual root
x = average response of root growth to a given temperature
$\overline{y}$ = average of y values
$\overline{x}$ = average of x values
b = coefficient of the regression equation (Barney 1951).

Increase in soil temperature influences heterotrophic respiration, thus affecting atmospheric CO_2. Enzyme activity and chemical kinetics get increased, thus increasing decomposition rates in the soil. This ultimately leads to decreased net ecosystem production (NEP). Elevated soil temperatures again trigger microbial activity and increase availability of N and also the net primary productivity (NPP). It also results in an increased activity of the roots and soil heterotrophic organisms, and these are thus responsible for the loss of carbon dioxide and methane from the soil. The soils often dry up owing to an increase in the soil temperature; thus various soil nutrients get immobilized. Higher temperature often results in an increase in the number of fine roots. This has been seen in *Pinus taeda* but not *Pinus ponderosa* (King et al. 1996). Root N concentration is found to increase in response to higher temperature (King et al. 1997; Kandeler et al. 2002; Wan et al; 2004), owing to diffusion and mineralization of N, which are elevated at increased temperature (BassiriRad et al. 1993; BassiriRad 2000). This higher N concentration sometimes results in mortality of the fine roots, thus affecting the soil N cycling (Zak et al. 2000).

2.1.2.4 Effect of pH Change on Rhizosphere Architecture

Whenever there is an imbalance of cation-anion uptake at the root-soil interface, the roots try to adjust the pH of the rhizosphere by releasing H^+ or OH^- ions. When a surplus of cations enter in comparison to anions, then to compensate for the extra positive charges entering the cell, H^+ ions are released into the apoplasm, thus increasing the pH of the cytosol. This type of pH adjustment occurs when the plant is treated with K_2SO_4 solution and an excess amount of K^+ ions enters the cell than the SO_4^{2-} ions (Haynes 1990; Hiatt 1967; Marschner 1995). Similarly, when an excess of anions enter the cell in comparison to cations, to compensate for the surplus negative charges entering, OH^- ions are released into the apoplasm, thereby decreasing the pH. This type of pH adjustment occurs when the plant is treated with $CaCl_2$ solution and an excess amount of Cl^- ions enters the cell than the Ca^{2+-} ions (Hiatt 1967; Haynes 1990; Marschner 1995). Plants like legumes, showing a dependence on atmospheric N_2, release excess positive charges in the form of H^+ since they are able to take up more cations than anions and thus increase the acidity by decreasing the pH.

Again, it is found that some portions of the root just behind the root apex may release H^+, while basal parts release OH^- ions into the rhizosphere, thus constructing a spatial variation along the root axes (Jaillard et al. 2002).

Organic acids, e.g. citric acid, malic acid and oxalic acid, play a major role in soil acidification (Jones and Brassington 1998; Jones et al. 2002). These acids are present in increased concentrations in the root cells. In few plant species, these acids are found to be present inside the vacuoles of root cells, while in others these acids are exuded into the rhizosphere.

It has been found that under various conditions of stresses, localized exudation of H^+ ions occurs to tackle the ion imbalance in the rhizosphere. Among these the most noted are shortage of iron (Fe) or phosphorus (P) or aluminium (Al)-induced toxicity in the form of Al^{3+} (Haynes 1990; Hinsinger 2001b). Al^{3+} toxicity decreases NO_3^- uptake, thus causing increased acidification of the rhizosphere. Hence nowadays, Al-resistant genotypes are being made by the scientists. These genotypes will be better suited to take up anions, thus decreasing acidity in the soil and alkalinizing the environment.

2.2 Conclusion

Thus it can be said that rhizosphere is one of richest biodiversity regions in the soil containing a variety of organisms. The very essence of rhizosphere lies at the basics of community and mutual interaction. Apart from that, there are many physical and chemical factors that play a vital role in shaping the rhizosphere into its complete structure. Most of the time, the biotic factors controlling the architecture of the rhizosphere are taken into account. But these physicochemical factors like bulk density, porosity and soil strength, light, temperature and pH also play a major role in modifying the rhizosphere architecture. Water uptake can change the water potential around the roots by affecting the microbial activities taking place in the rhizosphere as well as the radial movement of water particles. Usually soil temperature shows a positive correlation with the growth of roots. Change in pH in the rhizosphere affects the nutrient availability of the plant. Generally more roots are produced in response to increasing light intensity. With increase in CO_2 concentration, the existing C dynamics in the rhizosphere is altered resulting in an elevated C/N ratio of rhizodeposition. The random fluctuation of these factors is giving rise to loss of rhizospheric microflora, which in turn affects the plant which is in symbiotic relationship with the affected bacteria and fungi. If we want to save our planet, we need to save the trees, and for that we need to keep this rhizosphere intact. Especially with the event of global warming looming over us, the optimum values of each and every physical factor and how change of one factor can affect another are very much important. Presently it is our need to determine the range of change of these optimum values up to which a plant can tolerate, and above all it is also needed to determine how these physical factors cross-talk with each other; only then can one understand the language of rhizosphere. Still, lots of investigations need to be done to understand the proper architecture of the rhizosphere. Various short-term and long-term experiments need to be performed at the molecular level in the future to better understand the effects of the physical factors on the structure of the rhizosphere.

References

Alder NN et al (1996) Root and stem xylem embolism, stomatal conductance and leaf turgor in *Acer grandidentatum* populations along a soil moisture gradient. Oecologia 105:293–301

Allen MF, Klironomos JN, Treseder KK, Oechel WC (2005) Responses of soil biota to elevated CO2 in a chaparral ecosystem. Ecol Appl 15:1701–1711

Amellal N, Burtin G, Bartoli F, Heulin T (1998) Colonization of wheat roots by an exopolysaccharide-producing *Pantoea agglomerans* strain and its effects on rhizosphere soil aggregation. Appl Environ Microbiol 64:3740–3747

Barber SA (1995) Soil nutrient bioavailability: a mechanistic approach, 2nd edn. Wiley, New York, p 414

Barney CW (1951) Effects of soil temperature and light intensity on root growth of loblolly pine seedlings. Plant Physiol 26(1):146

BassiriRad H (2000) Kinetics of nutrient uptake by roots: responses to global change. New Phytol 147:155–169

BassiriRad H, Caldwell MM, Bilbrough C (1993) Effects of soil temperature and nitrogen status on kinetics of 15NO3 uptake by roots of field-grown Agropyron desertorum (Fisch. ex Link) Schult. New Phytol 123:485–489

Bhat KKS, Nye PH, Baldwin JP (1976) Diffusion of phosphate to plant roots in soil. IV. The concentration distance profile in the rhizosphere of roots with root hairs in a low-P soil. Plant Soil 44:63–72

Biswell HH (1935) Effect of environment upon the root habits of certain deciduous forest trees. Bot Gaz 96:676–708

Cotrufo MF, Gorissen A (1997) Elevated CO2 enhances below-ground C allocation in three perennial grass species at different levels of N availability. New Phytol 137:421–431

Crawford JW, Harris JA, Ritz K, Young IM (2005) Towards an evolutionary ecology of life in soil. Trends Ecol Evol 20:81–86

Curtis PS (1996) A meta-analysis of leaf gas exchange and nitrogen in trees grown under elevated carbon dioxide. Plant Cell Environ 19:127–137

Curtis TP, Sloan WT (2005) Exploring microbial diversity-A vast below. Science 309:1331–1333

Curtis PS, Wang XZ (1998) A meta-analysis of elevated CO2 effects on woody plant mass, form, and physiology. Oecologia 113:299–313

Curtis TP, Sloan WT, Scannell JW (2002) Estimating prokaryotic diversity and its limits. Proc Natl Acad Sci U S A 99:10494–10499

Czarnes S, Hiller S, Dexter AR, Hallett PD, Bartoli F (1999) Root:soil adhesion in the maize rhizosphere: the rheological approach. Plant Soil 211:69–86

Czarnes S, Hallett PD, Bengough AG, Young IM (2000) Root and microbial-derived mucilages affect soil structure and water transport. Eur J Soil Sci 51:435–443

Darrah PR (1993) The rhizosphere and plant nutrition: a quantitative approach. Plant Soil 155/156:1–20

Dexter AR (1987) Compression of soil around roots. Plant Soil 97:401–406

Diaz S, Grime JP, Harris J, McPherson E (1993) Evidence of a feedback mechanism limiting plant response to elevated carbon dioxide. Nature 364:616–617

Doussan C, Pagès L, Pierret A (2003) Soil exploration and resource acquisition by plant roots: an architectural and modelling point of view. Agronomie 23:419–431

Dunbabin VM, McDermott S, Bengough AG (2006) Upscaling from rhizosphere to whole root system: modelling the effects of phospholipid surfactants on water and nutrient uptake. Plant Soil 283(1–2):57–72

Ferguson JJ, Menge JA (1982) The influence of light intensity and artificially extended photoperiod upon infection and sporulation of Glomus fasciculatum on Sudan grass and on root exudation of Sudan grass. New Phytol 92(2):183–191

Gahoonia TS, Claassen N, Jungk A (1992) Mobilization of phosphate in different soils by ryegrass supplied with ammonium or nitrate. Plant Soil 140:241–248

Gamper H, Peter M, Jansa J, Luscher A, Hartwig UA, Leuchtmann A (2004) Arbuscular mycorrhizal fungi benefit from 7 years of free air CO2 enrichment in well-fertilized grass and legume monocultures. Glob Chang Biol 10:189–199

Gamper H, Hartwig UA, Leuchtmann A (2005) Mycorrhizas improve nitrogen nutrition of Trifolium repens after 8 yr of selection under elevated atmospheric CO2 partial pressure. New Phytol 167:531–542

Ge Z, Rubio G, Lynch JP (2000) The importance of root gravitropism for inter-root competition and phosphorus acquisition efficiency: results from a geometric simulation model. Plant Soil 218:159–171

Hacke U, Sauter JJ (1996) Drought-induced xylem dysfunction in petioles, branches and roots of *Populus balsamifera* and *Alnus glutinosa* (L.) Gaertn. Plant Physiol 111:413–417

Haig IT (1936) Factors controlling initial establishment of western white pine and associated species, Bulletin 41. Yale University, School Forestry, New Haven, p 41

Hartmann A, Rothballer M, Schmid M (2008) Lorenz Hiltner: a pioneer in rhizosphere microbial ecology and soil bacteriology research. Plant Soil 312:714

Hayman DS (1974) Plant growth responses to vesiculararbuscular mycorrhiza. New Phytol 73(1):71–80

Haynes RJ (1990) Active ion uptake and maintenance of cation– anion balance: a critical examination of their role in regulating rhizosphere pH. Plant Soil 126:247–264

Hendriks L, Claassen N, Jungk A (1981) Phosphatverarmung des wurzelnahen Bodens und Phosphataufnahme von Mais und Raps. Z Pflanzenern Bodenkd 144:486–499

Hiatt AJ (1967) Relationship of cell pH to organic acid change during ion uptake. Plant Physiol 42:294–298

Hiltner L (1904) Ueber neuere Erfahrungen und Probleme auf dem Gebiete der Bodenbakteriologie und unter besonderer BerUcksichtigung der Grundungung und Brache. Arb Deut Landw Gesell 98:5978

Hinsinger P (1998a) How do plant roots acquire mineral nutrients? Chemical processes involved in the rhizosphere. Adv Agron 64:225–265

Hinsinger P (1998b) Structure and function of the rhizosphere: mechanisms at the soil-root interface. Ocl-Oleaginoux Corps Gras Lipides 5(5):340–341

Hinsinger P (2001a) Bioavailability of soil inorganic P in the rhizosphere as affected by root-induced chemical changes: a review. Plant Soil 237(2):173–195

Hinsinger P (2001b) Bioavailability of trace elements as related to root-induced chemical changes in the rhizosphere. In: Gobran GR, Wenzel WW, Lombi E (eds) Trace elements in the rhizosphere. CRC Press LCC, Boca Raton, pp 25–41

Hinsinger P (2004) Nutrient availability and transport in the rhizosphere. In: Goodman RM (ed) Encyclopedia of plant and crop science. Marcel Dekker, Inc, New York, pp 1094–1097

Hinsinger P, Gobran GR, Gregory PJ, Wenzel WW (2005) Rhizosphere geometry and heterogeneity arising from root-mediated physical and chemical processes. New Phytol 168(2):293–303

Hinsinger P, Bengough AG, Vetterlein D, Young IM (2009) Rhizosphere: biophysics, biogeochemistry and ecological relevance. Plant Soil 321(1–2):117–152

Hu SJ, Firestone MK, Chapin FS (1999) Soil microbial feedbacks to atmospheric CO2 enrichment. Tree 14:433–437

Jaillard B, Plassard C, Hinsinger P (2002) Measurements of H+ fluxes and concentrations in the rhizosphere. In: Rengel Z (ed) Handbook of soil acidity. Marcel Dekker, New York

Jakobsen I, Abbott LK, Robson AD (1992) External hyphae of vesicular–arbuscular mycorrhizal fungi associated with *Trifolium subterraneum* L. 1. Spread of hyphae and phosphorus inflow into roots. New Phytol 120:371–380

Jones DL, Brassington DS (1998) Sorption of organic acids in acid soils and its implications in the rhizosphere. Eur J Soil Sci 49:447–455

Jones DL, Hinsinger P (2008) The rhizosphere: complex by design. Plant Soil 312:1–6

Jones DL, Dennis PG, Owen AG, van Hees PAW (2002) Organic acid behaviour in soils – misconceptions and knowledge gaps. Plant Soil 248:31–41

Jones DL, Hodge A, Kuzyakov Y (2004) Plant and mycorrhizal regulation of rhizodeposition. New Phytol 163:459–480

Jones DL, Nguyen C, Finlay RD (2009) Carbon flow in the rhizosphere: carbon trading at the soil-root interface. Plant Soil

Kandeler E, Tscherko D, Bardgett RD, Hobbs PJ, Lampichler C, Jones TH (2002) The response of soil microorganisms and roots to elevated CO2 and temperature in a terrestrial model ecosystem. Plant Soil 1998:251–262

King JS, Thomas RB, Strain BR (1996) Growth and carbon accumulation in root systems of *Pinus taeda and Pinus ponderosa* seedlings as affect by varying CO2, temperature and nitrogen. Tree Physiol 16:635–642

King JS, Thomas RS, Strain BR (1997) Morphology and tissue quality of seedling root systems of *Pinus taeda* and *Pinus ponderosa* as affected by varying CO2, temperature and nitrogen. Plant Soil 195:107–119

Klironomos JN, Rillig MC, Allen MF (1996) Below-ground microbial and microfaunal responses to Artemisia tridentata grown under elevated atmospheric CO2. Funct Eco 10:527–534

Kuchenbuch R, Jungk A (1982) A method for determining concentration profiles at the soil-root interface by thin slicing rhizospheric soil. Plant Soil 68:391–394

Li X-L, George E, Marschner H (1991) Extension of the phosphorus depletion zone in VA-mycorrhizal white clover in a calcareous soil. Plant Soil 136:41–48

Loosemore N, Straczek A, Hinsinger P, Jaillard B (2004) Zinc mobilization from a contaminated soil by three genotypes of tobacco as affected by soil and rhizosphere pH. Plant Soil 260:19–32

Lorenz SE, Hamon RE, McGrath SP (1994) Differences between soil solutions obtained from rhizosphere and non-rhizosphere soils by water displacement and soil centrifugation. Eur J Soil Sci 45:431–438

Lynch JM (1990) The rhizosphere. Wiley, New York

Lynch JM, Whipps JM (1990) Substrate flow in the rhizosphere. Plant Soil 129:1–10

Marschner H (1995) Mineral nutrition of higher plants, 2nd edn. Academic, London, 889 pp

McCully ME (1999) Roots in soil: unearthing the complexities of roots and their rhizospheres. Annu Rev Plant Physiol Plant Mol Biol 50:695–718

Michaud AM, Bravin MN, Galleguillos M, Hinsinger P (2007) Copper uptake and phytotoxicity as assessed in situ for durum wheat (Triticum turgidum durum L.) cultivated in Cu-contaminated, former vineyard soils. Plant Soil 298:99–111

Nye PH (1981) Changes of pH across the rhizosphere induced by roots. Plant Soil 61:7–26

Parrent JL, Morris WF, Vilgalys R (2006) CO2-enrichment and nutrient availability alter ectomycorrhizal fungal communities. Ecology 87:2278–2287

Paterson E, Hall JM, Rattray EAS, Griffiths BS, Ritz K, Killham K (1997) Effect of elevated CO2 on rhizosphere carbon flow and soil microbial processes. Glob Chang Biol 3:363–377

Pendall E, Bridgham S, Hanson PJ, Hungate B, Kicklighter DW, Johnson DW, Law BE, Luo Y, Megonigal JP, Olsrud M, Ryan MG (2004) Below-ground process responses to elevated CO2 and temperature: a discussion of observations, measurement methods, and models. New Phytol 162(2):311–322

Read DJ, Perez-Moreno J (2003) Mycorrhizas and nutrient cycling in ecosystems – a journey towards relevance. New Phytol 157:475–492

Read DB, Bengough AG, Gregory PJ, Crawford JW, Robinson D, Scrimgeour CM, Young IM, Zhang K, Zhang X (2003) Plant roots release phospholipid surfactants that modify the physical and chemical properties of soil. New Phytol 157:315–326

Rillig MC, Scow KM, Klironomos JN, Allen MF (1997) Microbial carbon-substrate utilization in the rhizosphere of Gutierrezia sarothrae grown in elevated atmospheric carbon dioxide. Soil Biol Biochem 29:1387–1394

Ryan PR, Delhaize E, Jones DL (2001) Function and mechanism of organic anion exudation from plant roots. Annu Rev Plant Physiol Plant Mol Biol 52:527–560

Sadowsky MJ, Schortemeyer M (1997) Soil microbial responses to increased concentrations of atmospheric CO2. Glob Chang Biol 3:217–224

Sanders IR, Streitwolf-Engel R, van der Heijden MGA, Boller T, Wiemken A (1998) Increased allocation to external hyphae of arbuscular mycorrhizal fungi under CO2 enrichment. Oecologia 117:496–503

Shirley HL (1936) Lethal high temperatures for conifers, and the cooling effect of transpiration. J Agric Res 53:239–258

Staddon PL, Heinemeyer A, Fitter AH (2002) Mycorrhizas and global environmental change: research at different scales. Plant Soil 244:253–261

Treseder KK, Allen MF (2000) Mycorrhizal fungi have a potential role in soil carbon storage under elevated CO2 and nitrogen deposition. New Phytol 147:189–200

Wan S, Norby RJ, Pregitzer KS, Ledford J, O'Neill EG (2004) CO2 enrichment and warming of the atmosphere enhance both productivity and mortality of maple tree fine roots. New Phytol 162

Winter AG, Meloh KA (1958) Untersuchungen über den Einfluss der endotrophen Mycorrhiza auf die Entwicklung von Zea mays L. Naturwissenschaften 45(13):319

Zak DR, Pregitzer KS, Curtis PS, Teeri JA, Fogel R, Randlett DL (1993) Elevated atmospheric CO2 and feedback between carbon and nitrogen cycles. Plant Soil 151:105–117

Zak DR, Pregitzer KS, Curtis PS, Holmes WE (2000) Atmospheric CO2 and the composition and function of soil microbial communities. Ecol Appl 10:47–59

Applications and Mechanisms of Plant Growth-Stimulating Rhizobacteria

3

Prem Chandra and Enespa Singh

Abstract

Plant growth-stimulating rhizobacteria (PGPR) are the symbiotic soil-dwelling bacteria existed at the outer part of the plant root and participate for growth and improvement of the crops. Various regulatory substances are secreted by these bacteria in the circumstances of rhizospheric regions. Normally, PGPR mechanisms simplify the growth of a plant by fixing the nitrogen from atmospheric regions, dissolved the phosphorus and other raw materials, siderophores assembly which liquefy the appropriated iron, or controlling the phytohormones levels at numerous phases of growth. When unplanned development of plant growth takes place, the activities of PGPR diminish or avoid the disastrous effect of one or more plant pathogens microbes in the form of biocontrol agents. Various researchers have been recognized to improve the fitness and proficiency of aquanaut's species of plants by using the growth-supporting rhizospheric bacteria under systematic and harassed circumstances. The advantageous rhizobacteria of the plant may reduce the comprehensive dependency on hazardous agronomic compounds which disrupt the agro-biota. This chapter emphasizes on the insight of the rhizospheric microbe which supports the growth of plant under the existing viewpoints. Conclusively, these favorable rhizospheric bacteria in various agro-biotas have been offered scientifically under normal and stress circumstances to focus on current developments with the objectives to improve forthcoming visions.

P. Chandra
Department of Environmental Microbiology, School for Environmental Sciences, Babasaheb Bhimrao Ambedkar (A Central) University, Lucknow 226025, Uttar Pradesh, India

E. Singh (✉)
Department of Botany, Dayanand P.G. College, Bachhrawan Raebareli 229301, Uttar Pradesh, India
e-mail: enespasingh@gmail.com

D.K. Choudhary et al. (eds.), *Plant-Microbe Interaction: An Approach to Sustainable Agriculture*, DOI 10.1007/978-981-10-2854-0_3

3.1 Introduction

The soil has one of the utmost essential raw materials for the existence of microbes directly or indirectly. Microbes elaborate the numerous biotic activities in the soil biota to make it energetic for nutrient turnover and biological for crop production (Chandler et al. 2008; Paul and Lade 2014). They stimulate the growth of plants through biological cycles of micro- and macronutrients in soils, producing plentiful plant growth substances, and protect the plants from plant pathogens by regulating or preventing them. They detoxify the polluted soil by sequestering contaminated heavy metals, biodegrading the xenobiotic components such as pesticides, as well as reducing the biotic or abiotic stress of plants, without deliberating the pathogenicity (Braud et al. 2009). Certainly, sufficient microbial colony is found surrounding the root of plants (rhizobacteria), which are responsible for the more adaptability in assembling, altering, and solubilizing the nutrients compared to those from bulk soils (Hayat et al. 2010). The integrated plant nutrient management system is more successful for cultivating the crop production between ecologists and agronomists using biological approaches. In this perspective, those types of rhizospheric bacteria which have original traits like heavy metal-detoxifying capabilities, salinity tolerance, pesticide removal, and biocontrol of plant pathogens and insects (Tank and Saraf 2010; Paul and Lade 2014; Mayak et al. 2004; Hynes et al. 2008; Berg et al. 2013), along with the natural plant improvement-promoting resources like plant hormones, nitrogenase activity, ammonia production, phosphate solubilization (Jahanian et al. 2012) and also essential for soil fertility, etc., should be discovered continuously at universe level by the superior assistance (Glick 2012; Bhardwaj et al. 2014). Therefore, the bio-inoculants of symbiotic and nonsymbiotic microbes are now being used globally to promote the plant growth and improvement under various stresses like heavy metals, insecticides herbicides, etc. (Berg et al. 2013; Chang et al. 2014; Oves et al. 2016). The present study in this chapter is an effort to define the mechanism and their theory of rhizospheric bacteria for plant growth improvement with recent updates. The newly updated examples of globally applicable rhizospheric bacteria in various agro-biotas have been offered to harvest the wide range of perspectives about their applicability.

3.2 Rhizosphere

The rhizospheric regions are a very modest setting where roots of connecting species and microbes compete for existence. The plant roots perform specific characters in the rhizospheric area providing mechanical supports to plant and uptake micro- and macronutrients with water content which depends on the production and releasing of metabolites (Brzostek and Finzi 2012; Feike et al. 2013). These metabolites are released by plant roots (such as oligosaccharides, α-aminoadipic acid, valeric acid, invertase, cytidine, pantothenate, etc.) and act as chemical attractants for dynamically metabolizing soil microbial populations.

Table 3.1 Various biochemical compounds secreted by plant roots of different plant species

Amino acids	Asparagines, cysteine, cystine, glycine, leucine, methionine, serine, valine, tryptophan, ornithine, histidine, arginine, α-aminoadipic acid, phenylalanine, β-alanine, α-Alanine, proline, homoserine, aspartate, glutamate, isoleucine, lysine, threonine
Organic acids	Acetic acid, pyruvic acid, malonic acid, citric acid, oxalic acid, succinic acid, butyric acid, aldonic acid, glycolic acid, malic acid, fumaric acid, aconitic acid, lactic acid, valeric acid, formic acid, glutaric acid, and tetronic acid
Sugars	Desoxyribose, raffinose, fructose, rhamnose, xylose, ribose, galactose, oligosaccharides, maltose, arabinose, and glucose
Enzymes	Invertase, amylase, protease, and acid/alkaline phosphatase
Vitamins	Riboflavin, niacin, pantothenate, thiamine, biotin
Nucleosides or purines	Cytidine, uridine, adenine, guanine
Gaseous molecules and inorganic ions	CO_2, H_2, HCO^{-3}, OH^-, H^+

Adapted from Dakora and Phillips (2002)

The biochemical released in the rhizospheric regions by plant roots known as root exudates. The chemical compounds of root change the physicochemical properties of the soil (Table 3.1) and also control the structure of microbial populations at the surface of plant root in the soil (Jung et al. 2003). These chemical compositions of the exudates are dependent upon the physiological status and the microbial and plant species (Kang et al. 2010). They also promote the plant growth by improving symbiotic interactions between plant and microbial community and inhibit the growth of antagonistic plant species (Ahemad and Kibret 2014). These microbial activities in the rhizospheric zone disturb the structure of root and the resources of available nutrients to plant species. The atmospheric carbon content is sequestered to the rhizospheric zone through root exudation by the photosynthesis (Hinsinger et al. 2009; Marschner et al. 2011). The rhizospheric zone of the soil is influenced by the biochemical components secreted by the plant roots and distress the activities of microorganisms. The soil particle strongly adheres to the root surface of rhizoplane. The root is the part of the system because various endophytic microbes have the capability to colonize in the internal part of root tissues (Barea et al. 2005). The root colonization is the zone of rhizoplane or root tissues of symbiotic microbes (Barea et al. 2005; Barros et al. 2014).

3.3 Plant Growth-Stimulating Rhizobacteria

The rhizospheric bacteria are soil-dwelling bacteria and have the capability of inhabiting at the vicinity of root environment (Kloepper 1994). Valuable root-inhabiting plant growth rhizospheric bacteria are defined on the basis of important characteristics; they have the capability to colonize the surface of the plant root, and they have the capability of proliferation and existence in microhabitats related with the superficial part of the root in comparisons with other microbes (Cleyet-Marcel et al. 2001). After sowing the seeds or propagation of plants in soils, then these bacteria promote the growth of the plants/crops directly by providing nutrients to plants or indirectly by decreasing the loss from soil-borne phytopathogens (Vessey 2003).

On the basis of biochemical compounds, rhizospheric activities classified as biofertilizers to provide the micro- and macronutrient promote the plant growth by using the plant hormones, biocontrol agents for monitoring the plant infections by production of antibiotics and antimicrobial metabolites, and rhizoremediators for degradation of organic pollutants and heavy metals (Somers et al. 2004). These substrates are capable of the production of plant hormones like gibberellic acid, indole acetic acid, ethylene, cytokinins, and symbiotic N_2 fixation (Haas and Defago 2005; Pérez-Miranda et al. 2007; Lugtenberg and Kamilova 2009; Kang et al. 2010; Laslo et al. 2012).

The rhizospheric zone has more nutrients compared to the loose soil because the root secretes various biochemical substances, like various organic acids, amino acids, sugars, and enzymes (Table 3.1), which provides the energy and micronutrient for the growth and development of microorganisms (Gray and Smith 2005). There are two types of rhizospheric bacteria that are found in this zone which is symbiotic and free-living (Khan 2005). On the basis of survival, these bacteria are divided into two groups, i.e., intracellular symbiotic bacteria (iPGPR) and extracellular free-living rhizobacteria (ePGPR). Symbiotic bacterial species like *Allorhizobium*, *Azorhizobium*, *Bradyrhizobium*, *Mesorhizobium*, and *Rhizobium* sp. reside inside root cells in specialized nodular forms (Viveros-Martinez et al. 2010; Figueiredo et al. 2011), and free-living rhizobacteria reside in the rhizospheric zone of the plant cells. The *Azotobacter*, *Arthrobacter*, *Azospirillum*, *Agrobacterium*, *Bacillus*, *Chromobacterium*, *Burkholderia*, *Caulobacter*, *Erwinia*, *Flavobacterium*, *Pseudomonas*, *Serratia*, and *Micrococcus* species of rhizobacteria are found in the rhizoplane regions or in the space between cells of the root cortex. These bacteria do not produce the nodules but still quicken the growth of plants (Gray and Smith 2005; Bhattacharyya and Jha 2012). The microbial population of rhizospheric regions shows the marvelous growth of plants in the presence of various actinomycetes (Merzaeva and Shirokikh 2006; Bhattacharyya and Jha 2012). These actinomycetes stimulate the growth of plants by constructing growth stimulators and are known as plant growth-promoting rhizobacteria. The various species of the actinomycetes, i.e., *Thermobifida*, *Streptomyces*, *Micromonospora*, and *Streptosporangium*, exhibited the potential against fungal pathogens of various roots as biocontrol agents (Franco-Correa et al. 2010; Bhattacharyya and Jha 2012).

3.4 Mechanisms of Plant Growth Promotion

Plant growth promotion increased in rhizospheric niche through the production of various biochemical substances by the entire population of microbes (Table 3.2) (Kloepper et al. 1980). Mostly, rhizospheric bacteria promote the growth of plants

Table 3.2 Plant growth-stimulating substances released by rhizospheric bacteria

PGPR	Plant growth-stimulating traits	References
Pseudomonas sp. and *Bacillus* sp.	Metal reclamation, antimicrobial activity	Oves et al. (2016)
Azotobacter sp., *Mesorhizobium* sp., *Pseudomonas* sp.	IAA, siderophore, antifungal activity, ammonia production, HCN, salt tolerant, production of harmonious solutes, plant hormones, genomic diversity, biocontrol potential	Shrivastava and Kumar (2015), Singh (2015)
Acinetobacter sp.	Production of ACC-deaminase	Chang et al. (2014)
Pseudomonas pseudoalcaligenes and *Bacillus pumilus*	Reduction of lipid peroxidation and superoxide dismutase activity	Jha and Subramanian (2014)
Pseudomonas putida, *Enterobacter cloacae*, *Serratia ficaria*, and *Pseudomonas fluorescens*	Improved sprouting rate, sprouting percentage, and index and enhanced the nutrient status	Nadeem et al. (2014)
Psychrobacter sp. SRS8, *Pseudomonas* sp. A3R3,	Heavy metal deployment, IAA, siderophores	Ma et al. (2011)
Acinetobacter sp.	IAA, phosphate solubilization, siderophores	Rokhbakhsh-Zamin et al. (2011)
Pseudomonas aeruginosa 4EA	Siderophores	Naik and Dubey (2011)
Bradyrhizobium sp. 750, *Pseudomonas* sp., *Ochrobactrum cytisi*	Heavy metal utilization	Dary et al. (2010)
Bacillus species PSB10	IAA, siderophores, HCN, ammonia production	Wani and Khan (2010)
Paenibacillus polymyxa	IAA, siderophores	Phi et al. (2010)
Stenotrophomonas maltophilia, *Rahnella aquatilis*	Nitrogenase activity, phosphate solubilization, IAA, ACC-deaminase	Mehnaz et al. (2010)
Ralstonia metallidurans	Siderophores	Braud et al. (2009)
Pseudomonas sp.	Phosphate solubilization, IAA, siderophore, HCN, biocontrol capacities	
Azospirillum amazonense	IAA, nitrogenase movement	Rodrigues et al. (2008)
Pseudomonas sp.	ACC-deaminase, IAA, siderophore	Poonguzhali et al. (2008)
Pseudomonas jessenii	ACC-deaminase, IAA, siderophore, heavy metal solubilization, phosphate solubilization	Rajkumar and Freitas (2008)
Bacillus sp., *Pseudomonas* sp., *Azotobacter* sp., *Rhizobium* sp.	IAA, ammonia production	Joseph et al. (2007)
Pseudomonas chlororaphis	Antifungal activity	Liu et al. (2007)

(continued)

Table 3.2 (continued)

PGPR	Plant growth-stimulating traits	References
Gluconacetobacter diazotrophicus	Zinc solubilization	Saravanan et al. (2007)
Xanthomonas sp. RJ3, *Azomonas* sp. RJ4, *Pseudomonas* sp. RJ10, *Bacillus* sp. RJ31	IAA	Sheng and Xia (2006)
Bacillus sp.	Phosphate solubilization	Canbolat et al. (2006)
Azotobacter chroococcum	Gibberellin, kinetin, IAA	Varma et al. (2001)

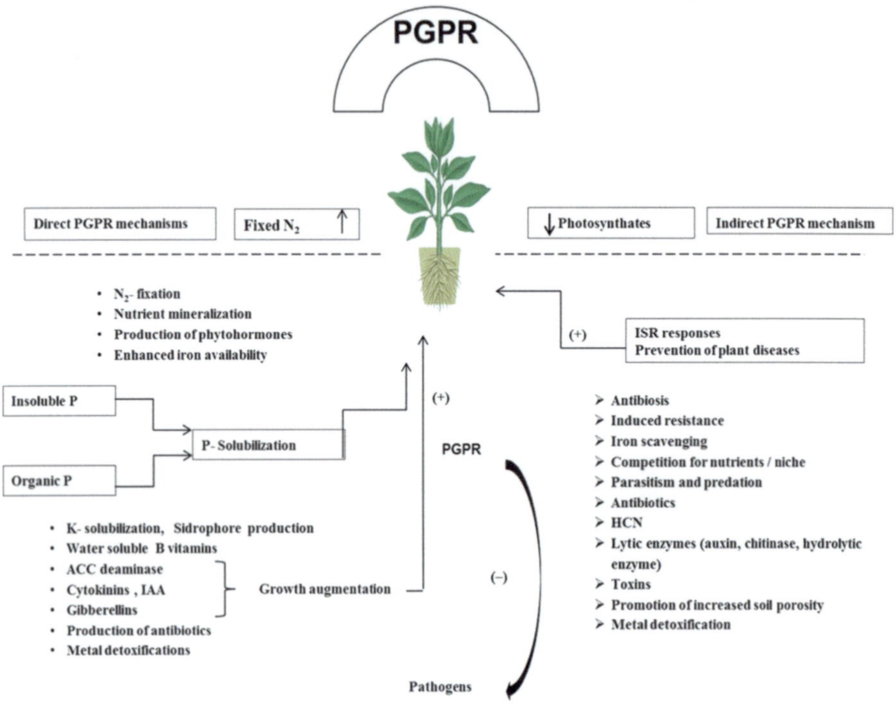

Fig. 3.1 Mechanism of plant growth promotion by rhizospheric bacteria

in a direct way by solubilizing phosphate minerals, releasing siderophores that dissolve the sequester irons, fixing the nitrogen of the atmosphere, and controlling the levels of phytohormones at various growth phases of the plant. The indirect mechanism of plant growth promotion arises when the rhizospheric bacteria reduce the injurious effect of microbial plant pathogens (Persello-Cartieaux et al. 2003; Glick 2012) (Fig. 3.1).

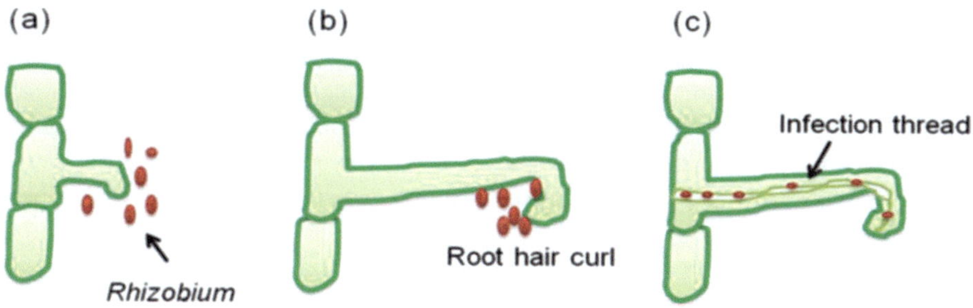

Fig. 3.2 Nodule formation method; (**a**) the interaction of bacterial-rich adhesion with host lectin; (**b**) exudation of nod factors by rhizobia causes root hair curling. (**c**) Penetration of root hair by bacteria which forms the contaminated filament and enters the cortical regions of the cells and forms bacteroid state, thereby nodules are molded (Adapted from Ahemad and Kibret (2014))

3.4.1 Direct Mechanisms

3.4.1.1 Nitrogen Fixation

Nitrogen is the essential macronutrient of plants and all vital components. These atmospheric N_2 are converted into utilizable forms to plants by biological nitrogen fixation (BNF). This amends nitrogen to ammonia by nitrogen-fixing bacteria using a nitrogenase enzyme (Kim and Rees 1994). Biological nitrogen fixation arises usually at insignificant temperatures by nitrogen-fixing bacteria, which are extensively circulated in nature (Raymond et al. 2004). Nitrogen can supply sufficient nutrients to increase production; it also leads to a global concern about environmental pollution resulting from extreme nitrate leaching (Dong et al. 2005). Its accessibility in soils may alter significantly at relatively short time intervals. Nitrogenase (*nif*) genes convoluted in the initiation of the iron–protein, Fe-Mo cofactor biosynthesis, electron transfer, and controlling genes required for the synthesis of enzymes. In both the process of symbiotic and nonsymbiotic classifications, nif genes are found (Kim and Rees 1994). Approximately two-thirds nitrogen is fixed worldwide by the BNF accounts. The remaining percent of nitrogen is synthesized industrially bases by the Haber–Bosch process (Rubio and Ludden 2008).

The nitrogen-fixing microbe generally characterized as symbiotic N_2-fixing bacteria forms a symbiotic relationship with the leguminous plants and includes the members of family Rhizobiaceae (Ahemad and Kibret 2014). And the nonleguminous plants are associated with the wider range of plants from eight families (e.g., *Frankia*) (Huss-Danell 1997; Vessey et al. 2004). The gram-negative soil-dwelling bacteria contain the unique capability to infect and establish a biological nitrogen-fixing symbiotic relationship with the roots of leguminous family. This type of symbiotic relationship forms a complex interaction between host and symbiont (Giordano and Hirsch 2004; Elmerich and Newton 2007). Finally, the nodulation takes place and the bacteria colonize as intercellular symbionts (Fig. 3.2). In the nonsymbiotic relationship, the nitrogen fixing takes place between the free-living bacteria and endophytes by *Azospirillum*, *Azotobacter*, *Azocarus*, *Gluconoacetobacter diazotrophicus*, and the cyanobacteria like *Nostoc*, *Anabaena*, etc. (Franche et al. 2009; Bhattacharyya and Jha 2012). In the nonleguminous plants, the diazotrophs fixed nitrogen and have the capability of establishing a non-obligate interaction with the host plants (Glick et al. 1999). However, diazotrophs provide only a little quantity of the fixed nitrogen that the bacterially associated host plant requires (Glick 2012). The nitrogenase enzyme which is coded by the nif gene involved in the nitrogen fixation process (Kim and Rees 1994).

The structure of nitrogenase enzyme was explained by two components, metalloenzyme which consists of dinitrogenase reductase is the iron–protein and dinitrogenase has a metallic cofactor.

The electrons transferred with high reducing influence by dinitrogenase reductase activity, while N_2 converted to NH_3 due to the production of these electrons by dinitrogenase. On the basis of the metallic cofactor, there are three different types of N-fixing systems that have been recognized which are iron-nitrogenase, vanadium nitrogenase, and Mo-nitrogenase. The existence of N_2-fixing system varies between different bacterial genera, which is carried out by the movement of the Mo-nitrogenase and found in all diazotrophs (Bishop and Jorerger 1990; Rubio and Ludden 2005; Newton 2007). The nitrogenase enzyme consisted of two metalloproteins and is purified from various sources. The first component is designated as MoFe protein and the second two known are as iron–protein (Hu et al. 2007; Newton 2007; Rubio and Ludden 2008). Mostly the compact association of *nif* genes is always defined. There are three structural genes *nifD*, *nifK*, and *nifH* that code for the Mo-nitrogenase polypeptides, for the Mo-protein subunits, and for the Fe protein, respectively. It is recognized that a core of nif genes (nifH, nifD, nifK, nifY, nifB, nifQ, nifE, nifN, nifX, nifU, nifS, nifV, nifW, nifZ) is compulsory for the synthesis of nitrogenase activity, and catalysis is preserved in all the diazotrophs. Various genes are responsible for the in vivo nitrogenase action based on the system; these code for the mechanisms of biological electron transport chains (the rnfABCDGEF cluster codes the ferredoxin, flavodoxin, and the NADH-ubiquinone oxidoreductase (NQR)) to nitrogenase, molybdenum endorsement, and homeostasis, including respiratory chains modified to oxygen situations at which the biological nitrogen fixation process can operate (Dixon and Kahn 2004; Pedrosa and Elmerich 2007; Glick 2012). Nitrogenase assemblage requires products of additional *nif* genes which are involved in the synthesis of FeMoCo (*nifB*, *nifQ*, *nifE*, *nifN*, *nifX*, *nifU*, *nifS*, *nifV*, *nifY*, also *nifH*) and also in association of iron–sulfur clusters (*nifS* and *nifU*) and the development of the nitrogenase mechanisms (*nifW and nifZ*) (Hu et al. 2007; Rubio and Ludden 2008).

3.4.1.2 Phosphate Solubilization

Phosphorus (P) is another vital macronutrient for growth and development of plant after nitrogen. It is plentifully accessible in soils in both inorganic and organic forms (Fig. 3.3) (Fernandez et al. 2007; Ahemad 2015). But the rate of absorption by plants is very slow. The less accessibility of phosphorus to plants is because,

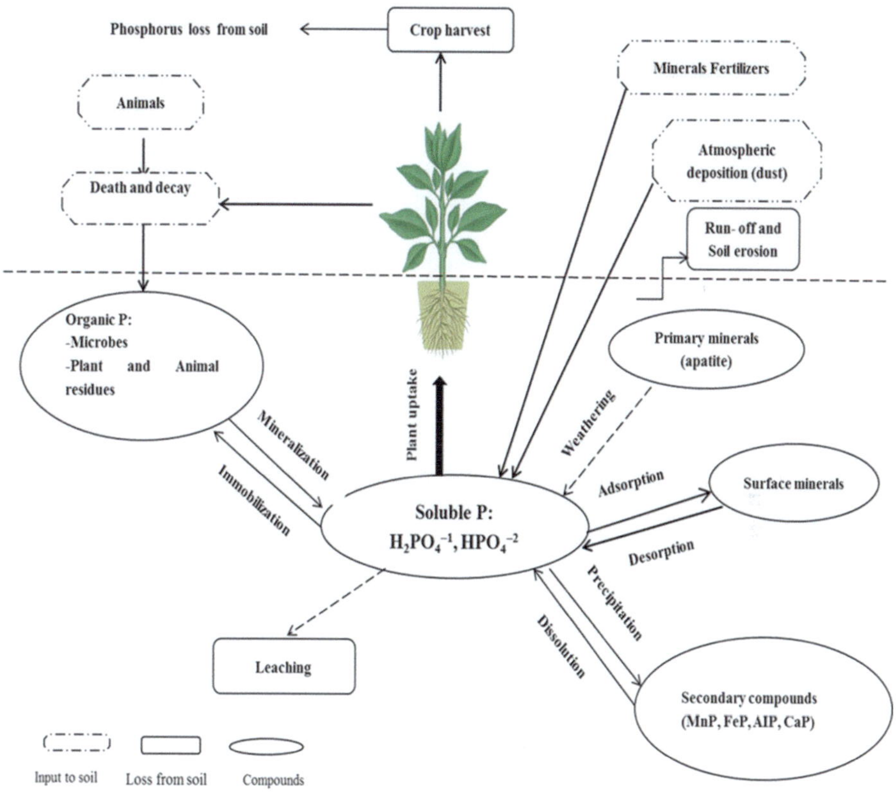

Fig. 3.3 Movement of phosphorus in soils

generally, the soil phosphorus is available in insoluble forms, while the plants uptake it in the forms of monobasic ($H_2PO_4^-$) and dibasic (HPO_4^{2-}) ions which are a soluble form of phosphorus (Bhattacharyya and Jha 2012).

The insoluble form of phosphorus is present in apatite which is an inorganic mineral or in the form of organic components such as inositol phosphate (soil phytate), phosphotriesterase, and phosphomonoesters (Glick 2012). Phosphatic fertilizers are applied for avoiding this type of shortage in soils. But a small amount of phosphorus is absorbed by the plants because most of the phosphatic fertilizers became precipitated (Mckenzie and Roberts 1990). Therefore, the regular application of phosphate fertilizer is objectionable for the ecological activity. In this context, the microbes have the capability of phosphate solubilization known as phosphate-solubilizing microorganisms. They may transfer the available forms of phosphorus to plants, therefore a sustainable substitute to chemical phosphatic fertilizers (Khan et al. 2006). The numerous phosphate-solubilizing microorganisms (PSMs) dwelling in the rhizospheric zone, considered as promising biofertilizers; subsequently they can supply the P to plants from various sources (Fig. 3.4) (Khan et al. 2006; Zaidi et al. 2009). Most substantial phosphate-solubilizing bacteria are

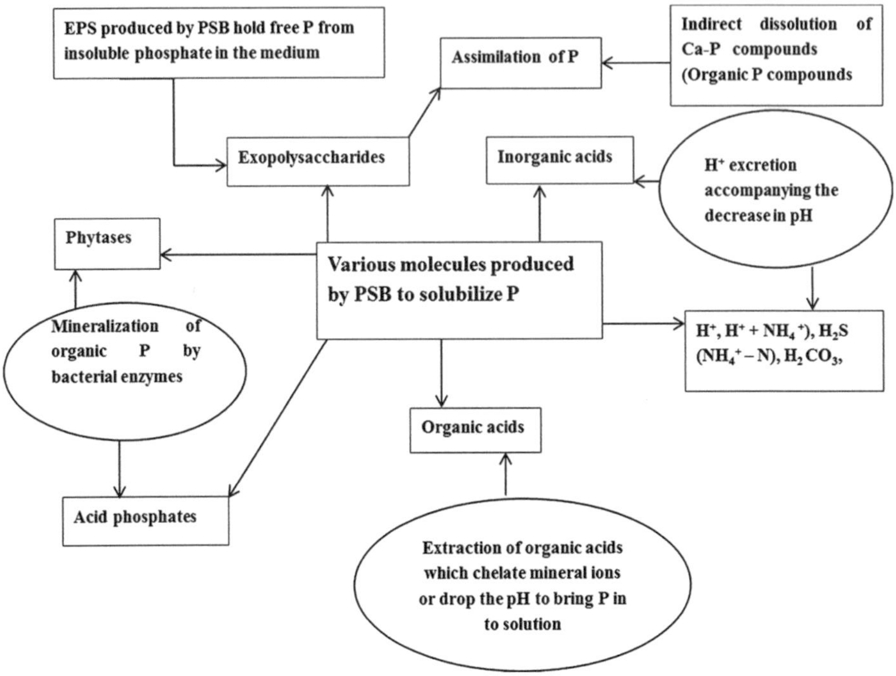

Fig. 3.4 PSB accountable for phosphate solubilization in the form of organic/inorganic substances in soils

reported such as *Azotobacter*, *Beijerinckia*, *Burkholderia*, *Bacillus*, *Enterobacter*, *Erwinia*, *Flavobacterium*, *Pseudomonas*, *Microbacterium*, *Serratia*, and *Rhizobium* (Bhattacharyya and Jha 2012). Usually, various soil-dwelling bacteria synthesized the low molecular weight organic acids which dissolved the inorganic phosphorus in soluble forms (Zaidi et al. 2009). On the other hand, the mineralization of organic phosphorus takes place by the production of various phosphatases, catalyzing the hydrolysis of phosphoric esters (Glick 2012). Significantly, the phosphate solubilization and mineralization can harmonize in the same strain of bacteria (Tao et al. 2008).

Under stress environment, the formation and performances of phosphate-solubilizing bacteria are harshly affected in soils (Ahemad and Khan 2013), but the advantageous effects of the inoculation with PSB used alone (Poonguzhali et al. 2008) or in amalgamation with other rhizospheric microbes have been reported (Zaidi and Khan 2005; Vikram and Hamzehzarghani 2008). Moreover providing the phosphorus to the plants, the phosphate-solubilizing bacteria enhance the plant growth by stimulating the efficiency of biological nitrogen fixation and augmenting the accessibility of other trace elements (Fe, Zn) and by producing essential plant growth-promoting substances (Ponmurugan and Gopi 2006; Mittal et al. 2008; Zaidi et al. 2009) (Table 3.2).

3.4.1.3 Siderophore Production

Iron is a dynamic micronutrient to plant growth, and it works as a cofactor of various enzymatic activities. It is also essential for major physiological procedures like N_2 fixation, photosynthesis, respiration, etc. (Sharma and Johri 2003). Iron is found mainly as Fe^{3+} in aerobic conditions and is prospectively to form insoluble hydroxides and oxyhydroxides. Thus it is normally inaccessible to both plants and microorganisms (Rajkumar et al. 2010). Siderophore complex iron (Fe^{3+}) is reduced to Fe^{2+} on the bacterial membrane which is further released *via* a gating mechanism into the cell from the siderophore (Ahemad and Kibret 2014). These are secreted to solubilize the iron from their circumstances, producing a complex of ferric siderophore that can transfer by diffusion and be returned to the cell surface (Beneduzi et al. 2012). These siderophores may be hydroxamates, phenol-catecholates, and carboxylates (Podile and Kishore 2006). The siderophores can be lesser peptidic particles containing various functional groups and side chains, which can deliver a high-affinity set of ligands to coordinate ferric ions (Crosa and Walsh 2002). Therefore, the siderophores work as solubilizing agents for Fe from minerals or organic compounds in limited conditions of iron (Indiragandhi et al. 2008). In the presence of heavy metals like Cr, Al, Cd, Cu, Ga, Pb, Zn, and radionuclides including U and Np, the siderophores form stable complexes that are of ecological concern (Neubauer et al. 2000; Ahemad and Kibret 2014). The bacterial siderophore sustenance to develop the strains is imposed on plants by rich soil levels of metals (Gamalero and Glick 2012). Plants absorb iron from bacterial siderophore using various mechanisms, for example, iron binding and discharge and absorption of siderophore-iron complexes directly or by a ligand interchange reaction (Ahemad and Kibret 2014).

Fluorescent *Pseudomonas* secreted yellow-green pigments and is characterized as pyoverdines which incandesce under UV light and utilized as siderophores (Agrawal et al. 2014). Additionally, the iron-pyoverdine complex production that has been detected in *P. fluorescens* C7 was taken up by *Arabidopsis thaliana* plant, inside the plant materials the iron content enhance the growth of plant (Vansuyt et al. 2007). The soil composition and various crop plants are also affected by siderophores activities which are effective pathogen-suppressive agents. Long-lasting specific suppression of *Fusarium oxysporum*-mediated wilt in flax and other susceptible crops by soil microbes (Janvier et al. 2007) and intercropping cultivation of corn and black-eyed pea against *Fusarium solani* CFF109 (Barros et al. 2014) are also examples of soil recollection. In both cases, suppression was recognized to a more diverse microbiome disturbed and sustained by diversified host disposal (Lapsansky et al. 2016). Furthermore, in the rhizosphere, the bacterial siderophores are commonly linked with biocontrol activities and not with the nutrition of plants (Vessey 2003). Latest researches confirmed the dominancy of soil-dwelling fungal pathogens through the discharge of iron-chelating siderophores by *Fluorescent pseudomonads*; adaptation is inaccessible to other microbes (Beneduzi et al. 2012).

3.4.1.4 Phytohormone Production

Plant growth-stimulating rhizobacteria (PGPR) produced plant hormones like indole acetic acid, gibberellins, cytokinins, auxins, and ethylene that can distress the cell propagation in the root way by excessive production of cross roots and hairs on root with a successive upturn of nutrients and uptake of water (Arora et al. 2013). Rhizospheric bacteria have the capability in the production of phytohormones which regulate the growth, improvement, and resistance responses of plants. Microbial production of the plant hormones such as auxin (indole-3-acetic acid/ indole acetic acid/IAA) recognized for a long time (e.g., cell division and differentiation), as well as rapid stimulation (e.g., increase in cell elongation) responses in plants (Egamberdiyeva 2007; Kaur et al 2016). It is described that eighty percent of microbes sequestered from rhizospheric regions from numerous crops have the capability to produce and discharge auxins as secondary metabolites (Patten and Glick 1996). Plant hormones such as ABA, IAA, and cytokinins were diligently linked to nitrogen signaling and provided insight that nitrogen and phytohormones signals were assimilated in order to change the morphological and physiological characters of plants (Kiba et al. 2011).

Mostly, the IAA which is secreted by rhizospheric bacteria interferes with the many plant innovative developments because the endogenous group of plant IAA may be improved by the attainment of IAA that has been secreted by soil bacteria (Spaepen et al. 2007a; Glick 2012) (Fig. 3.5). Indole acetic acid (IAA) promotes the seed and tuber propagation, with the increasing rate of xylem and root expansion; also affects the photosynthesis, pigment foundation, and biosynthesis of numerous metabolites; starts lateral and adventitious root formation; facilitates responses to light; and controls the procedures of vegetative growth, gravity, and fluorescence and resistance to hectic conditions (Spaepen and Vanderleyden 2007; Gupta et al. 2015). IAA has the capability of gene manifestation in numerous bacteria and acts as a mutual signaling molecule. The rhizobacterial IAA probably hampers the above biological processes of plants by modifying the plant's auxin pool. Similarly, the growth of root surface area and length increased by bacterial IAA, and thus it provides the plant greater access to soil nutrients (Glick 2012). Accordingly, the rhizobacterial IAA is recognized as an effective molecule in plant–microbe interactions, both in pathogenicity and plant growth promotion (Spaepen and Vanderleyden 2011; Glick 2014). Indole-3-acetamide that forms IAA through biosynthesis is stated for plant pathogenic bacteria *Pseudomonas syringae*, *Agrobacterium tumefaciens*, and *Erwinia herbicola* and saprophytic pseudomonads like *Pseudomonas putida* and *P. fluorescens*. Another amino acid like tryptophan is generally originated in root exudates and works as a precursor molecule for the biosynthesis of IAA in bacteria (Etesami et al. 2014). The independent pathway of tryptophan is more common in plants and also found in azospirilla and cyanobacteria. The biosynthesis of indole acetic acid by plant growth promoting rhizobacteria includes the production via indole-3- pyruvic acid and indole-3-acetic aldehyde, is the common apparatus in PGPRs like *Pseudomonas, Rhizobium, Bradyrhizobium, Agrobacterium, Enterobacter* and *Klebsiella* (Shilev 2013).

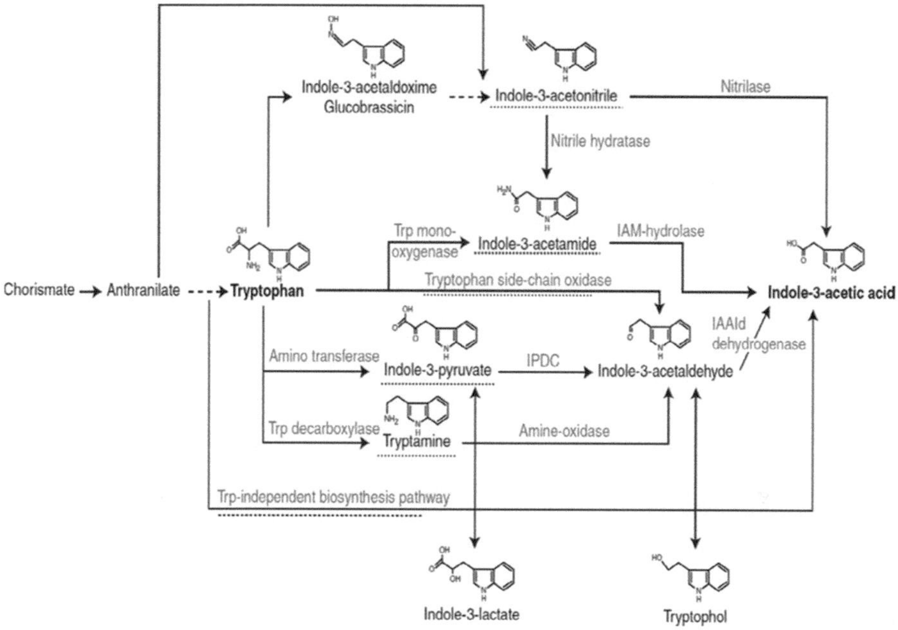

Fig. 3.5 Outline of several trails to the creation of IAA in bacteria. The intermediate mentioning to the name of the trail or the trail itself is highlighted with a dashed line. IAAld, indole-3-acetaldehyde; IAM, indole-3-acetamide; IPDC, indole-3-pyruvate decarboxylase; Trp, tryptophan (Adapted from Spaepen et al. (2007a)

3.4.1.5 1-Aminocyclopropane-1-Carboxylate (ACC) Deaminase

Ethylene has an extensive variety of biological activities that can distress the growth and improvement of plants; it is a key plant hormone (Khalid et al. 2006) and displays active role in root initiation, prevents root elongation, promotes lower wilting, motivates the seed germination, helps leaf abscission, stimulates fruit ripening, and activates the synthesis of other plant hormones (Kaur et al. 2016). It improves the growth of plants in various species such as *Arabidopsis thaliana* at a lower concentration, and generally it is known as senescence hormone because it inhibits the growth of the plant (Kaur et al. 2016). While, at higher concentrations, it encourages defoliation processes, it may reduce the production of the crop (Bhattacharyya and Jha 2012). An enzyme, 1-aminocyclopropane-1-carboxylate (ACC) deaminase, is reported in *Pseudomonas putida* bacterium and hydrolyzes ACC (precursor of ethylene) into ammonia and α- ketobutyrate (Zahir et al. 2003; Kang et al. 2010). The ACC is an essential enzyme for the production of ethylene and catalyzed by ACC oxidase. The several biochemical studies of ACC-deaminase designate that the substrate ACC is found essentially within plant tissues; the enzyme is not secreted by bacteria but is typically found in the cytoplasm (Glick 2014).

Pyridoxal 5-phosphate (vitamin B6) is an essential cofactor that is utilized by ACC-deaminase for enzymatic activity (Christen and Metzler 1985).

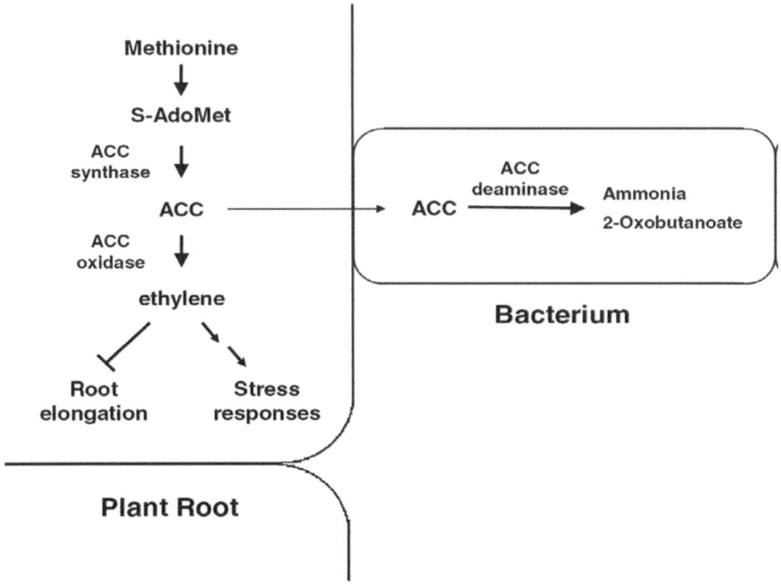

Fig. 3.6 A potential machinery of how stress controller bacteria diminish ethylene stages in the plant root by bacterial ACC-deaminase. ACC produced in plant tissues by ACC synthase is supposed to be transported from plant roots and be taken up by adjacent bacteria. After, the bacteria hydrolyze ACC to ammonia and 2-oxobutanoate. This ACC hydrolysis sustains ACC concentrations low in bacteria and permits continuous ACC relocation from plant roots to bacteria. Then, ethylene can be produced from ACC and then cause stress responses including growth inhibition. S-AdoMet: S-adenosyl-L-methionine; ACC: 1- aminocyclopropane-1-carboxylate (Adapted from Kang et al. (2010)

ACC-deaminase producers released various types of stress such as effects of plant pathogenic microbes like bacteria, viruses, and fungi and resistance to stress from salinity stress, oxidative stress, high temperature, hydrocarbons, heavy metals, radiation, wounding, insect predation, and water logging (Lugtenberg and Kamilova 2009). Consequently, ACC-deaminase-producing rhizospheric bacteria enhanced the growth of plants, rhizobial nodulation, mycorrhizal colonization, and N, P, and K uptake in various crops (Glick 2014). Several rhizospheric bacteria, for instance, *Achromobacter* sp., *Acinetobacter* sp., *Alcaligenes* sp., *Agrobacterium* sp., *Azospirillum* sp., *Burkholderia* sp., *Bacillus* sp., *Enterobacter* sp., *Ralstonia* sp., *Pseudomonas* sp., *Serratia* and *Rhizobium* sp., etc., have the capability for the production of ethylene (Zahir et al. 2010; Kang et al. 2010). These rhizospheric bacteria absorb the ethylene precursor ACC and convert it into 2-oxobutanoate and NH_3 (Arshad et al. 2007) (Fig. 3.6). The rhizospheric bacteria such as *Azotobacter* sp., *Rhizobium* sp., *Pantoea agglomerans*, *Rhodospirillum rubrum*, *Bacillus subtilis*, and *Pseudomonas fluorescens* also have the capability for the production cytokinins or gibberellins and enhanced the growth plants (Kang et al. 2010; Gupta et al. 2015).

3.4.2 Indirect Mechanisms

Biocontrol agents control the plant pathogens by microbial activity and have the eco-friendly approach (Lugtenberg and Kamilova 2009). The rhizospheric bacteria promote the plant growth by biocontrol agents through the indirect mechanisms (Glick 2012). Normally, the competition for nutrients, niche elimination, induced systemic resistance, and antifungal metabolites production are the main kinds of biocontrol activity in PGPR (Lugtenberg and Kamilova 2009). Various rhizospheric bacteria produced the antifungal metabolites such as HCN, 2,4-diacetylphloroglucinol, pyrrolnitrin, pyoluteorin, phenazines, viscosinamide, and tensin (Bhattacharyya and Jha 2012). Plant roots interaction of some rhizospheric bacteria can result in plant resistance against some pathogenic microbes such as bacteria, fungi, and viruses. It is known as induced systemic resistance (ISR) (Lugtenberg and Kamilova 2009). Moreover, the ISR mechanisms involve for ethylene signaling within the plant, and these hormones encourage the host plant's defense responses against several phytopathogens (Glick 2012). A lot of specific microbial mechanisms encourage ISR, like lipopolysaccharides (LPS), siderophores, flagella, homoserine lactones, 2, 4-diacetylphloroglucinol, cyclic lipopeptides, and volatiles like acetoin and 2,3-butanediol (Lugtenberg and Kamilova 2009).

3.4.2.1 Antibiosis

The biocontrol capabilities of *Pseudomonas* strains principally depend on antagonistic root settlement, initiation of collective conflict in plants, and manufacturing of antimicrobial antibiotics (Beneduzi et al. 2012). The antibiotic production machinery commonly linked with the rhizospheric bacteria to performance as hostile representatives against plant pathogens. There are six classes of antibiotic compounds, are better connected to the biocontrol of root diseases (Hass and Defago 2005), which are cyclic lipopeptides, pyrrolnitrin, phenazines, phloroglucinols, and pyoluteorin; all are found in the diffusible form and hydrogen cyanide (HCN) in volatile form (Beneduzi 2012; Nielsen and Sorensen 2003). Pyrrolnitrin antibiotic which is produced by *Pseudomonas fluorescens* BL915 strain has the capability to control the damage of *Rhizoctonia solani* throughout damping-off of cotton plants (Hill et al. 1994). Lipopeptide biosurfactants produced by some *Pseudomonas* and *Bacillus* species, which is used as biocontrol agent due to their possible progressive impact on functional associations with microorganisms like fungi, oomycetes, protozoa, nematodes, bacteria, and some plant species (Raaijmakers et al. 1999; de Bruijn et al. 2007). Pseudomonads also produced an effective and extensively studied antibiotic known as 2, 4-diacetylphloroglucinol (DAPG), which damages membrane to *Pythium* sp., produces the hindrance for zoospores formation in the oomycete (Bhattacharyya and Jha 2012), and acts as a biocontrol of bacterial canker of tomato (Lanteigne et al. 2012). Some *Pseudomonas* sp. also produced the phenazine, which possesses the redox activity, and it destroys the pathogenicity of plants, for instance, *F. oxysporum* and *Gaeumannomyces graminis* (Chin-A-Woeng et al. 2003). Phenazine −1-carboxamide is synthesized by *P. chlororaphis* PCL1391 bacterium isolated from roots of tomato plants, which has the capability to discharge the solvable form of iron from ferric oxides at neutral pH and contributes to iron utilization in soils (Hernandez et al. 2004; Haas and Defago 2005). Circulin, colistin, and polymyxin antibiotics are vibrant against phytopathogens formed by the popular *Bacillus* sp. (Maksimov et al. 2011). Excessively, use of antibiotic-producing rhizobacteria as biocontrol agents for plant growth promotion, various plant pathogens produced the induced systemic resistance (ISR) mechanism for particular antibiotics due to the augmented habit of these isolates. For the avoidance of this type of popularity, various researchers exploited those biocontrol isolates that produce more than one antibiotics (Lugtenberg and Kamilova 2009; Glick et al. 1999).

3.5 Applications of PGPR as Multifunctional Agents

Commercial application of PGPRs (inoculants) in crops to enhance the growth of plant and increase productivity or induce systemic resistance against pathogens or bring about mitigation of stress tolerance in any environments as prospective plans used for ecological farming. The PGPR affects the farming which differs from the laboratory, glasshouse, and field trials, because the soil has a volatile background and an envisioned product is sometimes difficult to attain. PGPR activity is very much affected when the climate fluctuations take place. But occasionally poor growth conditions in the field are to be predicted as the usual functioning of farming (Zaidi et al. 2009). Plant growth-supporting characters do not work freely from each other. But additionally, as it was recommended in the "additive hypothesis," that various tools, for instance, nitrogen fixation, siderophore biosynthesis, phosphate solubilization, IAA, ACC-deaminase, and antifungal activity are accountable for the promotion plant growth and improve the productivity. It is observed that after using the rhizospheric bacteria, the productivity of various crops increased in both organized forms of soil as well as the natural environment (Table 3.3). Due to the prevailing lack of enthusiasm worldwide to hold foods produced by genetically modified plants, advantageous as a means of promoting plant growth globally, it reduces the requirement of agrochemicals products. Moreover, PGPR machinery is freely available to planters globally (Gamalero et al. 2009).

3.6 Conclusion

A perfect ecological farming system as PGPR is one which protects the environment, increases the productivity of cereals, provides food to the whole inhabitants universally, and maintains and improves the health problem of human being. These microorganisms are found in rhizospheric regions, which includes rhizobacteria, and symbiotic fungus species; having the capability to bioremediation potentials of heavy metal, biodegradation of xenobiotic compounds regulates the wide range of plant pathogens as biocontrol and provides the phytohormones as well as nutrients. After this, they can be used as biofertilizers, biocontrol agents without any harmful effect to the environment and sustainable forming.

Table 3.3 The examples of several plant growth-stimulating rhizobacteria tested for crop types

Plant growth-promoting bacterial species	Crop plant	Conditions	Results of addition of bacteria to plants	References
Pseudomonas aeruginosa, Bacillus sp., *E. faecalis*	Cucumber, chickpea	In vitro and field experiments	Have the capability to control the bacterial and fungal root pathogens and improve the growth of plants significantly	Oves et al. (2016)
Bacillus sp., *Pseudomonas* sp., *Azotobacter* sp., *Mesorhizobium* sp.	Wheat, rice, maize, pea, green gram, peanut, potato,	In vitro conditions, pot experiments, field study,	The quantity of N, P, Fe, and Mn is increased by the bacterial inoculum in wheat shoots grown in normal and saline soil and enhanced the growth of pepper, canola, bean, tomato, and lettuce under salinity stress	Shrivastava and Kumar (2015) and Singh (2015)
Streptomyces sp. strain PGPA39	"Micro-Tom" tomato	Pot experiments	Improves the salinity and stimulates growth of "Micro-Tom" tomato plants	Palaniyandi et al. (2014)
Bacillus pumilis, Pseudomonas sp., and *Acinetobacter* sp.	Barley and oats	Field experiments	Assist the growth of barley and oats in salt affected soil	Chang et al. (2014)
Pseudomonas pseudoalcaligenes	Salt-sensitive rice GJ-17	Field experiments	Decrease superoxide dismutase activity and lipid peroxidation	Jha and Subramanian (2014)
Pseudomonas fluorescens, Pseudomonas putida, Enterobacter cloacae, Serratia ficaria	Wheat	Pot experiments, field experiments, and in vitro experiments	Increase the growth rate, growth percentage, index and status of nutrient in wheat crops	Nadeem et al. (2014)
Pseudomonas and *Rhizobium*	Mung bean (*Vigna radiata* L.)	In saline condition	Increase the osmotic stress tolerance	Ahmad et al. (2013)
Rhizobium phaseoli	*Vigna radiata* L.	Pots experiments	*Rhizobium* alleviated the adverse effects of salinity in the presence of tryptophan and improved the growth, biomass and grain yield, quantity of nodules, and grain concentration of nitrogen per plant significantly	Zahir et al. (2010)

(continued)

Table 3.3 (continued)

Plant growth-promoting bacterial species	Crop plant	Conditions	Results of addition of bacteria to plants	References
Paenibacillus polymyxa	Pepper	Gnotobiotic conditions	Improved the plant biomass considerably and stimulated induced systemic resistance against bacterial spot pathogen like *Xanthomonas axonopodis* pv. Vesicatoria untreated plants	Phi et al. (2010)
Pseudomonas putida strain R-168, *Pseudomonas fluorescens* DSM 50090, *Azospirillum brasilense* DSM 1690	Maize (*Zea mays* L.)	Fields experiments	Improved the growth of leaf area, dry weight of shoot, height, and seed weight and seed quantity per year	Gholami et al. (2009)
Ralstonia metallidurans and *Pseudomonas aeruginosa*	Maize	Pots experiments	Encouraged the plant growth, assisted soil metal utilization, and enhanced Cr and Pb uptake	Braud et al. (2009)
Pseudomonas sp.	Chickpea	Pots experiments	Upgraded the dry and fresh biomass of crops at 2 mM concentration of Ni	Tank and Saraf (2010)
Azospirillum amazonense	Rice (*Oryza sativa* L.)	Greenhouse	Grain dry matter accumulation (7–11.6 %), number of panicles (3–18.6 %), and nitrogen accumulation at grain maturation (3.5–18.5 %) increased	Rodrigues et al. (2008)
Pseudomonas sp.	Rice (*Oryza sativa*), maize (*Zea mays* L.)	In vitro experiments	Have the proficiency to control the phytopathogens which are obtained from maize crops	Lawongsa et al. (2008)
Pseudomonas aeruginosa strain MKRh3	Black gram	Pots experiments	Reduction of cadmium accumulation observed in plants, extensive rooting, and improved growth of plants	Ganesan (2008)
Azospirillum brasilense Sp245	Common bean (*Phaseolus vulgaris* L.)	Greenhouse experiment	Root development increased	Remans et al. (2008)

Bacillus sp. *Paenibacillus* sp.	Rice	Pots experiments	Shoot and root growth significantly induced	Beneduzi et al. (2008)
Pseudomonas tolaasii ACC23, *Alcaligenes* sp. ZN4, *Mycobacterium* sp. ACC14	*Brassica napus*	Pots experiments	Protected canola plant against the inhibitory effects of cadmium	Dell'Amico et al. (2008)
Bacillus sp.	Barley (*Hordeum vulgare*)	Greenhouse experiments	Shoot and root weight increased up to 34.7 % and 16.7 %, respectively	Canbolat et al. (2006)
Sinorhizobium sp. Pb002	*Brassica juncea*	Microcosms	Phytoextraction of lead efficiency increased by *Brassica juncea*	Di Gregorio et al. (2006)
Xanthomonas sp. RJ3, *Pseudomonas* sp. RJ10, *Bacillus* sp. RJ31	*Brassica napus*	Pots experiments	Increased cadmium accumulation with plant growth	Sheng and Xia (2006)
Pseudomonas jessenii PS06, *Mesorhizobium ciceri* C-2/2	*Cicer arietinum* (chickpea)	Greenhouse, fields	The co-inoculation treatment increased the seed yield (52 % greater than the uninoculated control treatment) and nodule fresh weight	Valverde et al. (2006)
Azotobacter chroococcum HKN-5, *Bacillus megaterium* HKP-1	*Brassica Juncea*	Greenhouse experiments	Plant protected from metal toxicity, enhanced plant growth	Wu et al. (2006)
Variovox paradoxus, *Rhodococcus* sp.,	*Brassica juncea*	In vitro experiments	Root elongation increased	Belimov et al. (2005)
Pseudomonas fluorescens PGPR1, PGPR2, PGPR4	Peanut (*Arachis hypogaea* L.)	Pots and fields experiments	Pod yield increased significantly, nodule dry weight and halum yield over the control	Dey et al. (2004)
Pseudomonas fluorescens Avm, *Rhizobium leguminosarum* bv *phaseoli* CPMex46	Alfalfa	Growth chamber	Cu and Fe translocation improved from root to shoot	Carrillo-Castaneda et al. (2003)
Enterobacter sakazakii 8MR5, *Pseudomonas* sp. 4MKS8, *Klebsiella oxytoca* 10MKR7	*Zea mays* L. (maize)	Pot experiments	Growth parameters increased by the inoculation	Babalola et al. (2003)

References

Agrawal PK, Agrawal S, Kundan R, Bhatt M (2014) Application and perspective of plant growth promoting rhizobacteria in the development of sustainable agriculture. Int J Curr Res 6:9044–9051

Ahemad M (2015) Phosphate-solubilizing bacteria-assisted phytoremediation of metalliferous soils: a review. Biotech 5:111–121

Ahemad M, Khan MS (2013) Pesticides as antagonists of rhizobia and the legume-*Rhizobium* symbiosis: a paradigmatic and mechanistic outlook. Biochem Mol Biol 1:63–75

Ahemad M, Kibret M (2014) Mechanisms and applications of plant growth promoting rhizobacteria: Current perspective. J King Saud Univ Sci 26:1–20

Ahmad M, Zahir A, Nazli F, Akram F, Arshad M, Khalid M (2013) The effectiveness of halotolerant, auxin producing *Pseudomonas* and *Rhizobium* strains to improve osmotic stress tolerance in mung bean (*Vigna radiata* L.). Braz J Microbiol 44(4):1341–1348

Arora NK, Tewari S, Singh S (2013) Multifaceted plant-associated microbes and their mechanisms diminish the concept of direct and indirect PGPRs. In: Arora NK (ed) Plant-microbe symbiosis: fundamentals and advances. Springer, New Delhi, pp 411–449

Arshad M, Saleem M, Hussain S (2007) Perspectives of bacterial ACC deaminase in phytoremediation. Trends Biotechnol 25:356–362

Babalola OO, Osir EO, Sanni A, Odhaimbo GD, Bulimo WD (2003) Amplification of 1-aminocy clopropane-1-carboxylic (ACC) deaminase from plant growth promoting rhizobacteria in Striga-infested soils. Afr J Biotechnol 2:157–160

Barea JM, Pozo MJ, Azcon R, Aguilar CA (2005) Microbial co-operation in the rhizosphere. J Exp Bot 56:1761–1778

Barros JA, Medeiros EV, Notaro KA, Moraes WS, Silva JM, Nascimento TCES, Moreira KA (2014) Different cover promote sandy soil suppressiveness to root rot disease of cassava caused by *Fusarium solani*. Afr J Microbiol Res 8:967–973

Belimov AA, Hontzeas N, Safronova VI, Demchinskaya SV, Piluzza G, Bullitta S, Glick BR (2005) Cadmium-tolerant plant growth promoting rhizobacteria associated with the roots of Indian mustard (*Brassica juncea* L. Czern.). Soil Biol Biochem 37:241–250

Beneduzi A, Peres D, Costa PBD, Zanettini MHB, Passaglia LMP (2008) Genitic and phenotypic diversity of plant-growth promoting bacilli isolated from wheae fields in Southern Brazil. Res Microbiol 159:244–250

Beneduzi A, Ambrosini A, Passaglia LMP (2012) Plant growth-promoting rhizobacteria (PGPR): their potential as antagonists and biocontrol agents. Genet Mol Biol 35(4):1044–1051

Berg G, Zachow C, Müller H, Phillips J, Tilcher R (2013) Next-generation bio-products sowing the seeds of success for sustainable agriculture. Agronomy 3:648–656

Bhardwaj D, Ansari MW, Sahoo RK, Tuteja N (2014) Biofertilizers function as a key player in sustainable agriculture by improving soil fertility, plant tolerance, and crop productivity. Microb Cell Factories 13:66

Bhattacharyya PN, Jha DK (2012) Plant growth-promoting rhizobacteria (PGPR): the emergence of agriculture. World J Microbiol Biotechnol 28:1327–1350

Bishop PE, Jorerger RD (1990) Genetics and molecular biology of an alternative nitrogen fixation system. Plant Mol Biol 41:109–125

Braud A, Jézéquel K, Bazot S, Lebeau T (2009) Enhanced phytoextraction of an agricultural Cr-, Hg- and Pb-contaminated soil by bioaugmentation with siderophores producing bacteria. Chemosphere 74:280–286

Brzostek ER, Finzi AC (2012) Seasonal variation in the temperature sensitivity of proteolytic enzyme activity in temperate forest soils. J Geophys Res Biogeosci 117:G01018

Canbolat MY, Bilen SC, Çakmakc R, Aydın A (2006) Effect of plant growth-promoting bacteria and soil compaction on barley seedling growth, nutrient uptake, soil properties and rhizosphere microflora. Biol Fertil Soils 42:350–357

Carrillo-Castaneda G, Munoz JJ, Peralta-Videa JR, Gomez E, Gardea-Torresdey JL (2003) Plant growth-promoting bacteria promote copper and iron translocation from root to shoot in alfalfa seedlings. J Plant Nutr 26:1801–1814

Chandler D, Davidson G, Grant WP, Greaves J, Tatchell GM (2008) Microbial biopesticides for integrated crop management: an assessment of environmental and regulatory sustainability. Trends Food Sci Technol 19:275–283

Chang P, Gerhardt KE, Yu HX-D, Xiao-Ming GBR, Gerwing PD, Greenberg BM (2014) Plant growth promoting bacteria facilitate the growth of barley and oats in salt impacted soil: implications for phytoremediation of saline soils. Int J Phytorem 16(11):1133–1147

Chin-A-Woeng TF, Bloemberg GV, Lugtenberg BJ (2003) Phenazines and their role in biocontrol by *Pseudomonas* bacteria. New Phytol 157:503–523

Christen P, Metzler DE (1985) Transaminases. Wiley, New York

Cleyet-Marcel JC, Larcher M, Bertrand H, Rapior S, Pinochet X (2001) Plant growth enhancement by rhizobacteria. In: MorotGaudry J-F (ed) Nitrogen assimilation by plants: physiological, biochemical, and molecular aspects. Science Publishers Inc., Plymouth, pp 185–197

Crosa JH, Walsh CT (2002) Genetics and assembly line enzymology of siderophore biosynthesis in bacteria. Microbiol Mol Biol Rev 66:223–249

Dakora FD, Phillips DA (2002) Root exudates as mediators of mineral acquisition in low-nutrient environments. Plant Soil 245:35–47

Dary M, Chamber-Pérez MA, Palomares AJ, Pajuelo E (2010) *In situ* phytostabilisation of heavy metal polluted soils using *Lupinus luteus* inoculated with metal resistant plant-growth promoting rhizobacteria. J Hazard Mater 177:323–330

de Bruijn I, de Kock MJD, Yang M, de Waard P, van Beek TA, Raaijmakers JM (2007) Genome-based discovery, structure prediction and functional analysis of cyclic lipopeptide antibiotics in *Pseudomonas* species. Mol Microbiol 63:417–428

Dell'Amico E, Cavalca L, Andreoni V (2008) Improvement of *Brassica napus* growth under cadmium stress by cadmium resistant rhizobacteria. Soil Biol Biochem 40:74–84

Dey R, Pal KK, Bhatt DM, Chauhan SM (2004) Growth promotion and yield enhancement of peanut (*Arachis hypogaea* L.) by application of plant growth-promoting rhizobacteria. Microbiol Res 159:371–394

Di Gregorio S, Barbafieri M, Lampis S, Sanangelantoni AM, Tassi E, Vallini G (2006) Combined application of Triton X-100 and *Sinorhizobium* sp. Pb002 inoculum for the improvement of lead phytoextraction by *Brassica juncea* in EDTA amended the soil. Chemosphere 63:293–299

Dixon R, Kahn D (2004) Genetic regulation of biological nitrogen fixation. Nat Rev Microbiol 2:621–631

Dong S, Neilsen D, Neilsen GH, Fuchigami LH (2005) Foliar N application reduces soil NO_3–N leaching loss in apple orchards. Plant Soil 268:357–366

Egamberdiyeva D (2007) The effect of plant growth promoting bacteria on growth and nutrient uptake of maize in two different soils. Appl Ecol 36:184–189

Elmerich C, Newton WE (2007) Associative and endophytic nitrogen-fixing bacteria and cyanobacterial associations. Springer, Dordrecht

Etesami H, Hosseini HM, Alikhani HA, Mohammadi L (2014) Bacterial biosynthesis of 1-Aminocyclopropane-1-Carboxylate (ACC) deaminase and Indole-3-Acetic Acid (IAA) as endophytic preferential selection traits by rice plant seedlings. J Plant Growth Regul 33(3):654–670

Feike A, Dijkstra YC, Elise P, Jack AM (2013) Rhizosphere priming: a nutrient perspective. Front Microbiol 4(216):1–8

Fernandez LA, Zalba P, Gromez MA, Sagardoy MA (2007) Phosphate-solubilization activity of bacterial strains in soil and their effect on soybean growth under greenhouse conditions. Biol Fertil Soil 43:803–805

Figueiredo MVB, Seldin L, Araujo FF, Mariano RLR (2011) Plant growth promoting rhizobacteria: fundamentals and applications. In: Maheshwari DK (ed) Plant growth and health promoting bacteria. Springer, Berlin/Heidelberg, pp 21–42

Franche C, Lindstrom K, Elmerich C (2009) Nitrogen fixing bacteria associated with leguminous and non-leguminous plants. Plan Soil 321:35–59

Franco-Correa M, Quintana A, Duque C, Suarez C, Rodríguez MX, Barea JM (2010) Evaluation of actinomycete strains for key traits related to plant growth promotion and mycorrhiza helping activities. Appl Soil Ecol 45:209–217

Gamalero E, Glick BR (2012) Plant growth-promoting bacteria, and metals phytoremediation. In: Anjum NA, Pereira ME, Ahmad I, Duarte AC, Umar S, Khan NA (eds) Phytotechnologies: remediation of environmental contaminants. CRC Press, Boca Raton, pp 361–376

Gamalero E, Berta G, Glick BR (2009) The use of microorganisms to facilitate the growth of plants in saline soils. In: Musarrat J, Khan MS, Zaidi A (eds) Microbial strategies for crop improvement. Springer, Berlin/Heidelberg

Ganesan V (2008) Rhizoremediation of cadmium soil using a cadmium-resistant plant growth-promoting rhizopseudomonad. Curr Microbiol 56:403–407

Gholami A, Shahsavani S, Nezarat S (2009) The effect of plant growth promoting rhizobacteria (PGPR) on germination, seedling growth, and yield of maize. Int J Biol Life Sci 1:35–40

Giordano W, Hirsch AM (2004) The expression of MaEXP1, a *Melilotus alba* expansin gene, is upregulated during the sweet clover-*Sinorhizobium meliloti* interaction. MPMI 17:613–622

Glick BR (2012) Plant growth-promoting bacteria: mechanisms and applications. Hindawi Publishing Corporation/Scientifica, Waterloo

Glick BR (2014) Bacteria with ACC deaminase can promote plant growth and help to feed the world. Microbiol Res 169:30–39

Glick BR, Patten CL, Holguin G, Penrose GM (1999) Biochemical and genetic mechanisms used by plant growth promoting bacteria. Imperial College Press, London

Gray EJ, Smith DL (2005) Intracellular and extracellular PGPR: commonalities and distinctions in the plant–bacterium signaling processes. Soil Biol Biochem 37:395–412

Gupta G, Parihar SS, Ahirwar NK, Snehi SK, Singh V (2015) Plant growth promoting rhizobacteria (PGPR): current and future prospects for development of sustainable agriculture. J Microb Biochem Technol 7:096–102

Haas D, Defago G (2005) Biological control of soil-borne pathogens by fluorescent pseudomonads. Nat Rev Microbiol 3:307–319

Hayat R, Ali S, Amara U, Khalid R, Ahmed I (2010) Soil beneficial bacteria and their role in plant growth promotion: a review. Ann Microbiol 60:579–598

Hernandez F, Jesus A, Jose JL, Mar P (2004) Role of tau protein in both physiological and pathological conditions. Physiol Rev 84:361–384

Hill DS, Stein JI, Torkewitz NR, Morse AM, Howell CR, Pachlatko JP, Becker JO, Ligon JM (1994) Cloning of genes involved in the synthesis of pyrrolnitrin from *Pseudomonas fluorescens* and role of pyrrolnitrin synthesis in biological control of plant disease. Appl Environ Microbiol 60:78–85

Hinsinger P, Bengough AG, Vetterlein D, Young IM (2009) Rhizosphere: biophysics, biogeochemistry, and ecological relevance. Plant Soil 321:117–152

Hu Y, Fay AW, Lee CC, Ribbe MW (2007) P-cluster maturation on nitrogenase MoFe protein. Proc Natl Acad Sci U S A 104:10424–10429

Huss-Danell K (1997) Actinorhizal symbioses and their N_2-fixation. Tansley review no. 93. New Phytol 136:375–405

Hynes RK, Leung GC, Hirkala DL, Nelson LM (2008) Isolation, selection, and characterization of beneficial rhizobacteria from pea, lentil, and chickpea are grown in Western Canada. Can J Microbiol 54:248–258

Indiragandhi P, Anandham R, Madhaiyan M, Sa TM (2008) Characterization of plant growth-promoting traits of bacteria isolated from larval guts of diamondback moth *Plutella xylostella* (Lepidoptera: Plutellidae). Curr Microbiol 56:327–333

Jahanian A, Chaichi MR, Rezaei K, Rezayazdi K, Khavazi K (2012) The effect of plant growth promoting rhizobacteria (pgpr) on germination and primary growth of artichoke (*Cynara scolymus*). Int J Agric Crop Sci 4:923–929

Janvier C, Villeneuve F, Alabouvette C, Edel-Hermann V, Mateille T, Steinberg C (2007) Soil health through soil disease suppression: which strategy from descriptors to indicators. Soil Biol Biochem 39:1–23

Jha Y, Subramanian RB (2014) PGPR regulate caspase-like activity, programmed cell death, and antioxidant enzyme activity in paddy under salinity. Physiol Mol Biol Plants 20(2):201–207

Joseph B, Patra RR, Lawrence R (2007) Characterization of plant growth promoting rhizobacteria associated with chickpea (*Cicer arietinum* L.). Int J Plant Prod 2:141–152

Jung C, Maeder V, Funk F, Frey B, Sticher H, Frossard E (2003) The release of phenols from *Lupinus albus* L. roots exposed to Cu and their possible role in Cu detoxification. Plant Soil 252:301–312

Kang BG, Kim WT, Yun HS, Chang SC (2010) Use of plant growth-promoting rhizobacteria to control stress responses of plant roots. Plant Biotechnol Rep 4:179–183

Kaur H, Kaur J, Gera R (2016) Plant growth promoting rhizobacteria: a boon to agriculture. Int J Cell Science Biotechnol 5:17–22

Khalid A, Akhtar MJ, Mahmood MH, Arshad M (2006) Effect of substrate-dependent microbial ethylene production on plant growth. Microbiology 75:231–236

Khan AG (2005) Role of soil microbes in the rhizospheres of plants growing on trace metal contaminated soils in phytoremediation. J Trace Elem Med Biol 18:355–364

Khan MS, Zaidi A, Wani PA (2006) Role of phosphate solubilizing microorganisms in sustainable agriculture – a review. Agron Sustain Dev 27:29–43

Kiba T, Kudo T, Kojim M, Sakakibara H (2011) Hormonal control of nitrogen acquisition: roles of auxin, abscisic acid, and cytokinin. J Exp Bot 62:1399–1409

Kim J, Rees DC (1994) Nitrogenase and biological nitrogen fixation. Biochemistry 33:389–397

Kloepper JW (1994) Plant growth-promoting rhizobacteria (other systems). In: Okon Y (ed) *Azospirillum*/Plant associations. CRC Press, Boca Raton, pp 111–118

Kloepper JW, Schroth MN, Miller TD (1980) Potato rhizosphere colonization by plant growth-promoting rhizobacteria increases plant development and yield. Phytopathology 70:1078–1082

Lanteigne C, Gadkar VJ, Wallon T, Novinscak A, Filion M (2012) Production of DAPG and HCN by *Pseudomonas* sp. LBUM300 contributes to the biological control of bacterial canker of tomato. Phytopathology 102:967–973

Lapsansky ER, Milroy AM, Andales MJ, Vivanco JM (2016) Soil memory as a potential mechanism for encouraging sustainable plant health and productivity. Curr Opin Biotechnol 38:137–142

Laslo É, György É, Mara G, Tamás E, Ábrahám B, Lányi S (2012) Screening of plant growth promoting rhizobacteria as potential microbial inoculants. Crop Prot 40:43–48

Lawongsa P, Boonkerd N, Wongkaew S, O'Gara F, Teaumroong N (2008) Molecular and phenotypic characterization of potential plant growth-promoting Pseudomonas from rice and maize rhizospheres. World J Microbiol Biotechnol 24:1877–1884

Liu H, He Y, Jiang H, Peng H, Huang X, Zhang X, Thomashow LS, Xu Y (2007) Characterization of a phenazine-producing strain *Pseudomonas chlororaphis* GP72 with broad-spectrum antifungal activity from green pepper rhizosphere. Curr Microbiol 54:302–306

Lugtenberg B, Kamilova F (2009) Plant–growth–promoting rhizobacteria. Annu Rev Microbiol 63:541–556

Ma Y, Rajkumar M, Vicente JA, Freitas H (2011) Inoculation of Ni-resistant plant growth promoting bacterium *Psychrobacter* sp. strain SRS8 for the improvement of nickel phytoextraction by energy crops. Int J Phytorem 13:126–139

Maksimov IV, Abizgil'dina RR, Pusenkova LI (2011) Plant growth promoting rhizobacteria as alternative to chemical crop protectors from pathogens (Review). Appl Biochem Microbiol 47:333–345

Marschner P, Crowley D, Rengel Z (2011) Rhizosphere interactions between microorganisms and plants govern iron and phosphorus acquisition along the root axis–model and research methods. Soil Biol Biochem 43:883–894

Mayak S, Tirosh T, Glick BR (2004) Plant growth-promoting bacteria confer resistance in tomato plants to salt stress. Plant Physiol Biochem 42:565–572

McKenzie RH, Roberts TL (1990) Soil, and fertilizers phosphorus update. In: Proceedings of Alberta soil science workshop proceedings, February, 20–22, Edmonton, pp 84–104

Mehnaz S, Baig DN, Lazarovits G (2010) Genetic and phenotypic diversity of plant growth promoting rhizobacteria isolated from sugarcane plants growing in Pakistan. J Microbiol Biotechnol 20:1614–1623

Merzaeva OV, Shirokikh IG (2006) Colonization of plant rhizosphere by actinomycetes of different genera. Microbiology 75:226–230

Mittal V, Singh O, Nayyar H, Kaur J, Tewari R (2008) Stimulatory effect of phosphate-solubilizing fungal strains (*Aspergillus awamori* and *Penicillium citrinum*) on the yield of chickpea (*Cicer arietinum* L. cv. GPF2). Soil Biol Biochem 40:718–727

Nadeem SM, Ahmad M, Zahir ZA, Javaid A, Ashraf M (2014) The role of mycorrhizae and plant growth promoting rhizobacteria (PGPR) in improving crop productivity under stressful environments. Biotechnol Adv 32:429–448

Naik MM, Dubey SK (2011) Lead-enhanced siderophore production and alteration in cell morphology in a Pb-resistant *Pseudomonas aeruginosa* strain 4EA. Curr Microbiol 62:409–14

Neubauer U, Furrer G, Kayser A, Schulin R (2000) Siderophores, NTA, and citrate: potential soil amendments to enhance heavy metal mobility in phytoremediation. Int J Phytorem 2:353–368

Newton WE (2007) Physiology, biochemistry and molecular biology of nitrogen fixation. In: Bothe H, Ferguson SJ, Newton WE (eds) Biology of the nitrogen cycle. Elsevier, Amsterdam, pp 109–130

Nielsen TH, Sørensen J (2003) Production of cyclic lipopeptides by *Pseudomonas fluorescens* strains in bulk soil and in the sugar beet rhizosphere. Appl Environ Microbiol 69:861–868

Oves M, Saghir Khan M, Huda Qari A, Nadeen Felemban M, Almeelbi T (2016) Heavy metals: biological importance and detoxification strategies. J Bioremed Biodeg 7:334

Palaniyandi SA, Damodharan K, Yang SH, Suh JW (2014) *Streptomyces* sp. strain PGPA39 alleviates salt stress and promotes the growth of 'Micro-Tom' tomato plants. J Appl Microbiol 117(3):766–773

Patten CL, Glick BR (1996) Bacterial biosynthesis of indole-3- acetic acid. Can J Microbiol 42:207–220

Paul D, Lade H (2014) Plant-growth-promoting rhizobacteria to improve crop growth in saline soils: a review. Agron Sustain Dev 34(4):737–752

Pedrosa FO, Elmerich C (2007) Regulation of nitrogen fixation and ammonium assimilation in associated and endophytic nitrogen-fixing bacteria. In: Elmerich C, Newton WE (eds) Associative and endophytic nitrogen-fixing bacteria and cyanobacterial associations. Springer, Dordrecht, pp 41–71

Pérez-Miranda S, Cabirol N, George-Téllez R, Zamudio-Rivera LS, Fernández FJ (2007) O-CAS, a fast and universal method for siderophore detection. J Microbiol Meth 70(1):127–131

Persello-Cartieaux F, Nussaume L, Robaglia C (2003) Tales from the underground: molecular plant-rhizobacteria interactions. Plant Cell Environ 26:189–199

Phi QT, Yu-Mi PS, Keyung-Jo R, Choong-Min P, Seung-Hwan K, Jong-Guk G, Sa-Youl G (2010) Assessment of root-associated *Paenibacillus polymyxa* groups on growth promotion and induced systemic resistance in pepper. J Microbiol Biotechnol 20:605–1613

Podile AR, Kishore GK (2006) Plant growth-promoting rhizobacteria. In: Gnanamanickam SS (ed) Plant associated bacteria. Springer, Dordrecht, pp 195–230

Ponmurugan P, Gopi C (2006) In vitro production of growth regulators and phosphate activity by phosphate solubilizing bacteria. Afr J Biotechnol 5:348–350

Poonguzhali S, Madhaiyan M, Sa T (2008) Isolation and identification of phosphate solubilizing bacteria from Chinese cabbage and their effect on growth and phosphorus utilization of plants. J Microbiol Biotechnol 18:773–777

Raaijmakers JM, Bonsall RF, Weller DM (1999) Effect of population density of *Pseudomonas fluorescens* on production of 2,4-diacetylphloroglucinol in the rhizosphere of wheat. Phytopathology 89:470–475

Rajkumar M, Freitas H (2008) Effects of inoculation of plant growth promoting bacteria on Ni uptake by Indian mustard. Bioresour Technol 99:3491–3498

Rajkumar M, Ae N, Prasad MNV, Freitas H (2010) Potential of siderophore-producing bacteria for improving heavy metal phytoextraction. Trends Biotechnol 28:142–149

Raymond J, Siefert JL, Staples CR, Blankenship RE (2004) The natural history of nitrogen fixation. Mol Biol Evol 21:541–554

Remans R, Beebe S, Blair M, Manrique G, Tovar E, Rao I, Croonenborghs A, Torres-Gutierrez R, El-Howeity M, Michiels J, Vanderleyden J (2008) Physiological and genetic analysis of root responsiveness to auxin-producing plant growth-promoting bacteria in common bean (*Phaseolus vulgaris* L.). Plant Soil 302:149–161

Rodrigues EP, Rodrigues LS, de Oliveira ALM, Baldani VLD, Teixeira KRS, Urquiaga S, Reis VM (2008) *Azospirillum amazonense* inoculation: effects on growth, yield and N_2 fixation of rice (*Oryza sativa* L.). Plant Soil 302:249–261

Rokhbakhsh-Zamin F, Sachdev D, Kazemi-Pour N, Engineer A, Pardesi KR, Zinjarde S, Dhakephalkar PK, Chopade BA (2011) Characterization of plant-growth-promoting traits of *Acinetobacter* species isolated from the rhizosphere of *Pennisetum glaucum*. J Microbiol Biotechnol 21:556–566

Rubio LM, Ludden PW (2005) Maturation of nitrogenase: a biochemical puzzle. J Bacteriol 187:405–414

Rubio LM, Ludden PW (2008) Biosynthesis of the iron-molybdenum cofactor of nitrogenase. Annu Rev Microbiol 62:93–111

Saravanan VS, Madhaiyan M, Thangaraju M (2007) Solubilization of zinc compounds by the diazotrophic, plant growth promoting bacterium *Gluconacetobacter diazotrophicus*. Chemosphere 66:1794–1798

Sharma A, Johri BN (2003) Growth promoting influence of siderophore-producing *Pseudomonas* strains GRP3A and PRS9 in maize (*Zea mays L.*) under iron limiting conditions. Microbiol Res 158:243–248

Sheng XF, Xia JJ (2006) Improvement of rape (*Brassica napus*) plant growth and cadmium uptake by cadmium-resistant bacteria. Chemosphere 64:1036–1042

Shilev S (2013) Soil rhizobacteria regulating the uptake of nutrients and undesirable elements by plants. In: Arora NK (ed) Plant-microbe symbiosis: fundamentals and advances. Springer, New Delhi, pp 147–150

Shrivastava P, Kumar R (2015) Soil salinity: a serious environmental issue and plant growth promoting bacteria as one of the tools for its alleviation. Saudi J Biol Sci 22:123–131

Singh BP (2015) Screening and characterization of plant growth promoting rhizobacteria (PGPR): an overview. Bull Environ Sci Res 4(1–2):1–14

Somers E, Vanderleyden J, Srinivasan M (2004) Rhizosphere bacterial signalling: a love parade beneath our feet. Crit Rev Microbiol 30:205–240

Spaepen S, Vanderleyden J (2011) Auxin and plant-microbe interactions. Cold Spring Harb Perspect Biol 3:1400–1438

Spaepen S, Vanderleyden J, Remans R (2007a) Indole- 3-acetic acid in microbial and microorganism-plant signalling. FEMS Microbiol Rev 31:425–448

Spaepen S, Versées W, Gocke D, Pohl M, Steyaert J, Vanderleyden J (2007b) Characterization of phenylpyruvate decarboxylase, involved in auxin production of *Azospirillum brasilense*. J Bacteriol 189:7626–7633

Tank N, Saraf M (2010) Salinity-resistant plant growth promoting rhizobacteria ameliorates sodium chloride stress on tomato plants. J Plant Interact 5:51–58

Tao GC, Tian SJ, Cai MY, Xie GH (2008) Phosphate solubilizing and -mineralizing abilities of bacteria isolated from. Pedosphere 18:515–523

Valverde A, Burgos A, Fiscella T, Rivas R, Velazquez E, Rodrıguez-Barrueco C, Cervantes E, Chamber M, Igual JM (2006) Differential effects of coinoculations with *Pseudomonas jessenii* PS06 (a phosphate-solubilizing bacterium) and *Mesorhizobium ciceri* C-2/2 strains on the growth and seed yield of chickpea under greenhouse and field conditions. Plant Soil 287:43–50

Vansuyt G, Robin A, Briat JF, Curie C, Lemanceau P (2007) Iron acquisition from Fe-pyoverdine by *Arabidopsis thaliana*. Mol Plant Microbe Interact 20:441–447

Varma A, Singh A, Sudha SN, Sharma J, Roy A, Kumari M, Rana D, Thakran S, Deka D, Bharti K, Franken P, Hurek T, Blechert O, Rexer K-H, Kost G, Hahn A, Hock B, Maier W, Walter M, Strack D, Kranner I (2001) *Piriformospora indica*: a cultivable mycorrhiza-like endosymbiotic fungus. In: Hock B (ed) Mycota IX. Springer, Berlin, pp 123–150

Vessey JK (2003) Plant growth promoting rhizobacteria as biofertilizers. Plant Soil 255:571–586

Vessey JK, Pawlowski K, Bergman B (2004) Root-based N_2- fixing symbioses: legumes, actinorhizal plants, Parasponia, and cycads. Plant Soil 266:205–230

Vikram A, Hamzehzarghani H (2008) Effect of phosphate solubilizing bacteria on nodulation and growth parameters of green gram (*Vigna radiate* L. Wilczec). Res J Microbiol 3:62–72

Viveros- Martinez O, Jorquera MA, Crowley DE, Gajardo G, Mora ML (2010) Mechanism and practical considerations involved in plant growth promotion by rhizobacteria. J Soil Sci Plant Nutr 10:293–319

Wani PA, Khan MS (2010) *Bacillus* species enhance growth parameters of chickpea (*Cicer arietinum* L.) in chromium stressed soils. Food Chem Toxicol 48:3262–3267

Wu CH, Wood TK, Mulchandani A, Chen W (2006) Engineering plant-microbe symbiosis for rhizoremediation of heavy metals. Appl Environ Microbiol 72:1129–1134

Zahir ZA, Arshad M, Frankenberger WT (2003) Plant growth promoting rhizobacteria: applications and perspectives in agriculture. Adv Agron 81:97–168

Zahir ZA, Shah MK, Naveed M, Akhter MJ (2010) Substrate-dependent auxin production by *Rhizobium phaseoli* improves the growth and yield of *Vigna radiata* L. under salt stress conditions. J Microbiol Biotechnol 20:1288–1294

Zaidi A, Khan MS (2005) Interactive effect of rhizospheric microorganisms on growth, yield and nutrient uptake of wheat. J Plant Nutr 28:2079–2092

Zaidi A, Khan MS, Ahemad M, Oves M (2009) Plant growth promotion by phosphate solubilizing bacteria. Acta Microbiol Immunol Hung 56:263–284

Microbial Ecology at Rhizosphere: Bioengineering and Future Prospective

4

Shyamalina Haldar and Sanghamitra Sengupta

Abstract

Rhizosphere, the interface between soil and plant roots, is a chemically complex environment which supports the development and growth of diverse microbial communities. Studies in rhizosphere science have undoubtedly improved our ability to steer the knowledge into technological applications in agricultural industry, ecological engineering, and nature restoration. In this chapter we provide a holistic perception of rhizosphere functioning with a highlight on the ecological drivers that promote colonization of coherent functional groups of microorganisms influencing plant life through several direct and indirect mechanisms. We also discuss how the activities of the indigenous microbes from rhizosphere may be exploited toward developing profitable techniques or methods in sustainable agriculture, biotechnology, and environmental management. In this context, we emphasize on the need for high degree of innovation and active collaboration between basic research and technology development wings for the best use of the knowledge in order to meet the increasing global demand for food, fiber, and bioenergy.

S. Haldar
Department of Microbiology, Goa University, Taleigao Plateau 403206, Goa, India

S. Sengupta (✉)
Department of Biochemistry, University of Calcutta,
35, Ballygunge Circular Road, 700019 Kolkata, India
e-mail: sanghamitrasg@yahoo.com

D.K. Choudhary et al. (eds.), *Plant-Microbe Interaction: An Approach to Sustainable Agriculture*, DOI 10.1007/978-981-10-2854-0_4

4.1 Introduction

Rhizosphere represents one of the most diverse habitats on our planet (Trabelsi and Mhamdi 2013). It is technically defined as the soil adhering to the root which is chemically enriched with the substances released from the plants and stimulate microbial growth and activities. Over the past two decades, a growing body of empirical research has shown that ecological and biochemical processes in the rhizosphere are mediated by intricate arrays of direct and indirect interactions occurring between the plants and residing microorganisms which cumulatively make the microenvironment unique, physically, chemically, and biologically (Fig. 4.1). Rhizosphere processes, at a global scale, utilize approximately half of the total energy fixed by photosynthesis in terrestrial ecosystems, contribute roughly 50 % of the total carbon dioxide emitted from terrestrial ecosystems, and mediate virtually all aspects of biogeochemical transformation, biomass turnover, and nutrient cycling (Hopkins et al. 2013). Plants and the rhizobiome together contribute to a significant extent, for the preservation of biodiversity and ecological sustainability of urban green infrastructures (Weyens et al. 2015). Consequently, there is a worldwide effort to comprehend and model rhizosphere functioning using multiscale information generated through genetics, genomics, metabolomics, and system biological approaches for effective translation of this knowledge for the upliftment of human health and living (Weyens et al. 2015). Fortunately for us, due to recent technological advances, the paradigm of microbiology has shifted toward understanding and

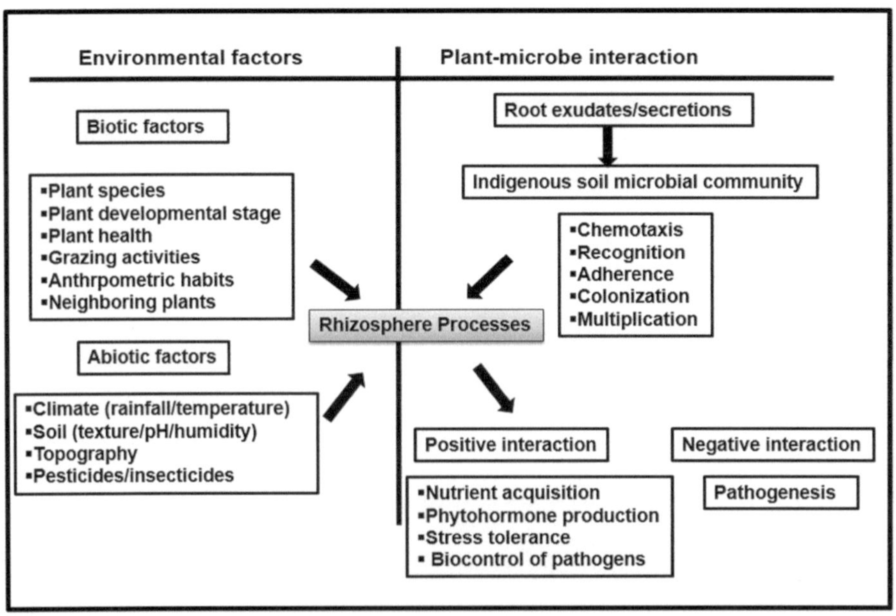

Fig. 4.1 Physical, chemical & biologial interactions affecting plant's rhizosphere

predicting the function of rhizo-consortia instead of a single dominant species. We are beginning to understand how plant systems interact with their environment, monitor biotic or abiotic stresses, and battle diseases in the field by modulating the associated microbial forum (Hill et al. 2013). Efforts have been initiated to engineer the knowledge into practical application such as production of sustainable food, fiber, and energy, maintenance of forest ecology and biodiversity, management of water resources, and alleviation of climate change (Mark et al. 2006; Jones 2008; Dessaux et al. 2016). In this book chapter, we summarize this worldwide endeavor by throwing light on the mechanistic and the ecological knowledge of the rhizosphere accumulated so far with a focus on agro-economy, biotechnology, and environmental management.

4.2 Microbial Ecology in the Rhizosphere

Following the colonization in the terrestrial ecosystem, "plants" as sessile organisms communicate with innumerable microorganisms residing in its ecosystem. The most significant part of this operates belowground, in general, and in the rhizosphere, in particular. Studies on rhizosphere date back to 1904, when Lorenz Hiltner (1904) observed that microbes are more abundant in rhizosphere than they are in distant soil environment. Since then a huge body of literature has accumulated and unequivocally demonstrated that this narrow zone of soil is a highly favorable site for microbial activities. It may contain up to 10^{11} microbial cells per gram root and more than 30,000 prokaryotic species (Egamberdieva et al. 2008; Mendes et al. 2011). Rhizosphere may be imagined as a cloud of microbes, vital for plant growth and survival, surrounding the plant root. Microbial proliferation at the plant roots and root–soil interface is supported by diverse varieties of root exudates (Badri et al. 2009). It is estimated that the collective genome of the rhizosphere microbial community is much larger than that of the plant itself and hence affectionately referred to as the plant's second genome (Berendsen et al. 2012). At present, the entire consortium of plants and associated microorganisms is perceived as a holobiont and is no longer recognized as "individual entities" (Dessaux et al. 2016).

4.2.1 Rhizosphere: A Platform for Microbial Growth

4.2.1.1 Rhizosphere as a Chemical Hotspot

Microbial colonization in rhizosphere is rooted to the phenomenon of "rhizosphere processes" which is collectively composed of various physicochemical and biological turnovers defined by the host plant through the uptake of minerals and water and subsequent release of nutrients and carbon dioxide, exudation, and secretion of an array of chemical compounds (Philippot et al. 2013). Significant advances in our knowledge about plant exudate chemistry and the impact of "rhizodeposition" on microbial growth at the root–soil interface have been made in the last few years (Smalla et al. 2001; Dunfield and Germida 2003; Nunan et al. 2005; Mougel et al.

2006). It may be surmised from these studies that many soil microorganisms remain dormant in the absence of organic input due to general carbon (C) limitation in normal soil (Owen et al. 2007). Growth of plants triggers an increase in the rate of turnover of soil organic matter (SOM) in the order of 20- to 30-folds around areas of exudation, and this pulse of organic input often fosters microbial activity and associated soil organic matter (SOM) turnover (Xiao et al. 2007). In contrast to bulk soil, rhizosphere is thus extremely nutrient rich due to the abundance of low molecular weight (LMW) compounds released during normal root cell metabolism (root exudates): both low and high molecular weight compounds that are synthesized for secretions (root secretions) and compounds released by plant cell lysis (Marschner 1995). Microorganisms thrive on these resources and form a plant-specific assemblage in the rhizosphere.

Mucilage secreted from the growing roots contains hydrated polysaccharides, organic acids, vitamins, and amino acids and can also bind to water molecules. This creates a well-hydrated environment supportive of microbial growth. Developing roots generally support fast-growing microorganisms like bacteria, whereas matured roots support slower-growing organisms such as fungi and actinomycetes; the latter produce less mucilage and fewer cell lysates due to the absence of border cells and emerging lateral roots and also leak less water due to the deposition of a water-impermeable layer around epidermal cells. Outward diffusion of nutrients and inward movement of salts and minerals during transpiration develop complex chemical gradients around the root and create a range of distinct microbial habitats. LMW carbon compounds such as sugar, organic acids, amino acids, and flavonoids are readily assimilated by microorganisms and play a primary role in regulating microbial community dynamics in the rhizosphere (Bais et al. 2006). Flavonoids, a diverse class of polyphenolic compounds, often serve as important chemical cues in mediating plant–microorganism interactions (reviewed by Shaw et al. 2006). Surfactant-active compounds such as carboxylic compounds in the root exudates have been found to increase the solubility of the heavy metals/toxic substances and make them bioavailable to root-colonizing microorganisms (Balseiro-Romero et al. 2014). Root volatiles include sulfur-containing compounds or the terpene (E)-b-caryophyllene which serve as foraging cues for parasitic entomopathogenic nematodes [EPNs] (Hiltpold and Turlings 2012; van Dam and Bouwmeester 2016). Organic phosphorus which is, in general, poorly available is solubilized through grazing by nematodes (Wenke et al. 2010). Phenolic compounds such as salicylic acid and gamma-aminobutyric acid (GABA) in the root exudates are suggested to send specific signals for soil bacteria, namely, *Sphingomonas*, *Methylobacterium*, *Frankineae*, *Variovorax*, *Micromonosporineae*, and *Skermanella* (Badri et al. 2013).

4.2.1.2 Factors Affecting Rhizo-Atmosphere

Much of our current understanding about rhizosphere incidences has emerged from studies on agricultural or horticultural crop plants: model species such as *Arabidopsis thaliana* (Bulgarelli et al. 2012) and *Medicago truncatula* (Kisiel and Kepczynska 2016) and a few noncultivated plant species such as arbuscular mycorrhizal associations (Bennett and Bever 2007). The excerpt from these findings points that plant genetic makeup unequivocally plays a dominant role in the selection of rhizobacterial community. This is not surprising because the variety and amount of the compounds synthesized and released by roots are mostly under the plant's physiological and genetic control (Costa et al. 2006; Berg and Smalla 2009; Badri et al. 2009). Corroboratively root microbiome of plants grown in the same soil has been found to differ between plant species (Curlango-Rivera et al. 2013; Bonito et al. 2014) and between ecotypes, chemotypes, and genotypes within species (Micallef et al. 2009; Hill et al. 2013; Bulgarelli et al. 2013). To be more specific, plant-specific variation in root exudation is regulated both quantitatively and qualitatively by the root system architecture (RSA) which is determined by the inherent genetic factors and varies across plant species (Badri and Vivanco 2009). Secretion of phytochemicals and proteins from roots is an important way for plants to respond to various environmental factors and stresses (Walker et al. 2004; Bais et al. 2004). Root structure additionally affects oxygen pressure and carbon and nitrogen availability which in turn influences nitrogen transformation by soil microorganisms (Blossfeld et al. 2011). Furthermore, root growth changes the physical and chemical properties of the soil, including the mineral and organic content, the water potential, the pH, and the salinity.

The type and condition of soil also influence the nature of rhizodeposits. Nutrient deficiency is a major factor enhancing the secretion of metabolites by plant roots (Rengel and Marschner 2005). Besides, the presence or absence of particular minerals or toxic metals affects the composition of root exudation. For example, citric, malic, and oxalic acids are secreted to detoxify aluminum (Wang et al. 2006). Secretion of phenolic compounds is increased in phosphorus-deficient soils (Khorassani et al. 2011), while secretion of flavanones and flavones is enhanced in nitrogen-limiting conditions (Schultze and Kondorosi 1998). Mineral deficiency enhances the production of elicitors that influence root exudation. For example, potassium deficiency increases jasmonic acid-mediated defense responses (Schachtman and Shin 2007). Hypoxia due to high soil moisture causes an increased anaerobic respiration rate resulting in accumulation of ethanol, lactic acid, and alanine in the rhizosphere (Rivoal and Hanson 1994). Low temperature and light reduce secretion of root exudates. For example, the exudation of tannins and phenolic compounds in *Vicia faba* was greatly reduced at 4 °C compared to the amounts secreted at 30 °C (Bekkara et al. 1998). The root exudation process follows diurnal rhythms with exudation increasing during light periods (Watt and Evans 1999). In the root exudates from *Alnus glutinosa* (L.), the flavonoid content has been found to be increased under light conditions (Hughes et al. 1999). Root exudation is even affected by neighboring plant species. Quantity of glucosinolates in the root exudates is increased when *Arabidopsis* plants are grown at a higher density causing a shift of the rhizobiota toward the glucosinolate-utilizing microorganisms (Wentzell and Kliebenstein 2008). Root-induced pH changes in the rhizosphere influence bioavailability of phosphate and copper by modulating adsorption and precipitation of ions and soil minerals, respectively, and thereby shape the microbial ecology in the root environment (Bravin et al. 2009).

4.2.2 Microbial Community Structure

In general, the microbes take the advantage of the nutrients that the plant provides as discussed above, and in effect, they assist their host plant in making more essential nutrients (reviewed in Mendes et al. 2013). Therefore, microorganisms typically represent the largest fraction of belowground biomass. One gram of soil is thought to constitute tens of thousands of microbial "species" (reviewed in Kent and Triplett 2002). Microbes in rhizosphere can be broadly classified as bacteria, fungi, nematodes, protozoa, and actinomycetes. Of these, bacteria and fungi are most well documented.

4.2.2.1 Bacteria

The ratio of the microbial population in the rhizosphere (R) to that in the bulk soil (S), i.e., R/S value, is >=20 for the bacteria, while that for fungi and actinomycetes are 10 and 2–3, respectively (Bagyaraj and Rangaswami 2005). The overall proportion of aerobic bacteria is relatively less in the rhizosphere because of low level of oxygen due to root respiration. Rhizosphere is programmed to recruit wide range of bacterial genera, beneficial to the plants by using the signals from the host. The beneficial bacteria are collectively termed as plant growth-promoting rhizobacteria (PGPRs). The most common genera of bacteria observed in the rhizosphere include *Acinetobacter, Agrobacterium, Alcaligenes, Arthrobacter, Azospirillum, Azotobacter, Bacillus, Cellulomonas, Flavobacterium, Micrococcus, Mycobacterium, Pseudomonas*, and *Rhizobium*. PGPRs may be categorized depending on their modes of beneficial action into biofertilizers, phytostimulators, biopesticides, and elicitors of tolerance to abiotic and biotic stresses (Bhattacharyya and Jha 2012; Bhardwaj et al. 2014; Perez-Montano et al. 2014; Yang et al. 2009). A list of common beneficial bacteria is given in Table 4.1.

Biofertilizers directly promote plant growth by endowing plants with nutrients such as nitrogen, phosphorus, and trace elements (iron) which otherwise would have remained inaccessible to the plants. They are composed of *Rhizobium* sp., *Pseudomonas fluorescens*, *Trichoderma* sp. (e.g., *Trichoderma asperellum* and *Trichoderma hamatum*), and *Allorhizobium* sp. Of these, root nodule symbiosis established by *Rhizobium* sp. with legumes for nitrogen fixation has been vividly investigated (reviewed in Wang et al. 2012). Besides nitrogen, iron is another important element required for the growth of plants. However, it is essentially unavailable in aerobic environments, as it tends to form insoluble hydroxides at biological pH (Guerinot 1994). Rhizobia (*Bradyrhizobium japonicum, Sinorhizobium meliloti, Rhizobium leguminosarum bv. Viciae, Rhizobium ciceri*), fluorescent pseudomonads, *Enterobacter*, *Burkholderia*, and streptomycetes are capable of producing LMW compounds called siderophores that complex with ferric iron and several membrane-bound/periplasmic proteins (Neilands 1995; Crosa and Walsh 2002). This iron sequestration helps these bacteria to establish themselves in the rhizosphere and also to provide plants with soluble iron. *Bradyrhizobium japonicum* and *Sinorhizobium meliloti*, on the other hand, also help the plants to take up the natural siderophores (ferrichromes) present in the soil directly by forming heterologous

Table 4.1 Mechanism of action of plant growth-promoting rhizobacteria (PGPR) and biocontrol agents (BCAs)

Microbial species	Mechanism of action	References
PGPR (biofertilizers and phytostimulators)		
Rhizobia	Nitrogen fixation; inorganic and organic nutrient solubilization; plant growth regulator (IAA, gibberellins, cytokinines) synthesis	Ferguson and Mathesius (2014)
Pseudomonads	IAA production; siderophore production and phosphate solubilization	Ajilogba and Babalola (2013)
Firmicutes		
Burkholderia		Farag et al. (2013) and Ghosh et al. (2016)
Azotobacter sp.	Cytokinine production; nitrogen fixation	Leaungvutiviroj et al. 2010
BCA (biopesticides)		
Rhizobium sp.	Disease suppression (antibiosis, competition for iron, enhancing plant defense mechanism)	Dutta et al. (2014)
Pseudomonads	Systemic resistance induction; antifungal volatile production; induced systemic tolerance to high temperature and salinity; stabilization of soil aggregates; quorum quenching (QQ)	Ajilogba and Babalola (2013) and Farag et al. (2013)
Firmicutes (*Bacillus* sp.)	Induced systemic resistance (ISR); antifungal volatile production; quorum quenching (QQ); induced systemic tolerance (IST) to high temperature and salinity	Shrivastava and Kumar (2015)
Burkholderia	Induced systemic tolerance for drought by producing ACC-deaminase	Onofre-Lemus et al. (2009)
Azotobacter sp.	Oxidative stress tolerance through production of abscisic acid (ABA) and degradation of reactive oxygen species (ROS)	Marsalek and Simek (1992)

PGPRs are also endowed with biocontrol properties, while BCA can also stimulate direct plant growth

siderophores (Powell et al. 1983). Likewise, a large proportion of phosphorus exists in insoluble forms in the soil. Phosphate-solubilizing bacteria like *Enterobacter*, *Pantoea*, *Pseudomonas*, *Klebsiella*, *Cedecea*, *Cronobacter*, *Bacillus*, *Rhodococcus*, *Arthrobacter*, *Chryseobacterium*, *Delftia*, *Gordonia*, *Phyllobacterium*, and *Serratia* identified from the rhizosphere of various plants have shown the ability to solubilize inorganic soil phosphates, such as $Ca_3(PO_4)_2$, $FePO_4$, and $AlPO_4$, by synthesizing organic acids (citric, gluconic, lactic, succinic, and propionic acids), siderophores, and hydroxyl ions (Chen et al. 2006; Sharma et al. 2013).

Rhizobium, *Pseudomonas*, *Bacillus*, *Azospirillum*, *Enterobacter*, *Azotobacter*, *Pantoea*, and *Streptomyces* are grouped as *phytostimulators* by the virtue of their ability to produce phytohormones like indole-3-acetic-acid (IAA), gibberellic acid (GA3), and cytokines which directly enhance plant growth by modulating root system architecture (Spaepen et al. 2007; Apine and Jadhav 2011; Duca et al. 2014). Coordination among the hormonal pathways is associated with overall development in plants. In addition, this is important for the maintenance of plasticity of plant morphogenesis essential for plants to respond to environmental cues (Hardtke et al. 2007). Gibberellins regulate cell elongation, cytokinins control cell proliferation, and auxin modulates both (Hardtke et al. 2007; Nakaya et al. 2002). Gibberellins act in collaboration with auxins to regulate the transition of root meristem cells from division to elongation (Hardtke et al. 2007). IAA and cytokinins regulate the root development by acting as a "control switch" between cell division and differentiation, thereby controlling the size of the organs (Barrada et al. 2015). Root morphogenesis and growth are basically controlled through a cross talk between these phytohormones where IAA increases the length and cytokinins maintain the length by inhibiting extra growth (De Vos et al. 2014).

Pseudomonas sp. (e.g., *P. fluorescens, P. cepacia, P. aeruginosa, and P. aureofaciens*) and *Bacillus* sp. (e.g., *Bacillus subtilis*) along with *Streptomyces* sp. are termed as *biopesticides* or *biocontrol agents* (BCAs) due to their role in inhibition of plant pathogens by producing hydrogen cyanide, 2,4-diacetylphloroglucinol, pyrrolnitrin, phenazine, oomycetes, and other compounds and thereby conferring protection against diseases (Kwak and Weller 2013; Haas and Keel 2003). BCA also inhibits the growth of pathogenic bacteria directly by quorum quenching. N-Acyl-homoserine lactone (AHL), an important molecule for cell to cell communication and used by several plant pathogens to establish virulence, is degraded by N-acyl-homoserine lactonase (AHL-lactonase), produced by *Bacillus* sp. Fluorescent pseudomonads can suppress the growth of pathogens by efficiently competing with them for the siderophores. Pyoverdines, produced by pseudomonads have a very high affinity toward iron in comparison to other microbes and thereby can limit the availability of iron for the pathogens in the rhizosphere (Cezard et al. 2015).

The other groups of rhizobacteria (*Burkholderia* sp., *Enterobacter* sp., *Rhizobium* sp.) assist plants to tolerate stress due to accumulation of reactive oxygen species and 1-aminocyclopropane-1-carboxylate (ACC). The latter molecule is an intermediate in ethylene biosynthesis. Ethylene is activated under nutrient stress and is harmful for the plants during post-harvest phases (Khan et al. 2015). *Bacillus subtilis* and *Achromobacter piechaudii* enhance salinity tolerance in plants growing in coastal regions, while *Paenibacillus polymyxa* and *Rhizobium tropici* have been reported to provide tolerance to drought stress in *Arabidopsis*, tomato and common bean (Mayak et al. 2004; Zhang et al. 2008; Timmusk and Wagner 1999; Yang et al. 2009).

4.2.2.2 Fungi

Both symbiotic and pathogenic fungi reside in the rhizosphere. One gram of rhizosphere soil harbors 10^5–10^6 organisms. Arbuscular mycorrhizal fungi (AMF), belonging to *Glomerales*, are one of the oldest groups of fungi that have been hypothesized to form the most primitive interaction with plant roots. This multitrophic symbiosis is present ubiquitously in the terrestrial plants, both from natural and agricultural ecosystems, including gymnosperms, ferns, angiosperms, halophytes, hydrophytes, and xerophytes. Therefore, the beneficial role of AMF in biofertilization and bioremediation has been widely explored (reviewed in Lee et al. 2013). The beneficial outcome of the root–AMF association is mutually determined by the microbial consortia and soil chemistry (Khan 2006). The positive effect of AMF on plant physiology is modulated in the presence of specific microbe–microbe interaction. For example, co-inoculation with mycorrhization helper bacteria (MHB) such as *Bacillus* sp. in nutrient-poor soils has been found to improve the mycorrhizal effects (Vivas et al. 2003). AMF improve the productivity, biomass and diversity of plants by mineral sequestration (phosphorus scavenging and nitrogen acquisition), nutrient acquisition, and increased tolerance of the plants to abiotic stresses [drought/salinity resistance] (Lenoir et al. 2016; Porras-Soriano et al. 2009). In addition, AMF alter the overall microbial activity in the soil by modulating the rhizodeposits and improving the soil quality through immobilization of heavy metals (Yang et al. 2015). The wide surface area of the extra-radical mycelium and the synthesized iron-containing protein, "glomalin," by AMF, cooperatively determine the beneficial activities of AMF. Glomalin also contributes to the sequestration of toxic elements to enhance survival rate of the plants in polluted soils (Khan 2006). Glomalin has a role in the stabilization of soil aggregates leading to an improved penetration of the soil by water and air and also rendering an enhanced resistance to soil toward erosion (Rillig and Mummey 2006). AMF may act as a biocontrol agent by protecting the host plant against biotic stresses such as plant-parasitic nematode (PPN) infection by induced systemic resistance (ISR) and direct competing with the nematodes for space and nutrients (Schouteden et al. 2015). Increased rate of root respiration and respiratory acclimation due to AMF colonization in tropical plants is also reported. This indicates a role of AMF in terrestrial organic carbon influx as well (Fahey et al. 2016).

A common mycorrhizal network (CMN) is sometimes shared among the plants. AMF in this instance contribute to the transfer, distribution, and partitioning of carbon, nitrogen, phosphorus, and water from source (resource-rich plants) to sink (resource-poor plants) (Walder et al. 2016). This net translocation offers multiple benefits to the plants under environmental stresses not only by providing an access to resources from other plants but aiding to a speedy revival from the periods of rigorous water stress and protecting against pathogens (Babikova et al. 2013). AMF can also act as a support system for seedling establishment and can influence plant invasion success (Klironomos 2002). Recent observations have shown that the co-occurrence of complementary diverse AMF species among the plants in a single region helps in increased plant to plant facilitation, an important ecological process practiced by the woody plants to establish themselves in semiarid areas (Montesinos-Navarro et al. 2012). However, species richness and genetic variation among AMF is the major factor influencing plant species diversity and ecosystem functioning (van der Heijden et al. 2006). Conversely, response to AMF inoculation is also dependent on the genotype of plants. But this plant genotype-dependent AMF colonization has been explored inadequately to date (Montes-Borrego et al. 2014). In natural systems, among the AMF, *Glomus* sp. has been most comprehensively studied for their pertinent role in plant productivity as biofertilizers and BCA (Labidi et al. 2015). Another widely studied AMF species in pot cultures is *Rhizophagus irregularis* (Tisserant et al. 2013). Nevertheless, recent advancements in "omics" techniques have helped to assess a huge number of both active and spore-forming AMF taxa from the soil, indicating that molecular diversity of soil AMF is enormous and thus might be manipulated toward sustainable agriculture and phytoremediation (Davison et al. 2012).

Trichoderma sp. is another important fungal BCA that combats with plant pathogens via multifarious mechanisms including competition, antibiosis, induced resistance, and parasitism and thereby offer an indirect beneficial effect on plant health (Zhang et al. 2015a, 2016). Many species of *Trichoderma* have been reported to exert direct effects on plant growth through solubilization of plant nutrients and/or better uptake of macro- and micronutrients and through production of plant growth factors (Li et al. 2015). However, our knowledge about *Trichoderma* sp.-mediated plant growth promotion is still limited.

4.2.2.3 Others

Although bacteria and fungi form the most significant members of the microbial consortia in rhizosphere, the other micro-/macroorganisms such as nematodes, protozoa, and oomycetes are also present. This latter group of species is mainly marked for their pathogenic invasion with the plants. Nevertheless, they too exert few microbial control properties to inhibit the growth of other pathogens like insects in the rhizosphere (Lacey et al. 2015).

The nematodes are complex eukaryotic invertebrate worms and mostly free-living which parasitize plants as well (reviewed in Kenney and Eleftherianos 2016). Among the nematodes, the genera *Steinernema* and *Heterorhabditis* have been identified as potent microbial control agents (MCA) that resist the growth of a number of pathogenic insects and pests in the rhizosphere. Beneficial EPNs are obligate parasites that destroy the insect very rapidly and hence are considered as one of the potent BCAs in several cropping systems (Ehlers 2003). Considerable progress toward the application of these EPNs in agriculture and pest management has been made in the past decade (Lacey et al. 2015). Secondary metabolites exuded from root tips play a dual role in attracting the nematodes so that the EPNs are selected over the pathogenic nematodes (Hiltpold et al. 2015). Studies have shown that insect herbivory at the roots could induce the secretion of volatile substances that attract EPN like *Heterorhabditis megidis* in many wild and cultivated plants, and thus this was adopted as an important plant defense mechanism against insects (Rasmann et al. 2005; Ali et al. 2010).

Protozoa, the unicellular animals, feed on microorganisms and form a ubiquitous group of rhizo-fauna. Protozoa by virtue of their grazing activities stimulate microbial decomposition and stimulate the release of organic matter and supplies plants with adequate nitrogen which otherwise would have remained limitedly accessible. Increased availability of nitrogen benefits AMF which transport it via the hyphae to the internal roots. The grazing also fosters rapid transportation of photosynthates from aboveground to belowground roots. In this process, the protozoa subsequently interact with AMF and strengthen the interaction by controlling the nutrient supply (Koller et al. 2013a, b).

Frankia, nitrogen-fixing *Actinobacteria* that form symbiotic association with actinorhizal plants, has a potential role in increasing soil fertility and thereby enhancing the plant productivity in degrading and nitrogen-limiting soils (Diagne et al. 2013). This actinorhizal association is very productive toward maintenance of soil stability and henceforth facilitates the establishment and development of subsequent plant communities in disturbed landforms (Gtari et al. 2012). *Frankia* indirectly influences the plant productivity mitigating the adverse effects of salinity, drought, and contamination of heavy metals in degraded lands. As a whole, this association considerably enhances the plant growth, nitrogen content in roots and shoots, overall biomass, and survival rate of the plants (Diagne et al. 2013).

4.3 Bioengineering: Turning New Knowledge into Useful Societal Benefit

Demand for food, fiber, fuel, and other amenities will continue to grow as a result of population growth and rising incomes. To meet up these mounting demands, the stress on the natural resources and environment is leading to the resource depletion and environment destruction. Sustainable intensification is proclaimed to be only alternative to overcome this problem (Gregory et al. 2013). The goal of sustainable intensification is to maximize agricultural output from existing farmland while minimizing pressure on the environment. To accomplish this, an integrative approach coalescing biological science with community ecology is needed (Reynolds et al. 2014). In this "omics" era of new molecular tools and biotechnological advances, the knowledge accrued from basic research is expected to contribute more meaningfully to the development of more sustainable systems of intensive production (Ryan et al. 2009).

4.3.1 Rhizosphere Engineering

"Rhizosphere engineering" refers to the manipulation of a plant's root and surrounding milieu with a view to create a "biased" environment that will specifically enhance the crop productivity and plant survival (Ryan et al. 2009). In nature, plants themselves can adapt to any unfavorable environment by developing a variety of strategies; one of them being the modulation of rhizosphere chemistry. The knowledge of plant's inherent mechanisms is basically applied for any kind of "rhizosphere engineering/management." The selection of appropriate crop species, soil amendments, introduction of beneficial microorganisms, and genetic modification of plant and microbial activities are the fundamental components of rhizosphere engineering (Ryan et al. 2009). The benefit of managing rhizosphere is multifold. It not only paves the way to increased production of food/fiber/fuel but also results in diminished dependence on agrochemicals through replacement of their functions with beneficial microbes, biodegradable biostimulants, or transgenic plants. As a consequence, the environmental and ecosystem integrities are preserved. Bioremediation is another aspect of rhizosphere engineering, which uses natural/genetically modified organisms/plants to degrade environmental pollutants and soil contaminants to restore the environmental and the ecological balance (Bisht et al. 2015). Rhizosphere engineering has emerged as an important tool to provide a cost-effective and environmentally sustainable "green technology" to address several global problems due to population growth. Research has confirmed a considerable progress in this field to date but still holds promise for further development (Ryan et al. 2009).

4.3.2 Rhizosphere Engineering and Agriculture

The goal of plant–microbiome engineering is to stimulate the wide spectrum of interaction among the phytomicrobiome toward overall enhanced beneficial outcome for the plant (Quiza et al. 2015). The two major aspects that are primarily taken care of are irrigation and application of fertilizers. The latter usually shifts the soil microbiome in and around the roots by altering soil pH. The acidic fertilizers (ammonium based) decrease the pH of the rhizosphere, while the basic fertilizers (nitrate based) enhance the alkalinity of the soil (Ryan et al. 2009). The practice of organic agriculture through input of organic fertilizers such as animal manure, biosolids, and composts is well established worldwide (Savka et al. 2002; Lim et al. 2015; Mazzola 2007). However, the lack of knowledge about the population of desired microorganisms in composts results in lack of reproducibility of the methods. Besides, this method adversely affects the soil acidity, salinity, and root colonization of certain species such as AMF. Moreover, biohazards due to toxic materials from biowaste and heavy metals used in compost composition cannot be overlooked (Quiza et al. 2015). To facilitate long-lasting modification in the rhizosphere, plant breeding and establishment of genetically modified (GM) organisms are alternative approaches for organic farming (Ryan et al. 2009). Genetic engineering has much to offer to bring about "new green revolution" in agriculture (Araus et al. 2014). Engineering of rhizosphere is mainly established through three potential means such as *plant-based methods*, *microbiome-based approach*, and *meta-organism-based techniques* (Quiza et al. 2015).

4.3.2.1 Plant-Based Methods

Plant-based strategies of rhizosphere engineering are achieved by either plant breeding (cultivar selection) or specific genetic modification of plant species. The basis for plant breeding is to develop and select cultivar lines that have the ability for (1) enhanced root exudation, (2) systemic resistance to disease and environmental stresses, and (3) increased rate of mutual symbiosis (Magalhaes et al. 2007; Campbell et al. 2002; Farrar et al. 2014). Genetically engineered plants are bestowed with a capacity of producing higher quantity of exudates that are highly specific for beneficial microorganisms (1) synthesizing quorum sensing/quorum quenching signal molecules, (2) altering soil organic anion efflux and transportation of the same through the roots, (3) modifying soil properties (pH, salinity), and (4) promoting disease suppressiveness in soils (Koyama et al. 2000; Gevaudant et al. 2007; Yang et al. 2007; Mazzola 2007; Savka et al. 2002). These are achieved by introduction of genes of interest in selective plants from either the same species [cisgenic] or different species [transgenic] or deleting the genes that might repress different physiological processes [subgenic] (Wang et al. 2014).

The first genetically modified crop plant was an antibiotic-resistant tobacco plant (Fraley et al. 1999). Introduction of foreign germplasm into crops has been achieved by traditional crop breeders by overcoming species barriers. Farmers have widely adopted this technology to produce GM crops (GMC)/biotech (Bt) crops. Breeding lines of GM cultivars have been well established for the food crops including rice, wheat, potato, egg plants, tomatoes, soybeans, apples, beans, melons, papaya, and plums and for fiber and fuel crops such as cotton and grass and commercially important plants like tobacco (GM Approval Databse-ISAAA.org; *www.isaaa.org, 2016* accessed). These crops have been modified for the traits including improved shelf life, stress resistance, herbicide resistance, pest resistance, disease suppression, production of useful goods such as biofuel or drugs, and ability to absorb toxins and for use in bioremediation of pollution. The recent aim of research is to develop locally important crop breeds for developing countries such as production of rice rich in vitamins and iron that may mitigate chronic malnutrition in Asian countries, virus-resistant sweet potato, insect-resistant cowpea, and brinjal in Africa (Bawa and Anilakumar 2013).

To date, most genetic modifications target the properties of aboveground parts of plants. However, recently root-specific modifications have been attempted in the plants like *Arabidopsis* (30 %), tobacco (14 %), rice (11 %), maize (8 %), *Medicago* (5 %), and potato and tomato (both 4 %) (Kabouw et al. 2012). This is possible due to accumulating knowledge on plant root properties and rhizosphere processes (Perez-Alfocea et al. 2011). Drought-resistant transgenic rice lines have been developed by introducing auxin-transporting genes with a root-specific promoter (Jeong et al. 2010), while in another study salt stress-resistant rice has been developed through introduction of *Arabidopsis* gene (AtHKT1) that is accountable for sequestering Na+ in roots (Plett et al. 2010). A decrease in nematode abundance in rhizosphere was recorded in GM rice and potato that constitutively expressed a proteinase inhibitor for nematode control (Kohli et al. 1998; Cowgill et al. 2002). Transgenic lines of potato and tobacco have been established with the property of "quorum quenching" by transforming with quorum sensing signal (NAHL) degrading/synthesizing genes from *Bacillus* sp. and *Yersinia enterocolitica* (Dong et al. 2001; Fray et al. 1999). The transgenic variety of potato could directly inactivate quorum sensing molecules and is tolerant to the pathogen *Pectobacterium*. The transgenic tobacco could synthesize bacterial quorum sensing signal molecules and complemented biocontrol ability of *Pseudomonas aureofaciens*, defective in NAHL synthesis. The genetic transformation of crops to produce insecticidal proteins from the soilborne beneficial bacterium, *Bacillus thuringiensis* (Bt), is now one of the most important elements of pest control management system. Insect-resistant Bt rice (*Oryza sativa*) lines, maize (sweet corn), and cotton have been developed leading to increase in the production of these crops with reduced pesticide application worldwide (Yang et al. 2011; Abbas et al. 2013; Blanco 2012). Transgenic lines of *Arabidopsis* (AVP1) and *Nicotiana tabacum* (PMA4) with modified H+−ATPase coding gene have been established with enhanced H+ efflux capabilities from roots, salinity resistance, phosphate mineralization, drought resistance, and auxin uptake (Yang et al. 2007; Gevaudant et al. 2007). Transgenic lines of *Medicago sativa, Brassica napus, Hordeum vulgare* (barley), sorghum, carrot, rice, tomato, and tobacco plants have also been established by transformation with genes encoding proteins for synthesis of citrate (citrate synthetase), extrusion of multidrug and toxic compound (MATE), and transport of malate (i.e., Al^{3}+-activated malate transporter gene or ALMT). The latter enhances the efflux of these anions from the roots and subsequently confers aluminum resistance and efficient phosphorus uptake (Koyama et al. 2000; Tesfaye et al. 2001; Delhaize et al. 2007). In recent times, the manipulation of regulatory genes (transcription factors or TFs) to establish stress-tolerant stable crops has emerged as an effective strategy. Transgenic rice, wheat, potato, apple, tobacco, sugarcane, alfalfa, and *Arabidopsis* with enhanced tolerance for drought, salinity, and cold have been developed with the capability of overexpressing TFs involved in regulating stress-responsive genes for abscisic acid (ABA)-dependent pathway or ABA-independent pathway (Wang et al. 2016a, b).

4.3.2.2 Microbiome-Based Approach

Microbiome-based approach involves either direct inoculation of individual microorganism or co-inoculation of mixed cultures of PGPR, AMF, ectomycorrhizal fungi (EMF), and endophytes to modulate crop productivity (Ping and Boland 2004; Ryan et al. 2009). PGPR with BCA promotes plant growth collaboratively through their abilities of biofertilization and phytostimulation through phosphate solubilization, siderophore production, nutrient and mineral uptake, and symbiosis for nitrogen assimilation: plant hormone production on one hand and disease suppression by inducing ISR in plants or through production of antifungal compounds (phenazines, pyoluteorin, and phloroglucinols), antibiotics (hydrogen cyanide, oligomycin, phenazine), and bacteriocins, on the other (Ping and Boland 2004; Paulin et al. 2009). Microbiome can also alter the plant metabolic profile toward producing better yield. However, this method of application of microbial inoculants into the soil requires the availability of cultured isolates and maintenance of their cultivability in soil (Quiza et al. 2015). An alternative strategy is to enhance plant performance through inoculation of recombinant microbial strains into the soil. The GM microorganisms (GMO) not only have an enhanced capability to specifically stimulate plant growth and kill pathogens, but they also stimulate the growth of members from indigenous soil community through transmission of genetic information by horizontal gene transfer [HGT] (Quiza et al. 2015). PGPR/BCA activities are also enhanced in GMO. The first report for GMO was *chiA*-introduced heterologous bacteria. This engineered species degrade chitin from fungal cell membrane to impart suppression of fungal infection. Engineered strains of *Escherichia coli* and *P. fluorescens* containing *chiA* could effectively control the infections caused by *Sclerotium rolfsii* in bean and *Fusarium oxysporum* f. sp. *redolens* and *Gaeumannomyces graminis* (*G. graminis*) var. *tritici* in wheat, respectively (Shapira et al. 1989). Transformation of ACC-deaminase gene, *acdS*, from *P. putida* into *P. fluorescens* CHA0 strains improved phytostimulation in canola seedlings and disease resistance in cucumber against *Pythium* sp. (Wang et al. 2000). A number of studies have shown that constitutive production or overproduction of antibiotics/antifungal compounds by engineered bacterial strains (*P. fluorescens* Q2–87; *P. fluorescens* BL915) and their subsequent application in field crops reduced the occurrence of diseases in plants (take-all disease and root rot) even more effectively than the wild types (Alsanius et al. 2002). The first study on the effect of inoculated microbes to inhibit pathogenesis in soil was performed by introducing diacetylphloroglucinol (DAPG) producing *P. fluorescens* strains to suppress the growth of *G. graminis* var. *tritici* (Ggt), the causative agent of take-all disease in wheat (Kwak and Weller 2013). Even pretreatment of soil with recombinant strains effectively decontaminated it, reducing the rate of disease outbreak (Timms-Wilson et al. 2000). In addition to application of wild-type and/or recombinant strains, disruption of indigenous microbial population through imposition of mechanical (tillage) or chemical (fungicides, antibiotics) disturbances and thereby introducing beneficial microorganisms in the rhizosphere are another method to establish exogenous communities and modulate the rhizosphere milieu (Bulluck and Ristaino 2002).

4.3.2.3 Meta-Organism-Based Techniques

Interdependence of plants and the microbes in the rhizosphere redefines plant and the rhizobiome collectively as a metabiome or holobiont (Lakshmanan et al. 2014). Therefore, a school of thought in rhizosphere engineering is in the favor of addressing both the partners together, instead of accounting on them separately. This approach is addressed in two ways: (1) crop rotation and (2) inoculation of GMC with GMO.

Crop rotation is a decade-old method that has been applied extensively worldwide. This approach basically involves culturing of plants in turns, so that the residual microorganisms and phytochemicals in the soil from one plant might be beneficial to the next, and thereby an associative rhizo-microbiome can be established (Quiza et al. 2015). Various reports have been documented for utilities of crop rotation. A study from North America showed the association of higher diversity of AMF and the antagonist species *Penicillium canescens* with two cultivars of chickpea (CDC Anna and CDC Amit, respectively). This in turn influenced the productivity of the soil that subsequently helped in the establishment of durum wheat in that same soil (Ellouze et al. 2013). Similarly, the alternate cropping of potato with alfalfa, white lupin, and oats promoted potato yield (Honeycutt 1998). This approach induces the formation of disease-suppressive soils. In addition, this improves organic carbon content, nutrient cycling, and physicochemical characteristics of soil, thereby promoting a diverse microbial community (Honeycutt 1998; Mazzola 2002, 2007).

Inoculation of genetically engineered plants with genetically engineered organisms basically stimulates the plant to exudate specific chemicals which can be degraded by the selected GMO, thereby causing a proliferation of a specific group of organisms. An example is "opine concept" where it has been observed that the transgenic plants (*Lotus corniculatus*) modified to produce opine which is a xenotopic compound produced from the *Agrobacterium tumefaciens*-induced tumor. Opine, in turn, selects opine-degrading bacteria over others that could maintain themselves at high concentrations, even after removal of the transgenic plants (Oger et al. 1997; Savka et al. 2002). A similar approach was adopted to study the interaction between rhizopine-synthesizing transgenic *Arabidopsis* and rhizopine-degrading transgenic bacteria, and rhizopine-degrading strains were favored in a rhizopine-rich environment (Gordon et al. 1996).

4.3.3 Rhizosphere Engineering in Bioremediation

The hindsight of urbanization, advancement, and development of technologies is increased accumulation of chemical/industrial/agricultural remnants which are often biohazardous. Therefore, there is also a need for "environment cleaning" to save our earth. Most of these pollutants are accumulated in soil and cannot be removed easily until the soil is excavated and treated at a particular site. However, this is arduous and expensive. Therefore, "bioremediation," i.e., the use of plants and their associated microbes to assimilate and degrade/stabilize/volatilize the pollutants, has become an attractive substitute (Eapen and D'Souza 2005; Pilon-Smits 2005). A huge body of literature suggests that bioremediation has been globally accepted as a cost-effective and environment-friendly alternative or complementary technology for conventional remediation (Clemens et al. 2002; Gisbert et al. 2003; Eapen and D'Souza 2005).

Microbial activity and plant intervention are both required for biodegradation of pollutants (Yergeau et al. 2014). Plants can directly take up the pollutants from the soil and degrade them to less bioavailable forms via precipitation in the rhizosphere or via phytase activity, a process known as phytodegradation (Newman and Reynolds 2004). In some cases, a part of the pollutants are lost into the atmosphere during transpiration through leaves. This is termed as phytovolatilization (Zhu and Rosen 2009). However, hydrophobic organic compounds which cannot be taken up by the plants are degraded by the rhizosphere microorganisms. Herbicides, trinitrotoluene (TNT), methyl tertiary butyl ether (MTBE), and trichloroethylene (TCE), which are mobile within plant tissues, are usually degraded by the plants directly, while polychlorinated biphenyls (PCBs), polycyclic aromatic hydrocarbons (PAHs), and other petroleum hydrocarbons are degraded by rhizosphere microorganisms (Harvey et al. 2002).

The process in which the plants stimulate soil microbes to degrade the pollutants is called phytostimulation or rhizodegradation. Besides participating in biodegradation, the microbes also stimulate the plants to take up contaminants from soil and to combat stresses developed due to accumulation of pollutants (Taghavi et al. 2005; Bell et al. 2015). It has been demonstrated that wild-type/genetically engineered microbial strains/microbial consortia accelerate the degradation of pollutants more effectively in the presence of plants (rhizoremediation) to bioaugmentation, the latter indicating the application of microbes in the soil without the plants. This is presumably because rhizosphere processes mediated cooperatively by plants and rhizobiota supplement the activity required for remediation, which cannot take place in bulk soil (Zhuang et al. 2007). Furthermore, plant roots provide greater surface area, transport the microbes to that proper depth of the soil where the contaminants are present, and induce soil aeration, which enhance oxidative degradation of recalcitrant compounds (Chaudhry et al. 2005). Rhizoremediation is thus better over bioaugmentation (Zhuang et al. 2007).

The process of rhizoremediation which occurs naturally may be manipulated by engineering suitable plant–microbe pairs, such as plant–PGPR or plant-contaminant degrading microbes (Bisht et al. 2015). Beneficial plant–microbe symbioses have been exploited for rhizoremediation of hazardous and xenobiotic compounds like PAHs, PCBs, and TCE by choosing the right type of plant cultivar with appropriate rhizobacteria or by mechanically injecting efficient rhizobacterial strains on plant seeds/roots (Narasimhan et al. 2003; Walton and Anderson 1990). A wide variety of plants including alfalfa, barley, grass, lupin, oat, pepper, pine, poplar, radish, rape, sugarbeat, wheat, willow, and corresponding rhizobacterial strains such as *Pseudomonas fluorescens*, *Burkholderia cepacia*, *Pseudomonas putida*, *Bacillus* sp., *Deinococcus* sp., *Kurthia* sp., *Micrococcus* sp., *Arthrobacter* sp., and *Actinomycetes* have been identified (Kuiper et al. 2001; Bisht et al. 2014).

However, it is to be remembered that rhizoremediation is the outcome of activities of an entire microbial consortium, rather than a particular species (Kuiper et al. 2004). Although there is no dearth of studies involving isolation and characterization of pollutant-degrading rhizobacterial strains, studies on specific plant–microbe pair selection for rhizoremediation system are still limited (Bisht et al. 2015). Nevertheless, attempts of rhizoengineering by modifying plants to increase their size/number and augmenting their competence for biodegradation have been widely undertaken (Kabouw et al. 2012; Zhang et al. 2015b). The process relies on the following elements: (i) enhancement of root biomass to foster accumulation of high quantity of contaminants so that stress tolerance toward accumulated substances develops, (ii) stimulation of secretion of enzymes that will mobilize and degrade the noxious waste, and (iii) modulation of root exudation to attract microbes which are capable of degrading specific pollutants (Bhargava et al. 2012; Abhilash et al. 2009; Lojkova et al. 2014).

For competent biodegradation, plants require the presence of membrane transporter proteins (MTPs) that will export inorganic metal ions from the soil to the root xylem. Transgenic *Arabidopsis* and tobacco plants with overexpressing genes encoding membrane transport proteins result in increased uptake and accumulation of inorganic pollutants and heavy metals like cadmium, calcium, nickel, lead, manganese, and zinc in the plant tissues (Arazi et al. 1999; Hirschi et al. 2000; Van der Zaal et al. 1999). Recombinant DNA technology has mainly been applied to existing hyperaccumulator plant species (*Thlaspi caerulescens*, a natural zinc–cadmium hyperaccumulator) and high biomass species (Pence et al. 2000). Protein engineering has been applied to model plant, *Arabidopsis*, to increase specificity of transport proteins for heavy metals (Rogers et al. 2000). Transformation of *Arabidopsis* with pea metallothionein-like gene *PsMTA* enhanced their capacity to chelate metal ions (Evans et al. 1992). Overexpression of glutathione synthetase and γ-glutamylcysteine synthetase in *Brassica juncea* (Indian mustard) enhanced cadmium tolerance and accumulation (Zhu et al. 1999). Iron fortification of rice seed by the soybean ferritin gene was also established (Goto et al. 1999). In another study, transfer of the yeast metallothionein gene (CUP1) caused remarkable improvement of heavy metal tolerance in GM plants (Thomas et al. 2003). Transgenic plants with bacterial pentaerythritol tetranitrate reductase and nitroreductase genes were reported to be more efficient in reductive transformation of TNT (French et al. 1999; Hannink et al. 2007). Even enhanced metabolism of halogenated hydrocarbons in transgenic plants containing mammalian cytochrome P450 2E1 was reported (Doty et al. 2000). Volatilization of heavy metals like mercury and arsenic in wild plants occurs in a very limited manner. However, introduction of a modified bacterial *merA* gene enhanced resistance and rate of reduction for mercury ions in *Arabidopsis* plants (Rugh et al. 1996).

Following AMF inoculation (either single or in combination with PGPR), stress tolerance for heavy metals like cadmium, cesium, iron, lead, trace elements (arsenic), PAHs, PCBs (petroleum), and their accumulation was enhanced in various plants including medicinal plants (*Cassia italica Mill*), food crops (sorghum, barley, oats, legumes, rice), flowering plants (sunflower), switch grass, rye grass, *Miscanthus* sp. via a variety of mechanisms including increasing chlorophyll content, endogenous hormone level, and protein content in the host plant and subsequently decreasing lipid oxidation, accumulation of ROS, and synthesis of ABA, peroxidase (PO), and superoxide dismutase (SOD) (Hashem et al. 2016; Huang et al. 2015; Arora et al. 2015; Mishra et al. 2015; de Melo et al. 2014; Xun et al. 2015; Chan et al. 2013; Arias et al. 2015; Cabral et al. 2015; Lu et al. 2014; Firmin et al. 2015).

However, to date, most studies have been performed under laboratory conditions. Due to various confounding factors, field trials appear to be more complex than anticipated. One field trial for selenium-resistant transgenic *Brassica juncea* (Indian mustard) was undertaken that showed enhanced Se accumulation in the field (Pilon-Smits et al. 1999; Zhu et al. 1999). The availability of genomic sequences of *Arabidopsis* and rice has led to the identification and manipulation of novel key genes and regulatory elements (transcription factors/tissue-specific transporters) for establishment of high biomass species for pollutant remediation and accelerated the pace of translational research and development of technology. Transgenic plants with modified hairy roots to enhance their capacity of absorbing inorganic chemicals have emerged as an attractive model system in the field of phytoremediation (Ibanez et al. 2015). Tailored transgenics is also emerging as a tool to study plant-site-specific or environment-specific gene expression and manipulation toward ecosystem management and environmental cleaning. Hyper-accumulation and increased tolerance for arsenic were established in plants by combining arsenate reductase and gamma-glutamylcysteine synthetase expression to combat arsenic toxicity (Dhankher et al. 2002).

4.3.4 Rhizoengineering for Industrial Application

Knowledge on plant–microbe interactions in the rhizosphere has unfolded diverse functionality of soil microorganisms in medicine and chemical industries in addition to agriculture. In this regard, isolation and purification of “enzymes” from various microbial strains suggest the potential use of microbes in biotechnological and/

or industrial processes. Presently, genetic engineering and protein engineering techniques have been applied to improve the production of enzymes both qualitatively and quantitatively (Gurung et al. 2013). Proteases and carbohydrases such as amylase and cellulase are the dominant enzyme groups isolated from the rhizosphere microorganisms because of their extensive use in dairy, detergent, textile, baking, and starch industries (Underkofler et al. 1958). Presently, hydrolases, which catalyze breakdown of molecules in the presence of water, find an extensive application in industries manufacturing food and beverages, cleaning supplies, clothing, paper products, transportation fuels, pharmaceuticals, and monitoring devices (Gurung et al. 2013).

The species under *Bacillus* genera (*Bacillus subtilis, Bacillus licheniformis, Bacillus stearothermophilus*, and *Bacillus amyloliquefaciens*) serve as the richest source of industrial amylases (Konsoula and Liakopoulou-Kyriakides 2007; Sokarda Slavić et al. 2016). Thermostable amylases isolated from *Bacillus subtilis, Bacillus licheniformis, Bacillus stearothermophilus*, and *Bacillus amyloliquefaciens* are widely used for starch degradation and production of crystalline sugar, dextrose syrup, and maltodextrins (Hua et al. 2014; Hwang et al. 1997). Amylases that can particularly function at halophilic environments have been isolated from halophilic bacteria such as *Chromohalobacter* sp., *Halobacillus* sp., *Haloarcula hispanica, Halomonas meridiana*, and *Bacillus dipsosauri* (Gupta et al. 2016; Kumar et al. 2012). Lipase derived from *Bacillus, Burkholderia* (*Achromobacter* sp.), *Pseudomonas, Enterococcus*, and *Arthrobacter* species are used in food, textile, detergent, cosmetic, biosensor, and medicine industries (Gurung et al. 2013). These enzymes have also found their use in therapeutics at a limited scale. Notable examples include streptokinase from *Streptomycetes*, urokinase from *Bacillus subtilis*, and glutaminase from *E. coli*, and these are used to treat thrombosis and leukemia, respectively (Banerjee et al. 2004; Zaitsev et al. 2010; Spiers and Wade 1976).

In addition, the microbes are genetically modified using genes/transcription factors from metabolic pathways or stress regulatory network to produce high quantities of metabolites like ethanol, N-butanol, glycerol, and mannitol which have wide applications as solvent, extractants, antifreeze, dye base, lubricants, detergents, pesticides, resins, explosives, plasticizers, synthetic fibers, brake fluids, and petroleum derivatives and also in medicine and food industry (reviewed in Jia et al. 2014).

The production of "biofuels" using plants forms another important application of rhizosphere biology. One of the important sources of biofuel is "biomass," i.e., deposition of free energy from photosynthesis. Usually, nonfood crops or residues are used as feedstock for biofuel production. Vegetative parts from sugarcane (*Saccharum* sp.), poplar, switch grass (*Panicum virgatum*), *Miscanthus* species (*Miscanthus x giganteus*), and *Erianthus* species (*Erianthus arundinaceus* Retz.) are utilized for this purpose (Furtado et al. 2014). Thus, any engineering event in plants that accelerate vegetative meristematic activity is advantageous for biofuel production. Genetic diversity among *Saccharum* sp., *Erianthus* sp., and *Miscanthus* sp. has been exploited in breeding programs targeting different genetic markers, growth factors (GFs), enzymes, and transcription factors to introduce disease resistance, adaptability feature, and biofuel traits (Zhu et al. 2014; De Souza et al. 2015).

Genetically modified sugarcane with high biomass and cellulose-degrading microbes has been used for biofuel production (reviewed in Arruda 2012). In another study, tobacco plants transformed with NAC family genes from *Lepidium latifolium* gave rise to increased production of a number of transcription factors that resulted in marked improvement of plant biomass indicating the future potential of NAC gene transgenesis in biofuel production (Singh et al. 2016). Attempts to manipulate cellulase and laccase production in *Arabidopsis*, maize, and rice have also been successful in providing a new direction toward production of lignocellulose-based biofuel (reviewed in Wang et al. 2015). Syngas, produced from lignocellulose, can be fermented to biofuels using acetogenic bacteria such as *Eubacterium limosum*, *Clostridium autoethanogenum*, or *Acetobacterium woodii* (Bertsch and Muller 2015). GM microorganisms with abilities to use hemicellulose-derived C5 sugars (pentoses) may also aid production of biofuel as pentose constitutes one-third of the lignocellulose component of biomass (Silva et al. 2010). In this respect, genetically engineered *Cyanobacteria* are worthy of mentioning as they are being largely used to convert CO_2 into various chemicals directly (Lai and Lan 2015).

4.4 Conclusion and Future Perspectives

Of the countless problems and challenges our globe is facing today, perhaps the most overwhelming is how to shape the "Fourth Industrial Revolution" that has been initiated in this century. New concept, information, and technologies from physical, digital, and biological worlds are propelling toward bringing about an altruistic societal change. Perhaps "biological science" has to offer the most important contribution in today's industrial revolution. Armed with genetic and protein engineering, the new era of synthetic biology integrates engineering to biological principles toward establishment of more systematic, efficient, robust, predictable, and scalable biological systems. The time is just ripe to harness the knowledge of rhizosphere biology with technology to yield fascinating results with beneficial impacts on mankind. Although there is monumental progress in understanding the existing plant–microbe coordination, in-depth knowledge is still missing in many parts. This lacuna needs to be bridged for maintenance of progress rate. Advances have been made utilizing "system approaches" to identify key molecular players (such as genes, RNAs, proteins, etc.) in plant–microbe cross talk associated with plant health and productivity. However, exigent issues still exist and need to be tackled with urgent priority. Last, but not the least, the plant performance needs to be investigated at a population scale. Therefore, multiscale mechanistic models that will link plant, microbes, and field ought to be developed taking care of influential environmental factors (Hill et al. 2013). Developing mixed genetic-ecophysiological models to amend the gap between genetic and environmental parameters is an important goal that may help in overcoming the constraints still prevailing while manipulating genes in recombinant species (Roose and Schnepf 2008). For example, implementation of "synthetic biology" for successful biotransformation has often proven cumbersome due to vulnerability of host organisms to intermittent and unpredictable environmental parameters (Jia et al. 2014). A robust knowledge of metabolic network and the mechanisms of systemic resistance operating in microorganisms at both cellular and community levels will provide solution to this problem. Besides, there is also a necessity of increasing public awareness and acceptance of genetically modified products (Adenle 2011; Kikulwe et al. 2011). This of course calls for coordinated participation of scientists and other professionals to spread the true information to the community.

References

Abbas HK, Zablotowicz RM, Weaver MA, Shier WT, Bruns HA, Bellaloui N, Accinelli C, Abel CA (2013) Implications of Bt traits on mycotoxin contamination in maize: overview and recent experimental results in southern United States. J Agric Food Chem 61(48):11759–11770. doi:10.1021/jf400754g

Abhilash PC, Jamil S, Singh N (2009) Transgenic plants for enhanced biodegradation and phytoremediation of organic xenobiotics. Biotechnol Adv 27(4):474–488. doi:10.1016/j.biotechadv.2009.04.002S0734-9750(09)00052-4 [pii]

Adenle AA (2011) Response to issues on GM agriculture in Africa: are transgenic crops safe? BMC Res Notes 4:388. doi:10.1186/1756-0500-4-3881756-0500-4-388 [pii]

Ajilogba CF, Babalola OO (2013) Integrated management strategies for tomato *Fusarium* wilt. Biocontrol Sci 18(3):117–127. doi:DN/JST.JSTAGE/bio/18.117 [pii]

Ali JG, Alborn HT, Stelinski LL (2010) Subterranean herbivore-induced volatiles released by citrus roots upon feeding by *Diaprepes abbreviatus* recruit entomopathogenic nematodes. J Chem Ecol 36(4):361–368. doi:10.1007/s10886-010-9773-7

Alsanius BW, Hultberg M, Englund JE (2002) Effect of lacZY-marking of the 2,4-diacetylphloroglucinol producing *Pseudomonas fluorescens*-strain 5-2/4 on its physiological performance and root colonization ability. Microbiol Res 157(1):39–45

Apine OA, Jadhav JP (2011) Optimization of medium for indole-3-acetic acid production using *Pantoea agglomerans* strain PVM. J Appl Microbiol 110(5):1235–1244. doi:10.1111/j.1365-2672.2011.04976.x

Araus JL, Li J, Parry MA, Wang J (2014) Phenotyping and other breeding approaches for a New Green Revolution. J Integr Plant Biol 56(5):422–424. doi:10.1111/jipb.12202

Arazi T, Sunkar R, Kaplan B, Fromm H (1999) A tobacco plasma membrane calmodulin-binding transporter confers Ni2+ tolerance and Pb2+ hypersensitivity in transgenic plants. Plant J 20(2):171–182. doi:tpj588 [pii]

Arias MS, Pena-Cabriales JJ, Alarcon A, Maldonado Vega M (2015) Enhanced Pb absorption by *Hordeum vulgare* L. and *Helianthus annuus* L. plants inoculated with an arbuscular mycorrhizal fungi consortium. Int J Phytoremediation 17(1–6):405–413. doi:10.1080/15226514.2014.898023

Arora K, Sharma S, Monti A (2015) Bio-remediation of Pb and Cd polluted soils by switchgrass: a case study in India. Int J Phytoremediation. doi:10.1080/15226514.2015.1131232

Arruda P (2012) Genetically modified sugarcane for bioenergy generation. Curr Opin Biotechnol 23(3):315–322. doi:10.1016/j.copbio.2011.10.012

Babikova Z, Gilbert L, Bruce TJ, Birkett M, Caulfield JC, Woodcock C, Pickett JA, Johnson D (2013) Underground signals carried through common mycelial networks warn neighbouring plants of aphid attack. Ecol Lett 16(7):835–843. doi:10.1111/ele.12115

Badri DV, Vivanco JM (2009) Regulation and function of root exudates. Plant Cell Environ 32(6):666–681. doi:10.1111/j.1365-3040.2008.01926.xPCE1926 [pii]

Badri DV, Weir TL, van der Lelie D, Vivanco JM (2009) Rhizosphere chemical dialogues: plant-microbe interactions. Curr Opin Biotechnol 20(6):642–650. doi:10.1016/j.copbio.2009.09.014 S0958-1669(09)00128-1 [pii]

Badri DV, Chaparro JM, Zhang R, Shen Q, Vivanco JM (2013) Application of natural blends of phytochemicals derived from the root exudates of Arabidopsis to the soil reveal that phenolic-related compounds predominantly modulate the soil microbiome. J Biol Chem 288(7):4502–4512. doi:10.1074/jbc.M112.433300M112.433300 [pii]

Bagyaraj DJ, Rangaswami G (2005) Microorganisms in soil. In: Agricultural microbiology, 2nd edn. Prentice Hall of India Private Limited, New Delhi, p 254

Bais HP, Park SW, Weir TL, Callaway RM, Vivanco JM (2004) How plants communicate using the underground information superhighway. Trends Plant Sci 9(1):26–32. doi:10.1016/j.tplants.2003.11.008 S1360-1385(03)00302-9 [pii]

Bais HP, Weir TL, Perry LG, Gilroy S, Vivanco JM (2006) The role of root exudates in rhizosphere interactions with plants and other organisms. Annu Rev Plant Biol 57:233–266. doi:10.1146/annurev.arplant.57.032905.105159

Balseiro-Romero M, Kidd PS, Monterroso C (2014) Influence of plant root exudates on the mobility of fuel volatile compounds in contaminated soils. Int J Phytoremediation 16(7–12):824–839. doi:10.1080/15226514.2013.856851

Banerjee A, Chisti Y, Banerjee UC (2004) Streptokinase – a clinically useful thrombolytic agent. Biotechnol Adv 22(4):287–307. doi:S0734975003001678 [pii]

Barrada A, Montane MH, Robaglia C, Menand B (2015) Spatial regulation of root growth: placing the plant TOR pathway in a developmental perspective. Int J Mol Sci 16(8):19671–19697. doi:10.3390/ijms160819671 ijms160819671 [pii]

Bawa AS, Anilakumar KR (2013) Genetically modified foods: safety, risks and public concerns-a review. J Food Sci Technol 50(6):1035–1046. doi:10.1007/s13197-012-0899-1 899 [pii]

Bekkara F, Jay M, Viricel MR, Rome S (1998) Distribution of phenolic compounds within seed and seedlings of two Vicia faba cvs differing in their see tannin content and study of their seed and root phenolic exudations. Plant Soil 203:27–36

Bell TH, Cloutier-Hurteau B, Al-Otaibi F, Turmel MC, Yergeau E, Courchesne F, St-Arnaud M (2015) Early rhizosphere microbiome composition is related to the growth and Zn uptake of willows introduced to a former landfill. Environ Microbiol 17(8):3025–3038. doi:10.1111/1462-2920.12900

Bennett AE, Bever JD (2007) Mycorrhizal species differentially alter plant growth and response to herbivory. Ecology 88(1):210–218

Berendsen RL, Pieterse CM, Bakker PA (2012) The rhizosphere microbiome and plant health. Trends Plant Sci 17(8):478–486. doi:10.1016/j.tplants.2012.04.001 S1360-1385(12)00079-9 [pii]

Berg G, Smalla K (2009) Plant species and soil type cooperatively shape the structure and function of microbial communities in the rhizosphere. FEMS Microbiol Ecol 68(1):1–13. doi:10.1111/j.1574-6941.2009.00654.x FEM654 [pii]

Bertsch J, Muller V (2015) Bioenergetic constraints for conversion of syngas to biofuels in acetogenic bacteria. Biotechnol Biofuels 8:210. doi:10.1186/s13068-015-0393-x 393 [pii]

Bhardwaj D, Ansari MW, Sahoo RK, Tuteja N (2014) Biofertilizers function as key player in sustainable agriculture by improving soil fertility, plant tolerance and crop productivity. Microb Cell Fact 13:66. doi:10.1186/1475-2859-13-66 1475-2859-13-66 [pii]

Bhargava A, Carmona FF, Bhargava M, Srivastava S (2012) Approaches for enhanced phytoextraction of heavy metals. J Environ Manage 105:103–120. doi:10.1016/j.jenvman.2012.04.002 S0301-4797(12)00183-1 [pii]

Bhattacharyya PN, Jha DK (2012) Plant growth-promoting rhizobacteria (PGPR): emergence in agriculture. World J Microbiol Biotechnol 28(4):1327–1350. doi:10.1007/s11274-011-0979-9

Bisht S, Pandey P, Aggarwal H et al (2014) Utilization of endophytic strain Bacillus sp. SBER3 for biodegradation of polyaromatic hydrocarbons (PAH) in soil model system. Eur J Soil Biol 60:67–76

Bisht S, Pandey P, Bhargava B, Sharma S, Kumar V, Sharma KD (2015) Bioremediation of polyaromatic hydrocarbons (PAHs) using rhizosphere technology. Braz J Microbiol 46(1):7–21. doi:10.1590/S1517-838246120131354 1517-8382-bjm-46-01-0007 [pii]

Blanco CA (2012) *Heliothis virescens* and Bt cotton in the United States. GM Crops Food 3(3):201–212. doi:10.4161/gmcr.21439 21439 [pii]

Blossfeld S, Suessmilch S, Le Marie CA, Kuhn AJ (2011) Exploration of key rhizosphere parameters in plant-MFCs. Commun Agric Appl Biol Sci 76(2):7–9

Bonito G, Reynolds H, Robeson MS 2nd, Nelson J, Hodkinson BP, Tuskan G, Schadt CW, Vilgalys R (2014) Plant host and soil origin influence fungal and bacterial assemblages in the roots of woody plants. Mol Ecol 23(13):3356–3370. doi:10.1111/mec.12821

Bravin MN, Tentscher P, Rose J, Hinsinger P (2009) Rhizosphere pH gradient controls copper availability in a strongly acidic soil. Environ Sci Technol 43(15):5686–5691

Bulgarelli D, Rott M, Schlaeppi K, Ver Loren van Themaat E, Ahmadinejad N, Assenza F, Rauf P, Huettel B, Reinhardt R, Schmelzer E, Peplies J, Gloeckner FO, Amann R, Eickhorst T, Schulze-Lefert P (2012) Revealing structure and assembly cues for Arabidopsis root-inhabiting bacterial microbiota. Nature 488(7409):91–95. doi:10.1038/nature11336 nature11336 [pii]

Bulgarelli D, Schlaeppi K, Spaepen S, Ver Loren van Themaat E, Schulze-Lefert P (2013) Structure and functions of the bacterial microbiota of plants. Annu Rev Plant Biol 64:807–838. doi:10.1146/annurev-arplant-050312-120106

Bulluck LR, Ristaino JB (2002) Effect of synthetic and organic soil fertility amendments on southern blight, soil microbial communities, and yield of processing tomatoes. Phytopathology 92(2):181–189. doi:10.1094/PHYTO.2002.92.2.181

Cabral L, Soares CR, Giachini AJ, Siqueira JO (2015) Arbuscular mycorrhizal fungi in phytoremediation of contaminated areas by trace elements: mechanisms and major benefits of their applications. World J Microbiol Biotechnol 31(11):1655–1664. doi:10.1007/s11274-015-1918-y [pii]

Campbell MA, Fitzgerald HA, Ronald PC (2002) Engineering pathogen resistance in crop plants. Transgenic Res 11(6):599–613

Cezard C, Farvacques N, Sonnet P (2015) Chemistry and biology of pyoverdines, *Pseudomonas* primary siderophores. Curr Med Chem 22(2):165–186. doi:CMC-EPUB-62749 [pii]

Chan WF, Li H, Wu FY, Wu SC, Wong MH (2013) Arsenic uptake in upland rice inoculated with a combination or single arbuscular mycorrhizal fungi. J Hazard Mater 262:1116–1122. doi:10.1016/j.jhazmat.2012.08.020 S0304-3894(12)00826-6 [pii]

Chaudhry Q, Blom-Zandstra M, Gupta S, Joner EJ (2005) Utilising the synergy between plants and rhizosphere microorganisms to enhance breakdown of organic pollutants in the environment. Environ Sci Pollut Res Int 12(1):34–48

Chen Y, Wang Y, Wu W, Lin Q, Xue S (2006) Impacts of chelate-assisted phytoremediation on microbial community composition in the rhizosphere of a copper accumulator and non-accumulator. Sci Total Environ 356(1–3):247–255. doi:10.1016/j.scitotenv.2005.04.028 S0048-9697(05)00245-7 [pii]

Clemens S, Palmgren MG, Kramer U (2002) A long way ahead: understanding and engineering plant metal accumulation. Trends Plant Sci 7(7):309–315. doi:S1360-1385(02)02295-1 [pii]

Costa R, Gotz M, Mrotzek N, Lottmann J, Berg G, Smalla K (2006) Effects of site and plant species on rhizosphere community structure as revealed by molecular analysis of microbial guilds. FEMS Microbiol Ecol 56(2):236–249. doi:10.1111/j.1574-6941.2005.00026.x FEM026 [pii]

Cowgill SE, Wright C, Atkinson HJ (2002) Transgenic potatoes with enhanced levels of nematode resistance do not have altered susceptibility to nontarget aphids. Mol Ecol 11(4):821–827. doi:1482 [pii]

Crosa JH, Walsh CT (2002) Genetics and assembly line enzymology of siderophore biosynthesis in bacteria. Microbiol Mol Biol Rev 66(2):223–249

Curlango-Rivera G, Huskey DA, Mostafa A, Kessler JO, Xiong Z, Hawes MC (2013) Intraspecies variation in cotton border cell production: rhizosphere microbiome implications. Am J Bot 100(9):1706–1712. doi:10.3732/ajb.1200607 ajb.1200607 [pii]

Davison J, Opik M, Zobel M, Vasar M, Metsis M, Moora M (2012) Communities of arbuscular mycorrhizal fungi detected in forest soil are spatially heterogeneous but do not vary throughout the growing season. PLoS One 7(8), e41938. doi:10.1371/journal.pone.0041938 PONE-D-12-12295 [pii]

de Melo RW, Schneider J, de Souza CE, Sousa SC, Guimaraes GL, de Souza MF (2014) Phytoprotective effect of arbuscular mycorrhizal fungi species against arsenic toxicity in tropical leguminous species. Int J Phytoremediation 16(7–12):840–858. doi:10.1080/15226514.2013.856852

De Souza AP, Alvim Kamei CL, Torres AF, Pattathil S, Hahn MG, Trindade LM, Buckeridge MS (2015) How cell wall complexity influences saccharification efficiency in *Miscanthus sinensis*. J Exp Bot 66(14):4351–4365. doi:10.1093/jxb/erv183 erv183 [pii]

De Vos D, Vissenberg K, Broeckhove J, Beemster GT (2014) Putting theory to the test: which regulatory mechanisms can drive realistic growth of a root? PLoS Comput Biol 10(10), e1003910. doi:10.1371/journal.pcbi.1003910 PCOMPBIOL-D-13-02201 [pii]

Delhaize E, Gruber BD, Ryan PR (2007) The roles of organic anion permeases in aluminium resistance and mineral nutrition. FEBS Lett 581(12):2255–2262. doi:10.1016/j.febslet.2007.03.057 S0014-5793(07)00311-0 [pii]

Dessaux Y, Grandclement C, Faure D (2016) Engineering the rhizosphere. Trends Plant Sci 21(3):266–278. doi:10.1016/j.tplants.2016.01.002 S1360-1385(16)00003-0 [pii]

Dhankher OP, Li Y, Rosen BP, Shi J, Salt D, Senecoff JF, Sashti NA, Meagher RB (2002) Engineering tolerance and hyperaccumulation of arsenic in plants by combining arsenate reductase and gamma-glutamylcysteine synthetase expression. Nat Biotechnol 20(11):1140–1145. doi:10.1038/nbt747 nbt747 [pii]

Diagne N, Arumugam K, Ngom M, Nambiar-Veetil M, Franche C, Narayanan KK, Laplaze L (2013) Use of *Frankia* and *Actinorhizal* plants for degraded lands reclamation. Biomed Res Int 2013:948258. doi:10.1155/2013/948258

Dong YH, Wang LH, Xu JL, Zhang HB, Zhang XF, Zhang LH (2001) Quenching quorum-sensing-dependent bacterial infection by an N-acyl homoserine lactonase. Nature 411(6839):813–817. doi:10.1038/35081101

Doty SL, Shang TQ, Wilson AM, Tangen J, Westergreen AD, Newman LA, Strand SE, Gordon MP (2000) Enhanced metabolism of halogenated hydrocarbons in transgenic plants containing mammalian cytochrome P450 2E1. Proc Natl Acad Sci U S A 97(12):6287–6291. doi:97/12/6287 [pii]

Duca D, Lorv J, Patten CL, Rose D, Glick BR (2014) Indole-3-acetic acid in plant-microbe interactions. Antonie Van Leeuwenhoek 106(1):85–125. doi:10.1007/s10482-013-0095-y

Dunfield KE, Germida JJ (2003) Seasonal changes in the rhizosphere microbial communities associated with field-grown genetically modified canola (*Brassica napus*). Appl Environ Microbiol 69(12):7310–7318

Dutta S, Morang P, Kumar SN, Dileep Kumar BS (2014) Two rhizobacterial strains, individually and in interactions with *Rhizobium sp.*, enhance fusarial wilt control, growth, and yield in pigeon pea. J Microbiol 52(9):778–784. doi:10.1007/s12275-014-3496-3

Eapen S, D'Souza SF (2005) Prospects of genetic engineering of plants for phytoremediation of toxic metals. Biotechnol Adv 23(2):97–114. doi:10.1016/j.biotechadv.2004.10.001 S0734-9750(04)00094-1 [pii]

Egamberdieva D, Kamilova F, Validov S, Gafurova L, Kucharova Z, Lugtenberg B (2008) High incidence of plant growth-stimulating bacteria associated with the rhizosphere of wheat grown on salinated soil in Uzbekistan. Environ Microbiol 10(1):1–9. doi:10.1111/j.1462-2920.2007.01424.x, EMI1424 [pii]

Ehlers RU (2003) Entomopathogenic nematodes in the European biocontrol market. Commun Agric Appl Biol Sci 68(4 Pt A):3–16

Ellouze W, Hamel C, Vujanovic V, Gan Y, Bouzid S, St-Arnaud M (2013) Chickpea genotypes shape the soil microbiome and affect the establishment of the subsequent durum wheat crop in the semi arid North American Great Plains. Soil Biol Biochem 63:129–141. doi:10.1016/j.soilbio.2013.04.001

Evans KM, Gatehouse JA, Lindsay WP, Shi J, Tommey AM, Robinson NJ (1992) Expression of the pea metallothionein-like gene PsMTA in *Escherichia coli* and *Arabidopsis thaliana* and analysis of trace metal ion accumulation: implications for PsMTA function. Plant Mol Biol 20(6):1019–1028

Fahey C, Winter K, Slot M, Kitajima K (2016) Influence of arbuscular mycorrhizal colonization on whole-plant respiration and thermal acclimation of tropical tree seedlings. Ecol Evol 6(3):859–870. doi:10.1002/ece3.1952 ECE31952 [pii]

Farag MA, Zhang H, Ryu CM (2013) Dynamic chemical communication between plants and bacteria through airborne signals: induced resistance by bacterial volatiles. J Chem Ecol 39(7):1007–1018. doi:10.1007/s10886-013-0317-9

Farrar K, Bryant D, Cope-Selby N (2014) Understanding and engineering beneficial plant-microbe interactions: plant growth promotion in energy crops. Plant Biotechnol J 12(9):1193–1206. doi:10.1111/pbi.12279

Ferguson BJ, Mathesius U (2014) Phytohormone regulation of legume-rhizobia interactions. J Chem Ecol 40(7):770–790. doi:10.1007/s10886-014-0472-7

Firmin S, Labidi S, Fontaine J, Laruelle F, Tisserant B, Nsanganwimana F, Pourrut B, Dalpe Y, Grandmougin A, Douay F, Shirali P, Verdin A, Lounes-Hadj Sahraoui A (2015) Arbuscular mycorrhizal fungal inoculation protects *Miscanthus x giganteus* against trace element toxicity in a highly metal-contaminated site. Sci Total Environ 527–528:91–99. doi:10.1016/j.scitotenv.2015.04.116 S0048-9697(15)30050-4 [pii]

Fraley RT, Rogers SG, Horsch RB, Sanders PR, Flick JS, Adams SP, Bittner ML, Brand LA, Fink CL, Fry JS, Galluppi GR, Goldberg SB, Hoffmann NL, Woo SC (1983) Expression of bacterial genes in plant cells. Proc Natl Acad Sci U S A 80(15):4803–4807

Fray RG, Throup JP, Daykin M, Wallace A, Williams P, Stewart GS, Grierson D (1999) Plants genetically modified to produce N-acylhomoserine lactones communicate with bacteria. Nat Biotechnol 17(10):1017–1020. doi:10.1038/13717

French CE, Rosser SJ, Davies GJ, Nicklin S, Bruce NC (1999) Biodegradation of explosives by transgenic plants expressing pentaerythritol tetranitrate reductase. Nat Biotechnol 17(5):491–494. doi:10.1038/8673

Furtado A, Lupoi JS, Hoang NV, Healey A, Singh S, Simmons BA, Henry RJ (2014) Modifying plants for biofuel and biomaterial production. Plant Biotechnol J 12(9):1246–1258. doi:10.1111/pbi.12300

Gevaudant F, Duby G, von Stedingk E, Zhao R, Morsomme P, Boutry M (2007) Expression of a constitutively activated plasma membrane H+−ATPase alters plant development and increases salt tolerance. Plant Physiol 144(4):1763–1776. doi:pp.107.103762 [pii] 10.1104/pp.107.103762

Ghosh R, Barman S, Mukherjee R, Mandal NC (2016) Role of phosphate solubilizing Burkholderia spp. for successful colonization and growth promotion of *Lycopodium cernuum* L. (*Lycopodiaceae*) in lateritic belt of Birbhum district of West Bengal, India. Microbiol Res 183:80–91. doi:10.1016/j.micres.2015.11.011 S0944-5013(15)30033-1 [pii]

Gisbert C, Ros R, De Haro A, Walker DJ, Pilar Bernal M, Serrano R, Navarro-Avino J (2003) A plant genetically modified that accumulates Pb is especially promising for phytoremediation. Biochem Biophys Res Commun 303(2):440–445. doi:S0006291X03003498 [pii]

GM Approval Databse-ISAAA.org (2016) http://www.isaaa.org. Accessed 5 Apr 2016

Gordon DM, Ryder MH, Heinrich K, Murphy PJ (1996) An experimental test of the rhizopine concept in *Rhizobium meliloti*. Appl Environ Microbiol 62(11):3991–3996

Goto F, Yoshihara T, Shigemoto N, Toki S, Takaiwa F (1999) Iron fortification of rice seed by the soybean ferritin gene. Nat Biotechnol 17(3):282–286. doi:10.1038/7029

Gregory PJ, Atkinson CJ, Bengough AG, Else MA, Fernandez-Fernandez F, Harrison RJ, Schmidt S (2013) Contributions of roots and rootstocks to sustainable, intensified crop production. J Exp Bot 64(5):1209–1222. doi:10.1093/jxb/ers385 ers385 [pii]

Gtari M, Ghodhbane-Gtari F, Nouioui I, Beauchemin N, Tisa LS (2012) Phylogenetic perspectives of nitrogen-fixing *Actinobacteria*. Arch Microbiol 194(1):3–11. doi:10.1007/s00203-011-0733-6

Guerinot ML (1994) Microbial iron transport. Annu Rev Microbiol 48:743–772. doi:10.1146/annurev.mi.48.100194.003523

Gupta S, Sharma P, Dev K, Sourirajan A (2016) Halophilic bacteria of Lunsu produce an array of industrially important enzymes with salt tolerant activity. Biochem Res Int 2016:9237418. doi:10.1155/2016/9237418

Gurung N, Ray S, Bose S, Rai V (2013) A broader view: microbial enzymes and their relevance in industries, medicine, and beyond. Biomed Res Int 2013:329121. doi:10.1155/2013/329121

Haas D, Keel C (2003) Regulation of antibiotic production in root-colonizing *Pseudomonas spp.* and relevance for biological control of plant disease. Annu Rev Phytopathol 41:117–153. doi:10.1146/annurev.phyto.41.052002.095656 052002.095656 [pii]

Hannink NK, Subramanian M, Rosser SJ, Basran A, Murray JA, Shanks JV, Bruce NC (2007) Enhanced transformation of tnt by tobacco plants expressing a bacterial nitroreductase. Int J Phytoremediation 9(5):385–401. doi:10.1080/15226510701603916

Hardtke CS, Dorcey E, Osmont KS, Sibout R (2007) Phytohormone collaboration: zooming in on auxin-brassinosteroid interactions. Trends Cell Biol 17(10):485–492. doi:10.1016/j.tcb.2007.08.003 S0962-8924(07)00191-2 [pii]

Harvey PJ, Campanella BF, Castro PM, Harms H, Lichtfouse E, Schaffner AR, Smrcek S, Werck-Reichhart D (2002) Phytoremediation of polyaromatic hydrocarbons, anilines and phenols. Environ Sci Pollut Res Int 9(1):29–47

Hashem A, Abd Allah EF, Alqarawi AA, Egamberdieva D (2016) Bioremediation of adverse impact of cadmium toxicity on *Cassia italica* Mill by arbuscular mycorrhizal fungi. Saudi J Biol Sci 23(1):39–47. doi:10.1016/j.sjbs.2015.11.007 S1319-562X(15)00277-6 [pii]

Hill K, Porco S, Lobet G, Zappala S, Mooney S, Draye X, Bennett MJ (2013) Root systems biology: integrative modeling across scales, from gene regulatory networks to the rhizosphere. Plant Physiol 163(4):1487–1503. doi:10.1104/pp.113.227215 pp.113.227215 [pii]

Hiltner L (1904) U¨ ber neuere Erfahrungen und Probleme auf dem Gebiete der Bodenbakteriologie unterbessonderer Ber ¨ ucksichtigung der Gr¨undung und Brache. Arb Dtsch Landwirtsch Ges Berl 98:59–78

Hiltpold I, Turlings TC (2012) Manipulation of chemically mediated interactions in agricultural soils to enhance the control of crop pests and to improve crop yield. J Chem Ecol 38(6):641–650. doi:10.1007/s10886-012-0131-9

Hiltpold I, Jaffuel G, Turlings TC (2015) The dual effects of root-cap exudates on nematodes: from quiescence in plant-parasitic nematodes to frenzy in entomopathogenic nematodes. J Exp Bot 66(2):603–611. doi:10.1093/jxb/eru345 eru345 [pii]

Hirschi KD, Korenkov VD, Wilganowski NL, Wagner GJ (2000) Expression of arabidopsis CAX2 in tobacco. Altered metal accumulation and increased manganese tolerance. Plant Physiol 124(1):125–133

Honeycutt CW (1998) Crop rotation impacts on potato protein. Plant Foods Hum Nutr 52(4):279–291

Hopkins F, Gonzalez-Meler MA, Flower CE, Lynch DJ, Czimczik C, Tang J, Subke JA (2013) Ecosystem-level controls on root-rhizosphere respiration. New Phytol 199(2):339–351

Hua H, Luo H, Bai Y, Wang K, Niu C, Huang H, Shi P, Wang C, Yang P, Yao B (2014) A thermostable glucoamylase from *Bispora sp.* MEY-1 with stability over a broad pH range and significant starch hydrolysis capacity. PLoS One 9(11), e113581. doi:10.1371/journal.pone.0113581 PONE-D-14-31239 [pii]

Huang RH, Yang HL, Huang W, Lu YM, Chen K (2015) Effects of *Funneliformis mosseae* on endogenous hormones and photosynthesis of *Sorghum haipense* under Cs stress. Ying Yong Sheng Tai Xue Bao 26(7):2146–2150

Hughes M, Donnelly C, Crozier A, Wheeler CT (1999) Effects of the exposure of roots Almus glutinosa to light on flavonoid and nodulation. Can J Bot 77:1311–1315

Hwang KY, Song HK, Chang C, Lee J, Lee SY, Kim KK, Choe S, Sweet RM, Suh SW (1997) Crystal structure of thermostable alpha-amylase from *Bacillus licheniformis* refined at 1.7 A resolution. Mol Cells 7(2):251–258

Ibanez S, Talano M, Ontanon O, Suman J, Medina MI, Macek T, Agostini E (2015) Transgenic plants and hairy roots: exploiting the potential of plant species to remediate contaminants. N Biotechnol. doi:S1871-6784(15)00267-8 [pii] 10.1016/j.nbt.2015.11.008

Jeong JS, Kim YS, Baek KH, Jung H, Ha SH, Do Choi Y, Kim M, Reuzeau C, Kim JK (2010) Root-specific expression of OsNAC10 improves drought tolerance and grain yield in rice under field drought conditions. Plant Physiol 153(1):185–197. doi:10.1104/pp.110.154773 pp.110.154773 [pii]

Jia H, Fan Y, Feng X, Li C (2014) Enhancing stress-resistance for efficient microbial biotransformations by synthetic biology. Front Bioeng Biotechnol 2:44. doi:10.3389/fbioe.2014.00044

Jones AG (2008) A theoretical quantitative genetic study of negative ecological interactions and extinction times in changing environments. BMC Evol Biol 8:119. doi:10.1186/1471-2148-8-1191471-2148-8-119[pii]

Kabouw P, van Dam NM, van der Putten WH, Biere A (2012) How genetic modification of roots affects rhizosphere processes and plant performance. J Exp Bot 63(9):3475–3483. doi:10.1093/jxb/err399 err399 [pii]

Kenney E, Eleftherianos I (2016) Entomopathogenic and plant pathogenic nematodes as opposing forces in agriculture. Int J Parasitol 46(1):13–19. doi:10.1016/j.ijpara.2015.09.005 S0020-7519(15)00260-X [pii]

Kent AD, Triplett EW (2002) Microbial communities and their interactions in soil and rhizosphere ecosystems. Annu Rev Microbiol 56:211–236. doi:10.1146/annurev.micro.56.012302.161120 012302.161120 [pii]

Khan AG (2006) Mycorrhizoremediation–an enhanced form of phytoremediation. J Zhejiang Univ Sci B 7(7):503–514. doi:10.1631/jzus.2006.B0503

Khan MI, Trivellini A, Fatma M, Masood A, Francini A, Iqbal N, Ferrante A, Khan NA (2015) Role of ethylene in responses of plants to nitrogen availability. Front Plant Sci 6:927. doi:10.3389/fpls.2015.00927

Khorassani R, Hettwer U, Ratzinger A, Steingrobe B, Karlovsky P, Claassen N (2011) Citramalic acid and salicylic acid in sugar beet root exudates solubilize soil phosphorus. BMC Plant Biol 11:121. doi:10.1186/1471-2229-11-121 1471-2229-11-121 [pii]

Kikulwe EM, Wesseler J, Falck-Zepeda J (2011) Attitudes, perceptions, and trust. Insights from a consumer survey regarding genetically modified banana in Uganda. Appetite 57(2):401–413. doi:10.1016/j.appet.2011.06.001 S0195-6663(11)00483-1 [pii]

Kisiel A, Kepczynska E (2016) *Medicago truncatula Gaertn.* as a model for understanding the mechanism of growth promotion by bacteria from rhizosphere and nodules of alfalfa. Planta. doi:10.1007/s00425-016-2469-7 10.1007/s00425-016-2469-7 [pii]

Klironomos JN (2002) Feedback with soil biota contributes to plant rarity and invasiveness in communities. Nature 417(6884):67–70. doi:10.1038/417067a 417067a [pii]

Kohli A, Leech M, Vain P, Laurie DA, Christou P (1998) Transgene organization in rice engineered through direct DNA transfer supports a two-phase integration mechanism mediated by the establishment of integration hot spots. Proc Natl Acad Sci U S A 95(12):7203–7208

Koller R, Robin C, Bonkowski M, Ruess L, Scheu S (2013a) Litter quality as driving factor for plant nutrition via grazing of protozoa on soil microorganisms. FEMS Microbiol Ecol 85(2):241–250. doi:10.1111/1574-6941.12113

Koller R, Rodriguez A, Robin C, Scheu S, Bonkowski M (2013b) Protozoa enhance foraging efficiency of arbuscular mycorrhizal fungi for mineral nitrogen from organic matter in soil to the benefit of host plants. New Phytol 199(1):203–211. doi:10.1111/nph.12249

Konsoula Z, Liakopoulou-Kyriakides M (2007) Co-production of alpha-amylase and beta-galactosidase by *Bacillus subtilis* in complex organic substrates. Bioresour Technol 98(1):150–157. doi:10.1016/j.biortech.2005.11.001 S0960-8524(05)00525-0 [pii]

Koyama H, Kawamura A, Kihara T, Hara T, Takita E, Shibata D (2000) Overexpression of mitochondrial citrate synthase in *Arabidopsis thaliana* improved growth on a phosphorus-limited soil. Plant Cell Physiol 41(9):1030–1037

Kuiper I, Bloemberg GV, Lugtenberg BJ (2001) Selection of a plant-bacterium pair as a novel tool for rhizostimulation of polycyclic aromatic hydrocarbon-degrading bacteria. Mol Plant Microbe Interact 14(10):1197–1205. doi:10.1094/MPMI.2001.14.10.1197

Kuiper I, Lagendijk EL, Bloemberg GV, Lugtenberg BJ (2004) Rhizoremediation: a beneficial plant-microbe interaction. Mol Plant Microbe Interact 17(1):6–15. doi:10.1094/MPMI.2004.17.1.6

Kumar S, Karan R, Kapoor S, Singh SP, Khare SK (2012) Screening and isolation of halophilic bacteria producing industrially important enzymes. Braz J Microbiol 43(4):1595–1603. doi:10.1590/S1517-83822012000400044 S1517-83822012000400044 [pii]

Kwak YS, Weller DM (2013) Take-all of wheat and natural disease suppression: a review. Plant Pathol J 29(2):125–135. doi:10.5423/PPJ.SI.07.2012.0112 ppj-29-125 [pii]

Labidi S, Jeddi FB, Tisserant B, Yousfi M, Sanaa M, Dalpé Y, Sahraoui AL (2015) Field application of mycorrhizal bio-inoculants affects the mineral uptake of a forage legume (*Hedysarum coronarium* L.) on a highly calcareous soil. Mycorrhiza 25(4):297–309. doi:10.1007/s00572-014-0609-0

Lacey LA, Grzywacz D, Shapiro-Ilan DI, Frutos R, Brownbridge M, Goettel MS (2015) Insect pathogens as biological control agents: back to the future. J Invertebr Pathol 132:1–41. doi:10.1016/j.jip.2015.07.009 S0022-2011(15)00134-2 [pii]

Lai MC, Lan EI (2015) Advances in metabolic engineering of *Cyanobacteria* for photosynthetic biochemical production. Metabolites 5(4):636–658. doi:10.3390/metabo5040636 metabo5040636 [pii]

Lakshmanan V, Selvaraj G, Bais HP (2014) Functional soil microbiome: belowground solutions to an aboveground problem. Plant Physiol 166(2):689–700. doi:10.1104/pp.114.245811 pp.114.245811 [pii]

Leaungvutiviroj C, Ruangphisarn P, Hansanimitkul P, Shinkawa H, Sasaki K (2010) Development of a new biofertilizer with a high capacity for N2 fixation, phosphate and potassium solubilization and auxin production. Biosci Biotechnol Biochem 74(5):1098–101. doi:10.1271/bbb.90898[pii]

Lee EH, Eo JK, Ka KH, Eom AH (2013) Diversity of arbuscular mycorrhizal fungi and their roles in ecosystems. Mycobiology 41(3):121–125. doi:10.5941/MYCO.2013.41.3.121

Lenoir I, Fontaine J, Lounes-Hadj Sahraoui A (2016) Arbuscular mycorrhizal fungal responses to abiotic stresses: a review. Phytochemistry 123:4–15. doi:10.1016/j.phytochem.2016.01.002 S0031-9422(16)30002-4 [pii]

Li RX, Cai F, Pang G, Shen QR, Li R, Chen W (2015) Solubilisation of phosphate and micronutrients by *Trichoderma harzianum* and its relationship with the promotion of tomato plant growth. PLoS One 10(6), e0130081. doi:10.1371/journal.pone.0130081 PONE-D-15-00619 [pii]

Lim SL, Wu TY, Lim PN, Shak KP (2015) The use of vermicompost in organic farming: overview, effects on soil and economics. J Sci Food Agric 95(6):1143–1156. doi:10.1002/jsfa.6849

Lojkova L, Vranova V, Rejsek K, Formanek P (2014) Natural occurrence of enantiomers of organic compounds versus phytoremediations: should research on phytoremediations be revisited? A mini-review. Chirality 26(1):1–20. doi:10.1002/chir.22255

Lu YF, Lu M, Peng F, Wan Y, Liao MH (2014) Remediation of polychlorinated biphenyl-contaminated soil by using a combination of ryegrass, arbuscular mycorrhizal fungi and earthworms. Chemosphere 106:44–50. doi:10.1016/j.chemosphere.2013.12.089 S0045-6535(14)00026-5 [pii]

Magalhaes JV, Liu J, Guimaraes CT, Lana UG, Alves VM, Wang YH, Schaffert RE, Hoekenga OA, Pineros MA, Shaff JE, Klein PE, Carneiro NP, Coelho CM, Trick HN, Kochian LV (2007) A gene in the multidrug and toxic compound extrusion (MATE) family confers aluminum tolerance in sorghum. Nat Genet 39(9):1156–1161. doi:10.1038/ng2074 ng2074 [pii]

Mark G, Morrissey JP, Higgins P, O'Gara F (2006) Molecular-based strategies to exploit *Pseudomonas* biocontrol strains for environmental biotechnology applications. FEMS Microbiol Ecol 56(2):167–177. doi:10.1111/j.1574-6941.2006.00056.x FEM056 [pii]

Marsalek B, Simek M (1992) Abscisic acid and its synthetic analog in relation to growth and nitrogenase activity of *Azotobacter chroococcum* and *Nostoc muscorum*. Folia Microbiol (Praha) 37(2):159–160

Marschner H (1995) Mineral nutrition of higher plants, 2nd edn. Academic, London

Mayak S, Tirosh T, Glick BR (2004) Plant growth-promoting bacteria confer resistance in tomato plants to salt stress. Plant Physiol Biochem 42(6):565–572. doi:10.1016/j.plaphy.2004.05.009 S0981-9428(04)00076-2 [pii]

Mazzola M (2002) Mechanisms of natural soil suppressiveness to soilborne diseases. Antonie Van Leeuwenhoek 81(1–4):557–564

Mazzola M (2007) Manipulation of rhizosphere bacterial communities to induce suppressive soils. J Nematol 39(3):213–220

Mendes R, Kruijt M, de Bruijn I, Dekkers E, van der Voort M, Schneider JH, Piceno YM, DeSantis TZ, Andersen GL, Bakker PA, Raaijmakers JM (2011) Deciphering the rhizosphere microbiome for disease-suppressive bacteria. Science 332(6033):1097–1100. doi:10.1126/science.1203980.science.1203980 [pii]

Mendes R, Garbeva P, Raaijmakers JM (2013) The rhizosphere microbiome: significance of plant beneficial, plant pathogenic, and human pathogenic microorganisms. FEMS Microbiol Rev 37(5):634–663. doi:10.1111/1574-6976.12028

Micallef SA, Channer S, Shiaris MP, Colon-Carmona A (2009) Plant age and genotype impact the progression of bacterial community succession in the Arabidopsis rhizosphere. Plant Signal Behav 4(8):777–780. doi:10.1093/jxb/erp053 9229 [pii]

Mishra V, Gupta A, Kaur P, Singh S, Singh N, Gehlot P, Singh J (2015) Synergistic effects of Arbuscular mycorrhizal fungi and plant growth promoting rhizobacteria in bioremediation of iron contaminated soils. Int J Phytoremediation. doi:10.1080/15226514.2015.1131231

Montes-Borrego M, Metsis M, Landa BB (2014) Arbuscular mycorrhizal fungi associated with the olive crop across the Andalusian landscape: factors driving community differentiation. PLoS One 9(5), e96397. doi:10.1371/journal.pone.0096397 PONE-D-13-49105 [pii]

Montesinos-Navarro A, Segarra-Moragues JG, Valiente-Banuet A, Verdu M (2012) Plant facilitation occurs between species differing in their associated arbuscular mycorrhizal fungi. New Phytol 196(3):835–844. doi:10.1111/j.1469-8137.2012.04290.x

Mougel C, Offre P, Ranjard L, Corberand T, Gamalero E, Robin C, Lemanceau P (2006) Dynamic of the genetic structure of bacterial and fungal communities at different developmental stages of *Medicago truncatula Gaertn*. cv. Jemalong line J5. New Phytol 170(1):165–175. doi:10.1111/j.1469-8137.2006.01650.x NPH1650 [pii]

Nakaya M, Tsukaya H, Murakami N, Kato M (2002) Brassinosteroids control the proliferation of leaf cells of Arabidopsis thaliana. Plant Cell Physiol 43(2):239–244

Narasimhan K, Basheer C, Bajic VB, Swarup S (2003) Enhancement of plant-microbe interactions using a rhizosphere metabolomics-driven approach and its application in the removal of polychlorinated biphenyls. Plant Physiol 132(1):146–153. doi:10.1104/pp.102.016295

Neilands JB (1995) Siderophores: structure and function of microbial iron transport compounds. J Biol Chem 270(45):26723–26726

Newman LA, Reynolds CM (2004) Phytodegradation of organic compounds. Curr Opin Biotechnol 15(3):225–230. doi:10.1016/j.copbio.2004.04.006 S0958166904000588 [pii]

Nunan N, Daniell TJ, Singh BK, Papert A, McNicol JW, Prosser JI (2005) Links between plant and rhizoplane bacterial communities in grassland soils, characterized using molecular techniques. Appl Environ Microbiol 71(11):6784–6792. doi:71/11/6784 [pii] 10.1128/AEM.71.11.6784-6792.2005

Oger P, Petit A, Dessaux Y (1997) Genetically engineered plants producing opines alter their biological environment. Nat Biotechnol 15(4):369–372. doi:10.1038/nbt0497-369

Onofre-Lemus J, Hernandez-Lucas I, Girard L, Caballero-Mellado J (2009) ACC (1-aminocyclopropane-1-carboxylate) deaminase activity, a widespread trait in *Burkholderia* species, and its growth-promoting effect on tomato plants. Appl Environ Microbiol 75(20):6581–6590. doi:10.1128/AEM.01240-09 AEM.01240-09 [pii]

Owen SM, Clark S, Pompe M, Semple KT (2007) Biogenic volatile organic compounds as potential carbon sources for microbial communities in soil from the rhizosphere of *Populus tremula*. FEMS Microbiol Lett 268(1):34–39. doi:10.1111/j.1574-6968.2006.00602.x FML602 [pii]

Paulin MM, Novinscak A, St-Arnaud M, Goyer C, DeCoste NJ, Prive JP, Owen J, Filion M (2009) Transcriptional activity of antifungal metabolite-encoding genes phlD and hcnBC in *Pseudomonas spp.* using qRT-PCR. FEMS Microbiol Ecol 68(2):212–222. doi:10.1111/j.1574-6941.2009.00669.x FEM669 [pii]

Pence NS, Larsen PB, Ebbs SD, Letham DL, Lasat MM, Garvin DF, Eide D, Kochian LV (2000) The molecular physiology of heavy metal transport in the Zn/Cd hyperaccumulator *Thlaspi caerulescens*. Proc Natl Acad Sci U S A 97(9):4956–4960, 97/9/4956 [pii]

Perez-Alfocea F, Ghanem ME, Gomez-Cadenas A, Dodd IC (2011) Omics of root-to-shoot signaling under salt stress and water deficit. OMICS 15(12):893–901. doi:10.1089/omi.2011.0092

Perez-Montano F, Alias-Villegas C, Bellogin RA, del Cerro P, Espuny MR, Jimenez-Guerrero I, Lopez-Baena FJ, Ollero FJ, Cubo T (2014) Plant growth promotion in cereal and leguminous agricultural important plants: from microorganism capacities to crop production. Microbiol Res 169(5–6):325–336. doi:10.1016/j.micres.2013.09.011 S0944-5013(13)00164-X [pii]

Philippot L, Raaijmakers JM, Lemanceau P, van der Putten WH (2013) Going back to the roots: the microbial ecology of the rhizosphere. Nat Rev Microbiol 11(11):789–799. doi:10.1038/nrmicro3109 mnrmicro3109 [pii]

Pilon-Smits E (2005) Phytoremediation. Annu Rev Plant Biol 56:15–39. doi:10.1146/annurev.arplant.56.032604.144214

Pilon-Smits EA, Hwang S, Mel Lytle C, Zhu Y, Tai JC, Bravo RC, Chen Y, Leustek T, Terry N (1999) Overexpression of ATP sulfurylase in indian mustard leads to increased selenate uptake, reduction, and tolerance. Plant Physiol 119(1):123–132

Ping L, Boland W (2004) Signals from the underground: bacterial volatiles promote growth in Arabidopsis. Trends Plant Sci 9(6):263–266. doi:10.1016/j.tplants.2004.04.008 S1360-1385(04)00105-0 [pii]

Plett D, Safwat G, Gilliham M, Skrumsager Moller I, Roy S, Shirley N, Jacobs A, Johnson A, Tester M (2010) Improved salinity tolerance of rice through cell type-specific expression of AtHKT1;1. PLoS One 5(9), e12571. doi:10.1371/journal.pone.0012571

Porras-Soriano A, Soriano-Martin ML, Porras-Piedra A, Azcon R (2009) Arbuscular mycorrhizal fungi increased growth, nutrient uptake and tolerance to salinity in olive trees under nursery conditions. J Plant Physiol 166(13):1350–1359. doi:10.1016/j.jplph.2009.02.010 S0176-1617(09)00080-7 [pii]

Powell PE, Szaniszlo PJ, Reid CP (1983) Confirmation of occurrence of hydroxamate siderophores in soil by a novel *Escherichia coli* bioassay. Appl Environ Microbiol 46(5):1080–1083

Quiza L, St-Arnaud M, Yergeau E (2015) Harnessing phytomicrobiome signaling for rhizosphere microbiome engineering. Front Plant Sci 6:507. doi:10.3389/fpls.2015.00507

Rasmann S, Kollner TG, Degenhardt J, Hiltpold I, Toepfer S, Kuhlmann U, Gershenzon J, Turlings TC (2005) Recruitment of entomopathogenic nematodes by insect-damaged maize roots. Nature 434(7034):732–737. doi:nature03451 [pii] 10.1038/nature03451

Rengel Z, Marschner P (2005) Nutrient availability and management in the rhizosphere: exploiting genotypic differences. New Phytol 168(2):305–312. doi:10.1111/j.1469-8137.2005.01558.x NPH1558 [pii]

Reynolds HL, Smith AA, Farmer JR (2014) Think globally, research locally: paradigms and place in agroecological research. Am J Bot 101(10):1631–1639. doi:10.3732/ajb.1400146 ajb.1400146 [pii]

Rillig MC, Mummey DL (2006) Mycorrhizas and soil structure. New Phytol 171(1):41–53. doi:10.1111/j.1469-8137.2006.01750.x NPH1750 [pii]

Rivoal J, Hanson AD (1994) Metabolic control of anaerobic glycolysis overexpression of lactate dehydrogenase in transgenic tomato roots supports the Davies-Roberts hypothesis and points to a critical role for lactate secretion. Plant Physiol 106:1179–1185

Rogers EE, Eide DJ, Guerinot ML (2000) Altered selectivity in an Arabidopsis metal transporter. Proc Natl Acad Sci U S A 97(22):12356–12360. doi:10.1073/pnas.210214197 210214197 [pii]

Roose T, Schnepf A (2008) Mathematical models of plant-soil interaction. Philos Trans A Math Phys Eng Sci 366(1885):4597–4611. doi:10.1098/rsta.2008.0198 2RT9X62Q81L87186 [pii]

Rugh CL, Wilde HD, Stack NM, Thompson DM, Summers AO, Meagher RB (1996) Mercuric ion reduction and resistance in transgenic Arabidopsis thaliana plants expressing a modified bacterial merA gene. Proc Natl Acad Sci U S A 93(8):3182–3187

Ryan P, Dessaux Y, Thomashow L, Weller D (2009) Rhizosphere engineering and management for sustainable agriculture. Plant Soil 321:363–383. doi:10.1007/s11104-009-0001-6

Savka MA, Dessaux Y, Oger P, Rossbach S (2002) Engineering bacterial competitiveness and persistence in the phytosphere. Mol Plant Microbe Interact 15(9):866–874. doi:10.1094/MPMI.2002.15.9.866

Schachtman DP, Shin R (2007) Nutrient sensing and signaling: NPKS. Annu Rev Plant Biol 58:47–69. doi:10.1146/annurev.arplant.58.032806.103750

Schouteden N, De Waele D, Panis B, Vos CM (2015) Arbuscular mycorrhizal fungi for the biocontrol of plant-parasitic nematodes: a review of the mechanisms involved. Front Microbiol 6:1280. doi:10.3389/fmicb.2015.01280

Schultze M, Kondorosi A (1998) Regulation of symbiotic root nodule development. Annu Rev Genet 32:33–57. doi:10.1146/annurev.genet.32.1.33

Shapira R, Ordentlich A, Chet I, Oppenheim AB (1989) Control of plant diseases by chitinase expressed from cloned DNA in *Escherichia coli*. Phytopathology 79:1246–1249

Sharma SB, Sayyed RZ, Trivedi MH, Gobi TA (2013) Phosphate solubilizing microbes: sustainable approach for managing phosphorus deficiency in agricultural soils. SpringerPlus 2:587. doi:10.1186/2193-1801-2-587 1439 [pii]

Shaw LJ, Morris P, Hooker JE (2006) Perception and modification of plant flavonoid signals by rhizosphere microorganisms. Environ Microbiol 8(11):1867–1880. doi:10.1111/j.1462-2920.2006.01141.x EMI1141 [pii]

Shrivastava P, Kumar R (2015) Soil salinity: a serious environmental issue and plant growth promoting bacteria as one of the tools for its alleviation. Saudi J Biol Sci 22(2):123–131. doi:10.1016/j.sjbs.2014.12.001 S1319-562X(14)00171-5 [pii]

Silva JP, Mussatto SI, Roberto IC (2010) The influence of initial xylose concentration, agitation, and aeration on ethanol production by Pichia stipitis from rice straw hemicellulosic hydrolysate. Appl Biochem Biotechnol 162(5):1306–1315. doi:10.1007/s12010-009-8867-6

Singh S, Grover A, Nasim M (2016) Biofuel potential of plants transformed genetically with NAC family genes. Front Plant Sci 7:22. doi:10.3389/fpls.2016.00022

Smalla K, Wieland G, Buchner A, Zock A, Parzy J, Kaiser S, Roskot N, Heuer H, Berg G (2001) Bulk and rhizosphere soil bacterial communities studied by denaturing gradient gel electrophoresis: plant-dependent enrichment and seasonal shifts revealed. Appl Environ Microbiol 67(10):4742–4751

Sokarda Slavić M, Pesic M, Vujcic Z, Bozic N (2016) Overcoming hydrolysis of raw corn starch under industrial conditions with *Bacillus licheniformis* ATCC 9945a alpha-amylase. Appl Microbiol Biotechnol 100(6):2709–2719. doi:10.1007/s00253-015-7101-4 10.1007/s00253-015-7101-4 [pii]

Spaepen S, Versees W, Gocke D, Pohl M, Steyaert J, Vanderleyden J (2007) Characterization of phenylpyruvate decarboxylase, involved in auxin production of *Azospirillum brasilense*. J Bacteriol 189(21):7626–7633. doi:10.1128/JB.00830-07 JB.00830-07 [pii]

Spiers AS, Wade HE (1976) Bacterial glutaminase in treatment of acute leukaemia. Br Med J 1(6021):1317–1319

Taghavi S, Barac T, Greenberg B, Borremans B, Vangronsveld J, van der Lelie D (2005) Horizontal gene transfer to endogenous endophytic bacteria from poplar improves phytoremediation of toluene. Appl Environ Microbiol 71(12):8500–8505. doi:71/12/8500 [pii] 10.1128/AEM.71.12.8500-8505.2005

Tesfaye M, Temple SJ, Allan DL, Vance CP, Samac DA (2001) Overexpression of malate dehydrogenase in transgenic alfalfa enhances organic acid synthesis and confers tolerance to aluminum. Plant Physiol 127(4):1836–1844

Thomas JC, Davies EC, Malick FK, Endreszl C, Williams CR, Abbas M, Petrella S, Swisher K, Perron M, Edwards R, Osenkowski P, Urbanczyk N, Wiesend WN, Murray KS (2003) Yeast metallothionein in transgenic tobacco promotes copper uptake from contaminated soils. Biotechnol Prog 19(2):273–280. doi:10.1021/bp025623q

Timms-Wilson TM, Ellis RJ, Renwick A, Rhodes DJ, Mavrodi DV, Weller DM, Thomashow LS, Bailey MJ (2000) Chromosomal insertion of phenazine-1-carboxylic acid biosynthetic pathway enhances efficacy of damping-off disease control by *Pseudomonas fluorescens*. Mol Plant Microbe Interact 13(12):1293–1300. doi:10.1094/MPMI.2000.13.12.1293

Timmusk S, Wagner EG (1999) The plant-growth-promoting rhizobacterium Paenibacillus polymyxa induces changes in Arabidopsis thaliana gene expression: a possible connection between biotic and abiotic stress responses. Mol Plant Microbe Interact 12(11):951–959. doi:10.1094/MPMI.1999.12.11.951

Tisserant E, Malbreil M, Kuo A, Kohler A, Symeonidi A, Balestrini R, Charron P, Duensing N, Frei dit Frey N, Gianinazzi-Pearson V, Gilbert LB, Handa Y, Herr JR, Hijri M, Koul R, Kawaguchi M, Krajinski F, Lammers PJ, Masclaux FG, Murat C, Morin E, Ndikumana S, Pagni M, Petitpierre D, Requena N, Rosikiewicz P, Riley R, Saito K, San Clemente H, Shapiro H, van Tuinen D, Becard G, Bonfante P, Paszkowski U, Shachar-Hill YY, Tuskan GA, Young JP, Sanders IR, Henrissat B, Rensing SA, Grigoriev IV, Corradi N, Roux C, Martin F (2013) Genome of an arbuscular mycorrhizal fungus provides insight into the oldest plant symbiosis. Proc Natl Acad Sci U S A 110(50):20117–20122. doi:10.1073/pnas.1313452110 1313452110 [pii]

Trabelsi D, Mhamdi R (2013) Microbial inoculants and their impact on soil microbial communities: a review. Biomed Res Int 2013:863240. doi:10.1155/2013/863240

Underkofler LA, Barton RR, Rennert SS (1958) Production of microbial enzymes and their applications. Appl Microbiol 6(3):212–221

van Dam NM, Bouwmeester HJ (2016) Metabolomics in the rhizosphere: tapping into belowground chemical communication. Trends Plant Sci 21(3):256–265. doi:10.1016/j.tplants.2016.01.008 S1360-1385(16)00009-1 [pii]

van der Heijden MG, Streitwolf-Engel R, Riedl R, Siegrist S, Neudecker A, Ineichen K, Boller T, Wiemken A, Sanders IR (2006) The mycorrhizal contribution to plant productivity, plant nutrition and soil structure in experimental grassland. New Phytol 172(4):739–752. doi:10.1111/j.1469-8137.2006.01862.x NPH1862 [pii]

van der Zaal BJ, Neuteboom LW, Pinas JE, Chardonnens AN, Schat H, Verkleij JA, Hooykaas PJ (1999) Overexpression of a novel Arabidopsis gene related to putative zinc-transporter genes from animals can lead to enhanced zinc resistance and accumulation. Plant Physiol 119(3):1047–1055

Vivas A, Voros I, Biro B, Campos E, Barea JM, Azcon R (2003) Symbiotic efficiency of autochthonous arbuscular mycorrhizal fungus (*G. mosseae*) and *Brevibacillus sp.* isolated from cadmium polluted soil under increasing cadmium levels. Environ Pollut 126(2):179–189, S0269749103003001957 [pii]

Walder F, Boller T, Wiemken A, Courty PE (2016) Regulation of plants' phosphate uptake in common mycorrhizal networks: role of intraradical fungal phosphate transporters. Plant Signal Behav 11(2), e1131372. doi:10.1080/15592324.2015.1131372

Walker TS, Bais HP, Deziel E, Schweizer HP, Rahme LG, Fall R, Vivanco JM (2004) *Pseudomonas aeruginosa*-plant root interactions. Pathogenicity, biofilm formation, and root exudation. Plant Physiol 134(1):320–331. doi:10.1104/pp.103.027888 pp.103.027888 [pii]

Walton BT, Anderson TA (1990) Microbial degradation of trichloroethylene in the rhizosphere: potential application to biological remediation of waste sites. Appl Environ Microbiol 56(4):1012–1016

Wang C, Knill E, Glick BR, Defago G (2000) Effect of transferring 1-aminocyclopropane-1-carboxylic acid (ACC) deaminase genes into *Pseudomonas fluorescens* strain CHA0 and its gacA derivative CHA96 on their growth-promoting and disease-suppressive capacities. Can J Microbiol 46(10):898–907

Wang P, Bi S, Wang S, Ding Q (2006) Variation of wheat root exudates under aluminum stress. J Agric Food Chem 54(26):10040–10046. doi:10.1021/jf061249o

Wang D, Yang S, Tang F, Zhu H (2012) Symbiosis specificity in the legume: rhizobial mutualism. Cell Microbiol 14(3):334–342. doi:10.1111/j.1462-5822.2011.01736.x

Wang Y, Cheng X, Shan Q, Zhang Y, Liu J, Gao C, Qiu JL (2014) Simultaneous editing of three homoeoalleles in hexaploid bread wheat confers heritable resistance to powdery mildew. Nat Biotechnol 32(9):947–951. doi:10.1038/nbt.2969 nbt.2969 [pii]

Wang J, Feng J, Jia W, Chang S, Li S, Li Y (2015) Lignin engineering through laccase modification: a promising field for energy plant improvement. Biotechnol Biofuels 15(8):145. doi:10.1186/s13068-015-0331-y

Wang J, Li Q, Mao X, Li A, Jing R (2016a) Wheat transcription factor TaAREB3 participates in drought and freezing tolerances in Arabidopsis. Int J Biol Sci 12(2):257–269. doi:10.7150/ijbs.13538 ijbsv12p0257 [pii]

Wang X, Wu D, Yang Q, Zeng J, Jin G, Chen ZH, Zhang G, Dai F (2016b) Identification of mild freezing shock response pathways in barley based on transcriptome profiling. Front Plant Sci 7:106. doi:10.3389/fpls.2016.00106

Watt M, Evans JR (1999) Linking development and determinacy with organic acid efflux from proteoid roots of white lupin grown with low phosphorus and ambient or elevated atmospheric CO_2 concentration. Plant Physiol 120(3):705–716

Wenke K, Kai M, Piechulla B (2010) Belowground volatiles facilitate interactions between plant roots and soil organisms. Planta 231(3):499–506. doi:10.1007/s00425-009-1076-2

Wentzell AM, Kliebenstein DJ (2008) Genotype, age, tissue, and environment regulate the structural outcome of glucosinolate activation. Plant Physiol 147(1):415–428. doi:10.1104/pp.107.115279 pp.107.115279 [pii]

Weyens N, Thijs S, Popek R, Witters N, Przybysz A, Espenshade J, Gawronska H, Vangronsveld J, Gawronski SW (2015) The role of plant-microbe interactions and their exploitation for phytoremediation of air pollutants. Int J Mol Sci 16(10):25576–25604. doi:10.3390/ijms161025576 ijms161025576 [pii]

Xiao C, Janssens IA, Liu P, Zhou Z, Sun OJ (2007) Irrigation and enhanced soil carbon input effects on below-ground carbon cycling in semiarid temperate grasslands. New Phytol 174(4):835–846. doi:10.1111/j.1469-8137.2007.02054.x NPH2054 [pii]

Xun F, Xie B, Liu S, Guo C (2015) Effect of plant growth-promoting bacteria (PGPR) and arbuscular mycorrhizal fungi (AMF) inoculation on oats in saline-alkali soil contaminated by petroleum to enhance phytoremediation. Environ Sci Pollut Res Int 22(1):598–608. doi:10.1007/s11356-014-3396-4

Yang H, Knapp J, Koirala P, Rajagopal D, Peer WA, Silbart LK, Murphy A, Gaxiola RA (2007) Enhanced phosphorus nutrition in monocots and dicots over-expressing a phosphorus-responsive type I H+−pyrophosphatase. Plant Biotechnol J 5(6):735–745. doi:10.1111/j.1467-7652.2007.00281.x PBI281 [pii]

Yang J, Kloepper JW, Ryu CM (2009) Rhizosphere bacteria help plants tolerate abiotic stress. Trends Plant Sci 14(1):1–4. doi:10.1016/j.tplants.2008.10.004 S1360-1385(08)00290-2 [pii]

Yang Z, Chen H, Tang W, Hua H, Lin Y (2011) Development and characterisation of transgenic rice expressing two *Bacillus thuringiensis* genes. Pest Manag Sci 67(4):414–422. doi:10.1002/ps.2079

Yang Y, Liang Y, Ghosh A, Song Y, Chen H, Tang M (2015) Assessment of arbuscular mycorrhizal fungi status and heavy metal accumulation characteristics of tree species in a lead-zinc mine area: potential applications for phytoremediation. Environ Sci Pollut Res Int 22(17):13179–13193. doi:10.1007/s11356-015-4521-8

Yergeau E, Sanschagrin S, Maynard C, St-Arnaud M, Greer CW (2014) Microbial expression profiles in the rhizosphere of willows depend on soil contamination. ISME J 8(2):344–358. doi:10.1038/ismej.2013.163 ismej2013163 [pii]

Zaitsev S, Spitzer D, Murciano JC, Ding BS, Tliba S, Kowalska MA, Marcos-Contreras OA, Kuo A, Stepanova V, Atkinson JP, Poncz M, Cines DB, Muzykantov VR (2010) Sustained thromboprophylaxis mediated by an RBC-targeted pro-urokinase zymogen activated at the site of clot formation. Blood 115(25):5241–5248. doi:10.1182/blood-2010-01-261610 blood-2010-01-261610 [pii]

Zhang H, Kim MS, Sun Y, Dowd SE, Shi H, Pare PW (2008) Soil bacteria confer plant salt tolerance by tissue-specific regulation of the sodium transporter HKT1. Mol Plant Microbe Interact 21(6):737–744. doi:10.1094/MPMI-21-6-0737

Zhang Y, Ruyter-Spira C, Bouwmeester HJ (2015a) Engineering the plant rhizosphere. Curr Opin Biotechnol 32:136–142. doi:10.1016/j.copbio.2014.12.006 S0958-1669(14)00221-3 [pii]

Zhang T, Chaturvedi V, Chaturvedi S (2015b) Novel *Trichoderma polysporum* strain for the biocontrol of *Pseudogymnoascus destructans*, the fungal etiologic agent of bat white nose syndrome. PLoS One 10(10), e0141316. doi:10.1371/journal.pone.0141316 PONE-D-14-51371 [pii]

Zhang F, Ge H, Guo N, Wang Y, Chen L, Ji X, Li C (2016) Biocontrol potential of *Trichoderma harzianum* isolate T-aloe against *Sclerotinia sclerotiorum* in soybean. Plant Physiol Biochem 100:64–74. doi:10.1016/j.plaphy.2015.12.017 S0981-9428(15)30195-9 [pii]

Zhu YG, Rosen BP (2009) Perspectives for genetic engineering for the phytoremediation of arsenic-contaminated environments: from imagination to reality? Curr Opin Biotechnol 20(2):220–224. doi:10.1016/j.copbio.2009.02.011 S0958-1669(09)00024-X [pii]

Zhu YL, Pilon-Smits EA, Tarun AS, Weber SU, Jouanin L, Terry N (1999) Cadmium tolerance and accumulation in Indian mustard is enhanced by overexpressing gamma-glutamylcysteine synthetase. Plant Physiol 121(4):1169–1178

Zhu JR, Zhou H, Pan YB, Lu X (2014) Genetic variability among the chloroplast genomes of sugarcane (*Saccharum spp*) and its wild progenitor species *Saccharum spontaneum* L. Genet Mol Res 13(2):3037–3047. doi:10.4238/2014.January.24.3 gmr3004 [pii]

Zhuang X, Chen J, Shim H, Bai Z (2007) New advances in plant growth-promoting rhizobacteria for bioremediation. Environ Int 33(3):406–413. doi:10.1016/j.envint.2006.12.005 S0160-4120(07)00003-7 [pii]

Mycorrhizosphere: The Extended Rhizosphere and Its Significance

5

P. Priyadharsini, K. Rojamala, R. Koshila Ravi, R. Muthuraja, K. Nagaraj, and T. Muthukumar

Abstract

Plant roots influence soil through the release of carbon-rich exudates and rhizodeposits. The soil region influenced by plant roots is termed as rhizosphere. A unique community of microorganisms thrives in the rhizosphere whose activities enable plants to acquire various resources from soil for their growth and survival. Most plants in natural and agricultural ecosystems are associated with mycorrhizal fungi, which act as interlink between two different environments, the root and the soil. Mycorrhizal fungi play an important role in plant uptake of nutrients and protecting plants against various abiotic and biotic stresses. Like roots, mycorrhizal fungal hyphae also release exudates containing carbon into the surrounding soil, the hyphosphere that contributes to the formation of microbial communities and aggregation of soil particles. The soil region influenced by the mycorrhizal roots is the mycorrhizosphere. A wide range of microorganisms like bacteria, fungi, protozoa, nematodes, arthropods, etc., inhabit the mycorrhizosphere. These microorganisms interact with each other and with the plant system either directly or indirectly. The activities in the mycorrhizosphere include stimulation in the activities and populations of microorganisms, changes in pH, nutrient release from organic matter and nutrient cycling, suppression of plant pathogens, mycorrhizal formation, and changes in soil structure. An understanding of the functional diversity of microorganisms inhabiting the mycorrhizosphere is necessary to optimize soil microbial technology for the benefit of plant growth and health. This chapter describes the concept of rhizosphere, hyphosphere, and mycorrhizosphere and the various activities in these regions.

P. Priyadharsini • K. Rojamala • R.K. Ravi • R. Muthuraja • K. Nagaraj
T. Muthukumar (✉)
Root and Soil Biology Laboratory, Department of Botany, Bharathiar University, Coimbatore 641046, Tamil Nadu, India
e-mail: tmkum@yahoo.com

D.K. Choudhary et al. (eds.), *Plant-Microbe Interaction: An Approach to Sustainable Agriculture*, DOI 10.1007/978-981-10-2854-0_5

5.1 Introduction

Roots, the hidden half of plants, form a linkage between the plant and the soil environment. The soil surrounding roots is inhabited by several fungi, bacteria, and other microflora and fauna (Linderman 1992). An intense effort is being carried out to understand and exploit the root system of plants in an endeavor to expand the yield probability of stable food crops to face the dual increase of population and global demand for food within the next 50 years (Zhang et al. 2010). The root system of the higher plants is interlinked with a diverse metabolically active microbial community together with the soil environment which has organic and inorganic matter (Fulekar and Pathak 2015). A unique habitat exists around the roots of the living plants. Hence, the population of microbes in soil surrounding the roots is greater than that of root-free soil. Further, these two regions of the soil have several qualitative and quantitative dissimilarities (Fulekar and Pathak 2015).

Mycorrhizal symbioses are associations formed between the roots of terrestrial plant species and soil fungi belonging to *Glomeromycota*, *Ascomycota*, and *Basidiomycota* (Smith and Read 2008). This is the most widespread type of symbiosis in the plant kingdom occurring in more than 80 % of the terrestrial vascular plant species and is believed to have facilitated colonization of terrestrial habitats by plants (Parniske 2008). During mycorrhization, the mycorrhizal fungi develop intimate relation with root cells on one side and soil on the other. The fungi grow on the root surface as runner hyphae or cover the roots in the form of mantle. It is fairly well established that mycorrhizal fungi extend the influence of roots into the non-root zones through their extraradical mycelium. The intraradical structures of mycorrhizal fungi within roots depending on the type of mycorrhizae include arbuscules, vesicles, Hartig net, hyphal coils/pelotons, and linear hyphae that may be inter- or intracellular (Smith and Read 2008).

Mycorrhizal association imparts a wide range of benefits to the plant species. These include plant growth promotion, nutrient uptake and translocation to the roots, and resistance to stresses induced by soilborne pathogens (Whipps 2004; Pozo and Azcon-Aguilar 2007), heavy metals (Vogel-Mikus et al. 2006; Dong et al. 2008), drought, and salinity (Feng et al. 2002; Miransari 2010). The prime benefit of mycorrhiza on plant growth is by providing large amounts of phosphorus (P) and nitrogen (N) to the host plant in nutrient-stressed soils. They also supply other mineral nutrients and help in dissolving and absorbing insoluble organic substances (non-glomeromycotean fungi) which in turn enhance plant growth. The extraradical mycelium of the mycorrhizal fungi facilitates the uptake of water and nutrients by plants as it extends its range into areas of the soil profile that is beyond the reach of the roots. Further, the smaller diameter of the fungal hyphae compared to roots enables their entry into soil pores that cannot be reached by roots (Kaiser et al. 2015). The extraradical mycelium of mycorrhizal fungi also acts as a pathway in the translocation and release of energy-rich molecules from the plant to the soil (Johnson et al. 2002). In addition, mycorrhizal symbiosis also enhances the efficiency of other plant-microbe symbiosis like nodulation and N_2-fixation in leguminous and actinorhizal plants.

The mycorrhizal fungi interact with other microorganisms both in the root and in the soil (Barea et al. 2002). The synergistic effect of mycorrhizal fungal association with plant growth-promoting rhizobacteria (PGPR), asymbiotic N_2-fixing bacteria, non-mycorrhizal fungi, and microfauna renders positive effects to plant growth and development (Sturz and Nowak 2000). Moreover, the microbial activity in the soil not only interferes with the plant root system, but it also stimulates the germination of mycorrhizal fungal propagules, mycelial growth, and mycorrhization of roots (Barea 2000). Though the mycorrhizal fungi could alter the plant root morphology, it also influences the activity and favors some organisms like bacteria to establish and flourish in the soil environment (Parniske 2008; Deveau et al. 2010).

In this chapter, we present an overview on the rhizosphere, hyposphere, and mycorrhizosphere. The changes in biological and soil properties in response to the development of mycorrhizosphere are also discussed.

5.2 The Rhizosphere Concept

Soil is one of the important frontiers of science, and the rhizosphere is the most dynamic part of that frontier where a multitude of biogeochemical activities occur, which influence many environmental and worldwide processes (McNear 2013; Haldar and Sengupta 2015). Rhizosphere is the narrow region of soil that is directly influenced by root secretions and rhizodeposits as well as the soil microorganisms associated with it (Fig. 5.1). This region can be differentiated from the bulk soil by the availability of the carbon compounds, water potential, and redox state, and it demarcates the distribution and activity of the widely diverse rhizosphere biota (Cardon and Whitbeck 2007). Being a resource-rich region, it is one of the most complex ecosystems on the earth (Pierret et al. 2007; Jones and Hinsinger 2008; Hinsinger et al. 2009). The rhizosphere constitutes about 10^{11} microbial cells per gram of root (Egamberdieva et al. 2008) and prokaryotic species that number around 30,000 which influence the productivity of plants (Mendes et al. 2011; 2013). The rhizosphere resembles a trade zone where symbionts or neighbor roots interact with the plant. Moreover, it acts as a preventive microbial buffer zone against pathogens (Baetz and Martinoia 2014).

The term rhizosphere includes three zones which are based on their relative proximity and influence from the plant root (Morgan et al. 2005). The endorhizosphere includes portions of the cortex and the endodermis, where the microbes and the cations occupy the apoplastic space, viz., the free space between the cells (Balandreau and Knowles 1978; Reinhold-Hurek et al. 2015). The rhizoplane or the root surface is the medial zone, which includes the root epidermis and mucilage (Nihorembere et al. 2011; Bulgarelli et al. 2012). The outermost zone is the ectorhizosphere which extends from the rhizoplane out into the region of bulk soil (Lynch and Whipps 1990; Badri and Vivanco 2009).

The plants are not only influenced by soil but also active microbial populations in the rhizosphere (Hiltner 1904). Recently, York et al. (2016) proposed the concept of a holistic rhizosphere that encompasses constituents like mucigel; modifications of soil

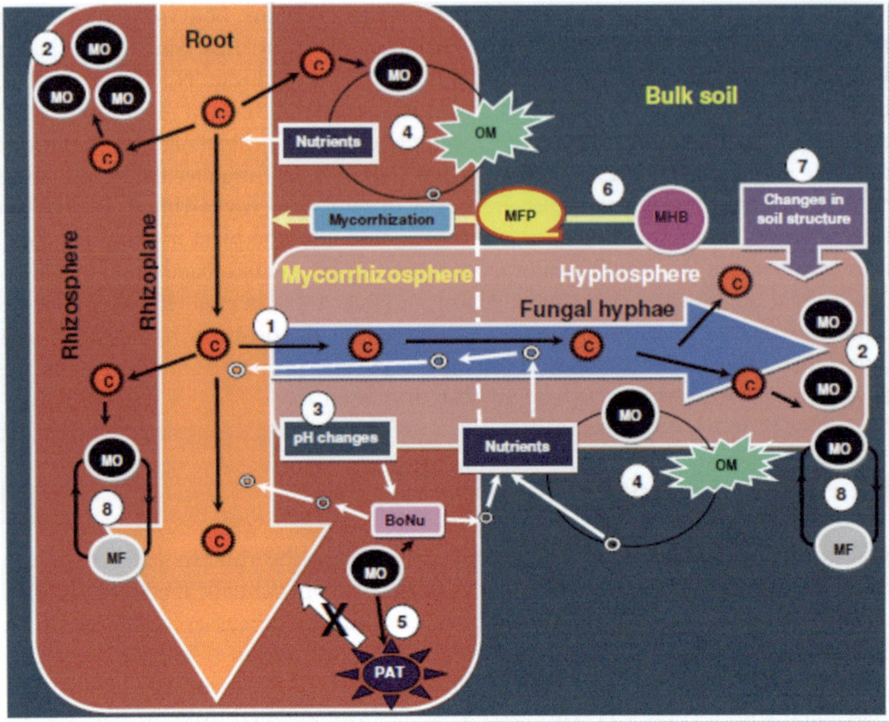

Fig. 5.1 Schematic representation of different types of activities occurring in the mycorrhizosphere. *1* Carbon transfer from root to rhizosphere, mycorrhizal fungal hyphae, and hyphosphere. *2* Increase in the activities and populations of microorganism. *3* Changes in pH resulting in the release of bound nutrients and their uptake. *4* Release of nutrients from organic matter by microorganisms and their uptake. *5* Control of plant pathogens. *6* Mycorrhizal fungal propagule germination and mycorrhization. *7* Changes in soil structure. *8* Interactions among microorganisms. *C* carbon, *MO* microorganisms, *OM* organic matter, *MHB* mycorrhiza helper bacteria, *MFP* mycorrhizal fungal propagule, *PAT* pathogens, *BoNu* bound nutrients, *MF* microfauna

structure; gradients of microbial communities, microorganisms; or a loss or gain of substances like water, root exudates, nutrients, gases, and other volatile products. However, the ingredients of the rhizosphere have been formed due to effectual processes. Hence, it is necessary to understand these processes for future considerations.

5.2.1 The Rhizosphere Effect

As the seeds germinate, they exude carbon compounds into soil which activates the microbial populations. This phenomenon is termed the rhizosphere effect (Morgan and Whipps 2001). Alternatively, it describes the development of soil microorganism communities as a result of the physical and chemical alterations of the soil by the products of excretions and organic debris of roots within a rhizosphere (McNear 2013).

Starkey (1938) defined the rhizosphere effect in terms of processes which occur at the root-soil interface of a plant inclusive of root exudation, microbial activity, genetic exchange, nutrient transformation, and gradient diffusion. In living plants, the organic carbon released by plant roots decomposes to carbon dioxide by a mechanism of rhizosphere priming effect (Kuzyakov 2002). Almost one-third to more than half of the total carbon assimilated by plants is allocated to belowground roots, of which 15–25 % is exuded into the soil resulting in a fast carbon turnover (Kuzyakov 2002). The intense microbial activity in response to carbon availability leads to intense competition for other nutrients in the rhizosphere (Hartmann et al. 2009; Haldar and Sengupta 2015). On the other hand, in a root-free bulk soil, all the nutrients except carbon are unlimited (Wardle 1992). Thus, the rhizosphere zone strongly differs from the root-free zones in physical, chemical, and biological properties (Hartmann et al. 2009).

5.2.2 Rhizodeposits

The substances secreted by the roots and its associated microorganisms into the rhizosphere are called rhizodeposits. Rhizodeposits have been classified depending on their chemical composition and mode of release or function (Rovira 1969). These compounds include: (1) Low-molecular mass compounds like monosaccharides, amino acids, organic acids, and water-soluble ions that are lost passively along a concentration gradient, (2) High-molecular mass compounds such as carbohydrates and proteins that act as signal molecules and lipids that are actively transported along an electrochemical gradient, (3) Insoluble mucilage composed of polygalacturonic acid and polysaccharides, (4) Secondary metabolites like antimicrobial compounds, flavonoids, and nematicides and (5) Remnants of the dead and lysed root cap and border cells (Marschner 1995; De-la-Peña et al. 2012; Weston et al. 2013; Zhang et al. 2014). The process of communication and interaction begins once the recipient organisms recognize the signaling phenomenon of the rhizodeposits. The nature and composition of root exudate can alter the microbial vitality and diversity of the soil, favoring growth of microorganisms that can benefit plant health and productivity, whereas, in others, root-exuded compounds prevent the growth of harmful microorganisms (Bais et al. 2006; Chaparro et al. 2012; Dutta et al. 2013; Li et al. 2013).

5.2.3 Effect of Rhizosphere Microbiome on Plant Growth and Health

The rhizosphere biome influences the composition and productivity, viz., biomass, of the natural plant communities (van der Heijden et al. 2006, 2008; Schnitzer et al. 2011). As a result, the species richness of the underground microbial communities can be considered as a factor to predict the aboveground plant diversity and productivity (Hooper et al. 2005; van der Heijden et al. 2008; Lau and Lennon 2011; Wagg et al. 2011).

5.2.4 Microbial Interactions in the Rhizosphere

5.2.4.1 Bacteria

The bacterial community in the rhizosphere promotes the production and germination of spores and hyphal growth of arbuscular mycorrhizal (AM) fungi. In addition to plant roots, spores (Bharadwaj et al. 2008; Cruz and Ishii 2012) and extraradical mycelium (Mansfeld-Giese et al. 2002) of AM fungi also associate predominantly with bacteria in the mycosphere. A bacterial community in the rhizosphere not only associates with the extraradical mycelium but also with spores of AM fungi. The association of bacteria with AM fungal spores is related to the size and surface roughness of the outer spore wall (Bharadwaj et al. 2008). Some bacterial taxa are exclusively restricted to a few mycorrhizal isolates, whereas others are extensively found in the mycosphere of several AM fungal taxa (Rillig et al. 2005). Bacterial association with AM fungal spores induces germination and establishment of mycorrhizal association under unfavorable conditions (Xavier and Germida 2003; Hildebrandt et al. 2006). This is often due to the bacterial secretion of volatile compounds, rupturing of the spore wall, and nutrient acquisition (Ruiz-Lozano and Bonfante 2000). Studies on AM fungal interactions with rhizosphere bacteria suggests that it may be either positive (Abdel-Fattah and Mohamedin 2000) or negative (Amora-Lazcano et al. 1998). Though AM fungal processes are enhanced by bacteria, some studies showed prohibitory activity of bacteria on AM fungal growth (Azcón 1989). This might be due to specificity in bacterial species and AM fungi. The AM fungi form a bridge between the root and soil (Bethlenfalvay and Schüepp 1994); in turn, the AM fungi affect the composition of bacterial communities in the rhizosphere (Linderman 1988; Paulitz and Lindennan 1991).

The fungi and bacteria in the rhizosphere are also involved in plant resistance to various types of stresses (Linderman 2000; Han and Lee 2005). The bacterial population in the rhizosphere mainly includes the beneficial associative N_2-fixing bacteria (Subba Rao et al. 1985), PGPR (Meyer and Linderman 1986), and phosphate-solubilizing bacteria (PSB) (Toro et al. 1997; Bonfante and Anca 2005). However, bacterial populations also vary under the influence of different plant and AM fungal species. The size and the composition of bacterial populations in the rhizosphere depend on the quantity of the root exudates (Azaizeh et al. 1995) and the competition for carbon source (Christensen and Jakobsen 1993). The carbon source is the major energy provider for various microbial communities, and its beneficial effect on plants has been well established. Mycorrhiza-associated bacteria also succeed to establish from AM fungal exudates (Toljander et al. 2007), which facilitate nutrition for both plant and fungal partners as well as protection from root pathogens (Larsen et al. 2015).

5.2.4.2 Fungi and Phytopathogens

In defense mechanism, mycorrhizal species directly or indirectly protect the host plant in the ecosystem. Such direct mechanisms include the production of physical structures (e.g., mantle by ectomycorrhizal fungi), secretion of toxic compounds against the pathogens, providing mechanical strength to the root system, and activating the host plant production of compounds like salicylic and jasmonic acids (Artursson et al. 2006; Finlay 2008). Indirect mechanisms include protection of the host plant by changing the microbial community, root exudates, and stimulation of suitable antagonistic microorganisms (Zarnea 1994; Zamfirache and Toma 2000; Miransari 2011). For example, the architecture of AM fungi-colonized roots is greatly modified. The mycorrhizal roots are highly branched, short and thick with reduced specific root length, resulting in conditions that are unfavorable for pathogenic microorganisms (Berta et al. 1993).

Studies have shown that rhizosphere bacteria could suppress plant pathogens (Berg and Hallmann 2006; Shehata et al. 2016). The rhizosphere fluorescent *Pseudomonas* strains produce the antibiotic 2,4-diacetylphloroglucinol (DAPG) that protects the plants against *Gaeumannomyces graminis* var. *tritici*. The bacterium produced significantly large amounts of DAPG in the presence of soluble carbon exuded by *Rhizophagus intraradices* and offered a sustainable strategy for plant protection (Siasou et al. 2009). The AM fungi and pathogen share common resources in the root system (Whipps 2004). But, competition in the endorhizosphere would arise when the carbon source from the host plant becomes scarce, resulting in the reduction in the colonization by AM fungi (Wehner et al. 2009). The intensity of the pathogenic effect on the host plant is reduced when multiple AM fungi species colonize the root system compared to colonization by an individual AM fungus (Jaiti et al. 2007). Bacteria associated with AM fungi enhance the plant resistance against pathogens through their antagonistic activity. For example, bacteria isolated from the spores of AM fungi inhibited the growth of *Ganoderma boninense*, which causes basal stem rot disease in oil palm (*Elaeis guineensis*) (Bakhtiar et al. 2010).

5.2.4.3 Microfauna

The rhizosphere contains microfauna like nematodes, protozoa, and arthropods. Most of these organisms are involved in the complex system of the food web that shares the plant resources (Pierret et al. 2007; Raaijmakers et al. 2009). Among these organisms, nematodes are free-living, eukaryotic invertebrates that feed on bacteria and fungi and some existing as plant parasites (Tiberius and Cătălin 2011). Nematodes cause diseases in plants by entering the root and establishing a stable feeding location within the root system (Badri et al. 2009). The interactions between mycophagous nematodes and mycorrhizal fungi result in the reduction of the extraradical hyphal production that can indirectly affect plant growth and yield (Giannakis and Sanders 1990; Khan 1993). However, to reduce the negative effect of nematode infestation, plants generally adapt various strategies like the association with mycorrhizal fungi, increased nutrient uptake, and structural and physiological changes in the root system (Schouteden et al. 2015). Even though AM fungi induce tolerance against adverse effects on host plants, several factors like host plant and AM fungal and nematode species determine the nature of interactions between AM fungi and nematodes (Hol and Cook 2005). Recently, Banuelos et al. (2014) found that roots of *Impatiens balsamina* inoculated with a consortium of AM fungi (11 species) reduced the root-knot disease caused by the nematode *Meloidogyne incognita* than the plant inoculated with *Glomus coronatum* alone.

However, the concentrations of antioxidant in shoots and phenolic compounds in roots were higher for AM fungal consortium inoculated plants and showed defense activity against the root-knot nematodes (Banuelos et al. 2014).

In addition to fungi, bacteria are also involved in the control of plant parasitic nematodes in soil. The nematophagous bacteria are differentiated based on their mode of activity and mostly belong to the genera *Bacillus*, *Pseudomonas*, and *Pasteuria* (Li et al. 2015). These bacteria have been isolated from soil, host tissues, and nematodes (Kerry 2000; Meyer 2003). The nematophagous bacteria affect nematodes through various mechanisms like producing toxins and antibiotics/enzymes, competing for nutrients, and inducing systemic resistance in plants (Tian et al. 2007). Some of the major rhizobacteria like *Azotobacter* and *Gluconacetobacter* also affect the plant parasitic nematodes. The antagonistic effect of bacteria against nematodes in the soil is due to the secretion of volatile compounds like ammonia and fatty acids which inhibit the juveniles of nematodes (Bansal and Bajaj 2003). A study by Bansal et al. (2005) in cotton showed that the antagonistic effect of *Gluconacetobacter diazotrophicus* (=*Acetobacter diazotrophicus*) on the root-knot nematode, *M. incognita*, was through suppression of egg hatching.

Abundance of bacterial grazers like the nematodes and protozoa significantly alters the bacterial community composition and their activities in the rhizosphere (Bonkowski 2004). Such changes in bacterial activities and populations are shown to significantly affect plant growth (Kreuzer et al. 2006; Mao et al. 2007). In a boreal forest, ectomycorrhizal fungus was shown to affect bacterial community composition, subsequently altering food resources for protozoa (Timonen et al. 2004). Both ectomycorrhizal fungi and protozoa can complement each other in rendering benefit for plants. For example, Bonkowski et al. (2001) showed that the protozoa increased the N availability to Norway spruce (*Picea abies*) seedlings, whereas the ectomycorrhizal fungus *Paxillus involutus* increased the availability of P. The excretion of N after the consumption of bacterial biomass by protozoans increases the N availability for direct or mycorrhizal mediated uptake by plants (Bonkowski 2004). In a microcosm study, Koller et al. (2013) showed that protozoa mobilized N by stimulating microbial activity in degradation of organic matter. The N released was transferred to the roots of *Plantago lanceolata* via hyphae of *R. intraradices*. Though different microorganisms in the rhizosphere complement each other from the plant's perspective, a competition for plant carbohydrates does exist between these microorganisms. A substantial reduction in the numbers of protozoa has been reported by Rønn et al. (2002) in AM-colonized pea plants. The presence of protozoa also affects root architecture and biomass in rice plants (Herdler et al. 2008). The influence of AM fungi on changes in the microbial community of the rhizosphere tends to vary with the growth phase of the plant. For example, in pea plants the presence of the AM fungus *R. intraradices* decreased the number of protozoa during late vegetative phase prior to flowering, but the negative effect on protozoa decreased during flowering and pod formation (Wamberg et al. 2003).

5.3 The Hyphosphere Concept

Generally, the amount of fungal hyphae in any given volume of the soil is enormous. There can be up to 20,000 km of fungal hyphae per cubic meter of soil (Moore et al. 2011). Ectomycorrhizal fungi can produce up to 800 m of hyphae per gram of soil and about 700–900 kg of mycelium per hectare in a humus-rich layer of the soil (van Elsas et al. 2007). Similarly, AM fungi can produce up to 100 m of hyphae per gram of soil (Miller et al. 1995). The extraradical mycelia of mycorrhizal fungi constitute around 20–30 % of total soil microbial biomass and have a powerful influence upon the biogeochemical cycling of nutrients and the composition and functioning of plant communities (Leake et al. 2004). Mycorrhizal mycelial networks in the soil are the most dynamic and functionally diverse components of the symbiosis, and they receive as much as 10 % or more of the net photosynthate from their host plant; part of which is exuded into the soil (Leake et al. 2004) (Fig. 5.1).

Like roots, the mycorrhizal hyphae also release compounds into the soil and play an important role in the microbial activity and nutrient dynamics of the soil (Jones et al. 2004). This holds well, especially for the ecto-, ericoid, and orchid mycorrhizal fungal mycelium which can release hydrolytic enzymes to acquire nutrients from organic sources (Chalot and Brun 1998) and other compounds (Sun et al. 1999). The highest concentration of glucose and trehalose and the lowest concentration of fructose, galactose, sucrose, raffinose, and mannitol were detected in hyposphere soil of olive trees (*Olea europaea*) colonized by *R. intraradices* (Mechri et al. 2014). The term hyphosphere was thus introduced to denote the soil region influenced by extraradical mycelium of the mycorrhizal fungus (Jones et al. 2004). In some cases, the activity and composition of microorganisms in the hyphosphere have been shown to affect the activities of AM fungi (Andrade et al. 1997; Filion et al. 1999).

The hyphosphere may be rich in carbon, but deficient in available nutrients like P for many microbes, which in turn may influence their activity of mineralizing the phytate-P of the soil and later adding the available P into microbial biomass P that is potentially available to AM fungal hyphae (Zhang et al. 2014). Plant genotype, development of the root system, quality and quantity of the root exudates, and plant's carbohydrate metabolism can influence microbial biomass and activity in the mycorrhizal hyphosphere (Marschner et al. 2001; Sood 2003; Hooker et al. 2007; Toljander et al. 2007). These have resulted in studies where colonization of plant roots by AM fungi has been shown to decrease (Wamberg et al. 2003; Cavagnaro et al. 2003), increase (Van Aarle et al. 2002; Albertsen et al. 2006), or have no effect (Olsson et al. 1996; Andrade et al. 1997) on the microbial biomass.

As per the reciprocal reward mechanism suggested for stabilizing the cooperation of mycorrhizal-plant symbiosis (Kiers et al. 2003; Hammer et al. 2011), the amount of carbon supplied to the soil by different fungi could vary significantly. For instance, when there is a one-to-one situation like an individual fungus colonizing the plant root system, the quantity of carbon that is provided by the plant to the fungus depends on the P contribution of its fungal partner and vice versa (Hammer et al. 2011). But, when situation involves many-to-many like different mycorrhizal fungi colonizing roots of different plant species in a community, the plants can detect and discriminate fungal partners with either excess or limited supply of carbohydrates. The fungal partners can also reciprocate by increasing the transfer of nutrients only to the roots that provide more carbohydrates (Kiers et al. 2011; Fellbaum et al. 2014). It can be seen that in mycorrhizal symbiosis, the plants and fungi have a choice to select between multiple potential partners. However, in the hyphosphere, AM fungi may receive P from different phosphate-solubilizing microorganisms (PSMs) including bacteria and fungi. It is possible that the choice of AM fungi is more limited for the PSMs because of scale and nonfilamentous growth issues, and thus each bacterium is probably dependent upon only a single AM fungal hypha for its carbon support. Thus, this may be expected to make the PSM more open to cooperative behavior (Zhang et al. 2016).

The mycorrhizal fungi aid plants in their forage for nutrients and water by extending its range into areas of soil that are not accessible by roots and to nutrient-rich soil hotspots through a large network of extraradical mycelium (Kaiser et al. 2015). Mycorrhizal hyphae also stimulate the surrounding soil microbes by release of carbon which is labile. This increases the availability of local nutrients in the hyphosphere (Hodge et al. 2010; Cheng et al. 2012; Jansa et al. 2013). Nevertheless, when the AM fungi pass on the plant photosynthates to the hyphosphere, it results in an increase in the availability of nutrients by stimulating the depolymerization of organic matter by soil microorganisms (Hodge et al. 2010; Jansa et al. 2013). This strategy is useful for AM fungi which do not have the ability to secrete extracellular enzymes to degrade complex organic compounds (Kaiser et al. 2015; Smith and Smith 2011).

5.3.1 Microbial Interactions in the Hyphosphere

Like rhizosphere, some bacteria are associated with the mycelium of both ectomycorrhizal and AM fungi in hyphosphere (Poole et al. 2001; Mansfeld-Giese et al. 2002; Naumann et al. 2010). The mycorrhizal fungal hyphae influence bacterial populations and their activity (Andrade et al. 1997). Although numerous studies exist on the rhizosphere colonization by bacterial populations, studies are limited to the distribution of bacterial populations in the hyphosphere (Ravnskov et al. 1999; Zhang et al. 2014). The AM fungal mycelium plays a crucial role in the transfer of carbon sources to the associated bacteria (Leake et al. 2006; Drigo et al. 2010). The AM fungi in the soil decompose the organic matter indirectly through the production of exudates, which stimulate the microbial communities in the hyphosphere that are involved in the decomposition of organic matter. The hyphal exudates not only promote microorganisms but also inhibit others (Toljander et al. 2007). The ectomycorrhizae harbor bacteria, which use the type III secretion system (T3SS) encoding for the attachment of infection needle (Warmink and van Elsas 2008). However, recently, sulfonate desulfurizing bacteria, *Gammaproteobacteria* and *Actinobacteria* with T3SS, were detected in the hyphosphere of AM fungi (Gahan and Schmalenberger 2015). But unlike the previous study where the system was used for encoding bacterial attachment, it helped plants to take up sulfur from unavailable forms.

Over the past decade, several attempts have been made to unravel the physical interactions that take place between AM fungi and bacteria in the hyphosphere (e.g., Bianciotto et al. 2001; Johansson et al. 2004). Ravnskov et al. (1999) studied the influence of *R. intraradices* on *Pseudomonas fluorescens* DF57 bacteria in the hyphosphere and rhizosphere soil. It was clear that the presence of AM did not induce P starvation response and it did not affect the metabolic activity of the bacterium. The authors also found that it could not use the hyphae as a carbon substrate. Proposals were made to conclude that *R. intraradices* can negatively affect the growth and survival of *P. fluorescens* DF57 both in the presence of roots, where the fungus can change the quality and quantity of root exudates and in the hyphosphere, where the microbes can interact directly.

Bacteria that were associated with the hyphosphere of the AM fungal species, namely, *Claroideoglomus etunicatum*, *R. intraradices*, and *Funneliformis mosseae*, were investigated by Andrade et al. (1997). The changes observed in the bacterial community in the hyphosphere were not related to the quantity of AM mycelium, but were due to qualitative effects like the composition of exudates of the fungal species which is very important for the composition and the proliferation of rhizobacteria (Johansson et al. 2004). The synergistic film surrounding the fungal hyphae acts as a highway for dispersal of bacteria in water-unsaturated media (Kohlmeier et al. 2005). Although this phenomenon is well demonstrated for non-mycorrhizal fungi like *Fusarium* and *Chaetomium* (Simon et al. 2015), it is yet to be demonstrated for mycorrhizal fungi.

5.4 The Mycorrhizosphere Concept

The concept of rhizosphere has been broadened to include the fungal component of the symbiosis, which has resulted in the term mycorrhizosphere (Rambelli 1973). The mycorrhizosphere is influenced both by the root and the mycelium of the mycorrhizal fungus (Fig. 5.1). Therefore, the term mycorrhizosphere is inclusive of the specific term hyphosphere that denotes exclusively the soil zone surrounding individual fungal hyphae that extends beyond the rhizosphere into the bulk soil (Johansson et al. 2004).

Mycorrhizosphere is the soil zone influenced by mycorrhizal roots, and it consists of two components: (1) the rhizosphere, a layer of soil surrounding the root system which is directly influenced by root and root hairs, and (2) the hyphosphere, a region where mycorrhizal fungal hyphae and soil interaction takes place (Marschner 1995). Both rhizosphere and hyphosphere influence several organisms, including saprotrophic fungi and bacteria (Meier et al. 2015). These microorganisms interact in the mycorrhizosphere and affect the abiotic and biotic factors (Rillig and Mummey 2006).

5.4.1 Significance of Mycorrhizosphere

Mycorrhizosphere communities play a significant role in sustainable agriculture (Johansson et al. 2004), plant fitness, soil quality (Barea et al. 2002), and nutrient cycling (Azcon-Aguilar and Barea 2015). The management of mycorrhizosphere with the focus on restoration of ecosystems, biological control of pathogenic roots, enhancing the quality of soil, phytoremediation of heavy metals in contaminated soil, and osmotic stress reduction was recently reviewed by Barea et al. (2013). The importance of mycorrhizosphere was also revealed in the sustainable, low-cost phytoremediation methods (Trotta et al. 2006), development of bioenergy crops (Philippot et al. 2013), carbon sequestration (Rees et al. 2005), mineral weathering (Frey-Klett et al. 2009; Koele et al. 2014), and pyrene degradation in soil (Li et al. 2008).

5.4.1.1 Role of Plant Growth-Promoting Rhizobacteria (PGPR)

PGPR are the rhizosphere bacteria involved in the plant growth and development through various direct and indirect processes (Gupta et al. 2000; Glick et al. 2007). They also act as biocontrol agents. PGPR facilitate plant growth by enhancing the uptake of nutrients (Adesemoye et al. 2009), solubilization of phosphate and other essential minerals (Ramachandran et al. 2007; Vyas and Gulati 2009), synthesizing plant growth regulators (Kannan and Surrendar 2009), production of siderophores (Idris et al. 2007; Ahmad et al. 2008), and suppression of disease-causing pathogens (Kloepper et al. 2004; Salimpour et al. 2010). PGPR are potential antagonists and are capable of producing hydrolytic enzymes which lyse the pathogenic fungal cells (Maksimov et al. 2011).

A significant increase in growth and yield has been reported in many important crops (Gray and Smith 2005; Peng et al. 2013). Some of the PGPR belong to the genera, *Azotobacter*, *Pseudomonas*, *Bacillus*, *Klebsiella*, *Arthrobacter*, *Azospirillum*, *Enterobacter*, *Azoarcus*, *Serratia*, and *Rhizobium* (Burdman et al. 2000). *Pseudomonas fluorescens* is present in the rhizosphere of many crop plants and is well known for its biocontrol activity (Costa et al. 2006; Ahmad et al. 2008). The suppression of root pathogens by symbiotic root-colonizing microorganisms is well documented (Rezzonico et al. 2007; Abbas-Zadeh et al. 2010). Okubara et al. (2010) demonstrated the biocontrol activity of *Pseudomonas fluorescens* against soilborne fungal pathogens of *Triticum aestivum* during the infection process. An isolate of *P. fluorescens* B16 isolated from the graminaceous plant roots colonized the roots of different plants and increased the height, flower and fruit number, and total weight of tomato plants (Minorsky 2008). PGPR have the ability to secrete metabolites such as antibiotics, fungal cell wall-degrading enzymes, some gaseous products, and siderophores (Idris et al. 2007). Further, they also produce plant growth regulators such as auxins, gibberellins, and cytokinins that facilitate nutrient release (Idrise et al. 2002) and provide resistance against environmental stresses (Ashraf et al. 2004). The IAA (indole-3-acetic acid) produced by bacteria enhances plant growth (Khare and Arora 2010). The production of siderophores, by *Pseudomonas fluorescens* and *Pseudomonas putida*, helps plants in their uptake of iron and deters the growth of plant pathogens (Santoyo et al. 2012).

Rhizosphere colonization by *Azospirillum* species enhances the plant growth due to its ability to fix N (Helman et al. 2011). They are also involved in the production of siderophores (Massenia Reis et al. 2011) and plant hormones like IAA and gibberellins (Martinez-Morales et al. 2003). Dual inoculation of PGPR (*Pseudomonas, Azotobacter, Azospirillum*) and AM fungi stimulates lycopene production and antioxidant activity in tomato (Kourosh et al. 2010). *Pseudomonas* sp. and *Azospirillum* sp. isolated from root cuttings of *Piper nigrum* resulted in significant phosphate solubilization (Ramachandran et al. 2007). Phosphate-solubilizing *Bacillus megaterium* isolated from the rhizosphere of tea was known to promote plant growth. Some PGPR contain ACC-deaminase enzyme which has the ability to scale down the ethylene levels, thereby promoting plant growth and development (Zahir et al. 2008). The role of PGPR with ACC-deaminase activity in phytoremediation (Cavalca et al. 2010) and biocontrol of the plant pathogens has also been demonstrated (Belimov et al. 2007).

5.4.1.2 Mycorrhization Helper Bacteria (MHB)

Bacteria which are beneficial to both arbuscular and ectomycorrhizal fungi are often called mycorrhiza or mycorrhization helper bacteria (Duponnois 2006; Frey-Klett et al. 2007). Some of the MHB strains include the Gram-positive bacteria (*Agrobacterium, Azospirillum, Azotobacter, Bradyrhizobium, Pseudomonas, Klebsiella*), Gram-negative bacteria (*Bacillus, Paenibacillus, Variovorax*), and Gram-positive actinomycetes (*Streptomycetes, Arthrobacter, Rhodococcus*) (Frey-Klett et al. 2007). Mycorrhization helper bacterium enhances the mycorrhiza formation mostly, ectomycorrhizal fungi by producing growth metabolites thus allowing fungal hyphae to colonize plant roots with large surface area (Bending et al. 2006).

Pseudomonas is one of the important groups of MHB that facilitate rapid colonization of the root by mycorrhizal fungi and stimulate the formation of lateral roots (Poole et al. 2001). MHB and the plant roots come in direct contact to promote mycorrhizal symbiosis (Aspray et al. 2006).

Actinomycetes are often free colonizers of plant roots, mycorrhizosphere and rhizospheres (Tarkka et al. 2008). Some actinomycetes are involved in the suppression of AM fungi due to its inhibitory effect in the rhizosphere. Root rot of apple caused by *Pythium, Dematophora*, and *Fusarium* is suppressed by MHB obtained from spores of AM fungi (Dohroo and Sharma 2012). MHB also induces AM fungal spore germination and mycelial growth due to the production of growth factors and by detoxifying the antagonistic compounds (Frey-Klett et al. 2007). *Streptomyces* sp. AcH505 produces metabolite auxofuran that suppresses the growth of pathogenic fungi and induces pre-symbiotic growth of *Amanita muscaria* and *Suillus bovinus* (Keller et al. 2006; Frey-Klett et al. 2007). One of the MHB isolates *Burkholderia* sp. EJP67 isolated from the ectomycorrhizal roots of *Pinus sylvestris* promoted the formation of long and short roots in the same host plant (Poole et al. (2001). On the whole, MHB has the ability to promote mycorrhization by different ectomycorrhizal fungi (Bending 2007).

5.4.1.3 Fungi

Other than PGPR and MHB, soil fungi also play a vital role in the mycorrhizosphere. The AM fungi also contribute to its role as biofertilizer, resistance against plant diseases, heavy metals, and various stress conditions (Hildebrandt et al. 2007). *Trichoderma* sp. secretes metabolites which act as a biological fungicide against the disease-causing fungi (Vinale et al. 2009). Other potential fungi known for biocontrol of soilborne diseases are *Trichothecium*, *Epicoccum*, and *Aspergillus*. Application of both *Trichoderma* sp., and mycorrhizal fungi improves plant growth by reducing the negative effect of pathogens (Ene and Alexandru 2008). The AM fungal association with plants decreases both the incidence and severity of disease caused by phytopathogens (Chakraborty et al. 2011).

5.4.1.4 Soil Aggregation

Soil aggregates that favours growth of plants and microbes consist of minerals, roots, fungal mycelium, bacteria, organic matter, and AM fungal derived glomalin (Rillig et al. 2007). These aggregates are formed by the adhesive action of oligosaccharides in glomalin with the help of iron or polyvalent cations (Rillig 2004). Glomalin is a fungal glycoprotein, insoluble in water, resistant to heat degradation, and found in almost all soils (Nichols and Wright 2004; Bai et al. 2009). The AM fungal mediated formation of aggregates in the soil enhances hyphal length, plant roots, and microbial communities in the mycorrhizosphere (Nichols 2008). The AM fungal hyphae may affect the soil aggregation directly by providing a skeletal structure to physically hold the mineral particles of the soil. The entangled hyphae also serve as a source for the organic and inorganic binding agents, and the soil microaggregates are transformed into macroaggregates (Miller and Jastrow 1992). The AM fungi are also capable of secreting considerable quantities of glomalin into the soil environment (Wright and Upadhyaya 1998; Wright 2000; Rillig et al. 2002, 2003), and in due course of time, a portion of these compounds may be reabsorbed by the mycorrhizal hyphae (Sun et al. 1999) similar to the reabsorption of exuded compounds by the roots (Jones and Darrah 1993). The soil aggregates are more stable in the hyphosphere when compared to the soils that are free from mycorrhizal hyphae. This clearly shows that the extraradical mycelium of mycorrhizal fungi can stabilize soil aggregates without any contributions from plant roots (Andrade et al. 1998). In addition, the fungi also play an important role in the improvement of soil structure as they provide reduced amount of carbon to the extrarhizosphere microflora of the hyphosphere (Tisdall and Oades 1982; Bagyaraj 1984). When the microbial biomass increases, the aggregate stability also increases (Lynch 1981). Moreover, the AM hyphal growth that is outside the rhizosphere helps the movement of organic nutrients from the plant to microorganisms (Jakobsen and Rosendahl 1990). Therefore, the hyphosphere is marked by severe bacterial colonization than the bulk soil without AM hyphae as revealed by the positive relationship between the length of hyphae and stability of the soil aggregates (Foster et al. 1983).

5.4.2 Molecular Mechanisms of Mycorrhizosphere

The molecular mechanisms could provide a better knowledge about the interactions occurring between plants and microbes in the mycorrhizosphere. Different types of biochemical and molecular mechanisms occur in the mycorrhizosphere. The mechanism involved in the uptake of nutrients and how mycorrhizosphere microbes defend plants against phytopathogens at the molecular level is discussed below.

5.4.2.1 Mechanism Involved in Biocontrol of Soilborne Diseases

The AM fungi have the ability to induce the establishment of rhizobacteria that deter the growth of pathogens in the mycorrhizosphere before they infect the plant roots (Lioussanne 2010). Plant diseases can be controlled by manipulation of indigenous or through introduction of antagonistic microbes or by management of resident soil microbes that can decrease the pathogen propagules responsible for causing diseases (Linderman 1992). Mycorrhizal plants are more resistant to infestation by soilborne pathogens, nematodes, and also root insects (Whipps 2004). A number of hypotheses have been put forward to explain the role AM fungi in controlling soilborne plant pathogens. Induced systematic resistance (ISR) is one of the mechanisms through jasmonic acid and ethylene production, in which rhizobacteria suppress diseases in plants. Systemic acquired resistance (SAR) is an induced defense mechanism through salicylic acid (SA) production, which protects the plants from pathogenic microbes (Van Loon et al. 1998). These ISR and SAR play an important role in plant defense mechanism through different signaling pathways that are interlinked with each other (Pozo et al. 2008). The plant defenses are preconditioned by an infection that results in tolerance to pathogens (Van Hulten et al. 2006; Beckers et al. 2009). Certain defense-related genes are activated by SA, known as pathogenesis-related proteins (Van Loon 2007). Another mechanism known as mycorrhiza-induced resistance (MIR) is also a well-known plant defense mechanism which has been demonstrated in rice against the blast fungus *Magnaporthe grisea* (Pozo et al. 2008; Campos-Soriano et al. 2012).

In the ISR, the AM colonization regulates the stimulated pathogenic symptoms in a systematic manner (Pozo and Azcon-Aguilar 2007). Many components have been isolated from AM-colonized plants which could control pathogenic activities. For example, high concentration of phenolic acids (Singh et al. 2004), few isoforms, superoxide dismutases, and peroxidases was also found (Garmendia et al. 2006). Isoforms of few enzymes such as chitosanases, β-1,3-glucanases, chitinases, and peroxidases have been identified in mycorrhizal roots (Pozo et al. 1996). In another mechanism, two strains of fluorescent pseudomonads proved to be an excellent biocontrol agent. They suppress other microbes by secretion of secondary metabolites (Srivastava et al. 2001; Kang et al. 2008).

5.4.2.2 Mechanism of Nutrient Exchange and Nutrient Cycling

Root-associated microbes involved in the nutrient cycling in plants are N_2-fixers, P mobilizers, and AM fungi. AM fungal interactions are responsible for the nutrient exchange between the plant and fungi. Moreover, saprophytic rhizobacteria also have the ability for N and P mobilization (Azcon-Aguilar and Barea 2015). Mostly the nutrient exchange takes place within root cortical cells having arbuscules. The extracellular hyphal network spreads into the surrounding soil and reaches the nutrient depletion zone and enhances the supply of inorganic nutrients mostly phosphate and nitrate (Smith et al. 2011). In return, the heterotrophic fungal partner receives photosynthates from the host plant (Smith and Smith 2011). This causes important changes in primary and secondary metabolism of plants (Harrison 1999). The N and P transport from the soil to plants occurs both via mycorrhizal pathway and direct pathway. AM roots absorb nutrients through root epidermis, root hairs, and AM fungal hyphae and transport them to root cortical cells. The N and P from the soil are transported to roots mostly by the mycorrhizal pathway (Smith and Smith 2011).

In N transport process, inorganic N is taken up by the extraradical mycelium of the mycorrhizal fungi and incorporated into arginine in the fungal cytoplasm. Next, it is transferred to the intraradical mycelium where it can associate with polyphosphate. The arginine is broken down in the intraradical mycelium prior to its translocation to the plant root. As a result, the fungal N reaches root as ammonium (Govindarajulu et al. 2005). Most of the fungal N that is transferred to the plants in the form of ammonia is acquired by the mycorrhizal fungi through the involvement of ammonium transporter gene like *GintAMT1* (Lopez-Pedrosa et al. 2006). In phosphate acquisition, the uptake of inorganic phosphate by AM fungi from soil requires high-affinity transporters which are present in the fungal extraradical mycelium (Benedetto et al. 2005). After the fungal uptake of inorganic phosphate, it is transported into the vacuoles and gets deposited in the form of polyphosphate chains. Further, polyphosphate chains are released by the fungus after moving into arbuscules and finally it reaches the cells of the root cortex. A number of AM fungi and plant phosphate transporters involved in P transport of mycorrhizal plants have been identified. For example, P transporter genes *GiPT* from *R. intraradices* (Maldonado-Mendoza et al. 2001); *ORYsa*, *Pht1*, *Pht11*, and *MtPT4* from rice (Paszkowski et al. 2002), and *Medicago truncatula* have been isolated and characterized (Harrison et al. 2002).

5.5 Conclusion

Non-mycorrhizal plants possess only rhizosphere, while the mycorrhizal plants have mycorrhizal mycelium that may contribute to around 75 % of the plant's absorptive surface. Mycorrhizospheres are unique and intricate systems whose composition and function determine the existence and sustainability of most terrestrial plant communities. The study of mycorrhizosphere is inherently difficult because of the intimacy of the plant roots and the fungal hyphae with the soil. This is further complicated by the mycorrhizal mycelial networks that interlink plant roots and stretch the conventional rhizosphere. Limited studies on the complicities of the mycorrhizosphere suggest that the microbial communities in the mycorrhizosphere have multilevel interactions among themselves and with the environment. Further studies using molecular tools as suggested by Timonen and Marschner (2006) would be of immense value in unrevealing the composition and physiological functions of the mycorrhizosphere.

References

Abbas-Zadeh PN, Saleh-Rastin H, Asadi-Rahmani K, Khavazi A, Soltani AR, Nejati S, Miransari M (2010) Plant growth-promoting activities of fluorescent pseudomonads, isolated from the Iranian soils. Acta Physiol Plant 32:281–288

Abdel-Fattah GM, Mohamedin AH (2000) Interactions between a vesicular-arbuscular mycorrhizal fungus (*Glomus intraradices*) and *Streptomyces coelicolor* and their effects on sorghum plants grown in soil amended with chitin of brawn scales. Biol Fertil Soils 32:401–409

Adesemoye AO, Torbert HA, Kloepper JW (2009) Plant growth promoting rhizobacteria allow reduced application rates of chemical fertilizers. Microb Ecol 58:921–929

Ahmad F, Ahmad I, Khan MS (2008) Screening of free-living rhizospheric bacteria for their multiple plant growth promoting activities. Microbiol Res 163:173–81

Albertsen A, Ravnskov S, Green H, Jensen DF, Larsen J (2006) Interactions between mycelium of the mycorrhizal fungus *Glomus intraradices* and other soil microorganisms as affected by organic matter. Soil Biol Biochem 38:1008–1014

Amora-Lazcano E, Vázquez MM, Azcón R (1998) Response of nitrogen-transforming microorganisms to arbuscular mycorrhizal fungi. Biol Fertil Soils 27:65–70

Andrade G, Mihara KL, Linderman RG, Bethlenfalvay GJ (1997) Bacteria from rhizosphere and hyphosphere soils of different arbuscular-mycorrhizal fungi. Plant Soil 192:71–79

Andrade G, Mihara KL, Linderman RG, Bethlenfalvay GJ (1998) Soil aggregation status and rhizobacteria in the mycorrhizosphere. Plant Soil 202:89–96

Artursson V, Finlay RD, Jansson JK (2006) Interactions between arbuscular mycorrhizal fungi and bacteria and their potential for stimulating plant growth. Environ Microbiol 8:1–10

Ashraf M, Hasnain S, Berge O, Mahamood T (2004) Inoculating wheat seedlings with exopolysaccharide-producing bacteria restricts sodium uptake and stimulates plant growth under salt-stress. Biol Fertil Soils 40:157–162

Aspray TJ, Jones EE, Whipps JM, Bending GD (2006) Importance of mycorrhization helper bacteria cell density and metabolite localization for the *Pinus sylvestris-Lactarius rufus* symbiosis. FEMS Microbiol Ecol 56:25–33

Azaizeh HA, Marschner A, Römheld V, Wittenmayer L (1995) Effects of a vesicular-arbuscular mycorrhizal fungus and other soil microorganisms on growth, mineral nutrient acquisition and root exudation of soil-grown maize plants. Mycorrhiza 5:321–327

Azcón R (1989) Selective interaction between free-living rhizosphere bacteria and vesicular-arbuscular mycorrhizal fungi. Soil Biol Biochem 21:639–644

Azcon-Aguilar C, Barea JM (2015) Nutrient cycling in the mycorrhizosphere. J Soil Sci Plant Nutr 25:372–396

Badri DV, Vivanco JM (2009) Regulation and function of root exudates. Plant Cell Environ 32:666–681

Badri DV, Quintana N, El Kassis EG, Kim HK, Choi YH, Suiyama A, Verpoorte R, Martinoia E, Manter DK, Vivanco JM (2009) An ABC transporter mutation alters root exudation of phytochemicals that provoke an overhaul of natural soil microbiota. Plant Physiol 151:2006–2017

Baetz U, Martinoia E (2014) Root exudates: the hidden part of plant defense. Trends Plant Sci 19:90–98

Bagyaraj DJ (1984) Biological interactions with VA mycorrhizal fungi. In: Powell CL, Bagyaraj DJ (eds) VA mycorrhiza. CRC, Boca Raton, pp 132–153

Bai CM, He XL, Tang HL, Shan BQ, Zhao LL (2009) Spatial distribution of arbuscular mycorrhizal fungi, glomalin and soil enzymes under the canopy of *Astragalus adsurgens* Pall. in the Mu Us sandland, China. Soil Biol Biochem 41:941–947

Bais H, Weir PTL, Perry LG, Gilroy S, Vivanco JM (2006) The role of root exudates in rhizosphere interactions with plants and other organisms. Annu Rev Plant Biol 57:233–66

Bakhtiar Y, Yahya S, Sumaryono W, Sinaga MS, Budi SW, Tajuddin T (2010) Isolation and identification of mycorrhizosphere bacteria and their antagonistic effects towards *Ganoderma boninense in vitro*. Microbiol Indones 4:96–102

Balandreau J, Knowles R (1978) The rhizosphere. In: Dommerques YR, Krupa SV (eds) Interactions between non-pathogenic soil microorganisms and plants. Elsevier, Amsterdam/Oxford/New York, pp 243–268

Bansal RK, Bajaj A (2003) Effect of volatile fatty acids on embryogenesis and hatching of Meloidogyne incognita eggs. Nematol Mediterr 31:135–140

Bansal RK, Dahiya RS, Narula N, Jain RK (2005) Management of *Meloidogyne incognita* in cotton, using strains of the bacterium *Gluconacetobacter diazotrophicus*. Nematol Mediterr 33:101–105

Banuelos J, Alarcón A, Larsen J, Cruz-Sánchez S, Trejo D (2014) Interactions between arbuscular mycorrhizal fungi and *Meloidogyne incognita* in the ornamental plant *Impatiens balsamina*. J Soil Sci Plant Nutr 14:63–74

Barea JM (2000) Rhizosphere and mycorrhiza of field crops. In: Toutant JP, Balazs E, Galante E, Lynch JM, Schepers JS, Werner D, Werry PA (eds) Biological resource management: connecting science and policy. INRA, Springer, Berlin, pp 110–125

Barea JM, Azcon R, Azcon-Aguilar C (2002) Mycorrhizosphere interactions to improve plant fitness and soil quality. Antonie Van Leeuwenhoek 81:343–351

Barea JM, Pozo MJ, Lopez-Raez JA, Aroca R, Ruiz-Lozano JM, Ferrol N, Azcon R, Azcon-Aguilar C (2013) Arbuscular mycorrhizas and their significance in promoting soil-plant systems sustainability against environmental stresses. In: Rodelas B, González-López J (eds) Beneficial plant-microbial interactions: ecology and applications. CRC Press, Boca Raton, pp 353–387

Beckers GJ, Jaskiewicz M, Liu Y, Underwood WR, He SY, Zhang S, Conrath U (2009) Mitogen -activated protein kinases 3 and 6 are required for full priming of stress responses in *Arabidopsis thaliana*. Plant Cell 21:944–953

Belimov AA, Dodd IC, Safronova VI, Hontzeas N, Davies WJ (2007) *Pseudomonas brassicacearum* strain Am3 containing 1-aminocyclopropane-1-carboxylate deaminase can show both pathogenic and growth promoting properties in its interaction with tomato. J Exp Bot 58:1485–1495

Bending G (2007) What are the mechanisms and specificity of mycorrhization helper bacteria? New Phytol 174:707–710

Bending GD, Aspray TJ, Whipps JM (2006) Significance of microbial interactions in the mycorrhizosphere. Adv Appl Microbiol 60:97–132

Benedetto A, Magurno F, Bonfante P, Lanfranco L (2005) Expression profiles of a phosphate transporter gene (GmosPT) from the endomycorrhizal fungus *Glomus mosseae*. Mycorrhiza 15:620–627

Berg G, Hallmann J (2006) Control of plant pathogenic fungi with bacterial endophytes. In: Barbara PD, Schulz JE, Boyle CJC, Sieber TN (eds) Microbial root endophytes. Springer, Berlin/Heidelberg, pp 53–69

Berta G, Fusconi A, Trotta A (1993) VA mycorrhizal infection and the morphology and function of root systems. Environ Exp Bot 33:159–173

Bethlenfalvay GJ, Schüepp H (1994) Arbuscular mycorrhizas and agrosystem stability. In: Gianinazzi S, Schüepp H (eds) Impact of arbuscular mycorrhizas on sustainable agriculture and natural ecosystems. Birkhäuser, Basel, pp 117–131

Bharadwaj DP, Lundquist P, Alstrom S (2008) Arbuscular mycorrhizal fungal spore-associated bacteria affect mycorrhizal colonization, plant growth and potato pathogens. Soil Biol Biochem 40:2494–2501

Bianciotto V, Andreotti S, Balestrini R, Bonfante P, Perotto S (2001) Mucoid mutants of the biocontrol strain *Pseudomonas fluorescens* CHA0 show increased ability in biofilm formation on mycorrhizal and nonmycorrhizal carrot roots. Mol Plant Microbe Interact 14:255–260

Bonfante P, Anca IA (2005) Plants, mycorrhizal fungi and bacteria: a network of interactions. Annu Rev Microbiol 63:363–383

Bonkowski M (2004) Protozoa and plant growth: the microbial loop in soil revisited. New Phytol 162:617–631

Bonkowski M, Jentschke G, Scheu S (2001) Contrasting effects of microbial partners in the rhizosphere: interactions between Norway Spruce seedlings (*Picea abies* Karst.), mycorrhiza (*Paxillus involutus* (Batsch) Fr.) and naked amoebae (protozoa). Appl Soil Ecol 18:193–204
Bulgarelli D, Rott M, Schlaeppi K, Themaat E Ver Loren van, Ahmadinejad N, Assenza F, Rauf P, Huettel B, Reinhardt R, Schmelzer E (2012) Revealing structure and assembly cues for *Arabidopsis* root-inhabiting bacterial microbiota. Nature 488:91–95
Burdman S, Okon Y, Jurkevitch E (2000) Surface characteristics of *Azospirillum brasilense* in relation to cell aggregation and attachment to plant roots. Crit Rev Microbiol 26:91–110
Campos-Soriano L, García-Martínez J, Segundo BS (2012) The arbuscular mycorrhizal symbiosis promotes the systemic induction of regulatory defense-related genes in rice leaves and confers resistance to pathogen infection. Mol Plant Pathol 13:579–592
Cardon ZG, Whitbeck JK (2007) The rhizosphere: an ecological perspective. Elsevier/Academic Press, New York
Cavagnaro TR, Smith FA, Ayling SM, Smith SE (2003) Growth and phosphorus nutrition of a *Paris* type arbuscular mycorrhizal symbiosis. New Phytol 157:127–134
Cavalca L, Zanchi R, Corsini A, Colombo M, Romagnoli C, Canzi E et al (2010) Arsenic-resistant bacteria associated with roots of the wild *Cirsium arvense* (L.) plant from an arsenic polluted soil, and screening of potential plant growth-promoting characteristics. Syst Appl Microbiol 33:154–164
Chakraborty A, Chakrabarti K, Chakraborty A, Ghosh S (2011) Effect of long-term fertilizers and manure application on microbial biomass and microbial activity of a tropical agricultural soil. Biol Fertil Soils 47:227–233
Chalot M, Brun A (1998) Physiology of organic nitrogen acquisition by ectomycorrhizal fungi and ectomycorrhizas. FEMS Microbiol Rev 22:21–44
Chaparro J, Sheflin A, Manter D, Vivanco J (2012) Manipulating the soil microbiome to increase soil health and plant fertility. Biol Fertil Soils 48:489–499
Cheng L, Booker F, Tu C, Burkey K, Zhou L, Shew H, Rufty TW, Hu S (2012) Arbuscular mycorrhizal fungi increase organic carbon decomposition under elevated CO_2. Science 2:1084–1087
Christensen H, Jakobsen I (1993) Reduction of bacterial growth by a vesicular-arbuscular mycorrhizal fungus in the rhizosphere of cucumber (*Cucumis sativus* L.). Biol Fertil Soils 15:253–258
Costa R, Newton CM, Raquel G, Peixoto S, Rumjanek N, Berg G, Leda CS, Hagler M, Smalla K (2006) Diversity and antagonistic potential of *Pseudomonas* sp. associated to the rhizosphere of maize grown in a subtropical organic farm. Soil Biol Biochem 38:2434–2447
Cruz AF, Ishii T (2012) Arbuscular mycorrhizal fungal spores host bacteria that affect nutrient biodynamics and biocontrol of soil-borne plant pathogens. Biol Open 1:52–57
De-la-Peña C, Badri D, Loyola-Vargas V (2012) Plant root secretions and their interactions with neighbors. In: Baluska F, Vivanco J (eds) Secretions and exudates in biological systems. Springer, Berlin, pp 1–26
Deveau A, Brulé C, Palin B, Champmartin D, Rubini P, Garbaye J, Sarniguet A, Frey-Klett P (2010) Role of fungal trehalose and bacterial thiamine in the improved survival and growth of the ectomycorrhizal fungus *Laccaria bicolor* S238N and the helper bacterium *Pseudomonas fluorescens* BBc6R8. Environ Microbiol Rep 2:560–568
Dohroo A, Sharma DR (2012) Role of plant growth promoting rhizobacteria, arbuscular mycorrhizal fungi and their helper bacteria on growth parameters and root rot of apple. WJST 2:35–38
Dong Y, Zhu YG, Smith FA, Wang Y, Chen B (2008) Arbuscular mycorrhiza enhanced arsenic resistance of both white clover (*Trifolium repens* Linn.) and ryegrass (*Lolium perenne* L.) plants in an arsenic contaminated soil. Environ Pollut 155:174–181
Drigo B, Pijl AS, Duyts H, Kielak A, Gamper HA, Houtekamer MJ et al (2010) Shifting carbon flow from roots into associated microbial communities in response to elevated atmospheric CO_2. Proc Natl Acad Sci U S A 107:10938–10942

Duponnois R (2006) Bacteria helping mycorrhiza development. In: Mukerji KG, Manoharachary C, Singh J (eds) Microbial activity in the rhizosphere. Springer, Berlin, pp 297–310

Dutta S, Rani TS, Podile AR (2013) Root exudate-induced alterations in *Bacillus cereus* cell wall contributes to root colonization and plant growth promotion. PLoS ONE 8, e78369

Egamberdieva D, Kamilova F, Validov SH, Gafurova L, Kucharova Z, Lugtenberg B (2008) High incidence of plant growth-stimulating bacteria associated with the rhizosphere of wheat grown on salinated soil in Uzbekistan. Environ Microbiol 10:1–9

Ene M, Alexandru M (2008) Microscopical examination of plant reaction in case of infection with *Trichoderma* and mycorrhizal fungi. Rome Biotechnol Lett 13:13–19

Fellbaum CR, Mensah JA, Cloos AJ, Strahan GE, Pfeffer PE, Kiers ET, Bucking H (2014) Fungal nutrient allocation in common mycorrhizal networks is regulated by the carbon source strength of individual host plants. New Phytol 203:646–656

Feng G, Zhang FS, Li XL, Tian CY, Tang C, Rengel Z (2002) Improved tolerance of maize plants to salt stress by arbuscular mycorrhiza is related to higher accumulation of soluble sugars in roots. Mycorrhiza 12:185–190

Filion M, St-Arnaud M, Fortin JA (1999) Direct interaction between the arbuscular mycorrhizal fungus *Glomus intraradices* and different rhizosphere microorganisms. New Phytol 141:525–533

Finlay RD (2008) Ecological aspects of mycorrhizal symbiosis: with special emphasis on the functional diversity of interactions involving the extraradical mycelium. J Exp Biol 59:1115–1126

Foster RC, Rovira AD, Cock TW (1983) Ultrastructure of the root-soil interface. American Phytopathological Society, St. Paul, pp 5–11

Frey-Klett P, Garbaye J, Tarkka M (2007) Tansley review: the mycorrhiza helper bacteria revisited. New Phytol 176:22–36

Frey-Klett P, Koele N, Turpault M-P, Hildebrand EE, Uroz S (2009) Interactions between mycorrhizal fungi and mycorrhizosphere bacteria during mineral weathering: budget analysis and bacterial quantification. Soil Biol Chem 41:1935–1942

Fulekar MH, Pathak B (2015) Rhizosphere: an innovative approach for remediation of contaminants. IJSER 6:291–303

Gahan J, Schmalenberger A (2015) Arbuscular mycorrhizal hyphae in grassland select for a diverse and abundant hyphospheric bacterial community involved in sulfonate desulfurization. Appl Soil Ecol 89:113–121

Garmendia I, Aguirreolea J, Goicoechea N (2006) Defence-related enzymes in pepper roots during interactions with arbuscular mycorrhizal fungi and/or *Verticillium dahliae*. Biocontrol 51:293–310

Giannakis N, Sanders FE (1990) Interactions between mycophagous nematodes mycorrhizal and other soil fungi. Agric Ecosyst Environ 29:163–167

Glick BR, Cheng Z, Czarny J, Duan J (2007) Promotion of plant growth by ACC deaminase-producing soil bacteria. Eur J Plant Pathol 119:329–39

Govindarajulu M, Pfeffer PE, Jin H, Abubaker J, Douds DD, Allen JW, Bucking H, Lammers PJ, Shachar-Hill Y (2005) Nitrogen transfer in the arbuscular mycorrhizal symbiosis. Nature 435:819–823

Gray EJ, Smith DL (2005) Intracellular and extracellular PGPR: commonalities and distinctions in the plant-bacterium signaling processes. Soil Biol Biochem 37:395–412

Gupta A, Gopal M, Tilak KV (2000) Mechanism of plant growth promotion by rhizobacteria. Indian J Exp Biol 38:856–862

Haldar S, Sengupta S (2015) Impact of plant development on the rhizobacterial population of *Arachis hypogaea*: a multifactorial analysis. J Basic Microbiol 55:922–928

Hammer EC, Pallon J, Wallander H, Olsson PA (2011) Tit for tat? A mycorrhizal fungus accumulates phosphorus under low plant carbon availability. FEMS Microbiol Ecol 76:236–244

Han HS, Lee KD (2005) Physiological responses of soybean-inoculation of *Bradyrhizobium japonicum* with PGPR in saline soil conditions. Res J Agric Biol Sci 1:216–221

Harrison M (1999) Biotrophic interfaces and nutrient transport in plant/fungal interfaces. J Exp Bot 50:1013–1022

Harrison MJ, Dewbre GR, Liu J (2002) A phosphate transporter from *Medicago truncatula* involved in the acquisition of phosphate released by arbuscular mycorrhizal fungi. Plant Cell 14:2413–2429

Hartmann A, Schmid M, van Tuinen D, Berg G (2009) Plant-driven selection of microbes. Plant Soil 321:235–257

Helman Y, Burdman S, Okon Y (2011) Plant growth promotion by rhizosphere bacteria through direct effects. In: Rosenberg E, Gophna U (eds) Beneficial microorganisms in multicellular life form. Springer, Heidelberg/Berlin, pp 89–103

Herdler S, Kreuzer K, Scheu S, Bonkowski M (2008) Interactions between arbuscular mycorrhizal fungi (*Glomus intraradices*, Glomeromycota) and amoebae (*Acanthamoeba castellanii*, Protozoa) in the rhizosphere of rice (*Oryza sativa*). Soil Biol Biochem 40:660–668

Hildebrandt U, Ouziad F, Marner FJ, Bothe H (2006) The bacterium *Paenibacillus validus* stimulates growth of the arbuscular mycorrhizal fungus *Glomus intraradices* up to the formation of fertile spores. FEMS Microbiol Lett 254:258–267

Hildebrandt U, Regvar MS, Bothe H (2007) Arbuscular mycorrhizal and heavy metal tolerance. Phytochemistry 68:139–146

Hiltner L (1904) Über neuere erfahrungen und probleme auf dem gebiet der bodenbakteriologie und unter besonderer berücksichtigung der gründüngung und brache. Arbeiten der Deutschen Landwirtschaftlichen Gesellschaft 98:59–78

Hinsinger P, Bengough A, Vetterlein D, Young I (2009) Rhizosphere: biophysics, biogeochemistry and ecological relevance. Plant Soil 321:117–152

Hodge A, Helgason T, Fitter AH (2010) Nutritional ecology of arbuscular mycorrhizal fungi. Fungal Ecol 3:267–273

Hol WHG, Cook R (2005) An overview of arbuscular mycorrhizal fungi-nematode interactions. Basic Appl Ecol 6:489–503

Hooker JE, Piatti P, Cheshire MV, Watson CS (2007) Polysaccharides and monosaccharides in the hyphosphere of the arbuscular mycorrhizal fungi *Glomus* E3 and *Glomus tenue*. Soil Biol Biochem 39:680–683

Hooper DU, Chapin FS, Ewel JJ, Hector A, Inchausti P, Lavorel S, Lawton JH, Lodge DM, Loreau M, Naeem S, Schmid B, Setala H, Symstad AJ, Vandermeer J, Wardle DA (2005) Effects of biodiversity on ecosystem functioning: a consensus of current knowledge. Ecol Monogr 75:3–35

Idris E, Iglesias D, Talon M, Borriss R (2007) Tryptophan-dependent production of indole-3-acetic acid (IAA) affects level of plant growth promotion by *Bacillus amyloliquefaciens* FZB42. Mol Plant Microbe Interact 20:619–626

Idrise EE, Makarewicz O, Farouk A, Rosner K, Greiner R, Bochow H, Richter T, Borriss T (2002) Extracellular phytase activity of *Bacillus amyloliquefaciens* FZB45 contributed to its plant growth promoting effect. Microbiology 148:2097–2109

Jaiti F, Meddich A, El Hadrami I (2007) Effectiveness of arbuscular mycorrhizal fungi in the protection of date palm (*Phoenix dactylifera* L.) against bayoud disease. Physiol Mol Plant Pathol 71:166–173

Jakobsen I, Rosendahl L (1990) Carbon flow into soil and external hyphae from roots of mycorrhizal cucumber plants. New Phytol 115:77–83

Jansa J, Bukovská P, Gryndler M (2013) Mycorrhizal hyphae as ecological niche for highly specialized hypersymbionts – or just soil free-riders? Front Plant Sci 4:134

Johansson JF, Paul LR, Finlay RD (2004) Microbial interactions in the mycorrhizosphere and their significance for sustainable agriculture. FEMS Microbiol Ecol 48:1–13

Johnson D, Leake JR, Ostle N, Ineson P, Read DJ (2002) *In situ* $^{13}CO_2$ pulse-labelling of upland grassland demonstrates that a rapid pathway of carbon flux from arbuscular mycorrhizal mycelia to the soil. New Phytol 153:327–334

Jones DL, Darrah PR (1993) Re-sorption of organic-compounds by roots of *Zea mays* L. and its consequences in the rhizosphere. 2. Experimental and model evidence for simultaneous exudation and re-sorption of soluble C compounds. Plant Soil 153:47–59

Jones D, Hinsinger P (2008) The rhizosphere: complex by design. Plant Soil 312:1–6

Jones DL, Hodge A, Kuzyakov Y (2004) Plant and mycorrhizal regulation of rhizodeposition. New Phytol 163:459–480

Kaiser C, Kilburn MR, Clode PL, Fuchslueger L, Koranda M, Cliff JB, Solaiman ZM, Murphy DV (2015) Exploring the transfer of recent plant photosynthates to soil microbes: mycorrhizal pathway vs direct root exudation. New Phytol 205:1537–1551

Kang SC, Pandey P, Khillon R, Maheshwari DK (2008) Process of phosphate solubilization by *Aspergillus* sp PS104 in soil amended medium. J Environ Biol 29:743–746

Kannan V, Surrendar R (2009) Synergistic effect of beneficial rhizosphere microflora in biocontrol and plant growth promotion. J Basic Microbiol 49:158–164

Keller S, Schneider K, Sussmuth RD (2006) Structure elucidation of auxofuran, a metabolite involved in stimulating growth of fly agaric, produced by the mycorrhiza helper bacterium *Streptomyces* AcH 505. J Anti biot 59:801–803

Kerry BR (2000) Rhizosphere interactions and exploitation of microbial agents for the biological control of plant-parasitic nematodes. Annu Rev Phytopathol 38:423–441

Khan MW (ed) (1993) Mechanisms of interactions between nematodes and other plant pathogens. In: Nematodes-interactions. Chapman & Hall, London, pp 175–202

Khare E, Arora NK (2010) Effect of indole-3-acetic acid (IAA) produced by *Pseudomonas aeruginosa* in suppression of charcoal rot disease of chickpea. Curr Microbiol 61:64–68

Kiers ET, Rousseau RA, West SA, Denison RF (2003) Host sanctions and the legume-rhizobium mutualism. Nature 425:78–81

Kiers ET, Duhamel M, Beesetty Y, Mensah JA, Franken O, Verbruggen E, Fellbaum CR, Kowalchuk GA, Hart MM, Bago A, Palmer TM, West SA, Vandenkoornhuyse P, Jansa J, Bücking H (2011) Reciprocal rewards stabilize cooperation in the mycorrhizal symbiosis. Science 333:880–882

Kloepper JN, Ryu CM, Zhang S (2004) Induced systemic resistance and promotion of plant growth by *Bacillus sp.* Phytopathology 94:1259–1266

Koele N, Dickie IA, Blum GD, Gleason JD, de Graaf L (2014) Ecological significance of mineral weathering in ectomycorrhizal and arbuscular mycorrhizal ecosystems from a field-based comparison. Soil Biol Chem 69:63–70

Kohlmeier S, Smits THM, Ford R, Keel C, Harms H, Wick LY (2005) Taking the fungal highway: mobilization of pollutant degrading bacteria by fungi. Environ Sci Technol 39:4640–4646

Koller R, Scheu S, Bonkowski M, Robin C (2013) Protozoa stimulate N uptake and growth of arbuscular mycorrhizal plants. Soil Biol Biochem 65:204–210

Kourosh O, Khavazi K, Moezzi A, Rajali F (2010) Influence of PGPR and AMF on antioxidant activity, lycopene and potassium contents in tomato. Afr J Agric Res 5:1108–1116

Kreuzer K, Adamczyk J, Iijima M, Wagner M, Scheu S, Bonkowski M (2006) Grazing of a common species of soil protozoa (*Acanthamoeba castellanii*) affects rhizosphere bacterial community composition and root architecture of rice (*Oryza sativa* L). Soil Biol Biochem 38:1665–1672

Kuzyakov Y (2002) Review: factors affecting rhizosphere priming effects. J Plant Nutr Soil Sci 165:382–396

Larsen J-LP, Nájera-Rincon M, González-Esquivel CE (2015) Biotic interactions in the rhizosphere in relation to plant and soil nutrient dynamics. J Soil Sci Plant Nutr 15:449–463

Lau JA, Lennon JT (2011) Evolutionary ecology of plant–microbe interactions: soil microbial structure alters selection on plant traits. New Phytol 192:215–224

Leake JR, Johnson D, Donnelly DP, Muckle GE, Boddy L, Read DJ (2004) Networks of power and influence: the role of mycorrhizal mycelium in controlling plant communities and agroecosystem functioning. Can J Bot 82:1016–1045

Leake JR, Ostle NJ, Rangel-Castro JI, Johnson D (2006) Carbon fluxes from plants through soil organisms determined by field $^{13}CO_2$ pulse-labelling in an upland grassland. Appl Soil Ecol 33:152–175

Li P, Gong Z, Fan S, He N (2008) Promotion of pyrene degradation in rhizosphere of alfalfa (*Medicago sativa* L). Chemosphere 71:1593–1598

Li XG, Zhang TL, Wang XX, Hua K, Zhao L, Han ZM (2013) The composition of root exudates from two different resistant peanut cultivars and their effects on the growth of soil-borne pathogen. Int J Biol Sci 9:164–173

Li J, Zou C, Xu J, Ji X, Niu X, Yang J, Huang X, Zhang K-Q (2015) Molecular mechanisms of nematode-nematophagus microbe interactions: basis for biological control of plant-parasitic nematodes. Annu Rev Phytopathol 53:67–95

Linderman RG (1988) Mycorrhizal interactions with the rhizosphere microflora: the mycorrhizosphere effect. Phytopathology 78:366–371

Linderman RG (1992) VA mycorrhizae and soil microbial interactions. In: Bethelenfalvay GJ, Linderman RG (eds) Mycorrhizae in sustainable agriculture. ASA Special Publication No. 54, Madison, WI, pp 45–70

Linderman RG (2000) Effects of mycorrhizas on plant tolerance to diseases. In: Kapulnick Y, Douds DD Jr (eds) Arbuscular mycorrhizas: physiology and function. Kluwer Academic Press, New York, pp 345–366

Lioussanne L (2010) The role of the arbuscular mycorrhiza-associated rhizobacteria in the biocontrol of soil borne phytopathogens. Span J Agric Res 8:51–61

Lopez-Pedrosa A, Gonzalez-Guerrero M, Valderas A, Azcon-Aguilar C, Ferrol N (2006) GintAMT1 encodes a functional high-affinity ammonium transporter that is expressed in the extraradical mycelium of *Glomus intraradices*. Fungal Genet Biol 43:102–110

Lynch JM (1981) Promotion and inhibition of soil aggregate stabilization by micro-organisms. J Gen Microbiol 126(37):1–375

Lynch JM, Whipps JM (1990) Substrate flow in the rhizosphere. Plant Soil 129:1–10

Maksimov IV, Abizgildina RR, Pusenkova LI (2011) Plant growth promoting rhizobacteria as alternative to chemical crop protectors from pathogens. Appl Biochem Microbiol 47:333–345

Maldonado-Mendoza IE, Dewbre GR, Harrison MJ (2001) A phosphate transporter gene from the extra-radical mycelium of an arbuscular mycorrhizal fungus *Glomus intraradices* is regulated in response to phosphate in the environment. Mol Plant Microbe Interact 14:1140–1148

Mansfeld-Giese K, Larsen J, Bodker L (2002) Bacterial populations associated with mycelium of the arbuscular mycorrhizal fungus *Glomus intraradices*. FEMS Microbiol Ecol 41:133–140

Mao XF, Hu F, Griffiths B, Chen XY, Liu MQ, Li HX (2007) Do bacterial-feeding nematodes stimulate root proliferation through hormonal effects? Soil Biol Biochem 39:1816–1819

Marschner H (1995) Mineral nutrition of higher plants, 2nd edn. Academic Press, London, p 889

Marschner P, Crowley DE, Ieberei R (2001) Arbuscular mycorrhizal infection changes the bacterial 16S rDNA community composition in the rhizosphere of maize. Mycorrhiza 11:297–302

Martinez-Morales LJ, Soto-Urzua L, Baca BE, Sanchez-Ahedo JA (2003) Indole-3-butyric acid (IBA) production in culture medium by wild strain *Azospirillum brasilense*. FEMS Microbiol Lett 228:167–173

Massenia Reis V, Regina dos Santos K, Teixeira K, Pedraza RO (2011) What is expected from the genus *Azospirillum* as a plant growth-promoting bacteria? In: Maheshwari DK (ed) Bacteria in agrobiology: plant growth responses. Springer, Berlin/Heildelberg, pp 123–138

McNear DH (2013) The rhizosphere-roots, soil and everything in between. Nat Educ Knowl 4:1

Mechri B, Manga AGB, Tekaya M, Attia F, Cheheb H, Meriem FB, Braham M, Boujnah D, Hammami M (2014) Changes in microbial communities and carbohydrate profiles induced by the mycorrhizal fungus (*Glomus intraradices*) in rhizosphere of olive trees (*Olea europaea* L.). Appl Soil Ecol 75:124–133

Meier IC, Pritchard SG, Brzostek ER, McCormack ML, Phillips RP (2015) The rhizosphere and hyphosphere differ in their impacts on carbon and nitrogen cycling in forests exposed to elevated CO_2. New Phytol 205:1167–1174

Mendes R, Kruijt M, de Bruijn I, Dekkers E, van der Voort M, Schneider JH, Piceno YM, DeSantis TZ, Andersen GL, Bakker PA, Raajimakers JM (2011) Deciphering the rhizosphere microbiome for disease-suppressive bacteria. Science 332:1097–1100

Mendes R, Garbeva P, Raaijmakers JM (2013) The rhizosphere microbiome: significance of plant beneficial, plant pathogenic, and human pathogenic microorganisms. FEMS Microbiol Rev 37:634–663

Meyer SLF (2003) United States Department of Agriculture – Agricultural Research Service research programs on microbes for management of plant-parasitic nematodes. Pest Manag Sci 59:665–670

Meyer RJ, Linderman RG (1986) Response of subterranean clover to dual inoculation with vesicular arbuscular mycorrhizal fungi and plant growth promoting rhizobacterium, *Pseudomonas putida*. Soil Biol Biochem 18:185–190

Miller RM, Jastrow JD (1992) The role of mycorrhizal fungi in soil conservation. In: Bethlenfalvay CJ, Linderman RG (eds) Mycorrhizae in sustainable agriculture. Crop Science Society Soil Science Society of America, Madison, pp 29–44

Miller RM, Reinhardt DR, Jastrow JD (1995) External hyphal production of vesicular-arbuscular mycorrhizal fungi in pasture and tallgrass prairie communities. Oecologia 103:17–23

Minorsky PV (2008) On the inside. Plant Physiol 146:323–324

Miransari M (2010) Contribution of arbuscular mycorrhizal symbiosis to plant growth under different types of soil stress. Plant Biol 12:563–569

Miransari M (2011) Interactions between arbuscular mycorrhizal fungi and soil bacteria. Appl Microbiol Biotechnol 89:917–930

Moore D, Robson GD, Trinci APJ (2011) 21st century guidebook to fungi. Cambridge University Press, New York

Morgan JAW, Whipps JM (2001) Methodological approaches to the study of rhizosphere carbon flow and microbial population dynamics. In: Pinton A, Varanini Z, Nannipieri P (eds) The rhizosphere. Biochemistry and organic substances at the soil-plant interface. Marcel Dekker, New York, pp 373–409

Morgan JA, Bending GD, White PJ (2005) Biological costs and benefits to plant-microbe interactions in the rhizosphere. J Exp Bot 56:1729–1739

Naumann M, Schüßler A, Bonfante P (2010) The obligate endobacteria of arbuscular mycorrhizal fungi are ancient heritable components related to the Mollicutes. ISMEJ 4:862–871

Nichols KA (2008) Indirect contributions of AM fungi and soil aggregation to plant growth and protection. In: Siddiqui ZA, Akhtar MS, Futai K (eds) Mycorrhizae: sustainable agriculture and forestry. Springer, Berlin, pp 177–194

Nichols KA, Wright SF (2004) Contributions of soil fungi to organic matter in agricultural soils. In: Magdoff F, Weil R (eds) Functions and management of soil organic matter in agroecosystems. CRC Press, Boca Raton, pp 179–198

Nihorembere V, Ongena M, Smargiass M (2011) Beneficial effect of the rhizosphere microbial community for plant growth and health. Biotechnol Agron Soc Environ 15:327–337

Okubara PA, Call DR, Wak YK, Skinner DZ (2010) Induction of defense gene homologues in wheat roots during interactions with *Pseudomonas fluorescens*. Biol Cont 55:118–125

Olsson PA, Baath E, Jakobsen I, Soderstrom B (1996) Soil bacteria respond to presence of roots but not to mycelium of arbuscular mycorrhizal fungi. Soil Biol Biochem 28:463–470

Parniske M (2008) Arbuscular mycorrhiza: the mother of plant root endosymbioses. Nat Rev Microbiol 6:763–775

Paszkowski U, Kroken S, Roux C, Briggs SP (2002) Rice phosphate transporters include an evolutionarily divergent gene specifically activated in arbuscular mycorrhizal symbiosis. Proc Natl Acad Sci U S A 99:13324–13329

Paulitz TC, Lindennan RG (1991) Mycorrhizal interactions with soil organisms. In: Aurora DK, Rai B, Mukerji KG, Knudsen G (eds) Handbook of applied mycology, soils and plants, vol 1. Marcel Dekker, New York, pp 77–129

Peng AP, Liu J, Gao YZ, Chen ZY (2013) Distribution of endophytic bacteria in *Alopecurus aequalis* Sobol and *Oxalis corniculata* L. from soils contaminated by polycyclic aromatic hydrocarbons. PLoS ONE 8:e83054

Philippot L, Raaijmakers JM, Lemanceau P, van der Putten WH (2013) Going back to the roots: the microbial ecology of the rhizosphere. Nat Rev Microbiol 11:789–799

Pierret A, Doussan C, Capowiez Y, Bastardie F, Pagès L (2007) Root functional architecture: a framework for modeling the interplay between roots and soil. Vadose Zone J 6:269–281

Poole EJ, Bending GD, Whipps JM, Read DJ (2001) Bacteria associated with *Pinus sylvestris*, *Lactarius rufus* ectomycorrhizas and their effects on mycorrhizal formation *in vitro*. New Phytol 151:743–751

Pozo MJ, Azcon-Aguilar C (2007) Unraveling mycorrhizal- induced resistance. Curr Opin Plant Biol 4:393–398

Pozo MJ, Dumas-Gaudot E, Slezack S, Cordier C, Asselin A, Gianinazzi S, Gianinazzi-Pearson V, Azcón-aguilar C, Barea JM (1996) Induction of new chitinase isoforms in tomato roots during interactions with *Glomus mosseae* and/or *Phytophthora nicotianae* var *parasitica*. Agronomie 16:689–697

Pozo MJ, Van Der Ent S, Van Loon LC, Pieterse CMJ (2008) Transcription factor MYC2 is involved in priming for enhanced defense during rhizobacteria-induced systemic resistance in *Arabidopsis thaliana*. New Phytol 180:511–523

Raaijmakers JM, Paulitz TC, Steinberg C, Alabouvette C, Moënne-Loccoz Y (2009) The rhizosphere: a playground and battlefield for soil borne pathogens and beneficial microorganisms. Plant Soil 321:341–361

Ramachandran K, Srinivasan V, Hamza S, Anadaraj M (2007) Phosphate solubilizing bacteria isolated from the rhizosphere soil and its growth promotion on black pepper (*Piper nigrum* L.) cutting. Plant Soil Sci 102:325–331

Rambelli A (1973) The rhizosphere of mycorrhizae. In: Marks GC, Kozlowski TT (eds) Ectomycorrhizae, their ecology and physiology. Academic Press, New York, pp 299–349

Ravnskov S, Nybroe O, Jakobsen I (1999) Influence of an arbuscular mycorrhizal fungus on *Pseudomonas fluorescens* DF57 in rhizosphere and hyphosphere soil. New Phytol 142:113–122

Rees RM, Bingham IJ, Baddeley JA, Watson CA (2005) The role of plants and land management in sequestering soil carbon in temperate arable and grassland ecosystems. Crop Soil Res 128:130–154

Reinhold-Hurek B, Bünger W, Burbano CS, Sabale M, Hurek T (2015) Roots shaping their microbiome: global hotspots for microbial activity. Annu Rev Phytopathol 53:403–424

Rezzonico F, Zala M, Keel C, Duffy B, Moenne-Loccoz Y, Defago G (2007) Is the ability of biocontrol fluorescent pseudomonads to produce the antifungal metabolite 2,4- diacetylphloroglucinol really synonymous with higher plant protection. New Phytol 173:861–872

Rillig MC (2004) Arbuscular mycorrhizae, glomalin and soil aggregation. Can J Soil Sci 84:355–363

Rillig MC, Mummey DL (2006) Mycorrhizas and soil structure. New Phytol 171:41–53

Rillig MC, Wright SF, Eviner V (2002) The role of arbuscular mycorrhizal fungi and glomalin in soil aggregation: comparing effects of five plant species. Plant Soil 238:325–333

Rillig MC, Ramsey PW, Morris S, Paul EA (2003) Glomalin, an arbuscular- mycorrhizal fungal soil protein, responds to land-use change. Plant Soil 253:293–299

Rillig MC, Lutgen ER, Ramsey PW, Klironomos JN, Gannon JE (2005) Microbiota accompanying different arbuscular mycorrhizal fungal isolates influence soil aggregation. Pedobiologia 49:251–259

Rillig MC, Caldwell BA, Wösten HAB, Sollins P (2007) Role of proteins in soil carbon and nitrogen storage: controls on persistence. Biogeochemistry 85:25–44

Rønn R, McCaig AE, Griffiths BS, Prosser JI (2002) Impact of protozoan grazing on bacterial community structure in soil microcosms. Appl Environ Microbiol 68:6094–6105

Rovira AD (1969) Plant root exudates. Biol Rev 35:35–57

Ruiz-Lozano JM, Bonfante P (2000) A *Burkholderia* strain living inside the arbuscular mycorrhizal fungus *Gigaspora margarita* possesses the *vacB* gene, which is involved in host cell colonization by bacteria. Microbiol Ecol 39:137–144

Salimpour S, Khavazi K, Nadian H, Besharati H, Miransari M (2010) Enhancing phosphorous availability to canola (*Brassica napus* L.) using P solubilizing and sulfur oxidizing bacteria. Aust J Crop Sci 4:330–334

Santoyo G, Orozco-Mosqueda MC, Govindappa M (2012) Mechanisms of biocontrol and plant growth-promoting activity in soil bacterial species of *Bacillus* and *Pseudomonas*: a review. Biocont Sci Technol 22:855–872

Schnitzer SA, Klironomos JN, HilleRisLambers J, Kinkel LL, Reich PB, Xiao K, Rillig MC, Sikes BA, Callaway RM, Mangan SA, van Nes EH, Scheffer M (2011) Soil microbes drive the classic plant diversity-productivity pattern. Ecology 92:296–303

Schouteden N, Waaele DD, Panis B, Vos CM (2015) Arbuscular mycorrhizal fungi for the biocontrol of plant-parasitic nematodes: a review of the mechanisms involved. Front Microbiol 6:1280

Shehata HR, Lyons EM, Jordan KS, Raizada MN (2016) Bacterial endophytes from wild and ancient maize are able to suppress the fungal pathogen *Sclerotinia homoeocarpa*. J Appl Microbiol 120:756–769

Siasou E, Standing D, Killham K, Johnson D (2009) Mycorrhizal fungi increase biocontrol potential of *Pseudomonas fluorescens*. Soil Biol Biochem 41:1341–1343

Simon A, Bindschedler S, Job D, Wick LY, Filippidou S, Kooli WM, Verrecchia EP, Junier P (2015) Exploiting the fungal highway: development of a novel tool for the *in situ* isolation of bacteria migrating along fungal mycelium. FEMS Microbiol Ecol 91:1–13

Singh DP, Srivastava JS, Bahadur A, Singh UP, Singh SK (2004) Arbuscular mycorrhizal fungi induced biochemical changes in pea (*Pisum sativum*) and their effect on powdery mildew (*Erysiphe pisi*). J Plant Dis Protect 111:266–272

Smith SE, Read DJ (2008) Mycorrhizal symbiosis, 3rd edn. Academic Press, Cambridge

Smith SE, Smith FA (2011) Roles of arbuscular mycorrhizas in plant nutrition and growth: new paradigms from cellular to ecosystem scales. Annu Rev Plant Biol 62:227–250

Smith SE, Jakobsen I, Gronlund M, Smith FA (2011) Roles of arbuscular mycorrhizas in plant phosphorus nutrition: interactions between pathways of phosphorus uptake in arbuscular mycorrhizal roots have important implications for understanding and manipulating plant phosphorus acquisition. Plant Physiol 156:1050–1057

Sood SG (2003) Chemotactic response of plant-growth promoting bacteria towards roots of vesicular-arbuscular mycorrhizal tomato plants. FEMS Microbiol Ecol 45:219–227

Srivastava AK, Singh T, Jana TK, Arora DK (2001) Induced resistance and charcoal rot in *Ciceri arietinum* (chickpea) by *Pseudomonas fluorescens*. Can J Bot 79:787–795

Starkey RL (1938) Some influences of the development of higher plants upon the microorganisms in the soil. VI. Microscopic examination of the rhizosphere. Soil Sci 45:207–249

Sturz AV, Nowak J (2000) Endophytic communities of rhizobacteria and the strategies required to create yield enhancing associations with crops. Appl Soil Ecol 15:183–190

Subba Rao NS, Tilak KVBR, Singh CS (1985) Synergistic effect of VAM and *Azospirillum brasilense* on the growth of barley in pots. Soil Biol Biochem 17: 119–121

Sun YP, Unestam T, Lucas SD, Johanson KJ, Kenne L, Finlay R (1999) Exudation-reabsorption in a mycorrhizal fungus, the dynamic interface for interaction with soil and soil microorganisms. Mycorrhiza 9:137–144

Tarkka M, Schrey S, Hampp R (2008) Plant associated micro-organisms. In: Nautiyal CS, Dion P (eds) Molecular mechanisms of plant and microbe coexistence. Springer, New York, pp 3–51

Tian B, Yang J, Zhang K-Q (2007) Bacteria used in the biological control of plant-parasitic nematodes: populations, mechanisms of action, and future prospects. FEMS Microbiol Ecol 61:197–213

Tiberius B, Cătălin T (2011) Interrelations between the mycorrhizal systems and soil organisms. J Plant Dev 18:55–69

Timonen S, Marschner P (2006) Mycorrhizosphere concept. In: Mukerji KG, Manoharachary C, Singh J (eds) Soil biology. Springer, Berlin, pp 155–172

Timonen S, Christensen S, Ekelund F (2004) Distribution of protozoa in scots pine mycorrhizosphere. Soil Biol Biochem 36:1087–1093

Tisdall JM, Oades JM (1982) Organic matter and waterstable aggregates in soils. J Soil Sci 33:141–163

Toljander JF, Lindahl BD, Paul LR, Elfstrand M, Finlay RD (2007) Influence of arbuscular mycorrhizal mycelial exudates on soil bacterial growth and community structure. FEMS Microbiol Ecol 61:295–304

Toro M, Azcón R, Barea JM (1997) Improvement of arbuscular mycorrhizal development by inoculation with phosphate solubilizing rhizobacteria to improve rock phosphate bioavailability (^{32}P) and nutrient cycling. Appl Environ Microbiol 63:4408–4412

Trotta A, Falaschi P, Cornara L, Minganti V, Fusconi A, Drava G, Berta G (2006) Arbuscular mycorrhizae increase the arsenic translocation factors in the As hyperaccumulating fern *Pteris vittata* L. Chenosphere 65:74–81

van Aarle IM, Olsson PA, Soderstrom B (2002) Arbuscular mycorrhizal fungi respond to the substrate pH of their extraradical mycelium by altered growth and root colonization. New Phytol 155:173–182

Van der Heijden MGA, Bakker R, Verwaal J, Scheublin TR, Rutten M, van Logtestijn R et al (2006) Symbiotic bacteria as a determinant of plant community structure and plant productivity in dune grassland. FEMS Microbiol Ecol 56:178–187

van der Heijden MGA, Bardgett RD, van Straalen NM (2008) The unseen majority: soil microbes as drivers of plant diversity and productivity in terrestrial ecosystems. Ecol Lett 11:296–310

Van Elsas JD, Torsvik V, Hartmann A, Ovreås L, Jansson JK (2007) The bacteria and archaea in soil. In: van Elsas JD, Jansson JK, Trevors JT (eds) Modern soil microbiology, 2nd edn. CRC Press, Boca Raton, pp 83–106

Van Hulten M, Pelser M, Van Loon LC, Pieterse CM, Ton J (2006) Costs and benefits of priming for defense in *Arabidopsis*. Proc Natl Acad Sci U S A 103:5602–5607

Van Loon LC (2007) Plant responses to plant growth-promoting rhizobacteria. Eur J Plant Pathol 119:243–254

Van Loon LC, Bakker PAHM, Pieterse CMJ (1998) Systemic resistance induced by rhizosphere bacteria. Annu Rev Phytopathol 36:453–483

Vinale F, Ghisalberti EL, Sivasithamparam K, Marra R, Ritieni A, Ferracane R, Woo S, Lorito M (2009) Factors affecting the production of *Trichoderma harzianum* secondary metabolites during the interaction with different plant pathogens. Lett Appl Microbiol 48:705–711

Vogel-Mikus K, Pongrac P, Kump P, Necemer M, Regvar M (2006) Colonisation of a Zn, Cd and Pb hyperaccumulator *Thlaspi praecox* Wulfen with indigenous arbuscular mycorrhizal fungal mixture induces changes in heavy metal and nutrient uptake. Environ Pollut 139:362–371

Vyas P, Gulati A (2009) Organic acid production *in vitro* and plant growth promotion in maize under controlled environment by phosphate-solubilizing fluorescent *Pseudomonas*. BMC Microbiol 9:174–189

Wagg C, Jansa J, Schmid B, van der Heijden MGA (2011) Belowground biodiversity effects of plant symbionts support aboveground productivity. Ecol Lett 14:1001–1009

Wamberg C, Christensen S, Jakobsen I, Muller AK, Sorensen SJ (2003) The mycorrhizal fungus (*Glomus intraradices*) affects microbial activity in the rhizosphere of pea plants (*Pisum sativum*). Soil Biol Biochem 35:1349–1357

Wardle DA (1992) A comparative assessment of factors which influence microbial growth carbon and nitrogen levels in soil. Biol Rev Camb Philos Soc 67:321–358

Warmink JA, Van Elsas JD (2008) Selection of bacterial populations in the mycosphere of *Laccaria proxima*: Is type III secretion involved? ISME J 2:887–900

Wehner J, Antunes PM, Powell JR, Mazukatow J, Rillig MC (2009) Plant pathogen protection by arbuscular mycorrhizas: a role for fungal diversity? Pedobiologia 53:197–201

Weston LA, Alsaadawi IS, Baerson SR (2013) Sorghum allelopathy: from ecosystem to molecule. J Chem Ecol 39:142–53

Whipps JM (2004) Prospects and limitations for mycorrhizas in biocontrol of root pathogens. Can J Bot 82:1198–1227

Wright SF (2000) A fluorescent antibody assay for hyphae and glomalin from arbuscular mycorrhizal fungi. Plant Soil 226:171–177

Wright SF, Upadhyaya A (1998) A survey of soils for aggregate stability and glomalin, a glycoprotein produced by hyphae of arbuscular mycorrhizal fungi. Plant Soil 198:97–107

Xavier LJC, Germida JJ (2003) Bacteria associated with *Glomus clarum* spores influence mycorrhizal activity. Soil Biol Biochem 35:471–478

York LM, Carminati A, Mooney SJ, Ritz K, Bennett MJ (2016) The holistic rhizosphere: integrating zones, processes, and semantics in the soil influenced by roots. J Exp Bot. doi:10.1093/jxb/erw108

Zahir ZA, Munir A, Asghar HN, Shahroona B, Arshad M (2008) Effectiveness of rhizobacteria containing ACC-deaminase for growth promotion of peas (*Pisum sativum*) under drought conditions. J Microbiol Biotechnol 18:958–63

Zamfirache M-M, Toma C (2000) Simbioza în lumea vie. Edit. Univ. Alexandru Ioan Cuza, Iaşi, pp 186–239

Zarnea G (1994) Tratat de microbiologie generală, vol 5. Bucureşti: Edit. Academiei Române, pp 367–391

Zhang FS, Shen JB, Zhang JL, Zuo YM, Li L, Chen XP (2010) Rhizosphere processes and management for improving nutrient use efficiency and crop productivity: implications for China. In: Sparks DL (ed) Advances in agronomy, vol 107. Academic Press, San Diego, pp 1–32

Zhang L, Fan J, Ding X, He X, Zhang F, Feng G (2014) Hyphosphere interactions between an arbuscular mycorrhizal fungus and a phosphate solubilizing bacterium promote phytate mineralization in soil. Soil Biol Biochem 74:177–183

Zhang L, Xu M, Liu Y, Zhang F, Hodge A, Feng G (2016) Carbon and phosphorus exchange may enable co-operation between an arbuscular mycorrhizal fungus and a phosphate-solubilizing bacterium. New Phytol 210:1022–1032

Arbuscular Mycorrhizae: Effect of Rhizosphere and Relation with Carbon Nutrition

6

Ibrahim Ortaş, Somayyeh Razzaghi, and Mazhar Rafique

Abstract

More than 90 % of terrestrial plants form symbiotic association with mycorrhizae which develop and promote cooperation belowground in rhizosphere. Mycorrhizal fungi produces spores in the soil and vegetative propagules in root fragments which respond to stimulation of root exudates in the rhizosphere. As a result, symbiotic relationship takes place where physiology and morphology of both participants rely on each other. Mycorrhizae are present in a range of horticultural, agricultural, forestry and other plant species. Along with mycorrhizae, other beneficial microbes also add in plant growth promotion, nutrient and uptake and stress tolerance either biotic or abiotic. The presence of bacteria in rhizosphere synchronizes with mycorrhizae termed as 'mycorrhizae helper bacteria' and increases plant growth by focusing on N and P in particular while micronutrients in general. Besides that, carbon has important structural and functional role in symbiotic association, because of mycorrhizal reliance on plants for food. Additionally, movement of C to the roots is an interesting area for exploration due to recent global focus on addressing climate change and carbon mitigation approaches particularly for sustainable agriculture. AM symbiosis can influence soil CO_2 emissions and soil in ecosystems dominated by mycorrhizal plants that

I. Ortaş (✉) • S. Razzaghi
Faculty of Agriculture, Department of Soil Science and Plant Nutrition, Cukurova University, 1150 Adana, Adana, Turkey
e-mail: iortas@cu.edu.tr; ortasibrahim@gmail.com

M. Rafique
Faculty of Agriculture, Department of Soil Science and Plant Nutrition, Cukurova University, 1150 Adana, Adana, Turkey

Faculty of Biological Sciences, Department of Plant Sciences, Quaid-i-Azam University, Islamabad 45320, Pakistan

D.K. Choudhary et al. (eds.), *Plant-Microbe Interaction: An Approach to Sustainable Agriculture*, DOI 10.1007/978-981-10-2854-0_6

contain 70 % more carbon per unit nitrogen than soil in ecosystems dominated by non-AM-associated plants. Absorption of CO_2 by mycorrhizae is contributing in climate change mitigation and translated as plant biomass production.

6.1 Introduction

The most widespread symbiotic association between microorganisms and plants is arbuscular mycorrhizal fungi (AMF), which is present in a range of horticultural, agricultural and forestry plants Marschner (2012). Different plant species are infected with indigenous AMF in their natural habitat (Ortas and Coskan 2016a). Mutualistic mycorrhizal fungi-root association has been known and being studied since 1885, when Frank for the first time gave the name *mycorrhiza* (*myco*, fungus; *rhiza*, root) to readily observable morphological complexes between fungi and tree roots. Mycorrhizal infection occurs in 83 % of dicotyledonous and 79 % of monocotyledonous plants (Trappe 1987).

According to Azcon-Aguilar and Barea (2015), beneficial microbes contribute in plant growth and increase nutrient uptake such as nitrogen and/or soluble phosphate. Understanding the mechanism of high N use efficiency by mycorrhizal/rhizobial plants and carbon allocation in a context of mutualistic system is critical for managing agricultural system for the ecosystem sustainability by microbial symbionts. Since there is significant effect of carbon on climate change and sustainability of agriculture, it is sound to explore the influence of beneficial organisms on carbon sequestration (Ortas et al. 2013). It has been suggested that soil microflora may have significant influence on the formation of mycorrhizal association. The results of Sutton and Sheppard (1976) showed that adding non-sterile soil leachate to a sterile soil increases biomass of AM hyphae.

Recently research groups on mycorrhiza have concentrated on the effect of mycorrhizal inoculum on nodulation, when both mycorrhizae and bacteria are inoculated together. It is an indication of additive and positive cooperation in between fungi and bacteria. Dual inoculation (of AMF and *Azotobacter*) had a synergetic effect on growth increase of the host plant. In rhizosphere, it is possible that some beneficial bacteria, such as symbiotic or free-living nitrogen fixer (Hamdia and Shaddad 2010), phosphate solubilizers and hormone producer organisms (Ratti et al. 2001), could develop cooperation with mycorrhizae. Besides that, very recently the role of mycorrhizae on CO_2 absorption is getting more attention because of continuously piling up of the atmospheric CO_2 concentration to affect climate. Since climate change is related with atmospheric CO_2, the role of mycorrhizae for plant growth promotion and biomass production through carbon absorption is significantly important.

6.2 Mycorrhizal Formation and Functions

Mycorrhizal fungi's characteristic structures are vesicles, arbuscules, hyphae (external and internal hyphae) and resting spores in the host root (Smith and Read 2008). One of the important structural features of mycorrhizae is hyphae which grow longitudinally between the cells of the root cortex. When fungal hyphae contact the root surface, penetration occurs through an aspersorium. After its development, fungus produces hyphae which penetrate between inner cortical cells (Berruti et al. 2013; Sieverding 1991; Smith and Read 2008). The AM root infection usually begins after hyphae extending from propagules (spores, hyphae and root fragments) penetrate into the host root from an entry point on root surface. The mycorrhizal fungi survive in the soil as resting spores, and when environmental conditions become favourable, they start to germinate. The spore formation is generally on the coarse external hyphae. These spores usually range from 50 to 600 μm (Sieverding 1991). Once infection is established, the developing fungi can produce inter- or intracellular vesicles (Rodriguez-Moran et al. 2015). Usually, vesicles are oval, round or lobe shaped and occur within or between cortical cells. They may contain lipid droplets which act as storage structures of fungus. Vesicle shape, wall structure content and their number can differ according to the fungal species forming mycorrhizae (Sieverding 1991; Smith and Read 2008). Arbuscules are intracellular, branched or tree-shaped structures of the symbiont and are formed by repeated branching. They are considered to function for the transfer of nutrients (Marschner 2012; Smith and Gianinazzipearson 1988).

The internal morphology of AMF can be easily observed on cleared and stained root samples under the microscope (Seok-Cho et al. 2007). The mycorrhizal fungus lives with host (plant partner) in a balanced close association. Mycorrhizal fungi can be seen in the soil as spores or as vegetative propagules in root fragments. Propagules of mycorrhizal fungi apparently respond to the stimulation of root exudate, and their hyphae and germ tubes grow and penetrate root epidermal cells. The colonization of the host tissue progresses, both internally and externally along the root surface. The formation depends on the association between host and fungi, the latter resulting in morphological and physiological changes which lead to the formation of different types of mycorrhizae. When mycorrhiza forms, symbiosis significantly changes the physiology and morphology of roots particularly and the whole plant generally (Bray et al. 2003; Wulf et al. 2003). In some plants such as onion and maize, there is a yellow pigmentation which accompanies root colonization. The physiological change is explained as the change has great impact on rhizospheric microorganisms, which alter permeability of the membranes. It is well understood that membrane permeability can alter the quantity and quality of root exudates and results in changed plant nutrient composition. The microbes in microsphere of mycorrhizal fungi may profoundly affect mycorrhizal functions, such as nutrient and water uptake. Moreover, mycorrhizospheric organisms and root exudates have significant influence on soil development as well. Mycorrhizal hyphae are normally supported by the host plants, but their biomass may be influenced by soil biotic and abiotic factors such as soil microorganisms.

6.3 Factors Affecting Mycorrhizal Association

Soil physical, chemical and biological factors affect mycorrhizal development. Water contents (Krishna et al. 2005), temperature (Zhang et al. 2016), light (Clark and St Clair 2011; Moratelli et al. 2007) soil type and their characteristics (Ortas and Coskan 2016b; Thougnon Islas et al. 2016) are illustrations of physical factors. Fitter et al. (2004) indicated that AM fungi respond directly to elevated soil temperature. Furthermore, examples of chemical factors are soil pH (Moon et al. 2016), phosphorus availability (da Silva et al. 2016), nitrogen forms (Smith and Read 2008), micronutrient levels (Hoffmann et al. 2009), salinity stress (Labidi et al. 2011; Ruiz-Lozano and Azcon 2000), organic matter content (Wang et al. 2015), excessive use of pesticides (Zocco et al. 2008), etc. Biological factors are based on host plant (Ocampo et al. 1980) interactions with other soil microorganisms such as pathogenic and competitive with other mycorrhizal fungi (Azcon-Aguilar and Barea 2015). These environmental conditions affect root colonization and fungal growth development in the cortex (Smith and Read 2008). It has been indicated that mean spore abundance was significantly different in cropped systems and soil management (Säle et al. 2015). Barea et al. (2011) and Burkle and Belote (2015) reported that disturbance of target semiarid ecosystem decreases density and diversity of mycorrhizal fungi population. In general, the sporulation of AMF is dependent on soil/plant nutrition, the ecophysiological status of hosts, climate, that is, previous precipitation and also sampling time.

Soil pH reflects the nutrient availability in soil through ion exchange process (Helgason and Fitter 2009). Varying soil pH can change species richness and community composition. For different mycorrhizal species, effect of soil pH on germination of mycorrhizal spores, hyphal growth from spores and hyphal growth from mycorrhizal roots may be different. Guo et al. (2012) have shown that soil medium with liming effect from pH 5.5 to 5.9 increased *G. mosseae* germination by 43–60 % with no further increase observed with addition of lime. The results of (Sivakumar 2013) showed positive correlation between the mycorrhizal spore abundance and soil pH moreover with root colonization. In general, soil with pH range 5.5–6.6, AMF is abundant (Sharma et al. 2009). Martensson et al. (2012) reported that the amount of AMF is very low in poor nutrient and drought-stressed habitat, and they also found that a high pH in the topsoil does not lead to higher AMF biomass.

The results of Alloush and Clark (2001) showed that mycorrhizal infection was strongly inhibited by Al and Mn. Similarly Lambais and Cardoso (1993) recorded that soil may have toxic concentration of Al and Mn to fungal growth, but it supported plant growth. It is believed that soil acidity is not an independent factor as pH itself may have little significant effect on spore germination and root colonization. Resting spores are thick-walled structures formed in the soil. Spore numbers are affected by several factors such as nutritional status of the host plant and soil moisture contents (Ortas and Coskan 2016a; Smith and Smith 2011; Smith and Read 2008).

The frequency of mycorrhizal infection is affected by nitrate (NO_3^-) and ammonium (NH_4^+) ions. According to Asghari and Cavagnaro (2012) and Valentine et al. (2002), the application of NO_3^- or NH_4^+ resulted in higher level of mycorrhizal infection. Mycorrhizae formation was decreased at high level of NH_4^+-N (Wallenda et al. 1996). The application of increasing amount of NO_3^- reduced the level of AM infection in lettuce roots when the root was inoculated with *G. mosseae* (Kohler et al. 2008). Wallenda et al. (1996) found that high NO_3^- levels can result in the low mycorrhizal formation of conifer seedlings, but with NH_4^+ supply, fungi formed abundant mycorrhizae (Correa et al. 2006). The results of Irshad et al. (2002) showed NO_3^- fertilizer is more inhibitory to AMF development than NH_4^+ fertilizer. Addition of NO_3^--N to the soil decreases AMF infection (van Diepen et al. 2013) and infectivity of mycorrhizal propagules (Cornejo et al. 2007). NH_4^+-N application magnified considerable morphological changes and showed plasticity of *G. intraradices* (Bago et al. 2004). The mechanism responsible for nitrogen inhibition of AM formation is not fully understood.

6.4 Carbon Relation with Mycorrhizae

Soil microorganisms are dependent upon plants for supply of energy mainly carbon. By this way, population of organisms is indirectly dependent on plant photosynthesis. A figure of soil-fungi-plant and their relationship with plant carbon nutrition is presented in Figs. 6.1 and 6.2. The relationship between mycorrhizal fungus and the host plant is bidirectional (Fig. 6.1) where both sides derive benefits. The fungi obtain its required carbon directly from the roots, and at the same time fungi supply inorganic minerals, especially phosphorus (P) from thc surrounding soil (Smith and

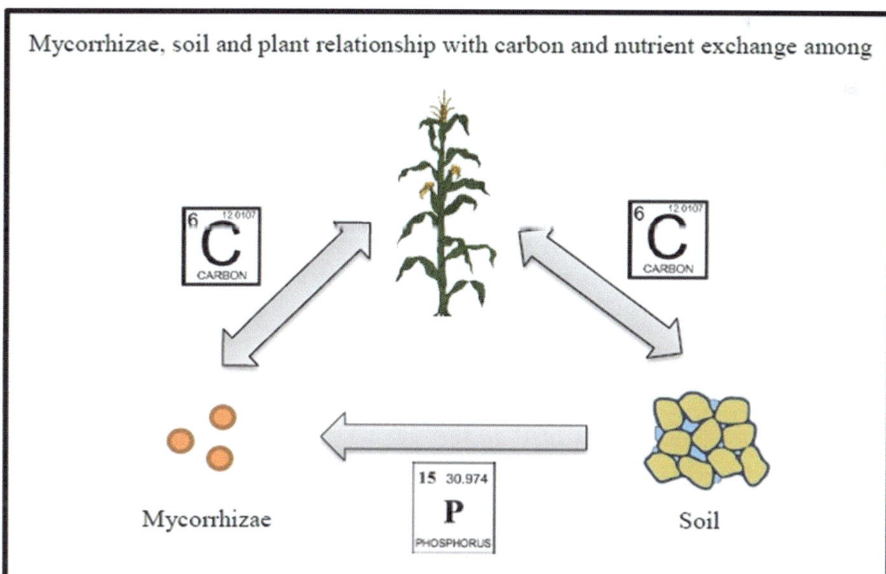

Fig. 6.1 Relationship between mycorrhizae-soil and plant-carbon-nutrient exchange (Ortas 1994)

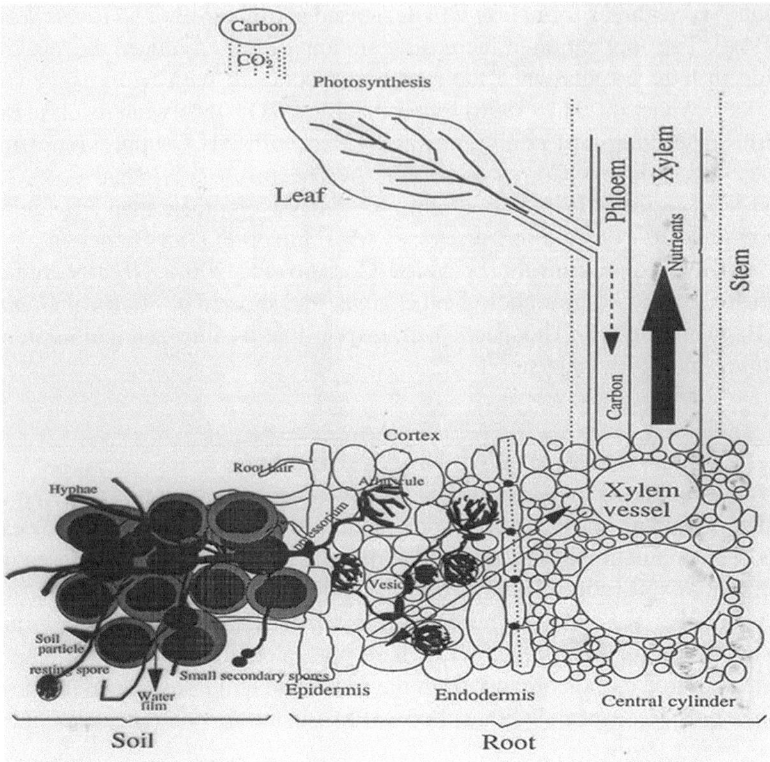

Fig. 6.2 Nutrient transportation from soil to plant tissue and phloem transportation from leaves to mycorrhizal hyphae (Ortas 1994)

Read 2008). Carbon (C) and P are the key nutrients required for mycorrhizal development and functions. It is well understood that AMF are completely dependent on soluble carbohydrates produced by the host plant for carbon. This carbon demand can inhibit plant growth in stress conditions of low light intensity, high level of root colonization (Asensio et al. 2012) and low soil temperature. Mycorrhizal fungi, because of their unique carbon system, can efficiently combine soil mineralization and nutrient uptake by plant roots (Mellado-Vazquez et al. 2016).

In rhizosphere, soil bacteria and fungi generally immobilize mineral nutrients, as carbon is consumed, and thereby compete with plant for nutrients. A useful indicator of plant material supply to soil is net primary production. Living plant roots supply a tremendous amount of C to the soil which can potentially be used by microorganisms. The C utilization by mycorrhizae becomes important when competing with other soil microorganisms.

It is generally accepted that root exudes sufficient quantity of organic compounds to support microorganism population in rhizosphere and support growth of certain microorganisms derived by root exudate quality. In rhizosphere, C losses by plant roots in complex associations of root and soil microorganisms contribute both positively and negatively which determine plant efficiency. Roots in non-sterile media support a large population of microorganisms on external surface (the rhizoplane) and in rhizospheric soil. The population density of bacteria and partly fungi in the rhizosphere depends on the amount of exudate, mucilage and sloughed-off cells which are carbon based. In soil-grown wheat plants under non-sterile conditions during a period of 3 to 8 weeks, 20–40 % of traslocated C from shoot to root was lost as organic carbon (OC) into the rhizosphere.

It has been suggested that enhanced plant growth and C flow below ground could increase C storage in soils, and it could be the missing sink (Ford et al. 2012). Increased C flow to the soil can be directly via plant roots or indirectly via soil organisms, of which mycorrhizae could potentially be a very important element. Indeed, C allocation to mycorrhizal fungi is often around 10 % of total fixed C (C allocated to belowground fractions such as roots and mycorrhizal hyphae accounted for an average of 10 %, with 4.3 % allocated to mycorrhizal hyphae) (Tome et al. 2015) and has been estimated to be high as 20 %. Cheng et al. (1996) showed that carbon availability index (CAI) and water soluble organic carbon (WSOC) were inversely related to the relative distance from root surface, with several times higher concentration in the rhizoplane soils. It is widely known that carbon availability in the rhizosphere is much higher than bulk soil where AMF could be responsible for carbon sequestration (Ortas et al. 2013).

Most of the carbon is utilized fairly and rapidly by rhizospheric microorganisms. A large turnover of OC by microorganism's activity in the rhizosphere has an important implication for both the carbon balance of plant and mineral nutrient relationships in the rhizosphere. At maturity, only a small fraction of the root-derived OC is retained in the root system.

Willis et al. (2013) indicated that the mechanisms involved in C transfer from plant to fungus are still not well understood. It has been estimated that mycorrhizal plants direct up to 20 % more photosynthate towards root system than non-mycorrhizal plants.

6.5 Mycorrhizae Affect Atmospheric CO_2 Absorption

Interaction between root and soil microorganisms controls nutrient availability and uptake by plants and influences soil greenhouse gas (GHG) emissions such as CO_2 and N_2O (Jackson et al. 2008). This symbiosis increases the uptake of soil nutrients in exchange of photoassimilated carbon compounds (Fellbaum et al. 2012). Mycorrhizal fungi of chlorophyllous plants absorb C compounds from their host. The role of AM symbiosis for plant and soil GHG emissions might be particularly important in ecologically managed systems. Several studies have reported higher CO_2 emissions in mycorrhizal plants than non-mycorrhizal. The results of Heinemeyer et al. (2006) showed that concentration of CO_2 flux is highest in the mycorrhizal treatments. It has been previously suggested that AM symbiosis can influence soil CO_2 emission either due to direct respiration of the fungi or indirect impacts on heterotrophic microorganisms (Cavagnaro et al. 2008). On the other hand, Tome et al. (2016) reported that mycorrhizal contribution to soil respiration (11 ± 6 %) was of similar magnitude to the roots (12 ± 4 %). However, respiration of SOM and mycorrhizae significantly increased in late summer and autumn terms; this is related with priming effect of roots on SOM degradation or to a stimulation of mycorrhizal respiration. The organic matter has key role in soil ecosystems (Lejon et al. 2007), while limited information is available for SOM effects on AMF, though it is well acknowledged that growth of AMF can be both increase and decrease (Cavagnaro 2014; Ravnskov et al. 2006) by soil organic amendments.

Fitter et al. (2004) indicated that under field studies, variation in vegetation due to environmental changes may play enormous role in determining AMF community structure. Elevated CO_2 could stimulate mycorrhizal colonization, since plants are fixing more C, and its availability to the fungus is increased. According to hypothesis, more C could flow in the soil via mycorrhizal hyphae. The amount of C translocated below ground by AM fungal structures varies between 4 and 20 % of the total C fixed by the plant (Smith and Read 2008). Several researchers have examined mycorrhizal colonization in conditions of elevated CO_2, which might stimulate ectomycorrhizal colonization in various species (Langley et al. 2003), but the evidence for arbuscular mycorrhizae is less clear. The effect of mycorrhizal type on soil carbon is dependent on the effects of net primary production, temperature, precipitation and soil clay content. Hence, the effect of mycorrhizal type on soil carbon content holds at the global scale.

The impact of ectomycorrhizal fungi (EMF) on C turnover in forest soils has been considered limited. Rineau et al. (2012) indicated, using global data sets, that soil in ecosystems dominated by mycorrhizae-associated plants contains 70 % more C per unit nitrogen. Although some climatic models estimated that increased C storage in temperate forest soils is because of increased photosynthetic C allocation to roots and symbiotic fungi (Clemmensen et al. 2013; Drigo et al. 2010), there is still a room for exact degree of C storage estimation. Mycorrhizosphere activity may also stimulate decomposition of previously recalcitrant SOM (Cheng et al. 2012). The saprotrophic and EMF species produce a range of hydrolytic and oxidative enzymes with a potential effect to break down C-containing compounds such as OM and mobilize nutrients from SOM (Norby et al. 2010) Although photosynthates are likely the primary source of C used by EMF in ideal condition (Wolfe et al. 2012), many studies suggest that fungi may directly (Vaario et al. 2012) or indirectly (Rineau et al. 2012) access SOM-C pools.

6.6 Mycorrhizal Development Influenced by Rhizospheric Organisms and Nitrogen

It has been previously reported that effect of AMF on P uptake and plant growth was more in sterilized soil in comparison to non-sterilized soil. It has been hypothesized that more plant growth and P uptake may result from more N mineralization especially the mineral NH_4^+-N flush after soil irradiation. According to Ortas et al. (1996), (Ortas and Rowell (2000) and Ortas et al. (2004), the amount of NH_4^+-N

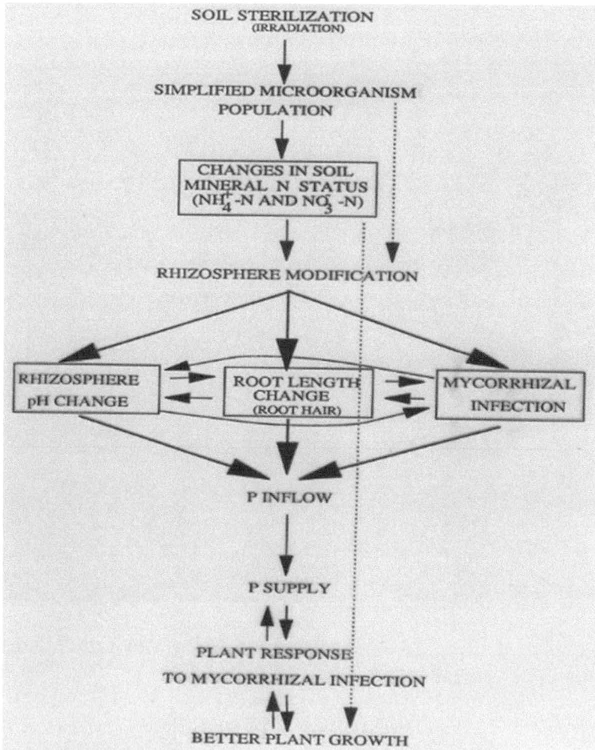

Fig. 6.3 The effect of soil sterilization on nutrient dynamics and P uptake

should be taken into account for plant growth and P uptake (Fig. 6.3). Phosphorus uptake mechanisms are related with nitrogen effect on rhizosphere pH (Hinsinger 2001; Isaac et al. 2012; Rubio et al. 2012). Under partial sterile soil conditions, N mainly comes after dead organisms and organic matter degradation.

So far numerous pot experiments have been carried out in partially sterilized soil to eliminate the effects of indigenous mycorrhizae on self-introduced mycorrhizal inoculum. The results of inoculation experiments are generally positive, because unwanted growth response in sterilized soil is related to increased level of available N and in particular N mineralized from the soil microbial biomass (Ortaş and Harris 1996) along with elimination of nutrient competition with other microorganisms. The effect of sterilization on soil properties (Gebremikael et al. 2015) and nutrient dynamics has been widely studied (Ortas and Rowell 2004; Ortas et al. 2004; Ortaş and Harris 1996).

The main aim of partial soil sterilization in mycorrhizal studies is to eliminate indigenous mycorrhizal spores and pathogenic microbial activity in the soil, but this procedure often alters the chemical and biological properties of the soil (Hassan et al. 2012). Under the greenhouse with sterile conditions, pathogen activities have been largely restricted, and plant growth especially root development was

Fig. 6.4 The effect of mycorrhizal inoculation on the maize growth under fumigated and unfumigated field conditions (Ortas unpublished photo)

maximized. AMF have been shown to affect root growth, root exudate, nutrient absorption and host physiological response to environmental stresses (Folli-Pereira et al. 2012; Liu et al. 2015; ZhongQun et al. 2007). Increasing the AMF in soil, P nutrition for root growth also enhances and expands the absorptive capacity of the root system for water and nutrients which influence cellular processes in root (Smith and Read 2008; Tischer et al. 2015).

The main procedure so far adopted is elimination of indigenous fungi from soil than reinoculated under controlled conditions before their effects on plant growth can be assessed by comparing mycorrhizal and non-mycorrhizal plants. As expressed by Hetrick et al. (1988) and Miransari et al. (2009)), the effect of other soil microorganisms eliminated during soil sterilization on plant growth or mycorrhizal growth response is generally not considered. The contribution of AMF to plant growth in non-sterile soil may be different from sterile one. Nevertheless, it is extremely difficult to evaluate the contribution of AMF on plant growth in non-sterile soil. Under the fumigated soil conditions, reinoculation of mycorrhizae increased maize growth (Fig. 6.4).

With soil sterilization, soil organisms are killed and organic matter mineralization releases sufficient nutrients such as NH_4^+-N and NO_3^--N. Partial soil sterilization can result in four–tenfold increase in NH_4^+-N level (Ortaş and Harris 1996; Tanaka et al. 2003). The contribution of soil partial sterilization to nutrient release may be explained as follows:

Table 6.1 Literature on the effect of partial sterilization on mineral nitrogen release from organic compounds and soil microorganisms

NH_4^+–N	NO_3^--N	References
ND	NO_3^--N	Bowen and Cawse (1964)
NH_4^+-N↑	ND	Salonius et al. (1967)
NH_4^+-N↑	NO_3^--N↓	Rovira and Bowen (1969)
NH_4^+-N↑	NO_3^--N↓	Singh and Kanehiro (1970)
NH_4^+-N↑	ND	Jenkinson et al. (1972)
NH_4^+-N↑	ND	Arunachalam et al. (1974)
NH_4^+-N↑	ND	Stribley et al. (1975)
NH_4^+-N↑	NO_3^--N	Jakobsen and Andersen (1982)
NH_4^+-N↑	NO_3^--N	Ramsay and Bawden (1983)
NH_4^+-N↑	NO_3^--N	Taufiaul and Habtem (1985)
NH_4^+-N↑	NO_3^--N	Speir et al. (1986)
NH_4^+-N↑	ND	Griffiths (1987)
NH_4^+-N↑	NO_3^--N↓	Kitt et al. (1988)
NH_4^+-N↑	NO_3^--N↓	Thompson (1990)
NH_4^+-N↑	NO_3^--N↑	Magnavacca and Sanchez (2003)
NH_4^+-N↑	NO_3^--N↑	Xiao et al. (2010)
NH_4^+-N↑	NO_3^--N↓	Gebremikael et al. (2015)
NH_4^+-N↓	NO_3^--N↑	Buchan et al. (2012, 2013)

↑ = Increase, ↓ = Decrease, ND = No Data

1. Some extractable nutrients come from decomposition or breakdown of organic matter present in soil as a direct result of irradiation treatment.
2. Several enzymes, including the urease (which produces ammonia as a decomposition product), are released after irradiation.
3. Some nitrogen is released from dead organisms (Ortaş and Harris 1996), and other nutrients are possibly released after the death of soil microorganisms (such as bacteria, fungi and actinomycetes).

Partial soil sterilization generally stimulates subsequent plant growth which is associated with an increased net mineralization of NH_4^+-N (Tanaka et al. 2003). The effects of partial soil sterilization on N-release reported in the literature are shown in Table 6.1.

The fertility of sterilized soil may be different than non-sterilized soil (Ortas 2003). According to Malkomes and Dietze (1998), partial soil sterilization drastically reduces the microbial population of soil with total eradication of certain microorganism groups. It is well known that N is one of the essential macronutrients and is required by plants in considerably large amount than P. According to Clemmensen et al. (2008) and (Tahovska et al. (2013) in different climates and neutral to slightly acid soils, the primary form of N available to plants is NH_4^+-N. The supply of nitrogen can influence rate of plant growth. Plant species differ in the form and amount of inorganic N uptake and its metabolism in roots (Azcon-Aguilar and Barea 2015). N supply to the host plant root influences, either directly or indirectly, susceptibility of the roots to infection, which can be influenced by rhizosphere pH change (Turnbull et al. 1995). Since N forms alter rhizospheric soil pH, two different N forms NH_4^+ and NO_3^- may affect mycorrhizal development in different ways as well (Ortas et al. 1996; Ortas et al. 2004).

It is widely accepted that plants can increase P uptake by chemically modifying the rhizosphere (Conversa et al. 2013; Marschner 2012; Ortas 1997). The main mechanism that has been suggested is pH alteration through excretion of H^+ and OH^-/HCO_3^-. pH is a major factor influencing the soil solution concentration of many plant nutrients, and plant-induced variation in pH affects the availability of many nutrients (Gao et al. 2012; Nietfeld and Prenzel 2015; Ortas and Rowell 2000; Valentinuzzi et al. 2015). The pH changes surrounding environment and infected roots, thereby affecting P availability. The intensity of AM root colonization, host plant P uptake and growth response to AM has been reported to be pH dependent (Baar et al. 2011; Zhu et al. 2007).

The local acidification around mycorrhizal-infected roots may be very important for P uptake. It was suggested that application of N and possibly P resulted in mycorrhizae making a significant contribution to the plant's P status. As mentioned above, because of utilization of NH_4^+-N by hyphae of mycorrhizal plants, this may have further consequences in the rhizosphere pH. Recent studies of Cely et al. (2016), Feitosa de Souza et al. (2016), Hall and Bell (2015), Zhou et al. (2016) and Zong et al. 2015) showed better understanding to the effect of mycorrhizal infection on P uptake. However, additional study, especially the effect of NH_4^+-N supply on P uptake with and without AMF, is required. It is also necessary to understand the relationship between P uptake and rhizosphere pH change (caused by NH_4^+-N) with VA inoculation an area which has received little attention in the past (Ortas 2012a). Gahoonia and Nielsen (2004) indicated that manipulation of rhizosphere pH through agronomic measures such as application of NH_4^+ or NO_3^- fertilizers may be more practical than breeding approaches.

6.7 Mycorrhizal Importance in Rhizospheric Soil

AMF are the largest symbiotic associations between plants and fungi which make significant contribution on physical, chemical and biological aspects of soil quality through AM fungal hyphae extending into the rhizosphere and thereby improving the absorption of nutrients especially P and micronutrients (Karandashov and Bucher 2005; Ortas 2003; Smith and Read 2008). Burkle and Belote (2015) results showed that the relationship between productivity and diversity varied among pioneer treatments and mycorrhizal amendments. This means that soil and crop management is related to the existence of mycorrhizae (Almaca and Ortas 2010; Ortas and Coskan 2016a). The establishment of mycorrhizae causes changes in the physiology of host plants. Like other soil microorganisms, AMF act as ecosystem engineers on roots and surface of the plants.

Fig. 6.5 Aggregate formation by the plant roots and AM mycelium (Ortaş 2008)

6.8 Mycorrhizae Effect on Soil Development Related with C Fixation

In an ecosystem, mycorrhizae actually play an important role across the rhizosphere and provide an organic link between the root and bulk soil. Moreover, mycorrhizae have significant effect on soil development. Aggregates encapsulate SOC and reduce rate of decomposition. Similarly, plant roots and AMF hyphae provide physical protection to soil C against microbial decomposers through aggregation (Leifheit et al. 2015; Ortas et al. 2013). AMF play a contributory effect on soil aggregate formation (Fig. 6.5) because of the symbiosis which significantly changes the root functioning (Espeland et al. 2013). The AMF symbiosis may also influence soil biogeochemical processes and GHG emissions through change in soil physical properties such as soil water holding capacity (Cavagnaro et al. 2006). Organic compounds and AMF hyphae are important in binding soil into macroaggregates and microaggregates (Singh et al. 2009). Thus, depletion of SOM and the degradation of soil structure can adversely affect soil fertility and crop productivity. Soil aggregation is one of the important soil characteristics that mediates many soil chemical, physical and biological properties and improves soil quality and sustainability (Ortas et al. 2013).

Graf and Frei (2013) reported that EMF increase water stable aggregates (WSA) along with promotion of plant growth. Therefore, mycorrhiza has a significant impact on soil resilience which is also an important component of soil quality. Several studies have reported that soil biology, especially mycorrhizal fungi, significantly influences soil fertility and soil quality.

6.9 Mycorrhizal Application for Plant Growth and Nutrient Acquisition

AMF influence soil functions such as C, N and P cycling to support plant growth and nutrition in the agro-ecosystem. Colonization by AMF ameliorates abiotic plant stress by enhancing plant nutrient uptake and delivering drought tolerance (Lehmann et al. 2014). The symbiosis influences plant water relation and drought resistance (Augé et al. 2015). AMF play a significant role in the establishment of plants in different environments by assisting in nutrient uptake enhancement along with stress tolerance such as drought and salt stress and even protecting them against soil pathogens (Azcon-Aguilar and Barea 2015).

There are good studies which explain that mycorrhizal plant has enhanced capability of plant root in acquiring nutrients from soil, particularly when the nutrient is poorly soluble and present in low concentration (Abrahao et al. 2014; Teste et al. 2014). For a given dry weight, mycorrhizal plants usually have higher P concentration in plant tissue than non-mycorrhizal plants (Zhang et al. 2014). Several crop plants absorb more P from low P soils when infected with AM fungi (Ortas 2003; Ortas et al. 2001). It has been found that mycorrhizal-infected roots can utilize rock P, whereas non-mycorrhizal roots cannot (Chinnusamy et al. 2006). How mycorrhizal plants obtain more P from soil than non-mycorrhizal plants is not yet fully understood. Several mechanisms have been proposed to define the AM effect on improving the absorption of available phosphate. Miranda et al. (2016) evaluated the effect of mycorrhizae and phosphorus (P) on forage peanut and reported that the seedlings grown in pots and fertilized with P, the extent of the response was higher for those inoculated with AMF. Moreover, Ortas et al. (2013) showed different mycorrhizal species significantly inoculated different plant roots and observed root colonization and P uptake. The species *G. clarum* was more efficient under conditions of low P availability for citrus seedlings (Ortas 2015).

Mycorrhizae may induce both quantitative and qualitative changes in plant P utilization (Smith and Read 2008). The amount of acid phosphatase present in AM hyphae (Cavagnaro 2014) and increased phosphatase activity of root surface as a result of infection (Guo et al. 2016) may liberate inorganic P from organic P sources, making P available for uptake. Alford et al. (2010) suggested that the roots of mycorrhizal plants may alter the rhizosphere chemistry by changing soil pH and produce exudates such as organic acids which may increase the availability of phosphorus by liberating phosphate ions in the soil (Rajkumar et al. 2012). There is still a wide research gap in understanding mechanism involved for increased P availability in the soil by mycorrhizal-infected roots. The low dry weight increment of experimentally inoculated plants in same conditions may be mycorrhizal, and all make greater demands on their host for carbon than the naturally released into the soil.

6.10 Soil P Influenced by Mycorrhizal Association

In agricultural and horticultural ecosystems, mycorrhizal colonization has been frequently observed less associated with high rates of P application (Elbon and Whalen 2015). The symbiosis of plant with AMF increases its efficiency in absorbing nutrients from the soil solution, especially the nutrients of low mobility such as phosphorus (Brito et al. 2013), which makes it possible to use phosphate fertilizers of low solubility in seedling production (Silva et al. 2016). Khade et al. (2010)

hypothesized that the infection is affected by P status of the plant rather than soil P levels. Graham et al. (1981) provided evidence that the roots of sorghum produce less root exudates in high P soils and hypothesized that colonization was affected by high soil P levels. It has been concluded that a large amount of P uptake can be explained by the increased surface area of hyphae alone (Sharif and Claassen 2011). In such case, plants have several mechanisms to employ for more P uptake, such as acidification of rhizosphere (Zahra et al. 2015) and excretion of organic acids (Palomo et al. 2006). Mobilization and solubilization of P are the principal chemical (soil pH change) and biological causes of increased nitrogen availability (Isaac et al. 2012). The soil organisms are also involved in mobilization of phosphate. Plant roots infected with AMF are known to have a higher phosphorus (P) absorption ability compared to non-mycorrhizal plants in P-deficient soils (Conversa et al. 2013; Smith and Smith 2011). Large inputs of soluble P, associated, for example, with application of superphosphate, can decrease mycorrhizal advantages by inhibiting the growth and activity of the vegetative mycelium (Greenhalgh et al. 1994). According to Feitosa de Souza et al. (2016), infection is affected by soil P as well as plant P concentration. As plants vary in their ability to absorb P and mycorrhizal fungi vary in their response to soil P, each plant-soil-AM symbiont system must be evaluated separately.

The soil P concentration is usually critical in mycorrhizal infection. Soil P must be sufficient for host plant growth and colonization of mycorrhizae. Very high and very low phosphorus levels may reduce mycorrhizal infection/colonization (Goncalves de Oliveira et al. 2015; Lirio Rondina et al. 2014). It is well established that infection by mycorrhizal fungi is significantly reduced at high soil P levels (Balzergue et al. 2013). The level of P in the plant has also been shown to influence the establishment of mycorrhizae with high levels inhibiting colonization; moreover it depends upon the root system (Yang et al. 2015). Ortas (2012a) showed that addition of P decreased AM infection in wheat under field conditions. The high concentration of soluble phosphate decreases AMF percentage (Table 6.2). Ortas (2012a) reported that with increasing P levels, mycorrhizal colonization significantly reduces. In *G. mosseae*-inoculated plants, when plant received no P fertilizer, root colonization was 90.8 % and with P treatment root colonization was 57.1 % (Table 6.2). Controlled plant roots had 3–10 % of colonization, but *G. etunicatum*-inoculated plants had 41–72 % and *G. mosseae*-inoculated plants had 30–75 % root colonization. Therefore, mycorrhizae formation, response to added P, host nutrient requirement and mycorrhizae responsiveness are all interrelated.

The results of mycorrhizal research have strongly suggested that infection does not change the size of labile pool, but the hyphae extend beyond P depletion zone and provide a well-distributed surface for absorbing phosphorus (Sharif and Claassen 2011). Rubio et al. (2012) demonstrated a field and greenhouse experiments and showed that different plant species have different effects on rhizosphere P depletion which is related with plant P demand. Plants uptake P from the soil solution at a much faster rate than they can diffuse to the root surface. Consequently, a P depletion zone develops around the absorbing organs (mycorrhizal hyphae or roots) of the plants (Marschner 2012).

Table 6.2 Effect of P, Zn and mycorrhizal inoculation on maize plant root infection (Ortas 2012a)

Mycorrhizal species	Treatments		Root infection (%)	
Control	P0		2.0	±1.00
	P1	**Zn0**	2.3	±2.50
	P2		2.7	±3.10
	P0		2.0	±2.00
	P1	**Zn1**	3.0	±3.00
	P2		3.0	±2.60
G. etunicatum	P0		93.7	±2.90
	P1	**Zn0**	87.9	±2.20
	P2		66.7	±13.6
	P0		89.1	±5.20
	P1	**Zn1**	78.0	±11.4
	P2		70.4	±9.40
G. mosseae	P0		90.8	±3.40
	P1	**Zn0**	91.4	±3.50
	P2		57.1	±35.4
	P0		82.4	±15.2
	P1	**Zn1**	79.7	±7.10
	P2		79.8	±9.40

Since the diffusion of P in soil is very slow, P absorption capacity of roots does not have great effect on rate of uptake. When plant roots are not able to absorb sufficient P for adequate growth, then plants employ other physiological variables of root on P absorption by increasing the amount of P available to diffuse the root surface interpreted (Isaac et al. 2012). The direct effect of soluble P on fungi metabolism, mainly by regulating enzymatic activities, is related to phosphate transfer to the host and has been recently discussed. Carrasco et al. (2011) showed higher levels of acid phosphate activity in the root and rhizosphere infected by *G. mosseae* and *G. geosporum* compared with control. Although increased phosphatase activity has been demonstrated in several mycorrhizal symbioses, plant roots and other microorganisms can also produce acid phosphatase. In this case, it is difficult to interpret contribution of mycorrhizae.

6.11 Effect of Mycorrhizal Infection on Nitrogen Uptake

The effect of N form and P rate application on total dry matter production on harvesting varied with mycorrhizal inoculation. Hoeksema et al. (2010) reported in a meta-analysis work that N-fertilization is an important predictor of plant response to mycorrhizal inoculation. AMF inoculation enhanced differences between N sources (Ortas and Rowell 2004). Sorghum plants infected with mycorrhizae had nearly three times more shoot dry matter yield than non-inoculated control plants

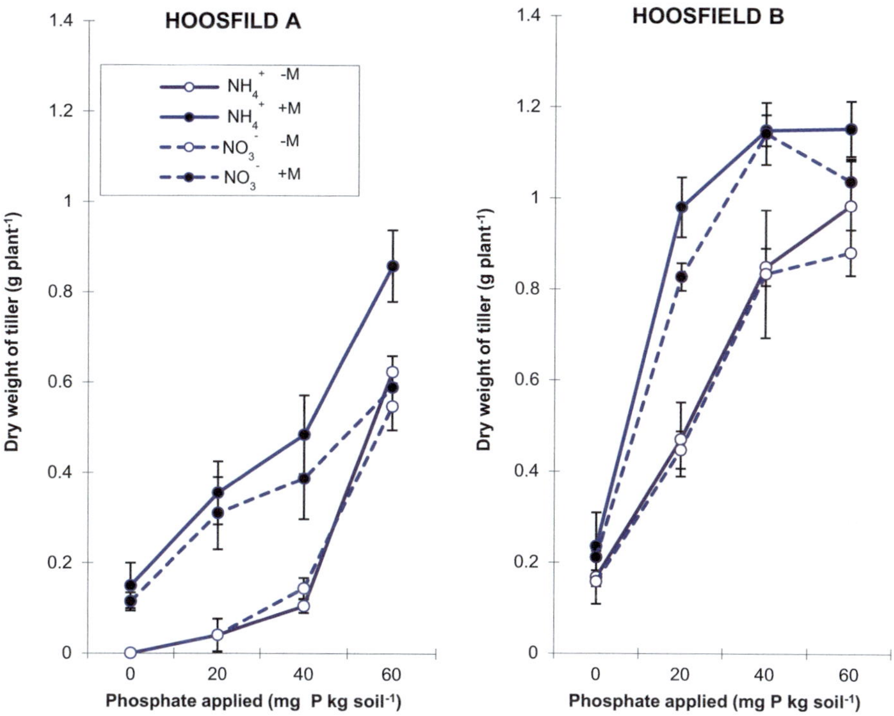

Fig. 6.6 Effect of nitrogen form, phosphate rate and AMF inoculum on dry weight of sorghum shoot at 40 days (+M AMF inoculum used, −M no inoculum) (Ortas et al. 1996)

(Fig. 6.6). Mycorrhizal inoculation with increased P application significantly enhanced tiller dry weight production (Ortas et al. 1996).

When the N source was $(NH_4)_2SO_4$, the specific absorption rate of N by mycorrhizal roots (nitrogen absorbed per g of root) was higher than that of non-mycorrhizal roots (Smith 1980). Similar results have been reported by Ortas et al. (1996) when soil was sterilized, as a result of more mineral NH_4^+-N, and the specific absorption rate of N was higher. Moreover, ectomycorrhizal and ericoid mycorrhizal fungi generally appear to prefer NH_4^+-N to NO_3^--N (Kosola et al. 2007; Kranabetter and MacKenzie 2010). Lundeberg (1970) has shown that most of the 27 ectotrophic mycorrhizal fungi grew better with NH_4^+-N than NO_3-N.

According to Azcon-Aguilar and Barea (2015), mycorrhizal infection stimulated growth of NH_4^+-N-fed plants more than that of NO_3^--N. Increasing availability of P by the rate of N fixation was related to AM fungus infection with dual application of mycorrhizal fungi and rhizobium. Barea et al. (1987) by using ^{15}N technique showed that both NH_4^+-N and NO_3^--N forms of N can be absorbed by AM hyphae and that growth enhancement of legumes by AM can be attributed to both enhanced N fixation as well as improved N uptake from the soil, especially with the NH_4^+-N forms (Tome et al. 2015). The release of NH_4^+ from nodules to the soil was immediately absorbed by the surrounding AM hyphae. N uptake by AMF has been reported to take place in the following situations:

1. Mycorrhizal fungi increase plant-absorbed NO_3^--N and NH_4^+-N from the growing substrate (Ortas et al. 1996).
2. Mycorrhizal fungi assimilate NH_4^+-N via glutamine syntheses, and this would have a significant influence on the function of external hyphae (Johansen et al. 1996).
3. Fungi directly uptake NH_4^+-N through the hyphae (Marschner and Dell 1994; Perez-Tienda et al. 2014). According to Javaid (2009) NH_4^+-N can be taken up by plant roots because it is relatively immobile compared to NO_3^--N in soil.
4. Mycorrhizal fungi increase N inflow of plant roots. N inflow was considerably increased when supplied as $(NH_4)_2SO_4$ (Ortas 1994).

6.12 Micronutrient Uptake Significantly Affected by Mycorrhizae

Since mycorrhizal hyphae can exploit more efficiently large volumes of soil, in the presence of AM symbiosis, more nutrients are taken up and transported specifically. In addition to P, AMF enhance the acquisition of other nutrients such as sulphur and potassium (K) (Ortas 2003) and immobile micronutrients, particularly Zn and Cu (Li and Christie 2001; Ortas 2012a). When P and Zn contents of leaves were compared with the recommended levels, entire P status of mycorrhizal plants was observed above the normal level, regardless of AM fungi. Watts-Williams et al. (2015) reported that up to 24 % of Zn in shoots of the AM plants was delivered via the AM pathway in soil with Zn concentration. In addition, non-mycorrhizal plants apparently suffered from Zn deficiency according to leaf analysis, whereas Zn status of inoculated plants was around an acceptable level. *G. intraradices* appeared to be more effective than *G. mosseae* in terms of Zn concentration of leaves. Wu and Zou (2009) and Ortas (2012b) showed that sole AMF inoculation significantly increased total dry weight, leaf P, K, Ca, Mg, Fe, Cu and Mn contents and root P, K, Ca, Fe, Cu and Zn contents of the seedlings, compared to the non-AMF control. The result of Balliu et al. (2015) indicates that AM fungi may increase the uptake of Fe to host plant.

References

Abrahao A, Lambers H, Sawaya ACHF, Mazzafera P, Oliveira RS (2014) Convergence of a specialized root trait in plants from nutrient-impoverished soils: phosphorus-acquisition strategy in a nonmycorrhizal cactus. Oecologia 176:345–355. doi:10.1007/s00442-014-3033-4

Alford ER, Pilon-Smits EAH, Paschke MW (2010) Metallophytes-a view from the rhizosphere. Plant Soil 337:33–50. doi:10.1007/s11104-010-0482-3

Alloush GA, Clark RB (2001) Maize response to phosphate rock and arbuscular mycorrhizal fungi in acidic soil. Commun Soil Sci Plant Anal 32:231–254. doi:10.1081/css-100103004

Almaca A, Ortas I (2010) Growth response of maize plants (Zea mays L.) to wheat and lentil pre-cropping and to indigenous mycorrhizae in field soil. Span J Agric Res 8:S131–S136
Arunachalam G, Oblisami G, Andrangaswami G (1974) Effect of gamma radiation on certain microbial properties of two soil types. Madras Agric J 61:992–996
Asensio D, Rapparini F, Penuelas J (2012) AM fungi root colonization increases the production of essential isoprenoids vs. nonessential isoprenoids especially under drought stress conditions or after jasmonic acid application. Phytochemistry 77:149–161. doi:10.1016/j.phytochem.2011.12.012
Asghari HR, Cavagnaro TR (2012) Arbuscular mycorrhizas reduce nitrogen loss via leaching. Plos One 7:151–155. doi:10.1371/journal.pone.0029825
Augé RM, Toler HD, Saxton AM (2015) Arbuscular mycorrhizal symbiosis alters stomatal conductance of host plants more under drought than under amply watered conditions: a meta-analysis. Mycorrhiza 25:13–24
Azcon-Aguilar C, Barea JM (2015) Nutrient cycling in the mycorrhizosphere. J Soil Sci Plant Nutr 15:372–396
Baar J, Paradi I, Lucassen ECHET, Hudson-Edwards KA, Redecker D, Roelofs JGM, Smolders AJP (2011) Molecular analysis of AMF diversity in aquatic macrophytes: a comparison of oligotrophic and utra-oligotrophic lakes. Aquat Bot 94:53–61. doi:10.1016/j.aquabot.2010.09.006
Bago B, Cano C, Azcon-Aguilar C, Samson J, Coughlan AP, Piche Y (2004) Differential morphogenesis of the extraradical mycelium of an arbuscular mycorrhizal fungus grown monoxenically on spatially heterogeneous culture media. Mycologia 96:452–462. doi:10.2307/3762165
Balliu A, Sallaku G, Rewald B (2015) AMF inoculation enhances growth and improves the nutrient uptake rates of transplanted, salt-stressed tomato seedlings. Sustainability 7:15967–15981. doi:10.3390/su71215799
Balzergue C, Chabaud M, Barker DG, Becard G, Rochange SF (2013) High phosphate reduces host ability to develop arbuscular mycorrhizal symbiosis without affecting root calcium spiking responses to the fungus. Front Plant Sci 4. doi: 10.3389/fpls.2013.00426
Barea J, Azcon-Aguilar C, Azcón R (1987) Vesicular-arbuscular mycorrhiza improve both symbiotic N2 fixation and N uptake from soil as assessed with a 15N technique under field conditions. New Phytol 106:717–725
Barea J, Palenzuela J, Cornejo P, Sánchez-Castro I, Navarro-Fernández C, Lopéz-García A, Estrada B, Azcón R, Ferrol N, Azcón-Aguilar C (2011) Ecological and functional roles of mycorrhizas in semi-arid ecosystems of Southeast Spain. J Arid Environ 75:1292–1301
Berruti A, Borriello R, Lumini E, Scariot V, Bianciotto V, Balestrini R (2013) Application of laser microdissection to identify the mycorrhizal fungi that establish arbuscules inside root cells. Front Plant Sci 4. doi: 10.3389/fpls.2013.00135
Bowen H, Cawse P (1964) Some effects of gamma radiation on the composition of the soil solution and soil organic matter. Soil Sci 98:358–361
Bray SR, Kitajima K, Sylvia DM (2003) Mycorrhizae differentially alter growth, physiology, and competitive ability of an invasive shrub. Ecol Appl 13:565–574. doi:10.1890/1051-0761(2003)013[0565:mdagpa]2.0.co;2
Brito I, Carvalho M, Goss MJ (2013) Soil and weed management for enhancing arbuscular mycorrhiza colonization of wheat. Soil Use Manag 29:540–546
Buchan D, Moeskops B, Ameloot N, De Neve S, Sleutel S (2012) Selective sterilisation of undisturbed soil cores by gamma irradiation: effects on free-living nematodes, microbial community and nitrogen dynamics. Soil Biol Biochem 47:10–13
Buchan D, Gebremikael MT, Ameloot N, Sleutel S, De Neve S (2013) The effect of free-living nematodes on nitrogen mineralisation in undisturbed and disturbed soil cores. Soil Biol Biochem 60:142–155
Burkle LA, Belote RT (2015) Soil mutualists modify priority effects on plant productivity, diversity, and composition. Appl Veg Sci 18:332–342. doi:10.1111/avsc.12149
Carrasco L, Azcon R, Kohler J, Roldan A, Caravaca F (2011) Comparative effects of native filamentous and arbuscular mycorrhizal fungi in the establishment of an autochthonous, leguminous shrub growing in a metal-contaminated soil. Sci Total Environ 409:1205–1209. doi:10.1016/j.scitotenv.2010.12.019

Cavagnaro TR (2014) Impacts of compost application on the formation and functioning of arbuscular mycorrhizas. Soil Biol Biochem 78:38–44. doi:10.1016/j.soilbio.2014.07.007
Cavagnaro T, Jackson L, Six J, Ferris H, Goyal S, Asami D, Scow K (2006) Arbuscular mycorrhizas, microbial communities, nutrient availability, and soil aggregates in organic tomato production. Plant Soil 282:209–225
Cavagnaro TR, Langley AJ, Jackson LE, Smukler SM, Koch GW (2008) Growth, nutrition, and soil respiration of a mycorrhiza-defective tomato mutant and its mycorrhizal wild-type progenitor. Funct Plant Biol 35:228–235
Cely MVT, de Oliveira AG, de Freitas VF, de Luca MB, Barazetti AR, dos Santos IMO, Gionco B, Garcia GV, Prete CEC, Andrade G (2016) Inoculant of arbuscular mycorrhizal fungi (Rhizophagus clarus) increase yield of soybean and cotton under field conditions. Front Microbiol 7. doi: 10.3389/fmicb.2016.00720
Cheng W, Zhang Q, Coleman DC, Carroll CR, Hoffman CA (1996) Is available carbon limiting microbial respiration in the rhizosphere? Soil Biol Biochem 28:1283–1288
Cheng L, Booker FL, Tu C, Burkey KO, Zhou L, Shew HD, Rufty TW, Hu S (2012) Arbuscular mycorrhizal fungi increase organic carbon decomposition under elevated CO2. Science 337:1084–1087
Chinnusamy M, Kaushik BD, Prasanna R (2006) Growth, nutritional, and yield parameters of wetland rice as influenced by microbial consortia under controlled conditions. J Plant Nutr 29:857–871. doi:10.1080/01904160600651803
Clark AL, St Clair SB (2011) Mycorrhizas and secondary succession in aspen-conifer forests: Light limitation differentially affects a dominant early and late successional species. For Ecol Manag 262:203–207. doi:10.1016/j.foreco.2011.03.024
Clemmensen KE, Sorensen PL, Michelsen A, Jonasson S, Stroem L (2008) Site-dependent N uptake from N-form mixtures by arctic plants, soil microbes and ectomycorrhizal fungi. Oecologia 155:771–783. doi:10.1007/s00442-008-0962-9
Clemmensen K, Bahr A, Ovaskainen O, Dahlberg A, Ekblad A, Wallander H, Stenlid J, Finlay R, Wardle D, Lindahl B (2013) Roots and associated fungi drive long-term carbon sequestration in boreal forest. Science 339:1615–1618
Conversa G, Lazzizera C, Bonasia A, Elia A (2013) Yield and phosphorus uptake of a processing tomato crop grown at different phosphorus levels in a calcareous soil as affected by mycorrhizal inoculation under field conditions. Biol Fertil Soils 49:691–703. doi:10.1007/s00374-012-0757-3
Cornejo P, Borie F, Rubio R, Azcon R (2007) Influence of nitrogen source on the viability, functionality and persistence of Glomus etunicatum fungal propagules in an Andisol. Appl Soil Ecol 35:423–431. doi:10.1016/j.apsoil.2006.06.006
Correa A, Strasser RJ, Martins-Loucao MA (2006) Are mycorrhiza always beneficial? Plant Soil 279:65–73. doi:10.1007/s11104-005-7460-1
da Silva EP, Freire Gomes VF, Mendes Filho PF, Tupinamb da Silva Junior JM, Lange Ness RL (2016) Development and mycorrhizal colonisation in embauba seedlings fertilised with natural phosphates and organic material. Revista Ciencia Agronomica 47:256–263. doi: 10.5935/1806-6690.20160030
Drigo B, Pijl AS, Duyts H, Kielak AM, Gamper HA, Houtekamer MJ, Boschker HT, Bodelier PL, Whiteley AS, van Veen JA (2010) Shifting carbon flow from roots into associated microbial communities in response to elevated atmospheric CO2. Proc Natl Acad Sci 107:10938–10942
Elbon A, Whalen JK (2015) Phosphorus supply to vegetable crops from arbuscular mycorrhizal fungi: a review. Biol Agric Hortic 31:73–90. doi:10.1080/01448765.2014.966147
Espeland E, Caesar AJ, Sainju UM, Lartey RT, Gaskin JF (2013) Effects of Agaricus lilaceps fairy rings on soil aggregation and microbial community structure in relation to growth stimulation of western wheatgrass (Pascopyrum smithii) in Eastern Montana rangeland. Microb Ecol 66:120–131
Feitosa de Souza TA, Rodriguez-Echeverria S, de Andrade LA, Freitas H (2016) Could biological invasion by Cryptostegia madagascariensis alter the composition of the arbuscular mycorrhizal fungal community in semi-arid Brazil? Acta Bot Bras 30:93–101. doi:10.1590/0102-33062015abb0190

Fellbaum CR, Gachomo EW, Beesetty Y, Choudhari S, Strahan GD, Pfeffer PE, Kiers ET, Bücking H (2012) Carbon availability triggers fungal nitrogen uptake and transport in arbuscular mycorrhizal symbiosis. Proc Natl Acad Sci 109:2666–2671

Fitter AH, Heinemeyer A, Husband R, Olsen E, Ridgway KP, Staddon PL (2004) Global environmental change and the biology of arbuscular mycorrhizas: gaps and challenges. Can J Bot 82:1133–1139. doi:10.1139/b04-045

Folli-Pereira MS, Meira-Haddad LSA, Soares Bazzolli DM, Megumi Kasuya MC (2012) Arbuscular mycorrhiza and plant tolerance to stress. Revista Brasileira De Ciencia Do Solo 36:1663–1679

Ford CR, McGee J, Scandellari F, Hobbie EA, Mitchell RJ (2012) Long- and short-term precipitation effects on soil CO2 efflux and total belowground carbon allocation. Agric For Meteorol 156:54–64. doi:10.1016/j.agrformet.2011.12.008

Gahoonia TS, Nielsen NE (2004) Root traits as tools for creating phosphorus efficient crop varieties. Plant Soil 260:47–57

Gao XP, Hoffland E, Stomph T, Grant CA, Zou CQ, Zhang FS (2012) Improving zinc bioavailability in transition from flooded to aerobic rice. A review. Agron Sust Dev 32:465–478. doi:10.1007/s13593-011-0053-x

Gebremikael MT, De Waele J, Buchan D, Soboksa GE, De Neve S (2015) The effect of varying gamma irradiation doses and soil moisture content on nematodes, the microbial communities and mineral nitrogen. Appl Soil Ecol 92:1–13. doi:10.1016/j.apsoil.2015.03.003

Goncalves de Oliveira JR, Matos e Silva E, Teixeira-Rios T, de Melo NF, Yano-Melo AM (2015) Response of an endangered tree species from Caatinga to mycorrhization and phosphorus fertilization. Acta Bot Bras 29:94–102. doi:10.1590/0102-33062014abb3420

Graf F, Frei M (2013) Soil aggregate stability related to soil density, root length, and mycorrhiza using site-specific Alnus incana and Melanogaster variegatus sl. Ecol Eng 57:314–323

Graham JH, Leonard RT, Menge JA (1981) Membrane-mediated decrease in root exudation responsible for phosphorus inhibition of vesicular-arbuscular mycorrhiza formation. Plant Physiol 68:548–552

Greenhalgh FC, Deboer RF, Merriman PR, Hepworth G, Keane PJ (1994) Control of phytophthora root-rot of irrigated subterranean clover with potassium phosphonate in Victoria, Australia. Plant Pathol 43:1009–1019. doi:10.1111/j.1365-3059.1994.tb01650.x

Griffiths B (1987) Growth of selected microorganisms and plants in soil sterilized by ethylene oxide or gamma-irradiation. Soil Biol Biochem 19:115–116

Guo YJ, Ni Y, Raman H, Wilson BAL, Ash GJ, Wang AS, Li GD (2012) Arbuscular mycorrhizal fungal diversity in perennial pastures; responses to long-term lime application. Plant Soil 351:389–403. doi:10.1007/s11104-011-0976-7

Guo Y, Du Q, Li G, Ni Y, Zhang Z, Ren W, Hou X (2016) Soil phosphorus fractions and arbuscular mycorrhizal fungi diversity following long-term grazing exclusion on semi-arid steppes in Inner Mongolia. Geoderma 269:79–90. doi:10.1016/j.geoderma.2016.01.039

Hall DJM, Bell RW (2015) Biochar and compost increase crop yields but the effect is short term on sandplain soils of Western Australia. Pedosphere 25:720–728

Hamdia MA, Shaddad MAK (2010) Salt tolerance of crop plants. J Stress Physiol Biochem 6:64–90

Hassan HM, Marschner P, McNeill A, Tang C (2012) Growth, P uptake in grain legumes and changes in rhizosphere soil P pools. Biol Fertil Soils 48:151–159. doi:10.1007/s00374-011-0612-y

Heinemeyer A, Ineson P, Ostle N, Fitter AH (2006) Respiration of the external mycelium in the arbuscular mycorrhizal symbiosis shows strong dependence on recent photosynthates and acclimation to temperature. New Phytol 171:159–170. doi:10.1111/j.1469-8137.2006.01730.x

Helgason T, Fitter AH (2009) Natural selection and the evolutionary ecology of the arbuscular mycorrhizal fungi (Phylum Glomeromycota). J Exp Bot: erp144.

Hetrick BAD, Kitt DG, Wilson GT (1988) Mycorrhizal dependence and growth habit of warm-season and cool-season tallgrass prairie plants. Can J Bot 66:1376–1380

Hinsinger P (2001) Bioavailability of soil inorganic P in the rhizosphere as affected by root-induced chemical changes: a review. Plant Soil 237:173–195. doi:10.1023/a:1013351617532

Hoeksema JD, Chaudhary VB, Gehring CA, Johnson NC, Karst J, Koide RT, Pringle A, Zabinski C, Bever JD, Moore JC, Wilson GWT, Klironomos JN, Umbanhowar J (2010) A meta-analysis of context-dependency in plant response to inoculation with mycorrhizal fungi. Ecol Lett 13:394–407. doi:10.1111/j.1461-0248.2009.01430.x

Hoffmann D, Vierheilig H, Riegler P, Schausberger P (2009) Arbuscular mycorrhizal symbiosis increases host plant acceptance and population growth rates of the two-spotted spider mite Tetranychus urticae. Oecologia 158:663–671. doi:10.1007/s00442-008-1179-7

Irshad M, Honna T, Eneji AE, Yamamoto S (2002) Wheat response to nitrogen source under saline conditions. J Plant Nutr 25:2603–2612. doi:10.1081/pln-120015525

Isaac ME, Hinsinger P, Harmand JM (2012) Nitrogen and phosphorus economy of a legume tree-cereal intercropping system under controlled conditions. Sci Total Environ 434:71–78. doi:10.1016/j.scitotenv.2011.12.071

Jackson LE, Burger M, Cavagnaro TR (2008) Roots, nitrogen transformations, and ecosystem services. Plant Biol 59:341

Jakobsen I, Andersen A (1982) Vesicular-arbuscular mycorrhiza and growth in barley: effects of irradiation and heating of soil. Soil Biol Biochem 14:171–178

Javaid A (2009) Arbuscular mycorrhizal mediated nutrition in plants. J Plant Nutr 32:1595–1618. doi:10.1080/01904160903150875

Jenkinson DS, Nowakowski TZ, Mitchell JDD (1972) Growth and uptake of nitrogen by wheat and regress in fumigated and irradiated soil. Plant Soil 36:149

Johansen A, Finlay RD, Olsson PA (1996) Nitrogen metabolism of external hyphae of the arbuscular mycorrhizal fungus Glomus intraradices. New Phytol 133:705–712. doi:10.1111/j.1469-8137.1996.tb01939.x

Karandashov V, Bucher M (2005) Symbiotic phosphate transport in arbuscular mycorrhizas. Trends Plant Sci 10:22–29. doi:10.1016/j.tplants.2004.12.003

Khade SW, Rodrigues BF, Sharma PK (2010) Symbiotic interactions between arbuscular mycorrhizal (AM) fungi and male papaya plants: Its status, role and implications. Plant Physiol Biochem 48:893–902. doi:10.1016/j.plaphy.2010.08.010

Kitt DG, Hetrick BD, Wilson GT (1988) Relationship of soil fertility to suppression of the growth response of mycorrhizal big bluestem in non-sterile soil. New Phytol 109:473–481

Kohler J, Hernandez JA, Caravaca F, Roldan A (2008) Plant-growth-promoting rhizobacteria and arbuscular mycorrhizal fungi modify alleviation biochemical mechanisms in water-stressed plants. Funct Plant Biol 35:141–151. doi:10.1071/fp07218

Kosola KR, Workmaster BAA, Spada PA (2007) Inoculation of cranberry (Vaccinium macrocarpon) with the ericoid mycorrhizal fungus Rhizoscyphus ericae increases nitrate influx. New Phytol 176:184–196. doi:10.1111/j.1469-8137.2007.02149.x

Kranabetter JM, MacKenzie WH (2010) Contrasts among mycorrhizal plant guilds in foliar nitrogen concentration and delta N-15 along productivity gradients of a Boreal Forest. Ecosystems 13:108–117. doi:10.1007/s10021-009-9304-y

Krishna H, Singh SK, Sharma RR, Khawale RN, Grover M, Patel VB (2005) Biochemical changes in micropropagated grape (Vitis vinifera L.) plantlets due to arbuscular-mycorrhizal fungi (AMF) inoculation during ex vitro acclimatization. Sci Hortic 106:554–567. doi:10.1016/j.scienta.2005.05.009

Labidi S, Calonne M, Ben Jeddi F, Debiane D, Rezgui S, Laruelle F, Tisserant B, Grandmougin-Ferjani A, Sahraoui AL-H (2011) Calcareous impact on arbuscular mycorrhizal fungus development and on lipid peroxidation in monoxenic roots. Phytochemistry 72:2335–2341. doi:10.1016/j.phytochem.2011.08.016

Lambais MR, Cardoso E (1993) Response of stylosanthes-guianensis to endomycorrhizal fungi inoculation as affected by lime and phosphorus applications. 2. nutrient-uptake. Plant Soil 150:109–116. doi:10.1007/bf00779181

Langley JA, Dijkstra P, Drake BG, Hungate BA (2003) Ectomycorrhizal colonization, biomass, and production in a regenerating scrub oak forest in response to elevated CO2. Ecosystems 6:424–430. doi:10.1007/s10021-002-0194-5

Lehmann A, Veresoglou SD, Leifheit EF, Rillig MC (2014) Arbuscular mycorrhizal influence on zinc nutrition in crop plants–A meta-analysis. Soil Biol Biochem 69:123–131

Leifheit EF, Verbruggen E, Rillig MC (2015) Arbuscular mycorrhizal fungi reduce decomposition of woody plant litter while increasing soil aggregation. Soil Biol Biochem 81:323–328. doi:10.1016/j.soilbio.2014.12.003

Lejon DP, Sebastia J, Lamy I, Chaussod R, Ranjard L (2007) Relationships between soil organic status and microbial community density and genetic structure in two agricultural soils submitted to various types of organic management. Microb Ecol 53:650–663

Li XL, Christie P (2001) Changes in soil solution Zn and pH and uptake of Zn by arbuscular mycorrhizal red clover in Zn-contaminated soil. Chemosphere 42:201–207. doi:10.1016/s0045-6535(00)00126-0

Lirio Rondina AB, Azevedo Marques Lescano LE, Alves RA, Matsuura EM, Nogueira MA, Zangaro W (2014) Arbuscular mycorrhizas increase survival, precocity and flowering of herbaceous and shrubby species of early stages of tropical succession in pot cultivation. J Trop Ecol 30:599–614. doi:10.1017/s0266467414000509

Liu Z, Li Y, Wang J, He X, Tian C (2015) Different respiration metabolism between mycorrhizal and non-mycorrhizal rice under low-temperature stress: a cry for help from the host. J Agric Sci 153:602–614. doi:10.1017/s0021859614000434

Lundeberg G (1970) Utilization of various nitrogen sources, in particular bound soil nitrogen, by mycorrhizal fungi. Studia Forestalia Suecica 79:1–95

Magnavacca C, Sanchez M (2003) Assessing nutrients availability of irradiated and non-irradiated blosolids for the agriculture re-use. In: International conference on wastewater sludge as a ResourceTrondheim, Norway 9, pp 65–73

Malkomes HP, Dietze T (1998) Effects of steaming and pesticides on soil microorganisms under laboratory conditions. II. Effects of partial sterilization and its combination with pesticides. Agribiological Research-Zeitschrift Fur Agrarbiologie Agrikulturchemie Okologie 51:155–165

Marschner P (2012) Marschner's mineral nutrition of higher plants. Academic, London

Marschner H, Dell B (1994) Nutrient uptake in mycorrhizal symbiosis. Plant Soil 159:89–102. doi:10.1007/bf00000098

Martensson LM, Schnoor TK, Olsson PA (2012) Allocation of carbon to mycorrhiza in the grasses Koeleria glauca and Corynephorus canescens in sandy grasslands. Appl Soil Ecol 54:55–62. doi:10.1016/j.apsoil.2011.12.006

Mellado-Vazquez PG, Lange M, Bachmann D, Gockele A, Karlowsky S, Milcu A, Piel C, Roscher C, Roy J, Gleixner G (2016) Plant diversity generates enhanced soil microbial access to recently photosynthesized carbon in the rhizosphere. Soil Biol Biochem 94:122–132. doi:10.1016/j.soilbio.2015.11.012

Miranda EM, Silva EMR, Saggin Júnior OJ (2016) Mycorrhizal inoculation and phosphate fertilizer in the production of seedlings of the forage peanut. Rev Ciênc Agron 47:240–246

Miransari M, Bahrami HA, Rejali F, Malakouti MJ (2009) Effects of soil compaction and arbuscular mycorrhiza on corn (Zea mays L.) nutrient uptake. Soil Tillage Res 103:282–290. doi:10.1016/j.still.2008.10.015

Moon JB, Wardrop DH, Bruns MAV, Miller RM, Naithani KJ (2016) Land-use and land-cover effects on soil microbial community abundance and composition in headwater riparian wetlands. Soil Biol Biochem 97:215–233. doi:10.1016/j.soilbio.2016.02.021

Moratelli EM, Costa MD, Lovato PE, Santos M, Paulilo MTS (2007) Efeito da disponibilidade de água e de luz na colonização micorrízica e no crescimento de Tabebuia avellanedae Lorentz ex Griseb. (Bignoniaceae). Revista Árvore 31:555–566. doi:10.1590/s0100-67622007000300021

Nietfeld H, Prenzel J (2015) Modeling the reactive ion dynamics in the rhizosphere of tree roots growing in acid soils. I. Rhizospheric distribution patterns and root uptake of M-b cations as affected by root-induced pH and Al dynamics. Ecol Model 307:48–65. doi:10.1016/j.ecolmodel.2015.02.011

Norby RJ, Warren JM, Iversen CM, Medlyn BE, McMurtrie RE (2010) CO2 enhancement of forest productivity constrained by limited nitrogen availability. Proc Natl Acad Sci 107:19368–19373

Ocampo J, Martin J, Hayman D (1980) Influence of plant interactions on vesicular-arbuscular mycorrhizal infections. I. Host and non-host plants grown together. New Phytol 84:27–35

Ortas I (1994) The effect of different forms and rates of nitrogen and different rates of phosphorus fertilizer on rhizosphere pH and P uptake in mycorrhizal and non-mycorrhizal Sorghum plants soil science. University of Reading, Reading

Ortas I (1997) Determination of the extent of rhizosphere soil. Commun Soil Sci Plant Anal 28:1767–1776. doi:10.1080/00103629709369914

Ortas I (2003) Effect of selected mycorrhizal inoculation on phosphorus sustainability in sterile and non-sterile soils in the Harran Plain in South Anatolia. J Plant Nutr 26:1–17. doi:10.1081/Pln-120016494

Ortas I (2012a) Do maize and pepper plants depend on mycorrhizae in terms of phosphorus and zinc uptake? J Plant Nutr 35:1639–1656. doi:10.1080/01904167.2012.698346

Ortas I (2012b) The effect of mycorrhizal fungal inoculation on plant yield, nutrient uptake and inoculation effectiveness under long-term field conditions. Field Crop Res 125:35–48. doi:10.1016/j.fcr.2011.08.005

Ortas I (2015) Comparative analyses of Turkey agricultural soils: Potential communities of indigenous and exotic mycorrhiza species' effect on maize (Zea mays L.) growth and nutrient uptakes. Eur J Soil Biol 69:79–87. doi:10.1016/j.ejsobi.2015.05.006

Ortaş İ (2008) The effect of mycorrhizal inoculation on forage and non-forage plant growth and nutrient uptake under field conditions. Options Méditerranéennes Série A: Séminaires Méditerranéens (CIHEAM)

Ortas I, Coskan A (2016a) Precipitation as the most affecting factor on soil-plant environment conditions affects the mycorrhizal spore numbers in three different ecological zones in Turkey. Acta Agric Scand Sect B Soil Plant Sci 66:369–378. doi:10.1080/09064710.2015.1132005

Ortas I, Coskan A (2016b) Precipitation as the most affecting factor on soil–plant environment conditions affects the mycorrhizal spore numbers in three different ecological zones in Turkey. Acta Agric Scand Sect B Soil Plant Sci 66:369–378. doi:10.1080/09064710.2015.1132005

Ortaş I, Harris PJ (1996) The effect of partial soil sterilization and seasonal change on soil degradation (N-mineralization and soil chemical properties). In: Kapur S (ed) 1st international conference on land degradation. Çukurova Universitey, Adana

Ortas I, Rowell DL (2000) Effect of pH on amount of phosphorus extracted by 10 mM calcium chloride from three rothamsted soils. Commun Soil Sci Plant Anal 31:2917–2923. doi:10.1080/00103620009370638

Ortas I, Rowell DL (2004) Effect of ammonium and nitrate on indigenous mycorrhizal infection, rhizosphere pH change, and phosphorus uptake by sorghum. Commun Soil Sci Plant Anal 35:1923–1944. doi:10.1081/lcss-200026820

Ortas I, Harris PJ, Rowell DL (1996) Enhanced uptake of phosphorus by mycorrhizal sorghum plants as influenced by forms of nitrogen. Plant Soil 184:255–264. doi:10.1007/bf00010454

Ortas I, Kaya Z, Cakmak I (2001) Influence of va-mycorrhiza inoculation on growth of maize and green pepper plants in phosphorus and zinc deficient soils. In: Horst W (ed) Plant nutrition-food security and sustainability of agro-ecosystems. Kluwer Akedmic Publishers, Dordrecht

Ortas I, Rowell DL, Harris PJ (2004) Effect of mycorrhizae and pH change at the root-soil interface on phosphorus uptake by Sorghum using a rhizocylinder technique. Commun Soil Sci Plant Anal 35:1061–1080. doi:10.1081/css-120030587

Ortas I, Akpinar C, Lal R (2013) Long-term impacts of organic and inorganic fertilizers on carbon sequestration in aggregates of an Entisol in mediterranean Turkey. Soil Sci 178:12–23. doi:10.1097/SS.0b013e3182838017

Palomo L, Claassenb N, Jones DL (2006) Differential mobilization of P in the maize rhizosphere by citric acid and potassium citrate. Soil Biol Biochem 38:683–692. doi:10.1016/j.soilbio.2005.06.019

Perez-Tienda J, Correa A, Azcon-Aguilar C, Ferrol N (2014) Transcriptional regulation of host NH4+ transporters and GS/GOGAT pathway in arbuscular mycorrhizal rice roots. Plant Physiol Biochem 75:1–8. doi:10.1016/j.plaphy.2013.11.029

Rajkumar M, Sandhya S, Prasad MNV, Freitas H (2012) Perspectives of plant-associated microbes in heavy metal phytoremediation. Biotechnol Adv 30:1562–1574. doi:10.1016/j.biotechadv.2012.04.011

Ramsay AJ, Bawden A (1983) Effects of sterilization and storage on respiration, nitrogen status and direct counts of soil bacteria using acridine orange. Soil Biol Biochem 15:263–268

Ratti N, Kumar S, Verma HN, Gautam SP (2001) Improvement in bioavailability of tricalcium phosphate to Cymbopogon martinii var. motia by rhizobacteria, AMF and Azospirillum inoculation. Microbiol Res 156:145–149. doi:10.1078/0944-5013-00095

Ravnskov S, Jensen B, Knudsen IMB, Bødker L, Funck Jensen D, Karliński L, Larsen J (2006) Soil inoculation with the biocontrol agent Clonostachys rosea and the mycorrhizal fungus Glomus intraradices results in mutual inhibition, plant growth promotion and alteration of soil microbial communities. Soil Biol Biochem 38:3453–3462. http://dx.doi.org/10.1016/j.soilbio.2006.06.003

Rineau F, Roth D, Shah F, Smits M, Johansson T, Canbäck B, Olsen PB, Persson P, Grell MN, Lindquist E (2012) The ectomycorrhizal fungus Paxillus involutus converts organic matter in plant litter using a trimmed brown-rot mechanism involving Fenton chemistry. Environ Microbiol 14:1477–1487

Rodriguez-Moran M, Navarro JM, Morte A (2015) Characterization of the Arum-type mycorrhiza in Citrus macrophylla wester rootstock under salt stress. In: SabaterMunoz B, Moreno P, Pena L, Navarro L (eds) Xii international citrus congress – International Society of Citriculture

Rovira A, Bowen G (1969) The use of radiation-sterilized soil to study the ammonium nutrition of wheat. Soil Res 7:57–65

Rubio G, Faggioli V, Scheiner JD, Gutierrez-Boem FH (2012) Rhizosphere phosphorus depletion by three crops differing in their phosphorus critical levels. J Plant Nutr Soil Sci 175. doi: 10.1002/jpln.201200307

Ruiz-Lozano JM, Azcon R (2000) Symbiotic efficiency and infectivity of an autochthonous arbuscular mycorrhizal Glomus sp from saline soils and Glomus deserticola under salinity. Mycorrhiza 10:137–143. doi:10.1007/s005720000075

Säle V, Aguilera P, Laczko E, Mäder P, Berner A, Zihlmann U, van der Heijden MG, Oehl F (2015) Impact of conservation tillage and organic farming on the diversity of arbuscular mycorrhizal fungi. Soil Biol Biochem 84:38–52

Salonius P, Robinson J, Chase F (1967) A comparison of autoclaved and gamma-irradiated soils as media for microbial colonization experiments. Plant Soil 27:239–248

Seok-Cho N, Kim D-H, Cho H-Y, Shin Y-S, Kim Y-C, Ohga S (2007) Identification of symbiotic arbuscular mycorrhizal fungi in Korean ginseng roots by 18S rDNA sequence. J Faculty Agric Kyushu Univ 52:265–274

Sharif M, Claassen N (2011) Action mechanisms of arbuscular mcorrhizal fungi in phosphorus uptake by Capsicum annuum L. Pedosphere 21:502–511

Sharma SD, Kumar P, Raj H, Bhardwaj SK (2009) Isolation of arbuscular mycorrhizal fungi and Azotobacter chroococcum from local litchi orchards and evaluation of their activity in the air-layers system. Sci Hortic 123:117–123. doi:10.1016/j.scienta.2009.07.019

Sieverding E (1991) Vesicular-arbuscular mycorrhiza management in tropical agro systems. Deutsche Gesellschaft für Technishe Zusammentarbeit (GTZ), Eschborn, Germany

Silva EP, Gomes VFF, Mendes Filho PF, Silva Júnior JMT, Ness RLL (2016) Development and mycorrhizal colonisation in embauba seedlings fertilised with natural phosphates and organic material. Rev Ciênc Agron 47:256–263

Singh B, Kanehiro Y (1970) Effects of gamma irradiation on the available nitrogen status of soils. J Sci Food Agric 21:61–64

Singh S, Mishra R, Singh A, Ghoshal N, Singh K (2009) Soil physicochemical properties in a grassland and agroecosystem receiving varying organic inputs. Soil Sci Soc Am J 73:1530–1538

Sivakumar N (2013) Effect of edaphic factors and seasonal variation on spore density and root colonization of arbuscular mycorrhizal fungi in sugarcane fields. Ann Microbiol 63:151–160. doi:10.1007/s13213-012-0455-2

Smith SS (1980) Mycorrhizas of autotrophic higher plants. Biol Rev 55:475–510

Smith SE, Gianinazzipearson V (1988) Physiological interactions between symbionts in vesicular-arbuscular mycorrhizal plants. Annu Rev Plant Physiol Plant Mol Biol 39:221–244. doi:10.1146/annurev.arplant.39.1.221

Smith SE, Read DJ (2008) Mycorrhizal symbiosis. Academic, San Diego

Smith FA, Smith SE (2011) What is the significance of the arbuscular mycorrhizal colonisation of many economically important crop plants? Plant Soil 348:63–79. doi:10.1007/s11104-011-0865-0

Speir T, Cowling J, Sparling G, West A, Corderoy D (1986) Effects of microwave radiation on the microbial biomass, phosphatase activity and levels of extractable N and P in a low fertility soil under pasture. Soil Biol Biochem 18:377–382

Stribley D, Read D, Hunt R (1975) The biology of mycorrhiza in the Ericaceae v. the effects of mycorrhizal infection, soil type and partial soil-sterilization (by gamma-irradiation) on growth of cranberry (Vaccinium macrocarpon ait.). New Phytol 75:119–130

Sutton JC, Sheppard BR (1976) Aggregation of sand-dune soil by endomycorrhizal fungi. Can J Bot 54:326–333

Tahovska K, Kana J, Barta J, Oulehle F, Richter A, Santruckova H (2013) Microbial N immobilization is of great importance in acidified mountain spruce forest soils. Soil Biol Biochem 59:58–71. doi:10.1016/j.soilbio.2012.12.015

Tanaka S, Kobayashi T, Iwasaki K, Yamane S, Maeda K, Sakurai K (2003) Properties and metabolic diversity of microbial communities in soils treated with steam sterilization compared with methyl bromide and chloropicrin fumigations. Soil Sci Plant Nutr 49:603–610

Taufiaul A, Habtem M (1985) Interaction of lacunae with glommus fasciculatumina typical oxisol. Lacunae Res Rep 6:89–97

Teste FP, Veneklaas EJ, Dixon KW, Lambers H (2014) Complementary plant nutrient-acquisition strategies promote growth of neighbour species. Funct Ecol 28:819–828. doi:10.1111/1365-2435.12270

Thompson J (1990) Soil sterilization methods to show VA-mycorrhizae aid P and Zn nutrition of wheat in vertisols. Soil Biol Biochem 22:229–240

Thougnon Islas AJ, Hernandez Guijarro K, Eyherabide M, Sainz Rozas HR, Echeverría HE, Covacevich F (2016) Can soil properties and agricultural land use affect arbuscular mycorrhizal fungal communities indigenous from the Argentinean Pampas soils? Appl Soil Ecol 101:47–56. http://dx.doi.org/10.1016/j.apsoil.2016.01.005

Tischer A, Werisch M, Doebbelin F, Camenzind T, Rillig MC, Potthast K, Hamer U (2015) Above- and belowground linkages of a nitrogen and phosphorus co-limited tropicalmountain pasture system-responses to nutrient enrichment. Plant Soil 391:333–352. doi:10.1007/s11104-015-2431-7

Tome E, Tagliavini M, Scandellari F (2015) Recently fixed carbon allocation in strawberry plants and concurrent inorganic nitrogen uptake through arbuscular mycorrhizal fungi. J Plant Physiol 179:83–89. doi:10.1016/j.jplph.2015.02.008

Tome E, Ventura M, Folegot S, Zanotelli D, Montagnani L, Mimmo T, Tonon G, Tagliavini M, Scandellari F (2016) Mycorrhizal contribution to soil respiration in an apple orchard. Appl Soil Ecol 101:165–173. doi:10.1016/j.apsoil.2016.01.016

Trappe JM (1987) Phylogenetic and ecologic aspects of mycotrophy in the angiosperms from an evolutionary standpoint. In: Safir GR (ed) Ecophysiology of VA mycorrhizal plants. CRC Press, Boca Raton

Turnbull MH, Goodall R, Stewart GR (1995) The impact of mycorrhizal colonization upon nitrogen source utilization and metabolism in seedlings of Eucalyptus grandis Hill ex Maiden and

Eucalyptus maculata Hook. Plant Cell Environ 18:1386–1394. doi:10.1111/j.1365-3040.1995.tb00199.x

Vaario L-M, Heinonsalo J, Spetz P, Pennanen T, Heinonen J, Tervahauta A, Fritze H (2012) The ectomycorrhizal fungus Tricholoma matsutake is a facultative saprotroph in vitro. Mycorrhiza 22:409–418

Valentine AJ, Osborne BA, Mitchell DT (2002) Form of inorganic nitrogen influences mycorrhizal colonisation and photosynthesis of cucumber. Sci Hortic 92:229–239. doi:10.1016/s0304-4238(01)00302-8

Valentinuzzi F, Mimmo T, Cesco S, Al Mamun S, Santner J, Hoefer C, Oburger E, Robinson B, Lehto N (2015) The effect of lime on the rhizosphere processes and elemental uptake of white lupin. Environ Exp Bot 118:85–94. doi:10.1016/j.envexpbot.2015.06.010

van Diepen LTA, Entwistle EM, Zak DR (2013) Chronic nitrogen deposition and the composition of active arbuscular mycorrhizal fungi. Appl Soil Ecol 72:62–68. doi:10.1016/j.apsoil.2013.05.012

Wallenda T, Schaeffer C, Einig W, Wingler A, Hampp R, Seith B, George E, Marschner H (1996) Effects of varied soil nitrogen supply on Norway spruce (Picea abies L Karst). 2. Carbon metabolism in needles and mycorrhizal roots. Plant Soil 186:361–369. doi:10.1007/bf02415531

Wang P, Wang Y, Shu B, Liu J-F, Xia R-X (2015) Relationships between arbuscular mycorrhizal symbiosis and soil fertility factors in Citrus orchards along an altitudinal gradient. Pedosphere 25:160–168

Watts-Williams SJ, Smith FA, McLaughlin MJ, Patti AF, Cavagnaro TR (2015) How important is the mycorrhizal pathway for plant Zn uptake? Plant Soil 390:157–166. doi:10.1007/s11104-014-2374-4

Willis A, Rodrigues BF, Harris PJC (2013) The ecology of arbuscular mycorrhizal fungi. Crit Rev Plant Sci 32:1–20. doi:10.1080/07352689.2012.683375

Wolfe BE, Tulloss RE, Pringle A (2012) The irreversible loss of a decomposition pathway marks the single origin of an ectomycorrhizal symbiosis. PLoS One 7, e39597

Wu QS, Zou YN (2009) Mycorrhizal influence on nutrient uptake of citrus exposed to drought stress. Philipp Agric Sci 92:33–38

Wulf A, Manthey K, Doll J, Perlick AM, Linke B, Bekel T, Meyer F, Franken P, Kuster H, Krajinski F (2003) Transcriptional changes in response to arbuscular mycorrhiza development in the model plant Medicago truncatula. Mol Plant-Microbe Interact 16:306–314. doi:10.1094/mpmi.2003.16.4.306

Xiao H, Griffiths B, Chen X, Liu M, Jiao J, Hu F, Li H (2010) Influence of bacterial-feeding nematodes on nitrification and the ammonia-oxidizing bacteria (AOB) community composition. Appl Soil Ecol 45:131–137

Yang H, Zhang Q, Dai Y, Liu Q, Tang J, Bian X, Chen X (2015) Effects of arbuscular mycorrhizal fungi on plant growth depend on root system: a meta-analysis. Plant Soil 389:361–374. doi:10.1007/s11104-014-2370-8

Zahra Z, Arshad M, Rafique R, Mahmood A, Habib A, Qazi IA, Khan SA (2015) Metallic nanoparticle (TiO2 and Fe3O4) application modifies rhizosphere phosphorus availability and uptake by Lactuca sativa. J Agric Food Chem 63:6876–6882. doi:10.1021/acs.jafc.5b01611

Zhang L, Fan J, Ding X, He X, Zhang F, Feng G (2014) Hyphosphere interactions between an arbuscular mycorrhizal fungus and a phosphate solubilizing bacterium promote phytate mineralization in soil. Soil Biol Biochem 74:177–183

Zhang T, Yang X, Guo R, Guo J (2016) Response of AM fungi spore population to elevated temperature and nitrogen addition and their influence on the plant community composition and productivity. Sci Rep 6. doi:10.1038/srep24749

ZhongQun H, ChaoXing H, ZhiBin Z, ZhiRong Z, HuaiSong W (2007) Changes of antioxidative enzymes and cell membrane osmosis in tomato colonized by arbuscular mycorrhizae under NaCl stress. Coll Surf B-Biointerf 59:128–133. doi:10.1016/j.colsurfb.2007.04.023

Zhou Y, Li X, Qin J, Liu H, Chen W, Niu Y, Ren A, Gao Y (2016) Effects of simultaneous infections of endophytic fungi and arbuscular mycorrhizal fungi on the growth of their shared host grass Achnatherum sibiricum under varying N and P supply. Fungal Ecol 20:56–65. doi:10.1016/j.funeco.2015.11.004

Zhu H-H, Yao Q, Sun X-T, Hu Y-L (2007) Colonization, ALP activity and plant growth promotion of native and exotic arbuscular mycorrhizal fungi at low pH. Soil Biol Biochem 39:942–950. doi:10.1016/j.soilbio.2006.11.006

Zocco D, Fontaine J, Lozanova E, Renard L, Bivort C, Durand R, Grandmougin-Ferjani A, Declerck S (2008) Effects of two sterol biosynthesis inhibitor fungicides (fenpropimorph and fenhexamid) on the development of an arbuscular mycorrhizal fungus. Mycol Res 112:592–601. doi:10.1016/j.mycres.2007.11.010

Zong K, Huang J, Nara K, Chen Y, Shen Z, Lian C (2015) Inoculation of ectomycorrhizal fungi contributes to the survival of tree seedlings in a copper mine tailing. J For Res 20:493–500. doi:10.1007/s10310-015-0506-1

Part II

Plant-Microbe Interaction Under Abiotic and Biotic Stress

7 Microbial-Mediated Amelioration of Plants Under Abiotic Stress: An Emphasis on Arid and Semiarid Climate

Amrita Kasotia, Ajit Varma, Narendra Tuteja, and Devendra Kumar Choudhary

Abstract

It is consensus that plant growth-promoting bacteria (PGPB) be studied extensively in the last two decades, but several of them are not fully investigated/ explored especially in arid and semiarid regions worldwide. They have been deployed as potent source of bioactive compounds useful in prospecting of sustainable agricultural. In the present scenario to meet food security, a number of different approaches have been employed to cultivate crops in salt- and drought-prone area. Hence, nowadays, the use of microbial inoculation to alleviate abiotic stress and amelioration of crops could be considered a more cost-effective eco-friendly approach. By keeping current approaches available for plant-microbe interaction, it is needed to pursue prospective research in this area. In the present chapter, authors will emphasize the role of benign PGPB in crop cultivation under stress through produced elicitors/determinants. It is very urgent need to explore this approach for sustainable agriculture grown under stress and

A. Kasotia
Department of Science, Faculty of Arts, Science & Commerce (FASC), Mody University of Science & Technology (MUST), Lakshmangarh, 332311 Sikar, Rajasthan, India

A. Varma • N. Tuteja
Amity Institute of Microbial Technology (AIMT),
Block 'E-3', 4th Floor, Amity University Campus, Sector-125, Gautam Buddha Nagar, 201313 Noida, Uttar Pradesh, India

D.K. Choudhary (✉)
Amity Institute of Microbial Technology (AIMT),
Block 'E-3', 4th Floor, Amity University Campus, Sector-125, Gautam Buddha Nagar, 201313 Noida, Uttar Pradesh, India
e-mail: dkchoudhary1@amity.edu

D.K. Choudhary et al. (eds.), *Plant-Microbe Interaction: An Approach to Sustainable Agriculture*, DOI 10.1007/978-981-10-2854-0_7

also to understand the mutual interactive activities belowground. Therefore, an exploitation of PGPB-plant interactions may be opted in the amelioration of plant health in arid and semiarid area.

7.1 Introduction

Among the total agricultural production around world, legumes encompass 25 % which include mainly pulses and oil seeds, namely, soybean and peanut. Based on the report generated by FAOSTAT (2012), India ranks first in worldwide pulse production and soybean ranks fourth. Abiotic and biotic stresses are major constraints for agriculture production worldwide. Therefore, an immediate and imperative integrated approach is required to avoid stresses and dissemination of the low-cost technologies in legume production (Reddy et al. 2013). In food web life does not exist without producers and, in natural resources wherein plants represent huge diversity in agroecosystems (AES), provides benign to detrimental metabolites. Among the benign are foods rich in proteins, feeds, and organic manures, and fix dinitrogen (N_2) improves soil structural characteristics and encourages beneficial microorganisms and the reclamation and revegetation of barren/degraded lands (Chaer et al. 2011). Based on these attributes, legumes are one of the most promising components of the Climate Smart Agriculture concept (FAO 2013). It finds major application as livestock forage and silage, grain, blooms, pharmaceutical/industrial, fallow/green manure, and human consumption as these are the good source of protein and rich in iron and vitamin B complex.

In India top legume producers constitute of Madhya Pradesh, Uttar Pradesh, Maharashtra, Rajasthan, Andhra Pradesh, Karnataka, Chhattisgarh, Gujarat, Jharkhand, and Bihar. Rajasthan ranks good enough in this list (Fig. 7.1a, b). The Thar Desert (Great Indian Desert) is a part of Rajasthan (constitutes 60 % of its area

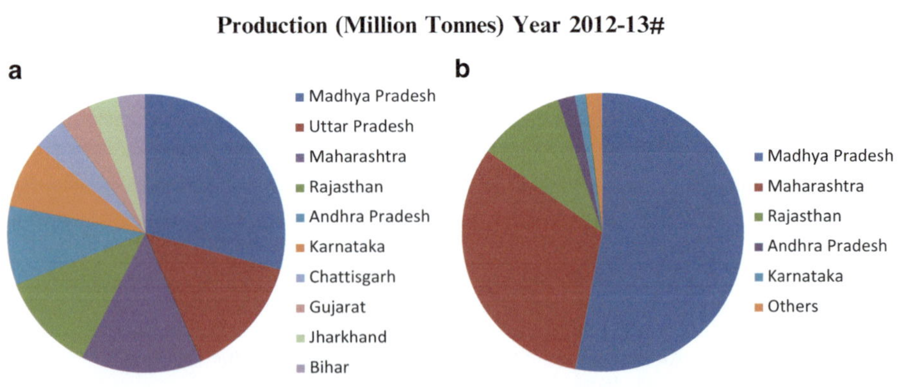

Fig. 7.1 (**a**) Top ten states in pulse production of India. (**b**) Major soybean-producing states of India. Fourth advance estimate (Source: DES, DAC, India)

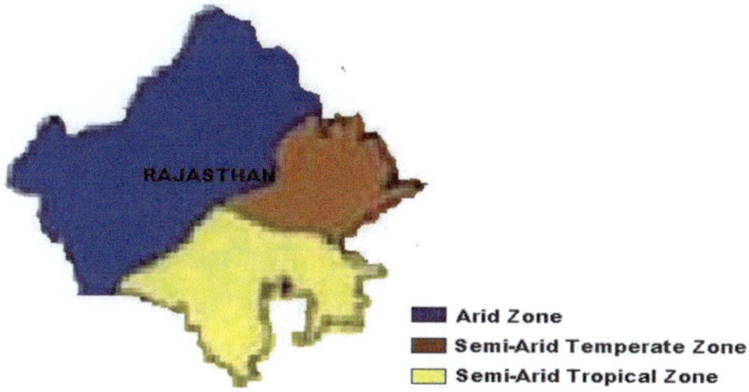

Fig 7.2 Different climatic zones of Rajasthan state (Source: http://www.nicra-icar.in/nicrarevised/index.php/component/content/article?layout=edit&id=195)

in Thar Desert). Geographically, Rajasthan lies between 23° 3′ to 30° 12′ longitude and 69° 30′ to 78° 17′ latitude. It occupies 342,239 km^2 land area which solely implies 10.41 % of the total land area of India. The Thar Desert lies between 24° to 28° N latitude and 68° to 71° E longitude, occupying an area of about 200,000 km^2. Its vegetation describes 911 wild species belonging to 780 genera and 154 families. Rajasthan consist of three climatic zones, namely, arid zone, semiarid temperate zone, and semiarid tropical zone (Fig. 7.2). Enduring flora of the Thar Desert (arid zone) involves tree and shrubs including cultivated leguminous plants, e.g., *Vigna* (*V. aconitifolia*, *V. mungo*, *V. radiata*, *V. unguiculata*, etc.), *Pisum sativum*, *Cicer arietinum*, *Trigonella foenumgraecum*, *Cajanus cajan*, *Cyamopsis tetragonoloba*, *Lens culinaris*, *Vicia faba*, *Phaseolus lunatus*, *Lablab purpureus*, *Canavalia ensiformis*, *Arachis hypogea*, etc. (http://dst.rajasthan.gov.in/), whereas semiarid zone is rich in *Glycine max*, *Arachis hypogaea*, *Cajanus cajan*, *Cicer arietinum*, *V. unguiculata*, etc. Legumes grown in these regions posses problem of abiotic stresses like salinity, alkalinity, high temperature, and drought, which lead to dehydration and osmotic stress in soil and thereby reduction in crop yields worldwide. Around 70 % of yield losses in major crops occur due to abiotic stress (da Silva et al. 2014).

The major limiting factors affecting the agricultural productivity worldwide are environmental stresses. Ecosystem of Rajasthan's Thar Desert is mainly affected with high temperature, salinity/alkalinity, low pH and several other abiotic factors. Apart from decreasing yield these introduces devastating impact on plant growth (Suzuki et al. 2014). High salinity and severe drought are the major constraints affecting the agricultural practices in Rajasthan. Out of this, soil degradation through salinization accounts the most wherein the main cause of salinization is irrigation. However, annual precipitation of rainfall (APRF) is poorly disseminated to make certain harvestable crops in arid and semiarid regions, resulting in gradual degradation (Singh et al. 2012). It has been reported that APRF affects approx. 50 % of irrigated areas worldwide and causes very stern threat to AES and leads to decline of natural resources (Gabrijel et al. 2009). In India, 8.4 Mha land is affected by soil salinity and alkalinity per se, of which about 5.5 Mha are waterlogged (Singh et al. 2012). And hence, over recent decades, soil salinization threatening environment health and sustainable development induced by human activities had developed sound land-use policies and planning actions for integrated land management to come in scenario (Zhang et al. 2011).

Soil salinization is considered as the occurrence of suspended inorganic ions that include Cl^-, SO_2^4, Mg^{++}, Ca^{++}, K^+, Na^+, HCO_3^-, and CO_3^{2-} in the aqueous phase of soil milieu. The change in soil salinity affects the survival of salt-sensitive plants so-called glycophytes, e.g., soybean. Soil with EC_e greater than 40 mM NaCl (4 dS/m) is considered saline (USDA Salinity Laboratory). Hence, increase in these limits leads to two major stresses for the plant osmotic and ionic stress. The occurrence of osmotic stress outside the plant root is the result of a rise in salt over threshold level which reflects hassles in H_2O uptake, cell growth, and expansion of lateral bud (Munns and Tester 2008). The ionic stress rose upon increase in toxic level of Na^+ that accumulates in leaf tissues over threshold level and causes leaf mortality with chlorosis/necrosis, whereby hindering cellular metabolic and enzyme activities (Chaves et al. 2009; Nawaz et al. 2013). To reduce salt-led phytotoxicity, the halophytes develop strategies to limit Na^+ uptake; further accumulation in shoot tissues is significant for survival (Zhang et al. 2008a, b).

According to crisis management plan (national) 2014, arid region of Rajasthan has shown drought efficiency of 2 in 5 years and semiarid region has 1 in 3 years. Drought has been considered as subtle peril of natural ecosystem and so-called creeping phenomenon and varied from one place to another. Land becomes dry when it gets light rain and sleet and leads to deep drought that cause noteworthy harm to the confined economy. Drought may also affect cropping system and threatens lasting erosion of AES enterprises (Kasotia and Choudhary 2014a, b). Water deficit caused by drought results in reduced turgor pressure of plant cells which thereby affects worth and measure of crop yield worldwide. It affects phenetic and genetic parameters of the plant and reflects reduction in cell division, enlargement, and differentiation including overall plant growth (Huang et al. 2012).

There is a cross talk between drought and salt stress as they eventually result in osmotic imbalance and lead to dehydration of the cell (Nakashima et al. 2014). This comprises three parameters: (1) restoration of ionic and osmotic equilibrium of the cell to develop homeostasis, (2) production of detoxification mechanisms to restore stress damage, and (3) induction of cell signaling to control cell division and metabolic pathways. Soil drying and salinization alter optimal supply of water, mineral nutrients, small organic molecules, proteins, and hormones in xylem (Pérez-Alfocea et al. 2011). Under stress condition plant cell implies signal transduction pathway that leads to production of secondary messengers, e.g., Ca^{+2}, ROS, and IMP. When plant possesses abiotic and biotic stresses, cytosolic level of calcium increases in the plant cell. Thereafter several simultaneous pathways are activated by calcium-interacting proteins (Kim et al. 2009). Mainly two stress responses are revealed by salinity stress, i.e., osmotic stress and ionic stress, whereas drought stress shows only osmotic stress (Huang et al. 2012). Osmotic stress produced by drought stress and salinity stress leads to ABA-dependant and ABA-independent signaling (Saibo et al. 2009), while ionic stress is alleviated by salt overly sensitive pathway (SOS pathway). Upon occurrence of salt stress, ion homeostasis of plant gets distressed that results in the rise of Na^+ and lack of K^+ in the cytoplasm. To mitigate such imbalance, ion transporters (plasma membrane Na^+/H^+ antiporter SOS1 and the high-affinity K^+ transporter 1 (HKT1)), located in the cell membrane, reflect exclusion of Na^+ entry into and exit out of cells and regulate Na^+/K^+ ration (Huang et al. 2012; Brini and Masmoudi 2012).

To alleviate such stressful conditions in plant, plant growth-promoting bacteria (PGPB) have been reported to implicate in the metabolism and growth of plants (Kang et al. 2014; Kasotia and Choudhary 2014b). In addition, PGPB that resides in the spermosphere (sphere that surround the seed) and rhizosphere (area around roots with 1–10 mm) enhance plant growth, after attaching to root surface. To alleviate abiotic stress, PGPB strains have been reported and include *Bacillus*, *Burkholderia*, *Acinetobacter*, *Enterobacter*, *Azospirillum*, *Beijerinckia*, *Rhizobium*, *Serratia*, *Erwinia*, *Flavobacterium*, *Alcaligenes*, etc. (Bharti et al. 2013). These microbes secrete bacterial AAC-deaminase, volatiles, antioxidants, cytokinin, IAA, and unknown metabolites in response to plant's ethylene, HKT1, ROS, and ABA under salt and drought stress (Yang et al. 2009). These microbial determinants result in "induced systemic tolerance (IST)" in plants, and further IST has been utilized to overcome the harmful effects of abiotic stress (Yang et al. 2009).

In higher plants, ethylene is produced under various abiotic stresses. It is a simple gaseous hydrocarbon that regulates many physiological processes, including root and shoot growth, seed germination, flower development, ripening of fruits, and senescence of plant organs. Under various abiotic stress conditions (salinity, chilling, drought, wounding, temperature, and heat), the level of C_2H_4 increases in plants (Li et al. 2013). Synthesis of ethylene mediated through L-methionine via the intermediates, S-adenosyl-L-methionine (SAM) and 1-aminocyclopropane-1-carboxylic acid (ACC) following Yang cycle (Yang and Hoffman 1984). The instant originator of C_2H_4 is ACC (Chen et al. 2013). It has been described that ACC-deaminase secreted by PGPB reduces the deleterious levels of ethylene. Bacteria use ACC as nitrogen source and degrade it to ammonia and α-ketobutyrate that are readily assimilated.[5] Salinity results in elevated levels of Na^+:K^+ which can be reduced by HKT1. HKT1 plays main physiological role in Na^+ homeostasis and thereby protects both mono- and dicotyledonous plants upon toxic level of Na^+ (Almeida et al. 2013). Among transporters, tt is one of Na^+ transporters that allows to transport Na^+ back to the soil by coupling to H^+ (Shi et al. 2002), transporters that avoid toxic level of Na^+ in the leaf tissues (Byrt et al. 2007), and antiporters that seize Na^+ in the vacuoles along with H^+-ATPase/H^+-PPase (Apse et al. 1999). Bacterial volatiles help the plant to regulate expression of HKT1 gene in maintaining low Na^+:K^+ ratio in plant (Zhang et al. 2008a, b).

Upon induction of salinity and drought, there is a rise in variety of ROS species which include radical (O_2^-, OH, HO_2, and RO) and non-radical forms (H_2O_2, and $^1O_2$) synthesized in plant cells (Gill and Tuteja 2010). To alleviate toxic level of ROS species, plant tissues per se contain several enzymatic (superoxide dismutase, catalase, glutathione reductase, peroxidase, etc.) and nonenzymatic (phenolic compounds, ascorbate, glutathione, carotenoid, and α-tocopherol) scavenging mechanisms (Jaleel et al. 2009; Gill and Tuteja 2010). The balance between the generation of ROSs and further the sequestration of antioxidants for ROSs gets disturbed under environmental stress conditions and leads to oxidative damage (Miller et al. 2010).

The induced activities resulted by PGPB detoxify plant cell by elevating antioxidant enzyme levels in plant cells (Kohler et al. 2008). The rise in ABA in plants showed a developmental process and allows an adaption to environmental stimuli in plants (Figueiredo et al. 2008; Fujita et al. 2011). Characteristically, it gets increased in roots, xylem sap, and shoots under osmotic stress (Albacete et al. 2008). For this cytokinin-producing bacteria are known to confer resistance (Nishiyama et al. 2011; Liu et al. 2013). Above all, PGPB secretes some more hormones such as IAA and GA that helps in the promotion of amplified root growth which leads to nutrient uptake in plants under stress (Kochar et al. 2011; Duca et al. 2014; Kang et al. 2014). They also act as signaling molecule in bacteria (Bashan and de-Bashan 2010). Nitrogen fixation via rhizobia-legume symbiosis is a well-known mechanism employed by PGPB to fix atmospheric nitrogen. PGPB convert atmospheric nitrogen to ammonia, a form that can be used up by plants (Franche et al. 2009). These bacteria contain enzyme complex nitrogenase that fixes atmospheric nitrogen to ammonia (Santi et al. 2013). Moreover PGPB influence soil fertility by solubilizing organic and precipitated phosphates in soil (Khan et al. 2009). PGPB excretes organic acids, namely, gluconic/citric acid, that dissolve calcium phosphates in the form of Pi and PO_4^{3-} (orthophosphate) and solubilize inorganic phosphate available largely in soil to bioavailable phosphorous. Besides, many phosphatase and cellulolytic enzymes are released for enzyme-labile soil organic phosphorous in favor of plant availability (Richardson and Simpson 2011).

To chelate iron in soil, PGPB also produce siderophore (Fe-III chelating agent) which can solubilize and sequester iron, whereby alleviating stress and allowing plant growth. Kintu et al. (2001) reported that microbially produced siderophores are of size <10,000 Da and showed the ability to chelate ferric ion as scavenging agent to fight against low iron stress (Kintu et al. 2001). Proteases secreted by PGPB break down complex proteins available in soil into plant-usable amino acids. They catalyze total hydrolysis of proteins to peptides and thereby function as degradative enzymes (Zhang et al. 2008a). In response to osmotic stress in soil, PGPB secretes compatible solutes which help them to adapt in external osmolarity (Paul and Nair 2008). Compatible solutes are low molecular weight hydrophilic molecular osmolytes including carbohydrates, amino acids, and their modified forms (Wood 2011). PGPB colonizes plant roots and alleviates the debilitating effects of salt stress (Paul 2013). Production of microbial EPS in soil under stress helps in removal of drought stress and whereby develops water retention capacity of soil (Sandhya et al. 2009). It is reported that EPS also binds to positively charged ions including Na^+ and therefore reduces the toxic level of Na^+ in soil and ameliorates plant growth (Nunkaew et al. 2014).

PGPB have been proven to be best eco-friendly remedy to accelerate the growth of plant in nutrient-deficient soil with respect to chemical fertilizers which are least available to plant. PGPB solubilizes nutrient and makes them available for uptake by plants (Choudhary 2011). There are some legumes like mung bean and soybean which are incapable of growing in drought and salt stresses as they may be devoid of mechanisms to survive in stressed conditions or due to unavailability of nutrient or increased secretion of ethylene hormone or decreased secretion of plant growth-promoting hormones.

Acknowledgments In the present review, some of the research has been partially supported by DBT and SERB grant no. BT/PR1231/AGR/021/340/2011 and SR/FT/LS-129/2012, respectively, to DKC. Authors would also like to acknowledge UGC-RGNF fellowship.

References

Albacete A, Ghanem ME, Martínez-Andújar C, Acosta M, Sánchez-Bravo J, Martínez V, Lutts S, Dodd IC, Pérez-Alfocea F (2008) Hormonal changes in relation to biomass partitioning and shoot growth impairment in salinized tomato (*Solanum lycopersicum* L.) plants. J Exp Bot 59:4119–4131

Almeida P, Katschnig D, de Boer AH (2013) HKT transporters—state of the art. Int J Mol Sci 14:20359–20385

Apse MP, Aharon GS, Snedden WA, Blumwald E (1999) Salt tolerance conferred by overexpression of a vacuolar Na^+/H^+ antiport in Arabidopsis. Science 285:1256–1258

Bashan Y, de-Bashan LE (2010) How the plant growth-promoting bacterium *Azospirillum* promotes plant growth-A critical assessment. In: Sparks DL (ed) Advances in agronomy, 108, Elsevier, Academic Press. Adv Agro 108:77–136

Bharti N, Yadav D, Barnawal D, Maji D, Kalra A (2013) Exiguobacterium oxidotolerans, a halotolerant plant growth promoting rhizobacteria, improves yield and content of secondary metabolites in *Bacopa monnieri* (L.) Pennell under primary and secondary salt stress. World J Microbiol Biotechnol 29:379–387

Brini F, Masmoudi K (2012) Ion transporters and abiotic stress tolerance in plants. ISRN Mol Biol. doi:10.5402/2012/927436

Byrt CS, Platten JD, Spielmeyer W, James RA, Lagudah ES, Dennis ES, Tester M, Munns R (2007) HKT1; 5-like cation transporters linked to Na^+ exclusion loci in wheat, Nax2 and Kna1. Plant Physiol 143:1918–1928

Chaer GM, Resende AS, de Balieiro FC, Boddey RM (2011) Nitrogen-fixing legume tree species for the reclamation of severely degraded lands in Brazil. Tree Physiol 31:139–149

Chaves MM, Flexas J, Pinheiro C (2009) Photosynthesis under drought and salt stress: regulation mechanisms from whole plant to cell. Ann Bot 103:551–560

Chen L, Dodd IC, Theobald JC, Belimov AA, Davies WJ (2013) The rhizobacterium *Variovorax paradoxus* 5C-2, containing ACC deaminase, promotes growth and development of Arabidopsis thaliana via an ethylene-dependent pathway. J Exp Bot. doi:10.1093/jxb/ert031

Choudhary DK (2011) Plant growth-promotion (PGP) activities and molecular characterization of rhizobacterial strains isolated from soybean (*Glycine max* L. Merril) plants against charcoal rot pathogen, *Macrophomina phaseolina*. Biotechnol Lett 33:2287–2295

da Silva GJ, Costa de Oliveira A (2014) Genes acting on transcriptional control during abiotic stress responses. Adv Agric. doi:10.1155/2014/587070

Duca D, Lorv J, Patten CL, Rose D, Glick BR (2014) Indole-3-acetic acid in plant-microbe interactions. Anton Leeuw 106:85–125

FAO (2013) Climate-smart agriculture sourcebook. FAO, Rome

Figueiredo MV, Burity HA, Martínez CR, Chanway CP (2008) Alleviation of drought stress in the common bean (*Phaseolus vulgaris* L.) by co-inoculation with *Paenibacillus polymyxa* and *Rhizobium tropici*. Appl Soil Ecol 40:182–188

Franche C, Lindström K, Elmerich C (2009) Nitrogen-fixing bacteria associated with leguminous and non-leguminous plants. Plant Soil 321:35–59

Fujita Y, Fujita M, Shinozaki K, Yamaguchi-Shinozaki K (2011) ABA-mediated transcriptional regulation in response to osmotic stress in plants. J Plant Res 124:509–525
Gabrijel O, Davor R, Zed R, Marija R, Monika Z (2009) Cadmium accumulation by muskmelon under salt stress in contaminated organic soil. Sci Total Environ 407:2175–2182
Gill SS, Tuteja N (2010) Reactive oxygen species and antioxidant machinery in abiotic stress tolerance in crop plants. Plant Physiol Biochem 48:909–930
Glick BR (2012) Plant growth-promoting bacteria: mechanisms and applications. Scientifica. doi:10.6064/2012/963401
Huang GT, Ma SL, Bai LP, Zhang L, Ma H, Jia P, Liu J, Zhong M, Guo ZF (2012) Signal transduction during cold, salt, and drought stresses in plants. Mol Biol Rep 39:969–987
Jaleel CA, Riadh K, Gopi R, Manivannan P, Inès J, Al-Juburi HJ, Chang-Xing Z, Hong-Bo S, Panneerselvam R (2009) Antioxidant defense responses: physiological plasticity in higher plants under abiotic constraints. Acta Physiol Plant 31:427–436
Kang SM, Khan AL, Waqas M, You YH, Kim JH, Kim JG, Hamayun M, Lee IJ (2014) Plant growth-promoting rhizobacteria reduce adverse effects of salinity and osmotic stress by regulating phytohormones and antioxidants in *Cucumis sativus*. J Plant Interact 9:673–682
Kasotia A, Choudhary DK (2014a) Induced inorganic phosphate solubilization through N-Methyl-N′-Nitro-N-Nitrosoguanidine treated mutants of *Pseudomonas koreensis* strain AK-1 (MTCC Number 12058) under polyethylene glycol. Proc Natl Acad Sci, India, Sect B Biol Sci 86:115–123
Kasotia A, Choudhary DK (2014b) Pseudomonas-mediated mitigation of salt stress and growth promotion in *Glycine max* L. Merrill Agric Res. doi:10.1007/s40003-014-0139-1
Khan AA, Jilani G, Akhtar MS, Naqvi SS, Rasheed M (2009) Phosphorus solubilizing bacteria: occurrence, mechanisms and their role in crop production. J Agric Biol Sci 1:48–58
Kim MC, Chung WS, Yun D-J, Cho MJ (2009) Calcium and calmodulin-mediated regulation of gene expression in plants. Mol Plant 2:13–21
Kintu K, Dave BP, Dube HC (2001) Detection and chemical characterization of siderophores produced by certain fungi. Indian J Microbiol 41:87–91
Kochar M, Upadhyay A, Srivastava S (2011) Indole-3-acetic acid biosynthesis in the biocontrol strain *Pseudomonas fluorescens* Psd and plant growth regulation by hormone overexpression. Res Microbiol 162:426–435
Kohler J, Hernandez JA, Caravaca F, Roldàn A (2008) Plant-growth-promoting rhizobacteria and arbuscular mycorrhizal fungi modify alleviation biochemical mechanisms in water-stressed plants. Funct Plant Biol 35:141–151
Li B, Sang T, He L, Sun J, Li J, Guo S (2013) Exogenous spermidine inhibits ethylene production in leaves of cucumber seedlings under NaCl stress. J Am Soc Hortic Sci 138:108–113
Liu F, Xing S, Ma H, Du Z, Ma B (2013) Cytokinin-producing, plant growth-promoting rhizobacteria that confer resistance to drought stress in *Platycladus orientalis* container seedlings. Appl Microbiol Biotechnol 97:9155–9164
López-Otín C, Overall CM (2002) Protease degradomics: a new challenge for proteomics. Nat Rev Mol Cell Biol 3:509–519
Miller GAD, Susuki N, Ciftci-Yilmaz S, Mittler RON (2010) Reactive oxygen species homeostasis and signalling during drought and salinity stresses. Plant Cell Environ 33:453–467
Munns R, Tester M (2008) Mechanisms of salinity tolerance. Annu Rev Plant Biol 59:651–681
Nakashima K, Yamaguchi-Shinozaki K, Shinozaki K (2014) The transcriptional regulatory network in the drought response and its crosstalk in abiotic stress responses including drought, cold, and heat. Front Plant Sci. doi:10.3389/fpls.2014.00170
Nawaz K, Hussain K, Majeed A, Khan F, Afghan S, Ali K (2013) Fatality of salt stress to plants: morphological, physiological and biochemical aspects. Afr J Biotechnol 9:5475–5480
Nishiyama R, Watanabe Y, Fujita Y, Le DT, Kojima M, Werner T, Vankovad R, Yamaguchi-Shinozakib K, Shinozakia K, Kakimoto T, Sakakibara H, Schmülling T, Tran LSP (2011) Analysis of cytokinin mutants and regulation of cytokinin metabolic genes reveals important

regulatory roles of cytokinins in drought, salt and abscisic acid responses, and abscisic acid biosynthesis. Plant Cell 23:2169–2183

Nunkaew T, Kantachote D, Nitoda T, Kanzaki H, Ritchie RJ (2014) Characterization of exopolymeric substances from selected *Rhodopseudomonas palustris* strains and their ability to adsorb sodium ions. Carbohydr Polym 115:334–341

Paul D (2013) Osmotic stress adaptations in rhizobacteria. J Basic Microbiol 53:101–110

Paul D, Nair S (2008) Stress adaptations in a plant growth promoting rhizobacterium (PGPR) with increasing salinity in the coastal agricultural soils. J Basic Microbiol 48:378–384

Pérez-Alfocea F, Ghanem ME, Gómez-Cadenas A, Dodd IC (2011) Omics of root-to-shoot signaling under salt stress and water deficit. Omics J Integr Biol 15:893–901

Reddy AA, Bantilan MCS, Mohan G (2013) Pulses production scenario: policy and technological options. Policy Brief 26, International Crop Research Institute for the Semi-Arid Tropics

Richardson AE, Simpson RJ (2011) Soil microorganisms mediating phosphorus availability update on microbial phosphorus. Plant Physiol 156:989–996

Saibo NJM, Lourenço T, Oliveira MM (2009) Transcription factors and regulation of photosynthetic and related metabolism under environmental stresses. Ann Bot 103:609–623

Sandhya V, Ali SZ, Grover M, Reddy G, Venkateswarlu B (2009) Alleviation of drought stress effects in sunflower seedlings by exopolysaccharides producing *Pseudomonas putida* strain P45. Biol Fertil Soils 46:17–26

Santi C, Bogusz D, Franche C (2013) Biological nitrogen fixation in non-legume plants. Ann Bot 111:743–767

Shi H, Lee BH, Wu SJ, Zhu JK (2002) Overexpression of a plasma membrane Na^+/H^+ antiporter gene improves salt tolerance in *Arabidopsis thaliana*. Nat Biotechnol 21:81–85

Singh A, Nath Panda S, Flugel WA, Krause P (2012) Waterlogging and farmland salinisation: causes and remedial measures in an irrigated semi-arid region of India. Irrig Drain 61:357–365

Suzuki N, Rivero RM, Shulaev V, Blumwald E, Mittler R (2014) Abiotic and biotic stress combinations. New Phytol 203:32–43

Wood JM (2011) Bacterial osmoregulation: a paradigm for the study of cellular homeostasis. Annu Rev Microbiol 65:215–238

Yang SF, Hoffman NE (1984) Ethylene biosynthesis and its regulation in higher plants. Annu Rev Plant Physiol Plant Mol Biol 35:155–189

Yang J, Kloepper JW, Ryu C-M (2009) Rhizosphere bacteria help plants tolerate abiotic stress. Trends Plant Sci 14:1–4

Zhang H, Kim MS, Sun Y, Dowd SE, Shi H, Paré PW (2008a) Soil bacteria confer plant salt tolerance by tissue-specific regulation of the sodium transporter HKT1. Mol Plant-Microbe Interact 21:737–744

Zhang H, Xie X, Kim MS, Kornyeyev DA, Holaday S, Pare PW (2008b) Soil bacteria augment Arabidopsis photosynthesis by decreasing glucose sensing and abscisic acid levels in planta. Plant J 65:264–273

Zhang T, Zeng S, Gao Y, Ouyang Z, Li B (2011) Assessing impact of land uses on land salinization in the Yellow River Delta, China using an integrated and spatial statistical model. Land Use Policy 28:857–866

Bacterial ACC-deaminase: An Eco-friendly Strategy to Cope Abiotic Stresses for Sustainable Agriculture

8

Sarita Kumari, Ajit Varma, Narendra Tuteja, and Devendra Kumar Choudhary

Abstract

Ethylene is the simplest unsaturated two-carbon gaseous plant hormone which regulates many physiological and developmental processes during plant growth at molecular level. High ethylene concentration, produced by virtually all higher plants under abiotic stresses such as drought, salinity, etc., acts as stress hormone which detrimentally affects the plant root growth, seed germination, and the whole plant growth. Under abiotic stresses like salinity and drought, the endogenous level of ethylene is enhanced substantially due to increased production of its immediate biochemical precursor, 1-aminocyclopropane-1-carboxylic acid (ACC), and it shows major contribution in stress ethylene production under such type of stress conditions. Plant growth-promoting bacteria (PGPB) around the plant root surface or in the roots having a potent enzyme ACC-deaminase maintain ethylene level in plants under adverse environmental conditions (severe drought and high salinity) by the enzymatic degradation of ACC into α-ketobutyrate and ammonia as a carbon and nitrogen source. The use of PGPB containing ACC-deaminase as a bio-inoculant is a most powerful technique in agricultural biotechnology for sustainable crop production in terms of decreasing the detrimental effect of high ethylene concentration and improving growth and development of plants under extreme environmental conditions. In this chapter we endeavor to explore current research on maintaining the physiological and molecular changes in the plants under diverse environmental conditions (drought and high salinity) by the use of PGPB having ACC-deaminase, mode of ACC-deaminase enzyme action, and severe effects of salinity and drought on growth of plant special due to ethylene evolution.

S. Kumari • A. Varma • N. Tuteja • D.K. Choudhary (✉)
Amity Institute of Microbial Technology (AIMT),
Block 'E-3', 4th Floor, Amity University Campus, Sector-125,
Gautam Buddha Nagar, 201303 Noida, Uttar Pradesh, India
e-mail: dkchoudhary1@amity.edu

D.K. Choudhary et al. (eds.), *Plant-Microbe Interaction: An Approach to Sustainable Agriculture*, DOI 10.1007/978-981-10-2854-0_8

8.1 Environmental Challenges

Several physiological and environmental factors that include availability of essential nutrients, physical and chemical nature of surrounding soil, and abiotic stresses (i.e., temperature, drought, salinity, flooding, etc.) reduce the plant growth. Apart from these physiological stresses, abiotic stresses negatively affect the productivity of agricultural crops mainly legume plants (*Glycine max* L., *Vigna radiata* L., *Cicer arietinum* L., etc.), and the population of beneficial microorganisms in rhizospheric soil is also adversely affected, which is a serious problem for the whole world. As an effect of increases in environmental destruction and population growth, the sufficient crop production to provide essential food for the world's people is a crucial challenge. To feed these entire populations, it is very necessary to improve the production of agricultural crops within the next few years. However, it is not an easy task to supply sufficient food to the growing worldwide population of people by using existing techniques; it will require some advanced and eco-friendly strategies and approaches. Many approaches have been applied such as the use of different chemicals including herbicides, insecticides, etc., and genetic engineering to solve this problem, but still these are not viable and will be effective only for a short period. To mitigate the deleterious effect of different abiotic stresses mainly drought and salinity, it is necessary to reevaluate these approaches to enhance the agricultural productivity and also improve soil fertility; that includes the interaction of plant roots with drought and salt-tolerant beneficial bacteria, which is an alternative eco-friendly and cost-effective approach to address this problem (Glick et al. 2013). When crops are exposed to environmental stress conditions, these beneficial soil microbes enhance the growth and yield of plants through direct and indirect mechanisms including inductions of expression of stress-responsive genes; accumulation of osmoprotectors such as proline, glycinebetaine, total soluble sugars, heat shock proteins, antioxidant enzymes, etc.; and suppression of disease caused by pathogenic microorganisms such as bacteria, fungi, viruses, and nematodes (Porcel and Ruiz-Lozano 2004; Nautiyal et al. 2008; Dimkpa et al. 2009; Mutava et al. 2015). In this present chapter, we focus on induced systemic tolerance (IST) in legume crops in response to drought and salt stress by the enzymatic mechanism of bacterial 1-aminocyclopropane-1-carboxylic acid (ACC)-deaminase, which cleaves ACC, the direct precursor of plant ethylene hormone into α-ketobutyrate and ammonia, and thereby lowers the accelerated level of ethylene in the plant tissue.

8.2 Ethylene Plant Stress Hormone

The detection of ethylene was recognized as a plant phytohormone which regulates the whole plant growth and development. The plant hormone ethylene which is found in the higher plants is an important regulator for several phases of plant growth and developmental process; this is also important for its role in plant responses under different abiotic and biotic stresses (Abeles and others 1992). Ethylene, which plays multiple roles in the regulation of plant biological processes, such as growth of plant roots, shoots, leaves, fruits, and flowers as well as plant development (Bleecker and Kende 2000; Binder 2008). The production of ethylene is highly regulated by the developmental process in plants, and normally it is required at low level (10–25 μg/L) for normal growth and functions of the plants. Ethylene is an inhibitor for plant growth but at very low concentration; it may promote plant growth in a wide range such as promoting lateral root formation in many plant species including *Arabidopsis* (Pierik et al. 2006). In the presence of a wide range of environmental stresses, like salinity (Mayak et al. 2004a; Nadeem et al. 2009), drought (Mayak et al. 2004b), temperature stress (Ghosh et al. 2003), metal stress (Belimov et al. 2009), etc., ethylene production may increase; this stress level of ethylene inhibits the root and shoot elongation, inhibiting root nodule formation, decreasing plant-microbe interaction, and inhibiting seed germination (Abeles et al. 1992). Hirsch and Fang (1994) have reported that in leguminous plants, more ethylene production adversely affects the nodule formation. At the molecular level, stress ethylene induces the gene expression of those responsible for fruit ripening (Lincoln and Fischer 1988). Many researchers experimentally provide evidence that the induced level of ethylene has been reduced by using ACC-deaminase-containing bacteria (Glick 2004; Mayak et al. 2004a, b; Cheng et al. 2007; Siddikee et al. 2011).

8.2.1 Ethylene Biosynthesis

Drought and salinity stress cause an imbalance in the production of endogenous ethylene and increased level of ethylene in the higher plants which is responsible for growth inhibition. The major discovery that made the ethylene biosynthesis through Yang cycle was the production of 1-aminocyclopropane-1-carboxylicacid (ACC) as the intermediate product by the conversion of methionine to S-adenosyl-L-methionine (SAM) and finally ethylene production in the higher plants (Adams and Yang 1979). In plants ACC acts as a precursor of plant ethylene biosynthesis pathway, which affects the eventual level of the plant hormone ethylene (Yang and Hoffman 1984). Ethylene biosynthetic pathway involved three enzymatic steps in higher plants (Fig. 8.1): (1) the conversion of methionine (Met) to SAM by SAM synthetase, (2) the conversion of SAM into ACC by ACC synthase (ACS), and (3) then the conversion of ACC into ethylene by ACC oxidase (ACO). For Met recycling, SAM is converted into 5-methylthioadenosine (MTA), and finally MTA is converted into 2-keto-4-methylthiobutyrate (KMTB), the immediate precursor of Met through many enzymatic steps of Yang cycle (Sauter et al. 2013; Li 1999). Adams and Yang (1979) reported that the discovery of SAM as an intermediate between methionine and ethylene was a tremendous progress in understanding the biosynthesis pathway of ethylene in the plant tissues. Murr and Yang (1975) also reported that the ethylene biosynthesis was started with the MTA, an intermediate product in the reaction of ACC production from SAM, and that MTA could be recycled back into methionine and maintained the methionine level in the plants. The steps of different enzymatic reactions in methionine cycle of plants, referred to

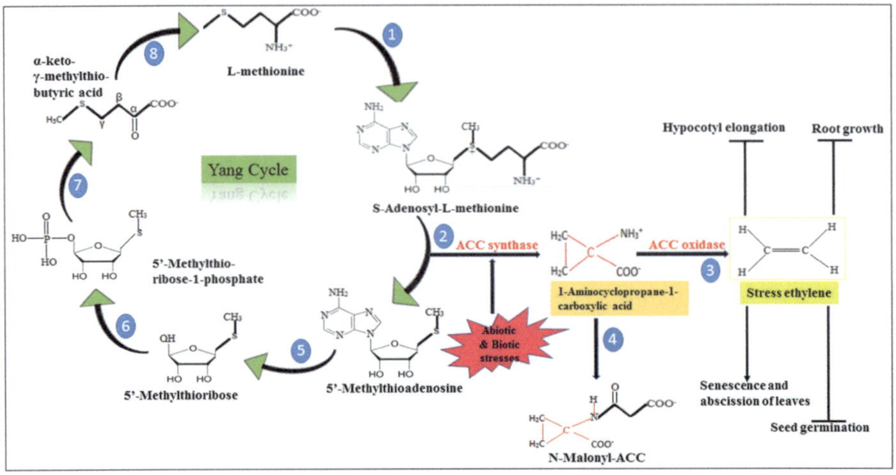

Fig. 8.1 A schematic presentation of ethylene biosynthetic pathway through L-methionine (an immediate precursor of ethylene synthesis) in plants and recycling of L-methionine (Source: Li 1999 with some changes)

as the Yang cycle, mainly showed the similarities between the plant pathway and the methionine salvage pathway in the prokaryotes, yeast, and mammalians (Fig. 8.1). Sauter et al. (2013) proposed a complete overview of the methionine and SAM metabolism in the biosynthesis of plant hormone ethylene in plants. Recently, Van de and Van der (2014) also documented that the ACC is an intermediate precursor between SAM and ethylene synthesis pathway.

In the whole process of ethylene biosynthesis, ACC is an immediate precursor of ethylene synthesis in higher plants. It is postulated that the higher plants release the large portion of ACC from roots in the surrounding environment, which is then taken up by PGPB, and ACC is hydrolyzed into α-ketobutyrate and ammonia by the help of bacterial ACC-deaminase enzyme (Glick et al. 1998; Penrose and Glick 2001).

8.3 Plant Growth-Promoting Bacteria

Soil is the largest favorable ecological niche for the microbes and their metabolic activities. Root zone area of plants contains huge microbial population and high metabolic activities. Instead of the high microbial population and metabolic activities in the rhizospheric area, these microbes occupy only 5 % area of the total space (Chakraborty et al. 2015). However, microbial activity or population is not uniform throughout the soil, but is highly concentrated in the region of the root surface area, known as the rhizosphere (Pinton et al. 2001; Chakraborty et al. 2015). Increased populations of microbes colonize the root zone of plants. Microbes present around the plant roots in higher concentration are the main reason of dense microbial population in the rhizosphere, making it a favorable habitat than bulk soil due to the presence of the higher level of nutrient availability including amino acid, sugars, organic acids, and flavonoids which are excreted from the roots of plants and are then used by the microbes in the soil (Dimkpa et al. 2009; Ashraf et al. 2013). More than 85 % of the total organic carbon in the rhizospheric soil can originate by excretion of root cells and tissues in the form of root exudates (Barber & Martin 1976). Gray and Smith (2005) also reported the soil conditions and the composition of different root exudates which play important roles in the specificity of plant-microbe interactions and microbial activities. Many literatures are repleted with reports describing the plant growth promotion under abiotic stresses in the occurrence of rhizospheric bacteria. Several approaches have been adopted in order to minimize the adverse effect of abiotic stresses including genetically modified crop, but the use of diverse species of rhizospheric microorganisms containing ACC-deaminase belonging to various taxonomic groups, including *Pseudomonas*, *Azospirillum*, *Azotobacter*, *Bacillus*, *Klebsiella*, *Enterobacter*, *Xanthomonas*, and *Serratia*, degrades the ACC into α-ketobutyrate and ammonia for use as carbon and nitrogen sources, which is shown to promote plant growth and is a well-known and sustainable approach for enhancing plant tolerance to abiotic stresses including drought and salinity (Duan et al. 2009; Yang et al. 2009; Egamberdieva 2009; Tilak et al. 2005). These microorganisms are generally termed as plant growth-promoting bacteria (PGPB) attached to plant tissue which provides a sink for ACC from the plant tissue and, thereby, reduces ethylene synthesis, promotes root clongation, and reduces the adverse effects of stress. ACC-deaminase-containing bacteria significantly decrease a lot of physiological damage in plants due to exposure to adverse environmental stresses including extreme high and low temperature, high soil salinity, water stresses (flooding and drought), metal stresses, and organic contaminants. PGPB are involved in the decomposition of organic matter as well as solubilization of nutrients and ions in the surrounding soil, which become easily available to the plant roots. PGPB, which live in association with the plant roots, elicit the largest influence on the plants, increasing their productivity and immune response and reducing disease caused by pathogens such as fungi, bacteria, and viruses (Kloepper et al. 2004). The exact mechanisms rf plant growth stimulation either of direct or indirect mechanism applying PGPB remain largely tentative, because of the difference between bacterial strains and most certainly its dependence on the various compounds such as auxin, ACC-deaminase, siderophore, osmolytes, and many others released by the different microorganisms. PGPB accelerate plant growth through either of direct or indirect mechanisms (Fig. 8.2).

The direct mechanism of plant growth promotion (PGP) by PGPB includes production of metabolites such as proline, betaine, total soluble sugars, etc. and volatile production, modulating plant hormone level such as IAA (indole-3-acetic acid), cytokinin, ABA (abscisic acid), salicylic acid (SA), and ethylene (Glick 2012; Vaishnav et al. 2015; Mutava et al. 2015) and facilitating the acquisition of nutrient

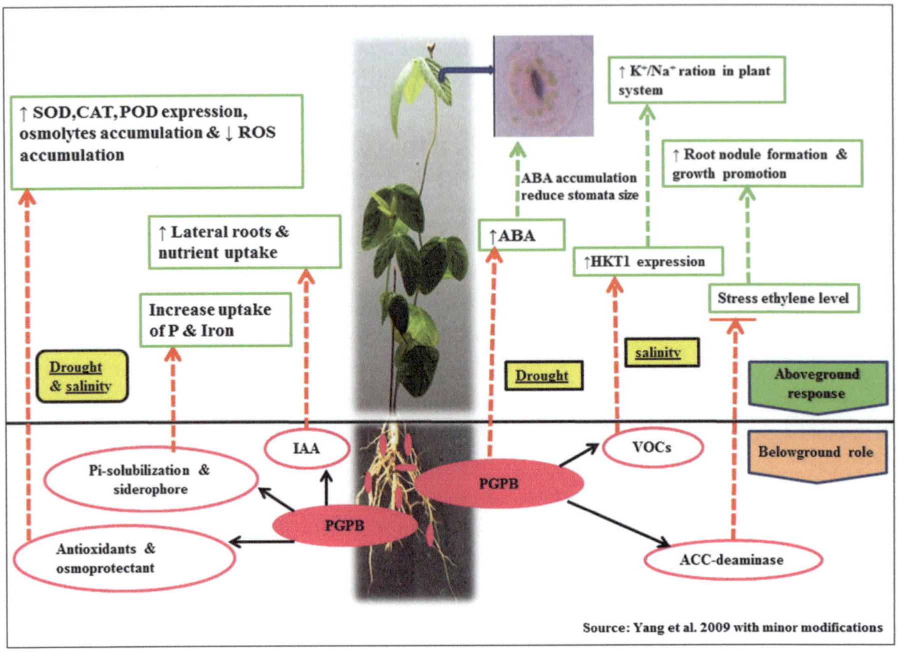

Fig. 8.2 PGPB-induced systemic tolerance against drought and salt stresses belowground (root) and aboveground (shoot and leaves). Solid black arrows indicate PGP traits; broken arrows indicate effects of PGPB on plants. PGPB strains, indicated in pin on the plant roots, having ACC-deaminase activity, suppressed the stress ethylene level by degradation of the ethylene precursor ACC and rescue normal plant growth under drought and salt stresses. Here, abscisic acid (ABA), 1-aminocyclopropane-1-carboxylate (ACC), high-affinity K^+ transporter 1 (HKT1), indole acetic acid (IAA), volatile organic compounds (VOCs), plant growth-promoting bacteria (PGPB), reactive oxygen species (ROS)

uptake from the surrounding environment including nitrogen, iron, auxin, and phosphate through nitrogen fixation, siderophore production, IAA synthesis, ACC-deaminase activity, and phosphate solubilization, respectively, through roots under environmental stress conditions (Yang et al. 2009; Kumari et al. 2015). In contrast, indirect mechanism of PGP by PGPR includes induced systemic resistance, antibiotic protection against pathogens, reduction of iron availability by sequestration of nutrients with siderophores, and synthesis of antifungal enzymes such as chitinase enzyme (Lucy et al. 2004; Dobbelaere and Okon 2007; Jain and Choudhary 2014). Hontzeas et al. (2004) have reported the ACC-deaminase-containing PGPB upregulate genes involved in plant growth and defense protein production while down-regulating plant genes involved in ethylene synthesis pathway. PGPB strains containing this enzyme may have a competitive advantage over other microorganisms in the rhizosphere because they can degrade the ACC into α-ketobutyrate and ammonia as carbon and nitrogen source and they help plants to overcome many of the harmful effects of abiotic stresses (Glick et al. 2007).

8.4 Bacterial Enzyme ACC-Deaminase

ACC-deaminase has been found only in microorganisms, they reduce the stress ethylene level in the plants by the use of ACC, and there are no microorganisms that produce ethylene through ACC. Glick (2012) described a scheme of the mechanism of action of ACC-deaminase to reduce the stress ethylene level by catalytic conversion of ethylene precursor ACC that includes cyclopropane ring fragmentation and deamination of ACC to form α-ketobutyrate and ammonia.

8.4.1 Biochemistry

The bacterial ACC-deaminase is a pyridoxal 5-phosphate (PLP)-dependent polymeric enzyme with a subunit molecular mass of approximately 35–42 kDa and was first studied in a soil bacterium *Pseudomonas* sp. strain ACP that degrades the ACC (Honma and shimomura 1978), and subsequently it was purified from *P. chlororaphis* 6G5, *P. putida* GR12-2, and *P. putida* UW4 (Klee et al. 1991; Jacobson et al. 1994; Hontzeas et al. 2004). Apart from above all these, the biochemical and physical characteristics of ACC-deaminase have been reported by numerous other researchers in different PGPB (Honma 1985; Hontzeas et al. 2004; Jia et al. 1999; Ose et al. 2003; Minami et al. 1998). The various biochemical studies of ACC-deaminase indicated that the substrate ACC is found largely within plant tissues; the enzyme ACC-deaminase is not secreted by bacteria but is typically found within the cytoplasm that did not have a particularly high affinity for ACC, ranging from 1.5 to 15 mM approximately (Glick et al. 2007). In this case the substrate ACC is exuded by the plants through roots and is then taken up by the PGPB containing ACC-deaminase. Glick (2005) reported the large differences in the level of ACC-deaminase activity in the different types of bacteria. ACC-deaminase activity is assayed by monitoring the concentration of α-ketobutyrate, the product of ACC hydrolysis. Organisms with high expression of ACC-deaminase activity typically bind nonspecifically to the variety of plant species, and those organisms express low level of ACC-deaminase activity and bind only to specific plant species, or they do not lower the overall stress ethylene level in plants (Glick 2005).

8.4.2 Prevalence of ACC-Deaminase Genes and Its Regulation

The ACC-deaminase activity has been reported in all three domains (Bacteria, Eukarya, and Archaea), but it is known to be present in the majority of different species of bacteria and fungi (Minami et al. 1998; Shah et al. 1998; Ma et al. 2003; Singh and Kashyap 2012; Nascimento et al. 2012). Recently many research studies reported that ACC-deaminase structural (*acdS*) gene has been found in a wide range of gram-negative bacteria such as *Achromobacter xylosoxidans*, *Rhizobium leguminosarum*, *Pseudomonas putida*, *Burkholderia phytofirmans*, etc. (Wand et al. 2001; Belimov et al. 2001; Hontzeas et al. 2004; Babalola et al. 2013; Duan et al. 2013);

gram-positive bacteria such as *Brevibacterium iodinum*, *Bacillus licheniformis*, *Zhihengliuela alba*, *Micrococcus* sp., *Brachybacterium saurashtrense*, *Brevibacterium casei*, etc. (Belimov et al. 2001; Siddikee et al. 2011; Timmusk et al. 2011; Jha et al. 2012); some fungi like *Penicillium citrinum*, *Trichoderma asperellum*, *Phytophthora sojae*, and *Issatchenkia occidentalis* (Jia et al. 1999; Palmer et al. 2007; Viterbo et al. 2010; Singh and Kashyap 2012); and different species of rhizobia such as *Mesorhizobium loti*, *Rhizobium leguminosarum*, *R. phaseoli*, etc. (Uchiumi et al. 2004; Duan et al. 2009; Ahmad et al. 2011). On the basis of existing literature, *acdS* genes from bacteria are being extensively used for the development of transgenic plants for tolerance toward abiotic stresses. Bacterial gene *acdS* is highly regulated and expressed differentially on the basis of many factors like the presence or absence of oxygen, concentration of ACC, and accumulation of product. Based on current literature, precise mechanisms of regulation of *acdS* are understood in few bacteria. Many of the *acdS* genes have been mainly regulated by the leucine-responsive regulatory protein (LrP) gene located approximately 50–100 base pairs upstream to the *acdS* gene and AcdB protein encoding glycerophosphoryl diester phosphodiesterase (for the binding with ACC) (Cheng et al. 2008; Duan et al. 2009). Glick et al. (2007) proposed a well-described model for the transcriptional regulation of *acdS* gene through LrP in bacteria; this protein has been encoded by the ACC-deaminase regulatory (*acdR*) gene. According to this model, in the presence of ACC, *acdR* gene encodes active octamer of LrP that interacts with AcdB protein and ACC; this complex further initiates transcription of *acdS* gene. Upon this interaction, *acdS* produced ACC-deaminase, and it hydrolyzes the ACC into ammonia and α-ketobutyrate (precursor of branched-chain amino acids such as leucine). When the concentration of leucine increases in the bacterial cell, it interacts with the active LrP octamer and forms inactive LrP dimer, which leads to shutting down the further transcription of *acdS* gene (Fig. 8.3).

In addition to above regulatory mechanism of *acdS* gene expression, it is also regulated by different regulatory proteins in different bacterial species such as *Burkholderia* sp. CCGE 1002 and *Burkholderia phymatum* STM 815, LrP, and σ70 promoter which are involved in the transcriptional regulation of *acdS* gene (Kaneko et al. 2002). In *Mesorhizobium loti* ACC-deaminase expression is regulated by *nifA2*, *nifA1*, and σ^{54} polymerase sigma recognition factor located upstream to *acdS* gene. The NifA2 protein (encoded by *nifA2)* interacts with σ^{54} polymerase sigma recognition factor favoring *acdS* gene transcription (Nukui et al. 2006), and *nifA1* also increases the *acdS* transcription; however the mechanism of regulation of *acdS* transcription through *nifA1* is not well understood.

8.4.3 Approach of ACC-Deaminase Action

The mode of action of PGPB containing ACC-deaminase was described by the originally proposed model (Glick et al. 1998); this model explains how ACC-deaminase-containing PGPB can lower plant ethylene level and in turn stimulate plant growth, especially under harsh conditions. ACC-deaminase-containing PGPB

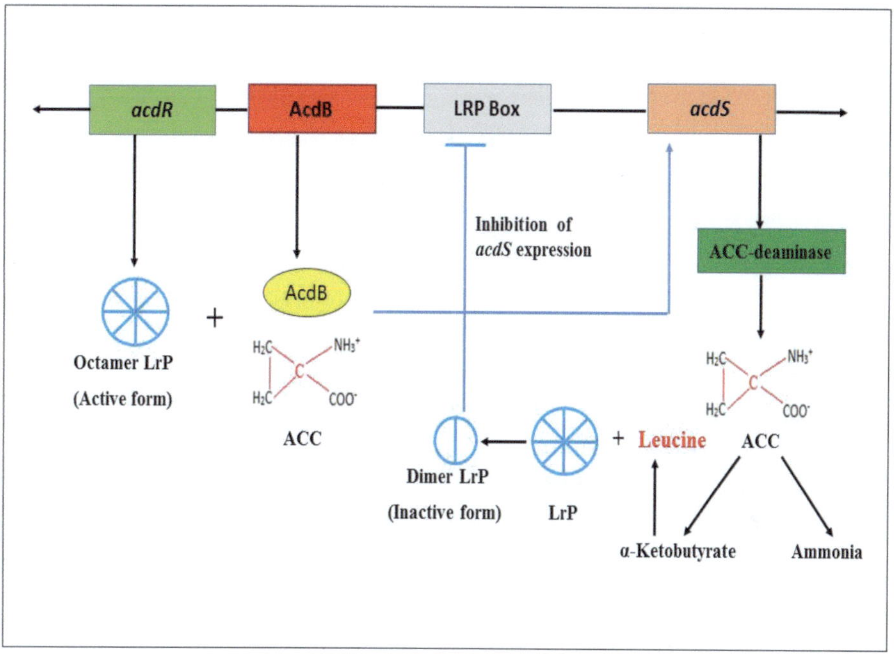

Fig. 8.3 A schematic presentation represents common mechanism of regulation of *acdS* gene transcription in bacteria (mainly *Pseudomonas putida* UW4). Abbreviations: *acdR* ACC-deaminase regulatory gene, *AcdB* encoding for glycerophosphoryl diester phosphodiesterase, *LrP* leucine-responsive protein, *acdS* deaminase structural gene (*acdS*) (Source: Glick et al. 2007)

first bind to the surface of either seed or plant roots in response to root exudates, although these bacteria may also be found on aerial parts of plants (leaves and flowers) or within a plant's internal tissues (Fig. 8.4).

Plants typically exude a huge fraction of their metabolically fixed different amino acids, sugars, and organic acids through their roots in the surrounding soil. Root exudates act as a bacterial food source which is the main reason that the numbers of bacteria present around the roots of plants (i.e., the rhizosphere) are 1000 times higher than in the bulk soil. In response to the presence of tryptophan and other photosynthetically fixed small molecules in the plant root exudates, the associated bacteria synthesize and secrete the phytohormone indole-3-acetic acid (IAA), some of IAA is taken up by the plant roots. IAA level in the plant trusses affects plant growth in different ways such as stimulation of the plant cell proliferation and cell elongation, and IAA can also induce the transcription of the plant enzyme ACC synthase that catalyzes the formation of ACC and stimulate the synthesis of ethylene in the plant. It can also act to loosen plant cell walls, thereby facilitating cell elongation and increasing the number of lateral roots and level of root exudation. Along with the IAA level in the plants also enhanced the plant ACC exuded from seeds or roots (Penrose et al. 2001), and this ACC may be taken up by the PGPB associated with these tissues and subsequently cleaved by ACC-deaminase (Stearns

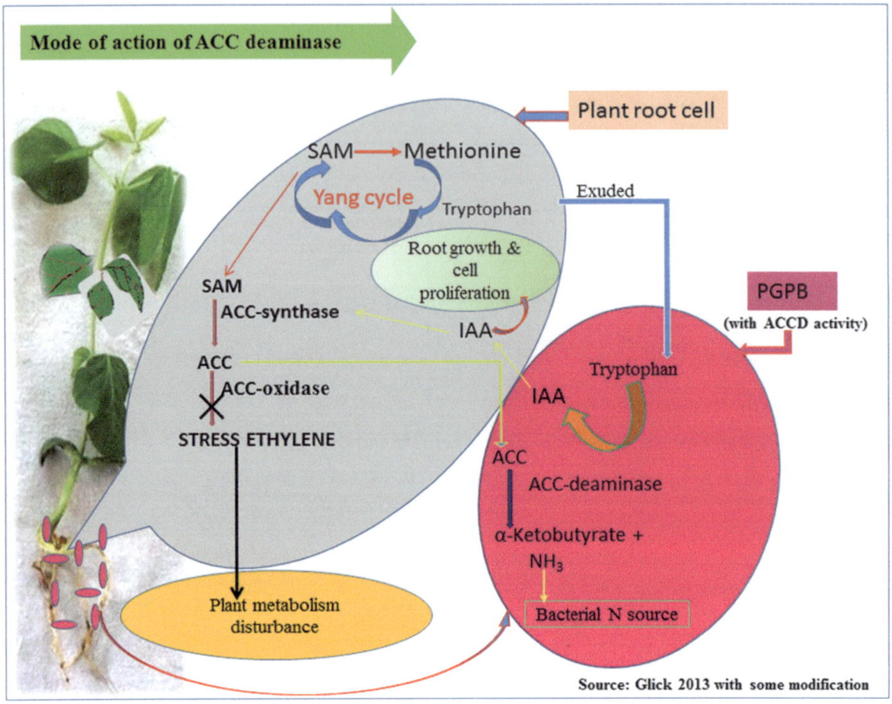

Fig. 8.4 Schematic presentation of how stress controller PGPB containing ACC-deaminase activity bound to plant root lower the stress ethylene concentration and prevent ethylene inhibition in plant's metabolism disturbance. Abbreviations: *SAM* S-adenosyl-L-methionine, *ACC* 1-aminocyclopropane-1-carboxylate, *IAA* indole acetic acid

et al. 2012; Penrose and Glick 2003). The net result of the cleavage of exuded ACC by bacterial ACC-deaminase is that lowering either the endogenous or the bacterial IAA-stimulated ACC level, the amount of ethylene in the plants is reduced. Subsequently, as a consequence of lowering plant ethylene levels, ACC-deaminase-containing PGPB can reduce a portion of the ethylene inhibition of plant growth and mitigate the adverse effect of a wide range of abiotic stresses. Plants which grow in association with ACC-deaminase-containing PGPB generally have longer roots and shoots and are more resistant to growth inhibition by a variety of ethylene-inducing stresses (Saleem et al. 2007).

8.5 ACC-Deaminase in Salinity and Drought Stress Amelioration

Glick et al. (2007) and Saleem et al. (2007) have reviewed that inoculation of plants with PGPB containing ACC-deaminase may lead to various subsequent physiological changes in plants. Belimov et al. (2001) and Penrose et al. (2001) also studied the ability of ACC-utilizing PGPB to improve plant growth inhibition caused by stress ethylene through decreased ACC content in plant tissue. Similarly many researchers also demonstrated the stimulation of root elongation and biomass production of different plant species by inoculations of PGPB containing ACC-deaminase activity, particularly when the plants were subjected to stressful growth conditions (Glick et al. 1998; Belimov et al. 2001; Safronova et al. 2006). Li et al. (2000) have documented that the ACC-deaminase-deficient mutated strain of *P. putida* UW4 simultaneously lost the ability to degrade the ACC and elongate roots in infected canola plants. Environmental stresses such as drought, salt, flooding, etc. cause overproduction of ethylene in the plant tissues which hamper plant growth by different ways like inhibition of root elongation, leaf senescence and abscission, early fruit ripping, and inhibition of legume root nodule formation (Gamalero and Glick 2012). Bacterial hydrolysis of ACC through ACC-deaminase leads to a decrease in plant stress ethylene level, which results in increased plant growth (Belimov et al. 2009; Glick et al. 2013). This property of bacterial ACC-deaminase activity and several other mechanisms of PGPB to alleviate abiotic stresses in plants are referred as "induced systemic tolerance" (IST) (Yang et al. 2009). Thus, the use of PGPB having ACC-deaminase activity is the most important and sustainable mechanism in agricultural sector to reduce the deleterious effect of adverse environmental stresses on the crop productivity (Table 8.1).

8.5.1 Salinity Stress

Soil salinity is a major abiotic environmental factor that adversely inhibits the growth of crops and reduced the yield. In these crops, legumes are mainly affected by a low level of salinity. Legumes also represent a very significant group of crops in agriculture ecosystem. Legume crops are the most important grains because they are rich sources of protein and oil in both human and animal diets. Furthermore, it plays a significant role in the maintenance of soil fertility, through its symbiotic association with rhizobia. Like other legumes, chickpea is very sensitive to salinity, which affects its growth and development. The excess salinity in the soil, due to the high concentration of Cl^- and Na^+ ions, affects the plant system and decreased the yield and quality of the crops (Shukla et al. 2012). Soil salinity induces production of reactive oxygen species (ROS) such as superoxide anion (O_2^-), singlet oxygen ($^1O_2$), and hydrogen peroxide (H_2O_2) and causes cellular damage in the plant system. To contract the adverse effect of these ROS, plant accumulates osmolytes (proline, glycinebetaine, sugars, etc.) resulting in the decrease of the adverse effect of oxidative damage (Qureshi et al. 2013a, b). To mitigate the deleterious effect of salinity, plant roots interact with PGPB containing ACC-deaminase (Table 8.1) which can decrease the high level of ethylene concentration, thereby alleviating the negative impact of salinity exerted on the plant growth and yield. PGPB promote plant growth by enhancing efficiency of water and nutrient uptake and maintain K^+/Na^+ ratio under salt stress (Mayak et al. 2004a). Many mechanisms of stress alleviation, such as lowering the ethylene concentration, production of phytohormones,

Table 8.1 List of plant growth-promoting bacteria containing ACC-deaminase activity reported for mitigating the adverse effect of abiotic stresses on different crops

PGPB	Plant	Stresses	PGPB-mediated mechanisms	References
Pseudomonas simiae	*Vigna radiata*	Drought	ACC-deaminase-mediated plant growth promotion under drought stress	Kumari et al. (2015)
Serratia spp. and *Mesorhizobium ciceri*	*Cicer arietinum* L.	Drought	Bacteria with ACC-deaminase played a pivotal role in plant growth and nodulation by lowering ethylene levels	Shahzad et al. (2014)
P. fluorescens ACC-5	*Pisum sativum*	Drought	Plant growth promotion under drought by the role of ACC-deaminase activity	Zahir et al. (2008)
Pseudomonas sp.	*P. sativum*	Drought	Alleviation of drought stress by decreased ethylene production	Arshad et al. (2008)
Achromobacter piechaudii	*Solanum lycopersicum*	Drought	Mitigation of drought stress through ACC-deaminase activity	Mayak et al. (2004b)
P. koraiensis	*Glycine max* L.	Salt	*Pseudomonas*-mediated salt tolerance in *Glycine max* L.	Kasotia et al. (2015)
PGPR strain A1 and A2	*V. radiata* L.	Salt	ACC-deaminase activity	Aamir et al. (2013)
Pseudomonas spp. and *Rhizobium*	*Lens culinaris*	Salt	Rhizobium with pseudomonas having ACC-deaminase would be a better approach for nodulation	Iqbal et al. (2012)
Pseudomonas strains	*V. radiata* L.	Salt	Improving nodule formation and inhibition of ethylene production	Ahmad et al. (2011)
Pseudomonas and *Serratia* sp.	*Triticum aestivum* L., *Lens culinaris*	Salt	Enhancing plant growth and yield	Zahir et al. (2009, 2010)
Pseudomonas, Flavobacterium, and *Enterobacter strains*	*Zea mays* L.	Salt	PGPB strains with ACC-deaminase activity significantly promote plant biomass, root and shoot length, cob and grain yield, etc.	Nadeem et al. (2007, 2010)

and regulation of nutrient uptake, are used by the PGPB containing ACC-deaminase activity to facilitate plant growth under salinity stress. PGPB-inoculated plants enhanced lateral root development due to higher indole-3-acetic acid (IAA) production and reduced the stress ethylene level in plants by the production of 1-aminocy clopropane-1-carboxylic acid (ACC)-deaminase enzyme activity (Senthil et al. 2009; Glick et al. 2007). Bacterial-inoculated plant root accumulated higher proline content compared to controlled plant root under salt stress, which may be due to higher uptake of nutrients, resulting in high biosynthesis rate (Vardharajula et al. 2011). The accumulation of proline in roots suggests an osmotic mechanism to keep a positive water potential for water entrance into the roots, leading to a lower stress damage in the plant (Porcel and Ruiz-Lozano 2004). In response to environmental stress such as salinity, reactive oxygen species (ROS) may generate in excess, which are extremely harmful to living organisms. The excess production of ROS can cause cell destruction by various pathways like peroxidation of lipids, oxidation of proteins, and nucleic acid damage finally leading to programmed cell death. The efficient removal of ROS requires the action of several antioxidant enzymatic reactions including catalase (CAT), peroxidase (POD), polyphenol oxidase (PPO), and superoxide dismutase (SOD) which can be considered as one mechanism of salt tolerance in plants (Sharma et al. 2012). Bacterial-inoculated plants showed less antioxidant enzyme activity under salinity stress suggesting that these plants are submitted to a lower oxidative stress under saline conditions (Porcel and Ruiz-Lozano 2004; Kohler et al. 2008; Kohler et al. 2009; Kumari et al. 2015). Recent studied related to the hydrolysis of ACC-deaminase by PGPB carrying ACC-deaminase activity diluted the detrimental effects of salinity by decreasing the ethylene level, thus improving the growth of plants (Glick et al. 1998; Cheng et al. 2007; Nadeem et al. 2009). Similarly, bacterial strain carrying ACC-deaminase has also been studied to decrease the adverse effects of soil salinity on the plants by the hydrolysis of salt stress-induced ACC concentration, regulating accelerated level of ethylene in response to stress in plants, and promote plant growth under saline condition (Belimov et al. 2009). Inoculation with PGPB containing ACC-deaminase has been reported to reduce the stress-induced ethylene-mediated negative effects on plants and boost plant growth particularly under stressed conditions (Glick 2005; Safronova et al. 2006).

8.5.2 Drought Stress

Water is a fundamental constituent of all life, about 90 % of the fresh weight in physiologically active plants. In most plants, if the water content goes down much below this level, many physiological processes are impaired. Water deficiency occurs when there is more water loss by evaporation than the amount taken up by roots; this is often referred to as "drought stress." Drought stress is common in many parts of the world, and more than 50 % of the globe is arid and semiarid. Soil water deficiency affects the water relations at whole plant level and finally makes plants more susceptible to other environmental stresses by decreasing the adequacy of defense mechanisms (Vardharajula et al. 2010) and can also adversely affect plant growth and yield, causing the most fatal economic losses in agricultural sector. The effects of drought stress can be counted as reduction in growth and simultaneous reduced dry weight, stomatal conductance decreases, and thus water loss from leaves and photosynthetic rate are reduced due to decreased intercellular CO_2 concentration (Mahajan and Tuteja 2005). Drought stress has been extensively associated with elevated release of endogenous ethylene in the plants which is responsible for growth inhibition (Mayak et al. 2004b; Arshad et al. 2008). Ethylene is a plant hormone that is involved in the regulation of many physiological responses (Arshad and Frankenberger 2002), and it was also regarded as a "stress hormone" because ethylene synthesis in plants is increased under a number of biotic and abiotic stresses. Stress ethylene production has been often coupled with reduced growth and premature senescence; therefore it may act as an indicator of plant susceptibility to stresses such as drought and heat (Wang et al. 2003; Belimov et al. 2009). PGPB containing ACC-deaminase might have decreased the drought stress-induced ethylene concentration in inoculated plants, which resulted in better plant growth under water stress conditions. Therefore, inoculation with PGPB containing ACC-deaminase could be helpful in eliminating the inhibitory effects of stress ethylene on the growth of plants under severe drought stress. Inoculation of plants with beneficial PGPB containing ACC-deaminase induced plant growth by root colonization and mitigates adverse effect of drought in arid or semiarid areas (Marulanda et al. 2007, 2009; Zahir et al. 2009; Vardharajula et al. 2011). Under adverse condition such as drought stress, bacterial cells accumulate various computable solutes such as proline and sugars that protect bacterial cells from degenerative process and improve survival under adverse environment. Drought stress can also create physicochemical and biological properties of soil unfavorable for soil microbial activities. *Pseudomonas* spp. can survive under stress conditions due to the production of EPS, which protects bacterial cells from drought stress by enhancing water preservation and regulating the diffusion of organic carbon sources (Vardharajula et al. 2009).Similarly other workers also reported that EPS-producing bacteria increased resistance to water stress and significant increase in root-adhering soil per root tissue (RAS/RT) ratio in plants (Alami et al. 2000). *Paenibacillus polymyxa* confers drought tolerance through the induction of drought-responsive gene, *early response to dehydration 15* (*ERD15*), in *Arabidopsis thaliana* (Timmusk and Wagner 1999). Researchers found the inoculation of *Pseudomonas*, *Bacillus*, and *Mycobacterium* strains significantly promotes the plant growth by increasing plant growth parameters in maize plants (Egamberdiyeva 2007) and alleviated the adverse effect of water deficiency on wheat (Kasim et al. 2013). Many other studies have reported on the ability of microbes under determined conditions in protecting plants from the deleterious effects of drought stress (Belimov et al. 2009; Arzanesh et al. 2011). Cohen et al. have reported the inoculation with *Azospirillum brasilense* increases the relative water content in the leaves due to the production of ABA hormone by the bacterial strain containing ACC-deaminase, thus reducing the ACC level in the plants. Inoculation with PGPB containing ACC-deaminase enhances root growth, which might be helpful in the uptake of relatively more water and nutrient uptake from deep soil under drought stress conditions (Dodd et al. 2005). The inoculation with PGPB containing ACC-deaminase was helpful in increasing water-use efficiency in peas under drought stress and confers resistance against drought stress in tomatoes and peppers (Zahir et al. 2008).

8.6 Conclusion and Future Prospects

In the present day, agricultural productivity is challenged by adverse environmental stresses. From the above discussion, it is clear that plants are exposed to numerous abiotic stresses such as drought and salinity in the environment which cause ethylene-induced inhibition of plant growth. Plants under both types of stresses are likely to face many conditions such as toxic level of ethylene, ionic imbalance, and more ROS production. There is now a consensus that PGPB with ACC-deaminase are able to mediate the enhanced tolerance to abiotic stresses in their host plants. Therefore, these PGPB have the potential to promote plant growth and productivity and reduce the detrimental effects of stress ethylene level in response to adverse environmental conditions. Therefore, it requires consideration of efficient PGPB with robust ACC-deaminase activity which can interact with plants and increase plant productivity under stress conditions. Moreover, this technique is also beneficial for the soil fertility and decreases the environmental hazards due to the use of chemical fertilizers which are harmful for the soil fertility as well as for animal health. Therefore, inoculation of plants with PGPB having ACC-deaminase gene could be very effective in facilitating plant growth both under normal and stressed environments. PGPB containing ACC-deaminase gene could have a positive benefit for tolerance to stresses in crops, where the environment and degree of stresses such as salinity, drought, high temperature, etc. are not always predictable, and development of *acdS* gene overexpressing transgenic plants to prevent these harsh environmental stresses. However, the expression of *acdS* gene has been reported in only few transgenic plants such as *Arabidopsis*, tomato, poplar, etc. Therefore, future research studies need to ascertain the mechanisms of regulation of *acdS* gene in bacteria and the development of other *acdS* transgenic plants to overcome the detrimental effects of stress-induced ethylene.

Acknowledgment The financial support for some of the research in this review has partially been supported by SERB and DBT grant no. SR/FT/LS-129/2012 and BT/PR1231/AGR/021/340/2011, respectively, to DKC.

References

Aamir M, Aslam A, Khan MY, Jamshaid MU, Ahmad M, Asghar HN, Zahir ZA (2013) Co-inoculation with rhizobium and plant growth promoting rhizobacteria (PGPR) for inducing salinity tolerance in mung bean under field condition of semi-arid climate. Asian J Agric Biol 1:17–22

Abeles FB, Morgan PW, Saltveit ME Jr (1992) Ethylene in plant biology, 2nd edn. Academic, New York

Adams DO, Yang SF (1979) Ethylene biosynthesis: identification of 1-aminocyclopropane-1-carboxylic acid as an intermediate in the conversion of methionine to ethylene. Proc Natl Acad Sci U S A 76:170–174

Ahmad M, Zahir ZA, Asghar HN, Asghar M (2011) Inducing salt tolerance in mung bean through co-inoculation with rhizobia and plant-growth-promoting rhizobacteria containing 1-aminocy clopropane-1-carboxylate-deaminase. Can J Microbiol 57:578–589

Alami Y, Achouak W, Marol C, Heulin T (2000) Rhizosphere soil aggregation and plant growth promotion of sunflowers by an exopolysaccharides producing *Rhizobium* sp. strain isolated from sunflower roots. Appl Environ Microbiol 66:3393–3398

Arshad M, Frankenberger WT (2002) Ethylene: agricultural sources and applications. Kluwer Academic Publishers, New York, p 342

Arshad M, Shaharoona B, Mahmood T (2008) Inoculation with *Pseudomonas* spp. containing ACC-deaminase partially eliminates the effects of drought stress on growth, yield, and ripening of pea (*Pisum sativum* L.). Pedosphere 18(5):611–620

Arzanesh M, Alikhani H, Khavazi K, Rahimian H, Miransari M (2011) Wheat (*Triticum aestivum* L.) growth enhancement by *Azospirillum* sp. under drought stress. World J Microbiol Biotechnol 27:197–205

Ashraf MA, Asif M, Zaheer A, Malik A, Ali Q, Rasool M (2013) Plant growth promoting rhizobacteria and sustainable agriculture: a review. Afr J Microbiol Res 7(9):704–709

Babalola OO, Tak HI, Ahmad F (2013) Advances in the application of plant growth promoting rhizobacteria in phytoremediation of heavy metals. Rev Environ Contam Toxicol 223:33–52

Barber DA, Martin JK (1976) The release of organic substances by cereal roots into soil. New Phytol 76:69–80

Belimov AA, Safronova VI, Sergeyeva TA, Egorova TN, Matveyeva VA, Tsyganov VE, Borosov AY, Tikhonovich IA, Kluge C, Preisfeld A, Dietz KJ, Stepanok VV (2001) Characterization of plant growth promoting rhizobacteria isolated from polluted soils and containing 1-aminocycl opropane-1-carboxylate deaminase. Can J Microbiol 47:242–252

Belimov AA, Dodd IC, Hontzeas N, Theobald JC, Safronova VI, Davies WJ (2009) Rhizosphere bacteria containing 1-aminocyclopropane-1-carboxylate deaminase increase yield of plants grow in drying soil via both local and systemic hormone signaling. New Phytol 181:413–423

Binder BM (2008) The ethylene receptors: complex perception for a simple gas. Plant Sci 75:8–17

Bleecker AB, Kende H (2000) Ethylene: a gaseous signal molecule in plants. Annu Rev Cell Dev Biol 16:1–18

Chakraborty U, Chakraborty B, Dey P, Chakraborty AP (2015) Role of microorganisms in alleviation of abiotic stresses for sustainable agriculture. In: Chakraborty U, Chakraborty B (eds) Abiotic stresses in crop plants. CABI, Wallingford/Boston. doi:10.1079/9781780643731.0232

Cheng Z, Park E, Glick BR (2007) 1-Aminocyclopropane-1-carboxylate deaminase from *Pseudomonas putida* UW4 facilitates the growth of canola in the presence of salt. Can J Microbiol 53:912–918

Cheng Z, Duncker BP, Mc Conkey BJ, Glick BR (2008) Transcriptional regulation of ACC deaminase gene expression in *Pseudomonas putida* UW4. Can J Microbiol 54:128–136

Dimkpa CO, Merten D, Svatos A, Buchel G, Kothe E (2009) Metal induced oxidative stress impacting plant growth in contaminated soil is alleviated by microbial siderophores. Soil Biol Biochem 41:154–162

Dobbelaere S, Okon Y (2007) The plant growth promoting effects and plant responses. In: Elmerich C, Newton WE (eds) Associative and endophytic nitrogen-fixing bacteria and cyanobacterial associations. Nitrogen fixation: origins, applications and research progress, vol 5. Springer, Heidelberg, pp 145–170

Dodd IC, Belimov AA, Sobeih WY, Safronova VI, Grierson D Davies WJ (2005) Will modifying plant ethylene status improve plant productivity in water limited environments? 4th International

Crop Science Congress. Available online: http://www.cropscience.org.au/icsc2004/poster/1/3/4/510-doddicref.htm. Accessed at 17 June 2007

Duan J, Müller KM, Charles TC, Vesely S, Glick BR (2009) 1-Aminocyclopropane-1-carboxylate (ACC) deaminase genes in *Rhizobia* from southern Saskatchewan. Microb Ecol 57:423–436

Duan J, Jiang W, Cheng Z, Heikkila JJ, Glick BR (2013) The complete genome sequence of the plant growth-promoting bacterium *Pseudomonas* sp. UW4. PLoS One 8(3), e58640

Egamberdieva D (2009) Alleviation of salt stress by plant growth regulators and IAA producing bacteria in wheat. Acta Physiol Plant 31:861–864

Egamberdiyeva D (2007) The effect of plant growth promoting bacteria on growth and nutrient uptake of maize in two different soils. Appl Soil Ecol 36:184–189

Gamalero E, Glick BR (2012) Plant growth-promoting bacteria and metal phytoremediation. In: Anjum NA, Pereira ME, Ahmad I, Duarte AC, Umar S, Khan NA (eds) Phytotechnologies: remediation of environmental contaminants. CRC Press, Boca Raton, pp 361–376

Ghosh S, Penterman JN, Little RD, Chavez R, Glick BR (2003) Three newly isolated plant growth promoting bacilli facilitate the seedling growth of canola, Brassica campestris. Plant Physiol Biochem 41:277–281

Glick BR (2004) Bacterial ACC deaminase and the alleviation of plant stress. Adv Appl Microbiol 56:291–312

Glick BR (2005) Modulation of plant ethylene levels by the bacterial enzyme ACC deaminase. FEMS Microbiol Lett 251:1–7

Glick BR (2012) Plant growth-promoting bacteria: mechanisms and applications. Scientifica. doi:10.6064/2012/963401

Glick BR, Penrose DM, Li J (1998) A model for the lowering of plant ethylene concentrations by plant growth promoting bacteria. J Theor Biol 190:63–68

Glick BR, Cheng Z, Czarny J, Duan J (2007) Promotion of plant growth by ACC deaminase-containing soil bacteria. Eur J Plant Pathol 119:329–339

Glick BR, Nascimento FX, Vicente CSL, Barbosa P, Espada M et al (2013) Evidence for the involvement of ACC deaminase from *Pseudomonas putida* UW4 in the biocontrol of pine wilt disease caused by *Bursaphelenchus xylophilus*. Biocontrol 58:427–433

Gray EJ, Smith DL (2005) Intracellular and extracellular PGPR: commonalities and distinctions in the plant–bacterium signaling processes. Soil Biol Biochem 37:395–412

Hirsch AM, Fang Y (1994) Plant hormones and nodulation: what's the connection? Plant Mol Biol 26:5–9

Honma M (1985) Chemically reactive sulfhydryl groups of 1-aminocyclopropane-1-carboxylate deaminase. Agric Biol Chem 49:567–571

Honma M, Shimomura T (1978) Metabolism of 1-aminocyclopropane-1-carboxylic acid. Agric Biol Chem 43:1825–1831

Hontzeas N, Zoidakis J, Glick BR, Abu-Omar MM (2004) Expression and characterization of1-aminocyclopropane-1-carboxylate deaminase from the rhizobacterium *Pseudomonas putida* UW4: a key enzyme in bacterial plant growth promotion. Biochim Biophys Acta 1703:11–19

Iqbal MA, Khalid M, Shahzad SM, Ahmad M, Soleman N, Akhtar N (2012) Integrated use of *Rhizobium leguminosarum*, plant growth promoting rhizobacteria and enriched compost for improving growth, nodulation and yield of lentil (*Lens culinaris* medik.). Chil J Agric Res 72:104–110

Jacobson CB, Pasternak JJ, Glick BR (1994) Partial purification and characterization of ACC deaminase from the plant growth-promoting rhizobacterium *Pseudomonas putida* GR12-2. Can J Microbiol 40:1019–1025

Jain S, Choudhary DK (2014) Induced defense-related proteins in soybean (Glycine max L. Merrill) plants by *Carnobacterium* sp. SJ-5 upon challenge inoculation of *Fusarium oxysporum*. Planta 239(5):1027–1040. doi:10.1007/s00425-014-2032-3

Jha B, Gontia I, Hartmann A (2012) The roots of the halophyte *Salicornia brachiata* are a source of new halotolerant diazotrophic bacteria with plant growth-promoting potential. Plant Soil 356:265–277

Jia YJ, Kakuta Y, Sugawara M, Igarashi T, Oki N, Kisaki M et al (1999) Synthesis and degradation of 1-aminocyclopropane-1-carboxylic acid by *Penicillium citrinum*. Biosci Biotechnol Biochem 63:542–549

Kaneko T, Nakamura Y, Sato S et al (2002) Complete genomic sequence of nitrogen-fixing symbiotic bacterium *Bradyrhizobium japonicum* USDA110. DNA Res 9:189–197

Kasim WA, Osman ME, Omar MN, Abd-Eldeim IA, Bejai S, Meijer J (2013) Control of drought stress in wheat using plant growth promoting bacteria. J Plant Growth Regul 32:122–130

Kasotia A, Varma A, Choudhary DK (2015) *Pseudomonas* mediated mitigation of salt stress and growth promotion in Glycine max. Agric Res 4(1):31–41

Klee HJ, Hayford MB, Kretzmer KA, Barry GF, Kishmore GM (1991) Control of ethylene synthesis by expression of a bacterial enzyme in transgenic tomato plants. Plant Cell 3:1187–1193

Kloepper JW, Ryu C-M, Zhang S (2004) Induced systemic resistance 1904 and promotion of plant growth by *Bacillus* spp. Phytopathology 94:1259–1266

Kohler J, Hernández JA, Caravacaa F, Roldána A (2008) Plant-growth-promoting rhizobacteria and arbuscular mycorrhizal fungi modify alleviation biochemical mechanisms in water-stressed plants. Funct Plant Biol 35:141–151

Kohler J, Hernández JA, Caravacaa F, Roldána A (2009) Induction of antioxidant enzymes is involved in the greater effectiveness of a PGPR versus AM fungi with respect to increasing the tolerance of lettuce to severe salt stress. Environ Exp Bot 65:245–252

Kumari S, Vaishnav A, Jain S, Varma A, Choudhary D K (2015) Bacterial-mediated induction of systemic tolerance to salinity with expression of stress alleviating enzymes in soybean (Glycine max L. Merrill). J Plant Growth Regul 34(3):558–573. doi:10.1007/s00344-015-9490

Li J (1999) Isolation, characterization and regulation of 1-aminocyclopropane-1-carboxylate deaminase genes from plant growth promoting rhizobacteria. Ph. D thesis, University of Waterloo, ON, Canada

Li J, Ovakim D, Charles TC, Glick BR (2000) An ACC deaminase minus mutant of Enterobacter cloacae UW4 no longer promotes root elongation. Curr Microbiol 41:101–105

Lincoln JE, Fischer RL (1988) Diverse mechanisms for the regulation of ethylene-inducible gene expression. Mol Gen Genet 212:71–75

Lucy M, Reed E, Glick BR (2004) Applications of free living plant growth-promoting rhizobacteria. A Van Leeuw 86:1–25

Ma W, Guinel FC, Glick BR (2003) *Rhizobium leguminosarum biovarviciae* 1-aminocyclopropan e-1-carboxylate deaminase promotes nodulation of pea plants. Appl Environ Microbiol 69:4396–4402

Mahajan S, Tuteja N (2005) Cold, salinity and drought stresses: an overview. Arch Biochem Biophys 444:139–158

Marulanda A, Porcel R, Barea JM, Azcón R (2007) Drought tolerance and antioxidant activities in lavender plants colonized by native drought-tolerant or drought-sensitive *Glomus* species. Microb Ecol 54:543–552

Marulanda A, Barea J-M, Azcón R (2009) Stimulation of plant growth and drought tolerance by native microorganisms (AM Fungi and Bacteria) from dry environments: mechanisms related to bacterial effectiveness. J Plant Growth Regul. doi:10.1007/s00344-009-9079-6

Mayak S, Tirosh T, Glick BR (2004a) Plant growth-promoting bacteria confer resistance in tomato plants to salt stress. Plant Physiol Biochem 42:565–572

Mayak S, Tirosh T, Glick BR (2004b) Plant growth-promoting bacteria that confer resistance to water stress in tomatoes and peppers. Plant Sci 166:525–530

Minami R, Uchiyama K, Murakami T, Kawai J, Mikami K, Yamada T (1998) Properties, sequence, and synthesis in Escherichia coli of 1-aminocyclopropane-1-carboxylate deaminase from *Hansenula saturnus*. J Biochem (Tokyo) 123:1112–1118

Murr DP, Yang SF (1975) Conversion of 5-methylthioadenosineto methionine by apple tissue. Phys Chem Chem Phys 14:1291–1292

Mutava RN, Prince SJ, Syed NH, Song L, Valliyodan B, Chen W, Nguyen HT (2015) Understanding abiotic stress tolerance mechanisms in soybean: a comparative evaluation of soybean response to drought and flooding stress. Plant Physiol Biochem 86:109–120

Nadeem SM, Zahir ZA, Naveed M, Arshad M (2007) Preliminary investigations on inducing salt tolerance in maize through inoculation with rhizobacteria containing ACC deaminase activity. Can J Microbiol 53:1141–1149

Nadeem SM, Zahir ZA, Naveed M, Arshad M (2009) Rhizobacteria containing ACC-deaminase confer salt tolerance in maize grown on salt-affected fields. Can J Microbiol 55:1302–1309

Nadeem SM, Zahir ZA, Naveed M, Ashraf M (2010) Microbial ACC deaminase: prospects and applications for inducing salt tolerance in plants. Crit Rev Plant Sci 29:360–393

Nascimento FX, Brígido C, Glick BR, Oliveira S (2012) ACC-deaminase genes are conserved between *Mesorhizobium* species able to nodulate the same host plant. FEMS Microbiol Lett 336:26–37

Nautiyal CS, Srivastava S, Chauhan PS (2008) Rhizosphere colonization: molecular determinants from plant-microbe coexistence perspective. In: Nautiyal CS, Dion P (eds) Molecular mechanisms of plant, microbe coexistence, Soil Biology Series. Springer, Berlin, pp 99–124

Nukui N, Minamisawa K, Ayabe SI, Aoki T (2006) Expression of the 1-aminocyclopropane-1-carboxylic acid deaminase gene requires symbiotic nitrogen-fixing regulator gene nifA2 in *Mesorhizobium loti* MAFF303099. Appl Environ Microbiol 72:4964–4969

Ose T, Fujino A, Yao M, Watanabe N, Honma M, Tanaka I (2003) Reaction intermediate structures of 1-aminocyclopropane-1-carboxylate deaminase. J Biol Chem 278:41069–41076

Palmer C, Golden K, Danniels L, Ahmad H (2007) ACC deaminase from Issatchenkia occidentalis. J Biol Sci 7:188–193

Penrose DM, Glick BR (2001) Levels of 1-aminocyclopropane-1-carboxylic acid (ACC) in exudates and extracts of canola seeds treated with plant growth-promoting bacteria. Can J Microbiol 47:368–372

Penrose DM, Glick BR (2003) Methods for isolating and characterizing ACC deaminase-containing plant growth-promoting rhizobacteria. Physiol Plant 118:10–15

Penrose DM, Moffatt BA, Glick BR (2001) Determination of 1-aminocyclopropane-1-carboxylic acid (ACC) to assess the effects of ACC deaminase-containing bacteria on roots of canola seedlings. Can J Microbiol 47:77–80

Pierik R, Tholen D, Poorter H, Visser EJW, Voesenek LACJ (2006) The Janus face of ethylene: growth inhibition and stimulation. Trends Plant Sci 11:176–183

Pinton R, Varanini Z, Nannipieri P (2001) The rhizosphere as a site of biochemical interactions among soil components, plants and microorganisms. In: Pinton R, Varanini Z, Nannipieri P (eds) The rhizosphere. Marcel Dekker, New York, pp 1–18

Porcel R, Ruiz-Lozano JM (2004) *Arbuscular mycorrhizal* influence on leaf water potential, solute accumulation, and oxidative stress in soybean plants subjected to drought stress. J Exp Bot 55(403):1743–1750

Qureshi MA, Shahzad H, Imran Z, Mushtaq M, Akhtar N, Ali MA, Mujeeb F (2013a) Potential of rhizobium species to enhance growth and fodder yield of maize in the presence and absence of tryptophan. J Anim Plant Sci 23(5):1448–1454

Qureshi MI, Abdin MZ, Ahmad J, Iqbal M (2013b) Effect of long-term salinity on cellular antioxidants, compatible solute and fatty acid profile of Sweet Annie (*Artemisia annua* L.). Phys Chem Chem Phys 95:215–223

Safronova VI, Stepanok VV, Engqvist GL, Alekseyev YV, Belimov AA (2006) Root-associated bacteria containing 1-aminocyclopropane-1-carboxylate deaminase improve growth and nutrient uptake by pea genotypes cultivated in cadmium supplemented soil. Biol Fertil Soils 42:267–272

Saleem M, Arshad M, Hussain S, Bhatti A (2007) Perspective of plant growth promoting rhizobacteria (PGPR) containing ACC-deaminase in stress agriculture. J Ind Microbiol Biotechnol 34:635–648

Sauter M, Moffatt BM, Saechao MC, Hell R, Wirtz M (2013) Methionine salvage and S-adenosylmethionine: essential links between sulfur, ethylene and polyamine biosynthesis. Biochem J 451:145–154

Senthil KM, Swarnlakshmi K, Govindasamy V, Lee YK, Annapurna K (2009) Biocontrol potential of soybean bacterial endophytes against charcoal rot fungus *Rhizoctonia bataticola*. Curr Microbiol 58:288–293

Shah S, Li J, Moffatt BA, Glick BR (1998) Isolation and characterization of ACC deaminase genes from two different plant growth promoting rhizobacteria. Can J Microbiol 44:833–843

Shahzad SM, Khalid A, Arif MS, Riaz M, Ashraf M, Iqbal Z, Yasmeen T (2014) Co-inoculation integrated with P-enriched compost improved nodulation and growth of Chickpea (*Cicer arietinum* L.) under irrigated and rainfed farming systems. Biol Fertil Soils 50:1–12

Sharma P, Jha AB, Dubey RS, Pessarakli M (2012) Reactive oxygen species, oxidative damage, and antioxidative defense mechanism in plants under stressful conditions. J Bot. doi:10.1155/2012/217037

Shukla PS, Agarwal PK, Jha B (2012) Improved salinity tolerance of *Arachis hypogaea* (L.) by the interaction of halotolerant plant growth promoting rhizobacteria. J Plant Growth Regul 31:195–206

Siddikee MA, Glick BR, Chauhan PS, Yim W-J, Sa T (2011) Enhancement of growth and salt tolerance of red pepper seedlings (*Capsicum annuum* L.) by regulating stress ethylene synthesis with halotolerant bacteria containing ACC deaminase activity. Plant Physiol Biochem 49:427–434

Singh N, Kashyap S (2012) In-silico identification and characterization of 1-aminocyclopropane-1-carboxylate deaminase from *Phytophthora sojae*. J Mol Model 18:4101–4111

Stearns JC, Woody OZ, McConkey BJ, Glick BR (2012) Effects of bacterial ACC deaminase on *Brassica napus* gene expression measured with an *Arabidopsis thaliana* microarray. Mol Plant-Microbe Interact 25:668–676

Tilak KVBR, Ranganayaki N, Pal KK, De R, Saxena AK, Nautiyal CS, Mittal S, Tripathi AK, Johri BN (2005) Diversity of plant growth and soil health supporting bacteria. Curr Sci 89:136–150

Timmusk S, Wagner EGH (1999) The plant growth-promoting rhizobacterium *Paenibacillus polymyxa* induces changes in *Arabidopsis thaliana* gene expression: a possible connection between biotic and abiotic stress responses. Mol Plant-Microbe Interact 12:951–959

Timmusk S, Paalme V, Pavlicek T, Bergquist J, Vangala A, Danilas T et al (2011) Bacterial distribution in the rhizosphere of wild barley under contrasting microclimates. PLoS One 6, e17968

Uchiumi T, Oowada T, Itakura M, Mitsui H, Nukui N, Dawadi P, Kaneko T, Tabatta S, Yokoyama T, Tejima T, Saeki K, Oomori H, Hayashi M, Maekawa T, Sriprang R, Murooka Y, Tajima S, Simomura K, Nomura M, Suzuki A, Shimoda S, Sioya K, Abe M, Minamisawa K (2004) Expression islands clustered on symbiosis island of *Mesorhizobium loti* genome. J Bacteriol 186:2439–2448

Vaishnav A, Kumari S, Jain S, Varma A, Choudhary DK (2015) Putative bacterial volatile-mediated growth in soybean (Glycine max L. Merrill) and expression of induced proteins under salt stress. J Appl Microbiol 119:539–551

Van de Poel B, Van Der Straeten D (2014) 1-aminocyclopropane-1- carboxylic acid (ACC) in plants: more than just the precursor of ethylene! Front Plant Sci 5:640. doi: 10.3389/fpls.2014.00640

Vardharajula S, Ali SKZ, Grover M, Reddy G, Bandi V (2009) Alleviation of drought stress effects in sunflower seedlings by the exopolysaccharides producing *Pseudomonas putida* strain GAP-P45. Biol Fertil Soils 46:17–26

Vardharajula S, Ali SZ, Grover M, Reddy G, Venkateswarlu B (2010) Effect of plant growth promoting *Pseudomonas* spp. on compatible solutes, antioxidant status and plant growth of maize under drought stress. Plant Growth Regul 62(1):21–30

Vardharajula S, Ali SA, Grover M, Reddy G, Bandi V (2011) Drought-tolerant plant growth promoting *Bacillus* spp.: effect on growth, osmolytes, and antioxidant status of maize under drought stress. J Plant Interact 6(1):1–14

Viterbo A, Landau U, Kim S, Chernin L, Chet I (2010) Characterization of ACC deaminase from the biocontrol and plant growth-promoting agent *Trichoderma asperellum* T203. FEMS Microbiol Lett 305:42–48

Wand C, Ramette A, Punjasamarnwong P, Zala M, Natsch A, Moenne-Loccoz Y, Defago G (2001) Cosmopolitan distribution of phlD-containing dicotyledonous crop associated biological control *Pseudomonas* of worldwide origin. FEMS Microbiol Ecol 37:105–116

Wang WX, Vinocur B, Altman A (2003) Plant responses to drought, salinity and extreme temperatures: towards genetic engineering for stress tolerance. Planta 218:1–14

Yang SF, Hoffman NE (1984) Ethylene biosynthesis and its regulation in higher plants. Annu Rev Plant Physiol 35:155–189

Yang J, Kloepper JW, Ryu CM (2009) Rhizosphere bacteria help plants tolerate abiotic stress. Trends Plant Sci 14:1–4

Zahir ZA, Munir A, Asghar HN, Arshad M, Shaharoona B (2008) Effectiveness of rhizobacteria containing ACC-deaminase for growth promotion of peas (*Pisum sativum*) under drought conditions. J Microbiol Biotechnol 18:958–963

Zahir ZA, Ghani U, Naveed M, Nadeem SM, Asghar HN (2009) Comparative effectiveness of *Pseudomonas* and *Serratia* sp. containing ACC deaminase for improving growth and yield of wheat (*Triticum aestivum* L.) under salt-stressed conditions. Arch Microbiol 191:415–424

Zahir ZA, Shah MK, Naveed M, Akhter MJ (2010) Substrate-dependent auxin production by *Rhizobium phaseoli* improves the growth and yield of *Vigna radiata* L. under salt stress conditions. J Microbiol Biotechnol 20:1288–1294

Increasing Phytoremediation Efficiency of Heavy Metal-Contaminated Soil Using PGPR for Sustainable Agriculture

9

Payman Abbaszadeh-Dahaji, Mahtab Omidvari, and Mansour Ghorbanpour

Abstract

Raising industrial activities and agricultural practices as well as other human anthropogenic actions adds a significant amount of heavy metals in soil and water, resulting in degradation of the environment. Some examples of the environmental concern metals are nickel, copper, arsenic, lead, cadmium, cobalt, and zinc. Due to their nonbiodegradable nature, toxic heavy metals accumulate in the environment and therefore contaminate the food chain. The presence of these hazardous metals further than the threshold limit exhibits a critical threat to the human health and total environment. Different physical, chemical, and biological procedures have been applied for the remediation of contaminants from the environment. Bioremediation is the application of biological remedy for cleanup and or mitigation of contaminants from the environment. This process is a cost-effective and worthwhile method for removal of heavy metal-contaminated soil compared to physicochemical remediation techniques which are expensive and deleterious for soil properties. Phytoremediation is defined as the direct use of appropriate living plants for removal, degradation, or sequester of contaminants from environments (atmosphere, hydrosphere, and lithosphere). The efficiency of phytoremediation depends on many factors like plant biomass yield, plant tolerance to metal toxicity, and heavy metal solubility or mobilization in the soil.

P. Abbaszadeh-Dahaji
Department of Agriculture, Faculty of Soil Science, University of Vali-e-Asr, Rafsanjan, Kerman, Iran

M. Omidvari
Department of Plant Pathology, University of Tehran, Karaj, Iran

M. Ghorbanpour (✉)
Department of Medicinal Plants, Faculty of Agriculture and Natural Resources, Arak University, 38156-8-8349 Arak, Iran
e-mail: m-ghorbanpour@araku.ac.ir

D.K. Choudhary et al. (eds.), *Plant-Microbe Interaction: An Approach to Sustainable Agriculture*, DOI 10.1007/978-981-10-2854-0_9

The success of the phytoremediation process can be attained through developing the association of hyperaccumularor plant species with microorganisms like heavy metal-resistant plant growth-promoting rhizobacteria.

9.1 Introduction

Developing industrialization and anthropogenic activities are the most common factors releasing toxic wastes into the soil, water, and air. This toxic chemical waste is classified as pollution. The consequent pollutants are dangerous to the environment and living things such as human beings, animals, plants, and microorganisms (Rajkumar et al. 2009). Inorganic chemical contaminants occur as natural elements in the biosphere, and result from man-made toxins, and several human activities such as industry, mining, agriculture, traffic, and military activities enhance their release into the environment, causing toxicity. Inorganic pollutants cannot be destroyed; however, they may be remediated through sequestration or stabilization inside the plant tissues. According to Pilon-Smits (2005), inorganic contaminants that can be phytoremediated comprised of plant macronutrients (e.g., nitrate and phosphate), plant trace elements (e.g., Fe, Cu, Cr, Mo, Mn, and Zn), and nonessential elements (e.g., Co, F, Cd, Hg, Pb, Se, and V), as well as radioactive isotopes (e.g., ^{238}U, ^{137}Cs, and ^{90}Sr). Heavy metals are one of the most important contaminants, which have a density ˃5 g cm^{-3} (Abdelatey et al. 2011). These toxic metals are the most important inorganic contaminants, which are very stable and progressively accumulate in the environment (Chaudhary and Khan 2015).

Soil contamination with heavy metals may take place by mining practices, discharge of industrial effluents, extensive use of synthetic fertilizers, pesticides, etc. and is of great environmental concern because of its harmful impacts on biological systems (Ma et al. 2013). High concentrations of heavy metals not only reduce activity of soil microorganisms and crop production but also intimidate human health because they may enter the food chain, enhancing the risk of toxicant transfer to food products (Boyd 2010). Several biological and physicochemical methods have been chosen for the removal of heavy metals, which is considered as a challenging work with regard to cost and technical intricacy (Sheoran et al. 2011; Wuana and Okieimen 2011). Physicochemical approaches comprise excavation, landfill, thermal treatment, leaching, as well as electro-reclamation. These procedures are quick but incompetent and expensive and cause negative impacts on different soil chemical, physical, and biological characteristics, resulting in secondary pollution (Glick 2010; Ali et al. 2013).

It has been suggested that the physicochemical approaches only alter the problem from one form to another, and can't entirely remediate the contaminants (Lambert et al. 2000). Bioremediation of contaminated sites has taken much attention globally as a procedure for enhancing soil quality through elimination of metals from soils (Marques et al. 2009). Phytoremediation is among these methods presenting significantly more advantageous than conventional technology of cleanup (Pilon-Smits 2005).

However, slow growth with low plant biomass in metal-polluted soils may restrict the efficiency of phytoremediation process (Li and Ramakrishna 2011; Ma et al. 2011). Furthermore, bioavailability of the metal in soil rhizosphere is believed to be another important factor that affects the successes of metal translocation and phytostabilization (Ma et al. 2011). In recent years, many chemical amendments, such as ethylenediaminetetraacetic acid (EDTA), have been applied to improve phytoextraction and/or phytostabilization process (Barrutia et al. 2010). However, chelators are known to be toxic to plants and to the most soil microorganisms (Evangelou et al. 2007). A promising option is the application of plant-microorganism interactions to enhance the effectiveness of phytoremediation, changing the bioavailability and mobilization of metals in soil environment (Glick 2010; Ma et al. 2011; Rajkumar et al. 2010). The efficiency of phytoremediation mainly depends on plant growth and a high content of heavy metal in the shoot parts of the plant (Petriccione et al. 2013). It has been acknowledged that plant growth-promoting rhizobacteria (PGPR) may improve the effectiveness of phytoremediation through increasing plant growth under harsh conditions and increasing heavy metal solubility by different mechanisms (Gadd 2004; DeBashan et al. 2012).

9.2 Phytoremediation

Phytoremediation may suggest a cost profitable, noninvasive, and secure alternative to standard soil-cleaning techniques through using specific shrubs, trees, and grass species to remove, immobilize, or even degrade hazardous materials from soil (Rajkumar et al. 2012).

Most recently, Arora et al. (2016) reported that in phytoremediation process, generally, two types of plants are utilized: (1) hyperaccumulators with a very high heavy metal accumulation potential and low biomass productivity and (2) non-hyperaccumulators, which have lower extraction capacity than hyperaccumulators, but whose total biomass yield is significantly higher and are fast-growing species.

Moreover, the type of plant species for phytoremediation is usually selected according to regional climate, root system, and the nature of the pollutants. It has been estimated that the approximate remediation depths for grasses, shrubs, and deep-rooting trees are 3, 10, and 20 ft, respectively (Chirakkara and Reddy 2015). Generally, there are six different processes for phytoremediation (Fig. 9.1) (Pilon-Smits 2005), but for heavy metal-polluted soils, four processes such as phytostabilization, phytoextraction, phytovolatilization, and rhizofiltration are more consequential (Laghlimi et al. 2015).

Phytoextraction, a subprocess of phytoremediation, can be defined as uptake of dangerous elements by roots from the soil and its translocation into harvestable biomass of plant (Ali et al. 2013). Phytostabilization can be defined as metal trapping from the rhizospheric soil through decreasing its availability or mobility in the environment (Arora et al. 2016). However, phytovolatilization is transpiration of certain contaminants from plant (Pilon-Smits 2005). Phytodegradation, also known as phytotransformation, is direct degradation of organic pollutants through plant

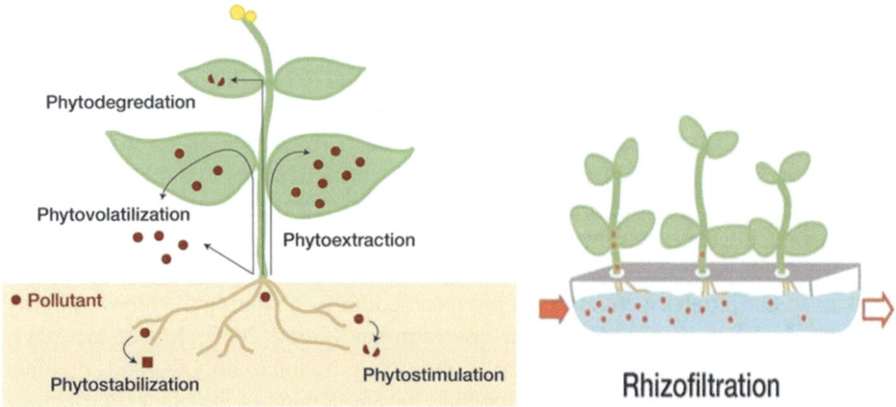

Fig. 9.1 Different processes of phytoremediation (Pilon-Smits 2005)

enzymatic activities (McCutcheon and Schnoor 2003). Phytostimulation, or rhizodegradation, is the process in which root-released materials increase microbial activity near the plant roots (Hutchinson et al. 2003). Rhizofiltration, however, involves filtering toxic pollutants by a mass of plant roots (Raskin et al. 1997). Phytoremediation, the same as other remediation procedures, contains many benefits and limitations.

The most important phytoremediation benefits include cost-effectiveness, potentially environmental friendly and long-term applicability to a variety of hazardous metals, as well as aesthetically pleasing. However, there are some limitations for phytoremediation. Duration of the process is too long, i.e., it is a slow process; therefore, it may take several years or even decades to clean up a polluted site, and it is only feasible to surface soils (Laghlimi et al. 2015). Phytoremediation process may also be limited through the bioavailability of toxic materials.

Multiprocess phytoremediation system (MPPS) has been proposed for an effective phytoremediation process. The process is based on combination of microbial, mechanical, and plant growth stages to improve biomass accumulation, especially plant below ground parts in the soil, and therefore increase the remediation kinetics. The processes utilized are land farming and inoculation with PGPR (Huang et al., 2005). Generally, plants with excellently high capacity for accumulating toxic metals often grow slowly and produce low biomass, especially when the metal content in the soil is high. However, there is a way to increase the efficiency of phytoremediation using PGPR, which are soil-living microorganisms that inhabit the rhizosphere. When PGPR are presented to a polluted site, they enhance the potential for plants that grow there to degrade toxic metals and to remobilize nutrients, preserve soil structure, detoxify synthetic chemicals, and suppress pathogens and pests; also, PGPR reduce the toxicity of heavy metals through altering their bioavailability in plants.

The plants subsequently supply the microorganisms with plant root secretions including proteins, carbohydrates, free amino acids, sugars, vitamins, hormones, mucilage, and alcohols, which are essential sources of their nutrition. The rhizosphere has high levels of easily degradable root-exuded compounds, attracting greater microbes than does bulk soil (Babalola 2010). According to Babalola (2010), efficiency of phytoremediation depends on the extent of soil pollution, bioavailability of metals in soil, and plants' ability to take up and accumulate metals as biomass (Babalola 2010). Several researchers have reported the positive effect of PGPR on bioavailability of heavy metals (Braud et al. 2006; Abou-Shanab et al. 2006). Generally, PGPR may improve phytoremediation efficiency of heavy metals by two mechanisms such as enhanced plant growth (abbaszadeh et al. 2010) and increased bioavailability of metals in soil (Whiting et al. 2001).

9.3 Plant Growth-Promoting Rhizobacteria (PGPR)

Among the rhizospheric microorganisms, PGPR believed to be worthy of special attention for heavy metal phytoremediation. PGPR are heterogeneous group of soil bacteria, categorized into two main parts, extracellular and intracellular, which are able to colonize plant root systems and enhance the plant growth (Pereira et al. 2015; Dimkpa et al. 2009a). The intracellular group bacteria are able to enter the plant as endophytic bacteria that create nodules, whereas extracellular PGPR exist in the rhizosphere, on the rhizoplane, or in the spaces between cells (apoplast) of the root cortex (Dimkpa et al. 2009a; Rajkumar et al. 2009). Since endophytic bacteria exist inside the plant, they could be more protected from different types of environmental stresses than rhizospheric bacteria (Rajkumar et al., 2009). PGPR may stimulate plant growth through various direct or indirect mechanisms such as production of growth substances (phytohormones), solubilization of insoluble elements, biocontrol of host plant pathogens, or enhancement of plant nutritional status (Abbaszadeh-dahaji et al. 2012; Glick et al. 1999).

These mechanisms comprise the synthesis of plant growth hormones such as auxin (Khakipour et al. 2008) and 1-aminocyclopropane-1-carboxyilic acid (ACC) deaminase, improved solubility of insoluble nutrients such as phosphorous and iron, fixation of atmospheric nitrogen, and control of the adverse impacts of pathogens on plant growth (Jalili et al. 2009). The major type of PGPR contains the strains of *Azospirillum*, *Azotobacter*, *Acetobacter*, *Burkholderia*, *Bacillus*, and *Pseudomonas* (Abbaszadeh et al. 2010).

9.4 Heavy Metal-Resistant PGPR

PGPR tolerance to the concentration of heavy metal is the most important limiting factor for using phytoremediation process. Metal-tolerant microorganisms have been often reported in the rhizosphere of hyperaccumulator plants growing in metal-contaminated soils exhibiting that these microorganisms have evolved a heavy metal tolerance and that they may play important roles in mobilization and/or immobilization of heavy metals through releasing different compounds such as organic acids or extracellular polymeric materials (Sessitsch et al. 2013). It has been suggested that plant inoculation with metal-resistant PGPR plays significant role in enhancing the efficiency of phytoremediation process (Ma et al. 2011; Rajkumar et al. 2012). Specific heavy metal-resistant PGPR may positively impact plants through improving plant tolerance to different environmental stresses, increasing root development and improving plant health and productivity (Glick 2010).

Plant growth-promoting and heavy metal-resistant activities are key attributes for bacteria used in metal phytoremediation; therefore, screening of effective PGPR and heavy metal-resistant bacteria should be taken into account as the primary work for cleaning up the heavy metal-polluted soil (Yu et al. 2014). Heavy metal-resistant PGPR may increase efficiency of phytoremediation by several mechanisms as follows:

A. Increasing the bioavailability of heavy metals for plant absorption.
B. Polluted soils are often poor in nutrient status; hence, PGPR enhance plants' nutrition and their growth and consequently phytoremediation efficiency.
C. Enhancement of plant growth through reducing plant stress generated by metal-polluted sites.

9.4.1 Enhanced Efficiency of Phytoremediation by Increasing the Bioavailability of Heavy Metals

Many factors such as soil nutrients, plant species type, pH, plant-associated microbial flora, and so on influence plant-microorganism interactions and thereby affect uptake of heavy metal by plants. However, as mentioned above heavy metal bioavailability in rhizosphere is considered to be a prominent factor determining the efficiency of phytoextraction and phytoremediation (Sessitsch et al. 2013). The content of bioavailable metals in the soil extremely affects the amount of metal accumulation in plants, because most parts of heavy metals are usually bound to different organic and inorganic compounds in contaminated soils and their phytoavailability are closely related to their chemical speciation (Ma et al. 2011). Rhizosphere plays an important role in phytoremediation process of heavy metal-polluted soils, in which microbial populations are believed to impact heavy metal mobility and availability to the plant by several pathways including release of chelating agents, solubilization activity, acidifcation, and redox changes (Bharti et al. 2014). According to the uptake by plants, heavy metals in soil may be categorized into three main classes, which comprise absorbable forms (e.g., free ions and chelating ions), exchangeable forms (bound to carbonates, organic matter, and iron-manganese oxides), and unavailable/or residual forms (Wei et al. 2008).

PGPR are among the soil microorganisms which are drawn in the plant interactions with metal-contaminated soil environments and require certain attention due to the fact that these may directly facilitate the phytoremediation process through changing bioavailability of metals by changes in synthesis of phytohormones (e.g., auxins), siderophores, and increased release of chelators (Ma et al. 2011). The use of heavy metal-solubilizing microorganisms is a promising procedure for improving bioavailability of heavy metals in contaminated soils. Plant-associated bacteria may be utilized to enhance phytoextraction activities through changing the solubility, availability, and translocation of toxic metals as well as nutrients, by decreasing soil pH and releasing chelating agents (Ma et al. 2011).

Production of metal chelating agents like siderophores by PGPR plays an important role in mobilization and accumulation of metals through a complexation reaction (Rajkumar et al. 2010; Gadd 2010). Braud et al. (2009) observed that Cr and Pb can be released into the soil solution after soil inoculation with *P. aeruginosa*. However, siderophores can constitute stable complexes with other environmental concern metals including Al, Cd, Cu, Ga, In, Pb, and Zn. The plant growth-promoting siderophore-producing rhizobacteria may enhance the phytoextraction rate that usually restricts the application of phytoremediation procedures (Braud et al. 2009). The importance of PGPR on Ni solubilization in soils was previously reported by Abou-Shanab et al. (2003).

According to Carrillo-Castaneda et al. (2003), siderophores secreted by PGPR play a key role in mobilization of soil metals. Some results emphasized the potential of plant inoculation with siderophore-producing bacteria to better enhance their phytoextraction efficiency (Pereira et al. 2015). It has been reported by Dimkpa et al. (2009b) that bacterial culture filtrates having hydroxamate siderophores produced via *S. tendae* F4 significantly improved uptake of Cd through the plant, when compared to the un-inoculated control. This study revealed that siderophores may help to mitigate toxicity of metals in bacteria while concurrently accelerating the uptake of such metals through plants. Soil beneficial rhizobacteria produce certain organic acids including gluconic, oxalic, and citric acids, which play essential roles in the solubility and mobilization of heavy metals. Ullah et al. (2015) reported that aforesaid organic acids play significant role in the adsorption reaction of heavy metals and improve their mobility for plants.

Moreover, Saravanan et al. (2007) displayed the production of gluconic acid derivative by *Gluconacetobacter diazotrophicus*, which contributes in the solubilization of Zn compounds.

Biosurfactants are believed to be integral significant metabolites produced by growth promoting that have the potential to increase mobilization and phytoremediation of toxic metals. The released biosurfactants by soil microorganisms create complexes with heavy metals at the soil interface, removing metals from the soil matrix and therefore accelerating solubility and bioavailability of metals and assisting phytoremediation process (Rajkumar et al. 2012). For example, biosurfactants produced through Pseudomonas aeruginosa BS2 would result in solubilization of Pb and Cd (Juwarkar et al. 2007). Furthermore, redox reactions are involved in mobilization of metals (Bolan et al. 2014).

According to (Gadd 2004), *Thiobacillus thiooxidans* (a promising sulfur-oxidizing bacteria) may acidify their surroundings through oxidation from ferrous iron to ferric iron and production of H_2SO_4 (Gadd 2004).

9.4.2 Improvement of Plant Nutrition, Root System, Metal Detoxification, and Biocontrol of Phytopathogens and Consequently Plant Growth and Phytoremediation Efficiency

PGPR in hosted plants not only trigger physiological changes but also alter the root architecture which may be monitored through variation in total root length and root tip (Bhattacharyya and Jha 2012). Phytohormones are very efficient on plant growth and development, and among them, IAA (indole-3-acetic acid) is one of the most important groups of growth regulators (Stepanova et al. 2008; Friml et al. 2003), often produced by PGPR (Mayak et al. 2004). PGPR generating IAA have been widely applied to enhance phytoremediation efficiency of metalliferous soil (Khan et al. 2009; Ma et al. 2011). Production of IAA is proposed as a key property of PGPR (Ahmad et al. 2008). IAA supplementation to soil may increase the uptake of metals in plant root surfaces. The presence and richness of strains producing high dose of IAA isolated from heavy metal-polluted soil in comparison to other soils implied that this PGP ability of the strains might aid in phytoremediating the soil (Yu et al. 2014). Marques et al. (2010) suggested that the correlation between IAA-producing rhizobacteria and corn (*Zea mays*) root biomass promotion was positive. Also, Khalid et al. (2004) reported that bacterial strains with ability to produce the highest amount of IAA increased the growth of wheat plant.

Enhancement of root growth is one of the main indicators by which the positive influence of PGPR is determined. Rapid establishment of plant root systems is useful for juvenile seedlings as it enhances their ability to attach themselves to the soil and to acquire water and different nutrients from their environment, increasing their chances of survival (Patten and Glick 2002).

The low level of iron uptake into plants cultivated in the presence of heavy metals may cause chlorosis because iron deficiency reduces both biosynthesis of chlorophyll and development of chloroplast (Ismande 1998). According to these ideas, the siderophore-overproducing bacteria could serve as potential source of iron for plants that grow under heavy metal stress conditions and hinder plants from chlorosis (yellowing of leaf tissue) through providing an adequate value of soluble iron to the plants (Suthersan 1999). PGPR's presence in heavy metal-polluted soil increases the uptake of iron in plants, which subsequently improve the chlorophyll contents and plant growth in the PGPR-treated plants (Kamran et al. 2015). Moreover, excessive value of accumulated heavy metals in plant tissues may cause changes in different critical processes of growth and have adverse effects on iron nutrition status. Under such conditions, the beneficial siderophore-producing rhizobacteria might offer an alternative biological rescue approach that is able to chelate Fe^{3+} and make it soluble to plants. The plant roots then could take up iron from siderophore-iron chelates possibly through various mechanisms including chelate degradation and release of iron, the direct uptake of siderophore-iron complexes, and/or a ligand exchange substitution (Rajkumar et al. 2010). Siderophores also stimulate biosynthesis of bacterial IAA through decreasing the deleterious impacts of heavy metals by chelation reaction (Ma et al. 2011). Rhizobacteria ability to convert insoluble forms from phosphorous to soluble forms is an important feature in PGPR for increasing plant productivity (Chen et al. 2006). However, under metal stress conditions, most heavy metal-tolerant PGPR strains can either change the insoluble phosphates into the soluble forms by the processes of acidification, chelation, exchange reactions, and production of organic acids (Chung et al. 2005).

Nitrogen-fixing rhizobacteria can beneficially influence on host plant growth through accelerating nitrogen availability. Therefore, they can act as an efficient biofertilizer which enhances plant growth and development (Kang et al. 2010). The PGPR can also contribute in decreasing phytotoxicity of metals through biosorption and bioaccumulation procedures. Because the bacterial cells (~1.0–1.5 mm^3) have very high-surface-area-to-volume ratio, they could take up more heavy metals than inorganic soil fractions (e.g., kaolinite and vermiculite) either through metabolism-independent or by a metabolism-dependent process, i.e., through passive or active pathways (Khan et al. 2007).

The ability of soil microorganisms to take up and accumulate toxic metals including Ag, Co, Cd, Cu, Zn, Mn, Pb, Ni, Hg, etc. has been previously reported. Various PGPR such as *Pseudomonas*, *Azospirillum*, *Azotobacter*, etc. and fungi including *Glomus* sp., *Gigaspora* sp., certain alga, and diatoms have such capability and are being investigated for their high biotechnological potential as effluent detoxification agents (Rajkumar et al. 2012). PGPR can also synthesize extracellular enzymes including cellulases, pectinases, proteases, and lipases. These bacterial characteristics will confer important benefits in the presence of phytopathogens, because their cell membrane and cell wall will be degraded through the exoenzymes and their negative impacts inhibited, resulting in enhancement of plant growth (Pereira et al. 2015).

9.4.3 Improvement of Plant Growth by Decreasing Plant Stress Generated by Metal-Contaminated Soils

Proline is a well-known α-amino acid, which is present in plants and microorganism (e.g., bacteria and fungi) experiencing abiotic stress conditions (John et al. 2009). Accumulation of proline also assists to maintain the structure of cell organelles and plays a key role in adjusting osmotic potential in plants under stress conditions (Kamran et al. 2015). Ahmad et al. (2006) reported that proline plays an essential role in the folding of proteins and molecular membrane, improving buffer cellular redox potential and protecting cell by scavenging reactive oxygen species (ROS) (Ahmad et al. 2006). Heavy metal stress induces accumulation of free proline in some plant species. Free proline enhances plant resistance to stress through osmoregulation process, which subsequently stabilizes protein synthesis, and protects enzymes against proteolytic degradation and denaturation (Modirroosta et al. 2014).

It has been reported that an increase in proline concentration was found in PGPR-inoculated wheat plants under heavy metal stress conditions (Janmohammadi et al. 2013). This process is regulated by PGPR-assisted tolerant plant species, which reduce the phytotoxic influence for several contaminates and are assumed to improve the metal absorption. According to Kamran et al. (2015), *Eruca sativa*-treated plants at higher Cd levels produce more proline contents, which verify the metal tolerance characteristics of this plant and role of PGPR to survive under heavy metal stress conditions.

The classical phytohormone, ethylene, has a prominent role in modulating the growth and metabolism of plants (Ping and Boland 2004) and seem to be involved in stress tolerance, disease resistance, plant-microorganism association, as well as plant nutrient cycle. Among its central role in triggering different physiological and biochemical variations in plants, the overproduction of ethylene may cause the reduction of primary root growth, lateral root initiation, and root hair formation (Ma et al. 2011); however, PGPR are potentially able to ameliorate the stress-mediated effect on plants via enzymatic hydrolysis of ACC (Glick et al. 2007). Bacterial ACC-deaminase enzyme is an inducible enzyme that plays a significant role in the regulation of ethylene, modifying growth and development of plants. It is acknowledged that inoculation with the ACC-deaminase-producing bacterial strains may ameliorate the stress-induced ethylene-mediated negative effect on plants (Glick 2005). Bacterial ACC-deaminase metabolizes the root's ACC into a-ketobutyrate and ammonia and inhibits ethylene production, which otherwise reduce plant growth by several mechanisms (Fig. 9.2).

The plant inoculation with bacteria possessing ACC-deaminase may have relatively extensive root system and therefore more growth and development because of less ethylene level and can better tolerate different biotic/abiotic stresses (Jalili et al. 2009). ACC-deaminase enzyme has been greatly reported in multiple PGPR such as *Agrobacterium genomovars*, *Alcaligenes*, *Azospirillum lipoferum*, *Bacillus*, *Burkholderia*, *Enterobacter*, *Methylobacterium fujisawaense*, *Pseudomonas*, *Ralstonia solanacearum*, *Rhizobium*, *Rhodococcus*, *Sinorhizobium meliloti*, and *Variovorax paradoxus* (Ghorbanpour and Hatami 2014; Ghorbanpour et al. 2013, 2016; Saleem et al. 2007).

Burd et al. (1998) reported that canola (*Brassica napus*) seedlings grown in the media supplemented with high concentrations of nickel produced lower quantities of ethylene when the canola seeds were treated with an ACC-deaminase-containing nickel-resistant bacteria. A positive correlation has been found between in vitro bacterial ACC-deaminase activity and their promoting effect on root elongation, suggesting that utilization of ACC is a prominent feature determining root growth enhancement. Soil bacteria offer promise as inoculants to enhance growth of *Brassica juncea* (a metal-accumulating plant) in the presence of toxic levels of Cd and are believed to be a useful approach for developing a successful phytoremediation strategy in contaminated soils (Belimov et al. 2005). It has been reported that

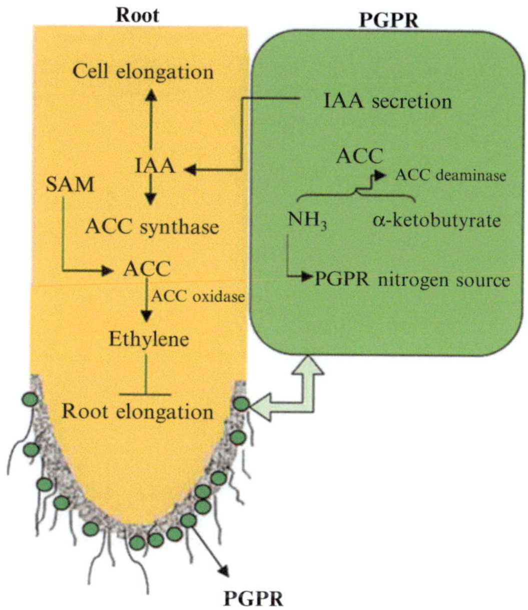

Fig. 9.2 Schematic diagram of the role of plant growth-promoting rhizobacteria (PGPR) on ethylene production and root growth inhibition. Abbreviations: *IAA* indoleacetic acid, *ACC* 1-aminocyclopropane-1-carboxylic acid, *SAM* S-adenosylmethionine (Ghorbanpour and Hatami 2014)

tobacco plants inoculated with *Pseudomonas putida* UW4 (containing ACC-deaminase enzyme activity) revealed better growth and accumulated a considerable value of metal from nickel-polluted site (Arshad et al. 2007). Furthermore, Sun et al. (2009) found that ACC-deaminase played influential for PGPR ability to enhance plant growth under heavy metal stress conditions. In a study, *S. plumbizincicola* inoculated with RC6b performed better in terms of growth in metal-polluted soils. The strain RC6b increased the root length, shoot length, fresh weight, and dry weight by 176 %, 27 %, 27 %, and 22 %, respectively, compared to non-inoculated plants. The increase in plant growth caused by *P. myrsinacearum* RC6b in metal-contaminated soils may be attributed to its ability to produce IAA, ACC-deaminase, and siderophores and solubilize P (Ma et al. 2013). Some researches in relation to the use of PGPR for increasing phytoremediation efficiency were listed in Table 9.1. Generally, the result of different researches and publications indicated that the use of PGPR could be effective and enhanced phytoremediation proficiency in heavy metal-contaminated sites.

Table 9.1 Some recent studies regarding the effects of PGPR on phytoremediation enhancement in heavy metal-contaminated soils

PGPR strains	Plant	Heavy metal	Role of PGPR	References
Azospirillum	*Panicum virgatum*	Pb and Cd	Bacteria increased the root length, branches, surface area, and root and shoot biomass	Arora et al. (2016)
Pseudomonas brassicacearum, *Rhizobium leguminosarum*	*Brassica juncea*	Zn	Plant growth increased	Adediran et al. (2016)
Streptomyces, *Azotobacter*, *Pseudomonas*, and *Paenibacillus*	*Pennisetum glaucum* and *Sorghum bicolor*	Fe	PGPR increased the extent of iron absorption	Mishra et al. (2016)
Pseudomonas putida	*Eruca sativa*	Cd	Inoculation with *P. putida* enhanced the Cd uptake potential of *E. sativa* and favors the healthy growth under Cd stress	Kamran et al. (2015)
Rhodococcus erythropolis EC34, *Achromobacter* sp. *1AP2*, and *Microbacterium* sp. *3ZP2*	*Trifolium repens*	Zn and Cd	Enhanced plant growth and the available and exchangeable metal concentrations in rhizosphere	Pereira et al. (2015)
Pseudomonas aeruginosa	*Amorpha fruticosa*	Cd, Cu, Ni, and Zn	Promoting plant growth and improving tolerance of the plant to heavy metals	Yu et al. (2015)
Bacillus licheniformis NCCP-59	*Oryza sativa*	Ni	Improved seed germination and plant growth	Jamil et al. (2014)
Pseudomonas fluorescence and *Pseudomonas aeruginosa*	*Prosopis juliflora*	F (fluoride)	Significantly increased the biomass and bioaccumulation factor	Vidyapith and Rajasthan, (2014)
Pseudomonas fluorescens	*Mirabilis jalapa*	Cd, Cr, Cu, Ni, and Zn	Significantly increased the biomass and heavy metal concentration	Petriccione et al. (2013)
Phyllobacterium myrsinacearum RC6b	*Sedum plumbizincicola*	Cd, Zn, and Pb	Increased growth, organ metal concentrations, and metal mobility in soil	Ma et al. (2013)

Bacillus safensis FO.036b and *Micrococcus roseus M2*	*Helianthus annuus*, *Amaranthus retroflexus*, *Medicago sativa*	Ni	Absorption of nickel increased	Moteshareزadeh and Savaghebi-Firoozabadi (2011)
Psychrobacter sp. SRS8	*Helianthus annuus*, *Ricinus communis*	Ni	Enhanced plant growth and Ni accretion in both plant species	Ma et al. (2011)
Agrobacterium radiobacter	*Populus deltoides*	As	Arsenic concentrations in roots, stems, and leaves were significantly increased	Wang et al. (2011)
Bradyrhizobium sp., *Pseudomonas* sp., *Ochrobactrum cytisi*	*Lupinus luteus*	Pb, Cu, and Cd	Increasing plant biomass	Dary et al. (2010)
Bacillus subtilis, *Bacillus cereus*, *Flavobacterium* sp., *Pseudomonas aeruginosa*	*Orychophragmus violaceus*	Zn	Increased shoot biomass and Zn accumulation.	He et al. (2010)
Bacillus subtilis, *B. cereus*, *Flavobacterium* sp., *Pseudomonas aeruginosa* (RS)	*Orychophragmus violaceus*	Zn	Increasing availability of water-soluble Zn in soil and Zn accumulation by plants	He et al. (2010)
Streptomyces tendae F4	*Sunflower* (*Helianthus annuus*)	Cd	Increasing cadmium availability and uptake	Dimkpa et al. (2009b)
Ralstonia metallidurans *Pseudomonas aeruginosa*	Maize	Cr and Pb	Increased metal uptake	Braud et al. (2009)
Pseudomonas fluorescens G10, *Microbacterium* sp. G16	Rape	Pb	Shoot Pb accumulation increased	Sheng et al. (2008)
Burkholderia cepacia	*Sedum alfredii*	Cd and Zn	Plant growth increased	Li et al. (2007)
Bacillus subtilis SJ-101	*Brassica juncea*	Ni	Approximately increased the accumulation of Ni; increased plant biomass	Zaidi et al. (2006)

References

Abbaszadeh P, Saleh-Rastin N, Asadi-Rahmani H, Khavazi K, Soltani A, Shoary-Nejati AR, Miransari M (2010) Plant growth-promoting activities of fluorescent pseudomonads, isolated from the Iranian soils. Acta Physiol Plant 32:281–288

Abbaszadeh-dahaji P, Savaghebi GR, Asadi-rahmani H, Rejali F, Farahbakhsh M, Moteshareh-zadeh B, Omidvari M, Lindstrom K (2012) Symbiotic effectiveness and plant growth promoting traits in some Rhizobium strains isolated from *Phaseolus vulgaris* L. Plant Growth Regul 68:361–370

Abdelatey LM, Khalil WK, Ali TH, Mahrous KF (2011) Heavy metal resistance and gene expression analysis of metal resistance genes in gram-positive and gram-negative bacteria present in egyptian soils. J Appl Sci Env San 6:201–211

Abou-Shanab RA, Angle JS, Delorme TA, Chaney RL, van Berkum P, Moawad H, Ghanem K, Ghozlan HA (2003) Rhizobacterial effects on nickel extraction from soil and uptake by *Alyssum murale*. New Phytol 158:219–224

Abou-Shanab RAI, Angle JS, Chaney RL (2006) Bacterial inoculants affecting nickel uptake by Alyssum murale from low, moderate and high Ni soils. Soil Biol Biochem 38:2882–2889

Adediran GA, Ngwenya BT, Mosselmans JFW, Heal KV, Harvie BA (2016) Mixed planting with a leguminous plant outperforms bacteria in promoting growth of a metal remediating plant through histidine synthesis. Int J Phytoremediation 18(7):720–729

Ahmad P, Sharma S, Srivastava PS (2006) Differential physio-biochemical responses of high yielding varieties of Mulberry (*Morus alba*) under alkalinity (Na_2CO_3) stress *in vitro*. J Plant Physiol Mol Biol 12:59–66

Ahmad F, Ahmad I, Khan M (2008) Screening of free-living rhizospheric bacteria for their multiple plant growth promoting activities. Microbiol Res 163:173–181

Ali H, Khan E, Sajad MA (2013) Phytoremediation of heavy metals concepts and applications. Chemosphere 91:869–881

Arora K, Sharma S, Monti A (2016) Bio-remediation of Pb and Cd polluted soils by switch grass: A case study in India. Int J Phytoremediation 18(7):704–709

Arshad M, Saleem M, Hussain S (2007) Perspectives of bacterial ACC deaminase in phytoremediation. Tren Biotech 25(8):356–361

Babalola OO (2010) Beneficial bacteria of agricultural importance. Biotechnol Lett 32(11):1559–1570

Barrutia O, Garbisu C, Hernandez-Allica J, Garcıa-Plazaola JI, Becerril JM (2010) Differences in EDTA-assisted metal phytoextraction between metallicolous and non-metallicolous accessions of *Rumex acetosa* L. Environ Pollut 58:1710–1715

Belimov AA, Hontzeas N, Safronova VI, Demchinskaya SV, Piluzza G, Bullitta S, Glick BR (2005) Cadmium-tolerant plant growth-promoting bacteria associated with the roots of Indian mustard (*Brassica juncea* L. Czern.). Soil Biol Biochem 37:241–250

Bharti RP, Shri Vastava A, Soni N, Tiwari A, More S, Ram Choudhary J (2014) Phytoremediation of heavy metal toxicity and role of soil in rhizobacteria. Int J Sci Res Pub 4(1):1–5

Bhattacharyya PN, Jha DK (2012) Plant growth-promoting rhizobacteria (PGPR): emergence in agriculture. World J Microbiol Biotechnol 28:1327–1350

Bolan N, Kunhikrishnan A, Thangarajan R, Kumpiene J, Park J, Makino T, Kirkham MB, Scheckel K (2014) Remediation of heavy metal(loid)s contaminated soils–to mobilize or to immobilize? J Hazard Mater 266:141–166

Boyd RS (2010) Heavy metal pollutants and chemical ecology: exploring new frontiers. J Chem Ecol 36(1):46–58

Braud A, Jezequel K, Vieille E, Tritter A, Lebeau T (2006) Changes in extractability of Cr and Pb in a polycontaminated soil after bioaugmentation with microbial producers of biosurfactants, organic acids and siderophores. Water Air Soil Pollut Focus 6:261–279

Braud A, Jezequel K, Bazot S, Lebeau T (2009) Enhanced phytoextraction of an agricultural Cr-, Hg and Pb-contaminated soil by bioaugmentation with siderophore producing bacteria. Chemosphere 74:280–286

Burd GI, Dixon DG, Glick BR (1998) A plant growth promoting bacterium that decreases nickel toxicity in plant seedlings. Appl Environ Microbiol 64:3663–3668

Carrillo-Castaneda G, Munoz JJ, Peralta-Videa JR, Gomez E, Gardea-Torresdey JL (2003) Plant growth promoting bacteria promote copper and iron translocation from root to shoot in alfalfa seedlings. J Plant Nutr 26:1801–1814

Chaudhary K, Khan S (2015) Review: plant microbe-interaction in heavy metal contaminated soils. Indian Res J Genet Biotech 7(2):235–240

Chen YP, Rekha PD, Arunshen AB, Lai WA, Young CC (2006) Phosphate solubilizing bacteria from subtropical soil and their tricalcium phosphate solubilizing abilities. Appl Soil Ecol 34:33–41

Chirakkara RA, Reddy KR (2015) Plant species identification for phytoremediation of mixed contaminated soils. J Hazard Toxic Radioact Waste 19(4):1–10

Chung H, Park M, Madhaiyan M, Seshadri S, Song J, Cho H, Sa T (2005) Isolation and characterization of phosphate solubilizing bacteria from the rhizosphere of crop plants of Korea. Soil Biol Biochem 37(10):1970–1974

Dary M, Chamber-Pérez MA, Palomares AJ, Pajuelo E (2010) In situ phytostabilisation of heavy metal polluted soils using *Lupinus luteus* inoculated with metal resistant plant growth promoting rhizobacteria. J Hazard Mater 177:323–330

DeBashan LE, Hernandez JP, Bashan Y (2012) The potential contribution of plant growth promoting bacteria to reduce environmental degradation. A comprehensive evaluation. Appl Soil Ecol 61:171–189

Dimkpa C, Weinand T, Asch F (2009a) Plant–rhizobacteria interactions alleviate abiotic stress conditions. Plant Cell Environ 32(12):1682–1694

Dimkpa CO, Merten D, Svatos A, Büchel G, Kothe E (2009b) Siderophores mediate reduced and increased uptake of cadmium by *Streptomyces tendae* F4 and sunflower (*Helianthus annuus*), respectively. J Appl Microbiol 107:1687–1696

Evangelou MWH, Bauer U, Ebel M, Schaeffer A (2007) The influence of EDDS and EDTA on the uptake of heavy metals of Cd and Cu from soil with tobacco *Nicotiana tabacum*. Chemosphere 68:345–353

Friml J, Vieten A, Sauer M, Weijers D, Schwarz H, Hamann T, Offringa R, Jurgens G (2003) Efflux-dependent auxin gradients establish the apical-basal axis of Arabidopsis. Nature 426:147–153

Gadd GM (2004) Microbial influence on metal mobility and application for bioremediation. Geoderma 122:109–119

Gadd GM (2010) Metals, minerals and microbes: geomicrobiology and bioremediation. Microbiology 156:609–643

Ghorbanpour M, Hatami M (2014) Biopriming of Salvia officinalis L. seed with plant growth promoting rhizobacteria (PGPRs) changes the invigoration and primary growth indices. J Biol Environ Sci 8:29–36

Ghorbanpour M, Hatami M, Khavazi K (2013) Role of plant growth promoting rhizobacteria on antioxidant enzyme activities and tropane alkaloid production of Hyoscyamus niger under water deficit stress. Turk J Biol 37:350–360

Ghorbanpour M, Hatami M, Kariman K, Abbaszadeh DP (2016) Phytochemical variations and enhanced efficiency of antioxidant and antimicrobial ingredients in Salvia officinalis as inoculated with different rhizobacteria. Chem Biodivers 13(3):319–330

Glick BR (2005) Modulation of plant ethylene levels by the bacterial enzyme ACC deaminase. FEMS Microbiol Lett 251:1–7

Glick BR (2010) Using soil bacteria to facilitate phytoremediation. Biotechnol Adv 28:367–487

Glick BR, Patten CL, Holguin G, Penrose DM (1999) Biochemical and genetic mechanisms used by plant growth promoting bacteria. Imperial College Press, London, p 267

Glick BR, Cheng Z, Czarny J, Duan J (2007) Promotion of plant growth by ACC deaminase producing soil bacteria. Eur J Plant Pathol 119(3):329–339

He CQ, Tan GE, Liang X, Du W, Chen YL, Zhi GY, Zhu Y (2010) Effect of Zn-tolerant bacterial strains on growth and Zn accumulation in *Orychophragmus violaceus*. Appl Soil Ecol 44:1–5

Huang XD, El-Alawi Y, Gurska J, Glick BR, Greenberg BM (2005) A multi-process phytoremediation system for decontamination of persistent total petroleum hydrocarbons (TPHs) from soils. Microchem J 81:139–147

Hutchinson SL, Schwab AP, Banks MK (2003) Biodegradation of petroleum hydrocarbons in the rhizosphere. In: McCutcheon SC, Schnoor JL (eds) Phytoremediation: transformation and control of contaminants Wiley, New York

Ismande J (1998) Iron, sulfur and chlorophyll deficiencies: a need for an integrative approach in plant physiology. Physiol Plant 103:139–144

Jalili F, Khavazi K, Pazira E, Nejati A, Asadi Rahmani H, Rasuli Sadaghiani H, Miransari M (2009) Isolation and characterization of ACC deaminase producing fluorescent pseudomonads, to alleviate salinity stress on canola (*Brassica napus* L.) growth. J Plant Physiol 166:667–674

Jamil M, Zeb S, Anees M, Roohi A, Ahmed I, Rehman S, Rha ES (2014) Role of Bacillus licheniformis in phytoremediationof nickel contaminated soil cultivated with rice. Int J Phytorem 16(6):554–571

Janmohammadi M, Bihamta M, Ghasemzadeh F (2013) Influence of rhizobacteria inoculation and lead stress on the physiological and biochemical attributes of wheat genotypes. Cercet Agron Moldova 46:49–67

John R, Ahmad P, Gadgil K, Sharma S (2009) Heavy metal toxicity: effect on plant growth, biochemical parameters and metal accumulation by *Brassica juncea* L. Int J Plant Prod 3:65–76

Juwarkar AA, Nair A, Dubey KV, Singh SK, Devotta S (2007) Biosurfactant technology for remediation of cadmium and lead contaminated soils. Chemosphere 10:1996–2002

Kamran MA, Syed JH, Eqani SAMAS, Munis MFH, Chaudhary HJ (2015) Effect of plant growth promoting rhizobacteria inoculation on cadmium (Cd) uptake by *Eruca sativa*. Environ Sci Pollut Res 22:9275–9283

Kang BG, Kim WT, Yun HS, Chang SC (2010) Use of plant growth-promoting rhizobacteria to control stress responses of plant roots. Plant Biotechnol Rep 4(3):179–183

Khakipour N, Khavazi K, Mojallali H, Pazira E, Asadirahmani H (2008) Production of auxin hormone by fluorescent pseudomonads. Am Eurasian J Agric Environ Sci 4:687–692

Khalid A, Arshad M, Zahir AA (2004) Screening plant growth promoting rhizobacteria for improving growth and yield of wheat. J Appl Microbiol 96:473–480

Khan MS, Zaidi A, Wani PA (2007) Role of phosphate-solubilizing microorganisms in sustainable agriculture a review. Agron Sustain Dev 27:29–43

Khan MS, Zaidi A, Wani PA, Oves M (2009) Role of plant growth promoting rhizobacteria in the remediation of metal contaminated soils. Environ Chem Lett 7:1–19

Laghlimi M, Baghdad B, El Hadi H, Bouabdli A (2015) Phytoremediation mechanisms of heavy metal contaminated soils: a review. Open J Ecol 5:375–388

Lambert M, Leven B, Green R (2000) New methods of cleaning up heavy metal in soils and water. Environmental science and technology briefs for citizens. Kansas State University, Manhattan

Li K, Ramakrishna W (2011) Effect of multiple metal resistant bacteria from contaminated lake sediments on metal accumulation and plant growth. J Hazard Mater 189:531–539

Li WC, Ye ZH, Wong MH (2007) Effects of bacteria on enhanced metal uptake of the Cd/Zn-hyperaccumulating plant, *Sedum alfredii*. J Exp Bot 58:4173–4182

Ma Y, Prasad MNV, Rajkumar M, Freitas H (2011) Plant growth promoting rhizobacteria and endophytes accelerate phytoremediation of metalliferous soils. Biotechnol Adv 29:248–258

Ma Y, Rajkumar M, Luo Y, Freitas H (2013) Phytoextraction of heavy metal polluted soils using *Sedum plumbizincicola* inoculated with metal mobilizing *Phyllobacterium myrsinacearum* RC6b. Chemosphere 93:1383–1392

Marques APGC, Rangel AOSS, Castro PML (2009) Remediation of heavy metal contaminated soils: phytoremediation as a potentially promising clean-up technology. Crit Rev Environ Sci Technol 39:622–654

Marques APGC, Pires C, Moreira H, Rangel AOSS, Castro PML (2010) Assessment of the plant growth promotion abilities of six bacterial isolates using *Zea mays* as indicator plant. Soil Biol Biochem 42:1229–1235

Mayak S, Tirosh T, Glick BR (2004) Plant growth-promoting bacteria confer resistance in tomato plants under salt stress. Plant Physiol Biochem 42:565–572

McCutcheon SC, Schnoor JL (2003) Over view of phytotransformation and control of wastes. In: McCutcheon SC, Schnoor JL (eds) Phytoremediation: transformation and control of contaminants, Wiley, NewYork, pp 3–58

Mishra V, Gupta A, Kaur P, Simranjeet S, Singh N, Gehlot P, Singh J (2016) Synergistic effects of Arbuscular mycorrhizal fungi and plant growth promoting rhizobacteria in bioremediation of iron contaminated soils. Int J Phytoremediation 18(7):704–709

Modirroosta S, Ardalan MM, Bayramzadeh V (2014) Impact of soil cadmium contamination on accumulation of cadmium and proline content of *Pinus sylvestris* L. Seedling Agric Sci Dev 3(2):167–172

Motesharezadeh B, Savaghebi-Firoozabadi GR (2011) Study of the increase in phytoremediation efficiency in a nickel polluted soil by the usage of native bacteria: *Bacillus safensis* FO.036b and *Micrococcus roseus* M2. Caspian J Env Sci 9(2):133–143

Patten CL, Glick BR (2002) Role of *Pseudomonas putida* indole-acetic acid in development of the host plant root system. Appl Environ Microbiol 18:3795–3801

Pereira SIA, Barbosa L, Castro PML (2015) Rhizobacteria isolated from a metal-polluted area enhance plant growth in zinc and cadmium-contaminated soil. Int J Environ Sci Technol 12:2127–2142

Petriccione M, Di Patre D, Ferrante P, Papa S, Bartoli G, Fioretto A, Scortichini M (2013) Effects of *Pseudomonas fluorescens* seed bioinoculation on heavy metal accumulation for *Mirabilis jalapa* phytoextraction in smelter-contaminated soil. Water Air Soil Pollut 224:1–17

Pilon-Smits E (2005) Phytoremediation. Annu Rev Plant Biol 56:15–39

Ping L, Boland W (2004) Signals from the underground: bacterial volatiles promote growth in Arabidopsis. Trends Plant Sci 9:263–266

Rajkumar M, Ae N, Freitas H (2009) Endophytic bacteria and their potential to enhance heavy metal phytoextraction. Chemosphere 77:153–160

Rajkumar M, Prasad MNV, Freitas H (2010) Potential of siderophore producing bacteria for improving heavy metal phytoextraction. Trends Biotechnol 128:142–149

Rajkumar M, Sandhya S, Prasad MNV, Freitas H (2012) Perspectives of plant-associated microbes in heavy metal phytoremediation. Biotechnol Adv 30:1562–1574

Raskin I, Smith RD, Salt DE (1997) Phytoremediation of metals: using plants to remove pollutants from the environment. Curr Opin Biotechnol 8:221–226

Saleem M, Arshad M, Hussain S, Bhatti AS (2007) Perspective of plant growth promoting rhizobacteria (PGPR) containing ACC deaminase in stress agriculture. J Ind Microbiol Biotechnol 34:635–648

Saravanan VS, Madhaiyan M, Thangaraju M (2007) Solubilization of zinc compounds by the diazotrophic, plant growth promoting bacterium *Gluconacetobacter diazotrophicus*. Chemosphere 66:1794–1798

Sessitsch A, Kuffner M, Kidd P, Vangronsveld J, Wenzel WW, Fallmann K, Puschenreiter M (2013) The role of plant-associated bacteria in the mobilization and phytoextraction of trace elements in contaminated soils. Soil Biol Biochem 60:182–194

Sheng XF, Xia JJ, Jiang CY, He LY, Qian M (2008) Characterization of heavy metal-resistant endophytic bacteria from rape (*Brassica napus*) roots and their potential in promoting the growth and lead accumulation of rape. Environ Pollut 156:1164–1170

Sheoran V, Sheoran A, Poonia P (2011) Role of hyperaccumulators in phytoextraction of metals from contaminated mining sites: a review. Crit Rev Environ Sci Technol 41:168–214

Stepanova AN, Robertson-Hoyt J, Yun J, Benavente LM, Xie DY, Dolezˇal K, Jürgens SG, Alonso JM (2008) TAA1-mediated auxin biosynthesis is essential for hormone crosstalk and plant development. Cell 133:177–191

Sun Y, Cheng Z, Glick B (2009) The presence of a 1-aminocyclopropane-1-carboxylate (ACC) deaminase deletion mutation alters the physiology of the endophytic plant growth-promoting bacterium *Burkholderia phytofirmans* PsJN. FEMS Microbiol Lett 296:31–36

Suthersan SS (1999) Phytoremediation: remediation engineering-design concepts. CRC Press LLC, Boca Raton

Ullah A, Mushtaq H, Ali H, Munis MFH, Javed MT, Chaudhary HJ (2015) Diazotrophs assisted phytoremediation of heavy metals: a novel approach. Environ Sci Pollut Res 22:2505–2710

Vidyapith B, Rajasthan BT (2014) Effect of plant growth promoting rhizobacteria (PGPR) on plant growth and flouride (f) uptake by F hyperaccumulator plant *prosopis juliflora*. Int J Recent Sci Res 5(11):1995–1999

Wang Q, Xiong D, Zhao P, Yu X, Tu B, Wang G (2011) Effect of applying an arsenic-resistant and plant growth promoting rhizobacterium to enhance soil arsenic phytoremediation by Populus deltoides LH05-17. J Appl Microbiol 111:1065–1074

Wei SH, Teixeira da Silva JA, Zhou QX (2008) Agro-improving method of phytoextracting heavy metal contaminated soil. J Hazard Mater 150:662–668

Whiting SN, deSouza MP, Norman T (2001) Rhizosphere bacteria mobilize Zn for hyperaccumulation by *Thlaspi caerulescens*. Environ Sci Technol 35:3144–3150

Wuana RA, Okieimen FE (2011) Heavy metals in contaminated soil: a sources review of, chemistry, risks and best available strategies for bioremediation. ISRN Ecology 2011:20

Yu X, Li Y, Zhang C, Liu H, Liu J, Zheng W, Kang X, Leng X, Zhao K, Gu Y, Zhang X, Xiang Q, Chen Q (2014) Culturable heavy metal-resistant and plant growth promoting bacteria in v-ti magnetite mine tailing soil from Panzhihua, China. Plos One 9(9):1–8

Yu Y, Zhang Y, Zhang Q, Zhang X, Meng X, Lu Z (2015) Improvement of heavy metal resistant bacteria on phytoremediation of reclaimed land using coal gangue. J Residuals Sci Tech 12(1):105–109

Zaidi S, Usmani S, Singh BR, Musarrat J (2006) Significance of *Bacillus subtilis* strain SJ-101 as a bioinoculant for concurrent plant growth promotion and nickel accumulation in *Brassica juncea*. Chemosphere 64:991–997

PGPR-Mediated Amelioration of Crops Under Salt Stress

10

Anukool Vaishnav, Ajit Varma, Narendra Tuteja, and Devendra Kumar Choudhary

Abstract

Soil salinity is a major abiotic factor which adversely affects the crop growth and productivity worldwide. Higher salt concentration caused ion imbalance and hyperosmotic stress which often lead to oxidative stress in plants. Soil salinization is mainly due to the poor irrigation management practices and natural causes. A total 20 % of the world's cultivated lands and almost half of all irrigated lands are affected by high salinity. This chapter begins by stressing the importance of research into plant salt tolerance. After a brief outline of salinity-induced damage to both agricultural yield and growth of plants, strategies which plants adopt to deal with salinity are discussed, and current biotechnological efforts towards producing salt-tolerant crops are summarized. Particular attention is paid towards the application of plant growth-promoting bacteria in agriculture system for producing salt stress-tolerant crops and a fundamental understanding towards the mechanisms of beneficial plant–microbe interaction in the presence of salt.

The global need for food production has never greater than it is today. This issue is a major concern for developing countries, where the population is expected to rise by 90 % and where food insecurity is a major subject. Hence, increasing human population and decreasing cultivable land are two threats for agricultural sustainability (Shahbaz and Ashraf 2013). The increasing demand of food may be meet either by bringing more area under cultivation or increasing the productivity from the land already under cultivation. The former one seems unlikely to happen because of increasing pollution, urbanization, industrialization, soil degradation and limited

A. Vaishnav • A. Varma • N. Tuteja • D.K. Choudhary (✉)
Amity Institute of Microbial Technology (AIMT),
Block 'E-3', 4th Floor, Amity University Campus, Sector-125, Gautam Buddha Nagar, 201313 Noida, Uttar Pradesh, India
e-mail: dkchoudhary1@amity.edu

D.K. Choudhary et al. (eds.), *Plant-Microbe Interaction: An Approach to Sustainable Agriculture*, DOI 10.1007/978-981-10-2854-0_10

water availability. Thus, to increase the agriculture productivity, the only possible environment-friendly approach is to improve the agricultural land already in use. However, it is a big challenge to increase the efficiency and sustainability of existing global agriculture system, because this system is regularly marked by several biotic and abiotic factors.

10.1 Stresses

Once a seed germinates, plants being immobile are destined to stay where they are. Thus, they experience heavy selection pressure in their environments. Plants have developed many traits that help them to evolve and succeed across the globe under different environmental regimes. This selective pressure can be divided into abiotic and biotic stresses. Biotic stresses occur when a plant is attacked by an unwanted organism that causes damage to them, may it be the attack by virus, bacteria and fungi or grazing by an insect or higher animal. Abiotic stresses include the effect of winds, extreme temperatures, soil salinity, drought and flood. Among these stresses soil salinity is one of the most serious abiotic stresses which majorly reduce cultivated land area, productivity and quality.

10.1.1 Soil Salinity

A soil is defined as saline when its electrical conductivity (EC) exceeds 4 dSm^{-1} (approximately 40 mM NaCl). This salinity level is critical to reduce yield of many crop plants (Jamil et al. 2011). Different types of salts, e.g. sodium chloride (NaCl), sodium sulphate (Na_2SO_4), sodium nitrate ($NaNO_3$), magnesium sulphate ($MgSO_4$), magnesium chloride ($MgCl_2$), potassium sulphate (K_2SO_4), calcium carbonate ($CaCO_3$), etc., are present in saline soil in which NaCl causes serious problems for higher plants. Increase in these salt limits leads to two major stresses for the plants: osmotic stress and ionic stress. The osmotic stress firstly comes in plants when salt concentrations increase outside the roots, which leads to reduction in water uptake and subsequently plant development. The ionic stress develops when Na^+ accumulation increases in plants particularly in leaves over threshold level which caused chlorosis in leaves and reduced photosynthesis and other metabolic activities (Munns and Tester 2008).

10.1.2 Reason for Soil Salinity

A saline soil possesses high concentration of Na^+, K^+, Ca^{2+}, Mg^{2+} and Cl^- salts, which comes from weathering of minerals, irrigation water or evaporation of shallow groundwater. Due to insufficient precipitation, ions could not leach from the soil profile resulting salts to accumulate in the soil. Rainfall contains seawater salts, mainly sodium chloride (10 mg/kg) that would affect the land by deposition of

10 kg/ha of salt during each 100 mm of rainfall per year. In arid and semiarid regions of the world, soils are becoming saline due to poor irrigation management. Soil salinity is regularly increasing, and it has been estimated that 20 % of total cultivated and 33 % of irrigated agricultural lands worldwide are salt affected. If it happens continuously, the cultivable land would be 50 % salinized by the year 2050.

10.2 Plant Physiology Under Salt Stress

In saline soil, water potential decreased in surrounding the root, and plants suffer from the osmotic stress and ionic effect of Na^+ and Cl^-. Accumulation of Na^+ plays a central role in reduction of plant growth and senescence during salinity. Therefore, cytoplasmic Na^+ concentration is regulated by the plants to tolerate salt stress. In general, the salt stress response of plants consists of ion homeostasis, osmotic adjustment and detoxification (Fig. 10.1).

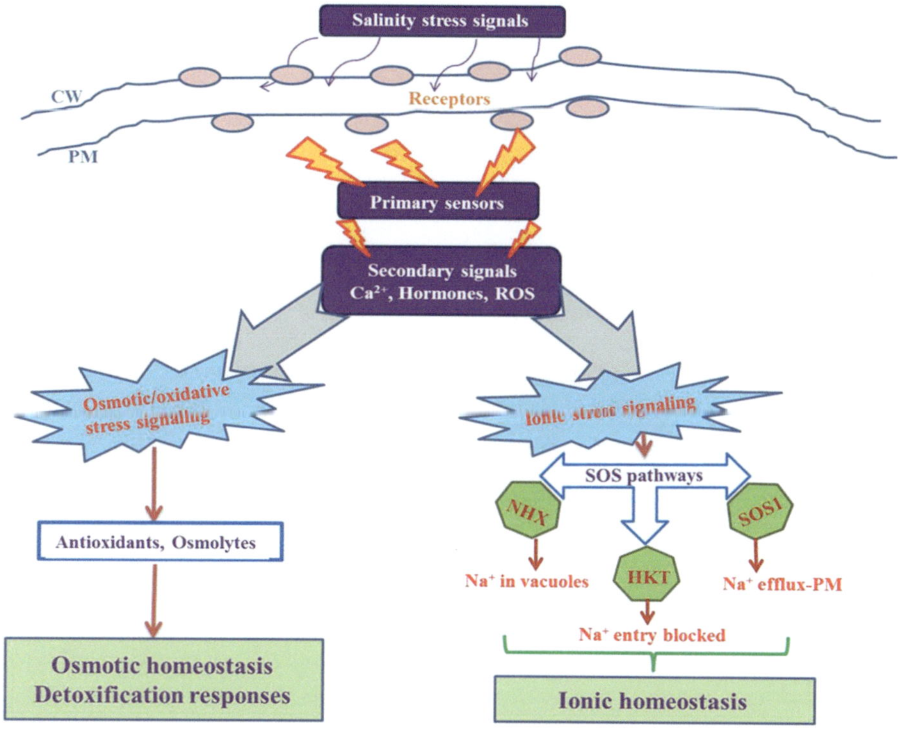

Fig. 10.1 Schematic pathway for the transduction of osmotic and ionic stress in plants

10.2.1 Ion Homeostasis

From an energetic viewpoint, it is preferable to control ion uptake initially rather than spending energy on detoxification and damage repair. Plant cells have mechanisms to buffer excess ions temporarily through the large, membrane-bound vacuoles. During salinity stress ion homeostasis of plant gets disturbed, resulting in excess toxic Na^+ in cytoplasm and deficiency into K^+. To regulate this, several ion transporters function to regulate Na^+ transport and its accumulation in plant tissues (Huang et al. 2012). The Na^+/H^+ antiporters catalyse the exchange of Na^+ for H^+ across the membranes. Plant Na^+/H^+ antiporters have been isolated from *Arabidopsis* (AtNHX1, SOS1; Gaxiola et al. 1999; Shi et al. 2002) and rice plants (Fukuda et al. 2004). *A. thaliana* plasma membrane Na^+/H^+ antiporter (*AtNHX1*), salt overly sensitive (*SOS1*) and the high-affinity K^+ transporter 1 (*HKT1*) genes were suggested to be essential for salt tolerance (Brini and Masmoudi 2012).

10.2.2 Osmotic Adjustment

Osmotic shock induced by a rapid increase of salinity triggers a fast and transient decrease in rates of leaf expansion and root elongation. Increased salt concentrations cause the water potential of the soil more negative than the root symplast, arising in tissue dehydration. The plant root must establish a water potential gradient so that water abounds into the plant from the soil. Plants can regulate their osmotic potentials within a certain range to indemnify for the low exterior water potential, and this is assembling osmotic adjustment. During salt stress, plants accumulate organic solutes together with primarily organic acids, nitrogen compounds and carbohydrates, e.g. malate, aspartate, glutamate, glycinebetaine, proline and sucrose, in the cytoplasm to take care of a low water potential within the cell. These solutes are involved in osmotic/oxidative stress management and protect macromolecules from damaging effects of increasing ionic strength of surrounding media of these stresses (Sharma et al. 2012). Polyols such as sorbitol, pinitol and mannitol also play a role as osmoprotectant. In addition, free amino acids have been reported to accumulate in plants subjected to salt stress. Proline accumulation occurs in larger amount in comparison to other amino acids and thus regulates N availability and osmotic adjustment. Proline being osmotically active contributes to stability of membrane and thus reduces the damaging effects of salt on membrane (Iqbal et al. 2014).

10.2.3 Detoxification

During salinity, the level of reactive oxygen species (ROS) is increased which creates oxidative stress in plants. Reactive oxygen species are consistently for metabolic pathways localized in altered cellular compartments. These are hydroxyl radicals ($OH^{\cdot}$), singlet oxygen ($^1O_2$), hydrogen peroxide (H_2O_2) and superoxide radical (O^{2-}) (Sharma et al. 2012). These radicals catalyse self-propagating autoxidation reactions that lead to the formation of other organic peroxides, which cause major damage to biological system. Higher ROS concentrations are responsible for plant cell death by causing lipid peroxidation, protein oxidation and harm to nucleic acid. To fight against the deleterious effects of reactive oxygen species, plants are endowed with several antioxidants and metabolites in different plant cell compartments. The enzymatic antioxidants include superoxide dismutase (SOD), catalase (CAT), guaiacol peroxidase (GPX) and enzymes of ascorbate glutathione (AsA-GSH) cycle such as ascorbate peroxidase (APX), monodehydroascorbate reductase (MDHAR), dehydroascorbate reductase (DHAR) and glutathione reductase (GR). The nonenzymatic antioxidants inside the cell are ascorbate (AsA), glutathione (GSH), carotenoids, tocopherols and phenolics (Gill and Tuteja 2010).

10.3 Approaches for Producing Salt-Tolerant Crops

Strategies for making tolerant plants to salinity and produce economically valuable species have been extensively studied for decades. A variety of strategies are used for improvement of crop tolerance, including traditionally (breeding), chemically (priming agents), molecular method (genetic engineering) and biologically (biofertilizers). Through traditional approaches, crops resistant/tolerant to salt stress have been bred, and the work is continuing. Direct selection based on higher yields on different location trials has been traditionally used for the development of tolerant/resistant varieties. This approach is time-consuming and labour intensive. In addition, strategic marker-assisted breeding is used in the development of tolerant cultivars that is further accelerated by development of molecular techniques such as molecular markers, gene mapping, QTL analysis and transgenics (Agarwal et al. 2013).

Genetic engineering (gene transfer) is the most progressive molecular approach which is used for the enhancement of tolerance level in plants. The variation between salt-tolerant (halophytes) and salt-sensitive (glycophytes) genotypes provides a genetic basis for engineering salt-tolerant crops. Several genes associated with salt response have been transferred into different plants to improve their tolerance against salt stress. These genes are involved in various types of activities during salt stress like compartmentalization of toxic ions in the vacuole, induction of antioxidant enzymes, synthesis of new proteins and accumulation of compatible solutes (Ashraf and Akram 2009). However, genetic engineering technique is not so successive due to its related ethical issues.

Salt-tolerant plants are also achieved via priming treatment with exogenous chemicals. These chemicals are natural products produced in plants at very low concentration, and when they are synthesized chemically and applied on plants, they start controlling downstream process. These chemicals include nitric oxide (NO), hydrogen peroxide (H_2O_2), sugars, hydrogen sulphide (H_2S), proline (Pro), glycinebetaine (GB), β-aminobutyric acid (BABA), jasmonates (JA), salicylic acid (SA), etc. (Ben Rejeb et al. 2013). However, these chemicals are

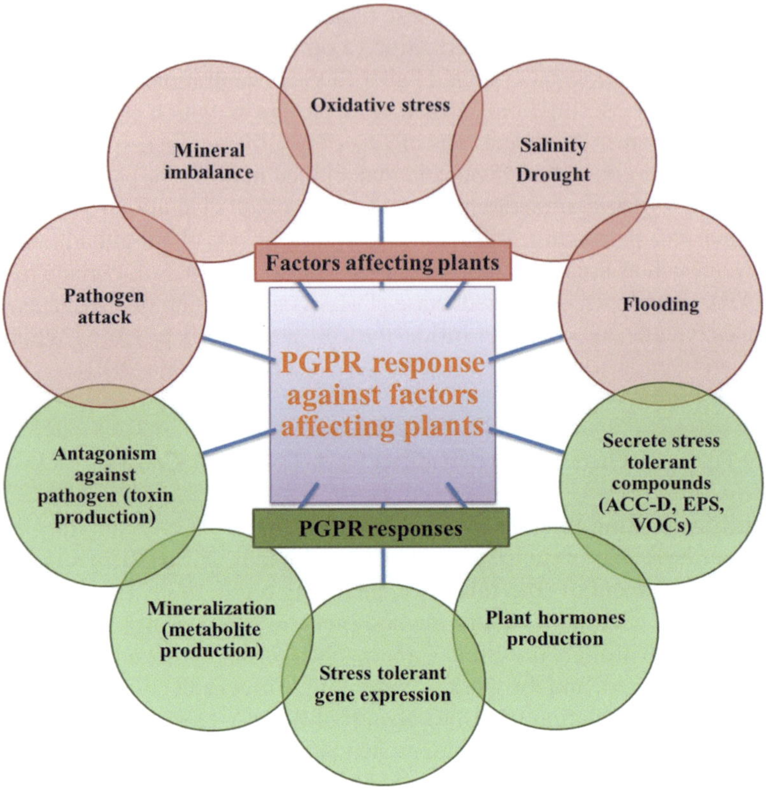

Fig. 10.2 Various types of PGPR responses against different factors affecting plant growth

cost-effective and also caused a number of long-term environmental problems. Hence, these chemicals are not suggested for easy and economical approaches for sustainable agriculture.

Alternatively, the use of plant growth-promoting rhizobacteria (PGPR) is the most promising approach to enhance crop production in saline regions. The beneficial relationship between PGPR and plants is an earlier theory, but the current concern is the application of these bacteria as inoculums in agriculture to mitigate stress conditions (Shrivastava and Kumar 2015). PGPR are bacteria which colonize in the rhizosphere/endorhizosphere of plants and promote growth of the plants through various direct and indirect mechanisms. Indirect mechanisms are related with plant pathogen inhibition. In this mechanism, bacteria secrete antibiotics and lytic enzymes that provide resistance against pathogen attack. In direct mechanisms, bacteria directly effect on plant itself and enhance growth by facilitating the nutrient uptake through mineral solubilization and phytohormone production (Fig. 10.2). Various salt-tolerant PGPR genera including *Rhizobium*, *Pseudomonas*, *Acetobacter*, *Bacillus*, *Serratia* and *Azospirillum* are being used and tested for plant growth promotion under salinity (Choudhary et al. 2015).

10.4 Induced Systemic Tolerance

Plants are sessile in nature; that's why they have evolved many adaptive strategies against a broad range of external factors. Adaptation to stresses has been suggested to be mediated by pre-existing or "memory" defences which lead to rapid and strong induction of first-line defence mechanisms upon subsequent exposure to stress (Pastor et al. 2013). The induction of "memory" defences in plants has been reported through priming with microbes and certain chemicals. Plant's perception for exogenous chemicals and microbe-associated molecular patterns (MAMPs) is able to induce response against abiotic stresses, providing tolerance in stress conditions. Such type of mechanism is known as "induced systemic tolerance (IST)".

10.5 PGPR-Mediated Induced Systemic Tolerance

Upon deployment of various mechanisms, it has been reported that IST reflects amelioration of plants under abiotic stress. Various traits of PGPR have been reported on amelioration of salinity stress, e.g. ACC-deaminase, exopolysaccharide, volatile production, Pi solubilization, indole-3-acetic acid (IAA) production, etc. (Fig. 10.3). Rhizosphere microbes are involved in altering hormonal root–shoot signalling in plants. Indole-3-acetic acid (IAA)-producing PGPR stimulate exudation of flavonoids by bean plants; regulate nodulation, nitrogen fixation and nutrient uptake; and relieved the negative effects of salt stress (Dodd and Pérez-Alfocea 2012). PGPR contain the enzyme 1-aminocyclopropane-1-carboxylic acid deaminase (ACC-D), which hydrolyzes the ethylene precursor ACC and reduces the level of ethylene during salt condition (Glick 2014; Choudhary et al. 2015). Microbial secreted exopolysaccharides (EPS) are reported to chelate excessive Na^+ and reduce their availability to plants during saline condition (Choudhary et al. 2015). PGPR are reported to enhance the nutrient uptake efficiency of plants by secreting enzymes in the soil and solubilizing bound nutrients such as phosphorus, potassium, zinc and iron. In recent years, volatile organic compounds (VOCs) were found as a novel way of signalling between PGPR and plants, in which VOCs from specific strains of bacteria enhanced plant growth by regulating different biological processes including hormone distribution, nutrient uptake, sodium homeostasis and biosynthesis of osmoprotectant (Liu and Zhang 2015). The ability of soil microbes to improve plant growth and alleviate negative effects of salinity is evaluated below.

10.5.1 Osmolyte Accumulation and Maintaining Water Homeostasis

Plant growth is firstly affected by osmotic stress and then recovers a little bit by accumulation of osmolytes during salt stress. Salinity creates water stress around the root due to accumulation of salt ions which decrease osmotic balance in plants. Water homeostasis and photosynthesis structures are critical to salt stress; hence,

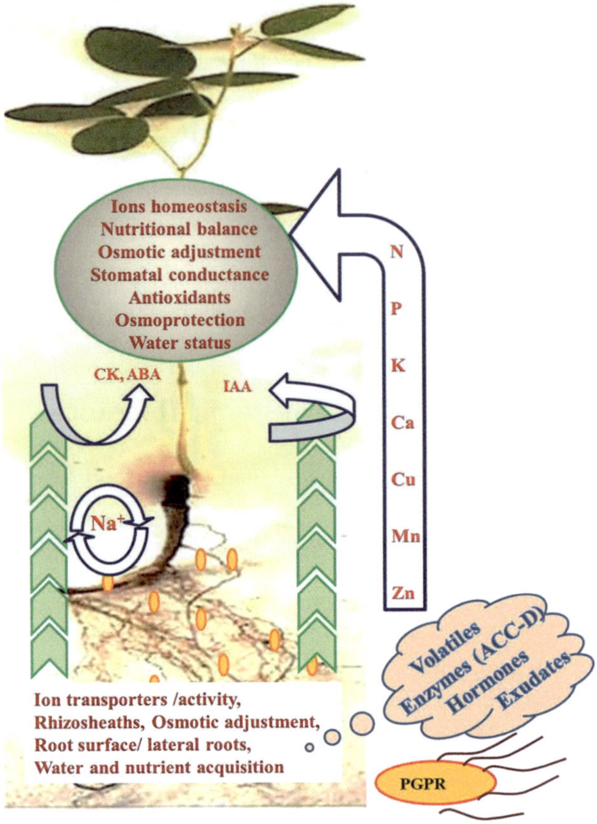

Fig. 10.3 Plant and PGPR interaction in response to salt stress

their maintenance is essential for alleviating the impact of salinity on plant growth (Iqbal et al. 2014). PGPR have the ability to improve plant–water relations by enhancing the accumulation of osmolytes in plants. Different bacterial genera such as *Burkholderia*, *Arthrobacter*, *Bacillus* and *Pseudomonas* are reported to enhance proline synthesis in abiotically stressed plants (Choudhary 2012). *Bacillus* sp.-inoculated plant root accumulated higher proline content in salt stress compared to non-inoculated plant root (Vardharajula et al. 2011). Kumari et al. (2015) suggested that higher proline content in roots maintain osmotic balance, leading to water entrance into the roots. Inoculation of *Rhizobium* and *Pseudomonas* bacterial strains in *Zea mays* was found to increase proline content along with relative water content (Bano and Fatima 2009). In addition, several reports determined the relation between proline accumulation and pyrroline-5-carboxylate synthase *(P5CS)* gene expression level in the presence of PGPR inoculation and suggested that bacterial treatment upregulated the *P5CS* gene expression in plant roots leading to accumulation of free proline content (Kim et al. 2007; Kumari et al. 2015). Similar to proline, the soluble sugars have been reported to increase in PGPR-inoculated plants during salt stress (Kumar et al. 2010; Shukla et al. 2012). Rhizobacteria are reported to maintain root hydraulic conductance which is suggested to assist the maintenance of plant–water status under saline environment (Marulanda et al. 2007). Relative water content is the best indicator of water stress which reduced during salt stress. Plants inoculated with PGPR have been reported to hydrate more than non-inoculated plants and enhanced photosynthesis activity and biomass content (Shukla et al. 2012; Kumari et al. 2015). Furthermore, Vardharajula et al. (2009) explored the role of bacterial exopolysaccharide in protection of plants from water stress. Authors examined that inoculation of *Pseudomonas putida* sp. GAP-P45 enhanced the survival rate of sunflower seedlings along with plant biomass and root-adhering soil under drought stress. The biofilm formation increased the percentage of stable soil aggregates which protect plants from water stress and improve soil health (Choudhary et al. 2015).

Plants keep protecting their photosynthesis activity and apparatus to ROS via the upregulation of antioxidative enzyme expression. Different *Rhizobium* strains (*Mesorhizobium ciceri*, *Mesorhizobium mediterraneum* and *Sinorhizobium medicae*) were studied on chickpea plant, in which *M. ciceri* strain was found potent to enhance plant tolerance by increasing POD enzyme activity under salt stress (Mhadhbi et al. 2004). *L. sativa* seedlings were inoculated with *P. mendocina* and affected by different levels of water stress. POD and CAT activities were increased in *P. mendocina*-inoculated seedlings in response to drought stress. *P. mendocina*-inoculated seedlings exhibited higher shoot and root biomass and relative water content compared with non-inoculated seedlings (Kohler et al. 2008). In another study, the effect of salt-tolerant *Bacillus amyloliquefaciens* NBRISN13 (SN13) inoculation was evaluated on rice plants exposed to salinity. SN13 increased plant growth as exposed by higher plant length, biomass and chlorophyll content and also enhanced salt tolerance by increasing proline content and upregulation of defence-related gene including *CAT* expression also (Nautiyal et al. 2013).

10.5.2 Ion Homeostasis

Na^+ and Cl^- accumulation in plant tissue is a crucial factor responsible for plant senescence and limiting growth. It is generally accepted that the exclusion of these ions is mostly related to salt tolerance mechanisms in glycophyte species. Rhizobacteria are reported to contribute in toxic ion homeostasis which improves plant growth and tolerance during salinity. These microbes can reduce the uptake of toxic ions by regulating ion transporter expression in plants and formation of rhizosheaths by producing exopolysaccharides (EPS) which work as a physical barrier around the roots. Rhizobacteria are also contributed in the macro-/micronutrient status in plants. These nutrients become more accessible to the plants due to some microbial activities in rhizosphere like Pi solubilization, organic acid excretion and siderophore production. These nutrients have been reported to reduce toxic ion accumulation, and the specific importance has been given to microbial-mediated enhancement of K^+/Na^+ ratios in plants (Shukla et al. 2012; Shkolnik-Inbar et al. 2013; Vaishnav et al. 2015). Ashraf et al. (2004) explained that *Aeromonas hydrophila/caviae* and *Bacillus* sp. decreased Na^+ accumulation in wheat plants by the excretion of EPS, which bind Na^+ in roots and prevent their transfer to leaves. In the same way, EPS producing *B. circulans* and *B. polymyxa* were found to enhance dry matter yield of root and shoot of wheat plants, K^+/Na^+ and Ca^{2+}/Na^+ ratio during salt stress. The effect may be attributed to the cation chelating capacity of EPS (Khodair et al. 2008). In another study, wheat rhizospheric bacteria were found to produce EPS which significantly decreased Na^+ uptake in plants under both nonsaline and saline conditions (Upadhyay et al. 2011). Moreover, Kumari et al. (2015) suggested that EPS producing bacterial strains enhanced K^+/Na^+ which maintained photosynthesis machinery in soybean plants under salt stress.

Another mechanism of PGPR-mediated ion homeostasis could be explained by the exposure of bacterial volatile organic compounds (VOCs) which are reported to modulate Na^+ homeostasis pathway in plants. These compounds have low molecular weight and are found as a novel way of signalling between two organisms. *Arabidopsis* plants exposed to *B. subtilis* GB03 VOCs were exhibited to tolerate salt stress than control plants. This exposure was found to decrease root *AtHKT1* expression in roots but upregulated it in the shoots which facilitate root-to-root Na^+ recirculation (Shkolnik-Inbar et al. 2013). In a study, Vaishnav et al. (2015) examined that *P. simiae* AU-mediated putative VOC blend enhanced the expression of vegetative storage protein (VSP) in soybean leaves, correlated with lower uptake of Na^+ ions by regulating sodium transporter activity under 100 mM NaCl stress.

10.5.3 Nutrient Acquisition

Plant growth and productivity is severely affected by inadequate supply and limitation of nutrients in the soil system. Availability and uptake of nutrients depend on several parameters of soil such as composition, pH, moisture, soil texture and microflora composition. Most of the nutrients are available in the range of 5–7 pH of soil. Salinity changes the pH of soil by which most of compounds bound to cations and anions to form a stable compound which further makes them less available in soil. PGPR solubilize these nutrients and make available to plants.

Phosphorus is a crucial macronutrient for plant growth and development. It is present in organic form (30–65 %) in the soil which is not assimilated by plants. The organic P in soils is present in the form of inositol phosphatases, phosphoesters, phosphodiesters and phosphotriesters (Sindhu et al. 2010). It is well-known facts that P chemical fertilizers which are added to soil have sparingly soluble nature which is completely not available to the plants leading to add large amount of fertilizers by farmers into the fields which later cause environmental problems. Phosphate-solubilizing bacteria (PSB) that belong to genera *Bacillus*, *Pseudomonas*, *Achromobacter*, *Alcaligenes*, *Brevibacterium*, *Corynebacterium*, *Serratia* and *Xanthomonas* are capable of hydrolyzing unavailable form of phosphorus in available form (Sindhu et al. 2010). In a study, a large number of fluorescent pseudomonad strains were screened for the solubilization of tricalcium phosphate on the basis of visible dissolution halos on Pikovskaya agar medium (Naik et al. 2008). These bacteria excrete low molecular weight organic acids such as gluconic acid, citric acid, succinic acid, propionic acid and lactic acids that mineralize and dissolve organic phosphate compounds and make available to plants in the form of inorganic phosphate (Choudhary 2012). Phosphatases are enzyme that can hydrolyze different form of phosphate, originating from the organic soil sources. Additionally, the production of hydrogen ions in rhizosphere environments alters the pH sufficiently to mobilize soil minerals (Khan et al. 2013). Salinity causes the depletion and precipitation of available phosphorus, and PSB have tendency to solubilize precipitated forms of phosphorus in hydroponic MS medium and enhanced the phosphorus content in plant system under NaCl stress (Shukla et al. 2012; Vaishnav et al. 2015). A PGPR *P. mendocina* has been observed to protect *Lactuca sativa* L. cv. Tafalla against different levels of salt stress. Kohler et al. (2008) reported that bacterial-inoculated plants exhibited higher phosphatase activity which released soluble phosphate from its insoluble compounds inside the plant cells and helped plants to tolerate salt stress.

Nitrogen is an essential component required for protein and nucleic acid synthesis and other nitrogen compounds, those that are considered as vital components of the living system. Soil microorganisms have the capacity to fix atmospheric and provide it to plants in the form of ammonia via nitrogen fixation process. PGPR can fix nitrogen by symbiotic or non-symbiotic mechanism. In symbiotic N_2 fixation, bacteria associated with host plant root by nodule formation and fixed nitrogen inside the cell that accounts nearly 65 % of the total biologically fixed nitrogen (Rajwar et al. 2013). Symbiotic N_2 fixation occurs in *Azotobacter* spp., *Bacillus* spp., *Beijerinckia* spp., etc., whereas non-symbiotic nitrogen fixation occurs through free living diazotrophs, *Azospirillum*, *Pseudomonas* and *Burkholderia*, those fixed nitrogen in the rhizosphere (Mia et al. 2013). Nitrogen fixation in legume plants is mediated mainly by associative bacteria *Rhizobium*. Salinity has a major effect on legume biology as it affects the diversity of rhizobia in soil and their interaction with legume plants. Salt stress mostly reduces number of nodules that result in reduction amount of nitrogen fixed. In the past few years, salt-tolerant PGPR which can tolerate higher levels of salts, up to 1.5–2.0 M NaCl, were co-inoculated with *Rhizobium* in legumes for growth enhancement and successful N_2 fixation (Divito and Sadras 2014). The co-inoculation of PGPR is a good strategy when *Rhizobium* is not so effective in saline environment. ACC-deaminase-containing PGPR have been observed to reduce the ethylene concentration which decreases nodulation efficiency in legumes under stress environment (Ahmad et al. 2011). *Pseudomonas* and *Rhizobium phaseoli* co-inoculation was observed very effective for enhancing nodulation process in mung bean plants under laboratory as well as field conditions affected by salinity stress (Ahmad et al. 2013). Similarly, co-inoculation of *Mesorhizobium* sp. with IAA-producing *Pseudomonas* has been found to increase nodulation in chickpea (Malik and Sindhu 2011). In another study, co-inoculation of *Pseudomonas* and *Rhizobium* was found to increase nodulation and nutrient uptake (Mishra et al. 2011). Furthermore, *Azospirillum* and *Rhizobium* consortia were found to enhance nodulation which increased tolerance in plants against unfavourable conditions (Bashan and de-Bashan 2010).

Iron is the fourth most abundant element required by most of the living organisms for growth. It plays a key role as cofactor for nearly 140 enzymes catalysing specific biochemical reactions and processes. Iron exists in the form of ferric state (Fe^{3+}) and produces insoluble hydroxides and oxyhydroxides which are not readily available to plants and microorganisms (Ma et al. 2011). In saline soil, the availability of ferric is further reduced due to decreased solubility from lower pH to higher pH (Thomine and Lanquar 2011). Organisms have employed various mechanisms to get available form of iron; among them siderophores have been best studied. Siderophores are iron-chelating agents and proved in different PGPR strains as an important attribute for plant growth promotion and phytopathogen protection (Scavino and Pedraza 2013). PGPR secrete siderophore in the rhizosphere, and then plant roots uptake iron from siderophore by either chelate degradation or direct uptake (Rajkumar et al. 2010). A great variability has been found in microbial siderophores such as peptidic siderophores, aminoalkane siderophore and citric acid-based siderophore (Budzikiewicz 2010). Siderophores have been concerned for both direct and indirect mechanism of plant growth by PGPR. Sharma and Johri (2003) reported that siderophore-producing *Pseudomonas* spp. strains GRP3A and PRS significantly increased maize seed germination and plant growth under iron-stressed condition and suggested application of these bacterial strains for crop productivity in calcareous soil system. Pandey et al. (2005) characterized *P. aeruginosa* GRC1-secreted siderophores. The purified siderophore was pyoverdine type with amino acid composition, when this bacterium applied in field trials, found to enhance the growth of *Brassica campestris*.

Potassium (K) is the third major essential nutrient for plant growth which involved in various metabolic processes in plants (Sindhu et al. 2010). Potassium is present in soil in the form of available (water soluble) and unavailable (micas, illite and orthoclase). The common components of potassium in the soil are feldspar and mica in 90–98 % (Sindhu et al. 2010). Potassium-solubilizing bacteria (KSB) are able to release K from its unavailable form. Meena et al. (2014) described the importance of KSB in K uptake efficiency by plants and reduction in the use of costly chemical fertilizers. Two KSB strains, KNP413 and KNP414, that possessed higher dissolution capacities of mineral K are widely used as potassium fertilizer in China (Hu et al. 2006). Three PGPR strains *B. mucilaginosus*, *Azotobacter chroococcum* and *Rhizobium* sp. were found to solubilize K from waste mica and enhanced its uptake in maize and wheat plants (Singh et al. 2010).

Sulphur (S) is the secondary essential macronutrient which has a crucial role in sulphur-containing amino acids, methionine and cysteine. Only 5 % of total soil S is available for plants in the form of sulphate (SO_4^{-2}), and the remaining 95 % is organically bound include pyrite (FeS_2), gypsum ($CaSO_4.2H_2O$) and epsonite ($MgSO_4.7H_2O$), which are unavailable for plants. Sulphur turnovers are reported by both biochemical and biological mineralization (Gharmakher et al. 2009). In biochemical mineralization, sulphate pools are hydrolyzed through enzymatic reactions, while the biological mineralization is driven by the soil microflora.

Sulphur-oxidizing bacteria are chemoautotrophic and photosynthetic bacteria which include *Beggiatoa*, *Chromatium*, *Chlorobium*, *Thiobacillus*, *Sulfolobus*, *Thiospira* and *Thiomicrospira*, *Bacillus*, *Pseudomonas*, *Arthrobacter* and *Flavobacterium*. Common PGPR species such as *Bacillus* and *Pseudomonas* have been reported to reduce sulphate to H_2S (Sindhu et al. 2010).

Zinc (Zn) is an essential micronutrient for plants, which plays several functions throughout the life of plants. It plays crucial roles in more than 100 enzymes which are involved in many types of functions in plants such as auxin synthesis, photochemical reactions of chlorophyll, stability of biological membranes and SOD and carbonic anhydrase enzymatic activity (Broadley et al. 2007). Plant growth, maturity, seed quality and yield are very much dependent upon Zn. Zinc is present in the soil as ZnS (sphalerite), and mineral ores include smithsonite ($ZnCO_3$), zincite (ZnO), zinkosite ($ZnSO_4$), franklinite ($ZnFe2O_4$) and hopeite [Zn_3 $(PO_4)^2$ $4H_2O$]. Zinc-solubilizing bacteria have been tested on different insoluble Zn ores (Abaid-Ullah et al. 2015). Tariq et al. (2007) have found that Zn-mobilizing bacteria enhanced Zn uptake in rice seedlings which had positive impact on plant growth and grain yield. In another study, *Serratia* sp. has been noted to more solubilize ZnO as compared to other insoluble ores and was able to significantly increased wheat yield under various climatic conditions (Abaid-Ullah et al. 2015).

10.5.4 Plant Hormones

The ability to produce plant hormones is a major property of many PGPR which directly influence plant growth. Among these hormones, IAA, gibberellic acid (GA), cytokinin (CK) and abscisic acid (ABA) may play a significant role in salt stress. These phytohormones alter metabolism and morphology, nutrient and water uptake efficiency and consequently larger and healthier plants.

10.5.4.1 Indole-3-Acetic Acid

Indole-3-acetic acid (IAA) is the best-studied compound involved in numerous plant mechanisms like cell division, differentiation, extension, apical dominance, gravitropism and phototropism (Korasick et al. 2013). Salinity was found to accumulate IAA in root which affects cell elongation and growth. It is also supposed to act as an inhibitor of cytokinin synthesis and their transport from root to shoot during stress condition leading to increased root elongation (Dodd et al. 2005). Most of PGPR adapted tryptophan-based pathway for IAA production either via indole-3-pyruvic acid (IPyA) or indole-3-acetamide formation (IAM) (Spaepen et al. 2007). Tryptophan is synthesized from chorismate in plant and secreted out from loosely bound root cells, then taken up by soil microbes. Phytohormone production in PGPR especially IAA was extensively studied in *Azospirillum* spp. during the last decade (Cassán et al. 2014). *Azospirillum brasilense* strain Cd was found to relieve the negative effects of 50 mM NaCl on *Phaseolus vulgaris* as exposed by higher branching of roots and flavonoid production in hydroponical condition (Dardanelli et al. 2008). An IAA-producing *Azospirillum brasilense*

Az39 strain was co-inoculated with *Bradyrhizobium japonicum* E109 and found to enhance germination and growth of corn and soybean (Cassán et al. 2009). Albacete et al. (2008) reported that plants inoculated with IAA-producing bacterial strains exhibited higher root and leaf growth which is considered as adaptive response of salinity. IAA-producing PGPR strains were also reported for enhancement of nutrient uptake efficiency under hydroponic conditions (Shukla et al. 2012). In another study, *B. subtilis* GB03-mediated VOCs were found to trigger *Arabidopsis* seedling growth by regulation of different RNA transcript involved in different metabolic processes. Further analysis confirmed that GB03 triggered growth promotion by regulating auxin homeostasis and cell wall loosening enzymes (Zhang et al. 2007). Similarly, bacterial VOCs were also reported to enhance expansin gene (*EXP2, EXP6* and *EXPA5*) expressions in *Nicotiana tabacum* and *Lactuca sativa* (Minerdi et al. 2011).

10.5.4.2 Abscisic Acid

Abscisic acid (ABA) is primarily known for abscission of leaves and shoot growth, but recent studies suggested that increased concentration of ABA is required for inhibiting excess ethylene production in plants during stress conditions. It is associated with phytohormone response to environmental stresses. Typically, stress condition increases ABA level triggering adaptive responses essential for survival (Pliego et al. 2011). During low water potential condition, ABA is produced in roots and then translocated in leaves, where it directly involves in stomatal closing to reduce transpiration activity and maintain water potential. It is also responsible for stimulation of root growth and emergence of lateral roots leading to enhancement of water uptake during drought condition. Furthermore, ABA is also involved in regulation of ion transport across the membrane and synthesis of specific proteins (Bashan and De-Bashan 2010). It was proposed that many PGPR produce ABA in vitro, and its production increases under osmotic stress (Dodd et al. 2010). The role of rhizobacteria on plant ABA status is conflicting; some reports explored that bacterial colonization prevented salinity-induced accumulation of ABA, while others were found that PGPR enhance accumulation of ABA, which may be responsible for survival in stress conditions. *P. putida* Rs-198-inoculated cotton seeds exhibited higher biomass accumulation and prevented ABA production in 10 % salinity, while uninoculated seedlings showed higher accumulation in foliar ABA concentration (Yao et al. 2010). In another report, Zhang et al. (2008b) proposed that *B. subtilis*-mediated VOCs were observed to promote photosynthesis capacity by decreasing transcription levels of ABA synthesis in aerial parts of *Arabidopsis* plants. Furthermore, *P. chlororaphis* O6 colonization was reported to decrease stomatal apertures in both wild-type and ABA-insensitive *Arabidopsis* mutant plants which suggested ABA-independent stomatal closure mechanism responsible for bacterial-mediated IST against drought and salt stress (Cho et al. 2012). In a recent study, ABA metabolizing rhizobacteria were observed to decrease ABA concentration *in planta* and alters plant growth during stress condition (Belimov et al. 2014).

10.5.4.3 Ethylene

Ethylene is known as stress hormone which synthesized in plants under stress conditions. Ethylene is also an inhibitor of rhizobial nodulation of legumes. However, during normal condition, ethylene production is minimal which regulate some physiological responses such as breaking of seed dormancy (Dodd et al. 2005). Ethylene is produced by Yang cycle in plants in which ACC is the precursor of ethylene biosynthesis that is converted into ethylene by ACC oxidase enzyme. ACC can be transported to particular stressed organ resulting in synthesis of ethylene in the affected tissue (Yoon and Kieber 2013). In salinity stress, increased foliar ethylene is correlated with Na^+ accumulation responsible for decreased growth of tomato (Mayak et al. 2004). In another study, 100 mM NaCl concentration enhanced ACC and Na^+ accumulation in the root, xylem sap and leaf correlated with onset of oxidative stress and decreased photosynthesis capacity suggesting ethylene role in foliar senescence (Albacete et al. 2008; Ghanem et al. 2008). Many soil bacteria contain ACC-deaminase (ACC-D) enzyme, which cleaves ACC to α-ketobutyrate and ammonia and thereby lowers the ethylene level in stressed plants. According to Glick et al.'s (1998) described model, ACC-D-producing bacteria attached to the root surface and take up ACC exuded from plant roots and then hydrolyze it through ACC-D mechanism. Hence, more ACC is exuded from roots to maintain equilibrium and finally ACC level reduced inside the cell. There are several reports that showed ACC-D containing PGPR can decrease salinity-induced growth inhibition (Ahmed and Farag 2011; Wu et al. 2012). *Achromobacter piechaudii* ARV8 which produced ACC-D was found to significantly increase plant weights and nutrient uptake efficiency in tomato seedlings under NaCl stress (Mayak et al. 2004). Shaharoona et al. (2006) reported that co-inoculation of ACC-D possessing PGPR with a *Rhizobium* strain *Bradyrhizobium* enhanced nodulation in mung bean by lowering ethylene production compared with *Bradyrhizobium* alone. ACC-deaminase-producing halotolerant bacterial strains *B. licheniformis* RS656, *Z. alba* RS111 and *Br. iodinum* RS16 have been reported to reduce ethylene production in red pepper plants at 150 mM NaCl stress. Bacterial-inoculated plants exhibited higher salt tolerance index and increased nutrient uptake as compared to non-inoculated plants which suggested amelioration of salt stress effect (Siddikee et al. 2011). Furthermore, plants treated with ACC-deaminase-producing PGPR strains exhibited higher root nodules, plant growth and yield under oxidative stress conditions (Roopa et al. 2012; Zafar-ul-Hye et al. 2013).

10.5.4.4 Cytokinin

Cytokinins are a group of purine-type phytohormone that regulate cell division, differentiation processes in meristematic tissues, chloroplast maturation, cell expansion and stomatal conductance of higher plants (Cassán et al. 2014). It is necessary for inducing root nodule organogenesis for nitrogen fixation (Kisiala et al. 2013). Auxin and CK ratio plays an important role in cell division and differentiation. Cytokinin production is a common PGP trait of rhizobacteria (Dodd et al. 2010). A PGPR strain *B. subtilis*, which produced CK, was found to enhance biomass content in the shoot of lettuce plants during drought stress considered root-to-shoot CK signalling (Arkhipova et al. 2007). Cytokinin-producing bacterium *B. megaterium* increased the level of CK in *A. thaliana* roots which predicted the role of CK signalling pathway in the plant growth promotion (Ortíz-Castro et al. 2008). Giraud et al. (2007) investigated that PGPR strain *Bradyrhizobium* has taken part in nodulation process with the help of CKs in the absence of nod factor in soybean plants.

10.5.4.5 Gibberellin

Gibberellins are diterpenoid acids synthesized by terpenoid pathway and involved in several developmental processes such as cell division and elongation, breaking seed dormancy. This is widely reported for starch hydrolysis mechanism during germination. Gibberellic acid stimulated the transcription level of α-amylase gene in seed embryo that is responsible in hydrolysis of starch into glucose (Richards et al. 2001). Inoculation of GA-producing *A. brasilense* enhanced germination of wheat and soybean and rapid plant growth at least partial to GA production under salt stress (Cassán et al. 2014). In another study, GA-producing *Promicromonospora* sp. SE188 was observed to improve tomato plant growth as exposed to higher shoot length and biomass (Kang et al. 2012). Kang et al. (2014) have reported that a GA-producing *P. putida* H-2-3 was found to significantly enhance plant length, weight and chlorophyll content in GA-deficient mutant soybean plants.

10.6 Selection, Characterization and Commercialization of PGPR Strains

For successful application of PGPR under saline regions, inoculants should be isolated from indigenous salt-affected soils. Certain PGPR lost their ability and failed to colonize with root system under salinity (Paul and Nair 2008). Under such condition, halotolerant bacteria as inoculums would be the most appropriate approach. In a study, five plant growth-promoting halotolerant bacteria were found to ameliorate salt stress (80, 160 and 320 mM NaCl) in wheat plants and increased root length up to 71.7 % (Ramadoss et al. 2013).

To commercialize any PGPR strain, different stages have been followed step by step which include isolation, screening, pot tests and field efficacy, formulation development, formulation viability, industrial linkages and quality control (Bhattacharyya and Jha, 2012). A potent PGPR strain is selected from diverse rhizospheric bacteria by screening on the basis of their ability to produce PGP activity and inhibit the growth of various phytopathogens and a positive interaction with the host plant. Pure cultures of PGPR strains are applied on seeds in in vitro glasshouse trials. Seeds are treated with pure and fresh bacterial suspension and then planted in soil for test. During the experiment, PGPR which are found potent for plant growth promotion and alleviate negative symptoms of stresses are selected for further field trials (Compant et al. 2005). Once isolates have been selected, then their characterization is done based on biochemical and molecular characteristics. Biochemical characterization is done according to Bergey's Manual of Determinative Bacteriology. In molecular characterization, DNA- and RNA-based homology testing, ribosomal protein profiling through MALDI and fatty acid profiling through GC-MS analysis are considered for identification (Bashan et al. 1993; Maiti et al. 2009; Bhattacharyya and Jha 2012). 16S ribosomal RNA is a component of the 30S small subunit of prokaryotic ribosome. Through evolution this region of the gene remained conserved and hence widely used to define molecular phylogeny and taxonomy of bacteria since the last decade (Sun et al. 2008).

The mass production of bacterial formulation is performed under liquid, semi-solid and solid state (Bhattacharyya and Jha 2012). The success and commercialization of PGPR formulations require a strong linkage between the research organizations and private industries, although it also depends on economical and viable market demand, longer shelf life, low capital costs and easy availability of career materials.

10.7 Conclusion

In saline conditions PGPR can induce tolerance mechanism in crop plants and promote plant growth and development. On the other hand, PGPR also improve soil fertility. In recent years a number of researchers have explained that plant–microbe interactions help to develop tolerance mechanisms in saline soil, but still need to understand details of their molecular and biochemical mechanism. Nevertheless, the microorganisms present in saline area or in the rhizosphere of halophytic plants may provide a valuable resource for improving the crop tolerance to salinity.

Acknowledgment The financial support for some of the research in this review has partially been supported by DBT and SERB grant no. BT/PR1231/AGR/021/340/2011 and SR/FT/LS-129/2012, respectively, to DKC.

References

Abaid-Ullah M, Hassan MN, Jamil M, Brader G, Shah MK, Sessitsch A, Hafeez FY (2015) Plant growth promoting rhizobacteria: an alternate way to improve yield and quality of wheat (*Triticum aestivum*). Int J Agric Biol 17:51–60

Agarwal DK, Billore SD, Sharma AN, Dupare BU, Srivastava SK (2013) Soybean: introduction, improvement, and utilization in India: problems and prospects. Agric Res 2(4):293–300

Ahmad M, Zahir ZA, Asghar HN, Asghar M (2011) Inducing salt tolerance in mung bean through coinoculation with rhizobia and plant-growth-promoting rhizobacteria containing 1-aminocyclopropane-1-carboxylate deaminase. Can J Microbiol 57(7):578–589

Ahmad M, Zahir ZA, Khalid M, Nazli F, Arshad M (2013) Efficacy of *Rhizobium* and *Pseudomonas* strains to improve physiology, ionic balance and quality of mung bean under salt-affected conditions on farmer's fields. Plant Physiol Biochem 63:170–176

Ahmed HMI, Farag MMA (2011) Alleviation of salinity stress in lettuce during germination by seed priming. J Plant Production Mansoura Univ 2(5):725–737

Albacete A, Ghanem ME, Martínez-Andújar C, Acosta M, Sánchez-Bravo J, Martínez V, Lutts S, Dodd IC, Pérez-Alfocea F (2008) Hormonal changes in relation to biomass partitioning and shoot growth impairment in salinized tomato (*Solanum lycopersicum* L.) plants. J Exp Bot 59(15):4119–4131

Arkhipova TN, Prinsen E, Veselov SU, Martinenko EV, Melentiev AI, Kudoyarova GR (2007) Cytokinin producing bacteria enhance plant growth in drying soil. Plant Soil 292(1-2):305–315

Ashraf M, Akram NA (2009) Improving salinity tolerance of plants through conventional breeding and genetic engineering: an analytical comparison. Biotechnol Adv 6:744–752

Ashraf M, Berge SH, Mahmood OT (2004) Inoculating wheat seedling with exopolysaccharides-producing bacteria restrict sodium uptake and stimulates plant growth under salt stress. Biol Fertil Soils 40:157–162

Bano A, Fatima M (2009) Salt tolerance in *Zea mays* (L.) following inoculation with *Rhizobium* and *Pseudomonas*. Biol Fertil Soils 45:405–413

Bashan Y, De-Bashan LE (2010) How the plant growth-promoting bacterium *Azospirillum* promotes plant growth: a critical assessment. Adv Agron 108:77–136

Bashan Y, Holguin G, Lifshitz R (1993) Isolation and characterization of plant growth-promoting rhizobacteria. In: Glick BR, Thompson JE (eds) Methods in plant molecular biology and biotechnology. CRC Press, Boca Raton, pp 331–345

Belimov AA, Dodd IC, Safronova VI, Dumova VA, Shaposhnikov AI, Ladatko AG, Davies WJ (2014) Abscisic acid metabolizing rhizobacteria decrease ABA concentrations in planta and alter plant growth. Plant Physiol Biochem 74:84–91

Ben Rejeb I, Atauri Miranda L, Cordier M, Mauch-Mani B (2013) Induced tolerance and priming for abiotic stress in plants. In: Gaur RK, Sharma P (eds) Molecular approaches in plant abiotic stress. CRC Press, Boca Raton

Bhattacharyya PN, Jha DK (2012) Plant growth-promoting rhizobacteria (PGPR): emergence in agriculture. World J Microbiol Biotechnol 28(4):1327–1350

Brini F, Masmoudi K (2012) Ion transporters and abiotic stress tolerance in plants. ISRN Mol Biol. doi:10.5402/2012/927436

Broadley MR, White PJ, Hammond JP, Zelko I, Lux A (2007) Zinc in plants. New Phytol 173(4):677–702

Budzikiewicz H (2010) Microbial siderophores. Springer, Vienna

Cassán F, Perrig D, Sgroy V, Masciarelli O, Penna C, Luna V (2009) *Azospirillum brasilense* Az39 and *Bradyrhizobium japonicum* E109, inoculated singly or in combination, promote seed germination and early seedling growth in corn (*Zea mays* L.) and soybean (*Glycine max* L.). Eur J Soil Biol 45(1):28–35

Cassán F, Vanderleyden J, Spaepen S (2014) Physiological and agronomical aspects of phytohormone production by model plant-growth-promoting rhizobacteria (PGPR) belonging to the genus *Azospirillum*. J Plant Growth Regul 33:440–459

Cho SM, Kang BR, Kim JJ, Kim YC (2012) Induced systemic drought and salt tolerance by *Pseudomonas chlororaphis* O6 root colonization is mediated by ABA-independent stomatal closure. Plant Pathol J 28(2):202–206

Choudhary DK (2012) Microbial rescue to plant under habitat-imposed abiotic and biotic stresses. Appl Microbiol Biotechnol 96(5):1137–1155

Choudhary DK, Kasotia A, Jain S, Vaishnav A, Kumari S, Sharma KP, Varma A (2015) Bacterial-mediated tolerance and resistance to plants under abiotic and biotic stresses. J Plant Growth Regul 35:276–300

Compant S, Duffy B, Jerzy N, Clement C, Barka EA (2005) Use of plant growth-promoting bacteria for biocontrol of plant diseases: principles, mechanisms of action, and future prospects. Appl Environ Microbiol 71(9):4951–4959

Dardanelli MS, Fernández de Córdoba FJ, Espuny MR, Rodríguez Carvajal MA, Soria Díaz ME, Gil Serrano AM, Okon Y, Megías M (2008) Effect of *Azospirillum brasilense* coinoculated with *Rhizobium* on *Phaseolus vulgaris* flavonoids and Nod factor production under salt stress. Soil Biol Biochem 40:2713–2721

Divito GA, Sadras VO (2014) How do phosphorus, potassium and sulphur affect plant growth and biological nitrogen fixation in crop and pasture legumes? A meta-analysis. Field Crop Res 156:161–171

Dodd IC, Belimov AA, Sobeih WY, Safronova VI, Grierson D, Davies WJ (2005) Will modifying plant ethylene status improve plant productivity in water limited environments? In: 4th International Crop Science Congress

Dodd IC, Pérez-Alfocea F (2012) Microbial amelioration of crop salinity stress. J Exp Bot 63:3415–3428

Dodd IC, Zinovkina NY, Safronova VI, Belimov AA (2010) Rhizobacterial mediation of plant hormone status. Ann Appl Biol 157(3):361–379

Fukuda A, Nakamura A, Tagiri A, Tanaka H, Miyao A, Hirochika H, Tanaka Y (2004) Function, intracellular localization and the importance in salt tolerance of a vacuolar Na^+/H^+ antiporter from rice. Plant Cell Physiol 45(2):146–159

Gaxiola RA, Rao R, Sherman A, Grisafi P, Alper SL, Fink GR (1999) The *Arabidopsis thaliana* proton transporters, AtNhx1 and Avp1, can function in cation detoxification in yeast. Proc Natl Acad Sci U S A 96(4):1480–1485

Ghanem ME, Albacete A, Martínez-Andújar C, Acosta M, Romero-Aranda R, Dodd IC, Lutts S, Pérez-Alfocea F (2008) Hormonal changes during salinity-induced leaf senescence in tomato (*Solanum lycopersicum* L.). J Exp Bot 59(11):3039–3050

Gharmakher HN, Machet JM, Beaudoin N, Recous S (2009) Estimation of sulfur mineralization and relationships with nitrogen and carbon in soils. Bio Fertil Soils 45(3):297–304

Gill SS, Tuteja N (2010) Reactive oxygen species and antioxidant machinery in abiotic stress tolerance in crop plants. Plant Physiol Biochem 48(12):909–930

Giraud E, Moulin L, Vallenet D, Barbe V, Cytryn E, Avarre JC, Jaubert M, Simon D, Cartieaux F, Prin Y, Bena G (2007) Legumes symbioses: absence of Nod genes in photosynthetic bradyrhizobia. Science 316(5829):1307–1312

Glick BR (2014) Bacteria with ACC deaminase can promote plant growth and help to feed the world. Microbiol Res 169(1):30–39

Glick BR, Penrose DM, Li J (1998) A model for the lowering of plant ethylene concentrations by plant growth-promoting bacteria. J Theor Biol 190(1):63–68

Hu X, Chen J, Guo J (2006) Two phosphate-and potassium-solubilizing bacteria isolated from Tianmu Mountain, Zhejiang, China. World J Microbiol Biotechnol 22(9):983–990

Huang GT, Ma SL, Bai LP, Zhang L, Ma H, Jia P, Liu J, Zhong M, Guo ZF (2012) Signal transduction during cold, salt, and drought stresses in plants. Mol Biol Rep 39:969–987

Iqbal N, Umar S, Nazar R (2014) Manipulating osmolytes for breeding salinity-tolerant plants. In: Ahmad P, Rasool S (eds) Emerging technologies and management of crop stress tolerance a sustainable approach. ISBN: 978-0-12-800875-1, Elsevier Inc., UK

Jamil A, Riaz S, Ashraf M, Foolad MR (2011) Gene expression profiling of plants under salt stress. Crit Rev Plant Sci 30(5):435–458

Kang S-M, Khan AL, Hamayun M, Hussain J, Joo G-J, You Y-H, Kim J-G, Lee I-J (2012) Gibberellin-producing *Promicromonospora* sp. SE188 improves *Solanum lycopersicum* plant growth and influences endogenous plant hormones. J Microbiol 50:902–909

Kang S-M, Radhakrishnan R, Khan AL, Kim M-J, Park J-M, Kim B-R, Shin D-H, Lee I-J (2014) Gibberellin secreting rhizobacterium, *Pseudomonas putida* H-2-3 modulates the hormonal and stress physiology of soybean to improve the plant growth under saline and drought conditions. Plant Physiol Biochem 84:115–124

Khan MS, Ahmad E, Zaidi A, Oves M (2013) Functional aspect of phosphate-solubilizing bacteria: importance in crop production In: Maheshwari DK et al (eds) Bacteria in agrobiology: crop productivity. doi:10.1007/978-3-642-37241-4_10

Khodair TA, Galal GF, El-Tayeb TS (2008) Effect of inoculating wheat seedlings with exopolysaccharide-producing bacteria in saline soil. J Appl Sci Res 4:2065–2070

Kim DW, Shibato J, Agrawal GK, Fujihara S, Iwahashi H, Kim DH (2007) Gene transcription in the leaves of rice undergoing salt-induced morphological changes (*Oryza sativa* L.). Mol Cell 24:45–59

Kisiala A, Laffont C, Emery RN, Frugier F (2013) Bioactive cytokinins are selectively secreted by *Sinorhizobium meliloti* nodulating and nonnodulating strains. Mol Plant Microbe In 26(10):1225–1231

Kohler J, Hernández JA, Caravacaa F, Roldána A (2008) Plant-growth-promoting rhizobacteria and arbuscular mycorrhizal fungi modify alleviation biochemical mechanisms in water-stressed plants. Funct Plant Biol 35:141–151

Korasick DA, Enders TA, Strader LC (2013) Auxin biosynthesis and storage forms. J Exp Bot 64(9):2541–2555

Kumar A, Sharma S, Mishra S (2010) Influence of arbuscular mycorrhizal (AM) fungi and salinity on seedling growth, solute accumulation, and mycorrhizal dependency of *Jatropha curcas* L. J Plant Growth Regul 29:297–306

Kumari S, Vaishnav A, Jain S, Varma A, Choudhary DK (2015) Bacterial-mediated induction of systemic tolerance to salinity with expression of stress alleviating enzymes in soybean (*Glycine max* L. Merrill). J Plant Growth Regul 34:558–573

Liu X-M, Zhang H (2015) The effects of bacterial volatile emissions on plant abiotic stress tolerance. Front Plant Sci 6:774

Ma Y, Prasad MN, Rajkumar M, Freitas H (2011) Plant growth promoting rhizobacteria and endophytes accelerate phytoremediation of metalliferous soils. Biotechnol Adv 29(2):248–258

Maiti B, Shekar M, Khusiramani R, Karunasagar I, Karunasagar I (2009) Evaluation of RAPD-PCR and protein profile analysis to differentiate *Vibrio harveyi* strains prevalent along the southwest coast of India. J Genet 88(3):273–279

Malik DK, Sindhu SS (2011) Production of indole acetic acid by *Pseudomonas* sp.: effect of coinoculation with *Mesorhizobium* sp. *Cicer* on nodulation and plant growth of chickpea (*Cicer arietinum*). Physiol Mol Biol Plants 17(1):25–32

Marulanda A, Porcel R, Barea JM, Azcón R (2007) Drought tolerance and antioxidant activities in lavender plants colonized by native drought-tolerant or drought-sensitive Glomus species. Microb Ecol 54:543–552

Mayak S, Tirosh T, Glick BR (2004) Plant growth-promoting bacteria confer resistance in tomato plants to salt stress. Plant Physiol Biochem 42:565–572

Meena VS, Maurya BR, Verma JP (2014) Does a rhizospheric microorganism enhance K+ availability in agricultural soils? Microbiol Res 169(5):337–347

Mhadhbi H, Jebara M, Limam F, Aouani ME (2004) Rhizobial strain involvement in plant growth, nodule protein composition and antioxidant enzyme activities of chickpea-rhizobia symbioses: modulation by salt stress. Plant Physiol Biochem 42(9):717–722

Mia MAB, Hossain MM, Zhamsuddin ZH, Islam MT (2013) Plant-associated bacteria in nitrogen nutrition in crops, with special reference to rice and banana. In: Maheshwari DK et al (eds) Bacteria in agrobiology: crop productivity. doi:10.1007/978-3-642-37241-4_10

Minerdi D, Bossi S, Maffei ME, Gullino ML, Garibaldi A (2011) *Fusarium oxysporum* and its bacterial consortium promote lettuce growth and expansin A5 gene expression through microbial volatile organic compound (MVOC) emission. FEMS Microbiol Ecol 76(2):342–351

Mishra PK, Bisht SC, Ruwari P, Joshi GK, Singh G, Bisht JK, Bhatt JC (2011) Bioassociative effect of cold tolerant *Pseudomonas* spp. and *Rhizobium leguminosarum*-PR1 on iron acquisition, nutrient uptake and growth of lentil (*Lens culinaris* L.). Eur J Soil Biol 47(1):35–43

Munns R, Tester M (2008) Mechanisms of salinity tolerance. Annu Rev Plant Physiol Plant Mol Biol 59:651–681

Naik PR, Raman G, Narayanan KB, Sakthivel N (2008) Assessment of genetic and functional diversity of phosphate solubilizing fluorescent pseudomonads isolated from rhizospheric soil. BMC Microbiol 8(1):230

Nautiyal CS, Srivastava S, Chauhan PS, Seem K, Mishra A, Sopory SK (2013) Plant growth-promoting bacteria *Bacillus amyloliquefaciens* NBRISN13 modulates gene expression profile of leaf and rhizosphere community in rice during salt stress. Plant Physiol Biochem 66:1–9

Ortíz-Castro R, Valencia-Cantero E, López-Bucio J (2008) Plant growth promotion by *Bacillus megaterium* involves cytokinin signaling. Plant Signal Behav 3(4):263–265

Pandey P, Kang SC, Gupta CP, Maheshwari DK (2005) Rhizosphere competent *Pseudomonas aeruginosa* GRC1 produces characteristic siderophore and enhances growth of Indian mustard (*Brassica campestris*). Curr Microbiol 51(5):303–309

Pastor V, Luna E, Mauch-Mani B, Ton J, Flors V (2013) Primed plants do not forget. Environ Exp Bot 94:46–56

Paul D, Nair S (2008) Stress adaptations in a plant growth promoting rhizobacterium (PGPR) with increasing salinity in the coastal agricultural soils. J Basic Microbiol 48:378–384

Pliego C, Kamilova F, Lugtenberg B (2011) Plant growth-promoting bacteria: fundamentals and exploitation. In: Bacteria in agrobiology: crop ecosystems, Springer, Berlin/Heidelberg, pp 295–343

Rajkumar M, Ae N, Prasad MN, Freitas H (2010) Potential of siderophore-producing bacteria for improving heavy metal phytoextraction. Trends Biotechnol 28(3):142–149

Rajwar A, Sahgal M, Johri BN (2013) Legume–rhizobia symbiosis and interactions in agroecosystems. In: Plant microbe symbiosis: fundamentals and advances. Springer, New Delhi, pp 233–265

Ramadoss D, Lakkineni VK, Bose P, Ali S, Annapurna K (2013) Mitigation of salt stress in wheat seedlings by halotolerant bacteria isolated from saline habitats. Springer Plus 2(6):1–7. http://dx.doi.org/10.1186/2193-1801-2-6

Richards DE, King KE, Ait-ali T, Harberd NP (2001) How gibberellin regulates plant growth and development: a molecular genetic analysis of gibberellin signaling. Annu Rev Plant Physiol Plant Mol Biol 52:67–88

Roopa B, Maya C, Makari HK (2012) Effect of different PGPR strain along with rhizobium on nodulation and chick pea productivity. Asian J Exp Biol Sci 3:424–426

Scavino AF, Pedraza RO (2013) The role of siderophores in plant growth-promoting bacteria. In: Maheshwari DK et al (eds) Bacteria in agrobiology: crop productivity. doi:10.1007/978-3-642-37241-4_11

Shaharoona B, Arshad M, Zahir ZA (2006) Effect of plant growth promoting rhizobacteria containing ACC-deaminase on maize (*Zea mays* L.) growth under axenic conditions and on nodulation in mung bean (*Vigna radiata* L.). Lett Appl Microbiol 42:155–159

Shahbaz M, Ashraf M (2013) Improving salinity tolerance in cereals. Crit Rev Plant Sci 32:237–249

Sharma A, Johri BN (2003) Growth promoting influence of siderophore-producing *Pseudomonas* strains GRP3A and PRS 9 in maize (*Zea mays* L.) under iron limiting conditions. Microbiol Res 158(3):243–248

Sharma P, Jha AB, Dubey RS, Pessarakli M (2012) Reactive oxygen species, oxidative damage, and antioxidative defense mechanism in plants under stressful conditions. J Bot. http://dx.doi.org/10.1155/2012/217037

Shi H, Xiong L, Stevenson B, Lu T, Zhu JK (2002) The Arabidopsis salt overly sensitive 4 mutants uncover a critical role for vitamin B6 in plant salt tolerance. Plant Cell 14(3):575–588

Shkolnik-Inbar D, Adler G, Bar-Zvi D (2013) ABI4 downregulates expression of the sodium transporter *HKT1* in *Arabidopsis* roots and affects salt tolerance. Plant J 73:993–1005

Shrivastava P, Kumar R (2015) Soil salinity: a serious environmental issue and plant growth promoting bacteria as one of the tools for its alleviation. Saudi J Biol Sci 22:123–131

Shukla PS, Agarwal PK, Jha B (2012) Improved salinity tolerance of *Arachis hypogaea* (L.) by the interaction of halotolerant plant growth promoting rhizobacteria. J Plant Growth Regul 31:195–206

Siddikee MA, Glick BR, Chauhan PS, Yim W, Sa T (2011) Enhancement of growth and salt tolerance of red pepper seedlings (*Capsicum annuum* L.) by regulating stress ethylene synthesis with halotolerant bacteria containing 1-aminocyclopropane-1-carboxylic acid deaminase activity. Plant Physiol Biochem 49:427–434

Sindhu SS, Dua S, Verma MK, Khandelwal A (2010) Growth promotion of legumes by inoculation of rhizosphere bacteria. In: Khan MS et al (eds) Microbes for legume improvement. doi:10.1007/978-3-211-99753-6_9

Singh G, Biswas DR, Marwaha TS (2010) Mobilization of potassium from waste mica by plant growth promoting rhizobacteria and its assimilation by maize (*Zea mays*) and wheat (*Triticum aestivum* L.): a hydroponics study under phytotron growth chamber. J Plant Nutr 33(8):1236–1251

Spaepen S, Vanderleyden J, Remans R (2007) Indole-3-acetic acid in microbial and microorganism-plant signaling. FEMS Microbiol Rev 31(4):425–448

Sun L, Qiu FB, Zhang XX, Dai X, Dong XZ, Song W (2008) Endophytic bacterial diversity in rice (*Oryza sativa* L.) roots estimated by 16S rDNA sequence analysis. Microb Ecol 55:415–424

Tariq M, Hameed S, Malik KA, Hafeez FY (2007) Plant root associated bacteria for zinc mobilization in rice. Pak J Bot 39(1):245

Thomine S, Lanquar V (2011) Iron transport and signaling in plants. In: Transporters and pumps in plant signaling, Springer, Berlin/Heidelberg, pp 99–131

Upadhyay SK, Singh JS, Singh DP (2011) Exopolysaccharide-producing plant growth-promoting rhizobacteria under salinity condition. Pedosphere 21(2):214–222

Vaishnav A, Kumari S, Jain S, Varma A, Choudhary DK (2015) Putative bacterial volatile-mediated growth in soybean (*Glycine max* L. Merrill) and expression of induced proteins under salt stress. J Appl Microbiol 119:539–551

Vardharajula S, Ali SKZ, Grover M, Reddy G, Bandi V (2009) Alleviation of drought stress effects in sunflower seedlings by the exopolysaccharides producing *Pseudomonas putida* strain GAP-P45. Biol Fertil Soils 46:17–26

Vardharajula S, Ali SA, Grover M, Reddy G, Bandi V (2011) Drought-tolerant plant growth promoting *Bacillus* spp.: effect on growth, osmolytes, and antioxidant status of maize under drought stress. J Plant Interact 6:1–14

Wu Z, Yue H, Lu J, Li C (2012) Characterization of rhizobacterial strain Rs-2 with ACC deaminase activity and its performance in promoting cotton growth under salinity stress. World J Microbiol Biotechnol 28(6):2383–2393

Yao L, Wu Z, Zheng Y, Kaleem I, Li C (2010) Growth promotion and protection against salt stress by *Pseudomonas putida* Rs-198 on cotton. Eur J Soil Biol 46(1):49–54

Yoon GM, Kieber JJ (2013) 1-Aminocyclopropane-1-carboxylic acid as a signalling molecule in plants. AoB Plants 5:plt017

Zafar-ul-Hye M, Ahmad M, Shahzad SM (2013) Synergistic effect of rhizobia and plant growth promoting rhizobacteria on the growth and nodulation of lentil seedlings under axenic conditions. Soil Environ 32:79–86

Zhang H, Kim MS, Krishnamachari V, Payton P, Sun Y, Grimson M, Farag MA, Ryu CM, Allen R, Melo SI, Paré PW (2007) Rhizobacterial volatile emissions regulate auxin homeostasis and cell expansion in *Arabidopsis*. Planta 226:839–851

Zhang H, Xie X, Kim MS, Kornyeyev DA, Holaday S, Paré PW (2008) Soil bacteria augment *Arabidopsis* photosynthesis by decreasing glucose sensing and abscisic acid levels in planta. Plant J 56:264–273

Plant–Microbe Interaction for the Removal of Heavy Metal from Contaminated Site

11

Asit Mandal, J.K. Thakur, Asha Sahu, Sudeshna Bhattacharjya, M.C. Manna, and Ashok K. Patra

Abstract

The diversity of microbes present in the rhizosphere plays a significant role in nutrient cycling and soil sustainability. Plant–microbe-modulated phytoremediation is a viable technology for the cleanup of contaminated environments. Several plants that were identified have various degrees of capacity to eliminate, degrade or detoxify, metabolize, or immobilize a wide range of soil contaminants. Plant-based remediation technologies are not yet commercialized because of its major limitation of slow process and restricted bioavailability of the contaminants, and it is greatly influenced by the climatic factors. The extensive use of plants can overcome most of the limitations by exploring the potential of microbe–plant–metal interaction. The biogeochemical process occurring in the root zone can influence on several rhizobacteria and mycorrhizae directly linked with microbial metabolite synthesis. Thus, a holistic approach of novel remediation technologies and understanding of plant–microbe–contaminant interaction would help for customizing phytoremediation process in relation to site-specific contamination. There is a huge challenge to remediation of contaminated sites by long-term accumulation of heavy metal. Unlike organic contaminants, metals are very much resistant to degradation, and in the long run, continuous accumulation may cause food chain contamination. It is very important to decontaminate the polluted sites in order to reach safe level of metal concentration below the threshold limit of toxicity. Recent studies revealed that phytoextraction, mainly the use of hyperaccumulator plants to extract toxic metals from the contaminated sites, has emerged as a cost-effective, eco-friendly cleanup technology. Novel, efficient microbes and their potential use in the plant rhizosphere could further enhance the phytoremediation for wider range of soil contaminants.

A. Mandal (✉) • J.K. Thakur • A. Sahu • S. Bhattacharjya • M.C. Manna • A.K. Patra
ICAR-Indian Institute of Soil Science, Nabi bagh, Berasia Road,
Nabibagh, Bhopal 462038, India
e-mail: asit.iari@gmail.com

D.K. Choudhary et al. (eds.), *Plant-Microbe Interaction: An Approach to Sustainable Agriculture*, DOI 10.1007/978-981-10-2854-0_11

11.1 Introduction

The era of industrial revolution and rapid urbanization caused various degree of soil contamination. The elevated levels of heavy metal at a long time in the soil are excessively absorbed by plant roots and translocated to aboveground parts, leading to impaired metabolism and reduced plant growth (Bingham et al. 1986). The severe soil contamination with various heavy metals tremendously hampered the soil biological function and soil fertility (McGrath et al. 1995) as well as food chain contamination (Richards et al. 2000). The contamination of the soil environment in the long run is considered as a potential threat to the soil ecosystem services. The soil contaminant bioavailability is highly influenced by various factors such as nature of pollutants, clay content, pH, moisture content, hydrogeology, microbial community dynamics, temperature, and redox potential (Dua et al. 2002). Thus, understanding the plant–microbe–heavy metal has received a great attention for the remediation of contaminated site. Biological means of remediation for the contaminated environment are a promising technique that offers the possibility to degrade or detoxify various contaminants by employing plants and microbes. The approaches of bioremediation are more economically viable, environment-friendly, and an aesthetically pleasing approach which is most widely used for the purpose of remediation of contaminated site. Developing sustainable remediation technologies by employing plant and microbes is a promising solution to reestablish the natural state of soil health (Jansen et al. 1994). However, introduction of numerous waste including toxic heavy metals into the soil leads to considerable loss of the microbial diversity, despite their vital role for the growth and survival of microbes at very low concentrations. The plants employing for cleanup of contaminated environments is quite old concept. More than 300 years ago, plants were used for the treatment of contaminated wastewater. During the nineteenth century, *Thlaspi caerulescens* and *Viola calaminaria* were reported as the first plant species to accumulate higher levels of metals in shoots (Baumann 1885). Several reports were available for the heavy metal accumulation plants like genus *Astragalus* which have a high potential to accumulate selenium up to 0.6 % in dry shoot biomass; some plants were indentified for Ni accumulator (1 %) in shoots (Minguzzi and Vergnano 1948), and *Thlaspi caerulescens* for high Zn accumulation (Rascio 1977). The plants used for phytoextraction of metals from the contaminated soil were developed and reintroduced by Utsunamyia (1980) and Chaney (1983). The first field trial for phytoextraction was conducted for Zn and Cd (Baker et al. 1991). Many plants that are classified as hyperaccumulator depend on type of metal and accumulation behavior from the soil. The diversity of plant rhizosphere microbes and mycorrhiza also play key role for the remediation of contaminated site with heavy metals. The key for successful bioremediation depends on the nature and bioavailability of pollutants. The comprehensive understanding is still required to learn the mechanisms and crucial factors influencing the plant–microbe–toxicant interaction in soils for the success of phytoremediation.

11.2 Rhizosphere Microbe-Assisted Phytoremediation

Phytoremediation involves the use of green plants to extract, sequester, degrade, and/or detoxify pollutants by means of biological processes (Wenzel et al. 1999) and has been reported to be an in situ, nonintrusive, cost-effective, ecologically benign, aesthetically pleasing, socially acceptable technology to remediate contaminated soils (Garbisu et al. 2002). It also helps to prevent landscape deterioration and enhances the diversity of soil microorganisms to maintain healthy ecosystems; hence, it is considered to be a more attractive technique than traditional approaches that are currently in use for heavy metal decontamination.

Phytoremediation process can be classified according to the method and nature of the soil pollutants (Salt et al. 1995). Various aspects of phytoremediation process in relation to organic and inorganic contaminants are depicted in the Fig. 11.1.

Phytoremediation techniques can be studied under different strategies such as: (a) Phytoextraction: It is the process by which plants absorb metal from the contaminated site and transfer it to aboveground parts of the plants. These plants have a high degree of potential to absorb and accumulate or translocate metals or metalloids to the aboveground biomass. (b) Phytostabilization: It involves restriction of the mobility of metals in the soil. The reduced mobility of the contaminants may be achieved by accumulation and absorption onto roots, or precipitation within the rhizosphere. (c) Phytostimulation: It is also called plant-assisted biodegradation. Phytostimulation is the process where root-induced microbial activity is capable of degrading the organic contaminants. (d) Phytovolatilization/rhizovolatilization: In this approach, plants take up contaminants from the soil and transformed it into volatile compounds into the atmosphere through transpiration. These methods are highly used for the metal(loid)s in the soil such as mercury (Hg), selenium (Se), and arsenic (As). (e) Phytodegradation: It is the process of enzymatic degradation of

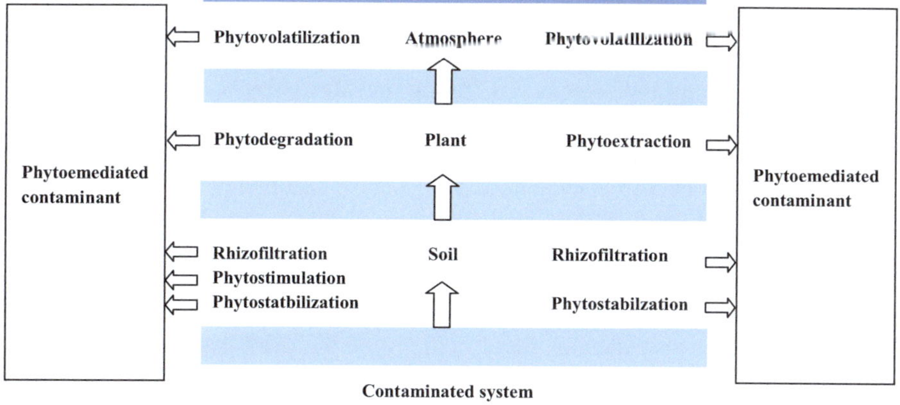

Fig. 11.1 Phytoremediation processes for organic and inorganic contaminants

complex organic molecules to simpler ones by means of enzymatic action or the incorporation of these molecules into plant tissues or into new plant material. (f) Rhizofiltration: It is primarily used to remediate aquatic systems with low levels of contaminant. It can be used for heavy metals such as lead (Pb), cadmium (Cd), copper (Cu), nickel (Ni), zinc (Zn), and chromium (Cr) which are generally retained within the roots and do not translocate to the shoots. This method can be explored in both terrestrial and aquatic plants for in situ or ex situ purposes.

11.2.1 Interactions in the Rhizosphere

Efficient phytoremediation techniques rely on the complex interactions among soil, contaminants, microbes, and plants.

11.2.1.1 Plant–Microbe Interactions

The interaction between plant roots and wide range of soil microbes, especially rhizospheric one, is the major determinants of the phytoremediation potential (Glick et al. 1995). Both the micropartner, i.e., plant-associated microbes and the host plant, control the functioning of associative plant–microbe symbioses in the contaminated soil. In plant bacterial symbiosis, plant provides specific carbon source to the bacteria inducing the bacteria to reduce the heavy metal phytotoxicity. Alternatively, in nonspecific association between plants and bacteria, plant metabolic processes stimulate the microbial community through root exudates, which in turn enable the microbes to degrade the contaminants in soil. Moreover, the adaptation capabilities of both the partners of associative symbiosis and the bioremediation potential of the microsymbiont play a vital role in minimizing the heavy metal toxicity.

11.2.1.2 Heavy Metal–Microbe Interactions

Rhizosphere microbes are empowered with different traits that can modify the solubility and bioavailability of the heavy metals in soil (Lasat 2002; McGrath et al. 2001; Whiting et al. 2001). Rhizobacteria may release different chelating substances by which acidification of the environment takes place through production of organic acid and changes the redox potential (Smith and Read 1997). Soil pH reduction mediated through *Sphingomonas macrogoltabidus*, *Microbacterium liquefaciens*, and *M. arabinogalactanolyticum* has been reported to enhance the Ni uptake in *Alyssum murale* grown in a serpentine soil (Abou-Shanab et al. 2003). An earlier study reported that the metal-polluted sites have negative impact on soil microbial diversity and microbial activities (Giller et al. 1998).

11.2.1.3 Plant–Bacteria–Soil Interactions

The soil condition also dictates the specificity of the plant–bacteria association. Different soil conditions regulate the bioavailability of soil contaminant such as composition of root exudate and levels of nutrient, influencing the bacterial metabolic activity as well as phytoremediation potential. Moreover, the requirements for heavy metals for bacterial metabolism may also govern whether the plant–bacteria interaction would be specific or nonspecific. Along with metal toxicity, there are several other factors that limit plant growth in the contaminated soils including harsh climatic conditions, poor soil structure, low water retention, and nutrient deficiency.

11.2.2 Rhizoremediation: Microorganism-Assisted Phytoremedation

Rhizoremediation is a subprocess of phytoremediation where plants along with their rhizospheric microorganisms are being used to enhance the efficiency of contaminant extraction (Jing et al. 2007). It is a beneficial association where the microorganisms enhance the bioavailability of the metals and the plants help in the extraction and removal of such compounds from soil (Chaudhry et al. 2005). It has positive role for both sides, where the plants supply nutrients to microorganisms, which, in turn, grow and proliferate, increasing the potential of degradation by the plant. However, there is a lack of studies about this synergism between plants and microorganisms facilitating phytoremediation (Kavamura and Esposito 2008). Some beneficial associations among plant and rhizospheric microbes that participated in the rhizoremediation are as follows:

11.2.2.1 Plant Growth-Promoting Rhizobacteria and Rhizoremediation

Plant growth-promoting rhizobacteria are generally known to promote the growth of the plants in the following manner:

1. Fix nitrogen from the atmosphere and deliver it to the plants.
2. Produce siderophores that can make complex with iron present in the soil and make available for assimilation to plant cells. Plants can easily take up the bacterial siderophore–iron complex and also through production of plant hormones like auxins, cytokinins, gibberellins, etc. which may stimulate the growth of the plant.
3. Solubilize mineral nutrients such as phosphorus through production of various organic acids, making them more easily available for plant growth.
4. Act as biocontrol agent.

Several experiments were conducted to examine the ability of a wide range of plants for heavy metal extraction and then to translocate those metals from roots to leaves and shoots. However, the potential of heavy metal removal is limited by slow plant growth and low biomass production by hyperaccumulator plants (Raskin and Ensley 2000). In this context, the use of plant growth-promoting rhizobacteria as adjuncts has been found to stimulate significant growth of plants even in the presence of higher concentration of heavy metals in soil (Zhuang et al. 2007; Glick 2010).

11.2.2.2 Endophytic Microorganisms and Rhizoremediation

Endophytic microorganisms can be defined as microbial colonizations in the internal tissues (root cortex or xylem) of plants without causing any symptoms of infection or negative impacts on their host (Schulz and Boyle 2006). Among the most predominant genera of culturable endophytes are *Pseudomonadaceae*, *Burkholderiaceae*, and *Enterobacteriaceae.* Endophytes play a very important role in phytoremediation especially in rhizoremediation. Idris et al. (2004) studied the endophytes and rhizobacteria with *Thlaspi goesingense*, a hyperaccumulator of Ni using both cultivation and cultivation-independent techniques. Results revealed that endophytes are generally culture independent and are more tolerant to higher concentration of Ni as compared to rhizobacteria. Though endophytes hold great promise for heavy metal remediation, the mechanisms by which endophytes enhanced metal accumulation are yet to be well understood. Furthermore, the application of culture-independent endophytes is quite a challenging task (Weyens et al. 2009).

11.2.2.3 Mycorrhizoremediation

Mycorrhizoremediation is an advanced phytoremediation strategy involving contribution from tripartite association among plant, mycorrhiza, and rhizobacteria. Mycorrhizae can be efficiently explored in the soil microsites that are not accessible for plant roots. They can further change the heavy metal bioavailability through competition with roots and other microorganisms for water and metal uptake, protection of roots from direct contact with the heavy metal via development of the ectomycorrhizal sheath, and restricted metal transport by increasing soil hydrophobicity (Lazcano et al. 2010). Ectomycorrhizal associations are reported to enable the host plant to withstand higher heavy metal toxicity. The structure of the fungal sheath, density, and surface area of the mycelium are key factors to determine the efficiency of an ectomycorrhizal association to resist/tolerate metal toxicity and to protect the host plant from pollutant contact (Hartley et al. 1997). Studies also reported increased uptake of metal(loid)s in the presence of arbuscular mycorrhizal fungi; however, there are some contradictory reports indicating negligible effect or decreased accumulation in plant tissues (Lazcano et al. 2010). The controversial results are difficult to interpret and could be attributed to the differential response under greenhouse experiment and field study.

11.2.3 Phytoextraction

Phytoextraction is a subprocess of phytoremediation where the pollutant-accumulating plants are being utilized for removal of heavy metals from contaminated soils by concentrating them in the aboveground biomass (Salt et al. 1998). The selection of plants for heavy metal phytoextraction should possess features like (a) potential tolerance to high levels of heavy metal concentration, (b) fast-growing plants for effective accumulation of heavy metal, (c) ready translocation of heavy metal in the aboveground biomass of plants, and (d) ease of harvest (Vangronsveld et al. 2009).

However, the success of phytoextraction depends upon factors such as bioavailability of heavy metal and the potential of the plant to intercept, take up, and accumulate the metals in shoots (Ernst 2000).

11.2.3.1 Role of Plant-Associated Rhizobacteria in Phytoextraction

To enhance the efficiency and rate of phytoextraction, the role of plant-associated rhizobacteria is highly beneficial. Microorganisms can increase plant uptake of heavy metal in the following way: (1) may increase the root surface area and root hair architecture, (2) enhance the metal bioavailability, and/or (3) increase the metal translocation from the rhizosphere to the plant shoot (Weyens et al. 2009). Further, improving the plant biomass production can influence the efficiency of trace element phytoextraction.

The plant-associated rhizobacteria metabolic performance may help develop new improved phytoremediation strategies. However, the dynamic and variable metabolic capacities of plant-associated rhizobacteria are still poorly highlighted. Plants stimulate the growth of rhizosphere microorganisms due to secretion of different organic molecules by their roots, which in turn improved the bacterial densities in the rhizosphere (Anderson and Coats 1995).

11.2.4 Bacterial Heavy Metal Resistance

The plant-associated rhizospheric bacteria have several benefits conferred to their hosts; the major qualification for protecting plants from heavy metals stress is resistance of the bacteria to heavy metals. Along with dynamic metabolic capacity of the bacteria, metal resistance operon is also important to empower the bacteria against heavy metal toxicity. Among the heterotrophic bacteria, members of the β-proteobacteria have the maximum levels of heavy metal resistance. *Alcaligenes eutrophus* is a potential member of this group. *A. eutrophus* CH34 species is the extensively reported that harbors two endogenous megaplasmids encoding genes for multiple heavy metal resistance. Plasmid pMOL28 is 180 kb and codes for resistance to various heavy metals such as cobalt, nickel, chromate, mercury, and thallium. Resistance genes are organized with the *chr, mer, and cnr* operons, coding for resistances to chromate, mercury, and both cobalt and nickel, respectively, (Mergeay et al. 1985; Taghavi et al. 1997). The plasmid from strain CH34 is pMOL30 (240 kb) responsible for resistance against some heavy metals. This plasmid also consists of organized operon out of which the *mer*, *cop*, and *pbr* operons encode resistance to heavy metal mercury, copper, and lead, respectively. The *czc* operon encodes for heavy metal cadmium, zinc, and cobalt resistance.

11.3 Plant–Microbe Association for Heavy Metal Transformation in Soil–Plant System

Rhizospheric microbes play an important role in improving phytoremediation process by changing the metal bioavailability through altering redox reactions, soil pH, or release of some chelators like siderophores, organic acids, biosurfactants, etc. (Zarei et al. 2010; Miransari 2011; Rajkumar et al. 2012) (Fig. 11.2).

Metabolites or reactions produced by plant-associated microbes have been reviewed and summarized in Table 11.1

11.3.1 Siderophores

Most plant-associated microorganisms can produce iron chelator siderophores at low levels of iron concentration in soil; however, siderophore can also form stable complex with other heavy metals such as Al, Cd, Cu, Ga, In, Pb, and Zn (Glick and Bashan 1997; Schalk et al. 2011) and cause solubilization of unavailable form of heavy metal to available form, thus improving efficacy of phytoextraction (Braud et al. 2009b; Rajkumar et al. 2010). Pyoverdine and pyochelin produced by *Pseudomonas aeruginosa* are responsible for enhancing the bioavailability of Cr and Pb in the rhizosphere of maize (Braud et al. 2009b). Similarly, siderophores produced by *Streptomyces tendae* F4 significantly enhanced uptake of Cd by sunflower plant (Dimkpa et al. 2009). Nevertheless, there are also contradictory reports

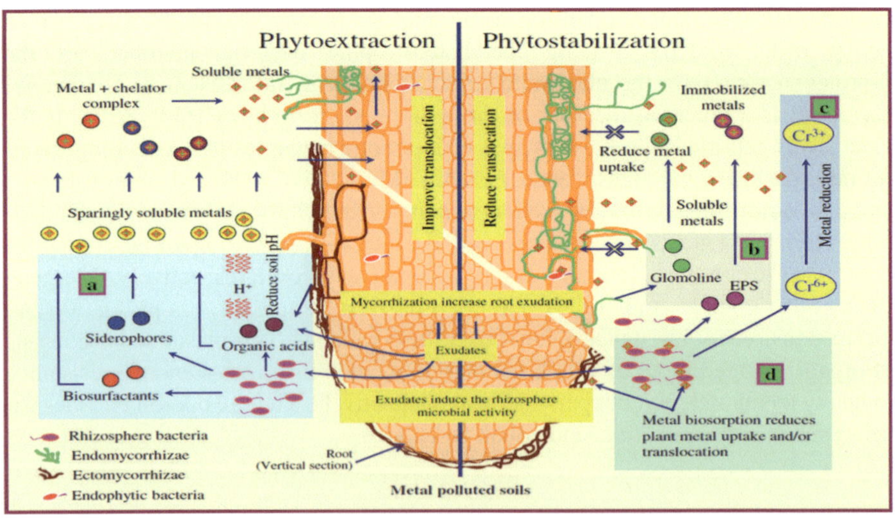

Fig. 11.2 Schematic representation of role of rhizospheric microbes for phytoremediation (**a**) by producing metal-mobilizing chelators, (**b**) by excreting metal-immobilizing metabolites, (**c**) by reducing metal reduction, and (**d**) by metal biosorption. EPS, extracellular polymeric substances (Source: Rajkumar et al. 2012)

Table 11.1 Potential of microbial metabolites/actions to mobilize/immobilize metals by plants

Metabolites or reactions	Microorganisms	Microbial potential	References
Siderophores			
Azoto chelin and azotobactin	*Azotobacter vinelandii*	Helps in Mo and V acquisition	Wichard et al. (2009)
Pyochelin	*Pseudomonas aeruginosa*	Chelates many metals like Cd^{2+}, Cr^{2+}, Al^{3+}, Mn^{2+}, Zn^{2+}	Braud et al. (2009a)
Desferrioxamine and coelichelin	*Streptomyces tendae*	Enhanced uptake of Cd and Fe by plants	Dimkpa et al. (2009)
Organic acids			
Oxalic acid, tartaric acid, formic acid, acetic acid, malic acid	*A. niger*, *Burkholderia cepacia*, *Beauveria caledonica*, *Oidiodendron maius*, *Pseudomonas fluorescens*, *Penicillium bilaiae*	Solubilized Zn, Ni, Fe, Pb, and Cd	Arwidsson et al. (2010), Li et al. (2010), and Hoberg et al. (2005)
Gluconic acid, 5-ketogluconic acid	*Gluconacetobacter diazotrophicus*, *Pseudomonas aeruginosa*	Solubilized ZnO, $ZnCO_3$, and $Zn_3(PO_4)_2$	Saravanan et al. (2007), and Fasim et al. (2002)
Biosurfactants			
Rhamnolipids, dirhamnolipid	*Pseudomonas aeruginosa*	Mobilized Cu, Cd, and Pb	Venkatesh and Vedaraman (2012), and Juwarkar et al. (2007)
Polymeric substances			
Polymeric substances (extracellular)	*Azotobacter* spp.	Immobilized Cd and Cr	Joshi and Juwarkar (2009)
Glomalin	*Glomus mosseae*	Immobilized Cu, Pb, and Cd	Gonzalez-Chavez et al. (2004)
Redox reaction			
Oxidation and reduction	*Streptomyces lividans* sp., *Rhodococcus* sp., *Acidithiobacillus thiooxidans*, *Leptospirillum ferrooxidans*	Increased the mobility of As, Cu, Cd, Hg, and Zn	Yang et al. (2012), Beolchini et al. (2009)

(Sinha and Mukherjee 2008; Tank and Saraf 2009; Kuffner et al. 2010) which generated the need to study the interaction of plant–siderophore-producing microorganisms–metals in the contaminated soils. Siderophore production by microbes is controlled by various factors, viz., iron availability, pH, nutrient status of soils, type, concentration of heavy metals, etc. Therefore, higher heavy metal concentration acts as stimuli to produce more siderophore by microbes. Findings of Braud et al. (2009a) revealed the fact that addition of heavy metals, Al, Cu, Ga, Mn, and Ni, in iron-limited succinate medium induced pyoverdine synthesis by P. *aeruginosa*.

Moreover, the presence of heavy metals such as Cu, Ni, and Cr stimulated pyoverdine synthesis even in the case of iron (Braud et al. 2010).

11.3.2 Organic Acids

Low molecular weight organic acids, synthesized by plant–microbe interaction, play an instrumental role in enhancing the bioavailability of the trace elements and metals in the soil mainly through formation of metal complex. Organic acids work as a ligand which form stable complex with the heavy metals. However, the stability of the complex is regulated by several factors, viz., number and the position of carboxyl groups in organic acids, form of heavy metals, and most importantly pH of the soil solution (Ryan et al. 2001). Different studies have reported that 5-ketogluconic acids and 2-gluconic acids are prime responsible for solubilizing and mobilizing of insoluble ZnO, $Zn_3(PO_4)_2$, and $ZnCO_3$. The bacterial strain involved in gluconic acid productions and Zn solubilization are reported to be *Gluconobacter diazotrophicus* and *Pseudomonas aeruginosa* (Fasim et al. 2002; Saravanan et al. 2007). Similarly, formic acid, succinic acid, oxalic acid, acetic acid, and tartaric acid produced by rhizospheric bacteria have been reported to solubilize Cd and Zn in the rhizosphere of *Sedum alfredii*, a hyperaccumulating plant (Li et al. 2010). Furthermore, organic acids secreted by plant-associated microbes expedite the absorption of Cu (Chen et al. 2005), Pb (Sheng et al. 2008), and Cd and Zn (Li et al. 2010) by plant root. Mycorrhizal fungi, especially ericoid mycorrhizal fungi (*Oidiodendron maius*) and other soil fungi (*Beauveria caledonica*), can also increase solubility of Zn from insoluble sources by releasing citric and malic acids. These organic acids either by chelation or by acidolysis process can increase the solubility and availability of Zn from insoluble ZnO, $Zn_3(PO_4)_2$, and pyromorphite (Martino et al. 2003; Fomina et al. 2005).

Although the role of organic acids seem promising, however, the factors governing the fate and the performance of the organic acids need to be considered for better understanding of their mechanisms. Moreover, the other root-mediated process such as contribution of root exudates and other metabolites in metal mobilization (Wenzel 2009) also need to be taken into account before describing the role of organic acids produced by plant–microbe interaction in heavy metal transformation and solubilization. In this respect, precise quantification of organic acids in rhizosphere and the genetic sequencing of responsible microbes could shed light in understanding organic acid dynamics between soil, plant, and microbe continuum.

11.3.3 Biosurfactants

Biosurfactants are amphiphilic molecules comprising of a nonpolar (hydrophobic) tail and a polar/ionic (hydrophilic) head. Biosurfactant produced by microbes can increase metal solubility and bioavailability through complex formation with heavy metals at the soil interface leading to desorption of metals from soil matrix. The potential of biosurfactant dirhamnolipid produced by *P. aeruginosa* in solubilizing and mobilizing Cd, Pb, and Cu has already been documented in earlier studies (Juwarkar et al. 2007; Venkatesh and Vedaraman 2012). In addition, biosurfactants produced by plant–microorganism association also show high promise for improving the metal (Cd) uptake by rape, maize, Sudan grass, and tomato plants, a desirable trait for plants to be used for phytoextraction. The biosurfactant released from *Bacillus* sp. J119 was capable of enhancing Cd uptake from soil artificially contaminated with different levels of Cd (0 and 50 mg kg^{-1} Sheng et al. (2008)). Hence, the knowledge regarding interactive effect of biosurfactant-producing microbes on plants will enrich our perception about the role of biosurfactant-producing microbes in heavy metal phytoremediation.

11.3.4 Polymeric Substances and Glycoprotein

Extracellular polymeric substance (EPS), mucopolysaccharides, and proteins produced by plant-associated microbes can form complex with heavy metals and reduce their mobility in soil. Joshi and Juwarkar (2009) reported that EPS produced by *Azotobacter* spp. could immobilize Cd and Cr through complex formation (15.2 mg g^{-1} of Cd and 21.9 mg g^{-1} of Cr) and reduce the uptake of Cd (−0.5) and Cr (−0.4) by *Triticum aestivum*. Arbuscular mycorrhizal fungi are also reported to produce glomalin which form complex with Cu, Pb, and Cd and extract approximately 4.3 mg Cu, 1.1 mg Pb, and 0.1 mg Cd per gram of glomalin from metal-polluted soils (Gonzalez-Chavez et al. 2004). Therefore, AMF with higher amount of glomalin secretion capacity could play an instrumental role in phytoextraction and phytostabilization effort.

11.3.5 Redox Transformation of Metal in Rhizosphere

Plant-associated microbes can change the mobility of heavy metals through redox transformation reactions. Oxidation of metals by rhizospheric microbes is particularly interesting from a phytoextraction point of view. For instance, Cu mobilization in contaminated soils and its uptake in plant tissue were enhanced in the presence of sulfur-oxidizing bacteria in the rhizosphere (Shi et al. 2011). This enhanced uptake of copper in the presence of sulfur-oxidizing bacteria was due to lowering of the soil pH as a result of conversion of reduced sulfur to sulfates. Potential of Fe-/S--oxidizing bacteria to enhance metal bioavailability in the soils through acidification reaction was also reported by Chen and Lin (2001).

Microbial reduction of heavy metals also sometime immobilizes the heavy metals in the rhizosphere. For example, decreased uptake of Cr by of 37 % in shoot and 56 % in root of green chili grown in Cr(VI)-contaminated soils upon inoculation with *Cellulosimicrobium cellulans* was reported by Chatterjee et al. (2009). This effect was brought about by microbial reduction of mobile and toxic Cr(VI) to non-toxic and immobile Cr(III) in the soil. Abou-Shanab et al. (2007) reported lower Cr translocation from root to shoots of water hyacinth as indicative of the Cr-reducing potential of rhizosphere microbes. Similarly, Di Gregorio et al. (2005) demonstrated the Se-reducing potential of *Stenotrophomonas maltophilia* isolated from the rhizosphere of *Astragalus bisulcatus*. This bacterium significantly reduced soluble and harmful Se(IV) to insoluble and unavailable Se(0), thereby reducing the uptake of Se by plant. These examples demonstrate mechanisms, by which metal-reducing microbes lock the metals within the rhizosphere soil and reflect the suitability of these microbes for phytostabilization applications.

Besides, the synergistic interaction of metal-oxidizing and metal-reducing microbes on heavy metal mobilization in contaminated soils has also been studied. Inoculation of Fe-reducing bacteria and the Fe-/S-oxidizing bacteria together significantly increased the mobility of Cu, Cd, Hg, and Zn by 90 %. This effect was attributed to the coupled and synergistic metabolism of oxidizing and reducing microbes Beolchini et al. (2009). Though these results open new perspectives for the bioremediation technology for metal mobilization, further investigations are needed to utilize such bacteria in phytoextraction process.

11.3.6 Biosorption

Through biosorption mechanism, the plant-associated microbes may also contribute in plant–metal uptake. Biosorption can be defined as the microbial adsorption of soluble/insoluble organic/inorganic contaminant by a metabolism-independent, passive or by a metabolism-dependent, active process (Ma et al. 2011). The biosorption process involves a solid phase (sorbent or biosorbent; biological material) and a liquid phase (solvent, normally water) containing a dissolved species to be sorbed (sorbate, metal ions). Higher affinity of the sorbent for the sorbate species (metals) is responsible for binding of metals on sorbent by different interactions. The process continues till equilibrium is established between the amount of solid-bound sorbate species and its portion remaining in the solution (Das et al. 2008). The efficiency of biosorption depends upon factors like initial metal concentration, pH, temperature, and biomass weight in solution. Several researchers have pointed out the restricted entry, reduced bioavailability, and lower metal uptake by plant due to biosorption. For instance, Madhaiyan et al. (2007) reported inoculation of metal-binding fungi *Magnaporthe oryzae* and bacteria *Burkholderia* sp. reduced Ni and Cd accumulation in roots and shoots of tomato. These effects of inoculation of *Trifolium repens* with *Brevibacillus* sp. B-I decreased the concentration of Zn in shoot tissues compared to respective uninoculated control due to the increased Zn biosorption by *Brevibacillus* sp. B-I Vivas et al. (2006).

The mycorrhizal fungi have also been reported to act as a filtration barrier against the translocation of heavy metals from plant roots to shoots. Experiments revealed that the inoculation of pine seedlings with *Scleroderma citrinum*, *Amanita muscaria*, and *Lactarius rufus* reduced translocation of Zn, Cd, or Pb from roots to

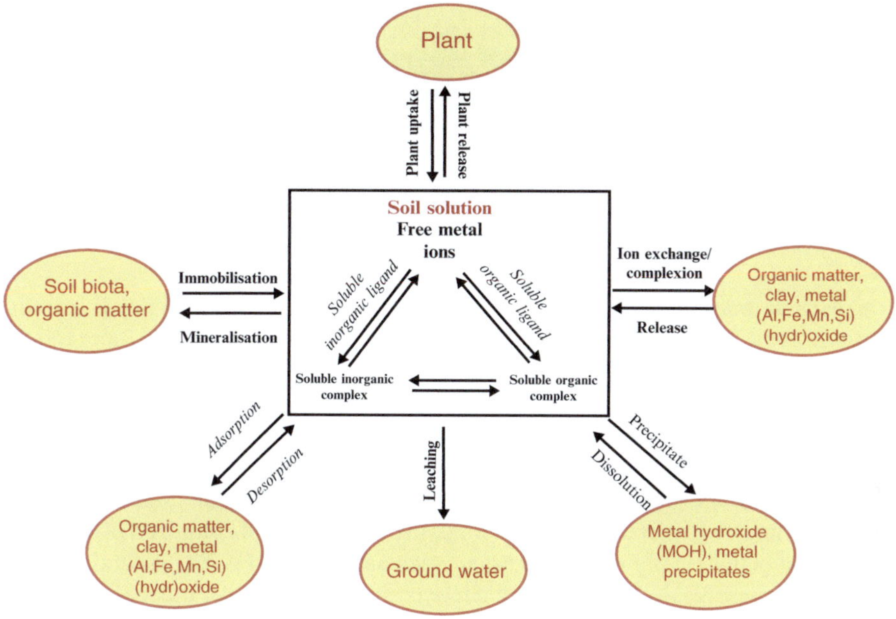

Fig. 11.3 Possible reactions involved in physical, chemical, and biological transformation of metal(loid)s in soil (Source: Seshadri et al. 2015)

shoots by increased metal biosorption in outer and inner components of the mycelium (Krupa and Kozdrój 2007). Large surface area of mycorrhizal fungi endows mycorrhizal fungi with a strong capacity for adsorbing heavy metals from soil. The fungal cell wall components (e.g., chitin, extracellular slime, etc.) and intracellular compounds (e.g., metallothioneins, P-rich amorphic material) may also immobilize/arrest the metals in the interior of plant roots (Meharg 2003). An exhaustive compilation of microbes for biosorption of heavy metals was made by Volesky and Holan (1995). Although inoculation of plants with metal-binding microbes could be a suitable approach for plant protection against heavy metals and phytostabilization of metal-polluted soils, many authors believe that the reduction in accumulation and translocation of metal in plants is not due to biosorption/bioaccumulation alone (Babu and Reddy 2011).

Plant-associated microorganisms differ in their ability to alter heavy metal bioavailability and its uptake by plants through metal-mobilizing/metal-immobilizing metabolites/processes. Colonization and survival of these microbes also greatly influence the quantity of metal accumulation in plants growing in metal-contaminated soils which in turn is governed by soil physicochemical–biological properties such as metal toxicity, indigenous microbial communities, adverse pH, nutrient deficiency, etc.

The general mechanisms involved in the transformation of metal(loid) ions in the soil lead to retention (mediated by sorption, precipitation, and complexation reactions) or loss (plant uptake, leaching, and volatilization) of heavy metal(loid)s (Fig. 11.3). Although most metal(loid)s do not undergo volatilization-related losses, some metal(loid)s such as As, Hg, and Se tend to form gaseous compounds (Bolan et al. 2013). A greater understanding of the microbiological (activity) and chemical (exudates) changes occurring in the rhizosphere would identify the mechanisms involved in the transformation of heavy metals in the contaminated soil.

11.4 Role of Mycorrhiza and PGPR for Heavy Metal Removal from Metal-Contaminated Site

Rhizosphere microbes have played a key role for nutrient cycling and soil sustainability. Arbuscular mycorrhizal fungi (AMF) are a group of endophytic fungi infecting the roots of majority of the terrestrial plants. This symbiosis association between mycorrhiza and host plants has very important role on the plant's growth and development through the acquisition of phosphorous and other essential mineral nutrients from the soil. Plant growth-promoting rhizobacteria (PGPR) is a group of bacteria that colonize plant roots and promote growth and yield (Wu et al. 2005). However, PGPR are known to increase root system uptake properties of colonized plants, thus facilitating better supply of plant nutrient such as N, P, and Fe. The potential application of mycorrhizal plants for land decontamination has several benefits such as increased plant biomass, plant phosphorus nutrition, and tolerance to heavy metal stress. Mycorrhizal species influences metal toxicity to plants through decreasing translocation of heavy metals and its concentration.

11.4.1 Role of Mycorrhiza for the Remediation of Contaminated Site

Remediators choose the applicable and suitable microbial species that are used as inoculants to plant growth promotion and bioremediation process. The arbuscular mycorrhizal (AM) fungi have several critical roles for improving the plant's resistance to various biotic and abiotic stresses (Harrier 2001). AM fungi also have great advantage to alleviate heavy metal toxicity of plants (Hildebrandt et al. 1999). AM fungi has significant role for improving the uptake of nutrient and water by host plants through their mycelial networks and protecting the host plants from heavy metal toxicity. Besides AM fungi, there are several other beneficial microorganisms in the rhizosphere that may also help for heavy metal tolerance to the plants. According to Khan et al. (2000), mycorrhizal species enhance the bioavailability of toxic metals by altering the microenvironment of the rhizosphere through decontamination. This AM fungi may improve the plant nutrient uptake in alkaline and calcareous soils of arid and semiarid regions in which the bioavailability of P and several cationic micronutrients is limited. The presence of carbonates in calcareous soils is also limiting water holding capacity. Furthermore, plant transpiration is significantly reduced with an increase in soil heavy metal concentration (Davari et al. 2010). It has been reported that heavy metals like Cd can affect the hydraulic conductivity of root by multiple mechanisms occurring on the apoplastic and/or the symplastic pathway (Shah et al. 2010). The ability of beneficial microorganisms to promote the growth of canola and tomato seedlings treated with toxic concentrations of various metal(loid)s such as As, Cd, Ni, Pb, Se, and Zn has been demonstrated. There have been few analytical studies available on AM fungi in the contaminated soils. While some workers highlighted that the external mycelium of the arbuscular mycorrhizae was the primary site for various heavy metal localization (Kaldorf et al. 1999; Turnau 1998), other reports emphasized the selective exclusion of toxic and nontoxic metals by adsorption onto chitinous cell wall structure (Zhou 1999), or onto extracellular glycoprotein called glomalin (Wright and Upadhyaya 1998), or intracellular crystallization. These mechanisms have great significance in reducing a plant's exposure to potentially toxic metals, which is called mycorrhizoremediation. Localization of Cu accumulation in the extraradical mycelium (ERM) of different AM fungi differed in their capacity for sorption of Cu which was directly related to the cation exchange capacity of ERM of AM fungi (Gonzalez-Chavez et al. 2002). Difference exists in accumulation and tolerance for different heavy metals among the species of AM fungi. Hence, mechanism involved in tolerance and accumulation of heavy metals require future research in order to explore the contribution of AM fungi in plant tolerance and its ecological significance in polluted soils.

11.4.2 Role of PGPR for the Remediation of Contaminated Site

Plant growth-promoting rhizobacteria (PGPR) colonize in the rhizosphere and improve plant growth through various mechanisms, such as plant nutrient uptake, suppressing harmful phytopathogens by producing antibiotics and siderophores or other bioactive compounds, phytohormone production, and fixation or solubilization of plant nutrient and making it available to the plants. Better colonization of rhizospheric microorganism increases stress endurance of a plant and improves the metal bioavailability. Many isolated strains of PGPR used to enhance crop yield and improve agriculture sustainability (Begonia et al. 2005). PGPR are known to increase root system uptake properties of colonized crops by facilitating ion nitrate adsorption, phosphate solubilization, and iron chelation (Islam et al. 2009). Maize seed inoculation with rhizobacteria such as *Pseudomonas cepacia*, *P. fluorescens*, and *Streptomyces aurantiacus* in combination with nitrogen increased 25 % more crop yield than the non-rhizobacterium-colonized control.

When Indian mustard (*Brassica juncea*) and canola (*Brassica campestris*) seeds grow in the presence of PGPR strain, the plants produce siderophores, and this plays an important role in the remediation of Ni-, Pb-, and Zn-contaminated site (Burd et al. 1998). According to Belimov et al. (2001), growth of *Brassica napus* plant is improved by inoculating recalcitrant PGPR through ACC-deaminase activity, and growth of barley plants is improved by biological nitrogen fixation and auxin production with PGPR inoculation in Cd-contaminated soil (Belimov and Dietz 2000). The rhizosphere is a type of microenvironment where microorganisms form a special type of communities with plant growth-promoting capabilities present to remove the toxic contaminants (Ma et al. 2009). Findings of Idris et al. 2004 confirmed that metal mobility and bioavailability to the plants are enhanced by rhizospheric bacteria by releasing chelating agents, acidification, phosphate solubilization, and redox changes.

Thus, interactions between plants and useful rhizosphere microbes can improve biomass production and accumulation of heavy metals. Growth of crop plant is promoted by PGPR which help in decreasing the plant stress related with phytoremediation methods (Reed and Glick 2005). Selection of highly potential microbial combination is a big challenge for developing phytoremediation strategies.

11.5 Conclusion

As an economic and green approach for decontamination of polluted soil and water, phytoremediation is an optimistic technology. Association of microbes has shown improved efficiency of phytoremediation in many cases. The capability of soil function is mostly regulated by the soil biological component. Plant–microbe interaction plays a critical role to remediate extensive contaminated sites and recover to health state of soil. Though the mechanism involved in reducing the load of contaminating metal through plant assisted by microbes is complex and involves several processes occurring simultaneously in a habitat, thorough understanding of processes will further improve the efficiency of phytoremediation by manipulating the interaction depending upon nature of pollutant, condition of microhabitat, concentration of contaminant, type of associated microbial community, etc. Further, identification of specific biomarker associated with the promising microbes for efficient microbe-assisted phytoremediation will further improve the remediation efficiency. Although promising response of inoculation of beneficial microbes particularly plant growth-promoting bacteria and/or mycorrhizae has been reported under laboratory conditions, the result under field condition showed limited effectiveness because of complexity of soil environment and competing microbes. Characterizing the physicochemical and biological features of target contaminated soils may be important for making successful microbe-assisted phytoremediation technology. The colonization and survival of inoculums in metal-contaminated soil is necessary to exhibit beneficial traits for improving the plant growth and overall phytoremediation process in metal-contaminated soils. Advancing the knowledge on identification of favorable soil condition, efficient microbes with multiple metal resistance/tolerance potential, survival, and compatibility with other microbes may be important to utilize the potential of inoculants for phytoremediation purpose. Identification of efficient microbes for bioaccumulation of heavy metal and understanding biochemical and molecular mechanisms of interaction of plant–microbe toxicant play a major role in the processes involved in phytoremediation.

References

Abou-Shanab RA, Angle JS, Delorme TA, Chaney RL, van Berkum P, Moawad H, Ghanem K, Ghozlan HA (2003) Rhizobacterial effects on nickel extraction from soil and uptake by Alyssum murale. New Phytol 158(1):219–224

Abou-Shanab RA, Angle JS, van Berkum P (2007) Chromate-tolerant bacteria for enhanced metal uptake by Eichhornia crassipes (Mart.). Int J Phytorem 9:91–105

Anderson TA, Coats JR (1995) Screening rhizosphere soil samples for the ability to mineralize elevated concentrations of atrazine and metolachlor. J Environ Sci Health B 30:473–484

Arwidsson Z, Johansson E, von Kronhelm T, Allard B, van Hees P (2010) Remediation of metal contaminated soil by organic metabolites from fungi—production of organic acids. Water Air Soil Pollut 205:215–226

Babu AG, Reddy S (2011) Dual inoculation of arbuscular mycorrhizal and phosphate solubilising fungi contributes in sustainable maintenance of plant health in fly ash ponds. Water Air Soil Pollut 219:3–10

Baker AJM, Reeves RD, McGrath SP (1991) In situ decontamination of heavy metal polluted soils using crops of metal-accumulating plants—a feasibility study. In: Hinchee RE, Olfenbuttel RF (eds) Situ bioreclamation. Butterworth Heinemann, Stoneham, pp 539–544

Baumann A (1885) Das Verhalten von Zinksatzen gegen Pflanzen und im Boden. Landwirtsch Vers-Statn 31:1–53

Begonia MT, Begonia GB, Ighoavodha M, Gilliard D (2005) Lead accumulation by tall fescue (Festucaarundinacea Schreb) grown on a lead contaminated soil. Int J Environ Res Public Health 2:228–233

Belimov AA, Dietz K-J (2000) Effect of associative bacteria on element composition of barley seedlings grown in solution culture at toxic cadmium concentrations. Microbiol Res 155:113–121

Belimov AA, Safronova VI, Sergeyeva TA, Egorova TN, Matveyeva VA, Tsyganov VE, Borisov AY, Tikhonovich IA, Kluge C, Preisfeld A, Dietz KJ, Stepanok VV (2001) Characterization of plant growth promoting rhizobacteria isolated from polluted soils and containing 1-aminocyclopropane-1- carboxylate deaminase. Can J Microbiol 47:242–252

Beolchini F, Dell'Anno A, Propris LD, Ubaldini S, Cerrone F, Danovaro R (2009) Auto- and heterotrophic acidophilic bacteria enhance the bioremediation efficiency of sediments contaminated by heavy metals. Chemosphere 74:1321–1326

Bingham FT, Pereyea FJ, Jarrell WM (1986) Metal toxicity to agricultural crops. Met Ions Biol Syst 20:119–156

Bolan N, Kunhikrishnan A, Gibb J (2013) Rhizoreduction of arsenate and chromate in Australian native grass, shrub and tree vegetation. Plant Soil 367:615–625

Braud A, Hannauer M, Milsin GLA, Schalk IJ (2009a) The Pseudomonas aeruginosa pyochelin-iron uptake pathway and its metal specificity. J Bacteriol 191:5317–5325

Braud A, Jézéquel K, Bazot S, Lebeau T (2009b) Enhanced phytoextraction of an agricultural Cr, Hg- and Pb-contaminated soil by bioaugmentation with siderophore producing bacteria. Chemosphere 74:280–286

Braud A, Geoffroy V, Hoegy F, Mislin GLA, Schalk IJ (2010) The siderophores pyoverdine and pyochelin are involved in Pseudomonas aeruginosa resistance against metals: another biological function of these two siderophores. Environ Microbiol Rep 2:419–425

Burd GI, Dixonand DG, Glick BR (1998) A plant growth promoting bacterium that decreases nickel toxicity in seedlings. Appl Environ Microbiol 64:3663–3668

Chaney RL (1983) Plant uptake of inorganic waste. In: Parr JE, Marsh PB, Kla JM (eds) Land treatment of hazardous waste. Noyes Data Corp, Park Ridge, pp 50–76

Chatterjee S, Sau GB, Mukherjee SK (2009) Plant growth promotion by a hexavalent chromium reducing bacterial strain, Cellulosimicrobium cellulans KUCr3. World J Microbiol Biotechnol 25:1829–1836

Chaudhry Q, Blom-Zandstra M, Gupta SK, Joner E (2005) Utilizing the synergy between plants and rhizosphere microorganisms to enhance breakdown of organic pollutants in the environment. Environ Sci Pollut Res 12:34–48

Chen SY, Lin JG (2001) Effect of substrate concentration on bioleaching of metal-contaminated sediment. J Hazard Mater 82:77–89

Chen YX, Wang YP, Lin Q, Luo YM (2005) Effect of copper-tolerant rhizosphere bacteria on mobility of copper in soil and copper accumulation by *Elsholtzia splendens*. Environ Int 31:861–866

Das N, Vimala R, Karthika P (2008) Biosorption of heavy metals – an overview. Indian J Biotechnol 7:159–169

Davari M, Homaee M, Khodaverdiloo H (2010) Modeling phytoremediation of Ni and Cd from contaminated soils using macroscopic transpiration reduction functions. J Sci Technol Agric Natural Resour Water Soil Sci 14:75–85
Di Gregorio S, Lampis S, Vallini G (2005) Selenite precipitation by a rhizospheric strain of Stenotrophomonas sp isolated from the root system of Astragalus bisulcatus: a biotechnological perspective. Environ Int 31:233–241
Dimkpa CO, Merten D, Svatoš A, Büchel G, Kothe E (2009) Siderophores mediate reduced and increased uptake of cadmium by Streptomyces tendae F4 and sunflower (Helianthus annuus), respectively. J Appl Microbiol 107:1687–1696
Dua M, Sethunathan N, Johri AK (2002) Biotechnology bioremediation success and limitations. Appl Microbiol Biotechnol 59:143–152
Ernst WHO (2000) Evolution of metal hyperaccumulation and phytoremediation hype. New Phytol 146:357–358
Fasim F, Ahmed N, Parsons R, Gadd GM (2002) Solubilization of zinc salts by a bacterium isolated from the air environment of a tannery. FEMS Microbiol Lett 213:1–6
Fomina MA, Alexander IJ, Colpaert JV, Gadd GM (2005) Solubilization of toxic metal minerals and metal tolerance of mycorrhizal fungi. Soil Biol Biochem 37:851–866
Garbisu C, Hernandez-Allica J, Barrutia O, Alkorta I, Becerril JM (2002) Phytoremediation: a technology using green plants to remove contaminants from polluted areas. Rev Environ Health 17:173–188
Giller KE, Witter E, McGrath SP (1998) Toxicity of heavy metals to microorganisms and microbial processes in agricultural soils. Soil Biol Biochem 30:1389–1414
Glick BR (2010) Using soil bacteria to facilitate phytoremediation. Biotechnol Adv 28:367–374
Glick BR, Bashan Y (1997) Genetic manipulation of plant growth-promoting bacteria to enhance biocontrol of phytopathogens. Biotechnol Adv 15:353–378
Glick BR, Karaturovic DM, Newell PC (1995) A novel procedure for rapid isolation of plant growth promoting pseudomonads. Can J Microbiol 41:533–536
Gonzalez-Chavez C, D'Haen J, Vangronsveld JJ, Dodd JC (2002) Copper sorption and accumulation by the extraradical mycelium of different Glomus spp. (arbuscular mycorrhizal fungi) isolated from the same polluted soil. Plant Soil 240:287–297
Gonzalez-Chavez MC, Carrillo-Gonzalez R, Wright SF, Nichols KA (2004) The role of glomalin, a protein produced by arbuscular mycorrhizal fungi, in sequestering potentially toxic elements. Environ Pollut 130:317–323
Harrier LA (2001) The arbuscular mycorrhizal symbiosis: a molecular review of the fungal dimension. J Exp Bot 52:469–478
Hartley J, Cairney JWG, Meharg AA (1997) Do ectomycorrhizal fungi exhibit adaptive tolerance to potentially toxic metals in the environment? Plant Soil 189:303–319
Hildebrandt U, Kaldorf M, Bothe H (1999) The zinc violet and its colonization by arbuscular mycorrhizal fungi. J Plant Physiol 154:709–717
Hoberg E, Marschner P, Lieberei R (2005) Organic acid exudation and pH changes by Gordonia sp. and Pseudomonas fluorescens grown with P adsorbed to goethite. Microbiol Res 160:177–187
Idris R, Trifonova R, Puschenreiter M, Wenzel WW, Sessitsch A (2004) Bacterial communities associated with flowering plants of the Ni hyperaccumulator Thlaspi goesingense. Appl Environ Microbiol 70:2667–2677
Islam MR, Madhaiyan M, Deka Boruah HP, Yim W, Lee G, Saravanan VS, Fu Q, Hu H, Sa T (2009) Characterization of plant growth-promoting traits of three-living diazotrophic bacteria and their inoculation effects on growth and nitrogen uptake of crop plants. J Microbiol Biotechnol 19:1213–1222
Jansen E, Michels M, van Til M, Doelman P (1994) Effects of heavy metals in soil on microbial diversity and activity as shown by the sensitivity-resistance index, an ecologically relevant parameter. Biol Fertil Soils 17:177–184
Jing Y, He Z, Yang X (2007) Role of soil rhizobacteria in phytoremediation of heavy metal contaminated soils. J Zhejiang Univ Sci B 8:192–207

Joshi PM, Juwarkar AA (2009) In vivo studies to elucidate the role of extracellular polymeric substances from Azotobacter in immobilization of heavy metals. Environ Sci Technol 43:5884–5889

Juwarkar AA, Nair A, Dubey KV, Singh SK, Devotta S (2007) Biosurfactant technology for remediation of cadmium and lead contaminated soils. Chemosphere 10:1996–2002

Kaldorf M, Kuhn M, Schroder WH, Hildebrandt U, Bothe H (1999) Selective element deposits in maize colonized by a heavy metal tolerance conferring arbuscular mycorrhizal fungus. J Plant Physiol 154:718–728

Kavamura NV, Esposito E (2008) Biotechnological strategies applied to the decontamination of soils polluted with heavy metals. Biotechnol Adv 28:61–69

Khan AG, Kuek C, Chaudhry TM, Khoo CS, Hayes WJ (2000) Role of plants, mycorrhizae and phytochelators in heavy metal contaminated land remediation. Chemosphere 41:197–207

Krupa P, Kozdrój J (2007) Ectomycorrhizal fungi and associated bacteria provide protection against heavy metals in inoculated pine (*Pinus sylvestris* L.) seedlings. Water Air Soil Pollut 182:83–90

Kuffner M, De Maria S, Puschenreiter M, Fallmann K, Wieshammer G, Gorfer M et al (2010) Culturable bacteria from Zn- and Cd accumulating Salix caprea with differential effects on plant growth and heavy metal availability. J Appl Microbiol 108:1471–1484

Lasat HA (2002) Phytoextraction of toxic metals: a review of biological mechanisms. J Environ Qual 31:109–120

Lazcano EA, Guerrero-Zuñiga LA, Rodriguez-Tovar A, Rodriguez-Dorantes A, Vasquez-Murrieta MS (2010) Rhizospheric plant-microbe interactions that enhance the remediation of contaminated soil. In: Méndez-Vilas A (ed) Current research, technology and education topics in applied microbiology and microbial biotechnology, Microbiology book series. Formatex Research Center, Barcelona, pp 251–256

Li WC, Ye ZH, Wong MH (2010) Metal mobilization and production of short-chain organic acids by rhizosphere bacteria associated with a Cd/Zn hyperaccumulating plant Sedum alfredii. Plant Soil 326:453–467

Ma Y, Rajkumar M, Freitas H (2009) Improvement of plant growth and nickel uptake by nickel resistant-plant-growth promoting bacteria. J Hazard Mater 166:1154–1161

Ma Y, Prasad MNV, Rajkumar M, Freitas H (2011) Plant growth promoting rhizobacteria and endophytes accelerate phytoremediation of metalliferous soils. Biotechnol Adv 29:248–258

Madhaiyan M, Poonguzhali S, Sa T (2007) Metal tolerating methylotrophic bacteria reduces nickel and cadmium toxicity and promotes plant growth of tomato (Lycopersicon esculentum L.). Chemosphere 69:220–228

Martino E, Perotto S, Parsons R, Gadd GM (2003) Solubilization of insoluble inorganic zinc compounds by ericoid mycorrhizal fungi derived from heavy metal polluted sites. Soil Biol Biochem 35:133–141

McGrath SP, Chaudri AM, Giller KE (1995) Long-term effects of metals in sewage sludge on soils, microorganisms and plants. J Ind Microbiol 14:94–104

McGrath SP, Zhao FJ, Lombi E (2001) Plant and rhizosphere processes involved in phytoremediation of metal-contaminated soils. Plant Soil 232(1–2):207–214

Meharg AA (2003) The mechanistic basis of interactions between mycorrhizal associations and toxic metal cations. Mycol Res 107:1253–1265

Mergeay M, Nies D, Schlegel HG, Gerits J, Charles P, Van Gijsegem F (1985) *Alcaligenes eutrophus* CH34 is a facultative chemolithotroph with plasmid-bound resistance to heavy metals. J Bacteriol 162(1):328–334

Minguzzi C, Vergnano O (1948) Il contento di nichel nelli ceneri di Alyssum bertlonii Desv. Atti della Societa Toscana di Science Naturali Mem Ser A 55:49–77

Miransari M (2011) Hyperaccumulators, arbuscular mycorrhizal fungi and stress of heavy metals. Biotechnol Adv 29:645–653

Rajkumar M, Ae N, Prasad MNV, Freitas H (2010) Potential of siderophore-producing bacteria for improving heavy metal phytoextraction. Trends Biotechnol 28:142–149

Rajkumar M, Sandhya S, Prasad MNV, Freitas H (2012) Perspectives of plant-associated microbes in heavy metal phytoremediation. Biotechnol Adv 30:1562–1574
Rascio W (1977) Metal accumulation by some plants growing on Zn mine deposits. Oikos 29:250–253
Raskin I, Ensley BD (2000) Phytoremediation of toxic metals: using plants to clean up the environment. Wiley, New York
Reed M, Glick B (2005) Growth of canola (*Brassica napus*) in the presence of plant growth-promoting bacteria and either copper or polycyclic aromatic hydrocarbons. Can J Microbiol 51:1061–1069
Richards B, Steenhus T, Peverly J, McBride M (2000) Effect of sludge-processing mode, soil texture and soil pH on metal mobility in undisturbed soil columns under accelerated loading. Environ Pollut 109:327–346
Ryan PR, Delhaize E, Jones DL (2001) Function and mechanism of organic anion exudation from plant roots. Annu Rev Plant Physiol Plant Mol Biol 52:527–560
Salt DE, Blaylock M, Kumar NP, Dushenkov V, Ensley BD, Chet I, Raskin I (1995) Phytoremediation: a novel strategy for the removal of toxic metals from the environment using plants. Biotechnology 13:468–474
Salt DE, Smith RD, Raskin L (1998) Phytoremediation. Annu Rev Plant Physiol Plant Mol Biol 49(1):643–668
Saravanan VS, Madhaiyan M, Thangaraju M (2007) Solubilization of zinc compounds by the diazotrophic, plant growth promoting bacterium Gluconacetobacter diazotrophicus. Chemosphere 66:1794–1798
Schalk IJ, Hannauer M, Braud A (2011) New roles for bacterial siderophores in metal transport and tolerance. Environ Microbiol 13:2844–2854
Schulz B, Boyle C (2006) What are endophytes? In: Schulz BJE, Boyle CJC, Sieber TN (eds) Microbial root endophytes. Springer, Berlin
Seshadri B, Bolan NS, Naidu R (2015) Rhizosphere-induced heavy metal(loid) transformation in relation to bioavailability and remediation. J Soil Sci Plant Nutr 15:524–548
Shah FR, Ahmad N, Masood KR, Peralta-Videa JR, Ahmad FD (2010) Heavy metal toxicity in plants. In: Ashraf M, Ozturk M, Ahmad MSA (eds) Plant adaptation and phytoremediation. Springer, Dordrecht/Heidelberg/London/New York, p 71
Sheng XF, Xia JJ, Jiang CY, He LY, Qian M (2008) Characterization of heavy metal-resistant endophytic bacteria from rape (*Brassica napus*) roots and their potential in promoting the growth and lead accumulation of rape. Environ Pollut 156:1164–1170
Shi JY, Lin HR, Yuan XF, Chen XC, Shen CF, Chen YX (2011) Enhancement of copper availability and microbial community changes in rice rhizospheres affected by sulfur. Molecules 16:1409–1417
Sinha S, Mukherjee SK (2008) Cadmium-induced siderophore production by a high Cd-resistant bacterial strain relieved Cd toxicity in plants through root colonization. Curr Microbiol 56:55–60
Smith SE, Read DJ (1997) Mycorrhizal symbiosis. Academic, San Diego
Taghavi S, Mergeay M, Nies D, van der Lelie D (1997) Alcaligenes eutrophus as a model system for bacterial interactions with heavy metals in the environment. Res Microbiol 148:536–551
Tank N, Saraf M (2009) Enhancement of plant growth and decontamination of nickel-spiked soil using PGPR. J Basic Microbiol 49:195–204
Turnau K (1998) Heavy metal content and localization in mycorrhizal Euphorbia cyparissias from zinc wastes in Southern Poland. Act Soc Bot Pol 67:105–113
Utsunamyia T (1980) Japanese Patent Application No. 55–72959
Vangronsveld J, Herzig R, Weyens N, Boulet J, Adriaensen K, Ruttens A, Thewys T, Vassilev A, Meers E, Nehnelajova E, van der Lelie D, Mench M (2009) Phytoremediation of contaminated soils and groundwater: lessons from the field. Environ Sci Pollut Res 16:765–794
Venkatesh NM, Vedaraman N (2012) Remediation of soil contaminated with copper using rhamnolipids produced from Pseudomonas aeruginosa MTCC 2297 using waste frying rice bran oil. Ann Microbiol 62:85–91

Vivas A, Biro B, Ruíz-Lozanoa JM, Azcon R (2006) Two bacterial strains isolated from a Zn-polluted soil enhance plant growth and mycorrhizal efficiency under Zn toxicity. Chemosphere 52:1523–1533

Volesky B, Holan ZR (1995) Biosorption of heavy metals. Biotechnol Prog 11:235–250

Wenzel WW (2009) Rhizosphere processes and management in plant-assisted bioremediation (phytoremediation) of soils. Plant Soil 321:385–408

Wenzel WW, Lombi E, Adriano DC (1999) Biochemical processes in the rhizosphere: role in phytoremediation of metal-polluted soils. In: Prasad MNV, Hagemeyer J (eds) Heavy metal stress in plants: from molecules to ecosystems. Springer, Heidelberg/Berlin/New York, pp 273–303

Weyens N, van der Lelie D, Taghavi S, Vangronsveld J (2009) Phytoremediation: plant-endophyte partnerships take the challenge. Curr Opin Biotechnol 20:248–254

Whiting SN, de Souza MP, Terry N (2001) Rhizosphere bacteria mobilize Zn for hyperaccumulation by Thlaspi caerulescens. Environ Sci Technol 35:3144–3150

Wichard T, Bellenger JP, Morel FM, Kraepiel AM (2009) Role of the siderophore azotobactin in the bacterial acquisition of nitrogenase metal cofactors. Environ Sci Technol 43:7218–7224

Wright SF, Upadhyaya A (1998) A survey of soils for aggregate stability and glomalin, a glycoprotein produced by hyphae of arbuscular mycorrhizal fungi. Plant Soil 198:97–107

Wu SC, Cao ZH, Li ZG, Cheung KC, Wong MH (2005) Effects of biofertilizer containing N-fixer, P and K solubilizers and AM fungi on maize growth: a greenhouse trial. Geoderma 125:155–166

Yang Q, Tu S, Wang G, Liao X, Yan X (2012) Effectiveness of applying arsenate reducing bacteria to enhance arsenic removal from polluted soils by Pteris vittata L. Int J Phytorem 14:89–99

Zarei M, Hempel S, Wubet T, Schäfer T, Savaghebi G, Jouzani GS, Nekouei MK, Buscot F (2010) Molecular diversity of arbuscular mycorrhizal fungi in relation to soil chemical properties and heavy metal contamination. Environ Pollut 158:2757–2765

Zhou JL (1999) Zn biosorption by Rhizopus arrhizus and other fungi. Appl Microbiol Biotechnol 51:686–693

Zhuang X, Chen J, Shim H, Bai Z (2007) New advances in plant growth-promoting rhizobacteria for bioremediation. Environ Int 33:406–413

Bacteria-Mediated Elicitation of Induced Resistance in Plants upon Fungal Phytopathogen

12

Shekhar Jain, Ajit Varma, Narendra Tuteja, and Devendra Kumar Choudhary

Abstract

Plants are sessile organism and primary producer of the ecosystem and communicate with above- and belowground communities that consist of benign/pathogenic microbes. Among these interactions, phytopathogenic fungi and oomycetes are the major causative agents of infectious crop plant diseases. To control these pathogens is extremely difficult, and a very small percentage of applied fungicides used for crop protection reach the target pathogen. To combat with such pathogen, higher level of resistance in addition to indigenous immune system is required which is elicited by plant growth-promoting bacteria (PGPB) in the form of induced systemic resistance in plants. Induced systemic resistance is prior activation of resistance in plants through PGPB via root priming that leads to defense-related protein activation which is independent of salicylic acid and dependent on jasmonic acid and ethylene. In case of it, nonexpressor of pathogenesis-related protein 1 (NPR1) plays the most important role by regulating hormonal defense signaling pathway leading to activation of pathogenesis-related and defense-related protein depending on the preceding signals. PGPB-elicited induced resistance showed that some of the bacterial determinants are responsible for the elicitation of induced systemic resistance (ISR). Although PGPB seem to actively suppress local host defense responses in the roots, it also produces elicitors that are responsible for the onset of systemic immunity. This chapter focuses on recent research study concerning the interaction between PGPB and plants under biotic stress condition.

S. Jain • A. Varma • N. Tuteja • D.K. Choudhary (✉)
Amity Institute of Microbial Technology (AIMT), Block 'E-3', 4th Floor, Amity University Campus, Sector-125, Gautam Buddha Nagar, 201313 Noida, Uttar Pradesh, India
e-mail: dkchoudhary1@amity.edu

D.K. Choudhary et al. (eds.), *Plant-Microbe Interaction: An Approach to Sustainable Agriculture*, DOI 10.1007/978-981-10-2854-0_12

12.1 Introduction

Plants are a source of nutrition for a vast biota in terrestrial environments and being sessile organisms continuously challenged by biotic and abiotic stresses. Fungal diseases caused by different phytopathogenic fungi and oomycetes are responsible for the major economic and social problems in the affected countries by severely decreasing the crop production (Vaishnav et al. 2014). To control these pathogens is extremely difficult, and a very small percentage of applied fungicides used for crop protection reach to the target pathogen. On the other hand, chemical control of diseases has negative effects on the environment such as a decrease in the biodiversity of soil microbiota, development of fungicide-resistant pathogens, and contamination of fruits and vegetables with chemicals that endanger the health of consumers (Bernard et al. 2012; Ludueña et al. 2012). In plant rhizosphere, the interactions of microorganisms with each other may be associative, competitive, mutualistic, or antagonistic. Some bacteria known as plant growth-promoting bacteria (PGPB) promote plant growth and increase the availability of essential nutrients through nutrient cycling activities, and some of them also induce resistance in the plants against plant pathogens (Wahyudi et al. 2011).

The extracellular products present in the rhizosphere and root-associated bacteria play an important role in inhibiting plant pathogens (Lugtenberg and Kamilova 2009). PGPB may colonize the rhizosphere and root surface and protect plants from various stresses. Biological control of plant disease by PGPB involves several mechanisms such as production of antifungal metabolites, cell wall-degrading enzymes, induced host resistance, and competition for nutrition and niches (Li et al. 2011). It is well documented that biological control agents based on PGPB are able to control plant diseases, increase plant growth, and improve resistance to environmental stresses, including drought and salt (Dodd and Perez-Alfocea 2012; Egamberdieva et al. 2013).

Plant growth-promoting bacteria have the ability to elicit changes in the physiology and induction of defenses in the host plant that leads to protection from the above- and belowground pathogenic communities involving organisms at different trophic levels (Pineda et al. 2010, 2013), and this defense elicitation mechanism is termed as induced systemic resistance (ISR) which expressed not only locally but also systemically (in the distal parts from the site of primary infection) against subsequent attack. A complex networking signaling pathway that involves salicylic acid (SA), jasmonic acid (JA), and ethylene (ET) as key signaling molecules regulates the induced resistance in plants (Glazebrook 2001; Thomma et al. 2001).

Upon getting affected by any pathogen infection or herbivore damage/PGPB, plants respond by activating defense machinery via activation of distinct sets of pathogenesis-related and defense-related genes preceded by accumulation of SA and JA. Defense reaction, mediated by the SA- and JA-dependent defense pathways, totally depends on the type of attacker encountered and can cross communicate with respect to plant protection (Pieterse et al. 2001; Choudhary et al. 2007; Jain et al. 2016). By keeping views of plant growth promotion under biotic stresses, the present chapter will unravel the mystification of mechanisms involved in plant defense including ISR and system acquired resistance (SAR) using sustainable development of plants.

12.2 Plant Immune System and Induced Resistance

Due to the nonhost resistance, majority of the phytopathogens cannot infect plants. The primary defense system of the plant contains layer-by-layer protection in the form of physical and chemical barriers such as the cell wall, waxes, hairs, antimicrobial enzymes, phytoanticipins, and secondary metabolites. Apart from these, plants also have heightened-level defense system that gets activated by signaling molecules, if primary defense is found deficient to overcome pathogens (Jain et al. 2016). Malinovsky et al. (2015) have described plant immune system in two broad types, namely, PAMP-triggered immunity (PTI) and effector-triggered immunity (ETI), based on the type of molecules recognized by the plant as indicator of pathogen attack (Fig. 12.1).

Microbe/pathogen-associated molecular patterns (MAMPs/ PAMPs), such as flagellin from bacterial flagella or chitin or different glucans present in fungal/ oomycete cell walls, are referred to as small molecular motifs/structures conserved within a class of microbes, hence characteristic of microbes, and required for the overall fitness of microbes, and these patterns act as "nonself" signals for the plants to activate basal/PAMP-triggered immunity (Newman et al. 2013). Apart from these, basal immunity can get activated by plants' ability to sense a compromised

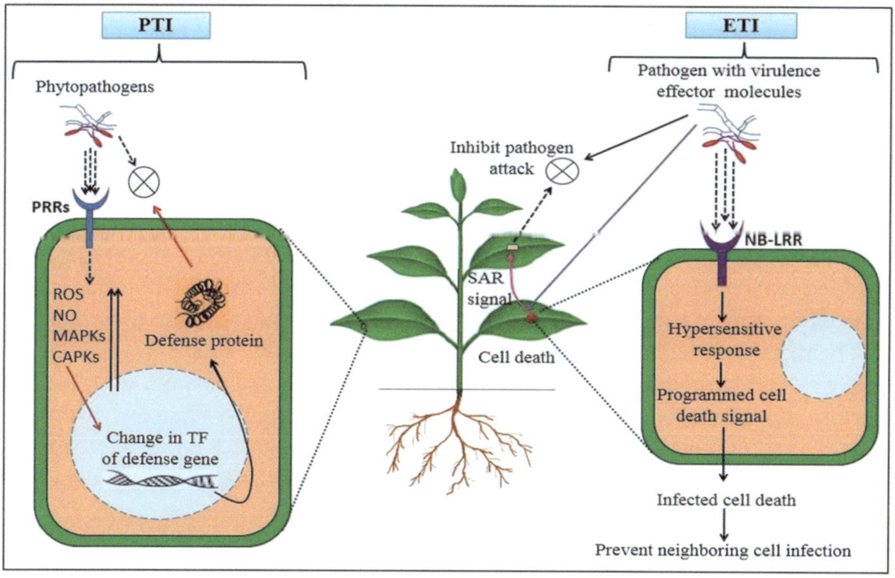

Fig. 12.1 Schematic representation of PAMP-triggered immunity (PTI) and effector-triggered immunity (ETI) in plants

"self" by detecting damage-associated molecular patterns (DAMPs) which are plant degradation products resulting from the action of invading pathogens or endogenous peptides, constitutively present or newly synthesized, that are released by the plants following pathogen attacks (Boller and Felix 2009). Recognition of DAMPs also triggers immune responses similar to the PTI response. Plasma membrane-localized pattern recognition receptors (PRRs), present on the plant cells, recognized these signature motifs (MAMP/PAMP) present on the invading pathogen leading to induction of a broad variety of defense responses through activation of a complex cascade of signaling events, including ion fluxes leading to plasma membrane depolarization, production of reactive oxygen species (ROS) and nitric oxide (NO), and activation of mitogen-activated protein kinases (MAPKs) and calcium-dependent protein kinases (CDPKs). All these signaling event activities lead to change in transcription factor (TF) activities to activate defense genes resulting in accumulation of different enzymes and stress-specific metabolites which kept most of the potential invader in check (Burketová et al. 2015). Apart from these, plants also possess a second line of defense in which it produces resistance (R) NB-LRR (nucleotide-binding–leucine-rich repeat) receptor proteins that recognize virulence effector molecules released by some of the powerful pathogen which can overcome the first line of defense either by suppressing PTI signaling or preventing detection by the host (Borges and Sandalio 2015). This second line of defense, induced due to effector molecule known as effector-triggered immunity (ETI) (Dodds and Rathjen 2010), is a manifestation of gene-for-gene resistance, which is often accompanied by a programmed cell death at the site of infection that prevents further access of biotrophic pathogens that flourish on living host tissue. Basically, ETI is coupled with hypersensitive response (HR), a strong local defense leading to programmed cell death at the site of infection (Dodds and Rathjen 2010). The onset of PTI and ETI often triggers an induced resistance in tissues distal from the site of infection and involves one or more long-distance signals that proliferate an enhanced defensive capacity in still undamaged plant parts (Shah and Zeier 2013). This well-characterized form of pathogen-induced resistance is commonly known as systemic acquired resistance (SAR) (Spoel and Dong 2012) and confers enhanced resistance against a broad spectrum of pathogens.

Likewise pathogen recognition system, plants also recognize herbivorous insects, most likely through a similar signaling concept (Howe and Jander 2008). Generally, induced resistance is an induced state of resistance in plants, triggered by biological/chemical inducers to protect plant parts against future attack by pathogenic microbes and herbivorous insects. Induced resistance takes place not even locally but also systemically (in distal plant parts that are spatially separated from the inducer) and confers an enhanced level of protection against a broad spectrum of attackers through a regulated network of interconnected signaling pathways in which plant hormones play a major regulatory role (Pieterse et al. 2012; Walters et al. 2013).

12.3 Signaling Events upon Fungal Pathogen

In the signaling cascade of plant defense system, the commencement of pathogen-induced SAR is triggered upon local activation of a PTI or ETI response due to limited primary infection with a pathogen and leads to long-lasting and broad-spectrum disease resistance to uninfected plant tissue against subsequent pathogen attack (Wendehenne et al. 2014; Gao et al. 2015). The establishment of SAR is coupled with increased levels of salicylic acid (SA) followed by regulated activation of a specific set of pathogenesis-related (PR) genes that encode PR proteins with antimicrobial activity (Van Loon et al. 2006). According to Conrath (2011), primary infection with pathogen is a vital step of SAR, and infected tissues are in an alert state that enables them to more rapidly and efficiently deal with both biotic and abiotic stresses. The defense alert expressed upon pathogen attack in the plant cell amplified and transferred from the site of infection by a system of mobile signals into distal (systemic) plant parts (Jain et al. 2016).

The use of SA transgenic and mutant plants for the research studies has revealed an essential role for this phytohormone in SAR (Loake and Grant 2007; Vlot et al. 2008a). The accumulation of SA in SAR had been proven by using *Arabidopsis* SA-non-accumulating mutant plant NahG which expressed the bacterial salicylate hydroxylase (*nahG*) gene responsible for conversion of SA into catechol. This type of plants cannot express SAR. SAR pathway is activated by SA, primary molecule for SAR, which further activates further signaling cascade to activate pathogenesis-related (PR) genes responsible for resistance against pathogen, which encode different pathogenesis-related proteins of families PR-2, PR-5, and PR-1, such as chitinases, β-1,3-glucanases, lipoxygenases, thaumatin-like proteins, antimicrobial peptides, etc. (Jain and Choudhary 2014). Upon elicitation of signal from SA accumulation, nonexpressor of PR genes-1 (*npr-1*) gets activated and encodes NPR1 which acts as a transcriptional coactivator of PR gene expression. Hence, as shown in Fig. 12.2, the overall sequence of the signaling event in SAR is in such a way that after recognition of pathogen, SA accumulation takes place which activates *npr-1* gene followed by activation of PR genes (Choudhary et al. 2016).

Methyl salicylate (MeSA), the volatile form of SA, is itself biologically inactive but in the systemic tissue gets hydrolyzed to SA by the MeSA esterase activity of SA-binding protein 2, and that's how it can act as long-distance mobile signal for SAR (Park et al. 2007; Vlot et al. 2008a, b). Being a volatile compound, MeSA can pass through by both air and vascular transport to intercede long-distance induction of resistance in distal leaves that lack a direct vascular connection to the attacked leaf and in neighboring plants (Heil and Ton 2008). In tobacco plant upon getting infection by tobacco mosaic virus, along with SA, ethylene (ET) perception is also required for the onset of SA-dependent SAR (Verberne et al. 2003). In addition, Truman et al. (2007) showed that the JA-signaling mutants *sgt1b* (suppressor of g2 allele of SKP1 1b), *opr3* (12-oxo-phytodienoate reductase 3), and *jin1* (jasmonate insensitive 1) failed to develop SAR upon leaf infiltration with an avirulent strain of

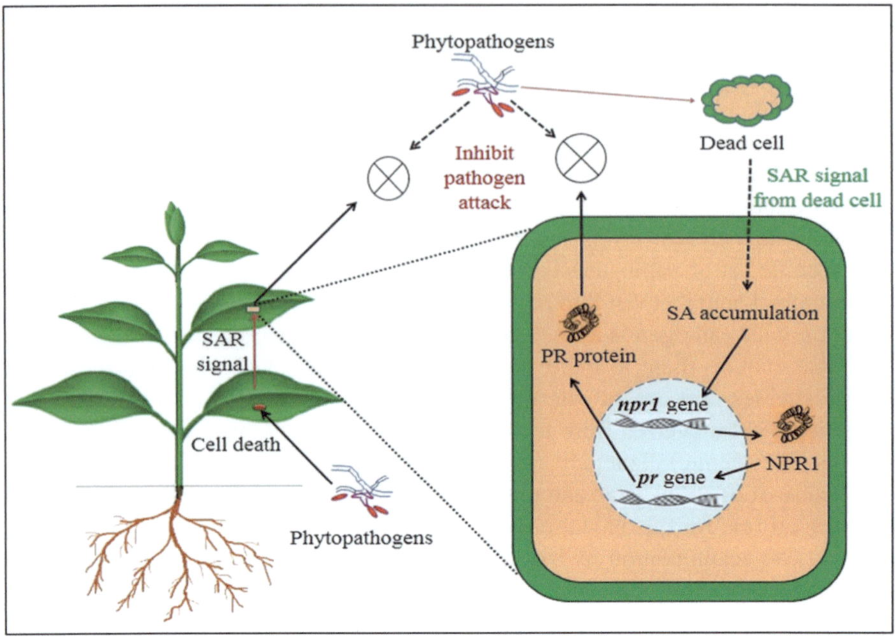

Fig. 12.2 Signal transduction pathway of pathogen-induced systemic acquired resistance (SAR) in plants

the pathogen *Pseudomonas syringae* pv. *tomato*, suggesting that JAs play a role in SAR as well. However, other JA-signaling mutants such as *jar1* (jasmonate resistant 1), *eds8* (enhanced disease susceptibility 8), and *coi1* (coronatine insensitive 1) were shown to develop normal levels of SAR (Attaran et al. 2009).

12.4 Defense Molecular Mechanisms

In comparison to SAR, ISR have a more assorted and composite route to establish a higher degree of prior resistance. In place of PR gene, defense-related gene activation takes place in ISR. Besides SA, the plant growth regulators JA and ET have been implicated in plant defense responses. ISR signal transduction pathway is independent of SA accumulation and totally dependent on JA and ET.

12.4.1 ISR: SA Independent but JA and ET Dependent

In contrast to plant defense system, two of the main signaling pathways, namely, ISR and SAR, confer a broad-spectrum disease resistance in systemic plant parts and look like similar, but actually they are regulated by different signal transduction pathways. The first facts in support of the differential regulation of SAR and

ISR came from studies on the plant growth-promoting rhizobacteria (PGPR) WCS417r. Pieterse et al. (1998) have used *Arabidopsis* SA mutant NahG plants that lack SA accumulation activity. NahG plants contain bacterial salicylate hydroxylase (*nahG*) gene responsible for conversion of SA into catechol. In research it was shown that this mutant plant also develops normal level of ISR after treatment of the root with ISR-inducing rhizobacterial strain *P. fluorescens* WCS417r against the challenge inoculation that confirms ISR independency over SA. After that, many research studies have been done in support of SA-independent ISR in *Arabidopsis* (Stein et al. 2008; Segarra et al. 2009) and other plant species, such as tobacco (Zhang et al. 2002), tomato (Hase et al. 2008), and rice (De Vleesschauwer et al. 2008). JA and ET are the central players in the regulation of ISR, and similar to SA mutant NahG plants, their role was also confirmed by using JA mutants such as *jar1*, *jin1*, *eds8*, and *coi1* and ET mutants such as *etr1* (ethylene response1) and *ein2* (ethylene insensitive 2). These plants were found unable to confer ISR upon against challenge inoculation that clears the dependency of ISR on JA and ET. Jasmonic acid and its different derivatives induce the expression of genes encoding defense-related proteins, such as thionins (Pieterse et al. 1998) and proteinase inhibitors, while ethylene is involved in the expression of the pathogen-inducible genes (van Wees et al. 1999).

The signaling cascade of the ISR is elicited by nonpathogenic rhizobacteria/PGPR and there is no need of initial infection as required in SAR. Upon receiving elicitation from PGPR, transient synthesis of JA and ET takes place, and the formation of phloem-mobile signal moves this signal in the direction of distal part of the plant, and after challenge inoculation, JA and ET responses activate *npr-1* gene expression, which encodes NPR1 followed by activation of defense-related gene. NPR1 is known as the master regulator of both defense pathways, as upon getting preceding signal, it activates the expression of either PR gene or defense-related gene for the establishment of SAR and ISR, respectively. Likewise MeSA, methyl jasmonate (MeJA) also works as a volatile signal for the distal part of the plant. Król et al. (2015) studied that tomato seed priming with MeJA is found to induce resistance to hemi-biotroph *Fusarium oxysporum* f.sp. *lycopersici*. Based on the preceding signal NPR1 get from either JA or ET or from both in concert expression of different defense-related genes will get express.

Saskia et al. (1999) have categorized different defense-related genes activated by JA and ET. Pathogen-inducible genes *Hel* (encoding a hevein-like protein) (Potter et al. 1993), *ChiB* (encoding a basic chitinase) (Samac et al. 1990), and *Pdf1.2* (encoding a plant defensin) (Penninckx et al. 1996) that code for the antifungal protein get induced by ET and JA (Thomma et al. 1998). Among the three, plant defensin proteins possess a wide range of activity that includes antifungal activity, antibacterial activity, proteinase inhibitory activity, and insect amylase inhibitory activity, and for its full expression, both ethylene and jasmonate are required, indicating that these hormonal signals act in concert (Penninckx et al. 1998). *Pal1* gene encodes for the phenylalanine ammonia-lyase, which plays an important regulatory role in the synthesis of phenylpropanoid such as lignin and of SA in *Arabidopsis* (Mauch-Mani and Slusarenko 1996), which has been also found to be induced by JA

(McConn et al. 1997). Along with these JA also protects plant from insect and herbivory. *Pin* gene which encoded for the proteinase inhibitor proteins was induced by JA in the tomato plant in case plant tissues get wounded by any intruder. This protein protects the plant against herbivory (Heitz et al. 1999). On another hand to combat against insect, it activates expression of the *Atvsp* gene (encoding vegetative storage protein) in *Arabidopsis* that possesses acid phosphate activity, and that's how it retards the development of insect and increases mortality rate. That's how, by triggering the activation of such a wide range of different defense-related genes, PGPR-elicited ISR help protect plant against a broad range of pathogens, insects, and herbivores (Berger et al. 1995).

12.4.2 NPR1: The Master Regulator of SAR and ISR

Even though both signal transduction pathways, ISR and SAR, vary from each other with respect to elicitor and signaling molecule, the defense regulatory protein NPR1 plays a key role in the regulation of both SA-dependent SAR and JA-/ET-dependent ISR (Dong 2004; Pieterse and Van Loon 2004), and that's why these signaling pathways are independent but overlapped due to requirement of NPR1 (van Wees et al. 2000). Research studies on the mutant *Arabidopsis npr1* plants were shown to be blocked in their ability to express ISR upon colonization of the roots by the PGPR WCS417r (Pieterse et al. 1998), *P. fluorescens* CHAO (Iavicoli et al. 2003), *P. fluorescens* 89B61 (Ryu et al. 2003), *P. putida* LSW17S (Ahn et al. 2007), *Serratia marcescens* 90–166, and *B. pumilus* SE34 (Ryu et al. 2003) upon challenge inoculation.

Based on the varied initiation site, that is, root in case of ISR, whereas leaves in SAR, it was recommended that these two responses may not compete for NPR1, but these are not independent, however, and may compete for NPR1 in leaves. In case of SAR, NPR1 works as transcriptional coactivator of SA-responsive PR gene expression (Kuai et al. 2015) while not in case of SA-independent ISR (Pieterse et al. 1996) that indicates a different role of NPR1 in ISR signaling pathway. Additive enhanced capacity in case of simultaneous activation of SAR and ISR suggests the roles of NPR1 are not mutually exclusive and it regulates and connects different hormone-dependent induced defense pathways by playing a junctional key role (Van Wees et al. 2000; Pieterse et al. 2009; Yang et al. 2015). SA signaling is clearly connected to a function of this regulatory protein in the nucleus; evidence is accumulating that the role of NPR1 in JA/ET signaling is connected to a cytosolic function of NPR1 (Dong 2004; Leon-Reyes et al. 2009).

12.5 Root Priming and Systemic Resistance

Root colonization, i.e., priming, is a critical step to establish ISR in plants. Although nonsymbiotic but mutualistic association between plant roots and PGPB is less well characterized, researchers have done well to resolve it (Zamioudis and Pieterse

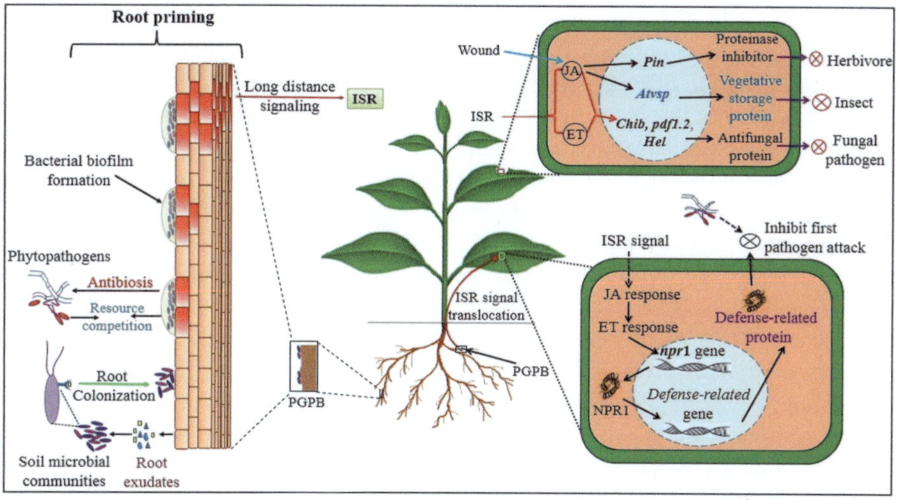

Fig. 12.3 Mechanism of root priming and induced systemic resistance signaling pathway in plants

2012). It was first confirmed by Van Loon et al. (1998) that colonization of plant roots by plant growth-promoting rhizobacteria (PGPR) protects aboveground plant tissues against different types of pathogens.

During the root priming phenomenon, different free-living PGPR get attracted toward root exudates secreted by plant roots. Then changes in the transcriptional program of the PGPR take place toward the traits involved in chemotaxis, root colonization, and energy metabolism (Fan et al. 2012; Mark et al. 2005). After root attachments, PGPB get enclosed within an extracellular matrix of self-produced polymeric substances, mainly exopolysaccharides (EPS) and mucilage, and form biofilm on the root surface. It is necessary for the colonization of roots by *B. subtilis* and was recently revealed to be stimulated by polysaccharides derived from host cell walls that function as signaling molecules for the expression of bacterial genes involved in matrix production (Beauregard et al. 2013).

In this matrix, the coordinated interpretation of the host and self-derived signals was done by bacterial cells to coordinate the production and release of compounds related to plant growth promotion, nutrition, and ISR. That's why this matrix can be considered as the mutualistic interface between the plant and bacteria through which they can exchange solutes and chemical information (Fig. 12.3). Due to the cell wall-degrading exoenzymes, such as cellulase and pectinase, PGPR endophytes commonly enter the root interior through cracks in the newly emerged lateral roots or utilize root hairs and the apical zone as entry points (Reinhold-Hurek and Hurek 2011).

Similar to pathogen-induced SAR, this PGPR-mediated ISR has been demonstrated in many plant species and has a wide range of effectiveness (Chen et al. 2014). Nonpathogenic *Pseudomonas* spp. and *Bacillus* spp. were found to be the most effective PGPR with respect to ISR (Van Loon and Bakker 2006). Maize plants treated with *P. putida* KT2440 are found to protect plant against fungal pathogen *Colletotrichum graminicola* (Planchamp et al. 2014). Even though both SAR and ISR work for the plant protection against different types of pathogens, their range of effectiveness is partly divergent.

Ton et al. (2002) have worked on *Arabidopsis thaliana*, and it was shown that SAR triggered by an avirulent strain of the bacterial leaf pathogen *P. syringae* pv. *tomato* and ISR elicited by the PGPR *P. fluorescens* WCS417r are equally effective against diseases caused by the fungal root pathogen *F. oxysporum* and the downy mildew pathogen *Hyaloperonospora arabidopsidis*. Over the last decade, it has become clear that many bacterial genera such as *Acinetobacter*, *Agrobacterium*, *Arthrobacter*, *Azospirillum*, *Bacillus*, *Bradyrhizobium*, *Frankia*, *Pseudomonas*, *Rhizobium*, *Serratia*, *Thiobacillus*, and many others can elicit induced systemic resistance to provide prior resistance to plants against fungal pathogens.

Along with these, the properties of PGPB to elicit a range of defense-responsive activities in plants such as activation of antioxidant status by reprogramming defense-related enzymes, modulation of quorum sensing phenomenon, and activation of phenylpropanoid pathway leading to phenolic production, lignin deposition, and transgenerational defense response in order to combat the pathogen challenge make it a powerful substitute of synthetic chemicals for the improvement of agroecosystem (Mishra et al. 2015).

12.6 Role of Bacterial Determinants

Research studies on the bacterial-elicited ISR have shown that some of the bacterial determinants are responsible for the elicitation of ISR. Although PGPB appear to actively repress local host defense responses in the roots, it also produces elicitors that are responsible for the onset of systemic immunity. Early reports on MAMPs and other elicitors of ISR-inducing PGPB pay attention on the contribution of lipopolysaccharides (LPS) and the iron-regulated metabolites pyoverdine and SA (De Vleesschauwer and Höfte 2009), but in the past years, many other bacterial determinants have been identified that elicit ISR, including antibiotics, flagella, *N*-acyl homoserine lactones, N-alkylated benzylamine, volatiles, exopolysaccharides, iron-regulated siderophores, and biosurfactants (De Vleesschauwer and Höfte 2009).

LPS is the major structural component of the outer membrane of gram-negative bacteria with highly conserved structure that consist of three components: a lipid A, a core oligosaccharide, and an O-antigen. Among the three, O-antigen of the LPS seems to be the moiety that triggers ISR in plants. Leeman et al. (1995) have shown the ISR-triggering capacity of two strains, *P. fluorescens* WCS374 and *P. fluorescens* WCS417, against *F. oxysporum* f.sp. *raphani* in the radish plant. The role of O-antigen in triggering ISR has been also proven by showing the inability to trigger ISR by the mutant of *P. fluorescens* strain WCS 417 that lacks the O-antigen side chain of the LPS. LPS of *P. fluorescens* strain WCS 417 have also found to be induced systemic resistance in carnation against *Fusarium* wilt caused by *F. oxysporum* f.sp. *dianthi* (Van Peer and Schippers 1992).

Biosurfactants are varied group of surface-active molecules/chemical compounds synthesized by microorganisms. These amphiphilic compounds more importantly cyclic lipopeptides in case of bacteria are produced on living surfaces or excreted extracellularly. Several plant-associated bacteria such as pathogenic and antagonistic *Pseudomonas* bacteria (Raaijmakers et al. 2006) and antagonistic *Bacillus* strains (Ongena and Jacques 2008) are reported to produce cyclic lipopeptides. Cyclic lipopeptides produced by *B. subtilis* include surfactin, iturin, and fengycin families and reported for ISR-mediated protective effect on bean plants against *Botrytis cinerea*, similar to the one induced by living cells of the strain *B. subtilis* S499 (Ongena et al. 2007). Raaijmakers et al. (2006) have classified *Pseudomonas* spp. that produced cyclic lipopeptides into four major groups, namely, the viscosin, amphisin, tolaasin, and syringomycin groups. The massitolide-producing *P. fluorescens* strain SS101 was effective in avoiding infection of tomato leaves by *Phytophthora infestans* and considerably reduced the extension of presented late blight lesions. A massitolide-negative mutant of *P. fluorescens* SS101 entirely lost the ability to induce systemic resistance. These results show that massitolide A is a bacterial determinant of ISR in tomato (Tran et al. 2007). Cell suspensions of *P. fluorescens* SS101 or massitolide A are also found to cause lysis of zoospores of oomycete pathogens (De Souza et al. 2003).

Another determinant, N-acyl-L-homoserine lactone (AHL) present in gram-negative bacteria, acts as signal molecules to control the expression of various functions in a cell density-dependent manner, and this phenomenon is termed as quorum sensing (Miller and Bassler 2001). *Serratia liquefaciens* MG1 produces two types of AHL molecules, namely, N-butanoyl and N-hexanoyl homoserine lactones. It can induce systemic resistance in the tomato plant against the fungal pathogen *Alternaria alternata*, while an AHL-negative mutant of *S. liquefaciens* MG1 slowed down the development of *A. alternata*-induced cell death, but infected plants showed no significant alterations in response to the fungal pathogen when compared with the non-inoculated control.

Inoculation with the *P. putida* strains IsoF having AHL-producing activity can also induce resistance against *A. alternata*. *S. liquefaciens* MG1 and pure N-hexanoyl homoserine lactone significantly increased free and conjugated SA levels in tomato leaves, while this increase was not observed for the AHL-negative mutant (De Vleesschauwer and Höfte 2009). Ongena et al. (2008) reported that *P. putida* BTP1 induces resistance in bean and tomato against *B. cinerea* and in cucumber against *P. aphanidermatum* and *Colletotrichum lagenarium*. They also showed that N, N-dimethyl, N-tetradecyl-N-benzylammonium (NABD) appears to be the bacterial determinant responsible for ISR. Besides this, pure benzylamine is also found to be effectual in triggering induced resistance in bean and cucumber that shows importance of the aromatic amino part for the biological activity of the entire molecule. According to Ahn et al. (2007), the aromatic phenol group present in thiamine is another inducer of systemic resistance in plants.

Siderophores are low molecular weight organic compounds and possess a very high and specific affinity to chelate iron (Boukhalfa and Crumbliss 2002). Different microorganisms produce a wide range of siderophore, but out of them, pseudobactines, also known as pyoverdine or fluorescein, is the most important.

It exhibited distinctive phenotypic trait of the rRNA homology group I species of the genus *Pseudomonas* (Visca et al. 2007). According to Compant et al. (2005), siderophores produced by different PGPB reduce the growth of pathogenic fungi through Fe^{3+} ion sequestering and showed heterologous siderophores produced by coinhabitant. Siderophores produced by fungi have lower affinity for ferric ion. Other than Fe^{3+} ion sequestering mediated protection, it also triggers immune response in plants (Höfte and Bakker 2007). A lot of research has been done on pseudobactines in the past decade which demonstrate its role in triggering resistance in plants. For example, pseudobactines produced by *P. putida* WCS358 were reported for its role in the suppression of *Ralstonia solanacearum* in *Eucalyptus urophylla* (Ran et al. 2005), *Erwinia carotovora* in tobacco (Van Loon et al. 2006), and *Botrytis cinerea* in tomato (Meziane et al. 2005). Earlier two strains of PGPB *Rhizobium meliloti*, RMP_3 and RMP_5, have been isolated by Arora et al. (2001) from *Mucuna pruriens* which produce siderophore and showed strong antagonism against pathogen *Macrophomina phaseolina*.

Antibiotics produced by PGPB also play a principal ISR elicitor task in plant defense. Finding of the characteristic of PGPB to produce antibiotics has made a significant increase in our knowledge about the biocontrol of disease. There is a wide range of antibiotics produced by gram-negative and gram-positive bacteria. Antibiotics produced by fluorescent *pseudomonads* comprise of 2,4-diacetylphloroglucinol (DAPG), pyoluteorin (PLT), pyrrolnitrin (PRN), phenazine-1-carboxyclic acid (PCA), 2-hydroxy phenazines, and phenazine-1-carboxamide (PCN) which have different structural configurations. Other than *Pseudomonas* a wide range of bacteria produces different types of antibiotic which target different pathogens and protect plant from respective diseases (Raaijmakers and Weller 1998; Weller et al. 2002; Fernando et al. 2005).

Diacetylphloroglucinol produced by *Pseudomonas* sp. is the most studied antibiotic among the listed one and most frequently reported in PGPB-mediated disease control. DAPG produced by *P. fluorescens* CHA0 is reported to induce resistance against oomycete *H. arabidopsidis* (Iavicoli et al. 2003) and the root-knot nematode *Meloidogyne javanica* (Siddiqui and Shaukat 2003). A wide range of bacterial strains have the ability to produce a broad range of antibiotics and help in suppression of diverse microbial competitors, e.g., *B. cereus* strain UW85 produced zwittermicin (Pal and Gardener 2006; Silo-Suh et al. 1994) and kanosamine (Milner et al. 1996). Upon study on *Arabidopsis* mutants and transgenic lines which implicated defense signaling pathways, it was find out that DAPG-induced resistance does not follow standard ISR pathway nor depend on the master regulator NPR1 or functional JAR1 protein but is regulated by *eir1* (ethylene-insensitive root-1) gene, which is ET insensitive in the roots only (De Vleesschauwer and Höfte 2009). Lack of ISR expression after exogenous exposure of DAPG on the *eir1* mutant recommended that an intact ET signaling pathway is required for the establishment of

DAPG-inducible resistance (De Vleesschauwer and Höfte 2009; Iavicoli et al. 2003). PCA is the another potent antibiotic with respect to plant protection having antagonistic activity coupled with the accumulation of toxic superoxide radicals in the target cells (Fernando et al. 2005). PCA produced by *P. fluorescens* 2–79 and *P. aureofaciens* 30–84 exhibited antagonism against *Gaeumannomyces graminis* var. *tritici* (Thomashow et al. 1990). Stem rot disease of canola caused by *Sclerotinia* is suppressed by activity of *P. chlororaphis* strain PA-23 (Zhang and Fernando 2004).

In the context of the plant defense, volatile organic compounds (VOCs) produced by PGPB-elicited plant growth promotion and induced systemic resistance provide a new insight in PGPB–plant interaction. Out of the different types of VOCs produced by bacteria with respect to plant defense, some of the most notable includes dodecane, 2-undecanone, 2-tridecanone, 2-tridecanol, tetramethylpyrazine, 2,3-butanediol and 3-hydroxy-2-butanone (acetoin), etc. Ryu et al. (2003) have done a lot of research on VOCs and reported that 2,3-butanediol and 3-hydroxy-2-butanone are the most important one with respect to their role in elicitation of ISR. Two bacterial strains, namely, *B. subtilis* GB03 and *B. amyloliquefaciens* IN937a, were found to consistently release 2,3-butanediol and 3-hydroxy-2-butanone. Priming of *A. thaliana* plants with these strains has shown significant resistance against the challenge inoculation of *Erwinia carotovora* subsp. *carotovora* SCC1. Priming activity of such VOCs to induce resistance against diseases is confirmed with genetically modified *Bacillus* strain which is unable to produce VOCs and found to be unable to elicit ISR (Ryu et al. 2003). Besides *Bacillus*, a number of strains of *P. fluorescens* were also reported for the production of VOCs and have shown more effectiveness in controlling root and seedling diseases (Landa et al. 2002).

12.7 Role of Defense Enzymes

By using the property of defense gene to get induced upon an appropriate stimulus/signal, through prior induction of plant's own defense mechanisms by application of a biological inducer, plants can be protected against invading pathogens. Different defense enzymes, namely, lipoxygenase (LOX), phenylalanine ammonia-lyase (PAL), peroxidase (POD), polyphenol oxidase (PPO), chitinase, and β-glucanase, get activated through prior treatment of plants with plant growth-promoting microbes and lead to plant protection against the biotic stress.

Peroxidases are expressed to restrain cell spreading of disease through generation of highly toxic environments by enormously producing ROS and reactive nitrogen species (RNS) or foundation of basic obstructions, for example, lignin and suberin deposition (Passardi et al. 2005; Cavalcanti et al. 2004), while PPO also plays an important role in defense against plant pathogens due to its reaction products and wound inducibility property (Mayer and Harel 1979; Chunhua et al. 2001). Sundaramoorthy et al. (2012) found the increased level of POD, PPO, and β-glucanase in the *Capsicum annum* L. treated with co-inoculation of two endophytic bacteria *B. subtilis*, namely, EPCO16 and EPC5, and rhizobacterium strain *P. fluorescence* PF1 after challenge inoculation with *F. solani*.

An elevated level of enzymatic activity of cell wall-bound PODs has been reported in different plants such as cucumber (Chen et al. 2000), soybean (Jain et al. 2013; Jain and Choudhary 2014), rice (Reimers et al. 1992), tomato (Mohan et al. 1993), and tobacco (Ahl Goy et al. 1992) against challenge inoculation. Research studies on different plants such as cucumber (Chen et al. 2000), banana (Thakker et al. 2007), tomato (Thipyapong and Steffens 1997), and poplar plant have been found with higher level of PPO upon pathogen infection. β-1,3-Glucan and chitin, polymer of N-acetylglucosamine (NAG), are the main components of fungal phytopathogen cell wall, and β-1,3-glucanase and chitinase play a straight role in plant protection by degrading these cell wall compounds, respectively. PAL and LOX are the other defense enzymes elicited by bacteria in plants against challenge inoculation. PAL regulates the critical steps in phenylpropanoid metabolism and plays an important role in lignin production which is an inducible defense mechanism used for protection against pathogen attack (Liang et al. 1989), while LOX is requisite for the synthesis of antifungal oxylipins, such as jasmonic acid (JA) that may act as signal factor for eliciting ISR in the plant (Creelman and Mullet 1997; Pieterse et al. 1998).

Numerous former studies on the plant–microbe interaction in the course of plant defense have found a significant role of PAL. Recently Ramamoorthy et al. (2002) found higher level of PAL and LOX in the roots of tomato plant treated with *P. fluorescens* Pf1 challenged inoculated with *F. oxysporum* f.sp. *lycopersici*. Daayf et al. (1997) have shown the role of PAL in the production of phenolics and phytoalexins in cucumber. PAL activity could be induced in plant–pathogen interactions and fungal elicitor treatment (Ramanathan et al. 2000). Chen et al. (2000) reported prominent level of PAL enzyme in the cucumber roots inoculated with *Pythium aphanidermatum* and treated with *P. corrugata*, but in later treatment, levels were deceased after challenge inoculation with *P. aphanidermatum*. De Meyer et al. (1999)) reported stimulation of PAL in bean roots and increased level of salicylic acid (SA) in leaves upon colonization of rhizosphere by *P. aeruginosa* 7NSK2.

12.8 Conclusion

Plants comprise an outstanding ecosystem for microorganisms that interact with plant tissues and cells with differing degrees of dependence. To attain practical agricultural applications, studies on the relationship between roots and microbiota are important. Among the bacterial strains that play important roles in the prevention of plant infectious diseases, many can promote plant growth by suppressing pathogenic microorganisms, synthesizing growth-stimulating plant hormones, and promoting increased plant disease resistance. New biotechnological products are currently being developed based on stimulation of the plant defense response and on the use of plant-beneficial bacteria for biological control of plant diseases and for plant growth promotion.

Acknowledgment Some of the research in the present review has partially been supported by DBT and SERB grant no. BT/PR1231/AGR/021/340/2011 and SR/FT/LS-129/2012, respectively, to DKC.

References

Ahl Goy P, Felix G, Metraux JP, Meins JR (1992) Resistance to disease in the hybrid *Nicotiana glutinosa*×*Nicotiana debneyi* is associated with high constitutive levels of β-1, 3-glucanase, chitinase, peroxidase and polyphenol oxidase. Physiol Mol Plant Pathol 41:11–21

Ahn I-P, Lee S-W, Suh S-C (2007) Rhizobacteria-induced priming in *Arabidopsis* is dependent on ethylene, jasmonic acid, and NPR1. Mol Plant-Microbe Interact 20:759–768

Arora NK, Kang SC, Maheshweri DK (2001) Isolation of siderophore producing strain of *Rhizobium meliloti* and their biocontrol potential against *Macrophomina phaseolina* that causes charcoal rot of groundnut. Curr Sci 81:673–677

Attaran E, Zeier TE, Griebel T, Zeier J (2009) Systemic acquired resistance in *Arabidopsis* is independent of methyl salicylate production and jasmonate signaling. Plant Cell 21:954–971

Beauregard PB, Chai YR, Vlamakis H, Losick R, Kolter R (2013) *Bacillus subtilis* biofilm induction by plant polysaccharides. Proc Natl Acad Sci U S A 110:E1621–E1630

Berger S, Bell E, Sadka A, Mullet JE (1995) *Arabidopsis thaliana AtVsp* is homologous to soybean *VspA* and *VspB*, genes encoding vegetative storage protein acid phosphatases, and is regulated similarly by methyl jasmonate, wounding, sugars, light and phosphate. Plant Mol Biol 27:933–942

Bernard E, Larkin RP, Tavantzis S, Erich MS, Alyokhin A, Sewell G, Lannan A, Gross SD (2012) Compost, rapeseed rotation, and biocontrol agents significantly impact soil microbial communities in organic and conventional potato production systems. App Soil Ecol 52:29–41

Boller T, Felix G (2009) A renaissance of elicitors: perception of microbe-associated molecular patterns and danger signals by pattern-recognition receptors. Annu Rev Plant Biol 60:379–406

Borges AA, Sandalio LM (2015) Induced resistance for plant defense. Front Plant Sci 6:109. doi:10.3389/fpls.2015.00109

Boukhalfa H, Crumbliss AL (2002) Chemical aspects of siderophore mediated iron transport. Biometals 15:325–339

Burketová L, Trdá L, Ott P, Valentova O (2015) Bio-based resistance inducers for sustainable plant protection against pathogens. Biotechnol Adv 33(6):994–1004. http://dx.doi.org/10.1016/j.biotechadv.2015.01.004

Cavalcanti FR, Oliveira JTA, Martins-Miranda AS, Viégas RA, Silveira JAG (2004) Superoxide dismutase, catalase and peroxidase activities do not confer protection against oxidative damage in salt stressed cowpea leaves. New Phytol 163:563–571

Chen C, Bélanger RR, Benhamou N, Paulitz T (2000) Defense enzymes induced in cucumber roots by treatment with plant growth promoting rhizobacteria (PGPR) and *Pythium aphanidermatum*. Physiol Mol Plant Pathol 56:13–23

Chen YC, Kidd BN, Carvalhais LC, Schenk PM (2014) Molecular defense responses in roots and the rhizosphere against *Fusarium oxysporum*. Plant Signal Behav 9(12):e977710. doi:10.4161/15592324.2014.977710

Choudhary DK, Prakash A, Johri BN (2007) Induced systemic resistance (ISR) in plants: mechanism of action. Indian J Microbiol 47:289–297

Choudhary DK, Kasotia A, Jain S, Vaishnav A, Kumari S, Sharma KP, Varma A (2016) Bacterial-mediated tolerance and resistance to plants under abiotic and biotic stresses. J Plant Growth Regul 35(1):276–300

Chunhua S, Ya D, Bingle X, Xiao L, Yonshu X, Qinguang L (2001) The purification and spectral properties of PPO I from *Nicotianan tababcum*. Plant Mol Biol 19:301–314

Compant S, Duffy B, Nowak J, Clément C, Ait Barka E (2005) Use of plant growth-promoting bacteria for biocontrol of plant diseases: principles, mechanisms of action, and future prospects. Appl Environ Microbiol 71:4951–4959

Conrath U (2011) Molecular aspects of defence priming. Trends Plant Sci 16:524–531

Creelman RA, Mullet JE (1997) Biosynthesis and action of jasmonates in plants. Annu Rev Plant Physiol Plant Mol Biol 48:355–381

Daayf F, Bel–Rhlid R, Bélanger RR (1997) Methyl ester of *p*-coumaric acid: a phytoalexin-like compound from long English cucumber leaves. J Chem Ecol 23:1517–1526

De Meyer G, Capieau K, Audenaert K, Buchala A, Metraux JP, Höfte M (1999) Nanogram amounts of salicylic acid produced by the rhizobacterium *Pseudomonas aeruginosa* 7NSK2 activate the systemic acquired resistance pathway in bean. Mol Plant Microbe Interact 12:450–458

De Souza JT, de Boer M, de Waard P, van Beek TA, Raaijmakers JM (2003) Biochemical, genetic, and zoosporicidal properties of cyclic lipopeptide surfactants produced by *Pseudomonas fluorescens*. Appl Environ Microbiol 69(12):7161–7172

De Vleesschauwer D, Djavaheri M, Bakker PAHM, Höfte M (2008) *Pseudomonas fluorescens* WCS374r-induced systemic resistance in rice against *Magnaporthe oryzae* is based on pseudobactin-mediated priming for a salicylic acid-repressible multifaceted defense response. Plant Physiol 148:1996–2012

Dodd IC, Perez-Alfocea F (2012) Microbial alleviation of crop salinity. J Exp Bot 63:3415–3428

Dodds PN, Rathjen JP (2010) Plant immunity: towards an integrated view of plant-pathogen interactions. Nat Rev Genet 11:539–548

Dong X (2004) NPR1, all things considered. Curr Opin Plant Biol 7:547–552

Egamberdieva D, Jabborova D, Wirth S (2013) Alleviation of salt stress in legumes by co-inoculation with Pseudomonas and Rhizobium. In: Plant microbe symbiosis: fundamentals and advances. Springer, New Delhi, pp 291–303

Fan B, Carvalhais L, Becker A, Fedoseyenko D, Von Wiren N, Borriss R (2012) Transcriptomic profiling of *Bacillus amyloliquefaciens* FZB42 in response to maize root exudates. BMC Microbiol 12:116

Fernando WD, Ramarathnam R, Krishnamoorthy AS, Savchuk SC (2005) Identification and use of potential bacterial organic antifungal volatiles in biocontrol. Soil Biol Biochem 37(5):955–964

Gao Q-M, Zhu S, Kachroo P, Kachroo A (2015) Signal regulators of systemic acquired resistance. Front Plant Sci 6:228

Glazebrook J (2001) Genes controlling expression of defense responses in *Arabidopsis*: 2001 status. Curr Opin Plant Biol 4:301–308

Hase S, Takahashi S, Takenaka S, Nakaho K, Arie T, Seo S, Ohashi Y, Takahashi H (2008) Involvement of jasmonic acid signalling in bacterial wilt disease resistance induced by biocontrol agent *Pythium oligandrum* in tomato. Plant Pathol 57:870–876

Heil M, Ton J (2008) Long-distance signalling in plant defense. Trends Plant Sci 13:264–272

Heitz T, Geoffroy P, Fritig B, Legrand M (1999) The PR-6 family: proteinase inhibitors in plant-microbe and plant-insects interactions. In: Datta SK, Muthukrishnan S (eds) Pathogenesis-related proteins in plants. CRC Press, Boca Raton, pp 131–155

Höfte M, Bakker PAHM (2007) Competition for iron and induced systemic resistance by siderophores of plant growth-promoting rhizobacteria. In: Varma A, Chincholkar SB (eds) Microbial siderophores. Springer, Berlin/Heidelberg, pp 121–134

Howe GA, Jander G (2008) Plant immunity to insect herbivores. Annu Rev Plant Biol 59:41–66

Iavicoli A, Boutet E, Buchala A, Métraux J-P (2003) Induced systemic resistance in *Arabidopsis thaliana* in response to root inoculation with *Pseudomonas fluorescens* CHA0. Mol Plant-Microbe Interact 16:851–858

Jain S, Choudhary DK (2014) Induced defense-related proteins in soybean (*Glycine max* L. Merrill) plants by *Carnobacterium* sp. SJ-5 upon challenge inoculation of *Fusarium oxysporum*. Planta 239(5):1027–1040. doi:10.1007/s00425-014-2032-3

Jain S, Vaishnav A, Kasotia A, Kumari S, Gaur RK, Choudhary DK (2013) Rhizobacterium-mediated growth promotion and expression of stress enzymes in *Glycine max* L. Merrill against Fusarium wilt upon challenge inoculation. World J Microbiol Biotechnol. doi:10.1007/s11274-013-1455-5

Jain S, Varma A, Tuteja N, Choudhary DK (2016) Plant growth-promoting microbial-mediated induced systemic resistance in plants: induction, mechanism, and expression. In: Microbial-mediated induced systemic resistance in plants. Springer, Singapore, pp 213–226

Król P, Igielski R, Pollmann S, Kępczyńska E (2015) Priming of seeds with methyl jasmonate induced resistance to hemi-biotroph *Fusarium oxysporum* f.sp. *lycopersici* in tomato via 12-oxo-phytodienoic acid, salicylic acid, and flavonol accumulation. J Plant Physiol 179:122–132

Kuai X, MacLeod BJ, Després C (2015) Integrating data on the *Arabidopsis* NPR1/NPR3/NPR4 salicylic acid receptors; a differentiating argument. Front Plant Sci 6:235. doi:10.3389/fpls.2015.00235

Landa BB, Mavrodi OV, Raaijmakers JM, Gardene BBM, Thomashow LS, Weller DM (2002) Differential ability of genotypes of 2,4-diacetylphloroglucinol-producing *Pseudomonas fluorescens* strains to colonize the roots of pea plants. Appl Environ Microbiol 68:3226–3237

Leeman M, Van Pelt JA, Den Ouden FM, Heinsbroek M, Bakker PAHM, Schippers B (1995) Induction of systemic resistance against *Fusarium* wilt of radish by lipopolysaccharides of *Pseudomonas fluorescens*. Phys Chem Chem Phys 85:1021–1027

Leon-Reyes A, Spoel SH, De Lange ES, Abe H, Kobayashi M, Tsuda S, Millenaar FF, Welschen RAM, Ritsema T, Pieterse CMJ (2009) Ethylene modulates the role of NPR1 in cross-talk between salicylate and jasmonate signaling. Plant Physiol 149:1797–1809

Li R, Zhang H, Liu W, Zheng X (2011) Biocontrol of postharvest gray and blue mold decay of apples with *Rhodotorula mucilaginosa* and possible mechanisms of action. Int J Food Microbiol 146(2):151–156

Liang X, Dron M, Schmid J, Dixon R, Lamb C (1989) Developmental and environmental regulation of a phenylalanine ammonia-lyase-b-glucuronidase gene fusion in transgenic tobacco plants. Proc Natl Acad Sci 86:9284–9288

Loake G, Grant M (2007) Salicylic acid in plant defence-the players and protagonists. Curr Opin Plant Biol 10:466–472

Luduen˜a LM, Taurian T, Tonelli ML, Angelini JG, Anzuay MS, Valetti L, Munoz V, Fabra AI (2012) Biocontrol bacterial communities associated with diseased peanut (*Arachis hypogaea* L.) plants. Eur J Soil Biol 53:48–55

Lugtenberg B, Kamilova F (2009) Plant-growth-promoting rhizobacteria. Annu Rev Microbiol 63:541–556

Malinovsky FG, Fangel JU, Willats WG (2015) The role of the cell wall in plant immunity. Plant cell wall in pathogenesis, parasitism and symbiosis 13:38

Mark GL, Dow JM, Kiely PD, Higgins H, Haynes J, Baysse C, Abbas A, Foley T, Franks A, Morrissey J, O'Gara F (2005) Transcriptome profiling of bacterial responses to root exudates identifies genes involved in microbe-plant interactions. Proc Natl Acad Sci U S A 102:17454–17459

Mauch-Mani B, Slusarenko AJ (1996) Production of salicylic acid precursors is a major function of phenylalanine ammonialyase in the resistance of *Arabidopsis* to *Peronospora parasitica*. Plant Cell 8:203–212

Mayer AM, Harel E (1979) Polyphenol oxidases in plants. Phys Chem Chem Phys 18:193–215

McConn M, Creelman RA, Bell E, Mullet JE, Browse J (1997) Jasmonate is essential for insect defense in *Arabidopsis*. Proc Natl Acad Sci U S A 94:5473–5477

Meziane H, Van der Sluis I, Van Loon LC, Ho¨fte M, Bakker PAHM (2005) Determinants of *Pseudomonas putida* WCS358 involved in inducing systemic resistance in plants. Mol Plant Path 6:177–185

Miller MB, Bassler BL (2001) Quorum sensing in bacteria. Annu Rev Microbiol 55(1):165–199

Milner JL, Silo-Suh L, Lee JC, He HY, Clardy J, Handelsman J (1996) Production of kanosamine by *Bacillus cereus* UW85. Appl Environ Microbiol 62:3061–3065

Mishra S, Singh A, Keswani C, Saxena A, Sarma BK, Singh HB (2015) Harnessing plant-microbe interactions for enhanced protection against phytopathogens. In: Arora N (ed) Plant microbes symbiosis: applied facets. Springer, New Delhi, pp 111–125

Mohan R, Vijayan P, Kolattukudy PE (1993) Developmental and tissue specific expression of a tomato anionic peroxidase (tap1) gene by a minimal promoter with wound and pathogen induction by an additional 5'-flanking region. Plant Mol Biol 22:475–490

Newman MA, Sundelin T, Nielsen JT, Erbs G (2013) MAMP (microbe-associated molecular pattern) triggered immunity in plants. Front Plant Sci 4:139

Ongena M, Jacques P (2008) *Bacillus* lipopeptides: versatile weapons for plant disease biocontrol. Trends Microbiol 16(3):115–125

Ongena M, Jourdan E, Adam A, Paquot M, Brans A, Joris B, Arpigny JL, Thonart P (2007) Surfactin and fengycin lipopeptides of *Bacillus subtilis* as elicitors of induced systemic resistance in plants. Environ Microbiol 9(4):1084–1090

Ongena M, Jourdan E, Adam A, Schäfer M, Budzikiewicz H, Thonart P (2008) Amino acids, iron, and growth rate as key factors influencing production of the *Pseudomonas putida* BTP1 benzylamine derivative involved in systemic resistance induction in different plants. Microb Ecol 55(2):280–292

Pal KK, Gardener BM (2006) Biological control of plant pathogens. Plant Healt Instruct. doi:10.1094/PHI-A-2006-1117-02

Park S-W, Kaimoyo E, Kumar D, Mosher S, Klessig DF (2007) Methyl salicylate is a critical mobile signal for plant systemic acquired resistance. Science 318:113–116

Passardi F, Cosio C, Penel C, Dunand C (2005) Peroxidases have more functions than a Swiss army knife. Plant Cell Rep 24:255–265

Penninckx IAMA, Eggermont K, Terras FRG, Thomma BPHJ, De Samblanx GW, Buchala A, Métraux J-P, Manners JM, Broekaert WF (1996) Pathogen-induced systemic activation of a plant defensin gene in *Arabidopsis* follows a salicylic acid-independent pathway. Plant Cell 8:2309–2323

Penninckx IA, Thomma BP, Buchala A, Métraux JP, Broekaert WF (1998) Concomitant activation of jasmonate and ethylene response pathways is required for induction of a plant defensin gene in *Arabidopsis*. Plant Cell 10(12):2103–2113

Pieterse CMJ, Van Loon LC (2004) NPR1: the spider in the web of induced resistance signaling pathways. Curr Opin Plant Biol 7:456–464

Pieterse CMJ, Van Wees SCM, Hoffland E, Van Pelt JA, Van Loon LC (1996) Systemic resistance in *Arabidopsis* induced by biocontrol bacteria is independent of salicylic acid accumulation and pathogenesis-related gene expression. Plant Cell 8:1225–1237

Pieterse CMJ, van Wees SCM, van Pelt JA, Knoester M, Laan R, Gerrits H, Weisbeek PJ, van Loon LC (1998) A novel signalling pathway controlling induced systemic resistance in *Arabidopsis*. Plant Cell 10:1571–1580

Pieterse CMJ, Ton J, Van Loon LC (2001) Cross-talk between plant defence signalling pathways: boost or burden? AgBiotechNet 3:1–8

Pieterse CMJ, Leon-Reyes A, Van der Ent S, Van Wees SCM (2009) Networking by small-molecule hormones in plant immunity. Nat Chem Biol 5:308–316

Pieterse CMJ, Van der Does D, Zamioudis C, Leon-Reyes A, Van Wees SCM (2012) Hormonal modulation of plant immunity. Annu Rev Cell Dev Biol 28:489–521

Pineda A, Zheng SJ, Van Loon JJA, Pieterse CMJ, Dicke M (2010) Helping plants to deal with insects: the role of beneficial soil-borne microbes. Trends Plant Sci 15:507–514

Pineda A, Soler R, Weldegergis BT, Shimwela MM, VAN Loon JJ, Dicke M (2013) Non-pathogenic rhizobacteria interfere with the attraction of parasitoids to aphid-induced plant volatiles via Jasmonic acid signaling. Plant Cell Environ 36:393–404

Planchamp C, Glauser G, Mauch-Mani B (2014) Root inoculation with *Pseudomonas putida* KT2440 induces transcriptional and metabolic changes and systemic resistance in maize plants. Front Plant Sci 5:719

Potter S, Uknes S, Lawton K, Winter AM, Chandler D, DiMaio J, Novitzky R, Ward E, Ryals J (1993) Regulation of a hevein-like gene in *Arabidopsis*. Mol Plant-Microbe Interact 6:680–685

Raaijmakers JM, Weller DM (1998) Natural plant protection by 2, 4-diacetyl-phloroglucinol-producing *Pseudomonas* spp. in take-all decline soils. Mol Plant-Microbe Interact 11:144–152

Raaijmakers JM, de Bruijn I, de Kock MJ (2006) Cyclic lipopeptide production by plant-associated *Pseudomonas* spp.: diversity, activity, biosynthesis, and regulation. Mol Plant-Microbe Interact 19(7):699–710

Ramamoorthy V, Raguchander T, Samiyappan R (2002) Induction of defense-related proteins in tomato roots treated with *Pseudomonas fluorescens* Pf1 and *Fusarium oxysporum* f. sp. *lycopersici*. Plant Soil 239:55–68

Ramanathan A, Samiyappan R, Vidhyasekaran P (2000) Induction of defense mechanisms in green gram leaves and suspension cultured cells *by Macrophomina phaseolina* and its elicitors. J Plant Dis Protect 107:245–257

Ran LX, Li ZN, Wu GJ, Van Loon LC, Bakker PAHM (2005) Induction of systemic resistance against bacterial wilt in *Eucalyptus urophylla* by fluorescent *Pseudomonas* spp. Eur J Plant Pathol 113:59–70

Reimers PJ, Guo A, Leach JE (1992) Increased activity of a cationic peroxidase associated with an incompatible interaction between *Xanthomonas oryzae* pv. *oryzae* and rice (*Oryza sativa*). Plant Physiol 99:1044–1050

Reinhold-Hurek B, Hurek T (2011) Living inside plants: bacterial endophytes. Curr Opin Plant Biol 14:435–443

Ryu C-M, Hu C-H, Reddy MS, Kloepper JW (2003) Different signaling pathways of induced resistance by rhizobacteria in *Arabidopsis thaliana* against two pathovars of *Pseudomonas syringae*. New Phytol 160:413–420

Samac DA, Hironaka CM, Yallaly PE, Shah DM (1990) Isolation and characterization of the genes encoding basic and acidic chitinase in *Arabidopsis thaliana*. Plant Physiol 93:907–914

Segarra G, Van der Ent S, Trillas I, Pieterse CMJ (2009) MYB72, a node of convergence in induced systemic resistance triggered by a fungal and a bacterial beneficial microbe. Plant Biol 11:90–96

Shah J, Zeier J (2013) Long-distance communication and signal amplification in systemic acquired resistance. Front Plant Sci 4:30

Siddiqui MA, Shaukat SS (2003) Suppression of root-knot disease by *Pseudomonas fluorescens* CHA0 in tomato: importance of bacterial secondary metabolite, 2, 4-diacetylpholoroglucinol. Soil Biol Biochem 35:1615–1623

Silo-Suh LA, Lethbridge BJ, Raffel SJ, He H, Clardy J, Handelsman J (1994) Biological activities of two fungistatic antibiotics produced by *Bacillus cereus* UW85. Appl Environ Microbiol 60:2023–2030

Spoel SH, Dong X (2012) How do plants achieve immunity? Defence without specialized immune cells. Nat Rev Immunol 12:89–100

Stein E, Molitor A, Kogel K-H, Waller F (2008) Systemic resistance in Arabidopsis conferred by the mycorrhizal fungus *Piriformospora indica* requires Jasmonic acid signaling and the cytoplasmic function of NPR1. Plant Cell Physiol 49:1747–1751

Sundaramoorthy S, Raguchander T, Ragupathi N, Samiyappan R (2012) Combinatorial effect of endophytic and plant growth promoting rhizobacteria against wilt disease of Capsicum annum L. caused by *Fusarium solani*. Biol Control 60(1):59–67

Thakker JN, Patel N, Kothari IL (2007) *Fusarium oxysporum* derived elicitor-induced changes in enzymes of banana leaves against wilt disease. J Mycol Plant Pathol 37:510–513

Thipyapong P, Steffens JC (1997) Tomato polyphenol oxidase (Differential response of the polyphenol oxidase F promotor to injuries and wound signals). Plant Physiol 115:409–418

Thomashow LS, Weller DM, Bonsall RF, Pierson LS (1990) Production of the antibiotic phenazine-1-carboxylic acid by fluorescent *Pseudomonas* Species in the rhizosphere of wheat. Appl Environ Microbiol 56:908–912

Thomma BPHJ, Eggermont K, Penninckx IAMA, Mauch-Mani B, Cammue BPA, Broekaert WF (1998) Separate jasmonate-dependent and salicylic acid-dependent defense response pathways in *Arabidopsis* are essential for resistance to distinct microbial pathogens. Proc Natl Acad Sci U S A 95:15107–15111

Thomma BPHJ, Penninckx IAMA, Cammue BPA, Broekaert WF (2001) The complexity of disease signaling in *Arabidopsis*. Curr Opin Immunol 13:63–68

Ton J, Van Pelt JA, Van Loon LC, Pieterse CMJ (2002) Differential effectiveness of salicylate-dependent and jasmonate/ethylene-dependent induced resistance in *Arabidopsis*. Mol Plant-Microbe Interact 15:27–34

Tran H, Ficke A, Asiimwe T, Höfte M, Raaijmakers JM (2007) Role of the cyclic lipopeptide massetolide A in biological control of *Phytophthora infestans* and in colonization of tomato plants by *Pseudomonas fluorescens*. New Phytol 175:731–742

Truman W, Bennett MH, Kubigsteltig I, Turnbull C, Grant M (2007) *Arabidopsis* systemic immunity uses conserved defense signaling pathways and is mediated by jasmonates. Proc Natl Acad Sci U S A 104:1075–1080

Vaishnav A, Jain S, Kasotia A, Kumari S, Gaur RK, Choudhary DK (2014) Molecular mechanism of benign microbe-elicited alleviation of biotic and abiotic stresses for plants. In: Approaches to plant stress and their management. Springer, New Delhi, pp 281–295

Van Loon LC, Bakker PAHM (2006) Root-associated bacteria inducing systemic resistance. In: Gnanamanickam SS (ed) Plant-associated bacteria. Springer, Dordrecht, pp 269–316

Van Loon LC, Bakker PAHM, Pieterse CMJ (1998) Systemic resistance induced by rhizosphere bacteria. Annu Rev Phytopathol 36:453–483

Van Loon LC, Rep M, Pieterse CMJ (2006) Significance of inducible defense related proteins in infected plants. Annu Rev Phytopathol 44:135–162

Van Peer R, Schippers B (1992) Lipopolysaccharides of plant growth promoting *Pseudomonas* spp. strain WCS417r induce resistance in carnation to *Fusarium* wilt. Neth J Plant Pathol 98:129–139

van Wees SCM, Luijendijk M, Smoorenburg I, Leendert C, van Loon, Pieterse CMJ (1999) Rhizobacteria-mediated induced systemic resistance (ISR) in *Arabidopsis* is not associated with a direct effect on expression of known defense-related genes but stimulates the expression of the jasmonate-inducible gene *Atvsp* upon challenge. Plant Mol Biol 41:537–549

Van Wees SCM, de Swart EAM, van Pelt JA, van Loon LC, Pieterse CMJ (2000) Enhancement of induced disease resistance by simultaneous activation of salicylate- and jasmonate dependent defense pathways in *Arabidopsis thaliana*. Proc Natl Acad Sci U S A 97:8711–8716

Verberne MC, Hoekstra J, Bol JF, Linthorst HJM (2003) Signaling of systemic acquired resistance in tobacco depends on ethylene perception. Plant J 35:27–32

Visca P, Imperi F, Lamont IL (2007) Pyoverdine siderophores: from biogenesis to biosignificance. Trends Microbiol 15:22–30

Vleesschauwer D, Ho¨Fte M (2009) Rhizobacteria-induced systemic resistance. Adv Bot Res 51:223–281

Vlot AC, Klessig DF, Park SW (2008a) Systemic acquired resistance: the elusive signal(s). Curr Opin Plant Biol 11:436–442

Vlot AC, Liu P-P, Cameron RK, Park S-W, Yang Y, Kumar D, Zhou F, Padukkavidana T, Gustafsson C, Pichersky E, Klessig DF (2008b) Identification of likely orthologs of tobacco salicylic acid-binding protein 2 and their role in systemic acquired resistance in *Arabidopsis thaliana*. Plant J 56:445–456

Wahyudi AT, Astuti RP, Widyawati A, Meryandini A, Nawangsih AA (2011) Characterization of *Bacillus* sp. strains isolated from rhizosphere of soybean plants for their use as potential plant growth for promoting Rhizobacteria. J Microbiol Antimicrobial 3(2):34–40

Walters DR, Ratsep J, Havis ND (2013) Controlling crop diseases using induced resistance: challenges for the future. J Exp Bot 64:1263–1280

Weller DM, Raaijmakers JM, McSpadden Gardener BB, Thomashow LS (2002) Microbial populations responsible for specific soil suppressiveness to plant pathogens. Annu Rev Plant Physiol Plant Mol Biol 40:309–348

Wendehenne D, Gao QM, Kachroo A, Kachroo P (2014) Free radical-mediated systemic immunity in plants. Curr Opin Plant Biol 20:127–134

Yang YX, Ahammed GJ, Wu C, Fan SY, Zhou YH (2015) Crosstalk among jasmonate, salicylate and ethylene signaling pathways in plant disease and immune responses. Curr Protein Pept Sci 16(5):450–461

Zamioudis C, Pieterse C (2012) Modulation of host immunity by beneficial microbes. Mol Plant Microbe Interact 25:139–150

Zhang Y, Fernando WGD (2004) Presence of biosynthetic genes for phenazine-1-carboxylic acid and 2,4-diacetylphloroglucinol and pyrrolnitrin in *Pseudomonas chlororaphis* strain PA-23. Can J Plant Pathol 6:430–431

Zhang S, Moyne A-L, Reddy MS, Kloepper JW (2002) The role of salicylic acid in induced systemic resistance elicited by plant growth-promoting rhizobacteria against blue mold of tobacco. Biol Control 25:288–296

Essential Oils as Antimicrobial Agents Against Some Important Plant Pathogenic Bacteria and Fungi

13

Bachir Raho Ghalem

Abstract

Plant diseases impact negatively on human well-being through agricultural and economic loss and also have consequences for biodiversity conservation. They are caused by some pathogens like bacteria, fungi, nematodes and viruses. Bacteria and fungi are the most common cause of many diseases of plants. The use of antibiotics for the control of plant diseases is limited due to the possibility to the production of some pathogen populations resistant to fungicides and pathogen populations resistant to antimicrobial agents and the ability to the transfer of responsible resistant genes to human and animal pathogenic microbes. In addition, these chemical compounds can cause undesirable effects on environment due to their slow biodegradation and several serious side effects on mammalian health associated to toxic residues in agricultural products. There is, therefore, a need to develop alternative control agents to pathogenic bacterial and fungal diseases in plants. Essential oils are a concentrated hydrophobic liquid containing volatile aroma compounds derived from the different parts of the plants. They were previously known to possess many biological activities such as antifungal and antibacterial properties. In addition, the potential effectiveness of essential oils against many plant pathogenic bacteria and fungi has been verified by many authors. This review discusses the susceptibility of most important ten bacterial and fungal plant pathogens towards different essential oils and their constituents, which have been reported in scientific references.

B.R. Ghalem (✉)
Department of Biology, University of Mascara, Mascara, Algeria
e-mail: bachir_raho@yahoo.fr

D.K. Choudhary et al. (eds.), *Plant-Microbe Interaction: An Approach to Sustainable Agriculture*, DOI 10.1007/978-981-10-2854-0_13

13.1 Introduction

Plants are continuously confronted in their environment with a wide range of potential pests and pathogens that include insects, nematodes, viruses, bacteria, fungi and oomycetes. While many pathogens evolved to infect aerial parts of the plant (leaves, stems, flowers and fruits), others target belowground organs (roots and tubers) (Hajek 2004; Yadeta and Thomma 2013). From plant diseases, fungal and bacterial plant pathogens cause major production and economic losses in agriculture by reducing quality and yield of particular crops or limiting their production in areas with environmental conditions conducive for disease development (Obradovic et al. 2008; Sankaran et al. 2010; Tajane and Janwe. 2014). In addition to fungal diseases which account for many of disease problems due to their prevalence and the amount of loss caused in plant production, there are an important number of bacterial diseases that are extremely destructive, and they are a threat to crops (Obradovic et al. 2008). Despite that chemical control reduces the effects of most fungal and bacterial disease effectively and extensively, these are not always desirable and present a danger to the health of humans, animals and the environment if they are used in excessive and improper manner (Shabana et al. 2008; El-Mohamedy and Aboelfetoh 2014). In addition, many farmers particularly in developing countries cannot use them for their high costs and lack of knowledge about its application (Dhlamini et al. 2005). Therefore, extensive studies for biopesticides that are easily biodegradable and safe to the environment to control fungal – and bacterial – diseases of crops have been carried out during the last two decades (Al-Askar 2012). Hence, aromatic plants that have been used in traditional medicine for their antimicrobial activities since ancient times could represent a promising solution and wise tool (Ismail et al. 2012; Pawar 2013). Studies done previously have confirmed that many plant essential oils exhibited antimicrobial effect on Gram-positive and Gram-negative bacteria, fungi and phytopathogens. The aim of this paper is to provide an overview of the published data on essential oils that have been reported to be effective against the most important bacteria, fungi and phytopathogens.

13.2 Brief History of Plant Essential Oils as Antimicrobial Agents

Since ancient times, herbs and spices have enjoyed a rich tradition of use both for their flavour-enhancement characteristics and for their medicinal properties (Kaefer and Milner 2008). The earliest records of the valuable properties of medicinal plants were by the Sumerians (6000 BC), followed by Chinese and Greek. A Chinese (4000 BC) wrote the first book about herbal plants (Kaliora and Kountouri 2012). Around 1550 BC, the ancient Egyptians used these substances as preservative agents for food conservation and as embalming agent to preserve their deceased pharaohs (Davidson et al. 1983). EO production appeared in the East (India and Persia) more than 2000 years ago and was improved in the ninth century by Arabs (Tajkarimi et al. 2010). Essential oils started to be manufactured by chemists after the thirteenth century AD, and their pharmacological effects are described in pharmacopoeias. After their use only in London, they disseminated consequently to the rest of Europe in the fourteenth century (Burt 2004). In the sixteenth century, Paracelsus von Hohenheim used the term of "essential oil" for the first time to the component of a drug as "quinta essential" (Guenther 1950). Many researches in the nineteenth century on plant products have been focused on the antimicrobial properties of herbs, spices and their constituents; the interest in the properties of these compounds continues to grow (Zaika 1988). In 1881, la Croix carried out the first bactericidal experiment of EOs (Boyle 1955), followed by Chamberland in 1887 who tested over 100 essential oils against spores of *Bacillus anthracis* and found that the vapour of the cinnamon oil was lethal to the spores. Hoffman and Evans were among the first researchers to describe the preservative effect of some spices such as cinnamon, cloves, mustard, allspice, nutmeg, ginger, black pepper and cayenne pepper. They demonstrated that cinnamon, cloves and mustard were most effective, while ginger, black pepper and cayenne pepper were least effective (Arora 2003). The utilization of essential oil pharmaceutical preparations had been declined by the middle of the twentieth century, and its role had been limited almost totally to be used in food flavourings, cosmetics and perfumes (Edris 2007).

In the last two decades, many studies have been conducted with a large number of essential oils from different plants in order to investigate their antimicrobial properties on plant pathogenic microorganisms (Vasinauskiene et al. 2006; Tabanca et al. 2007; Ozturk and Ercisli 2007; Kowalska and Smolinska 2008; Kotan et al. 2010; Dadasoglu et al. 2011; Kokoskova ct al. 2011; Silva et al. 2012; Adebayo et al. 2013; Islam et al. 2013; Kotan et al 2014; Alamshahi and Nezhad 2015).

13.3 Chemical Composition of Essential Oils

Essential oils are volatile, natural, complex mixtures of compounds characterized by a strong odour and are formed by aromatic plants as secondary metabolites (Bakkali et al. 2008). There are many different methods of essential oil extraction from plants: water or steam distillation, solvent extraction, expression under pressure and supercritical fluid and subcritical water extraction (Edris 2007). They have a complex composition, containing from a few dozen to several hundred constituents (Miguel 2010). These compounds are mainly terpenes and terpenoids, aromatic (phenolic) components and in a lower extent aliphatic (alkanes and alkenes) compounds (Bayala et al. 2014).

Terpenes are known as the large group of hydrocarbons made up of isoprene units (C_5H_8). They are synthesized in the cytoplasm of plant cells via the mevalonic acid pathway starting from acetyl CoA. They have a hydrocarbon backbone which can be rearranged into cyclic structures (monocyclic or bicyclic structures) by cycles. Monoterpenes ($C_{10}H_{16}$) and sesquiterpenes ($C_{15}H_{24}$) are usually the main terpenes, but longer chains such as diterpenes ($C_{20}H_{32}$), triterpenes ($C_{30}H_{40}$), etc., also exist. Examples of terpenes include ρ-cymene, limonene, terpinene, sabinene and pinene (Hyldgaard et al. 2012).

The monoterpenes are formed from the coupling of two isoprene units. They are the most representative molecules constituting 90 % of the essential oils and allow a great variety of structures. The following chemical classes are included as monoterpenes such as carbures, alcohols, aldehydes, ketones, esters, ethers, peroxides and phenols. The sesquiterpenes are formed from the assembly of three isoprene units. The extension of the chain increases the number of cyclizations which allows a great variety of structures. The structure and function of sesquiterpenes are similar to those of the monoterpenes which also include carbures, alcohols, ketones and epoxide (Bajpai et al. 2011).

Most terpenes do not possess high inherent antimicrobial activity of ρ-cymene, one of the most important components of thyme essential oil, and do not show antimicrobial activity against many Gram-negative pathogens (Bagamboula et al. 2004). Other terpenes, such as limonene, α-pinene, β-pinene, γ-terpinene, δ-3-carene, (+)-sabinene and α-terpinene, showed a very low or no antimicrobial activity against 25 genera of bacteria (Dorman and Deans 2000). These in vitro tests indicate that terpenes show ineffective antimicrobial activity when used as singular compounds (Nazzaro et al. 2013).

Terpenoids are terpenes that undergo biochemical modifications via enzymes that add oxygen molecules and move or remove methyl groups. Terpenoids can be subdivided into alcohols, esters, aldehydes, ketones, ethers, phenols and epoxides.

Examples of terpenoids are thymol, carvacrol, linalool, linalyl acetate, citronellal, piperitone, menthol and geraniol (Hyldgaard et al. 2012). The antimicrobial or antifungal mode of action of essential oil may be due to terpenoids. These compounds are highly lipophilic and are of low molecular weight which disrupt the cell membrane, cause the cell death and are also effective in the inhibition of sporulation and germination of food spoilage fungi (Tian et al. 2011). For example, the ethanol-soluble fraction of purple prairie clover yields a terpenoid called petalostemumol, which produces significant activity against *Bacillus subtilis* and *Staphylococcus aureus* while lesser activity against Gram-negative bacteria as well as *Candida albicans* (Cowan 1999; Ciocan and Bra 2007).

The other chemical classes of typical constituents in the essential oils are some aromatic compounds which are derived from phenylpropane and occur less frequently than all the terpenes mentioned above. The most common aromatic compounds are aldehydes such as cinnamaldehyde, alcohols (e.g., cinnamic alcohol), various phenols and methoxy and methylenedioxy derivatives. Nitrogenous or sulphur components such as glucosinolates or isothiocyanate derivatives can occur in some specific essential oils. However, these last classes of compounds are really less frequent in comparison with mono- and sesquiterpenes and their derivatives (Bertoli et al. 2011).

13.4 Mechanism of Action of Essential Oils

The mechanisms by which essential oils inhibit bacteria involve different modes of action; one of the well documented is membrane disruption by the lipophilic components (Dreger and Wielgus 2013). They cause lipid partitioning of bacterial cell membranes and mitochondria, disturbing the cell structures and rendering them more permeable. Extensive leakage from bacterial cells or the exit of critical molecules and ions will lead to death (Božović et al. 2015). Another possibility of action is inhibition of production of amylase and protease which stops the toxin production and electron flow and results in coagulation of the cell content (Djilani and Dicko 2012). Antifungal actions are quite similar to those described for bacteria (Djilani and Dicko 2012). According to Freiesleben and Jäger (2014), the antifungal agent can destroy the fungus by one of these six mechanisms:

Inhibition of cell wall formation: The fungal cell wall primarily consists of β-glucans. If the synthesis of these compounds is inhibited, the cell wall integrity will disrupt.

Cell membrane disruption: The ergosterols are essential for the cell membrane. If these sterols are bound by antifungal drugs, or the synthesis of them is inhibited by ergosterol biosynthesis inhibitors, the cell membrane integrity will disrupt. Thereby the membrane becomes leaky.

Dysfunction of the fungal mitochondria: Inhibition of the mitochondrial electron transport will result in reduction in mitochondrial membrane potential. The inhibition can occur via inhibition of the proton pumps in the respiratory chain, leading to reduction in ATP production and subsequent cell death.

Inhibition of cell division: Inhibition of cell division can happen via inhibition of microtubule polymerization, thereby inhibiting the formation of the mitotic spindle.

Inhibition of RNA/DNA synthesis or protein synthesis: If the antifungal agent enters the cell, for instance, via active transport on ATPases, and interferes with the RNA, it can cause faulty RNA synthesis and inhibition of DNA transcription. Inhibition of protein synthesis is also a known antifungal target.

Inhibition of efflux pumps: Efflux pumps are present in all living cells, and their function is to transport toxic substances out of the cell. This transport often includes transport of accumulated drug out of the fungal cell. Overexpression of efflux pumps can lead to drug resistance. By inhibiting the efflux pumps, it is believed that drug resistance can be reduced.

The antimicrobial activity of several essential oils has been attributed to the presence of specific terpenoid and phenolic compounds (Villa and Veiga-Crespo 2013; Zengin and Baysal 2014), as well as the chemical constituents and functional groups contained in the essential oil, the proportions in which they are present and the interactions between them (Dorman and Deans 2000). Interactions between these components may lead to antagonistic, additive or synergistic effects. Some studies have showed that whole essential oils usually have a greater antibacterial activity than the major components mixed, suggesting that the minor components are critical to the synergistic activity, though antagonistic and additive effects have also been observed (Davidson and Parish 1989; Gill et al. 2002; Mourey and Canillac 2002).

13.5 Review of the Susceptibility of Ten Fungal Plant Pathogens to Different EOs

Dean et al. (2012) published the top ten fungal plant pathogen list, based on scientific/economic importance. The top ten list includes, in rank order, (1) *Magnaporthe oryzae*, (2) *Botrytis cinerea*, (3) *Puccinia* spp., (4) *Fusarium graminearum*, (5) *Fusarium oxysporum*, (6) *Blumeria graminis*, (7) *Mycosphaerella graminicola*, (8) *Colletotrichum* spp., (9) *Ustilago maydis* and (10) *Melampsora lini*.

13.5.1 *Magnaporthe oryzae*

Magnaporthe oryzae was inhibited by essential oils of *Gliomastix murorum* and *Pichia guilliermondii* at 0.84 mg/mL and 1.56 mg/mL, respectively (Zhao et al. 2009); methanol extract of *Myristica fragrans* Houttyn (nutmeg) seeds (Cho et al. 2007); hydro-distilled roots and rhizomes of two valerianaceous species, *Nardostachys chinensis* and *Valeriana officinalis* (Wang et al. 2010a); EOs of the leaves of *Ocimum gratissimum*, *Chromolaena odorata* and *Cymbopogon citratus*; seeds of *Eugenia aromatica* and *Piper guineense*; nuts of *Garcinia kola* (Olufolaji et al. 2015); *Corymbia citriodora* and *Cymbopogon nardus* essential oils (Aguiar et al. 2014); the essential oil of star anise (*Illicium verum* Hook. f.) fruit (Huang et al. 2010); EOs of *Callistemon lanceolatus* DC leaves at 3000 ppm (Misra et al. 1997); EOs from *Piper nigrum* at 500 ppm and *Coriandrum sativum* oils at 1000 ppm concentration (Sukanya et al. 2011); liquid extract of *Ruta graveolens* (Reis et al. 2015); the root bark essential oil of *Periploca sepium* Bunge (Asclepiadaceae/Apocynaceae) (Wang et al. 2010b); hydro-distilled essential oil from flowering shoots of *Tanacetum annuum* (Greche et al. 2000); and the aqueous extracts of processed *Coffee arabica*, *Nicotiana tabacum*, *Aloe vera* and *Chrysanthemum coccineum* (Hubert et al. 2015). Extracts of three different plants, evaluated by food poisoning method, had the following relative inhibitory effects on *Magnaporthe oryzae*: garlic (*Allium sativum* L.) > neem (*Azadirachta indica* L.) > *Calotropis* (*Calotropis procera* L.) (Jamal-U-Ddin et al. 2012).

13.5.2 *Botrytis cinerea*

Antifungal activity of essential oils or extracts on *Botrytis cinerea* has been reported by several researchers. In this regard, vapours of thyme, oregano and lemongrass and their respective major components showed complete growth inhibition of *Botrytis cinerea* as reported by Plotto et al. (2003). Also, Arrebola et al. (2010) indicated that thyme and lemongrass oils caused over 50 % and 25 % inhibition of radial mycelium growth in the presence of lemon and oregano essential oils at concentration of 17 μl/ml and 0.02 μl/ml, respectively (Vitoratos et al. 2013). Jaspers et al. (2001) reported that thyme oil at concentration of 0.33 % reduced significantly *B. cinerea* sporulation on artificially induced necrotic leaf lesion. In this respect,

Tzortzakis and Economakis (2007) reported that lemongrass oil at 25 ppm could inhibit *B. cinerea* spore production, and at 500 ppm, the highest oil concentration employed, fungal sporulation was completely inhibited. However, complete inhibition of *B. cinerea* to the black caraway and fennel oils at concentrations of 400 and 600 μLL^{-1}, respectively, in vivo and black caraway, fennel and peppermint oils at all applied concentrations in vivo on plum fruits has been shown (Aminifard and Mohammadi 2013). Daferera et al. (2003) observed that the growth of *B. cinerea* was completely inhibited by oregano, thyme, dictamnus and marjoram essential oils at relatively low concentrations (85–300 μg/ml). Moreover, Hammam et al. (2011) reported that *Viola odorata* L. essential oils exhibited strong antifungal activity against *B. cinerea* based on the inhibition zone and minimal inhibitory concentration values. Doğu and Zobar (2014) tested the antifungal effect of seven different plant essential oils and observed that thyme, mint and rosemary oils were found more effective to *B. cinerea*, but sage, grapeseed, ozone and basil oils showed varying effects. The effect of several essential oils on the growth of *B. cinerea* was investigated by Mohammadi et al. (2014) where this fungi was completely inhibited by the essential oil of black caraway at 400 $\mu g\ l^{-1}$. Recently, Şesan et al (2015) tested nine different plant extracts on *Botrytis cinerea* and found high inhibition of *Hyssopus officinalis* (at 20, 10 and 5 %), *Satureja hortensis*, *Allium sativum*, *Tagetes patula* (at 20 and 10 %) and *Mentha* (at 20 %), and a moderate anti-*Botrytis* activity (efficiency between 35.7 and 65.7 %) has been noticed for *Mentha* (at 10 and 5 %), *Satureja hortensis*, *Allium sativum* and *Tagetes patula* (at 5 %) extracts.

13.5.3 *Puccinia* spp.

To date very few researches have been conducted on the antifungal activities of the plant essential oils or extracts on the three *Puccinia* spp. which rust diseases occur on wheat (*Puccinia graminis* Pers. f. sp. *tritici* (*Pgt*), *Puccinia striiformis* f. sp. *tritici* (*Pst*) and *Puccinia triticina* Eriks (*Pt*)). Somaya and El-Sharkawy (2014) assayed the effect of chamomile, thyme, cumin, basil, eucalyptus and garlic essential oil on wheat rust disease at seeding and adult stage of two susceptible wheat cultivars (Morocco and Sids-1) under greenhouse and field conditions in 2013/2014 growing season and found that the addition of these essential oils decreased leaf rust severity (%), significantly, increased both spike weight (g), grains weight/spike (g) and 1000 kernel weight (g).

13.5.4 *Fusarium graminearum*

A variety of antifungal activities on *Fusarium graminearum* has been shown by *Eucalyptus camaldulensis* essential oils (Mehani et al. 2014); cinnamon, clary sage and marjoram essential oils (Gömöri et al. 2013); water-distilled EOs of mint (*Mentha spicata* var. *crispa* L.) and the commercially essential oil of cinnamon (*Cinnamomum verum*) (Aromax Ltd., Hungary) (Horváth et al. 2013); *Ocimum sanctum* L. essential oil at 1250 μg/mL (minimum inhibitory concentration) and 1800 μg/mL (minimum fungicidal concentration) (Kalagatur et al. 2015); *Zataria multiflora*, *Satureja hortensis* essential oils, thymol and carvacrol at 16, 31.5, 70 and 15 μl/100 ml, respectively, in PDA media and at 16, 30, 70 and 20 μl/100 ml, respectively, in PDB media (Lahooji et al. 2010); oregano, cinnamon, lemongrass, clove and palmarosa essential oils at two concentrations (500 and 1000 mg kg^{-1}), at different water activity (aw) (0.95 and 0.995) and temperature (20 and 30 ° C) levels (Velluti et al. 2004); EOs from seed of asafoetida at 0.15 and 0.3 % (Mostafa et al. 2013); hydro-distilled aerial parts of *Echinophora platyloba* (Hashemi et al. 2016); *O. vulgare* essential oil (Marín et al. 2004); *Allium fistulosum* L., *Allium sativum* L. and *Allium cepa* L. oils (Benmeddour et al. 2015); rice, oat and wheat crude protein extracts (Pagnussatt et al. 2013); hydro-distilled *Artemisia afra*, *Conyza scabrida*, *Helichrysum foetidum*, *Leucosidea sericea*, *Mentha piperita* and *Pelargonium graveolens oils* (Samie and Nefefe 2012); *Mentha piperita* (peppermint) and *Salvia officinalis* L. (sage) oils (Tomescu et al. 2015); essential oil of *Artemisia sieberi* Besser (Amir et al. 2013); hydro-distilled of *Zataria multiflora*, *Thymus vulgaris* and *Thymus kotschyanus* (Amini et al. 2012); *Thymus vulgaris*, *Satureja hortensis*, *Anethum graveolens*, *Mentha sativa* and *Capsicum annuum* essential oils (Hoseiniyeh et al. 2012); and pure eugenol and carvacrol and clove EO (Cardiet et al. 2012).

13.5.5 *Fusarium oxysporum*

A significant antifungal effect was observed with *Cinnamomum zeylanicum*, *Thymus vulgaris* and *Syzygium aromaticum* oils which had a total inhibition at 100, 150, 200, 250 and 300 ppm. *Teloxys ambrosioides*, *Mentha piperita* and *Citrus aurantifolia* oils exhibited a dose-dependent inhibition on mycelial growth of *Fusarium oxysporum* to increase the dose of 100 at 300 ppm (Barrera-Necha et al. 2009). In another study, the volatile essential oils of *Cinnamomum zeylanicum* and *Syzygium aromaticum* showed good effects in controlling the Panama disease caused by *Fusarium oxysporum* f. sp. *cubense* in both 1000 and 2000 ppm (Monteiro et al. 2013).

Istianto and Emilda (2011) studied the inhibitory effect of *Cymbopogon nardus*, *Eugenia aromatica*, *Pogostemon cablin* and *Vitiveria zizanoides* essential oils on the mycelial growth of *Fusarium oxysporum* f. sp. *cubense* (*Foc*), and the results showed that *E. aromatica* oil provided the strongest suppression of *Foc* mycelial growth, mainly when used at a volume of 9 and 18 μl. Lima et al (2010) studied the growth and survival of *Fusarium oxysporum* S. and *Thanatephorus cucumeris* F. in the presence of essential oil from leaves of *Hedychium coronarium*, which showed inhibitory effect on the in vitro growth of *F. oxysporum.* Arango et al. (2011) reported the fungicide effect of *Eucalyptus tereticornis* essential oil on the pathogenic fungus *Fusarium oxysporum.* Guzmán-Guzmán et al. (2003) analysed the inhibitory effect of various concentrations of mint, eucalyptus, laurel, clove, sweet marjoram, rosemary, origanum, thyme, cinnamon, pepper and grapefruit essential oils (1,250, 2,500 and 3,750 ppm) on the growth of *Fusarium oxysporum* f. sp. *phaseoli* and reported that it is possible to control *Fusarium oxysporum* f. sp. *phaseoli* with essential oils. Manganyi (2013) and Manganyi et al. (2015) also noticed that clove and thyme essential oils inhibited the mycelial growth of *Fusarium oxysporum*. Recently, a study carried out by La Torre et al.(2016) analysed the action of clove oil, thyme oil, rosemary oil and their major components in controlling *Fusarium* wilt in tomato, which showed that the clove oil and its major component eugenol were the most effective, while rosemary oil gives the lowest inhibitory activity on *Fusarium oxysporum* f. sp. *lycopersici*.

13.5.6 *Blumeria graminis*

The effects of plant extracts on *Blumeria graminis* have been studied by a very large number of researchers in different parts of the world. Haugaard et al. (2002) reported that the whole-plant bioassays for testing possible effects of mycelial extract from *Bipolaris oryzae*, *Pythium ultimum* and *Rhizopus stolonifer* showed that the mycelial extracts strongly reduced the numbers of *Blumeria graminis* colonies formed on the leaves and that the few colonies that developed appeared small and with reduced conidial production. Terzi et al. (2007) observed the inhibitory action of *Melaleuca alternifolia* essential oil and its principal components (terpinen-4-ol, γ-terpinen and 1,8-cineole). Terpinen-4-ol was the most effective. Hafez (2008) tested the antifungal effect of black seed (*Nigella sativa*) oil, rapeseed (*Brassica napus*) oil and paraffin oil and observed great reduction of the disease severity of barley powdery mildew (*Blumeria graminis* f. sp. *hordei*) from 63.4 % (control) to 9.4 % (black seed oil), 16 % (rapeseed oil) and 16.4 % (paraffin oil). The results obtained after the treatment of infected barley (*Blumeria graminis*) leaves with different concentrations of *Artemisia herba-alba* essential oil (0.25 and 0.5 % of oil) showed that this essential oil had a strong toxic effect against the hyphal growth and spore germination (Morcia et al. 2015). Recently, Věchet and Šerá (2015) analysed the antifungal activity of extracts from plants (oak, giant knotweed, curcuma and ginger) on powdery mildew (*Blumeria graminis* f. sp. *tritici*) attack on winter wheat and found that the two plant extracts, giant knotweed (*Polygonum sachalinense*) and turmeric spice (curcuma, *C. longa*), showed the best results.

13.5.7 *Mycosphaerella graminicola*

Recently, in a study by Matusinsky et al.(2015), the antifungal activity of essential oils extracted from *Pimpinella anisum*, *Thymus vulgaris*, *Pelargonium odoratissimum*, *Rosmarinus officinalis* and *Foeniculum vulgare* against five fungi, including *M. graminicola*, was studied using agar dilution method. All essential oils used in this experiment affected the growth of these fungi. Ultimately, the best antifungal activity (on the basis of inhibitory effect) was demonstrated by *Thymus vulgaris*. A screening of the level of inhibitory activity of essential oils from aerial parts of

Acantholippia deserticola, *Artemisia proceriformis*, *Achillea micrantha* and *Libanotis buchtormensis* tested by Sampietro et al. (2015) on *M. graminicola* showed a moderate antimicrobial activity of *A. deserticola*, *A. micrantha* and *L. buchtormensis* oils on these phytopathogenic fungi with MIC_{100} ranged between 0.5 and 1.5 mg/ml. Scher et al. (2004) showed also antifungal activity of the dichloromethane and a methanol extract of liverwort *Bazzania trilobata* against *M. graminicola*. Deweer et al.(2013) reported that dill seed essential oils used crude or with DMSO are more efficient on S6 – sensitive strain of *Zymoseptoria tritici* (teleomorph: *M. graminicola*) – at 350 mg/L than on R1187, resistant strain, at 1000 mg/L but with Tween 80; the essential oil effectiveness is the same on both strains (300 mg/L).

13.5.8 *Colletotrichum* spp.

Colletotrichum is one of the most common and important genera of plant pathogenic fungi. Virtually every crop grown throughout the world is susceptible to one or more species of *Colletotrichum*. Many essential oils have been reported as effective compounds against *Colletotrichum* sp.: the oil isolated by hydrodistillation of basil, rosemary and cinnamon on *C. musae* isolated from banana (Idris et al. 2015); hydro-distilled essential oils of *Zanthoxylum monophyllum Z. rhoifolium* and *Z. fagara* oils on *Colletotrichum acutatum* (Prieto et al. 2011); oregano and thyme essential oils, among 56 EOs investigated on *C. acutatum* and *C. gloeosporioides* (Grahovac et al. 2012); crude extracts of *Acorus calamus* L., *Stemona curtisii* HK.f., *Stemona tuberose* L., *Memmea siamensis* Kost, *Eugenia caryophyllus* and an eugenol essential oil (Thobunluepop et al. 2009); and *Amomum cardamomum*, *Asarum sieboldii*, *Illicium verum*, *Juniperus chinensis*, *Myristica fragrans* and *Schizonepeta tenuifolia* oils on *C. gloeosporioides* (Sun et al. 2007).

13.5.9 *Ustilago maydis*

Steam distillate from leaves of *Cymbopogon citratus* completely inhibited the growth of *U. maydis*, and hot water extracts from fresh leaves of *Ocimum gratissimum* and *Chromolaena odorata* and dry fruits of *Xylopia aethiopica* reduced radial growth by 37–57 %. A hot water extract from dry fruits of *Monodora myristica* was ineffective as a fungi toxicant (Awuah 1989). Maize oil was more effective against *Ustilago maydis* followed by soybean and sunflower oils in controlling the disease (Moursy et al. 2001). The plant oils of eucalyptus, clove, cinnamon, peppermint and anise at 750 and 1000 ppm and clove and anise oils only at 500 ppm caused 100 % inhibition of the in vitro *Ustilago maydis* growth (El-Fiki et al. 2003).

13.5.10 *Melampsora lini*

Melampsora lini (Ehrenb.) Desm., the fungal pathogen responsible for rust disease on flax and linseed (*Linum usitatissimum* L.), is of interest for both economic and scientific reasons. It can cause severe losses in seed yield as well as reducing fibre quality in flax plants grown for linen production (Lawrence et al. 2007).

In my research in literature, I have not found any work on the antifungal activity of essential oils on this fungus, which constitutes a very important field of research to discover natural antifungal components against this species.

13.6 Review of the Susceptibility of Ten Bacterial Plant Pathogens to Different EOs

The top bacterial species have been listed based on their scientific and economic importance in plant disease: (1) *Pseudomonas syringae* pathovars, (2) *Ralstonia solanacearum*, (3) *Agrobacterium tumefaciens*, (4) *Xanthomonas oryzae* pv. *oryzae*, (5) *X. campestris* pathovars, (6) *X. axonopodis* pathovars, (7) *Erwinia amylovora*, (8) *Xylella fastidiosa*, (9) *Dickeya* (former *Erwinia*) (*dadantii* and *solani*) and (10) *Pectobacterium* (former *Erwinia*) carotovorum (and *Pectobacterium atrosepticum*) (Mansfield et al. 2012).

13.6.1 *Pseudomonas syringae*

A large number of investigations have been performed on the antifungal activities of essential oils against *Pseudomonas syringae* pv. Kokoskova et al. (2011) examined the antimicrobial effects of five aromatic herb species of the family Lamiaceae against plant pathogenic (*Erwinia amylovora* and *Pseudomonas syringae* pv. *syringae*) and saprophytic (*Pseudomonas fluorescens*, *Pantoea dispersa* and *P. agglomerans*). Plant essential oils from *M. officinalis* and *M. arvensis* were significantly more effective against *P. syringae* pv. *syringae*. The steam-distilled essential oils from oregano, sweet flag, caraway, peppermint, common, fern leaf and willow-leaved yarrow field accessions were investigated against the growth of phytopathogenic bacteria, *Erwinia carotovora* subsp. *carotovora*, *Xanthomonas vesicatoria*, *Pseudomonas marginalis* pv. *marginalis*, *P. syringae* pv. *syringae*, *P. syringae* pv. *tomato* and *Bacillus* sp. by Vasinauskienė et al. (2006). *P. syringae* pv. *syringae* was sensitive to oregano and willow-leaved yarrow essential oils. Poswal and Witbooi (2012) evaluated essential oils of *Artemisia afra*, *Eriocephalus punctulatus*, *Mentha piperita*, *Lavandula angustifolia* and *Lippia javanica* on the growth of *Pseudomonas syringae* pv. *syringae*. Essential oils from *L. angustifolia* and *A. afra* (indigenous to South Africa) were the most effective in inhibiting the growth of *P. syringae* pv. *syringae*. Antifungal properties (IŞcan et al. 2002) of the *Mentha piperita* oils were also investigated against 21 human and plant pathogenic microorganisms. Peppermint oils showed stronger inhibition (MIC 0.07–2.5 mg mL^{-1}) against

Pseudomonas syringae pv. *phaseolicola, P. syringae* pv. *tomato* and *Pseudomonas syringae* pv. *syringae.* Essential oils extracted by hydrodistillation from fruits of *Cuminum cyminum* L. and *Carum carvi* L. were also investigated on Gram-positive and Gram-negative bacteria. Among them, *P. syringae* pv. *phaseolicola, P. syringae* pv. *pisi, P. syringae* pv. *syringae, P. syringae* pv. *aptata, P. syringae* pv. *apii, P. syringae* pv. *atrofaciens, P. syringae* pv. *lachrymans, P. syringae* pv. *maculicola, P. syringae* pv. *tomato* and *P. syringae* pv. *glycinea* were sensitive (MIQ 910–7360 (μg)) (Iacobellis et al. 2005). Similar activity was also verified on the same organisms for essential oils that were extracted from fruits of *Coriandrum sativum* L. and *Foeniculum vulgare* Miller. *C. sativum* oil inhibits *P. syringae* pv. *phaseolicola, P. syringae* pv. *pisi, P. syringae* pv. *syringae, P. syringae* pv. *aptata, P. syringae* pv. *atrofaciens, P. syringae* pv. *maculicola, P. syringae* pv. *tomato* and *P. syringae* pv. *glycinea*, but *F. vulgare* oil inhibits only *P. syringae* pv. *atrofaciens* and *P. syringae* pv. *glycinea* (Cantore et al. 2004). Karaman et al. (2003) reported low inhibition of *Juniperus oxycedrus* extracts *Pseudomonas syringae* pvs. (8–10 mm). In recent screening study, Gormez et al. (2015) evaluated antibacterial activities of *Satureja hortensis* and *Calamintha nepeta* oils against 20 phytopathogenic bacteria causing serious crop loss. *C. nepeta* oil was most active on *P. syringae* pv. *syringae, P. syringae* pv. *phaseolicola, P. syringae* pv. *pisi, P. syringae* pv. *tabaci* and *P. syringae* pv. tomato.

13.6.2 *Ralstonia solanacearum*

Many plant EOs have demonstrated antibacterial activity against *Ralstonia solanacearum.* The efficacy of thymol, palmarosa oil and lemongrass oil against *R. solanacearum* was investigated by Pradhanang et al. (2003). The results showed that the tomato seedlings transplanted in soil treated with 700 mg/l of thymol, palmarosa oil and lemongrass oil were free from bacterial wilt, and 100 % of plants in thymol treatments were free of *R. solanacearum.* Huang and Laksman (2010) observed antibacterial activity of clove oil against seven different species of plant pathogenic bacteria. Both Gram-positive and Gram-negative bacteria tested were sensitive to clove essential oil (0.1 and 0.5 %), *R. solanacearum* being the most sensitive one. In another investigation performed by Paret et al. (2010), palmarosa (*Cymbopogon martini*), lemongrass (*C. citratus*) and eucalyptus (*Eucalyptus globulus*) oils evaluated for their efficacy against *R. solanacearum.* The experiments revealed that 0.04 % of palmarosa and lemongrass oils reduced the growth of the bacterium compared with control, and at 0.07 and 0.14 %, they showed complete inhibition of bacterial growth. However, eucalyptus oil treatments at 0.04 and 0.07 % had bacteriostatic effects on the cells. Wagura et al. (2011) reported that extract and essential oils derived from leaves of *Ocimum gratissimum* at concentration of 0.4, 0.2, 0.1, 0.05 and 0.025 mg ml^{-1} exhibited highly significant ($p < 0.0001$) differences on their effects against the growth of *R. solanacearum*, the causal agent of bacterial wilt of potato. Nezhad et al. (2012) reported on the antibacterial potential of essential oils of *Coriandrum sativum*, *Thymus vulgaris*, *Cumimum cymimum*, *Rosmarinus officinalis* and *Eucalyptus globulus* against *Pectobacterium carotovorum*, *Ralstonia solanacearum* and *Escherichia coli*. Results showed that the most active essential oils against tested bacteria were thyme oil with the inhibition zone of 34.8 mm against *R. solanacearum* and the MIC of 1 μl/ml. In another experiment reported by Hosseinzadeh et al. (2013), antibacterial activity of essential oil from *Cinnamomum zeylanicum*, *Thymus vulgaris*, *Lavandula angustifolia* and *Eucalyptus camaldulensis* was also tested against *R. solanacearum*. The results of in vitro assay indicated that the sub-bactericidal concentrations of essential oils applied in this study suppressed *R. solanacearum* pathogenicity and virulence factors. Oboo et al. (2014a) investigated the antibacterial activity of *Rosmarinus officinalis*, *Ocimum suave*, *Tarchonanthus camphorates*, *Lantana trifolia*, *Lippia javanica* and *Lippia ukambensis* oil against *R. solanacearum*. Results demonstrated that essential oils extracted from *O. suave*, *L. javanica* and *T. camphorates* possess antibacterial activity that is effective in the control of *R. solanacearum* at 24 °C, 28 °C and 32 °C. The inhibitory effect of essential oil from *Lippia javanica*, *Ocimum suave* and *Tarchonanthus camphoratus* against *R. solanacearum* was investigated by the same team (Oboo et al. 2014b). Treatment with the three plants, *T. camphorates*, *L. javanica* and *O. suave*, reduced the bacterial wilt disease caused by *R. solanacearum* by 38, 21 % and more than 90 %, respectively. Alamshahi and Nezhad (2015) evaluated the antibacterial effects of the essential oils extracted from *Coriandrum sativum*, *Cuminum cyminum*, *Rosmarinus officinalis* and *Eucalyptus globulus* against *R. solanacearum*. Of all the essential oils tested, the treatment by thyme essential oil caused significant reduction in bacterial wilt incidence on potato by 44 %. The antibacterial activity of essential oil from the leaves of *Macleaya cordata* was evaluated for the effect on the growth of several pathogens (Li and Yu 2015). The results showed that bacteria had different sensitivities to essential oil of *M. cordata* with diameters of inhibition zones ranging from 8.5± 0.6 to 18.6± 1.9 mm and MIC values from 125 to 500 μg/ml. Among these bacteria, *R. solanacearum* was the most sensitive to essential oil. Screening for antibacterial oils was done on nine essential plant oils, *viz*. spearmint (*Spearmint viridis*), neem (*Azadirachta indica*), marigold (*Tagetes erecta*), castor (*Ricinus communis*), calamus (*Acorus calamus*), olive (*Olea europaea*), turpentine (*Syncarpia glomulifera*), eucalyptus (*Eucalyptus macrorhyncha*) and garlic (*Allium sativum*), against growth of tomato, brinjal and capsicum isolates of *Ralstonia solanacearum* (Sood and Pardeep 2015). Neem oil was found most effective to all the three isolates of *R. solanacearum* followed by eucalyptus oil and spearmint oil.

13.6.3 *Agrobacterium tumefaciens*

Agrobacterium tumefaciens was inhibited by *Thymbra spicata* essential oil with MBC of 328 mg/ml (Basim et al. 2000); cinnamon, clove, chenopodium, caraway, rosemary and thyme oils and their constituents thymol, chlorothymol and carvacrol

(El-Zemity et al. 2008); *Heracleum persicum* oils and extracts and essential oils of *Cinnamomum zeylanicum* with MIC ranged between 0.2 and 50 mg/mL (Noudeh et al. 2010); and leaves essential oils of *Syzygium cumini* L. (12 mm, MIC< 250 µg/mL) and *Cupressus sempervirens* L. (8 mm, MIC 500 µg/mL) (Elansary et al. 2012).

13.6.4 *Xanthomonas oryzae*

A variety of antimicrobial effects on *Xanthomonas oryzae* has been shown by leaf extract of *Datura metel* (Kagade et al. 2004); *Cymbopogon citratus*, *Monodora myristica*, *Ocimum gratissimum*, *Thymus vulgaris* and *Zingiber officinale* oils (Nguefack et al. 2005); essential oil of *Metasequoia glyptostroboides* at 125 µg/ml concentration (Bajpai et al. 2010a) and *Cleistocalyx operculatus* at 62.5 µg/ml at 62.5 µg/ml concentration (Bajpai et al. 2010b); essential oil and compounds of limonin and imperatorin or *Poncirus trifoliata* (Rahman et al. 2014); essential oil of *Ocimum ciliatum* Hornem (Moghaddam et al. 2014); essential oil of neem and lemon (Singh et al. 2015); garlic bulb (*Allium sativum*), tamarind fruit (*Tamarindus indica*), gooseberry fruit (*Phyllanthus emblica*), green mango (*Mangifera indica*) and lemon juice (*Citrus aurantifolia*) extracts; and palmarosa (*Cymbopogon martinii*) lemongrass (*Cymbopogon flexuosus*), cinnamon (*Cinnamomum zeylanicum*) and vetiver (*Chrysopogon zizanioides*) oils (Raji et al. 2016).

13.6.5 *Xanthomonas campestris*

The below components have been reported as effective compounds against *Xanthomonas campestris*: *Mentha citrata* essential oils at 10^{-1} dilution (Maiti et al. 1985), herb and seed essential oils of *Coriandrum sativum* L. (Minija and Thoppil 2001), coriander and hyssop essential oil (Kizil et al. 2005), eugenol (Cantore et al. 2009), hydro-distilled essential oil of seeds of *Citrullus colocynthis* (Mehr et al. 2012) and *Cinnamomum cassia*, *Cinnamomum zeylanicum*, *Syzygium aromaticum*, *Thymus vulgaris*, *Laurus nobilis*, *Salvia sclarea*, *Boswellia carterii*, *Rosmarinus officinalis*, *Ocimum basilicum*, *Calendula officinalis* and *Cassia tora* essential oils (Chudasama and Thaker 2012).

13.6.6 *Xanthomonas axonopodis*

Thymbra spicata essential oils were bactericide at 323 mg/ml on *Xanthomonas axonopodis* pv. *vesicatoria* (Basim et al. 2000). *Mentha arvensis* and *Ocimum sanctum* oils showed the highest inhibition zone (17 and 12 mm, respectively) against *Xanthomonas axonopodis* pv. *malvacearum* (Thakare et al. 2003). The essential oil of *Rosa damascena* petals had inhibitory effect against *Xanthomonas axonopodis* spp. *vesicatoria* (Basim and Basim 2003). The pure carvacrol and thymol showed the highest inhibition zone (85 mm), and MIC value was 3.125 µl/ml on *X.*

axonopodis. Also, *Thymus canoviridis*, *Satureja hortensis*, *Melissa officinalis inodora*, *Helichrysum plicatum*, *Thymus haussknechtii*, *Thymus sipyleus* and *Thymus sipyleus rosulans* essential oils were the most active on *X. axonopodis* showing an inhibition zone of 22–46.3 mm and a MIC of 25–200 μl/ml (Kotan et al. 2007). Hydro-distilled essential oils from *Origanum acutidens*, *O. rotundifolium* and *O. vulgare* showed a wide spectrum of antibacterial activity on *X. axonopodis* pathovars (*malvacearum, vesicatoria, campestris, vitians and pelargoni*). It was also shown that carvacrol, thymol and other main components such as terpinen-4-ol and linalool possess antimicrobial activity (Dadasoglu et al. 2011). Leaf solvent (methanolic, ethanolic, petroleum ether and water) extracts of *Juniperus communis* L. and *Vitex negundo* show inhibitory effect on linear growth of *Xanthomonas axonopodis* pv. *punicae*.

Among these plant extracts, methanolic and ethanolic extracts at 300 ppm were more effective than that of other extracts for both plants (Digvijay et al. 2014). *Satureja hortensis* and *Calamintha nepeta* oils showed a strong antimicrobial effect against *Xanthomonas axonopodis* pv. *campestris* (Gormez et al. 2015).

13.6.7 *Erwinia amylovora*

It has been shown that *Erwinia amylovora* is sensitive to essential oils from *Melissa officinalis*, *Mentha arvensis*, *Origanum compactum*, *O. vulgare*, *Thymus vulgaris*, *Eugenia caryophyllata*, *Mentha pulegium* and *Nepeta cataria* (Kokoškova and Pavela 2007); *Citrus maxima* essential oil (Măruţescu et al. 2009); *Satureja montana* spp. *montana* L. (inhibition zone 25 mm) and *S. adamovici* (MIC= MBC= 0.09 μlml^{-1}) essential oils (Mihajilov-Krstev et al. 2010); *Satureja hortensis* L. and *Thymus vulgaris* L. essential oils and their major constituents thymol and carvacrol (Karami-Osboo et al. 2010); *Thymus vulgaris*, *Origanum compactum*, *Origanum vulgare*, *Nepeta cataria*, *Mentha arvensis* and *Mellissa officinalis* essential oils (Kokoskova et al. 2011); *Pelargonium odoratissimum*, *Salvia officinalis* and *Tagetes patula* oils (Chiriac and Ulea 2012); and sage and clove oils (Mikiciński et al. 2012).

13.6.8 *Xylella fastidiosa*

Ribeiro et al.(2008) investigated the antibacterial activity of a number of flavonoids, coumarins, alkaloids, dihydrocinnamic acid derivative, anacardic acid, triterpenes and limonoids on the growth of *Xylella fastidiosa*. Their experiments showed that sesquiterpenoid components were more effective than monoterpenoid components of the leaf oil. These results revealed that azadirachtin A was the most active and hesperidin showed a moderate activity. Screening of essential oil from 17 plant species against *Xylella fastidiosa* causing citrus variegated chlorosis (CVC) disease was carried by Massuco et al. (2013)). The essential oil of sandal proved bactericidal on *Xylella fastidiosa* at 125 μg/mL concentration followed by the oil of *Salvia sclarea*, cinnamon, cedar, patchouli and myrrh with concentrations ranged between

250 μg/mL and 500 μg/mL. The inhibitory effect of 12 phenolic compounds, representing phenolic acid, coumarin, stilbene and flavonoid, against *Xylella fastidiosa* was investigated by Maddox et al. (2010). They found that catechol, caffeic acid and resveratrol showed strong anti-*Xylela* activities.

13.6.9 *Dickeya* (*dadantii and solani*)

Several studies have demonstrated antimicrobial activity of essential oils against *Dickeya* spp. (*dadantii* and *solani*). Stefanova et al. (2005) found that *Erwinia chrysanthemi* is sensitive to the extracts of the two oregano species (*Hyptis suaveolens* and *Coleus amboinicus*) at 1 %, with inhibition areas of 20.6 mm. *Artemisia santonicum* and *Artemisia spicigera* essential oils showed a weak activity on *E. chrysanthemi* (Kordali et al. 2005). Sqalli et al. (2009) have reported the inhibition of *Erwinia chrysanthemi* (*Dickeya dadantii*) by the aqueous and ethanolic extracts and the essential oil of *Thymus pallidus* Batt. Paradza et al. (2012) demonstrated that the botanical extract of neem (*Azadirachta indica* A. Juss.) leaf and garlic (*Allium sativum* L.) inhibited the growth of *Dickeya dadantii* at concentration of 10 and 25 % (w/v). Among 100 plant essential oils tested for their antibacterial activity on *E. chrysanthemi*, *Cinnamomum cassia*, *Cinnamomum zeylanicum* and *Citrus bigaradia*, EOs were very active on these phytopathogenic bacteria (Chudasama and Thaker 2012). Ethanol and methanol extracts of *Juniperus squamata* were found effective by showing a mark zone of inhibition, MIC and MBC on *Erwinia chrysanthemi* (21 mm, 31.25 μg/ml and 125 μg/ml for ethanol extract) (16 mm, 250 μg/ml and 500 μg/ml for methanol extract) (Sati and Kumar 2015). In another research, Sati et al (2015) reported that *Berberis aristata*, *Chenopodium ambrosioides* and *Tinospora cordifolia* extract displayed a weak antibacterial activity on *E. chrysanthemi.* Essential oils from *Eugenia caryophyllata*, *Lavandula angustifolia*, *L. latifolia*, *Melaleuca quinquenervia*, *Melissa officinalis*, *Mentha pulegium*, *Origanum majorana*, *Pelargonium graveolens*, *P. roseum*, *Rosmarinus officinalis*, *Salvia officinalis*, *Thuja occidentalis* and *Thymus mastichina* (Kokošková and Pavela 2005), essential oil from *Salvia mukerjeei* (Mohan et al. 2011) and essential oil from *Thymus bleicherianus*, *Thymus algeriensis* and *Thymus zygis* (Zayyad et al. 2014) were active against *E. chrysanthemi*. Recently, Sledz et al (2015) have investigated the effect of caffeine *Dickeya solani* and found 18.3 mM as MIC and 80.0 mM as MBC.

13.6.10 *Pectobacterium carotovorum* (and *Pectobacterium atrosepticum*)

Essential oils of many medicinal plants have been used for evaluation of their antibacterial on *Pectobacterium carotovorum* and *Pectobacterium atrosepticum* by many workers. The antibacterial activity of the aqueous extracts of 20 plants against

Erwinia carotovora subsp. *carotovora* syn. *Pectobacterium carotovorum* subsp. *carotovorum* was studied by (Bhardwaj and Laura 2008). The strongest inhibitory effect was of the leaf extracts of *Camellia sinensis* and bark extracts of *Acacia arabica* and *Acacia catechu.* The inhibitory effect against tested bacteria was also shown by leaf extracts of *Azadirachta indica*, root extracts of *Asparagus racemosus*, seed extracts of *Acacia farnesiana* and fruit extracts of *Aegle marmelos.*

The extract of neem (*Azadirachta indica A. Juss*) leaf and garlic (*Allium sativum* L.) cloves inhibited *Pectobacterium atrosepticum* at a concentration of 10 and 25 % (w/v) (Paradza et al. 2012). Recently, Ikeura and Kobayashi (2015) reported that essential oils of coriander (*Coriandrum sativum* L.) inhibited 40.94 % of *P. carotovorum.*

Acetone extracts of *Olea europaea* L. leaves and methanol extracts of *Salvia officinalis* and *Olea europaea* L. leaves with 400 mg/ml (Zaidi-Yahiaoui et al. 2008), essential oil of *Cymbopogon citratus* (Jeong et al. 2009) and the extracts of *Artemisia kermanensis*, *Lavandula officinalis*, *Rosmarinus officinalis* and *Eucalyptus caesia* are reported to have inhibitory effect on *Pectobacterium carotovorum* (Mehrorosh et al. 2014). Thyme oil (Rojas Fernández et al. 2014; Alamshahi and Nezhad 2015) was active against *P. carotovorum.* In another study, the hydro-distilled essential oils of *Coriandrum sativum*, *Thymus vulgaris*, *Cuminum cyminum*, *Rosmarinus officinalis* and *Eucalyptus globulus* showed a weak antibacterial activity on *P. carotovorum* at high concentrations (Nezhad et al. 2012).

The inhibitory effect of caffeine against *P. atrosepticum* and *P. carotovorum* was investigated by Sledz et al. (2015). They found that caffeine was able to significantly inhibit bacterial growth. The MIC and MBC values for growth inhibition were 8.7 and 100.0 mM for *P. atrosepticum* and 9.0 and 100.0 mM for *P. carotovorum.*

13.7 Conclusion

Different studies have demonstrated the effectiveness of essential oils or their active compounds on a range of plant pathogenic bacteria and fungi responsible for pre- and postharvest diseases. Also, because of the increasing demand for effective, safe, natural products, quantitative data on plant oils and extracts and the resurgence of interest in natural control of plant infectious bacterial and fungal pathogens are required and could lead to a new antimicrobial agent, which could support the use of the plant to treat various infective diseases. Nonetheless, plant essential oils have several important benefits; they are superior for disease control, effective at very low dosages of even less than one gallon per acre, excellent in spreading and sticking properties on leaf surfaces and at low cost and have little or no toxicity to man and animals and have much lower level of risk to the environment than with current synthetic pesticides.

References

Adebayo O, Dang T, Bélanger A et al (2013) Antifungal studies of selected essential oils and a commercial formulation against *Botrytis Cinerea*. J Food Res 2:217–226

Aguiar RWS, Ootani MA, Ascencio SD et al (2014) Fumigant antifungal activity of *Corymbia citriodora* and *Cymbopogon nardus* essential oils and citronellal against three fungal species. Sci World J ID492138:1–8

Alamshahi L, Nezhad H (2015) Effect of essential oils of five medicinal plants on two microbial diseases of potato and tomato under laboratory and field condition. IJAIR 4(2):390–395

Al-Askar AAA (2012) In vitro antifungal activity of three Saudi plant extracts against some phytopathogenic fungi. J Plant Protect Res 52(4):458–462

Amini M, Safaie N, Salmani MJ et al (2012) Antifungal activity of three medicinal plant essential oils against some phytopathogenic fungi. Trakia J Sci 10(1):1–8

Aminifard MH, Mohammadi S (2013) Essential oils to control *Botrytis cinerea* in vitro and in vivo on plum fruits. J Sci Food Agric 93:348–353

Amir YS, Maryam N, Saeed M et al (2013) The new formulation of nanoencapsulated essential oil as green antifungal against *Fusarium graminearum*. Jundishapur J Microbiol Spec Edn, p 9–9. 2/3p

Arango WM, Ruíz JMA, Jaramillo CAP (2011) Fungicidal activity of *Eucalyptus tereticornis* essential oil on the pathogenic fungus *Fusarium oxysporum*. Rev Cubana Farm 45(2):264–274

Arora DK (2003) Fungal biotechnology in agricultural, food, and environmental applications. CRC Press, New York, p 294

Arrebola E, Sivakumar D, Bacigalupo R et al (2010) Combined application of antagonist Bacillus amyloliquefaciens and essential oils for the control of peach postharvest diseases. Crop Prot 29(4):369–377

Awuah RT (1989) Fungitoxic effects of extracts from some West African plants. Ann Appl Biol 115:451–453

Bagamboula CF, Uyttendaele M, Debevere J (2004) Inhibitory effect of thyme and basil essential oils, carvacrol, thymol, estragole, linalool and p-cymene towards *Shigella sonnei* and *S. flexneri*. Food Microbiol 21:33–42

Bajpai VK, Cho MJ, Kang SC (2010a) Control of plant pathogenic bacteria of *Xanthomonas* spp. by the essential oil and extracts of *Metasequoia glyptostroboides* Miki ex Hu *In vitro* and *In vivo*. J Phytopathol 158:479–486

Bajpai VK, Dung NT, Suh HJ et al (2010b) Antibacterial activity of essential oil and extracts of *Cleistocalyx operculatus* buds against the bacteria of *Xanthomonas* spp. J Am Oil Chem Soc 87:1341–1349

Bajpai VK, Kang S, Xu H et al (2011) Potential roles of essential oils on controlling plant pathogenic bacteria *Xanthomonas* species: a review. Plant Pathol J 27:207–224

Bakkali F, Averbeck S, Averbeck D et al (2008) Biological effects of essential oils – a review. Food Chem Toxicol 46:446–475

Barrera-Necha LL, Garduno-Pizana C, Garcia-Barrera LJ (2009) In vitro antifungal activity of essential oils and their compounds on mycelial growth of *Fusarium oxysporum* f. sp. gladioli (Massey) Snyder and Hansen. Plant Pathol J 8(1):17–21

Basim E, Basim H (2003) Antibacterial activity of *Rosa damascena* essential oil. Fitoterapia 74(4):394–396

Basim H, Yegen O, Zeller W (2000) Antibacterial effect of essential oil of *Thymbra spicata* L. var. *spicata* on some plant pathogenic bacteria. J Plant Dis Protect 107(3):279–284

Bayala B, Bassole IH, Scifo R et al (2014) Anticancer activity of essential oils and their chemical components – a review. Am J Cancer Res 4(6):591–607

Benmeddour T, Laouar H, Benabdi AA et al (2015) Evaluation of antibacterial and antifungal activity of extracts from three species of the genus *Allium*: *A. cepa*, *fistulosum* and *sativum* grown in agricultural area of Doussen (wilaya of Biskra). Courrier du Savoir 19:09–14

Bertoli A, Cirak C, da Silva JAT (2011) *Hypericum* species as sources of valuable essential oils. Med Arom Plant Sci Biotechnology 5 (Spec Issue 1): 29–47

Bhardwaj SK, Laura JS (2008) Antibacterial activity of some plant extracts against pathogenic bacteria *Erwinia carotovora* subsp. *carotovora*. Potato J 35(1–2):72–77

Boyle W (1955) Spices and essential oils as preservatives. Am Perfum Essent Oil Rev 66:25–28

Božović M, Pirolli A, Ragno R (2015) *Mentha suaveolens* Ehrh. (*Lamiaceae*) essential oil and its main constituent piperitenone oxide: biological activities and chemistry. Molecules 20(5):8605–8633

Burt S (2004) Essential oils: their antibacterial properties and potential applications in foods – a review. Int J Food Microbiol 94:223–253

Cantore PLO, Iacobellis NS, Marco A et al (2004) Antibacterial activity of *Coriandrum sativum* L. and *Foeniculum vulgare* Miller Var. *vulgare* (Miller) essential oils. J Agric Food Chem 52:7862–7866

Cantore PL, Shanmugaiah V, Iacobellis SN (2009) Antibacterial activity of essential oil components and their potential use in seed disinfection. J Agric Food Chem 57:9454–9461

Cardiet G, Fuzeau B, Barreau C et al (2012) Contact and fumigant toxicity of some essential oil constituents against a grain insect pest *Sitophilus oryzae* and two fungi, *Aspergillus westerdijkiae* and *Fusarium graminearum*. J Pest Sci 85(3):351–358

Chiriac IP, Ulea E (2012) In vitro susceptibility of *Erwinia amylovora* (burrill) winslow *et al.* strain isolated from pear to several plant extracts and different pesticides. Cercetări Agronomice în Moldova XLV 1(149):65–71

Cho JY, Choi GJ, Son SW et al (2007) Isolation and antifungal activity of lignans from *Myristica fragrans* against various plant pathogenic fungi. Pest Manag Sci 63(9):935–940

Chudasama KS, Thaker VS (2012) Screening of potential antimicrobial compounds against *Xanthomonas campestris* from 100 essential oils of aromatic plants used in India: an eco-friendly approach. Archiv Phytopathol Plant Protect 45(7):783–795

Ciocan ID, Bra II (2007) Plant products as antimicrobial agents. Genet Biol Mol 8:151–156

Cowan MM (1999) Plant products as antimicrobial agents. Clin Microbiol Rev 12:564–582

Dadasoglu F, Aydin T, Kotan R et al (2011) Antibacterial activities of extracts and essential oils of three *Origanum* species against plant pathogenic bacteria and their potential use as seed disinfectants. J Plant Pathol 93(2):271–282

Daferera DJ, Ziogas BN, Polissiou MG (2003) The effectiveness of plant essential oils in the growth of *Botrytis cinerea*, *Fusarium* sp. and *Clavibacter michiganensis* subsp. *michiganensis*. Crop Prot 22:39–44

Davidson PM, Parish ME (1989) Methods for testing the efficacy of food antimicrobials. Food Technol 43:148–155

Davidson PM, Post SL, Brannen AL et al (1983) Naturally occurring and miscellaneous food antimicrobials. In: Branen AL, Davidson PM (eds) Antimicrobials in foods. Mercel Dekker Inc, New York, pp 371–419

Dean R, Van Kan JL, Pretorius ZA et al (2012) The top 10 fungal pathogens in molecular plant pathology. Mol Plant Pathol 13(4):414–430

Deweer C, Yaguiyan A, Muchembled J et al (2013) *In vitro* evaluation of dill seed essential oil antifungal activities to control *Zymoseptoria tritici*. Commun Agric Appl Biol Sci 78(3):489–495

Dhlamini Z, Spillane C, Moss JP et al (2005) FAO status of research and application of crop biotechnologies in developing countries. Preliminary assessment. Food and Agriculture Organization of the United Nations, Rome, p 25

Digvijay BS, Kumar P (2014) Antibacterial activity of *Juniperus communis* L. and *Vitex negundo* against *Xanthomonas axonopodis* pv *punicae in vitro*. World J Pharm Res 3(10):1489–1500

Djilani A, Dicko A (2012) The therapeutic benefits of essential oils. In Bouayed J (ed) Nutrition, well-being and health. InTech, pp 155–178

Doğu DM, Zobar D (2014) Effects of some plant essential oils against *Botrytis cinerea* and *Tetranychus urticae* on Grapevine. Turkish J Agric Nat Sci Spec Issue 1:1268–1273

Dorman HJD, Deans SG (2000) Antimicrobial agents from plants: antibacterial activity of plant volatile oils. J Appl Microbiol 88:308–316

Dreger M, Wielgus K (2013) Application of essential oils as natural cosmetic preservatives. Herba Pol 59(4):142–156

Edris AE (2007) Pharmaceutical and therapeutic potentials of essential oils and their individual volatile constituents: a review. Phytother Res 21(4):308–323

Elansary HO, Salem MZM, Ashmawy NA et al (2012) Chemical composition, antibacterial and antioxidant activities of leaves essential oils from *Syzygium cumini* L., *Cupressus sempervirens* L. and *Lantana camara* L. from Egypt. J Agric Sci 4(10):144–152

El-Fiki AII, Fahmy ZM, Mohamed FG et al (2003) Studies on the control of corn common smut disease [Ustilago maydis]. Egypt J Appl Sci 18(11B):433–453

El-Mohamedy RSR, Aboelfetoh MA (2014) Evaluation of antifungal activity of *Moringa oleifera* extracts as natural fungicide against some plant pathogenic fungi in-vitro. J Agric Technol 10(4):963–982

El-Zemity SR, Radwan MA, Shady AEMM et al (2008) Antibacterial screening of some essential oils, monoterpenoids and novel N-methyl carbamates based on monoterpenoids against *Agrobacterium tumefaciens* and *Erwinia carotovora*. Arch Phytopathol Plant Protect 41(6):451–461

Freiesleben SH, Jäger AK (2014) Correlation between plant secondary metabolites and their antifungal mechanisms–a review. Med Aromat Plants 3:2, 6 pages

Gill AO, Delaquis P, Russo P et al (2002) Evaluation of antilisterial action of cilantro oil on vacuum packed ham. Int J Food Microbiol 3:83–92

Gömöri C, Farkas EN, Kerekes EB et al (2013) Evaluation of five essential oils for the control of food spoilage and mycotoxin producing fungi. Acta Biologica Szegediensis 57(2):113–116

Gormez A, Bozari S, Yanmis D et al (2015) Chemical composition and antibacterial activity of essential oils of two species of *Lamiaceae* against phytopathogenic bacteria. Pol J Microbiol 64(2):121–127

Grahovac M, Hrustić J, Tanović B et al (2012) In vitro effects of essential oils on *Colletotrichum* spp. Agric For 57(11)4: 7–15

Greche H, Hajjaji N, Ismaili-Alaoui M et al (2000) Chemical composition and antifungal properties of the essential oil of *Tanacetum annuum*. J Essent Oil Res 12(1):122–124

Guenther E (1950) The essential oils. van Nostrand Co., Inc., New York

Guzmán-Guzmán A, Navarrete-Maya R, Navarrete-Maya J (2003) In vitro control of *Fusarium oxysporum* f sp. *phaseoli* with vegetal essential oils. XLVI Report Bean Improv Cooper 46:221–222

Hafez YM (2008) Effectiveness of the antifungal black seed oil against powdery mildews of cucumber (*Podosphaera xanthii*) and barley (*Blumeria graminis* f.sp. *hordei*). Acta Biologica Szegediensis 52(1):17–25

Hajek AE (2004) Natural enemies: an introduction to biological control. Cambridge University Press, Cambridge, UK, 261 p

Hammami I, Kamoun N, Rebai A (2011) Biocontrol of *Botrytis cinerea* with essential oil and methanol extract of *Viola odorata* L. flowers. Archiv Appl Sci Res 3(5):44–51

Hashemi M, Ehsani A, Afshari A et al (2016) Chemical composition and antifungal effect of *Echinophora platyloba* essential oil against *Aspergillus flavus*, *Penicillium expansum* and *Fusarium graminearum*. J Chem Health Risks 6(2):91–97

Haugaard H, Collinge DB, Lyngkjær MF (2002) Mechanisms involved in control of *Blumeria graminis* f. sp. *hordei* in barley treated with mycelial extracts from cultured fungi. Plant Pathol 51:612–620

Horváth A, Kovács B, Nagy G (2013) Application of mint and cinnamon against Fusarium disease of winter wheat. Episteme: Czasopismo Naukowo-Kulturalne 18(3):297–304

Hoseiniyeh FSH, Mirabolfathy M, Rezaie DH et al (2012) Effect of five essential oils on zearalenone production and growth of *Fusarium graminearum*. Pests Plant Dis 80(1):81–91(Persian)

Hosseinzadeh S, Shams-Bakhsh M, Hosseinzadeh E (2013) Effects of sub-bactericidal concentration of plant essential oils on pathogenicity factors of *Ralstonia solanacearum*. Arch Phytopathol Plant Protect 46(6):643–655

Huang Q, Laksman DK (2010) Effect of clove oil on plant pathogenic bacteria and bacterial wilt of tomato and geranium. J Plant Pathol 92:701–707

Huang Y, Zhao J, Zhou L et al (2010) Antifungal activity of the essential oil of *Illicium verum* fruit and its main component trans-anethole. Molecules 15(11):7558–7569

Hubert J, Mabagala RB, Mamiro DP (2015) Efficacy of selected plant extracts against *Pyricularia grisea*, causal agent of rice blast disease. Am J Plant Sci 6:602–611

Hyldgaard M, Mygind T, Rikke LM (2012) Essential oils in food preservation: mode of action, synergies, and interactions with food matrix components. Front Microbiol 3:1–24

Iacobellis NS, Cantore PL, Capasso F et al (2005) Antibacterial activity of *Cuminum cyminum* L. and *Carum carvi* L. essential oils. J Agric Food Chem 53:57–61

Idris FM, Ibrahim AM, Forsido SF (2015) Essential oils to control *Colletotrichum musae in vitro* and *in vivo* on banana fruits. Am-Eur J Agric Environ Sci 15(3):291–302

Ikeura H, Kobayashi F (2015) Antimicrobial and antifungal activity of volatile extracts of 10 herb species against *Glomerella cingulata*. J Agric Sci 7(9):77–84

IŞcan G, Kirimer N, Kürkcüoğlu M et al (2002) Antimicrobial screening of Mentha piperita essential oils. J Agric Food Chem 50(14):3943–3946

Islam MT, AhnSY JSM et al (2013) Isolation of antibacterial compounds from hairy vetch (*Vicia villosa*) against grapevine crown gall pathogen. Hortic Environ Biotechnol 54:338–345

Ismail MM, Essam TM, Mohamed AF et al (2012) Screening for the antimicrobial activities of alcoholic and aqueous extracts of some common spices in Egypt. Int J Microbiol Res 3(3):200–207

Istianto M, Emilda D (2011) Preliminary study of the activity of some essential oils against *Fusarium oxysporum* f. sp. *cubense*. J Fruit Ornam Plant Res 19(2):111–121

Jamal-U-Ddin H, Mubeen LA, Mumtaz AP et al (2012) In vitro evaluation of fungicides, plant extracts and bio control agents against rice blast pathogen *Magnaporthe oryzae* Couch. Pak J Bot 44(5):1775–1778

Jaspers MV, Walter M, Frampton CM et al (2001) Control of *Botrytis cinerea* in grape using thyme oil. Australas Plant Pathol 30(1):21–25

Jeong MR, Park PB, Kim DH et al (2009) Essential oil prepared from *Cymbopogon citratus* exerted an antimicrobial activity against plant pathogenic and medical microorganisms. Mycobiology 37:48–52

Kaefer CM, Milner JA (2008) The role of herbs and spices in cancer prevention. J Nutr Biochem 19(6):347–361

Kagade S, Marimuthu T, Thayumanavan B et al (2004) Antimicrobial activity and induction of resistance in rice by leaf extract of *Datura metel* against *Rhizoctonia solani* and *Xanthomonas oryzae* pv. *oryzae*. Physiol Mol Plant Pathol 65:91–100

Kalagatur NK, Mudili V, Siddaiah C et al (2015) Antagonistic activity of Ocimum sanctum L. essential oil on growth and zearalenone production by Fusarium graminearum in maize grains. Front Microbiol 6:892

Kaliora AC, Kountouri AM (2012) Chemopreventive activity of Mediterranean medicinal plants. In: Georgakilas A (ed) Cancer prevention-from mechanisms to translational benefits. Tech Publishers, pp 261–283

Karaman I, Sahin F, Güllüce M et al (2003) Antimicrobial activity of aqueous and methanol extracts of *Juniperus oxycedrus* L. J Ethnopharmacol 85:231–235

Karami-Osboo R, Khodaverdi M, Ali-Akbari F (2010) Antibacterial effect of effective compounds of *Satureja hortensis* and *Thymus vulgaris* essential oils against *Erwinia amylovora*. J Agric Sci Tech 12:35–45

Kizil S, Uyar F, Sağir A (2005) Antibacterial activities of some essential oils against plant pathogens. Asian J Plant Sci 4(3):225–228

Kokošková B, Pavela R (2005) Effectivity of essential oils against pectinolytic *Erwinia chrysanthemi* and *Pseudomonas marginalis*. In: Proceedings of the 1st international symposium on biological control of bacterial plant diseases, Seeheim, Darmstadt, Germany, 23–26 October 2005

Kokoškova B, Pavela R (2007) Effectiveness of plant essential oils on the growth of *Erwinia amylovora*, the causal agent of Fire Blight Disease. Pest Technol 1(1):76–80

Kokoskova B, Pouvova D, Pavela R (2011) Effectiveness of plant essential oils against *Erwinia amylovora*, Pseudomonas syringae pv. syringae and associated saprophytic bacteria on/in host plants. Journal of Plant Pathology 93(1):133–139

Kordali S, Kotan R, Mavi A et al (2005) Determination of the chemical composition and antioxidant activity of the essential oil of *Artemisia dracunculus* and of the antifungal and antibacterial activities of Turkish *Artemisia absinthium, A. dracunculus, Artemisia santonicum*, and *Artemisia spicigera* essential oils. J Agric Food Chem 53(24):9452–9458

Kotan R, Dadasoglu F, Kordali S et al (2007) Antibacterial activity of essential oils extracted from some medicinal plants, carvacrol and thymol on Xanthomonas axonopodis pv. vesicatoria (Doidge) Dye causes bacterial spot disease on pepper and tomato. J Agric Technol 3:299–306

Kotan R, Cakir A, Dadasoglu F et al (2010) Antibacterial activities of essential oils and extracts of Turkish *Achillea, Satureja* and *Thymus* species against plant pathogenic bacteria. J Sci Food Agric 90:145–160

Kotan R, Cakir A, Ozer H et al (2014) Antibacterial effects of *Origanum onites* against phytopathogenic bacteria: possible use of the extracts from protection of disease caused by some phytopathogenic bacteria. Sci Hortic 172:210–220

Kowalska B, Smolinska U (2008) The effect of selected plant materials and extract on the development of bacterial disease on onion. Vegetable Crops Res Bull 68:33–45

La Torre A, Caradonia F, Matere A et al (2016) Using plant essential oils to control Fusarium wilt in tomato plants. Eur J Plant Pathol 144(3):487–496

Lahooji A, Mirabolfathy M, Karami Osboo R (2010) Effect of *Zataria multiflora* and *Satureja hortensis* essential oils, thymol and carvacrol on growth of *Fusarium graminearum* isolates and deoxynivalenol production. Iran J Plant Path 46(1):11–13

Lawrence GJ, Dodds PN, Ellis JG (2007) Pathogen profile: rust of flax and linseed caused by *Melampsora lini*. Mol Plant Pathol 8:349–364

Li CM, Yu JP (2015) Chemical composition, antimicrobial activity and mechanism of action of essential oil from the leaves of Macleaya Cordata (Willd). R Br J Food Safety 35:227–236

Lima RA, Santos MRA, Fernandes CF et al (2010) Antifungal activity of *Hedychium coronarium* J. essential oil against *Fusarium oxysporum* S. and *Thanatephorus cucumeris* F. V Simposio Iberoamericano de plantas Medicinais, 18 a 20 de Outubro, Itajaí-SC, Brazil

Maddox CE, Laur LM, Tian L (2010) Antibacterial activity of phenolic compounds against the phytopathogen *Xylella fastidiosa*. Curr Microbiol 60(1):53–58

Maiti D, Kole CR, Sen C (1985) Antimicrobial efficacy of some essential oils. J Plant Dis Protect 92(1):64–68

Manganyi M (2013) Antimicrobial activity of essential oils against *Fusarium oxysporum* isolates and their biofilms. Magister of Technology in Pharmaceutical Sciences. Tshwane University of Technology

Manganyi MC, Regnier T, Olivier EI (2015) Antimicrobial activities of selected essential oils against *Fusarium oxysporum* isolates and their biofilms. S Afr J Bot 99:115–121

Mansfield J, Genin S, Magori S et al (2012) Top 10 plant pathogenic bacteria in molecular plant pathology. Mol Plant Pathol 13:614–629

Marín S, Velluti A, Ramos AJ et al (2004) Effect of essential oils on zearalenone and deoxynivalenol production by *Fusarium graminearum* in non-sterilized maize grain. Food Microbiol 21:313–318

Măruţescu L, Saviuc C, Oprea E et al (2009) In vitro susceptibility of Erwinia amylovora (Burrill) Winslow et al. to Citrus maxima essential oil. Roum Arch Microbiol Immunol 68(4):223–227

Massuco JE, Abrao F, Alves JA et al (2013) Evaluation of the antibacterial action of essential oils against *Xylella fastidiosa*. In: 9th international congress of pharmaceutical sciences, 20–23 November 2013, Ribeirão Preto, Brazil.

Matusinsky P, Zouhar M, Pavela R et al (2015) Antifungal effect of five essential oils against important pathogenic fungi of cereals. Ind Crop Prod 67:208–215
Mehani M, Salhi N, Valeria T et al (2014) Antifungal effect of essential oil of *Eucalyptus camaldulensis* plant on *Fusarium graminearum* and *Fusarium sporotrichioides*. Int J Curr Res 6(12):10795–10797
Mehr ZS, Sanadgol N, Ghasemi LV (2012) Effects of essential oil extracted from *Citrullus colocynthis* (CCT) seeds on growth of phytopathogenic bacteria. Afr J Microbiol Res 6(36):6572–6575
Mehrorosh H, Gavanji S, Larki B et al (2014) Essential oil composition and antimicrobial screening of some Iranian herbal plants on *Pectobacterium carotovorum*. Global NEST J X(X):XX–XX
Miguel MG (2010) Antioxidant and anti-inflammatory activities of essential oils: a short review. Molecules 15:9252–9287
Mihajilov-Krstev T, Radnović D, Kitić D (2010) Antimicrobial activity of *Satureja* L. essential oils against phytopathogenic bacteria *Erwinia amylovora*. Biologica Nyssana 1(1–2):95–98
Mikiciński A, Sobiczewski P, Berczyński S (2012) Efficacy of fungicides and essential oils against bacterial diseases of fruit trees. J Plant Protect Res 52(4):468–471
Minija J, Thoppil JE (2001) Volatile oil constitution and microbicidal activities of essential oils of *Coriandrum sativum* L. J Nat Remed 1(2):147–150
Misra D, Misra M, Tewari SN (1997) Toxic effect of volatiles from Callistemon lanceolatus on six fungal pathogens of rice. Indian Phytopathol 50(1):103–105
Moghaddam M, Alymanesh MR, Mehdizadeh L et al (2014) Chemical composition and antibacterial activity of essential oil of *Ocimum ciliatum*, as a new source of methyl chavicol, against ten phytopathogens. Ind Crop Prod 59:144–148
Mohammadi S, Aroiee H, Aminifard MH et al (2014) Effect of fungicidal essential oils against *Botrytis cinerea* and *Rhizopus stolonifer* rot fungus in vitro conditions. Arch Phytopathol Plant Protect 47(13):1603–1610
Mohan L, Negi A, Melkani AB et al (2011) Chemical composition and antibacterial activity of essential oil from Salvia mukerjeei. Nat Prod Commun 6(12):1949–1952
Monteiro FP, Ferreira LC, Silva JL et al (2013) Influence of plant extracts and essential oils against panama disease (*Fusarium oxysporum* f. sp. *cubense*) in banana seedlings. J Agric Sci 4:63–74
Morcia C, Mehani M, Salhi N et al (2015) On the role of natural compounds in mycotoxigenic fungi control. In: Méndez-Vilas A (ed) The battle against microbial pathogens: Basic science, technological advances and educational programs, 193–198
Mostafa Z, Soheil P, Mahdi J et al (2013) Antifungal effects of *asafoetida* seed essential oil on in vitro growth of five species of plant pathogenic fungi. Int Res J Appl Basic Sci 4(5):1159–1162
Mourey A, Canillac N (2002) Anti-Listeria monocytogenes activity of essential oils components of conifers. Food Control 13:289–292
Moursy MA, Mansour IM, Bekheet FM (2001) Effect of plant oils on *Ustilago maydis* infecting maize and teosinte. J Agric Sci Mansoura Univ 26(9):3559–3568
Nazzaro F, Fratianni F, De Martino L et al (2013) Effect of essential oils on pathogenic bacteria. Pharmaceuticals 6(12):1451–1474
Nezhad MH, Alam L, Panjehkeh N (2012) Biocontrol efficiency of medicinal plants against *Pectobacterium Carotovorum*, *Ralstonia Solanacearum* and *Escherichia Coli*. Open Conf Proc J 3(Suppl 1-M8):46–51
Nguefack J, Somda I, Mortensen CN et al (2005) Evaluation of five essential oils from aromatic plants of Cameroon for controlling seed-borne bacteria of rice (*Oryza sativa* L.). Seed Sci Technol 33:397–407
Noudeh GD, Sharififar F, Noodeh AD et al (2010) Antitumor and antibacterial activity of four fractions from *Heracleum persicum* Desf. and *Cinnamomum zeylanicum* Blume. J Med Plant Res 4(21):2176–2180

Oboo H, Muia AW, Kinyua ZM (2014a) Effect of essential oil plant extracts on in vitro growth of *Ralstonia solanacearum*. Egerton J Sci Technol 14:141–160

Oboo H, Muia AW, Kinyua ZM (2014b) Effect of selected essential oil plants on bacterial wilt disease development in potatoes. J Appl Biosci 78:6666–6674

Obradovic A, Jones JB, Balogh B et al (2008) Integrated management of tomato bacterial spot. In: Ciancio A, Mukerji KG (eds) Integrated management of diseases caused by fungi, phytoplasma and bacteria. Springer, Dordrecht, pp 211–223

Olufolaji DB, Adeosun BO, Onasanya RO (2015) In vitro investigation on antifungal activity of some plant extracts against *Pyricularia oryzae*. Nig J Biotechnol 29:38–43

Ozturk S, Ercisli S (2007) Antibacterial activity and chemical constitutions of *Ziziphora clinopodioides*. Food Control 18:535–540

Pagnussatt FA, Bretanha CC, Kupski L et al (2013) Promising antifungal effect of rice (*Oryza sativa* L.), oat (*Avena sativa* L.) and wheat (*Triticum aestivum* L.) extracts. J Appl Biotechnol 1(1):37–44

Paradza VM, Icishahayo D, Ngadze E (2012) Efficacy of botanical extracts from garlic and neem on controlling potato soft rot pathogens. UNISWA J Agric 16:1–10

Paret ML, Cabos R, Kratky BA et al (2010) Effect of plant essential oils on *Ralstonia solanacearum* race 4 and bacterial wilt of edible ginger. Plant Dis 94:521–527

Pawar BT (2013) Antifungal activity of some fruit extracts against seedborne pathogenic fungi. Adv Biores 4(3):95–97

Plotto A, Roberts DD, Roberts RG (2003) Evaluation of plant essential oils as natural postharvest disease control of tomato (*Lycopersicon esculentum*). Acta Horticulturae (ISHS) 628:737–745

Poswal MAT, Witbooi W (2012) Antibacterial properties of essential oils on *Pseudomonas Syringae* pv. *syringae* and *Pseudomonas solanacearum*. In: Rudolph K, Burr TJ, Mansfield J, Stead DE, Vivian A, Kietzell JV (eds) *Pseudomonas Syringae* Pathovars and related pathogens volume 9 of the series Developments in Plant Pathology. Springer, pp 606–610

Pradhanang PM, Momol MT, Olson SM et al (2003) Effects of plant essential oils on *Ralstonia solanacearum* population density and bacterial wilt incidence in tomato. Plant Dis 87:423–427

Prieto JA, Patiño OJ, Delgado WA et al (2011) Chemical composition, insecticidal, and antifungal activities of fruit essential oils of three Colombian *Zanthoxylum* species. Chil J Agric Res 71(1):73–82

Rahman A, Islam R, Al-Reza SM et al (2014) In vitro control of plant pathogenic *Xanthomonas* spp. using *Poncirus trifoliata* rafin. EXCLI J 13:1104–1110

Raji P, Sumiya KV, Dhanya S et al (2016) Inhibitory effect of plant extracts and plant oils on *Xanthomonas oryzae* pv *oryzae*, the bacterial blight pathogen of rice. Int J Appl Nat Sci 52:71–76

Reis KB, Côrtes MVCB, Martins FS et al (2015) Characterization of rue extract and its potential for controlling rice blast. Pesq Agropec Bras Brasília 50(12):1121–1130

Ribeiro AB, Abdelnur PV, Garcia CF et al (2008) Chemical characterization of *Citrus sinensis* grafted on *C. limonia* and the effect of some isolated compounds on the growth of *Xylella fastidiosa*. J Agric Food Chem 56(17):7815–7822

Rojas Fernández MM, López MC, Pérez YS et al (2014) Antibacterial activity of essential oils against *Pectobacterium carotovorum* subsp. *carotovorum*. Rev Protección Veg 29(3):197–203

Samie A, Nefefe T (2012) Antifungal activities of essential oils from Southern African medicinal plants against five *Fusarium* species. J Med Plants Res 6(3):465–478

Sampietro DA, Lizarraga EF, Ibatayev ZA et al (2015) Chemical composition and antimicrobial activity of essential oils from *Acantholippia deserticola*, *Artemisia proceriformis*, *Achillea micrantha* and *Libanotis buchtormensis* against phytopathogenic bacteria and fungi. Nat Prod Res 24:1–6

Sankaran S, Mishra A, Ehsani R et al (2010) A review of advanced techniques for detecting plant diseases. Comput Electron Agric 72(1):1–13

Sati SC, Kumar P (2015) Assessment of himalayan juniper, Juniperus squamata buch– ham ex d don for phytochemical screening and antimicrobial potential against some infection causing pathogens. World J Pharmaceut Res 4:998–1011

Sati SC, Takuli P, Kumar P et al (2015) Antibacterial activity of three medicinal plants of Kumaun Himalaya against some pathogenic bacteria. Int J Pharma Sci Res 16(11):1361–1368

Scher JM, Speakman JB, Zapp J et al (2004) Bioactivity guided isolation of antifungal compounds from the liverwort *Bazzania trilobata* (L.) S.F. Gray. Phytochemistry 65(18):2583–2588

Şesan TE, Enache E, Iacomi BM et al (2015) Antifungal activity of some plant extracts against *Botrytis cinerea* Pers. in the blackcurrant crop (*Ribes nigrum* L.). Acta Sci Pol Hortorum Cultus 14(1):29–43

Shabana YM, Abdel-Fattah GM, Ismail AE et al (2008) Control of brown spot pathogen of rice (*Bipolaris oryzae*) using some phenolic antioxidants. Braz J Microbiol 39(3):438–444

Silva CL, Souza EB, Felix KCS et al (2012) Essential oils and plant extracts on the control of soft rot of crispy lettuce. Hortic Bras 30(4):632–638

Singh R, Yadav RS, Javeria S (2015) Management of bacterial leaf blight of Basmati rice caused by *Xanthomonas oryzae* pv. *oryzae* with some available antibiotics and plant products. Int J Innov Appl Res 3(11):1–6

Sledz W, Los E, Paczek A et al (2015) Antibacterial activity of caffeine against plant pathogenic bacteria. Acta Biochim Pol 62(3):605–612

Somaya T, El-Sharkawy HHA (2014) Effect of some plant essential oils against wheat leaf rust caused by *Puccinia triticina* f. sp. *tritici*. Egypt J Biol Pest Control 24:211–216

Sood PAK, Pardeep K (2015) Evaluation of essential oils against *Ralstonia solanacearum* causing bacterial wilt of solanaceous crops. Plant Dis Res 30(1):67–72

Sqalli H, El Ouarti A, Farah A et al (2009) Antibacterial activity of Thymus pallidus Batt. and determination of the chemical composition of its essential oil. Acta Bot Gallica 156(2):303–310

Stefanova M, Rizo SG, Romeu C et al (2005) In vitro bactericidal activity of essential oils from *Hyptis suaveolens* (L.) Poit and *Coleus amboinicus* (lour). In: Proceedings of the 1st international symposium on biological control of bacterial plant diseases, Seeheim/Darmstadt, Germany, 23–26 October, pp 69–71

Sukanya SL, Yamini D, Fathima SK (2011) Eco-friendly management of *Pyricularia oryzae* – the causal agent of blast of paddy. Curr Bot 2(8):46–49

Sun Og L, Park IK, Choi GJ et al (2007) Fumigant activity of essential oils and components of *Illicium verum* and *Schizonepeta tenuifolia* against *Botrytis cinerea* and *Colletotrichum gloeosporioides*. J Microbiol Biotechnol 17(9):1568–1572

Tabanca N, Demirci B, Crockett SL et al (2007) Chemical composition and antifungal activity of *Arnica longifolia*, *Aster hesperius*, and *Chrysothamnus nauseosus* essential oils. J Agric Food Chem 55:8430–8435

Tajane V, Janwe NJ (2014) Ayurvedic plants disease identification using CBIR. IJIRCCE 2(6):4794–4801

Tajkarimi MM, Ibrahim SA, Cliver DO (2010) Antimicrobial herb and spice compounds in food. Food Control 21:1199–1218

Terzi V, Morcia C, Faccioli P et al (2007) *In vitro* antifungal activity of the tea tree (*Melaleuca alternifolia*) essential oil and its major components against plant pathogens. Lett Appl Microbiol 44:613–618

Thakare AR, Wankhade SG, Somani RB et al (2003) Growth inhibition in *Rhizoctonia bataticola* and *Xanthomonas axonopodis* pv. *malvacearum* by herbal oils. J Spices Aromat Crops 12(1):83–85

Thobunluepop P, Jatisatienr C, Pawelzik E et al (2009) In vitro screening of the antifungal activity of plant extracts as fungicides against rice seed borne fungi. Acta Hortic 837:223–228

Tian J, Ban B, Zeng H et al (2011) Chemical composition and antifungal activity of essential oil from *Cicuta virosa* L. var. *latisecta* Celak. Int J Food Microbiol 145:464–470

Tomescu A, Sumalan RM, Pop G et al (2015) Chemical composition and protective antifungal activity of *Mentha Piperita L.* and *Salvia Officinalis L.* essential oils against *Fusarium Graminearum Spp*. Rev Chim (Bucharest) 66(7):1027–1030

Tzortzakis NG, Economakis CD (2007) Antifungal activity of lemongrass (*Cymbopogon citrates* L.) essential oil against key postharvest pathogen. Innov Food Sci Emerg Technol 8(2):253–258

Vasinauskiene M, Radusiene J, Zitikaite I et al (2006) Antibacterial activities of essential oils from aromatic and medicinal plants against growth of phytopathogenic bacteria. Agron Res 4:437–440

Věchet L, Šerá B (2015) Effectiveness of both synthetic compounds and biological extracts against powdery mildew (*Blumeria graminis* f. sp. *tritici*) on winter wheat. Agrociencia 9(1):77–85

Velluti A, Sanchis V, Ramos AJ et al (2004) Impact of essential oils on growth rate, zearalenone and deoxynivalenol production by *Fusarium graminearum* under different temperature and water activity conditions in maize grain. J Appl Microbiol 96(4):716–724

Villa TG, Veiga-Crespo P (2013) Antimicrobial compounds: current strategies and new alternatives. Springer, Heidelberg, 61 p

Vitoratos A, Bilalis D, Karkanis A et al (2013) Antifungal activity of plant essential oils against *Botrytis cinerea*, *Penicillium italicum* and *Penicillium digitatum*. Not Bot Hortic Agrobo 41:86–92

Wagura AG, Kimenju JW, Gichimu BM (2011) Comparative antibacterial effects of raw extracts and essential oils of *Ocimum gratissimum* L. against *Ralstonia solanacearum* (Smith). Int J Plant Pathol 2(3):144–152

Wang J, Liu H, Zhao J et al (2010a) Antimicrobial and antioxidant activities of the root bark essential oil of *Periploca sepium* and its main component 2-Hydroxy-4-methoxybenzaldehyde. Molecules 15:5807–5817

Wang J, Zhao J, Liu H et al (2010b) Chemical analysis and biological activity of the essential oils of two valerianaceous species from China: *Nardostachys chinensis* and *Valeriana officinalis*. Molecules 15(9):6411–6422

Yadeta KA, J Thomma BP (2013) The xylem as battleground for plant hosts and vascular wilt pathogens. Front Plant Sci 4:97

Zaidi-Yahiaoui R, Zaidi F, Bechar S (2008) Effect of Olea europaea L and Salvia officinalis leaves phenolics extracts on potato tubers soft rot disease. Planta Med 74:PE9.r. doi:10.1055/s-0028-1084708

Zaika LL (1988) Spices and herbs: their antimicrobial activity and its determination. J Food Saf 9:97–118

Zayyad N, Farah A, Bahhou J (2014) Chemical analysis and antibacterial activity of essential oils from three species of Thymus: *Thymus zygis, T. algeriensis* and *T. bleicherianus*. Bulletin de la Société Royale des Sciences de Liège 83:118–132

Zengin H, Baysal AH (2014) Antibacterial and antioxidant activity of essential oil terpenes against pathogenic and spoilage-forming bacteria and cell structure-activity relationships evaluated by SEM microscopy. Molecules 19:17773–17798

Zhao J, Shan T, Huang Y et al (2009) Chemical composition and in vitro antimicrobial activity of the volatile oils from *Gliomastix murorum* and *Pichia guilliermondii*, two endophytic fungi in *Paris polyphylla* var. *yunnanensis*. Nat Prod Commun 4(11):1491–1496

Halophilic Bacteria: Potential Bioinoculants for Sustainable Agriculture and Environment Management Under Salt Stress

14

Anjney Sharma, Anukool Vaishnav, Hena Jamali, Anchal Kumar Srivastava, Anil Kumar Saxena, and Alok Kumar Srivastava

Abstract

Salinity is one of the most critical environmental constraints which cause soil degradation and hampering agricultural production throughout the world. In the present time, a total 831 million hectares of land is affected by salinity. The salinity affects the processes in plant life from its germination to maturation stage. Regulation of phytohormones, root/shoot development, nutrient uptake, and photosynthesis are severely affected by salt stress and ultimately reduce agricultural productions. The loss of agriculture production due to salinization is one of the major constraints to feed to the growing population. High salt levels in the soil limit its agroecological potential and represent a considerable ecological and socioeconomic threat to sustainable development. In this context, the use of halophilic bacteria has been gained a great interest in eco-friendly and sustainable agriculture approach with emphasis on plant growth promotion in salt stress. This chapter paid attention to the use of halophilic bacteria in agriculture system toward producing salt stress-tolerant crops and an understanding the mechanisms of plant and halophilic bacterial interaction. Halophilic bacteria help plants to cope with salinity by supporting them in the restoration of essential activities such as nutrient uptake efficiency, ROS scavenging, and phytohormone production. The second part of this chapter describes different enzymatic potentials of halophilic bacteria and their uses in food processing, industrial bioconversions,

A. Sharma • H. Jamali • A.K. Srivastava • A.K. Saxena • A.K. Srivastava (✉)
National Bureau of Agriculturally Important Microorganisms,
Mau Nath Bhanjan, Uttar Pradesh 275103, India
e-mail: aloksrivastva@gmail.com

A. Vaishnav
Amity Institute of Microbial Technology (AIMT), Block 'E-3', 4th Floor, Amity University Campus, Sector-125, Gautam Buddha Nagar, 201313 Noida, Uttar Pradesh, India

D.K. Choudhary et al. (eds.), *Plant-Microbe Interaction: An Approach to Sustainable Agriculture*, DOI 10.1007/978-981-10-2854-0_14

and bioremediation. After that, a brief outline of characterization of halophilic bacteria by phenotypic, biochemical, chemotaxonomy, and molecular methods is discussed. The exploitation of halophilic bacteria in agriculture is required for environment and human welfare.

14.1 Introduction

A major challenge for present agriculture is to cope with the increasing demand for food production for constantly rising world population. This increasing demand for food production is paralleled by continuous losses of arable land due to enhanced soil destruction and erosion. Soil salinity is one of the most critical environmental constraints that affect more than 831 Mha of land throughout the world equating to more than 6 % of the world's total land (FAO 2015) and hampering plant growth and development. Salinity not only decreases the agricultural crop production but also affects the associated ecological balance of the area by changing soil physicochemical properties. The negative impacts of soil salinity include low agricultural crop production and low economic returns due to high cost of cultivation, reclamation, and management. High salt levels in the agriculture soil limit its agroecological potential and represent a considerable socioeconomic and ecological threat to sustainable development. Different region-wise distributions (world level) of salt-affected soil are presented in Table 14.1.

Accumulation of salts over long periods of time in arid and semiarid zones is the main factor behind development of the salt-affected land (Bui 2013). Recent report of ICAR-Central Soil Salinity Research Institute (2012) exhibits that India covers 1.7 Mha saline, 3.8 Mha sodic–saline, and 1.2 Mha coastal saline soil, i.e., a total of 6.7 Mha area of the country is saline (Fig. 14.1 and Table 14.2).

Soil salinity is the concentration of dissolved different mineral salts present in the soils and waters, in which the electrical conductivity (EC) of the saturation extract (ECe) in the root zone exceeds >4 dS m^{-1} (40 mM NaCl approximately) at 25 °C and has exchangeable sodium of 15 %. The dissolved mineral salts consist of

Table 14.1 Worldwide distribution of salt-affected soil (in Mha and %)

	Total area	Saline soil		Sodic soil	
Regions	(Mha)	Percent (%)	Area (Mha)	Percent (%)	Area (Mha)
Africa	1899	2.0	39	1.8	34
Asia, Australia, and the Pacific	3107	6.3	195	8.0	249
Europe	2011	0.3	07	3.6	73
Latin America	2039	3.0	61	2.5	51
Near East	1802	5.1	92	0.8	14
North America	1924	0.2	05	0.8	15
Total area	**12,781**	**3.1**	**397**	**3.4**	**434**

Source: FAO Land and Plant Nutrition management Service

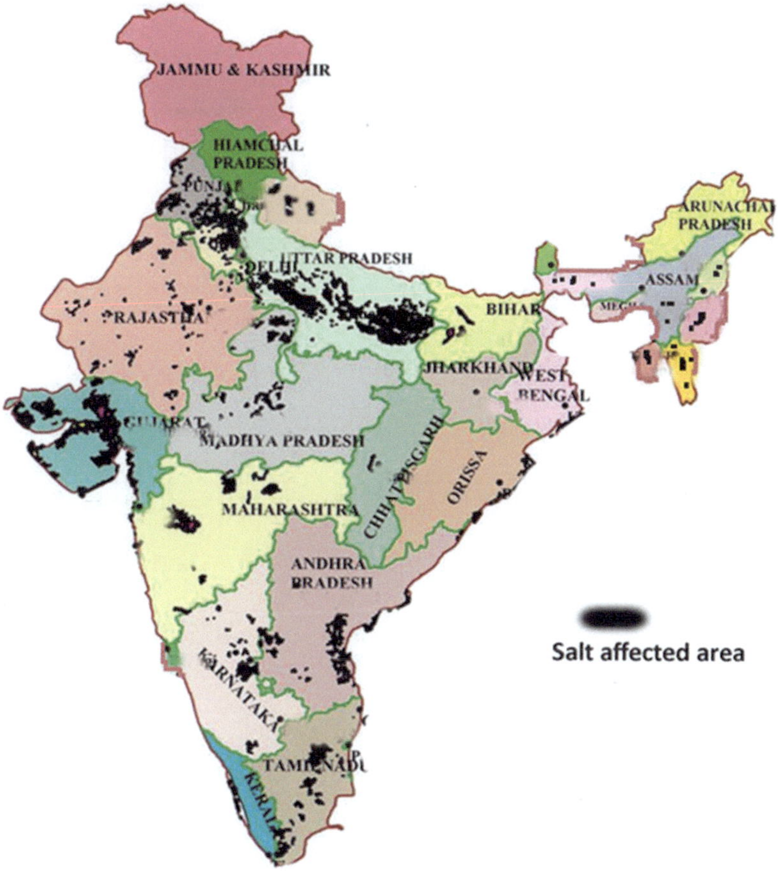

Fig. 14.1 Distribution of saline areas in India (*black color*) (CSSRI 2012)

the electrolytes of major cations (Na^+, Ca^{2+}, Mg^{2+}, and K^+) and major anions (Cl^-, SO_4^{2-}, HCO^{3-},CO_3^{2-}, and NO_3^-). Excess of these salts in the soil solution adversely affects every aspect of plant physiology and metabolisms through osmotic stress in an early growth phase and ionic stress at later growth stage (Rojas-Tapias et al. 2012). The physiological processes affected by salt acclimation include ionic toxicity, osmotic stress, nutrient deficiency, and changes in oxygen-scavenging enzymes (Munns and Tester 2008; Daneshmand et al. 2009). Specific ion effects cause direct toxicity, or, alternatively, the competitive absorption or insolubility of ions may affect the plant's metabolisms by nutritional imbalance. Under salt stress conditions, excessive uptake of Na^+ results in a drastic decline in K^+ uptake. This is because Na^+ ions compete with K^+ for binding sites essential for various cellular functions (Rus et al. 2001). Potassium ion is involved in activation of enzymes, stomatal movements, and protein synthesis (Wang et al. 2013; Ahmad et al. 2014). Higher salinity reduces the K^+/Na^+ ratio, disturbs ionic balance of the cytoplasm, and ultimately affects plant growth and productivity.

Table 14.2 Major salt-affected area in India

State	Saline soils (ha)	Sodic soils (ha)	Coastal saline soil (ha)	Total (ha)
Andhra Pradesh	0	196,609	77,598	274,207
Andaman and Nicobar Islands	0	0	77,000	77,000
Bihar	47,301	105,852	0	153,153
Gujarat	1,218,255	541,430	462,315	2,222,000
Haryana	49,157	183,399	0	232,556
Jammu and Kashmir	0	17,500	0	17,500
Karnataka	1307	148,136	586	150,029
Kerala	0	0	20,000	20,000
Maharashtra	177,093	422,670	6996	606,759
Madhya Pradesh	0	139,720	0	139,720
Orissa	0	0	147,138	147,138
Punjab	0	151,717	0	151,717
Rajasthan	195,571	179,371	0	374,942
Tamil Nadu	0	354,784	13,231	368,015
Uttar Pradesh	21,989	1,346,971	0	1,368,960
West Bengal	0	0	441,272	441,272
Total	**1,710,673**	**3,788,159**	**1,246,136**	**6,744,968**

Source: CSSRI, Karnal, India (2012)

The salinity stress constraint is most acute for agriculture and developing countries like India. A sustainable management practice is a major challenge for successful remediation of salt-degraded areas. In addition, sustainable management practices in agriculture are one of the potentially important factors to meet our future agricultural needs, something that conventional agriculture will not be able to do. Synthetic fertilizers, development of genetically modified and salinity-tolerant varieties, resource management practices, etc. are cost-intensive and give negative effects on human and environment. Recently, the use of halophilic beneficial microorganisms gained interest in eco-friendly and sustainable agriculture (Paul and Lade 2014), which can help crops to cope with salinity stress. Since microorganisms are an integral part of any ecosystem, interest has been renewed in the nature and properties of microbes that play a major role in nutrient cycling in salt ecosystems. The ability of halotolerant bacteria to grow in a broad range of NaCl (0–33 %) makes them one of the suitable organisms for its interaction with plants under saline condition (Oren 2008). Hence, it was hypothesized that the bacteria with PGP activities from naturally saline habitats could help to ameliorate saline stress effect on plants.

The mechanism for growth and survival of halophilic bacteria in environments with elevated osmolarity has been studied well (Grover et al. 2011). The members of *Halobacteraceae* normally posses the compatibility of high salt concentrations within the cell cytoplasm because of extensive structural and enzymatic modifications. Whereas, other prokaryotes have evolved the mechanism to accumulate a specific group of molecules of low molecular mass, termed compatible solutes, as a general mechanism to cope with environments of elevated osmolarity. They balance the external osmolarity with high intracellular concentrations of these osmolytes and protect cellular processes (Sleator and Hill 2001). The role of these microorganisms is well reported in plant growth promotion, nutrient management, and disease control (Dodd and Pérez-Alfocea 2012). The rhizobacteria play a significant role in stress alleviation in the crops grown in saline soils due to their unique properties of tolerance to extremities, their interaction with crop plants, and their potential deployment methods. Several recent studies have demonstrated that these beneficial halotolerant bacteria colonize the rhizosphere/endorhizosphere of plants and promote plant growth and ameliorate the salinity stress in crop plants through various direct and indirect mechanisms such as phosphate solubilization, secretion of various phytohormones, and production of antioxidant enzymes and siderophores (Alizadeh and Parsaeimehr 2011; Chakraborty et al. 2011; Kohler et al. 2009). Furthermore, the halotolerant plant growth-promoting bacteria (PGPB) have reduced the negative effects of saline stress by increasing the relative water content in leaves and enhancing photosynthetic pigment production under stress conditions (Saghafi et al. 2013).

The studies about the utilization of halotolerant and halophilic plant growth-promoting bacteria in mitigation of the deleterious effects of salt stress have been performed using *Bacillus*, *Pseudomonas*, *Rhizobium*, *Azospirillum*, and *Planococcus* (Mayak et al. 2004; Egamberdiyeva 2005; Sapsirisopa et al. 2009; Rajput et al. 2013; Sharma et al. 2015). The effects have been demonstrated well in wheat seedlings (Nabti et al. 2012; Ramadoss et al. 2013), tomato plants (Tank and Saraf 2010; Shen et al. 2012), and maize and soybean plants by application with osmotolerant rhizobacteria (Naz et al. 2009; Vaishnav et al. 2015). The improvement of nutrient elements was recorded in sunflower under high salinity (Shirmardi et al. 2010) and the increase in growth of barley by the inoculation with novel halotolerant rhizobacteria (Cardinale et al. 2014).

14.2 Halophilic Bacteria

Halophilic bacteria are salt-loving organisms inhabiting various environments with the capacity to balance the osmotic pressure of the environment. The halophilic microorganisms are highly diverse in nature and belong to three domains of life, i.e., Archaea, Bacteria, and Eukarya. They inhabit salt ponds, soda lakes, and even rock salt crystals as dormant cells or as biopolymers. These prokaryotic organisms are likely evolutionary adaptations of more conventional bacterial forms rather than a more fundamental group or branch on the evolutionary tree (Woese 1993). A distinction must be made between "tolerance for salt" and "requirement for salt" (Larsen 1986). Halotolerant microorganisms do not specifically require salt, more than the usual concentration of 100–200 mM NaCl which is needed by all microorganisms including non-halotolerant, but they could grow up to ~1.25 M NaCl. Within the group of halotolerant microorganisms, a distinction can be made between those for which growth rate is decreased by the addition of any salt and those for which the growth rate reaches an optimum with the addition of some salt. There are several different definitions and classifications for halophiles. The definition proposed by Kushner (1993) is widely accepted (Oren 2008). The halophiles have been classified as (1) slight halophiles, able to grow optimally between 1 and 3 % (0.2–0.5 M) NaCl; (2) moderate halophiles, growing optimally in media with 3–15 % (0.5–2.5 M) NaCl; and (3) extreme halophiles, able to grow optimally in media with 15–30 % (2.5–5.2 M) NaCl. There are aerobic as well as anaerobic halophiles; heterotrophic, phototrophic, and chemoautotrophic types are found within halophilic bacteria. The unique properties of the halophilic microorganisms make them valuable resources in the development of novel biotechnological processes and industrial applications, e.g., proteases and amylases in detergent industry, biosurfactant production, poly-beta hydroxyalkanoate and exopolysaccharide as biodegradable plastic, biopolymers in oil recovery, bioremediation of contaminated hypersaline brines, etc. (Kanekar et al. 2012).

14.3 Taxonomy of Halophilic Bacteria

The halophilic bacteria belong to the order *Halobacteriales*, which contains only a single family, the *Halobacteriaceae* previously. After publication of *Bergey's Manual of Systematic Bacteriology*, a total number of 19 genera and 57 validated species have been recognized. The first halophilic bacterial species *Halanaerobium praevalens* was isolated from the sediments of the Great Salt Lake (Utah), and after characterization it was placed in the family *Bacteroidaceae* as a genus with uncertain affiliation (Zeikus et al. 1983). Later, new families and different orders of halophilic bacteria were proposed time to time on the basis of 16S rRNA sequence similarity and membrane lipid profiling. The most commonly recognized genera and species of the halobacteria are listed in Table 14.3.

Most of the halophilic bacteria contained ether-linked phosphoglycerides (C20C20; diphytanyl isoprenoids) and methyl ester-linked phosphatidyl glycerol phosphate (C20C25; phytanyl-sesterterpanyl isoprenoids). These polar lipid compositions have proved to be remarkably consistent in different halophilic bacterial taxonomy as compared to 16S rRNA gene sequence. The 16S rRNA gene sequence of halophilic bacteria showed most distantly related species in phylogenetic tree with 83.2 % similarity. The methanogens are their closest relatives with less than 80 % similarity (Aljohny 2015).

14.4 Mechanisms for Adaptation of Halophilic Bacteria in Saline Environments

Adaptation of microorganisms to such saline environments is a complex multilevel regulatory process in which different types of genes may be involved for protecting organisms against the lethal effects of dehydration (Srivastava et al.

Table 14.3 Introduced halophilic bacterial species since 2010

Halobacterial species	Salt concentration range for growth (%)	Reference
Bacillus halochares	6–23	Pappa et al. (2010)
Marinimicrobium haloxylanilyticum	2–22	Moller et al. (2010)
Marinobacterium lutimaris	1–10	Kim et al. (2010)
Virgibacillus byunsanensis	8	Yoon et al. (2010)
Halanaerocella petrolearia	6–26	Gales et al. (2011)
Kangiella spongicola	2–15	Ahn et al. (2011)
Salisediminibacterium halotolerans	3–30	Jiang et al. (2011)
Amphibacillus cookie	6–26	Pugin et al. (2012)
Arhodomonas recens	2–25	Saralov et al. (2012)
Desulfohalophilus alkaliarsenatis	12.5–33	Blum et al. (2012)
Fodinibius salinus	10–15	Wang et al. (2012)
Halanaerobacter jeridensis	6–30	Mezghani et al. (2012)
Halobellus salinus	15–30	Cui et al. (2012)
Natribacillus halophilus	7–23	Echigo et al. (2012)
Salinibacter iranicus	12–30	Makhdoumi-Kakhki et al. (2012)
Alkalibacterium gilvum	0–17.5	Ishikawa et al. (2013)
Halomicroarcula pellucida	20–30	Echigo et al. (2013)
Halanaerobium sehlinense	5–30	Abdeljabbar et al. (2013)
Limimonas halophila	15–30	Amoozegar et al. (2013a)
Saliterribacillus persicus	0.5–22.5	Amoozegar et al. (2013b)
Aquibacillus halophilus	0.5–20	Amoozegar et al. (2014)
Bacillus daqingensis	0–16	Wang et al. (2014)
Halomonas huangheensis	1–20	Miao et al. (2014)
Oceanicola flagellatus	0–21	Liu and Yang (2014)
Oceanobacillus aidingensis	0–21	Liu and Yang (2014)
Spiribacter salinus	10–25	León et al. (2014)

2008). The optimum metabolic processes like enzymatic activities and membrane stability occur at high salinity in certain halophilic bacterial species (Oren 1999), whereas other microorganisms develop different adaptation mechanisms to combat the stress such as the salt-in and compatible-solute strategy and exopolysaccharide production.

The cytoplasm is exposed to high ionic strength to achieve osmotic equilibrium by maintaining a cytoplasmic salt concentration similar to that of the surrounding media. Microorganisms that grow optimally in the presence of extremely high salinities (up to 5 M NaCl), accumulate intracellular potassium and chloride ions in concentrations higher than the external NaCl concentration to maintain a turgor pressure. This is called "salt-in" strategy found in *Halobacteriales* (Archaea) and *Halanaerobiales* (anaerobic halophilic bacteria) (Hanelt and Muller 2013).

Compatible-solute strategy is a more flexible strategy, mostly found in halotolerant as well as moderately halophilic microorganisms that grow over a wide range of salinities (typically 0.5–3 M NaCl) (Roessler and Muller 2001). In this strategy, "the low-salt-in" strategy depends on the accumulation of high concentrations of organic compatible solutes. Compatible solutes are small, mainly neutral but polar compounds, which are highly soluble in water and do not interfere with the cellular metabolism. Such solutes include glutamate and proline (amino acids), peptides, and N-acetylated amino acids (amino acid derivatives), glycine, betaine, and carnitine (quaternary amines), sucrose and trehalose (sugars), and ectoines (tetrahydropyrimidines) (Paul and Lade 2014). Paul and Nair (2008) reported that *Pseudomonas fluorescens* MSP-393, a PGPR strain, as a means of salt tolerance, de novo-synthesized the osmolytes, alanine, glycine, glutamic acid, serine, threonine, and aspartic acid in their cytosol. The uptake or synthesis of compatible solutes retains a cytoplasm iso-osmotic with or slightly hyperosmotic compared to its surroundings. In addition to their well-studied function as osmoprotectants, compatible solutes also have protein-stabilizing properties that support the correct folding of polypeptides under denaturing conditions both in vitro and in vivo (Street et al. 2006). Compatible solutes confer the changes in structure of solvent and/or elusive changes in protein's dynamic properties, not by structural changes in the protein itself (Lamosa et al. 2003),but it also helps in protein–DNA interaction (Kurz 2008).

Certain Gram-negative bacteria survive under abiotic stress conditions through exopolysaccharide (EPS) production, which protects bacteria from hydric stress and fluctuations in water potential by enhancing water retention and regulating the diffusion of carbon sources in microbial environment (Sandhya et al. 2009a, b). In addition, exopolysaccharides possess unique water holding and cementing properties, thus play a vital role in the formation and stabilization of soil aggregates and regulation of nutrients and water flow across plant roots through biofilm formation (Bezzate et al. 2000).

14.5 Role of Halophilic Bacteria in Plant Growth Promotion Under Salt Stress

Any living organisms under stressful condition opt either fight or flight strategy. Since plants are sessile, they cannot run away from adverse conditions, so they fight back; their tolerance capacity, growth, and production can be increased with the help of several mechanistic actions of salt-tolerant PGPRs as shown in Fig. 14.2. Bacteria that help plants overcome the negative effects of abiotic stress are endowed with certain specialized functional traits. Previous studies reported that plant growth promotion and amelioration of salinity stress in crop plants by salt-tolerant bacteria could involve different mechanisms such as secretion of various phytohormones, ACC-deaminase activity, phosphate solubilization, antioxidant enzymes, and siderophore production (Chakraborty et al. 2011). The use of halotolerant PGPB possessing the traits of PGP under saline stress is becoming prevalent worldwide to achieve sustainable agriculture along with soil reclamation through phytoremediation as well

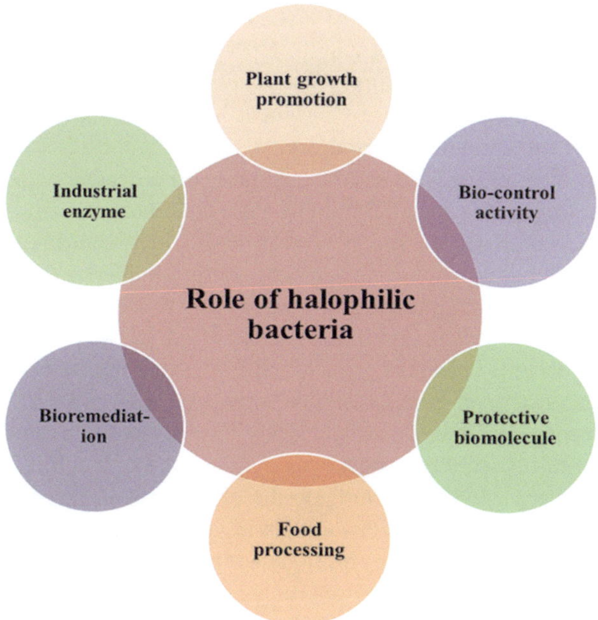

Fig. 14.2 Application of halophilic bacteria in different environmental aspects

as bioremediation (Tank and Saraf 2010). The ameliorative effects of PGPR on plant growth under saline conditions have been shown for various plant species, such as tomato, pepper, canola, bean, lettuce, soybean, and mung bean (Barassi et al. 2009; Kang et al. 2009; Egamberdieva 2009; Vaishnav et al. 2013; Kumari et al. 2016). Kohler et al. (2006) demonstrated the beneficial effect of PGPR *Pseudomonas mendocina* strains on stabilization of soil aggregate. The three PGPR isolates *Pseudomonas alcaligenes* PsA15, *Bacillus polymyxa* BcP26, and *Mycobacterium phlei* MbP18 were able to tolerate high temperatures and salt concentrations and thus confer on them potential competitive advantage to survive in arid and saline soils such as calcisol (Egamberdiyeva 2007). Heidari et al. (2011) also reported increase in plant growth of *Ocimum basilicum* with increased auxin and protein contents under drought stress conditions when inoculated by *Pseudomonas* sp. Two strains of *Azotobacter* sp. have shown increased growth of *Zea mays* under high NaCl concentration (Rojas-Tapias et al. 2012); the experiment revealed a significant restoration of plant biomass (length and weight), exclusion of Na^+ and K^+, improvement of chlorophyll and polyphenol contents, and maintenance of nitrogen fixation and phosphate solubilization activities under saline stress conditions. In another study Egamberdieva et al. (2013) demonstrated that the colonization of *G. officinalis* root tips by *Rhizobium* cells increased almost twofold under saline conditions when the plants were inoculated besides *Rhizobium* with *Pseudomonas* strains. This combined inoculation could also enhance formation of nodules on legumes grown in salinated potting soil. Vaishnav et al. (2015) reported that salt-tolerant *Pseudomonas simiae* strain AU have growth promotion attributes at 100 mM NaCl on soybean seedling and showed significant improvement of shoot/root length, K^+/Na^+ ratio, and P content of soybean seedling after 10 days of germination. Upadhyay et al. (2011) studied the effect of PGPR on growth and oxidative stress in wheat in saline soil and reported that co-inoculation of *B. subtilis* and *Arthrobacter* sp. alleviates the adverse effects of soil salinity with an increase in proline content, total soluble sugars, and dry biomass. Similar reports were given by Shukla et al. (2012) in peanuts (*Arachis hypogaea*) where an increase of NaCl stress inhibited growth which could be alleviated after inoculation with the diazotrophic rhizosphere bacterium *Brachybacterium saurastrense* and other halotolerant isolates from the halophyte *Salicornia* (Jha et al. 2012). Most recently, Ramadoss et al. (2013) studied the effect of five plant growth-promoting halotolerant bacteria on wheat growth and found that inoculation of those halotolerant bacterial strains to ameliorate salt stress (80, 160 and 320 mM) in wheat seedlings produced an increase in root length of 71.7 % in comparison with uninoculated positive controls. Another study was conducted on the utilization of marine bacteria as salt-tolerant PGPB to mitigate the effect of stress on inoculated plants (Kim et al. 2014). Maziah et al. 2009 and Das et al. 2011 performed studies even on trees such as banana and mangrove forest growing under hard environmental conditions caused by salinity. Halophilic bacteria are also reported for biological control activity. The biological control referred as microbial activity to control plants diseases. The different genera of halophilic bacteria like *Virgibacillus*, *Terribacillus*, *Halomonas*, *Halobacillus*, *Planococcus*, *Staphylococcus*, *Marinococcus*, *Salinococcus, and Halovibrio* have been identified in biocontrol activity by producing chitinase and β-1,3-glucanase enzyme activity (Sadfi-Zouaoui et al. 2007).

14.6 Mechanism of Plant Growth Promotion by Halophilic Bacteria

14.6.1 The Role of Bacterial Phytohormones

Phytohormones are naturally occurring, organic substances which influence physiological processes of plants at low concentrations. These hormones affect differentiation and development of plant growth through the regulation of diverse processes. The plant growth hormones of microbial origin in the vicinity of plant roots could evoke a physiological response in the host plant. Production of indole acetic acid, cytokinins (CK), gibberellins, abscisic acid (ABA), and other growth regulators produced by halotolerant PGPB apparently supports the rooting with increased root length, surface area, and number of root tips; ultimately it leads to enhanced uptake of nutrients and thereby improves plant health under stress conditions (Egamberdieva and Kucharova 2009; Jha et al. 2013). Furthermore, Jha and Subramanian (2013) showed clearly the direct and potential effect of some osmotolerant bacteria on germination of paddy seeds under saline conditions. Kumari et al. (2015) reported that two IAA-producing bacterial strains *Bacillus* and *Pseudomonas* enhanced soybean growth under 100 mM NaCl stress by enhancing antioxidant enzyme activity and lowering lipid peroxidation.

14.6.2 Aminocyclopropane-1-Carboxylate (ACC) Deaminase

Ethylene is a volatile phytohormone and plays an important role in plant growth regulation at very low concentrations such as development of different vegetative plant parts, nodulation, or rooting of cuttings (Davis 2004) and also involved in the transduction of a signal for the recognition of salt stress (Selvakumar et al. 2012). The overproduction of ethylene in response to abiotic stresses leads to inhibition of root growth and, consequently, growth of the plant. Chemical inhibitors of ethylene synthesis, such as cobalt ions and aminoethoxyvinylglycine, are often used to overcome the problems associated with salt stress. However, these chemicals are not only expensive, but they have harmful effects on the environment (Dodd 2009). Halotolerant PGPB contains aminocyclopropane-1-carboxylate (ACC) deaminase which hydrolyzes ACC into ammonia and α-ketobutyrate, thereby lowering the level of ethylene in stressed plants. In the presence of 1-aminocyclopropane-1-carboxylate deaminase-producing bacteria, plant 1-aminocyclopropane-1-carboxylate is sequestered and degraded by bacterial cells to supply nitrogen and energy (Mayak et al. 2004), facilitating plant growth under the salinity stress condition (Nadeem et al. 2010; Aamir et al. 2013). ACC-deaminase has been widely reported in numerous microbial species of Gram-negative and Gram-positive halotolerant bacterial strains that belong to different bacterial genera, i.e., *Bacillus*, *Brevibacterium*, *Planococcus*, *Zhihengliuella*, *Halomonas*, *Exiguobacterium*, *Oceanimonas*, *Corynebacterium*, *Arthrobacter*, and *Micrococcus*, that were originally isolated from saline environments and have a real potential to enhance plant growth under saline stress via 1-aminocyclopropane-1-carboxylate deaminase activity (Siddikee et al. 2010; Hussain et al. 2013). In a recent study, an overproducing ACC-deaminase mutant bacterial strain *Pseudomonas simiae* AU5 was found most prominent to alleviate salt stress in mung bean plants as compared to wild strain AU. *P. simiae* AU5-inoculated plants showed lower level of ethylene hormone and salt-induced membrane injury (Kumari et al. 2016).

14.6.3 Phosphate Solubilization

Phosphorous is the major nutrient for plant growth as it is an integral part of different biochemicals like nucleic acids, nucleotides, phospholipids, and phosphoproteins. In most cases, salinity decreased P accumulation in plant, which developed P deficiency symptoms (Parida and Das 2005). Phosphorus exists in two forms in soil, as organic and inorganic phosphate, and, like other nutrient elements such as potassium, iron, zinc, and copper, possesses limited mobility in the soil (Hayat et al. 2010). The conversion of insoluble phosphate compounds (both organic and inorganic) in a form accessible to the plant is an important trait of PGPB strains. This is achieved through the acidification, chelation, ion-exchange reactions, and production of low-molecular-weight organic acids such as gluconic acids. Halotolerant PGPBs have been proved to be vital for circulation of plant nutrients in many ways, thereby reducing the need for chemical fertilizers. Apart from phytohormones and

ACC-deaminase activity, many strains of bacteria can affect plant growth directly by solubilizing inorganic phosphate, improving nutrient uptake, and mineralizing organic phosphate (Ogut et al. 2010). Solubilization of phosphate in the rhizosphere is the most common mode of action implicated in PGPR that increase the nutrient availability to the host plant (Rashid et al. 2004). These rhizobacteria are critical for the transfer of P from poorly available forms and are important for maintaining P in readily available pools. Diby et al. (2005) reported enhanced nutrient mobilization in the rhizosphere of black pepper and significant uptake of nitrogen (N) and phosphorus (P) in the PGPR-treated black pepper vines that resulted in root proliferation and enhanced plant growth. Strains of rhizobacteria that have efficient phosphate-solubilizing ability even under high saline (60 g L^{-1} NaCl) conditions have been reported (Upadhyay et al. 2011). *Pseudomonas* inoculation had favorable effect on salt tolerance of *Zea mays* L., under NaCl stress (Bano and Fatima 2009). Baldani et al. (2000) inoculated phosphate-solubilizing bacteria, *Herbaspirillum seropedicae* and *Burkholderia* sp., to the soil and showed that these bacteria increased the weight of crop 1.5–21 % over uninoculated controls under saline conditions. *Azospirillum*-inoculated lettuce seeds had better germination and vegetative growth than non-inoculated controls after being exposed to NaCl (Barassi et al. 2009). Dardanelli et al. (2008) reported as salt stress affects nodulation during *Phaseolus–Rhizobium* interaction that a secondary inoculation of the salt-stressed plants with *Azospirillum* caused an extended exudation of plant flavonoids compared to *Rhizobium* alone. This co-inoculation of plants with different bacterial strains contributed to relieving of the abiotic stress. Vaishnav et al. (2015) proposed that *P. simiae*-mediated volatile compounds enhanced vegetative storage protein (VSP) expression, which is responsible for acid phosphatase activity and enhanced P uptake in soybean plants under 100 mM NaCl stress.

14.6.3.1 Antioxidative Activity

During salt stress, the ROS level is high which causes oxidative damage to biomolecules such as lipids and proteins and finally leads to plant death (Del Rio et al. 2003). Plants cope up with these ROS through their antioxidant machinery. Major antioxidative enzymes SOD, APX, and CAT and nonenzymatic antioxidants such as ascorbic acid and glutathione participate in ROS-scavenging mechanism (Miller et al. 2010). PGPRs such as *S. proteamaculans* and *Rhizobium leguminosarum* are reported to produce antioxidant enzymes such as superoxide dismutase (SOD), peroxidase (POX), and catalase (CAT) and nonenzymatic antioxidants such as ascorbate, glutathione, and tocopherol. Ruiz-Lozano et al. (2001) reported that mycorrhizal-inoculated lettuce plants exhibited higher superoxide dismutase (SOD) activity under drought stress, and this was correlated to plant protection. Stress resistance in plants has been related to more effective antioxidant systems (Bor et al. 2003). A recent study reports that salt-tolerant bacteria *P. simiae* AU enhanced peroxidase and CAT gene expression in soybean plants when inoculated under 100 mM NaCl stress (Vaishnav et al. 2016). Reduction in the activity of antioxidant enzymes was also observed in bacterial inoculated plants. Five potential drought-tolerant plant growth-promoting *Pseudomonas* spp. strains were found to alleviate drought stress effects in maize plants. Inoculated plants showed significantly lower activity of antioxidant enzyme plants as compared to uninoculated plants (Sandhya et al. 2010). Omar et al. (2009) reported that catalase and peroxidase activities were increased in non-inoculated plants during salinity, whereas *Azospirillum brasilense*-inoculated plants exhibited lower enzyme activity and significantly ameliorated the deleterious effects of salinity.

14.6.4 Siderophore Producers

Iron is an essential micronutrient of plants as it serves as a cofactor of many enzymes with redox activity. A large portion of iron in soils is present in highly insoluble form of ferric hydroxide; thus iron acts as a limiting factor for plant growth even in iron-rich soils. Its availability to the organism is very limited due to the rapid oxidation of ferrous (Fe^{++}) to ferric (Fe^{+++}) state. Ferric ion is highly insoluble under physiological conditions and makes its acquisition by microorganisms a considerable challenge (Neilands 1995). Microorganisms have evolved specialized mechanisms for the assimilation of iron, including the production of low-molecular-weight iron-chelating compounds known as siderophores, which transport this element into their cells. Siderophores have been implicated for both direct and indirect enhancement of plant growth by rhizospheric microorganisms (Neilands 1981). Siderophores provide an advantage in the survival of both plants and bacteria because they mediate competition that results in exclusions of fungal pathogens and other microbial competitors in the rhizosphere by a reduction in the availability of iron for their survival (Masalha et al. 2000; Wang et al. 2000).

14.7 Other Applications of Halophilic Bacteria

14.7.1 As a Source of Industrially Important Enzymes

Under extreme salt or temperature conditions, the stability and characterization of industrially important enzymes of microbial origin is an important factor. A few enzymes of halophilic bacteria make them desired candidates for industry because of its stability under high ionic circumstance during process. In addition, mostly these halophilic bacterial enzymes not only remain active and stable in high-salt environments but are also thermotolerant and alkaliphilic. Thus, halobacteria have attracted much attention in recent years due to beneficial production of halophilic exoenzymes that can be used in diverse fields of biotechnology. The most well-investigated haloezymes are hydrolases such as amylases (Amoozegar et al. 2003), lipases, esterases, xylanases, chitinases, proteases, cellulases, nucleases, etc. (Oren 2010; Moreno et al. 2013). The use of these enzymes in industrial products and processes has been an accepted technology for several decades because they are safe, environmental-friendly biological molecules that make a substantial contribution to the environmental sustainability of industrial processes.

14.7.1.1 Amylases

Amylases catalyze the hydrolysis of starch and their related saccharides and are used widely in several fields of biotechnology and are one of the most commonly used industrial enzymes. The best-known enzymes of this group are α-amylase, β-amylase, and glucoamylase. Halophilic α-amylases have received more attention due to their ability to remain active in the presence of high salt concentrations. α-Amylases could be used in food, pharmaceutical, biofuel, fermentation, paper, detergent, and textile industries (Kadziola et al. 1998; Machius et al. 1995; Souza 2010). The amylases produced by halophilic bacteria such as *Micrococcus varians* subsp. *halophilus* has two protein components of 86 and 60 kDa molecular mass (Kamekura 1986) with optimal activity at 4.5–6 % NaCl and pH 6–7. Similarly, an extracellular α-amylase from *Halomonas meridiana* exhibited maximal activity at pH 7.0, 37 °C and 10 % (w/v) NaCl, respectively (Coronado et al. 2000a, b). Another extracellular α-amylase isolated from a haloalkaliphilic bacterium was active up to 4 M salt, with optimal activity at 2 M salt, pH 10.0–11.0 and 50 °C (Pandey and Singh 2012). The amylases have significantly applied for treatment of wastewater. The extracellular amylase production from *Halomonas meridian* was highest at 5 % salt concentration with maximal activity at pH 7; its amylase gene, AmyH, has also been isolated.

14.7.1.2 Cellulases

Cellulases are one of the important enzymes for biomedical science, paper, agriculture, food and laundry industries (Zhang et al. 2012). The cellulolytic enzymes hydrolyze the β-1,4-D-glucosidic linkages in cellulose, lichenin, and cereal β-D-glucans. The enzyme has been categorized in three main groups, viz., endocellulase, exocellulase, and β-glucosidase (ShaoMin and Guang 2013; Karnchanatat et al. 2008), which completely hydrolyze the β-1,4-D-glycosidic bonds of cellulose to form glucose by acting together (Bhat and Bhat 1997; Bhat 2000). Huang et al. (2010) identified a novel endoglucanase from halophilic bacterium *Halomonas* sp. S66-4, cloned in *E. coli*, the purified recombinant enzyme, which showed the highest activity (4.9 U/mg) at pH 5 and 6 % NaCl. Shivanand et al. (2013) also reported production of cellulases from *Halomonas* sp. PS47 at 6 % NaCl. The maximum activity was at pH 7.1 and 50 °C. A novel salt-tolerant endo-β-1,4-glucanase Cel5A was also identified from *Vibrio* sp. G21; it has a cellulose-binding domain and a catalytic domain of glycosyl hydrolase (Gao et al. 2010).

14.7.1.3 Xylanases

Xylan is the second most abundant hemicellulose in nature after cellulose, which together with lignin, cellulose, pectin, and other polysaccharides constitutes the major components of plant cell walls and maintains cell wall integrity. Xylan is generally insoluble in nature but enzymatically degradation of it converts into useful products like xylose, xylitol, and ethanol. Biodegradation of these xylans, i.e., hydrolysis of 1,4-*β*-D-xylosidic linkages, involves catalysis by the action of two major xylanolytic enzymes, endoxylanase (1,4-β-D-xylan xylanohydrolase) and β-xylosidase (1,4-β-D-xylan xylohydrolase). There has been resurgence in interest in microbial xylanases due to their numerous uses in industrial applications, such as biobleaching of pulp and most notably the conversion of lignocellulosic materials into fermentable substrates for production of economical and environmental-friendly biofuels (Oksanen et al. 2000). Pioneer work on characterization and purification of halotolerant endoxylanases was done by Wejse et al. (2003) from a novel halophilic bacterium, strain CL8, which had highest sequence similarity with *Oceanospirillum linum* and *Marinobacter* sp. str. CAB. Xylobiose and xylotriose are the major products of these enzymes. Though the optimal activity of these enzymes was at IM NaCl, it remained stable at 5 M NaCl. Sustainable xylanase activity was demonstrated up to 30% NaCl from *Gracibacillus* sp. TSGPVG at 60 °C (Giridhar and Chandra 2010). This kind of stability would be essential for the extracellular activity in a high-salt environment in industry.

14.7.1.4 Lipases

Lipases (triacylglycerol acylhydrolases EC 3.1.1.3) catalyze esterification, trans-esterification, and aminolysis and have considerable physiological significance and industrial potential (Babu et al. 2008). The characterization of salt stable lipases from halophilic source has been a growing interest nowadays. Lipases have emerged as one of the leading biocatalysts with proven potential of contribution to the under-exploited lipid industry with several applications: esterification, interesterification, transesterification fat hydrolysis, and organic biosynthesis during production of drugs in the pharmaceutical industry. These lipases could be used for the hydrolysis of milk fat in the dairy industry, for the removal of subcutaneous fat in the leather industry, for the removal of impurities from raw cotton in the paper industry, and as additives in detergents (Gomes and Steiner 2004). A moderately halophilic *Salinivibrio* sp. SA-2 produces extracellular lipase; its maximum activity was reported at pH 7.5 and 50 °C. The enzyme remained active in presence of 17% NaCl (Amoozegar et al. 2008). Similarly, an intracellular lipase enzyme from *Salicola marasensis* LipL shows maximum activity at 1 M NaCl; however, it could tolerate up to 4 M NaCl with 6 mM of betaine (Moreno et al. 2013). *Halobacillus* sp. strain LY5 from the saline soil in Yuncheng, China, produced extracellular esterase of molecular mass of 96 kDa. Its optimum enzyme activity was found at 10% (w/v) NaCl at pH 10.0 at 50 °C (Li et al. 2012). In another study, a novel moderate halophile *Marinobacter lipolyticus* isolated from a hypersaline habitat exhibited lipolytic activity optimally at 7.5% NaCl (Martín et al. 2003).

14.7.1.5 Proteases

Bacterial proteases are one among the largest studied groups of hydrolytic enzymes with diverse applications in industrial and biotechnological fields (Joshi et al. 2008; Singh et al. 2012). Extensive research has been done on the extremophile proteases because of their utility in food industries, detergents, laundry, wool quality improvement, and waste treatment (Vijayaraghavan et al. 2012). Recently, proteases in the pharmaceutical industry and bioremediation process have attracted more attention. A moderately halophilic *Geomicrobium* sp. EMB2 produces an extracellular protease stable at 20% salts, 75% organic solvents, 2.0% detergents, and 1.0%

surfactants (Karan et al. 2011). *Pseudomonas* sp. strain A-14 also produces extracellular protease which has its optimal activity at pH 8 at 18 % NaCl (Van Qua et al. 1981). *Chromohalobacter* sp. strain TVSP101 produces halothermophilic protease with maximum activity at 4.5 M NaCl at pH 8 (Vidyasagar et al. 2007). Similarly, the extracellular proteases produced by *Halobacillus blutaparonensis* were stable up to 20 % NaCl concentration and organic solvents (Santos et al. 2013).

14.7.1.6 Chitinase

Chitinase is one of the most potential enzymes of biocontrol in agriculture and environmental investigations (Duo-Chuan 2006). Chitinases hydrolyze the chitin from the cell wall of fungi and are used to produce protoplasts; they are also used during production of biologically active oligosaccharides (Bhattacharya et al. 2007). The bacterial strains *Arhodomonas* HCh2 and *Saccharospirillum* HCh1 isolated from hypersaline lakes in Russia had optimal growth on chitin at 1.5–1 M NaCl and growth range 0.5–3.25 M NaCl (Sorokin and Kolganova 2013). *Virgibacillus marismortui*, a moderately halophilic bacterium isolated from shallow salt lakes, had the ability to produce chitinase (in absence of salt as well as in presence of high salinity, 25–30 % NaCl w/v). Such strains can be significant for biocontrol purposes (Essghaier et al. 2012).

14.7.2 Bioremediation of Polluted Environments

Generally, the wastes coming into the environment have high salt concentration; hence the use of halophilic bacteria could be a promising alternative in waste treatment and management. Industrial processes like tannery or food processing produce large volumes of saline wastewater that cannot be treated by conventional methods due to low efficiency. Application of halophilic bacteria can improve the removal efficiency of COD from saline wastewater. Tannery wastewater-adapted bacteria *Pseudomonas aeruginosa*, *Bacillus flexus*, *Exiguobacterium homiense*, and *Staphylococcus aureus* showed 80 % decrease in COD at 8 % salinity (Sivaprakasam et al. 2008). Introducing halophilic bacteria in textile effluents is a prominent approach for treatment of synthetic dyes, where other microorganisms are not able to degrade. In a study, *Halomonas* sp. strain IP8 showed decolorization of dye from 50 to 20 mg/L, during 16–24 h at 1–1.5 M NaCl salt concentration, at temperature range of 25–45 °C (Pourbabaee et al. 2011). Presence of high salt concentrations in heavy metal and hydrocarbon contamination sites also arises great demand of halophilic microorganisms for biotreatment of these sites.

14.7.3 Halophiles in Food Biotechnology

The use of halophilic bacteria has a number of advantages in relation to the production of salt-containing food. Halophilic fermentation gives taste, aroma, and flavor to food, and acetate production during fermentation protects food from contamination. Protease secretion from halophilic bacteria in fermented product plays an important role in lowering the fermentation time. Halophilic fermentative bacteria are used to produce a wide variety of food products, notably fermented fish, shrimp, meat, fruits, and vegetables (pickles), Asian fish and meat sauces, rice noodles and flours, and Indonesian soy sauce. The major species of the genera *Lactobacillus*, *Halobacterium*, *Halococcus*, *Bacillus*, *Pediococcus*, and *Tetragenococcus* are involved in food production (Aljohny 2015).

14.8 Tools for Characterization of Halophilic Bacteria

Characterization of different halobacterial strains from saline environments could be done based on the following three different approaches: (1) morphological and biochemical characterization (phenotypic), (2) chemotaxonomic characterization (chemotypic), and (3) molecular or genomic characterization (genotypic). It is difficult to classify all bacterial species based on only phenotypic characteristics; therefore, polyphasic approach employing all phenotypic, chemotypic, and genotypic characteristics is suggested. A complete integrated information would allow a confident classification and a reliable grouping of the organism.

14.8.1 Phenotypic and Biochemical Characterization

The phenotypic characterization of microorganisms is done by morphological, physiological, and biochemical properties of the microorganism (de Vos et al. 2009). Traditionally, colony morphology (color, dimensions, form) and microscopic appearance of the cells (shape, endospore, flagella, inclusion bodies); characteristics of the organism on different growth substrates; growth range of microorganisms on different conditions of salt, pH, and temperature; susceptibility toward different kinds of antimicrobial agents; etc. are measured in phenotypic characterization. Even if cell wall composition is analyzed, the Gram reaction is still a valuable diagnostic character. Biochemical tests in bacterial identification include the relationship with oxygen, fermentation reactions, carbon utilization, and nitrogen metabolism. A Biolog system has been developed for determination of carbon utilization pattern in different bacterial communities (Garland and Mills 1991). In this culture-dependent technique, a 96-well Biolog microtiter plate containing 95 different carbon sources and one control well per plate with growth medium are used with the redox dye tetrazolium salt. The color changes in tetrazolium salt because of bacterial metabolic actions on the substrate. Other tests may be performed as appropriate, depending on the bacterial strains studied (Heritage et al. 1996; Rodríguez-Díaz et al. 2008). However, reproducibility of results from phenotypic tests between different laboratories is a major concern, and only standardized procedure should be used during execution of experiment. Other major disadvantage with phenotypic methods is the conditional nature of gene expression wherein the same organism might show different phenotypic characters in different environmental conditions.

Therefore, phenotypic data must be compared with similar set of data from type strain of closely related organism(s). Miniaturized versions of traditional biochemical tests are available for taxonomical studies and mostly contain a battery of dehydrated reagents. Addition of a standardized inoculum initiates the reaction (growth, production of enzymatic activity, etc.). The results are interpreted as recommended by the manufacturer and are readily accessible with a minimal input of time. The phenotypic fingerprinting system API 50CH has the highest rate of correct identification; it uses forty-nine different carbohydrates and one negative control to identify different bacterial genera including *Bacillus* (Logan and Berkeley 1984), *Paenibacillus*, and *Pseudomonas* species (Barr et al. 1989).

14.8.2 Chemotaxonomic Characterization

Chemotaxonomy is the method of biological identification and classification based on similarities in the structure of certain compounds, i.e., cellular fatty acid among the organisms being classified. In cellular fatty acid analysis, chemical and physical techniques are employed to elucidate the chemical composition of whole bacterial cells and/or their individual cellular components in order to produce a chemical signature or profile of taxonomic significance. One successful and commercialized chemotaxonomic approach for obtaining bacterial fatty acid profiles is based on the fatty acid methyl ester (FAME) analysis by gas chromatography (GC) (Fang et al. 2001). In microorganisms, PLFAs are found exclusively in cell membranes and not in other parts of the cell such as storage products. This method provides information on the microbial community composition based on groupings of fatty acids (Ibekwe and Kennedy 1998). Fatty acids make up a relatively constant proportion of the cell biomass, and signature fatty acids exist that can differentiate major taxonomic groups within a community. Therefore, a change in the fatty acid profile would represent a change in the microbial population. However, the fatty acid composition of microorganisms does not change by plasmid loss or gain or by simple mutations. Fatty acid profiles showing variability in double-bond position, chain length, and substituent groups are perfectly suitable for taxon description and also for comparative studies of profiles that have been obtained under similar growth conditions (Suzuki et al. 1993). The automated MIDI Sherlock Microbial Identification System identifies microorganisms based on unique FAME patterns of known strains (Whittaker et al. 2003).

14.8.3 Molecular Characterization

Because of the inherent limitations of conventional phenotyping methods for detecting microorganism strains within culture-dependent techniques, as well as their mechanisms of resistance, molecular techniques that complement the information provided by these methods have been developed (Perez et al. 2007). The application of molecular biological methods to study the identification, diversity, and ecology of microorganisms in natural environments has been practiced since the mid-1980s. Methods that use this approach are directed toward DNA or RNA molecules and comprise measurements of DNA relatedness over the entire genome; comparisons of restriction patterns, especially ribotyping; and comparative analyses of sequences of homologous genes. DNA–DNA relatedness and ribotyping are best suited for the identification of closely related species or strains within a single species. Presently, a direct comparison of rRNA sequences is probably the most powerful tool for the identification of many bacteria. Indeed, rRNA genes (rDNA) are present in all bacterial species, are truly homologous in all organisms, are easily sequenced, and now offer a large and ever-increasing database of sequences and allow the identification of cultured as well as uncultured bacteria.

Studies of microbial isolation, identification, and characterization have always been intimately entwined. Comparative analysis of rRNA sequences not only provided the phylogenetic framework which was lacking in microbial diversity but also allowed the development of tools to address this vast microbial wealth. The ubiquity of rRNA molecules (small subunit 5S, 16S, large subunit 23S) in all cellular life forms and comparative analysis of their sequences can be universally applied to infer relationships among organisms. Among the three rRNA molecules, 16S rRNA gene (1500 bp) is the most commonly used marker. It has a universal distribution, highly conserved nature, fundamental role of ribosome in protein synthesis, no horizontal transfer, and its rate of evolution which represents an appropriate level of variation between organisms (Stackebrandt and Goebel 1994). The 16S rRNA molecule comprises highly conserved sequence domains interspersed with more variable regions.

The most commonly used form of comparative rRNA sequence analysis involves the construction of phylogenetic trees. Ribosomal RNA sequence analyses have been greatly facilitated by the availability of an excellent, indispensable, curated database of rRNA sequences (the ribosomal database project, RDP-II) (Maidak et al. 2001; Cole et al. 2007). Sequences can be retrieved from these databases for comparative phylogenetic analysis of the microbial species. The sequence-comparing tools such as BLAST and CLUSTAL X are used to align the 16S rRNA gene sequence in which after alignment the relatedness between bacterial species can be scrutinized by the construction of phylogenetic trees or dendrograms. The phylogenetic tree ascertains the identity to the genus and its nearest neighbors. At present, by correlation with experimental data obtained in the comparison of total genomic DNA (DNA–DNA hybridization), it is accepted that a similarity below 98.7–99 % on the 16S rRNA gene sequences of two bacterial strains is sufficient to consider them as belonging to different species.

Besides ribosomal genes, other structural, metabolic, or housekeeping gene sequences have also been used to differentiate bacterial strains to species or subspecies level. Among them, highly conserved housekeeping or other protein-encoding genes such as *rpo*B (the RNA polymerase β-subunit-encoding gene), *rpo*D, *gyr*B (gyrase subunit β-gene), *rec*A (encoding a protein involved in repairing damaged DNA in the SOS regulon), and multilocus sequence analysis (MLSA) (Carro et al.

2012; Jacques et al. 2012) are more informative than that of 16S rDNA because of their size and conserved and alternating variable regions and are used for taxonomic and phylogenetic studies, especially when it is suspected that the tested strain may be a new species (Meintanis et al. 2006).

Currently, there are several genetic fingerprinting techniques that can be used to characterize bacterial communities or single bacterial isolates which include amplified ribosomal DNA restriction analysis (ARDRA), enterobacterial repetitive intergenic consensus–polymerase chain reaction (ERIC–PCR), and rapid fragment length polymorphism (RFLP). The genetic fingerprinting of microbial communities provides a pattern or profile of the community diversity, based upon the physical separation of unique nucleic acid sequences (Meyer et al. 2007). The use of molecular methods for study of genetic diversity primarily the sensitive and accurate PCR-based genotyping methods enables differentiation among closely related bacterial strains and the detection of higher bacterial diversity than previously considered (Tan et al. 2001).

At present, temperature gradient gel electrophoresis (TGGE) and denaturing gradient gel electrophoresis (DGGE) are predominantly being used. These are based on the direct extraction of DNA or RNA from soil; the amplification of this DNA is done by PCR, followed by electrophoretic separation in a temperature gradient for the former, or by using chemical denaturing substances for the latter. These techniques allow the separation of DNA fragments of exactly the same length but with different sequences, based on their melting properties. Each of the methods described above possesses its own distinctive advantages and disadvantages. Generally, the more selective the method, the less able it is to detect global changes in communities and vice versa. These tools can provide an estimate of the rhizosphere diversity in the soil.

14.9 Conclusion

Salinity is one of the most critical factors which have many detrimental effects on agriculture and the environment. Many resident microflora in this ecology perform all functions of life for survival of their own and associated biological entities. Halophilic bacteria have evolved the capacity to function under so-called unusual conditions. The studies on halophilic bacteria and their metabolites have clearly demonstrated its potential for wide agricultural, industrial, and environmental applications. The successful restoration of plant growth under saline environment after inoculation with halophilic bacteria provides insight for a better alternative to improve crop growth and yield in saline soils. Additionally, halophilic bacteria are also involved in production of haloenzymes, bioremediation, and biodegradation of effluents from saline-based industries. Understanding and exploitation of the beneficial characters of halophilic microorganisms would provide better tool kits for sustainable agricultural and industrial productivity and monitor and regulate anthropogenic detrimental activities that affect biological and environmental health.

References

Aamir M, Aslam A, Khan MY et al (2013) Co-inoculation with *Rhizobium* and plant growth promoting rhizobacteria (PGPR) for inducing salinity tolerance in mung bean under field condition of semi-arid climate. Asian J Agric Biol 1:17–22

Abdeljabbar H, Cayol JL, Hania WB, Boudabous A, Sadfi N, Fardeau ML (2013) *Halanaerobium sehlinense* sp. nov., an extremely halophilic, fermentative, strictly anaerobic bacterium from sediments of the hypersaline lake Sehline Sebkha. Int J Syst Evol Microbiol 63(6):2069–2074

Ahmad P, Ashraf M, Azooz MM, Rasool S, Akram NA (2014) Potassium starvation-induced oxidative stress and antioxidant defense responses in *Brassica juncea*. J Plant Interact 9(1):460–467

Ahn J, Park JW, McConnell JA, Ahn YB, Häggblom MM (2011) *Kangiella spongicola* sp. nov., a halophilic marine bacterium isolated from the sponge Chondrilla nucula. Int J Syst Evol Microbiol 61(4):961–964

Alizadeh O, Parsaeimehr A (2011) The influence of plant growth promoting rhizobacteria (PGPR) on the reduction of abiotic stresses in crops. Extrem Life Biospeol Astrol 3:93–99

Aljohny BO (2015) Halophilic bacterium – a review of new studies. Biosci Biotechnol Res Asia 12(3):2061–2069

Amoozegar MA, Malekzadeh F, Malik KA (2003). Production of amylase by newly isolated moderate halophile, *Halobacillus* sp. strain MA-2. J Microbiol Methods 52(3):353–359

Amoozegar MA, Salehghamari E, Khajeh K, Kabiri M, Naddaf S (2008) Production of an extracellular thermohalophilic lipase from a moderately halophilic bacterium, *Salinivibrio* sp. strain SA-2. J Basic Microbiol 48(3):160–167

Amoozegar MA, Makhdoumi-Kakhki A, Ramezani M, Nikou MM, Fazeli SAS, Schumann P, Ventosa A (2013a) *Limimonas halophila* gen. nov., sp. nov., an extremely halophilic bacterium in the family Rhodospirillaceae. Int J Syst Evol Microbiol 63(4):1562–1567

Amoozegar MA, Bagheri M, Didari M, Fazeli SAS, Schumann P, Sánchez-Porro C, Ventosa A (2013b) *Saliterribacillus persicus* gen. nov., sp. nov., a moderately halophilic bacterium isolated from a hypersaline lake. Int J Syst Evol Microbiol 63(1):345–351

Amoozegar MA, Bagheri M, Makhdoumi-Kakhki A, Didari M, Schumann P, Nikou MM, Sánchez-Porro C, Ventosa A (2014) *Aliicoccus persicus* gen. nov., sp. nov., a halophilic member of the Firmicutes isolated from a hypersaline lake. Int J Syst Evol Microbiol. doi:10.1099/ijs.0.058545-0

Babu J, Pramod WR, George T (2008) Cold active microbial lipases: some hot issues and recent developments. Biotechnol Adv 26:457–470

Baldani JL, Reis VM, Baldani VLD, Dobereiner J (2000) A brief story of nitrogen fixation in sugarcane – reasons for success in Brazil. Funct Plant Biol 29:417–423

Bano A, Fatima M (2009) Salt tolerance in *Zea mays* L. following inoculation with *Rhizobium* and *Pseudomonas*. Biol Fertil Soils 45:405–413

Barassi CA, Ayrault G, Creus CM, Sueldo RJ, Sobrero MT (2009) Seed inoculation with *Azospirillum mitigates* NaCl effects on lettuce. Sci Hortic 109:8–14

Barr JG, Emmerson AM, Hogg GM, Smyth E (1989) API-20NE and sensititre autoidentification systems for identifying *Pseudomonas* spp. J Clin Pathol 42:1113–1114

Bezzate S, Aymerich S, Chambert R, Czarnes S, Berge O, Heulin T (2000) Disruption of the *Paenibacillus polymyxa* levansucrase gene impairs its ability to aggregate soil in the wheat rhizosphere. Environ Microbiol 2:333–342

Bhat M (2000) Cellulases and related enzymes in biotechnology. Biotechnol Adv 18:355–383

Bhat M, Bhat S (1997) Cellulose degrading enzymes and their potential industrial applications. Biotechnol Adv 15:583–620

Bhattacharya D, Nagpure A, Gupta RK (2007) Bacterial chitinases: properties and potential. Crit Rev Biotechnol 27:21–28

Blum JS, Kulp TR, Han S, Lanoil B, Saltikov CW, Stolz JF, Miller LG, Oremland RS (2012) *Desulfohalophilus alkaliarsenatis* gen. nov., sp. nov., an extremely halophilic sulfate-and arsenate- respiring bacterium from Searles Lake, California. Extremophiles 16(5):727–742

Bor M, Ozdemir F, Turkan I (2003) The effect of salt stress on lipid peroxidation and antioxidants in leaves of sugar beet *Beta vulgaris* L. and wild beet *Beta maritime* L. Plant Sci 164:77–84

Bui EN (2013) Soil salinity: a neglected factor in plant ecology and biogeography. J Arid Environ 92:14–25

Cardinale M, Ratering S, Suarez C, Montoya AMZ, Geissler-Plaum R, Schnell S (2014) Modulation rhizosphere microbiota to mitigate salt stress of barley plants. In: Microbial ecology and application of inoculants in biocontrol – Indo-German Workshop. Book of Abstracts, IARI, New Delhi, pp 35–37

Carro L, Spröer C, Alonso P, Trujillo ME (2012) Diversity of *Micromonospora* strains isolated from nitrogen fixing nodules and rhizosphere of *Pisum sativum* analyzed by multilocus sequence analysis. Syst Appl Microbiol 35(2):73–80

Chakraborty U, Roy S, Chakraborty AP, Dey P, Chakraborty B (2011) Plant growth promotion and amelioration of salinity stress in crop plants by salt-tolerant bacterium. Recent Res Sci Technol 3:67–70

Cole JR, Chai B, Farris RJ, Wang Q (2007) The ribosomal database project (RDP-II): introducing my RDP space and quality controlled public data. Nucleic Acids Res 35(SI):D169–D172

Coronado MJ, Vargas C, Hofemeister J, Ventosa A, Nieto JJ (2000a) Production and biochemical characterization of an α-amylase from the moderate halophile *Halomonas meridiana*. FEMS Microbiol Lett 183(1):67–71

Coronado MJ, Vargas C, Mellado E, Tegos G, Drainas C, Nieto J, Ventosa A (2000b) The α-amylase gene amyH of the moderate halophile *Halomonas meridiana* : cloning and molecular characterization. Microbiology 146(4):861–868

CSSRI (Central Soil Salinity Research Institute) (2012) Computerized database on salt effected soils in India. Indian Council of Agricultural Research, Ministry of Agriculture, Government of India, New Delhi

Cui HL, Yang X, Zhou YG, Liu HC, Zhou PJ, Dyall-Smith ML (2012) Halobellus limi sp. nov. and *Halobellus salinus* sp. nov., isolated from two marine solar salterns. Int J Syst Evol Microbiol 62(6):1307–1313

Daneshmand F, Mohammad JA, Khosrow MK (2009) Effect of acetylsalicylic acid (Aspirin) on salt and osmotic stress tolerance in *Solanum bulbocastanum* in vitro: enzymatic antioxidants. Am Eurasian J Agric Environ Sci 6:92–99

Dardanelli MS, Fernández de Córdoba FJ, Rosario Espuny M, Rodríguez Carvajal MA, Soria Díaz ME, et al (2008) Effect of *Azospirillum brasilense* coinoculated with *Rhizobium* on *Phaseolus vulgaris flavonoids* and Nod factor production under salt stress. Soil Biol Biochem 40:2713–2721

Das S, De M, Ray R, Ganguly D, Jana TK, De TK (2011) Salt tolerant culturable microbes accessible in the soil of the Sundarbans mangrove forest, India. Open J Ecol 1:35–40

Davis PJ (2004) Nature, occurrence and functions. In: Davis PJ (ed) Plant hormones biosynthesis, signal transduction, action, vol 1. Kluwer, Dordrecht, pp 1–15

de Vos P, Garrity G, Jones D, Krieg NR, Ludwig W, Rainey FA, Schleifer KH, Whitman WB (2009) Bergey's manual of systematic bacteriology, vol 3: The Firmicutes, 2nd edn. XXVI, 1450 p. 393

Del Rio LA, Corpas FJ, Sandalio LM, Palma JM, Barroso JB (2003) Plant peroxisomes, reactive oxygen metabolism and nitric oxide. IUBMB Life 55(2):71–81

Diby P, Sarma YR, Srinivasan V, Anandaraj M (2005) *Pseudomonas fluorescense* mediated vigour in black pepper (*Piper nigrum* L.) under green house cultivation. Ann Microbiol 55(3):171–174

Dodd IC (2009) Rhizosphere manipulations tomaximize 'crop per drop' during deficit irrigation. J Exp Bot 60(9):2454–2459

Dodd IC, Pérez-Alfocea F (2012) Microbial amelioration of crop salinity stress. J Exp Bot 8:1–14

Duo-Chuan L (2006) Review of fungal chitinases. Mycopathologia 161(6):345–360

Echigo A, Minegishi H, Shimane Y, Kamekura M, Usami R (2012) *Natribacillus halophilus* gen. nov., sp. nov., a moderately halophilic and alkalitolerant bacterium isolated from soil. Int J Syst Evol Microbiol 62(2):289–294

Echigo A, Minegishi H, Shimane Y, Kamekura M, Itoh T, Usami R (2013) *Halomicroarcula pellucida* gen. nov., sp. nov., a non-pigmented, transparent-colony-forming, halophilic archaeon isolated from solar salt. Int J Syst Evol Microbiol 63(10):3556–3562

Egamberdieva D (2009) Alleviation of salt stress by plant growth regulators and IAA producing bacteria in wheat. Acta Physiol Plant 31:861–864

Egamberdieva D, Kucharova Z (2009) Selection for root colonizing bacteria stimulating wheat growth in saline soils. Biol Fertil Soils 45:563–571

Egamberdieva D, Jabborova D, Wirth S (2013) Alleviation of salt stress in legumes by co-inoculation with *Pseudomonas* and *Rhizobium*. In: Arora NK (ed) Plant microbe symbiosis: fundamentals and advances. Springer, New Delhi, pp 291–303

Egamberdiyeva D (2005) Characterization of *Pseudomonas* species isolated from the rhizosphere of plants grown in szernozem soil, semi-arid region of Uzbekistan. Sci World J 5:501–509

Egamberdiyeva D (2007) The effect of plant growth promoting bacteria on growth and nutrient uptake of maize in two different soils. Appl Soil Ecol 36(2):184–189

Essghaier B, Hedi A, Bejji M, Jijakli H, Boudabous A, Sadfi -Zouaoui N (2012) Characterization of a novel chitinase from a moderately halophilic bacterium, *Virgibacillus marismortui* strain M3-23. Ann Microbiol 62:835–841

Fang C, Radosevich M, Fuhrmann JJ (2001) Characterization of rhizosphere microbial community structure in five similar grass species using FAME and BIOLOG analyses. Soil Biol Biochem 33(4):679–682

Food and Agricultural Organization (2015) Land and plant nutrition management service. Available from: www.fao.org/ag/agl/agll/spush/.pdf

Gales G, Chehider N, Joulian C, Battaglia-Brunet F, Cayol JL, Postec A, Borgomano J, Neria-Gonzalez I, Lomans B, Ollivier B (2011) Characterization of Halanaerocella petrolearia gen. nov., sp. nov., a new anaerobic moderately halophilic fermentative bacterium isolated from a deep subsurface hypersaline oil reservoir. Extremophiles 15(5):565–571

Gao Z, Ruan L, Chen X, Zhang Y, Xu X (2010) A novel salt-tolerant endo-β-1, 4-glucanase Cel5A in Vibrio sp. G21 isolated from mangrove soil. Appl Microbiol Biotechnol 87(4):1373–1382

Garland JL, Mills AL (1991) Classification and characterization of heterotrophic microbial communities on the basis of patterns of community-level-sole-carbon-source-utilization. Appl Environ Microbiol 57:2351–2359

Giridhar PV, Chandra T (2010) Production of novel halo-alkali-thermo-stable xylanase by a newly isolated moderately halophilic and alkali-tolerant *Gracilibacillus* sp. TSCPVG. Process Biochem 45(10):1730–1737

Gomes J, Steiner W (2004) The biocatalytic potential of extremophiles and extremozymes. Food Technol Biotechnol 42:223–235

Grover M, Ali SZ, Sandhya V, Rasul A, Venkateswarlu B (2011) Role of microorganisms in adaptation of agriculture crops to abiotic stresses. World J Microbiol Biotechnol 27(5):1231–1240

Hanelt I, Muller V (2013) Molecular mechanisms of adaptation of the moderately halophilic bacterium Halobacillus halophilus to its environment. Life 3:234–243

Hayat R, Ali S, Amara U, Khalid R, Ahmed I (2010) Soil beneficial bacteria and their role in plant growth promotion: a review. Ann Microbiol 60:579–598

Heidari M, Mosavinik SM, Golpayegani A (2011) Plant growth promoting rhizobacteria (PGPR) effect on physiological parameters and mineral uptake in basil (*Ociumum basilicm* L.) under water stress. ARPN J Agric Biol Sci 6:6–11

Heritage J, Evans EGV, Killington RA (1996) Introductory microbiology. Cambridge University Press, England, 234 p

Huang X, Shao Z, Hong Y, Lin L, Li C, Huang F, Wang H, Liu Z (2010) Cel8H, a novel endoglucanase from the halophilic bacterium *Halomonas* sp. S66-4: molecular cloning, heterogonous expression, and biochemical characterization. J Microbiol 48(3):318–324

Hussain MI, Asghar HN, Arshad M, Shahbaz M (2013) Screening of multi-traits rhizobacteria to improve maize growth under axenic conditions. J Anim Plant Sci 23:514–520

Ibekwe AM, Kennedy AC (1998) Phospholipid fatty acid profiles and carbon utilization pattern for analysis of microbial community structure under field and green house conditions. FEMS Microbiol Ecol 26:151–163

Ishikawa M, Yamasato K, Kodama K, Yasuda H, Matsuyama M, Okamoto-Kainuma A, Koizumi Y (2013) *Alkalibacterium gilvum* sp. nov., slightly halophilic and alkaliphilic lactic acid bacterium isolated from soft and semi-hard cheeses. Int J Syst Evol Microbiol 63(4):1471–1478

Jacques MA, Durand K, Orgeur G, Balidas S, Fricot C, Bonneau S, Quillévéré A, Audusseau C, Olivier V, Grimault V, Mathis R (2012) Phylogenetic analysis and polyphasic characterization of *Clavibacter michiganensis* strains isolated from tomato seeds reveal that nonpathogenic strains are distinct from *C. michiganensis* subsp. *michiganensis*. Appl Environ Microbiol 78(23):8388–8402

Jha Y, Subramanian RB (2013) Paddy plants inoculated with PGPR show better growth physiology and nutrient content under saline conditions. Chil J Agric Res 73:213–219

Jha B, Gontia I, Hartmann A (2012) The roots of the halophyte *Salicornia brachiata* are source of new halotolerant diazotrophic bacteria with plant growth promoting potential. Plant Soil 356:265–277

Jha M, Chourasia S, Sinha S (2013) Microbial consortium for sustainable rice production. Agroecol Sustain Food Syst 37(3):340–362

Jiang F, Cao S-J, Li ZH, Fan H, Li HF, Liu WJ, Yuan HL (2011) *Salisediminibacterium halotolerans* gen. nov., sp. nov., a halophilic bacterium isolated from Xiarinaoer soda lake sediment in Inner Mongolia, China. Int J Syst Evol Microbiol. doi:10.1099/ijs.0.034488-034480

Joshi RH, Dodia MS, Singh SP (2008) Optimization of culture parameters for production of commercially valuable alkaline protease from a haloalkaliphilic bacterium isolated from sea water. Biotechnol Bioprocess Eng 13:552–559

Kadziola A, Søgaard M, Svensson B, Haser R (1998) Molecular structure of a barley α-amylase-inhibitor complex: implications for starch binding and catalysis. J Mol Biol 278:205–217

Kamekura M (1986) Production and function of enzymes of eubacterial halophiles. FEMS Microbiol Rev 2(1–2):145–150

Kanekar PP, Kanekar SP, Kelkar AS, Dhakephalkar PK (2012) Halophiles – taxonomy, diversity, physiology and applications. In: Satyanarayana T, Johri BN, Prakash A (eds) Microorganisms in environmental management: microbes and environment. Springer, Dordrecht, pp 1–34

Kang SM, Joo GJ, Hamayun M, Na CI, Shin DH, Kim HY, Hong JK, Lee IJ (2009) Gibberellin production and phosphate solubilization by newly isolated strain of *Acinetobacter calcoaceticus* and its effect on plant growth. Biotechnol Lett 31:277–281

Karan R, Singh S, Kapoor S, Khare S (2011) A novel organic solvent tolerant protease from a newly isolated *Geomicrobium* sp. EMB2 (MTCC 10310): production optimization by response surface methodology. New Biotechnol 28:136–145

Karnchanatat A, Petsom A, Sangvanich P, Piapukiew J, Whalley AJ, Reynolds CD, Gadd GM, Sihanonth P (2008) A novel thermostable endoglucanase from the wood-decaying fungus Daldinia eschscholzii (Ehrenb.: Fr.) Rehm. Enzym Microb Technol 42:404–413

Kim JM, Lee SH, Jung JY, Jeon CO (2010) *Marinobacterium lutimaris* sp. nov., isolated from a tidal flat. Int J Syst Evol Microbiol 60(8):1828–1831

Kim K, Jang YJ, Lee SM, Oh BT, Chae JC, Lee KJ (2014) Alleviation of salt stress by *Enterobacter* sp EJ01 in tomato and Arabidopsis is accompanied by Up-regulation of conserved salinity responsive factors in plants. Mol Cells 37:109–117

Kohler J, Caravaca F, Carrasco L, Roldan A (2006) Contribution of *Pseudomonas mendocina* and *Glomus intraradices* to aggregates stabilization and promotion of biological properties in rhizosphere soil of lettuce plants under field conditions. Soil Use Manag 22:298–304

Kohler J, Caravaca F, Alguacil MM, Roldan A (2009) Elevated CO_2 increases the effect of an *arbuscular mycorrhizal* fungus and a plant-growth promoting rhizobacterium on structural stability of a semiarid agricultural soil under drought conditions. Soil Biol Biochem 41:1710–1716

Kumari S, Vaishnav A, Jain S, Varma A, Choudhary DK (2015) Bacterial-mediated induction of systemic tolerance to salinity with expression of stress alleviating enzymes in Soybean (*Glycine max* L. Merrill). J Plant Growth Regul 34(3):558–573

Kumari S, Vaishnav A, Jain S, Varma A, Choudhary DK (2016) Induced drought tolerance through wild and mutant bacterial strain *Pseudomonas simiae* in mung bean (*Vigna radiata* L.). World J Microbiol Biotechnol 32(1):1–10

Kurz M (2008) Compatible solute influence on nucleic acids: many questions but few answers. Saline Syst 4(1):1

Kushner D (1993) Growth and nutrition of halophilic bacteria. In: Russell HV, Hechstein LI (eds) The biology of halophilic bacteria, vol 1. CRC Press, Boca Raton, pp 87–103

Lamosa P, Turner DL, Ventura R, Maycock C, Santos H (2003) Protein stabilization by compatible solutes. Eur J Biochem 270(23):4606–4614

Larsen H (1986) FEMS Microbiol Rev 39:3–7

León MJ, Fernández AB, Ghai R, Sánchez-Porro C, Rodriguez-Valera F, Ventosa A (2014) From metagenomics to pure culture: isolation and characterization of the moderately Halophilic bacterium *Spiribacter salinus* gen. nov., sp. nov. Appl Environ Microbiol 80(13):3850–3857

Li X, Yu HY, Lin YF (2012) Purification and characterization of an extracellular esterase from a moderately halophilic bacterium, *Halobacillus* sp. strain LY 5. Afr J Biotechnol 11:6327–6334

Liu W, Yang SS (2014) *Oceanobacillus aidingensis* sp. nov., a moderately halophilic bacterium. Antonie Van Leeuwenhoek 105(5):801–808

Logan NA, Berkeley RCW (1984) Identification of *Bacillus* strains using the API system. J Gen Microbiol 130:1871–1882

Machius M, Wiegand G, Huber R (1995) Crystal structure of calcium-depleted *Bacillus licheniformis* α-amylase at 2.2 Å resolution. J Mol Biol 246:545–559

Maidak BL, Cole JR, Lilburn TG, Parker CT, Saxman PR, Stredwick JM, Garrity GM, Li B, Olsen GJ, Pramanik S, Schmidt TM, Tiedje JM (2001) The RDP II (Ribosomal Database Project) continues. Nucleic Acids Res 28:173–174

Makhdoumi-Kakhki A, Amoozegar MA, Ventosa A (2012) *Salinibacter iranicus* sp. nov. and *Salinibacter luteus* sp. nov., isolated from a salt lake, and emended descriptions of the genus Salinibacter and of Salinibacter ruber. Int J Syst Evol Microbiol 62(7):1521–1527

Martín S, Márquez M, Sánchez-Porro C, Mellado E, Arahal D, Ventosa A (2003) *Marinobacter lipolyticus* sp. nov., a novel moderate halophile with lipolytic activity. Int J Syst Evol Microbiol 53:1383–1387

Masalha J, Kosegarten H, Elmaci Ö, Mengel K (2000) The central role of microbial activity for iron acquisition in maize and sunflower. Biol Fertil Soils 30:33–439

Mayak S, Tirosh T, Glick BR (2004) Plant growth-promoting bacteria that confer resistance in tomato to salt stress. Plant Physiol Biochem 42:565–572

Maziah M, Zuraida AR, Halimi M, Zulkifli S, Sreeramanan S (2009) Responses of banana plantlets to rhizobacteria inoculation under salt stress condition. Am-Euras J Sustain Agric 3:290–305

Meintanis C, Chalkou KI, Kormas KA, Lymperopoulou DS, Katsifas EA, Hatzinikolaou DG, Karagouni AD (2006) Application of rpoB sequences similarity analysis, REP-PCR and BOX-PCR for the differentiation of species within the genus *Geobacillus*. Lett Appl Microbiol 46:395–401

Meyer M, Stenzel U, Myles S, Prufer K, Hofreiter M (2007) Targeted high-throughput sequencing of tagged nucleic acid samples. Nucleic Acids Res 35:97

Mezghani M, Alazard D, Karray F, Cayol JL, Joseph M, Postec A, Fardeau ML, Tholozan JL, Sayadi S (2012) *Halanaerobacter jeridensis* sp. nov., isolated from a hypersaline lake. Int J Syst Evol Microbiol 62(8):1970–1973

Miao C, Jia F, Wan Y, Zhang W, Lin M, Jin W (2014) *Halomonas huangheensis* sp. nov., a moderately halophilic bacterium isolated from a saline–alkali soil. Int J Syst Evol Microbiol 64(3):915–920

Miller G, Susuki N, Ciftci-Yilmaz S, Mittler R (2010) Reactive oxygen species homeostasis and signalling during drought and salinity stresses. Plant Cell Environ 33:453–467

Moller MF, Kjeldsen KU, Ingvorsen K (2010) *Marinimicrobium haloxylanilyticum* sp. nov., a new moderately halophilic, polysaccharide-degrading bacterium isolated from Great Salt Lake, Utah. Antonie Van Leeuwenhoek 98(4):553–565

Moreno ML, Perez D, Garcia MT, Mellado E (2013) Halophilic bacteria as a source of novel hydrolytic enzymes. Life 3:38–51

Munns R, Tester M (2008) Mechanisms of salinity tolerance. Annu Rev Plant Physiol 59:651–681

Nabti E, Sahnoune M, Ghoul M, Fischer D, Rothballer M, Schmid M, Hartmann A (2012) Enhancement and restoration of growth of durum wheat (*Triticum durum*, varwaha) on saline soil by using *Azospirillum brasilense* NH and marine alga Ulva lactuca. In: Krueger D, Meyer H (eds) Algae, ecology, economic uses and environmental impact marine biology. Nova Sciences, New York, pp 29–52

Nadeem SM, Zahir ZA, Naveed M, Asghar HN, Arshad M (2010) Rhizobacteria capable of producing ACC-deaminase may mitigate salt stress in wheat. Soil Sci Soc Am J 74(2):533–542

Naz I, Bano A, Ul-Hassan T (2009) Isolation of phytohormones producing plant growth promoting rhizobacteria from weeds growing in Khewra salt range, Pakistan, and their implication in providing salt tolerance to *Glycine max* (L.). Afr J Biotechnol 8:5762–5766

Neilands JB (1981) Iron adsorption and transport in microorganisms. Annu Rev Nutr 1:27–46

Neilands JB (1995) Siderophores: structure and function of microbial iron transport compounds. J Biol Chem 270:26723–26726

Ogut M, Er F, Kandemir N (2010) Phosphate solubilization potentials of soil *Acinetobacter* strains. Biol Fertil Soils 46(7):707–715

Oksanen T, Pere J, Paavilainen L, Buchert J, Viikari L (2000) Treatment of recycled kraft pulps with *Trichoderma reesei* hemicellulases and cellulases. J Biotechnol 78(1):39–48

Omar MNA, Osman MEH, Kasim WA, Abd El-Daim IA (2009) Improvement of salt tolerance mechanisms of barley cultivated under salt stress using *Azospirillum brasiliense*. Tasks Veg Sci 44:133–147

Oren M (1999) Thermophilic and halophilic extremophiles. Curr Opin Microbiol 2:265–269

Oren A (2008) Microbial life at high salt concentrations: phylogenetic and metabolic diversity. Saline Syst 4:2

Oren A (2010) Industrial and environmental applications of halophilic microorganisms. Environ Technol 3(1):825–834

Pandey S, Singh SP (2012) Organic solvent tolerance of an α-amylase from haloalkaliphilic bacteria as a function of pH, temperature, and salt concentrations. Appl Biochem Biotechnol 166:1747–1757

Pappa A, Sánchez-Porro C, Lazoura P, Kallimanis A, Perisynakis A, Ventosa A, Drainas C, Koukkou A (2010) *Bacillus halochares* sp. nov., a halophilic bacterium isolated from a solar saltern. Int J Syst Evol Microbiol 60(6):1432–1436

Parida AK, Das AB (2005) Salt tolerance and salinity effects on plants: a review. Ecotoxicol Environ Saf 60:324–349

Paul D, Lade H (2014) Plant-growth-promoting rhizobacteria to improve crop growth in saline soils: a review. Agron Sustain Dev 34(4):737–752

Paul D, Nair S (2008) Stress adaptations in a plant growth promoting rhizobacterium (PGPR) with increasing salinity in the coastal agricultural soils. J Basic Microbiol 48(5):378–384

Perez E, Miguel S, Marıa MB, Yarzabal LA (2007) Isolation and characterization of mineral phosphate-solubilizing bacteria naturally colonizing a limonitic crust in the south-eastern Venezuelan region. Soil Biol Biochem 39(11):2905–2914

Pourbabaee AA, Bostani S, Amozzegar MA, Naddaf R (2011) Decolorization of cibacron black w-55 under Alkaline conditions by new strain of Halomonas sp. isolated from textile effl uent. Iran J Chem Chem Eng 30(4):63–70

Pugin B, Blamey JM, Baxter BK, Wiegel J (2012) *Amphibacillus cookii* sp. nov., a facultatively aerobic, spore-forming, moderately halophilic, alkalithermotolerant bacterium. Int J Syst Evol Microbiol 62(9):2090–2096

Rajput L, Imran A, Mubeen F, Hafeez FY (2013) Salt-tolerant PGPR strain *Planococcus rifi* etoensis promotes the growth and yield of wheat (*Triticum aestivum* L.) cultivated in saline soil. Pak J Bot 45:1955–1962

Ramadoss D, Lakkineni VK, Bose P, Ali S, Annapurna K (2013) Mitigation of salt stress in wheat seedlings by halotolerant bacteria isolated from saline habitats. Springer Plus 2:1–7

Rashid N, Imanaka H, Fukui T, Atomi H, Imanaka T (2004) Presence of a novel phosphopentomutase and a 2-deoxyribose 5-phosphate aldolase reveals a metabolic link between pentoses and central carbon metabolism in the hyperthermophilic *Archaeon thermococcus kodakaraensis*. J Bacteriol 186(13):4185–4191

Rodríguez-Díaz M, Rodelas B, Pozo C, Martínez-Toledo MV, González-López J (2008) A review on the taxonomy and possible screening traits of plant growth promoting rhizobacteria. In: Ahmad I, Pichtel J, Hayat S (eds) Plant-bacteria interactions: strategies and techniques to promote plant growth. Wiley, Chichester

Roessler M, Muller V (2001) Chloride dependence of glycine betaine transport in *Halobacillus halophilus*. FEBS Lett 489:125–128

Rojas-Tapias D, Moreno-Galván A, Pardo-Díaz S, Obando M, Rivera D, Bonilla R (2012) Effect of inoculation with plant growth-promoting bacteria (PGPB) on amelioration of saline stress in maize (*Zea mays*). Appl Soil Ecol 61:264–272

Ruiz-Lozano JM, Collados C, Barea JM, Azcon R (2001) Cloning of cDNAs encoding SODs from lettuce plants which show differential regulation by *Arbuscular mycorrhizal* symbiosis and by drought stress. J Exp Bot 52:2241–2242

Rus A, Yokoi S, Sharkhuu A, Reddy M, Lee B, Matsumoto TK et al (2001) AtHKT1 is a salt tolerance determinant that controls Na1 entry into plant roots. Proc Natl Acad Sci U S A 98:14150–14155

Sadfi-Zouaoui N, Essghaier B, Hannachi I, Hajlaoui MR, Boudabous A (2007) First report on the use of moderately halophilic bacteria against stem canker of greenhouse tomatoes caused by *Botrytis cinerea*. Ann Microbiol 57(3):337–339

Saghafi K, Ahmadi J, Asgharzadeh A, Bakhtiari S (2013) The effect of microbial inoculants on physiological responses of two wheat cultivars under salt stress. Int J Adv Biol Biomed Res 4:421–431

Sandhya V, Ali SZ, Grover M, Kishore N, Venkateswarlu B (2009a) *Pseudomonas* sp. strain P45 protects sunflowers seedlings from drought stress through improved soil structure. J Oilseed Res 26:600–601

Sandhya V, Ali SZ, Grover M, Reddy G, Venkateswarlu B (2009b) Alleviation of drought stress effects in sunflower seedlings by exopolysaccharides producing *Pseudomonas putida* strain P45. Biol Fertil Soil 46:17–26

Sandhya V, Ali SZ, Grover M, Reddy G, Venkateswarlu B (2010) Effect of plant growth promoting *Pseudomonas* spp. on compatible solutes, antioxidant status and plant growth of maize under drought stress. Plant Growth Regul 62:21–30

Santos AF, Valle RS, Pacheco CA, Alvarez VM, Seldin L, Santos AL (2013) Extracellular proteases of *Halobacillus blutaparonensis* strain M9, a new moderately halophilic bacterium. Braz J Microbiol 44:1299–1304

Sapsirisopa S, Chookietwattana K, Maneewan K, khaengkhan P (2009) Effect of salt-tolerant *Bacillus inoculum* on rice KDML 105 cultivated in saline soil. Asian J Food Agro-Ind (Special Issue):S69–S74

Saralov A, Kuznetsov B, Reutskikh E, Baslerov R, Panteleeva A, Suzina N (2012) Arhodomonas recens sp. nov., a halophilic alkane-utilizing hydrogen-oxidizing bacterium from the brines of flotation enrichment of potassium minerals. Microbiology 81(5):582–588

Selvakumar G, Panneerselvam P, Ganeshamurthy AN (2012) Bacterial mediated alleviation of abiotic stress in crops. In: Maheshwari DK (ed) Bacteria in agrobiology: stress management. Springer, Berlin, pp 205–223

ShaoMin Y, Guang W (2013) Secretory pathway of cellulase: a mini-review. Biotechnol Biofuels. doi:10.1186/1754-6834-6-177

Sharma A, Singh P, Kumar S, Kashyap PL, Srivastava AK, Chakdar H, Singh RN, Kaushik R, Saxena AK, Sharma AK (2015) Deciphering diversity of salt-tolerant Bacilli from saline soils of eastern Indo-Gangetic Plains of India. Geomicrobiology 32:70–180

Shen M, Kang YJ, Wang HL, Zhang XS, Zhao QX (2012) Effect of plant growth promoting rhizobacteria (PGPRs) on plant growth, yield, and quality of tomato (*Lycopersicon esculentum Mill*) under simulated seawater irrigation. J Gen Appl Microbiol 58:253–262

Shirmardi M, Savaghebi GR, Khavazi K, Akbarzadeh A, Farahbakhsh M, Rejali F, Sadat A (2010) Effect of microbial inoculants on uptake of nutrient elements in two cultivars of sunflowers (*Helianthus annuus* L). Not Sci Biol 2:57–66

Shivanand P, Mugeraya G, Kumar A (2013) Utilization of renewable agricultural residues for the production of extracellular halostable cellulase from newly isolated *Halomonas* sp. strain PS47. Ann Microbiol 63(4):1257–1263

Shukla PS, Agarwal PK, Jha B (2012) Improved salinity tolerance of *Arachis hypogeae* (L.) by the interaction of halotolerant plant-growth promoting rhizobacteria. J Plant Growth Regul 31:195–206

Siddikee MA, Chauhan PS, Anandham R, Han GH, Sa T (2010) Isolation, characterization and use for plant growth promotion under salt stress, of ACC deaminase-producing halotolerant bacteria derived from coastal soil. J Microbiol Biotechnol 20:1577–1584

Singh SP, Raval VH, Purohit MK, Pandey S, Thumar JT, Gohel SD, Akbari VG, Rawal CM (2012) Haloalkaliphilc bacteria and actinobacteria from the saline habitats: new opportunities for biocatalysis and bioremediation. In: Satyanarayana T, Johri BN, Prakash A (eds) Microorganisms in environmental management: microbes and environment. Springer, New York/Dordrecht (Library of Congress Control Number: 2011944027), pp 415–429

Sivaprakasam S, Mahadevan S, Sekar S, Rajakumar S (2008) Biological treatment of tannery wastewater by using salt-tolerant bacterial strains. Microb Cell Fact 7:15

Sleator RD, Hill C (2001) Bacterial osmoadaptation: the role of osmolytes in bacterial stress and virulence. FEMS Microbiol Rev 26:49–71

Sorokin D, Kolganova T (2013) Bacterial chitin utilization at halophilic conditions. Extremophiles 18:243–248

Souza PM (2010) Application of microbial α-amylase in industry – a review. Braz J Microbiol 41:850–861

Srivastava S, Yadav A, Seem K, Mishra S, Chaudhary V, Srivastava CS (2008) Effect of high temperature on *Pseudomonas putida* NBRI0987 biofilm formation and expression of stress sigma factor RpoS. Curr Microbiol 56(4):453–457

Stackebrandt E, Goebel BM (1994) Taxonomic note: a place for DNA-DNA reassociation and 16S rRNA sequence analysis in the present species definition in bacteriology. Int J Syst Bacteriol 44:846–849

Street TO, Bolen DW, Rose GD (2006) A molecular mechanism for osmolyte-induced protein stability. Proc Natl Acad Sci U S A 103(38):13997–14002

Suzuki K, Goodfellow M, O'Donnell AG (1993) Cell envelopes and classification. In: Goodfellow M, O'Donnell AG (eds) Handbook of new bacterial systematics. Academic, London, pp 195–250

Tan Z, Hurek T, Vinuesa P, Müller P, Ladha JK, Reinhold-Hurek B (2001) Specific detection of *Bradyrhizobium* and *Rhizobium* strains colonizing Rice (Oryza sativa) roots by 16S-23S ribosomal DNA intergenic spacer-targeted PCR. Appl Environ Microbiol 67(8):3655–3664

Tank N, Saraf M (2010) Salinity-resistant plant growth promoting rhizobacteria ameliorates sodium chloride stress on tomato plants. J Plant Interact 5:51–58

Upadhyay SK, Singh JS, Singh DP (2011) Exo-polysaccharide-producing plant growth-promoting rhizobacteria salinity condition. Pedosphere 21:214–222

Vaishnav A, Jain S, Kasotia A, Kumari S, Gaur RK, Choudhary DK (2013) Effect of nitric oxide signaling in bacterial treated soybean plant under salt stress. Arch Microbiol 195:171–177

Vaishnav A, Kumari S, Jain S, Varma A, Choudhary DK (2015) Putative bacterial volatile-mediated growth in soybean (*Glycine max* L. Merrill) and expression of induced proteins under salt stress. J Appl Microbiol 119(2):539–551

Vaishnav A, Kumari S, Jain S, Varma A, Tuteja N, Choudhary DK (2016) PGPR-mediated expression of salt tolerance gene in soybean through volatiles under sodium nitroprusside. J Basic Microbiol 56:1–15

Van Qua D, Simidu U, Taga N (1981) Purifi cation and some properties of halophilic protease produced by a moderately halophilic marine *Pseudomonas* sp. Can J Microbiol 27:505–510

Vidyasagar M, Prakash S, Jayalakshmi S, Sreeramulu K (2007) Optimization of culture conditions for the production of halothermophilic protease from halophilic bacterium *Chromohalobacter* sp. TVSP101. World J Microbiol Biotechnol 23:655–662

Vijayaraghavan P, Jebamalar TRJ, Vincent SGP (2012) Biosynthesis optimization and purification of a solvent stable alkaline serine protease from *Halobacterium* sp. Ann Microbiol 62:403–410

Wang C, Knill E, Glick BR, Defago G (2000) Effect of transferring 1-aminocyclopropane-1-carboxylic acid (ACC) deaminase genes into *Pseudomonas fluorescens* strain CHAO and its gacA derivative CHA96 on their growth-promoting and disease-suppressive capacities. Can J Microbiol 46:898–907

Wang YX, Liu JH, Xiao W, Zhang XX, Li YQ, Lai YH, Ji KY, Wen ML, Cui XL (2012) *Fodinibius salinus* gen. nov., sp. nov., a moderately halophilic bacterium isolated from a salt mine. Int J Syst Evol Microbiol 62(2):390–396

Wang M, Zheng Q, Shen Q, Guo S (2013) The critical role of potassium in plant stress response. Int J Mol Sci 14(4):7370–7390

Wang S, Sun L, Wei D, Zhou B, Zhang J, Gu X, Zhang L, Liu Y, Li Y, Guo W (2014) *Bacillus daqingensis* sp. nov., a halophilic, alkaliphilic bacterium isolated from saline-sodic soil in Daqing, China. J Microbiol 52(7):548–553

Wejse PL, Ingvorsen K, Mortensen KK (2003) Purification and characterisation of two extremely halotolerant xylanases from a novel halophilic bacterium. Extremophiles 7(5):423–431

Whittaker P, Mossoba MM, Al-Khaldi S, Fry FS, Dunkel VC, Tall BD, Yurawecz MP (2003) Identification of foodborne bacteria by infrared spectroscopy using cellular fatty acid methyl esters. J Microbiol Methods 55(3):709–716

Woese C (1993) The Archaea: their history and significance. In Kates M, Kushner D, Matheson A (eds) The biochemistry of Archaea (Archaebacteria). Elsevier, Amsterdam, pp vii–xxix

Yoon JH, Kang SJ, Jung YT, Lee KC, Oh HW, Oh TK (2010) *Virgibacillus byunsanensis* sp. nov., isolated from a marine solar saltern. Int J Syst Evol Microbiol 60(2):291–295

Zeikus JG, Hegge PW, Thompson TE, Phelps TJ, Langworthy TA (1983) Isolation and description of *Haloanaerobium praevalens* gen. nov. and sp. nov. J Biotechnol 152:114–124

Zhang G, Li S, Xue Y, Mao L, Ma Y (2012) Effects of salts on activity of halophilic cellulase with glucomannanase activity isolated from alkaliphilic and halophilic *Bacillus* sp. BG-CS10. Extremophiles 16:35–43

Abiotic Stress Mitigation Through Plant-Growth-Promoting Rhizobacteria

15

Palika Sharma, Veena Khanna, and Suman Kumari

Abstract

Abiotic and biotic stresses highly impacts production of principal crops all around the world. Due to climate change, extreme abiotic factors like high and low temperatures, droughts, salinity, osmotic stress, heavy rains, floods and frost damages are posing grave threats to crop production. There is a dire need to mitigate these stresses, so in order to cope with such impacts, microorganisms can be employed as best alternatives to chemical inputs by exploiting their unique properties of tolerance to extreme environments, their ubiquity, their genetic diversity and their interaction with crop plants and by developing methods for their successful employment in agriculture production. Plant-growth-promoting rhizobacteria (PGPRs) mitigate abiotic stresses on plants most effectively through degradation of ACC, the ethylene precursor by bacterial ACC-deaminase and through biofilm and exopolysaccharide production. Alleviation of environmental stresses in crop plants using these microorganisms opens new and emerging applications in sustainable agriculture.

15.1 Introduction

Agriculture is considered to be one of the most vulnerable sectors to climate change. Elevation in abiotic and biotic stresses has become major cause for stagnation of productivity in principal crops. Exposure of plants to a large number of different environmental stresses like flooding, drought, extremes of pH and temperature, high salt, heavy metals and various pathogens affect plant growth and productivity.

P. Sharma (✉) • V. Khanna • S. Kumari
Department of Plant Breeding and Genetics, Punjab Agricultural University,
Ludhiana 141004, Punjab, India
e-mail: ps.batish@gmail.com

D.K. Choudhary et al. (eds.), *Plant-Microbe Interaction: An Approach to Sustainable Agriculture*, DOI 10.1007/978-981-10-2854-0_15

Under these environmental stresses, plants synthesize raised levels of the plant hormone ethylene which halts growth and proliferation of plant to greater extent, until the stress is mitigated by lowering ethylene level (Gamalero and Glick 2012). The decline in yield of wheat and paddy in various parts of South Asia has been evidenced due to increased osmotic stress and temperature and reduced rainfall. The average temperature in the Indian subcontinent is likely to rise to 5.8 °C by 2100 (Grover et al. 2011). Besides elevated temperature, droughts, increased CO_2, heavy rainfall, flooding, cold and heat waves, and other huge natural disasters causing severe economic losses are being witnessed globally. These events cause serious negative impacts on crop growth and yields and impose severe pressure on our land and water resources. The major abiotic stress in India is drought or osmotic stress due to high temperatures, soil salinity/alkalinity, low pH, and metal toxicity affecting about two-thirds area, leading to formation of the arid and semiarid regions. Nearly 11 m ha area is affected by salinity, a chemical stress and another 16 m ha by water logging, a physical stress (Grover 2010). It is a major challenge to develop efficient, low-price, and easily adaptable methods for the abiotic stress mitigation. Globally, extensive study is being conducted, to develop strategies to deal with abiotic stresses by developing heat- and drought-tolerant crop varieties, shifting the crop calendars, resource management practices, etc. (Venkateswarlu and Shanker 2009). An unanticipated amplification in agricultural practices aimed to improve the crop production at an unprecedented rate has exploited the cost-intensive technologies and strategies which are unfavorable for the sustainability of soil health (Kumar et al. 2010). The ill-advised exceeding use of agro-chemicals in agricultural land is posing grave threats to the soil fertility. In this context, there should be a paradigm shift toward eco-friendly strategies to mitigate abiotic stress and enhance crop yields. Recently, it has been indicated that some microorganisms can also help crops to tolerate environmental stresses and promote plant growth through nutrient management and biocontrol. This beneficial group of bacteria colonizing plant's rhizosphere/endorhizosphere promotes plant growth through varied direct and indirect mechanisms (Shahzad et al. 2014). However, recently, the microbes alleviating biotic and abiotic stresses has attained great importance. The concept of PGPR-eliciting tolerance to abiotic stresses has been reviewed recently (Yang et al. 2004). The present review compiles the recent work on the role of rhizobacteria aiding crops to tolerate various abiotic stresses due to climate change like heat, salinity, drought, chilling injury, and waterlogging. Plant-growth-promoting rhizobacteria (PGPRs) with ACC-deaminase activity facilitate the proliferation of plants under stressed conditions.

15.1.1 Plant-Growth-Promoting Rhizobacteria

There exists diverse group of bacteria in the soil which are associated with the roots of all higher plants. These efficient bacteria compete in the rhizosphere resulting in plant–microbe interactions which could be positive, neutral, or negative (Shahzad et al. 2014). Bacteria colonizing plant roots aggressively are able to stimulate plant growth through various mechanisms and are referred as plant-growth-promoting rhizobacteria (PGPR) (Kloepper et al. 1986).

Plant-growth-promoting rhizobacteria are the beneficial group of rhizobacteria that are known to enhance plant growth via direct and indirect means and are viable options for meeting demand of sustainable agriculture alternate to chemical inputs which are hazardous to living forms and impose harmful impact on environment. PGPR acts as (1) biofertilizers (enhancing nutrient (N, P, K, Zn, Fe, etc.) availability to plant), (2) phytohormone producers, (3) rhizoremediators (degrading organic pollutants), and (4) biocontrol agents (Antoun and Pre 'vost 2005). These PGPRs can be extracellular (ePGPR), existing in the rhizosphere, on the rhizoplane, or in the spaces between root cortex cells, or can be intracellular (iPGPR) residing in root cells or in specialized nodular structures (Sundaramoorthy and Balabaskar 2012). *Arthrobacter*, *Agrobacterium*, *Azotobacter*, *Azospirillum*, *Bacillus*, *Burkholderia*, *Pseudomonas*, and *Serratia* are some examples of ePGPR, whereas iPGPRs are *Azorhizobium*, *Bradyrhizobium*, and *Mesorhizobium* (Bhattacharyya and Jha 2012). PGPR can promote plant growth mainly via following means:

1. ACC-deaminase production to lower ethylene levels in plant roots
2. Phytohormone production like indoleacetic acid, gibberellic acid, cytokinins, and ethylene
3. Symbiotic nitrogen fixation
4. Exhibiting antagonistic activity against phytopathogens through siderophores, b-1,3-glucanase, chitinases, antibiotics, fluorescent pigment, and cyanide production
5. Solubilization of mineral phosphates and other nutrients (Shahzad et al. 2014)

These PGPRs when inoculated with crops not only promote their growth and yield but also maintain soil fertility; thus, PGPR as biofertilizer is an eco-friendly approach.

15.2 PGPR-Mediated Stress Tolerance Mechanisms

PGPRs use various mechanisms to protect plants from abiotic stresses which pose grave threats to agricultural production (Fig. 15.1).

15.2.1 Ethylene Biosynthesis and Role in Plant Physiology

Ethylene is a gaseous plant hormone which is produced by almost all plants and imposes different effects on plant growth depending on its concentration in root tissues. Ethylene at low levels plays an active role in seed germination, tissue differentiation, anthocyanin synthesis, root and shoot primordia formation, root elongation, lateral bud development, flowering initiation, opening and senescence of flower, pollination, ripening and degreening of fruit, and the production of volatile organic

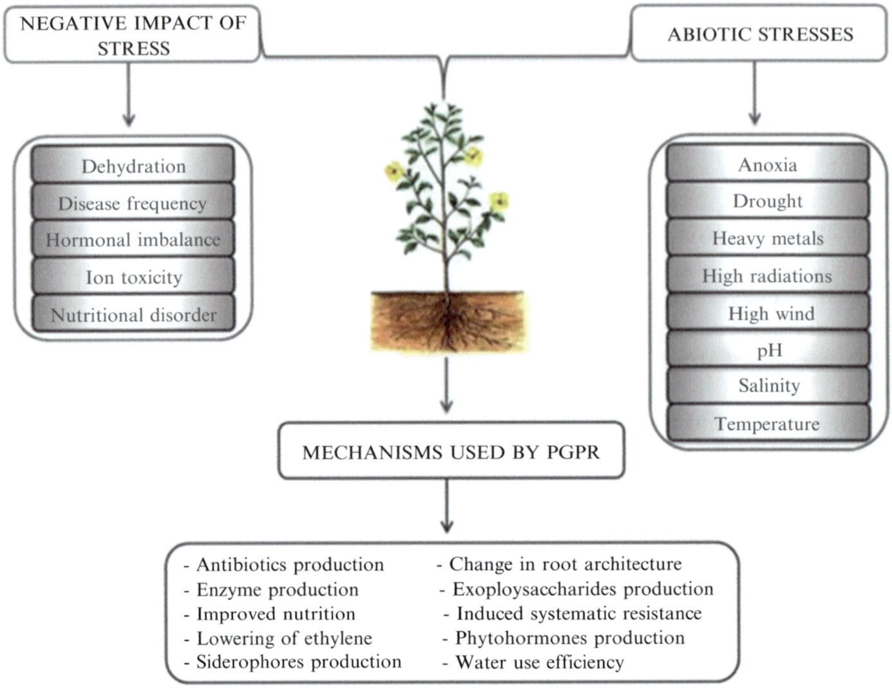

Fig. 15.1 Mechanisms used by PGPR for abiotic stress mitigation

compounds, which imparts aroma to fruits and is also involved in plant–microbial symbiotic interactions that are important for establishment of the legume–*Rhizobium* association (Lynch and Brown 1997). At high concentrations, ethylene is usually deleterious to plant growth and health as it leads to defoliation, inhibition of stem and root growth, as well as premature senescence and also causes decreased vegetative period, which ultimately reduces crop performance. The classical "triple" response in etiolated dicot seedlings due to ethylene is a renowned example of ethylene as stress hormone. This effect includes three distinct morphological changes in the seedling shape, inhibition of stem elongation, increased stem diameter, and horizontal growth (Khalid et al. 2006).

Ethylene is generated by most of the plant tissues. Synthesis of this hormone begins biologically with S-adenosylmethionine (SAM) compound (Fig. 15.2) that acts as a precursor in many other pathways and is, therefore, present in abundance within plant tissues. The ethylene pathways along with the Yang cycle (Yang and Hoffman 1984) initiate with the enzyme ACC-synthase that converts SAM to 1-aminocyclopropane-1-carboxylic acid (ACC) and 5 V methylthioadenosine (MTA). It has been considered to be the foremost step in the ethylene biosynthetic pathway, whereas the extremely labile ACC-synthase enzyme has been shown to be rate limiting which rises proportionally to ethylene levels within the plant tissue.

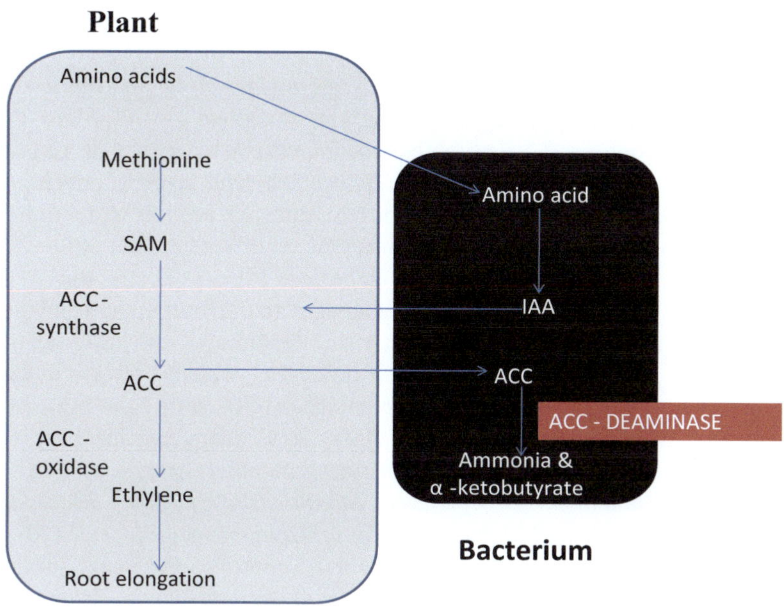

Fig. 15.2 Ethylene biosynthesis and degradation by ACC-deaminase

The next step is the conversion of ACC to ethylene by ACC-oxidase, which is present in most tissues at very low levels. A huge increase in ethylene production is activated due to ACC-synthase and/or ACC-oxidase when ethylene is applied exogenously to plant (Wang et al. 2002). Stimulation of ethylene by IAA also occurs in etiolated pea seedlings, via a rapid increase in the buildup of ACC-oxidase (Peck and Kende 1995).

Ethylene is also known as "stress" hormone, and its accelerated production is associated with both biotic and abiotic stresses (Arshad et al. 2008). When "stress" ethylene synthesis increases, it causes senescence in response to stress in the plant, leading to physiological changes in cells near to the site of stress. As a consequences of different types of environmental stress, viz., chilling, drought, flooding, pathogens, and heavy metal toxicity, plants respond by synthesizing 1-aminocyclopropane-1-carboxy late (ACC), a precursor for ethylene (Glick et al. 2007). An increased ethylene concentration in the root zone is also known to inhibit nodulation and subsequently nitrogen fixation in lentil plants; PGPR can help overcome these deleterious effects.

15.2.2 ACC-Deaminase Production

A pyridoxal phosphate-dependent enzyme, 1-aminocyclopropane-1-carboxylate deaminase (ACCD) is widespread in various bacterial and fungal species. Ethylene production is accelerated endogenously in response to abiotic and biotic stresses via elevated levels of the ethylene precursor ACC which has adverse affects on the root and plant growth (Shrivastava and Kumar 2013). Owing to ACCD activity, certain plant-associated bacteria help plant to proliferate under abiotic and biotic stresses by lowering the "stress ethylene" level which inhibits plant growth. Many PGPRs have 1-aminocyclopropane-1-carboxylate deaminase activity which can cleave ACC to α-ketobutyrate and ammonia to reduce ethylene level in developing or stressed plants. acdS is a gene which encodes this enzyme, and it is under tight regulation and regulated differentially under different environmental conditions (Singh et al. 2015). Hontzeas et al. (2005) have elaborated the crystal structure of ACC-deaminase from *Pseudomonas putida* UW4 along with the biochemical and thermodynamic properties. One of the characteristic features of all ACC-deaminase enzymes is their low affinity toward the substrate ACC, which is always in the millimolar range. In the past few years, a large number of bacteria have been isolated encoding ACC-deaminase activity (Glick 2005). ACC-deaminase has been widely reported and extensively studied in numerous microbial species of PGPR like *Agrobacterium genomovars*, *Azospirillum*, *Bacillus*, *Burkholderia*, *Enterobacter*, *Pseudomonas*, *Rhizobium*, *Rhodococcus*, *Sinorhizobium meliloti*, etc. (Ali et al. 2014). ACC-deaminase active PGPR strains have improved the maize production under dual stress conditions, i.e., drought and soil salinity/sodicity (Zafar-Ul-Hye et al. 2014). Plants inoculated with ACC-deaminase-positive rhizobacteria are highly resistant to the injurious effects of the stress ethylene produced under flooding (Grichko and Glick 2001), drought (Creus et al. 2004), and high salt concentrations (Nadeem et al. 2007).

15.2.3 EPS Production

Extracellular exopolysaccharide (EPS) accumulation is commonly observed feature in many bacteria. Bacterial EPS production in saline soil can prevent osmotic stress in plants, and it also shields microorganisms from water stress by increasing water retention and by regulation of organic carbon source permeability. Role of EPS producing root-colonizing bacteria in improving plant growth has been variously reported (Ali et al. 2014). EPS aids the microorganisms to attach and colonize the plant roots irreversibly involving a network of fibrillar material that results in firm gripping of bacteria to the root surface. Sandhya et al. (2009) have reported that exopolysaccharide secretion by PGPR forms an organo-mineral sheath around microbial cell which enables specific bacteria to survive under stressed conditions such as drought, and it also improves drought tolerance in plants through osmotic and intracellular adjustment. The type of polysaccharide determines water retention capacity, but water retention capacity by EPS may exceed 70 g water per g polysaccharide. A study revealed that inoculation of barley with exopolysaccharide-producing PGPR exhibited extended drought tolerance in comparison to uninoculated control (Timmusk 2003).

15.2.4 Biofilm Production

Most bacteria in natural environments persist as "biofilm" communities where cells are encased in an extracellular polymeric matrix. Biofilms established on varied surfaces like root zones and soil particles and cement soil particles form aggregates which improve crop production and physiochemical properties of soil (Qurashi and Sabri 2012). Under stressed conditions, bacteria persist in the form of biofilm communities for their better survival, as an extracellular matrix provides an infinite range of macromolecules. The dense biofilm matrix regulates diffusion of bioactive substances and nutritional secretions by rhizobacteria which therefore remain concentrated at the root surface in order to affect plant growth. Timmusk (2003) suggested that the major components of biofilm in the model bacterium *Bacillus subtilis* are polysaccharides and a Tas A protein, and when these components get mutated, they pose severe effects on biofilm production. The sugars in biofilms can be divided into simple sugars (monosaccharides, oligosaccharides, polysaccharides) and complex sugars: all of which can play various roles in host–microbe interactions (Vu et al. 2009). The biofilm improves water retention and enhances soil aggregation and microbial biomass which in turn stimulates root exudation under stress. Hence, there is a great advantage of a slimy layer of extracellular matrix produced in the rhizosphere, especially under stressful conditions; it contributes to mechanical stability of the biofilm and interacts with other macromolecules and micromolecules, creating a microenvironment within the biofilm (Timmusk 2003).

15.2.5 Nutrient Deficiency Tolerance

Plants require various macro (N, P, and K) and micro (Zn, Mn, and Fe) nutrients for their growth and metabolism; thus, their deficiency in plants may lead to reduction in crop yields to greater extent. The deficiency of macro- and micronutrient is the major factor contributing not only to yield plateaus but also to declining crop production, shrinking profits, and environmental footprint (Velu et al. 2013).

15.2.6 Biological Nitrogen Fixation

Nitrogen (N) ranks first among the major plant nutrients, yet its low availability to plants due to the high losses by emission or leaching is a limiting factor in agricultural ecosystems. Some microorganisms are capable of making atmospheric N available to plants through biological nitrogen fixation (BNF) which is of great importance (Martínez-Viveros et al. 2010). For agricultural sustainability advancement, an increase in the utilization of BNF as a major source of nitrogen for plants is required. PGPRs with biological nitrogen fixation (BNF) property include symbiotic nitrogen fixers (*Rhizobium* in leguminous plants and *Frankia* in nonleguminous trees) and nonsymbiotic N_2-fixers (*Azospirillum*, *Azotobacter*, *Gluconacetobacter*, *Achromobacter*, *Azoarcus*, *Acetobacter diazotrophicus*, *Bacillus*, *Klebsiella*, and

Pseudomonas). Nonsymbiotic nitrogen fixation has a great agronomic significance. Free-living diazotrophs carrying out nonsymbiotic BNF can promote growth in nonleguminous plants. Studies by Antoun et al. (1998) have revealed that N-fixers, free-living bacteria, as well as symbiotic *Rhizobium* strains can stimulate the growth of radish which is a nonlegume.

15.2.7 Phosphorus Solubilization

Element phosphorus (P) ranks second after nitrogen among mineral nutrients which also limits the growth of crops. Phosphorus as an essential mineral nutrient participates in numerous metabolic processes such as transfer of energy, respiration, biosynthesis of macromolecules, and signal transduction (Khan and Joergensen 2009). The P content in soil (organic plus inorganic) is 100–400 g/ha which usually exceeds plant requirements; however its bioavailability to plant is one of the major constraints in limited plant growth. Even if phosphorus is added to soil as phosphatic fertilizer, it may get fixed and precipitate in calcareous and alkaline soils and become unavailable to plant. Phosphate anions due to their extreme reactivity get precipitated with cations such as Ca^{2+}, Mg^{2+}, Fe^{3+}, and Al^{2+} and become immobilized. On the other hand, much of this P is in mineral form and is available to plants only at gradual basis (Richardson et al. 2009).

In order to meet sustainable agricultural demands, the use of special microorganisms as inoculum for mobilization of a large pool of soil phosphorus is one of the useful strategies to improve crop yields. A group of PGPRs referred to as phosphate-solubilizing bacteria (PSB) are employed as solubilizers of insoluble inorganic P compounds at frequent basis through production of low-molecular-weight organic acids such as gluconic and keto-gluconic acids, and genera belonging to *Azotobacte*r, *Bacillus*, *Enterobacter*, *Klebsiella*, *Pseudomonas*, and *Rhizobium* are the potent phosphate solubilizers (Verma et al. 2013). In addition, inorganic acids such as carbonic acid, sulfuric acid, and nitric acid produced by PSB could also facilitate the solubilization of P and Zn compounds.

The PSB constitute 1–50 % whereas phosphorous solubilizing fungi (PSF) contribute only 0.1–0.5 % in P solubilization potential among the whole microbial population in soil (Chen et al. 2006). The microorganisms involved in P solubilization can enhance plant growth also by enhancing biological nitrogen fixation efficiency and the availability of other trace elements such as Fe, Zinc, etc. (Gyaneshwar et al. 2002).

15.2.8 Siderophore Production

To satisfy nutritional requirements of iron, microorganisms are blessed with a special mechanism that assimilates iron via secretion of low-molecular-weight (500–1000 Da) chelators having greater affinity for iron which are termed as siderophores and are generally produced in response to iron starvation (Sarode et al. 2009). Strains of *Pseudomonas*, *Enterobacter*, *Azotobacter*, *Gluconacetobacter*, *Bacillus*, and *Rhodococcus* genera are known to produce siderophores classified as catecholates (phenolates), carboxylates, and hydroxamates by the ligands used to chelate the ferric iron (Saharan and Nehra 2011). Under aerated environment at physiological pH, the unstable reduced ferrous (Fe^{2+}) form is readily oxidized to the ferric (Fe^{3+}) form, which is unavailable to biological systems, and its concentration is less than optimal for bacteria which necessitates special mechanisms for iron acquisition; thus, secreted siderophores by microorganisms solubilize iron by forming a complex ferric siderophore (Sharma and Johri 2003). Although some siderophores are known to chelate other ions along with iron, many siderophores chelate iron not only for microbial nutrition but also for microbial infection and the antagonism of PGPR against plant pathogen by sequestering Fe^{3+} in root area (Beneduzi et al. 2012); thus, siderophores have been implicated for both direct and indirect mechanisms of plant growth promotion.

Siderophores such as salicylic acid and pyoverdine are known to induce systemic resistance. Further, most of the catechols are derivatives of 2,3-dihydroxy benzoic acid (DHBA) and consist of 2,3-DHBA and one or more amino acid residues. *Pseudomonas fluorescens* is known to secrete pyoverdines which are yellow-greenish fluorescent siderophores involved in the transport of iron into the cell (Meyer et al. 2002). Siderophore enterobactin secreted by *Escherichia coli*, bacillibactin from *Bacillus subtilis* and *Bacillus anthracis*, and vibriobactin produced by *Vibrio cholera* are some of the examples of catecholates. Ferrichromes produced by *Ustilago sphaerogena*, desferrioxamine B (deferoxamine) by *Streptomyces pilosus* and *Streptomyces coelicolor*, and desferrioxamine E by *Streptomyces coelicolor* belong to the hydroxamate class of siderophores (Sarode et al. 2009).

In recent years, considerable interest has been paid to rhizobacteria, which are aggressive root colonizers and produce siderophores. The role of microbial siderophores in N-fixation has also been implicated. Indirect mode of plant growth promotion is the ability of siderophore to protect from heavy metal toxicity (Glick 2005). Thus plants are benefitted in a number of ways, i.e., by direct uptake of iron, suppression of proliferation of fungal pathogens, improved N-fixation, and prevention from heavy metal toxicity.

15.2.9 Zinc Solubilization

Zinc is an imperative micronutrient necessary at low concentrations (5–100 mg kg^{-1}) in plant tissues for healthy growth and reproduction and acts as a cofactor in many enzymes. Zn deficiency is currently listed as a major risk factor for both plant and human health globally. As a result of prevalent Zn deficiency, the production of cereal crops suffers twin problems of low food production and Zn malnutrition in the population using cereals as their staple diet (Vaid et al. 2013). In plants, its deficiency results in reduction in membrane integrity and synthesis of starch, protein, growth hormones, nucleotides, chlorophyll, and cytochromes and also leads to development of susceptibility to heat stress. Zn influences basic life processes of plant, such as (a) nitrogen metabolism–nitrogen uptake and protein quality, (b) photosynthesis–chlorophyll synthesis and carbon anhydrase activity, (c) abiotic stress tolerance, and (d) rate of protein synthesis and protein content (Potarzycki and Grzebisz 2010). On the other hand, dietary deficiency of zinc (Zn) leads to human health complications including impairments in the immune system together with incidence of infectious diseases such as severe acute malnutrition, diarrhea, and pneumonia which affects more than two billion people worldwide (WHO 2012).

Like phosphorus zinc is also present in soil in insoluble form that so plants are unable to utilize it. The problem of Zn deficiency in crops is attributed to its lesser solubility in soils instead of its low total amount. Solubilization of zinc through PGPR can be achieved by various mechanisms, including excretion of metabolites such as organic acids and chelating agents or through proton extrusion (Ramesh et al. 2014). Rhizobacterial genera belonging to spp. *Pseudomonas* and *Bacillus* have been reported to solubilize zinc through proton extrusion, chelating ligands, and oxidoreductive systems established on the cell surface and membranes (Goteti et al. 2013). Reduction in pH through organic acid production by microbial isolates is regarded as one of the major mechanisms of Zn solubilization. Solubilization of Zn compounds using soil bacteria has been reported by Fasim et al. (2002).

15.2.10 Phytohormone Production

PGPRs are known to secrete phytohormones, viz., auxins, cytokinins, gibberellins, and ethylene. The naturally occurring auxin is indoleacetic acid (IAA) which exists in abundance and has the ability to control many aspects of plant growth and development such as differentiation of vascular tissues, apical dominance, root elongation, initiation of lateral roots, and fruit setting and ripening (Maheswari et al. 2013). Gibberellins (GA) are plant-growth-promoting hormones that are engaged in the germination of seeds, seedling emergence, stem and leaf growth, induction of flowering, regulation of vegetative and reproductive bud dormancy, and fruit growth (Maheswari et al. 2013). The phytohormones produced by rhizospheric and endophytic bacteria (*Enterobacter*, *Pseudomonas*, *Stenotrophomonas*) play a key role in modifying root morphology in plants exposed to drought, salinity, high temperature, and toxicity of heavy metals (Spaepen and Vanderleyden 2010). Plants inoculated with IAA producers have resulted in stimulation of seed germination which accelerates root growth and also leads to modification in root architecture to increase the root biomass even under stressed conditions. IAA-producing bacteria are considered as potential plant growth promoters as they increase the root surface and create a larger infection area for colonization of potential diazotrophs (Molla et al. 2001).

Ethylene, a phytohormone, is produced almost in all plants and is known to mediate several responses to environmental and developmental signals in plants. Arshad and Frankenberger (1998) have shown that ethylene when exudated by the roots exhibits involvement in plant growth. The fate of rhizobial infection in legume root hairs is regulated by the levels of ethylene in the underlying plant cortex; a low level of ethylene allows proper deposition of infection thread, whereas a higher level of the hormone induces abortion of the infection thread by inducing cross-linking of its matrix glycoproteins (Ma et al. 2003). Certain free-living rhizobacteria with ACC-deaminase activity promote nodulation in plant roots by endogenously regulating the biosynthesis of ethylene.

15.2.11 Induced Systemic Tolerance

Plant-growth-promoting rhizobacteria are known to alleviate the impact of abiotic stresses on plants effectively through induced systemic tolerance (IST), via (a) cytokinin production by bacteria, (b) antioxidants, and (c) enzyme ACC-deaminase degrading ethylene precursor ACC. The terminology, induced systemic tolerance, has been proposed for PGPR-induced physical and chemical changes resulting in increased tolerance to abiotic stress. Another term, "induced systemic resistance," (ISR) refers to a process involving physical or chemical changes related to plant defense by PGPRs. PGPR eliciting ISR has been reported to suppress plant diseases caused by a range of phytopathogens both in the greenhouse and field (Kloepper et al. 2004). However, few reports on PGPR as elicitors of tolerance to abiotic stresses, such as drought, salt, and nutrient deficiency, have also been published. More recently, the subject of PGPR-eliciting tolerance to heavy metal toxicity has also been reviewed. The term "induced systemic tolerancc" (IST) is proposed here for physical and chemical changes induced in plants by PGPR resulting in enhanced tolerance to abiotic stress only whereas biotic stress is excluded from IST because conceptually it is part of biological control and induced resistance (Yang et al. 2004).

PGPR strain, *Achromobacter piechaudii* ARV8, producing ACC-deaminase, conferred IST to drought stress in pepper and tomato (Mayak et al. 2004). Under stressed environment, the stress hormone ethylene endogenously regulates plant homeostasis resulting in decreased root and shoot growth. However, breaking down of the ethylene precursor ACC by bacterial enzyme ACC-deaminase releases plant stress and rescues normal plant growth (Glick 2005).

15.2.12 Proline Accumulation in Plants Under Stress Conditions

A proteinogenic amino acid, proline is an indispensable component for primary metabolism which regulates plant development and also acts as a signaling molecule. However, accumulation of proline is known to influence stress tolerance in various ways. It has also been reported that proline can work as a molecular chaperone in order to protect protein integrity and to activate different enzymes. An antioxidant feature has been attributed to proline which suggests its ROS scavenging activity and its role as a singlet oxygen quencher (Matysik et al. 2002). Accumulation of proline could be due to de novo synthesis or decreased degradation or both. Many studies have revealed that under different stress conditions, an increase in proline

Table 15.1 Bacteria-mediated abiotic stress tolerance in plants

	Bacteria inoculated	Plant species	References
Osmotic stress PEG 6000	*Bacillus cereus*	Lentil (*Lens culinaris* Medikus)	Sharma et al. (2015)
Osmotic stress PEG 6000	*Bacillus* sp. and *Pseudomonas* sp.	Chickpea (*Cicer arietinum*)	Sharma et al. (2013)
Salinity	*Achromobacter piechaudii*	*Tomato* (*L. esculentum*)	Mayak et al. (2004)
Salinity	*Pseudomonas fluorescens*	Maize (*Zea mays*)	Nadeem et al. (2007)
Salinity	*B. amyloliquefaciens*	Wheat (*T. aestivum*)	Ashraf et al. (2004)
Salinity	*Piriformospora indica*	*Barley*	Waller et al. (2005)
Drought	*Azospirillum*	Wheat (*T. aestivum*)	Creus et al. (2004)
Drought	*Pseudomonas* sp.	Pea (*Phaseolus vulgaris*)	Arshad et al. (2008)
Drought	*P. polymyxa*	Bean (*P. vulgaris*)	Figueiredo et al. (2008)
Flooding	*Pseudomonas putida*	Tomato (*L. esculentum*)	Grichko and Glick (2001)
Temperature – heat	*Pseudomonas* sp. AMK-P6	Sorghum	Ali et al. (2009)
Temperature – cold	*P. putida*	*Canola*	Chang et al. (2007)
Nutrient deficiency	*Bacillus polymyxa*, *Pseudomonas alcaligenes*	Maize (*Z. mays*)	Egamberdiyeva (2007)

content in higher plants has been reported (Yang et al. 2004). It was first discovered in bacteria that proline functions as an osmoprotective agent, where a correlation between accumulated proline and salt tolerance has long been demonstrated. Numerous reviews have emphasized on protective function of proline which get accumulated in plants under stressed conditions (Verbruggen and Hermans 2008). However, the correlation between proline accumulation and abiotic stress tolerance in plants is not always apparent.

The potential of PGPR for enhancing plant growth and yield was also tested under different abiotic stresses, i.e., water shortage, high and low temperature, salinity, and under-deficiency nutrients (Table 15.1).

15.3 Future Prospects

Tremendous progress has been attained worldwide, in the field of PGPR biofertilizer technology as they are very effective plant growth promoters and have potential to alleviate various environmental stresses, enrich soil fertility and food nutritional quality, and enhance the agricultural production. The use of PGPR as stress mitigators, biofertilizers, biocontrol agents, and biofortifiers is an efficient alternative to the use of chemicals for sustainable crop cultivation at global level. Thus, present and future progression in understanding of diversity of PGPR, their ability to colonize plant roots, and their mode of action, formulation, and application can lead to their development as reliable components in the management of sustainable agriculture.

References

Ali Sk Z, Sandhya V, Grover M, Kishore N, Rao LV, Venkateswarlu B (2009) *Pseudomonas* sp. strain AKM-P6 enhances tolerance of sorghum seedlings to elevated temperatures. Biol Fertil Soil 46:45–55

Ali SZ, Sandhya V, Rao LV (2014) Isolation and characterization of drought-tolerant ACC deaminase and exopolysaccharide-producing fluorescent. Ann Microbiol 64(2):493–502

Antoun H, Pre'vost D D (2005) Ecology of plant growth promoting rhizobacteria. In: Siddiqui ZA (ed) PGPR: biocontrol and biofertilization. Springer, Dordrecht, pp 1–38

Antoun H, Beauchamp CJ, Goussard N, Chabot R, Lalande R (1998) Potential of Rhizobium and Bradyrhizobium species as plant growth promoting rhizobacteria on non-legumes: effect on radishes (*Raphanus sativus* L.). Plant Soil 204:57–67

Arshad M, Frankenberger WT (1998) Plant growth substances in the rhizosphere: microbial production and functions. *Adv*. Bloemberg and Lugtenberg (2001) Molecular basis of plant growth promotion and biocontrol by rhizobacteria. Plant Biol 4:343–350

Arshad M, Shaharoona B, Mahmood T (2008) Inoculation with plant growth promoting rhizobacteria containing ACC-deaminase partially eliminates the effects of water stress on growth, yield and ripening of *Pisum sativum* L. Pedosphere 18:611–620

Ashraf M, Berge SH, Mahmood OT (2004) Inoculating wheat seedling with exopolysaccharide-producing bacteria restricts sodium uptake and stimulates plant growth under salt stress. Biol Fertil Soils 40:157–162

Beneduzi A, Ambrosini A, Passaglia LMP (2012) Plant growth-promoting rhizobacteria (PGPR): their potential as antagonists and biocontrol agents. Genet Mol Biol 35(4):1044–1051

Bhattacharyya NP, Jha DK (2012) Plant growth promoting rhizobacteria (PGPR): emergence in agriculture. World J Microbiol Biotechnol 28:1327–1350

Chang WS, van de Mortel M, Nielsen L, de Guzman GN, Li X, Halverson LJ (2007) Alginate production by *Pseudomonas putida* creates a hydrated microenvironment and contributes to biofilm architecture and stress tolerance under water-limiting conditions. J Bacteriol 189:8290–8299

Chen YP, Rekha PD, Arunshen AB, Lai WA, Young CC (2006) Phosphate solubilizing bacteria from subtropical soil and their tricalcium phosphate solubilizing abilities. Appl Soil Ecol 34:33–41

Creus CM, Sueldo RJ, Barassi CA (2004) Water relations and yield in Azospirillum-inoculated wheat exposed to drought in the field. Can J Bot 82(2):273–281

Egamberdiyeva D (2007) The effect of plant growth promoting bacteria on growth and nutrient uptake of maize in two different soils. Appl Soil Ecol 36:184–189

Fasim F, Ahmed N, Parsons R, Gadd GM (2002) Solubilization of zinc salts by the bacterium isolated by the air environment of tannery. FEMS Microbiol Lett 213:1–6

Figueiredo MVB, Burity HA, Martinez CR, Chanway CP (2008) Alleviation of drought stress in common bean (Phaseolus vulgaris L.) by co-inoculation with *Paenibacillus polymyxa* and *Rhizobium tropici*. Appl Soil Ecol 40:182–188

Gamalero E, Glick BR (2012) Ethylene and abiotic stress tolerance in plants. In: Ahmad P, Prasad MNV (eds) Environmental adaptations and stress tolerance of plants in the era of climate change. Springer, New York, pp 395–412

Glick BR (2005) Modulation of plant ethylene levels by the bacterial enzyme ACC deaminase. FEMS Microbiol Lett 251:1–7

Glick BR, Todorovie B, Czarny J, Cheng Z, Duan J (2007) Promotion of plant growth by bacterial ACC deaminase. Crit Rev Plant Sci 26:227–242

Goteti PK, Emmanuel LDA, Desai S, Shaik MHA (2013) Prospective zinc solubilising bacteria for enhanced nutrient uptake and growth promotion in Maize (*Zea mays* L.). Int J Microbiol. http://dx.doi.org/10.1155/2013/869697

Grichko VP, Glick BR (2001) Amelioration of flooding stress by ACC deaminase-containing plant growth-promoting bacteria. Plant Physiol Biotechnol 39:11–17

Grover M (2010) Role of microorganisms in adaptation of agriculture crops to abiotic stresses. W J Microbiol Biotechnol 27(5):1231–1240

Grover M, Ali Z, Sandhya V, Rasul A, Venkateswarlu B (2011) W J Microbiol Biotechnol 27:1231–1240. doi:10.1007/s11274-010-0572-7

Gyaneshwar P, Kumar GN, Parekh LJ, Poole PS (2002) Role of soil micro-organisms in improving P Nutrition of plants. Plant Soil 245:83–93

Hontzeas N, Richardson AO, Belimov A, Safronova V, Abu-Omar MM, Glick BR (2005) Evidence for horizontal transfer of 1-aminocyclopropane-1-carboxylate deaminase genes. Appl Environ Microbiol 71:7556–7558

Khalid A, Akhtar MJ, Mahmood MJ, Arshad M (2006) Effect of substrate-dependent microbial ethylene production on plant growth. Microbiology 75:231–236

Khan KS, Joergensen RG (2009) Changes in microbial biomass and P fractions in biogenic household waste compost amended with inorganic P fertilizers. Bioresour Technol 100:303–309

Kloepper JW, Scher FM, Laliberte M, Tipping B (1986) Emergence-promoting rhizobacteria: description and implications for agricultures. NATO ASI Ser A Life Sci 117:155–164

Kloepper JW et al (2004) Induced systemic resistance and promotion of plant growth by Bacillus species. Phytopathology 94:1259–1266

Kumar H, Bajpai VK, Dubey RC, Maheshwari DK, Kang SC (2010) Wilt disease management and enhancement of growth and yield of *Cajanus cajan* (L) var. Manak by bacterial combinations amended with chemical fertilizer. Crop Prot 29:591–598

Lynch JP, Brown KM (1997) Ethylene and plant responses to nutritional stress. Physiol Plant 100:613–619

Ma W, Charles TC, Glick BR (2003) The *Rhizobium leguminosarum* bv.*viciae* ACC deaminase protein promotes the nodulation of pea plants. Appl Environ Microbiol 69:4396–4402

Maheswari UT, Anbukkarasi K, Hemalatha T, Chendrayan K (2013) Studies on phytohormone producing ability of indigenous endophytic bacteria isolated from tropical legume crops. Int J Curr Microbiol App Sci 2:127–136

Martínez-Viveros O, Jorquera MA, Crowley DE, Gajardo G, Mora ML (2010) Mechanisms and practical considerations involved in plant growth promotion by rhizobacteria. J Soil Sci Plant Nutr 10:293–319

Matysik J, Alia BB, Mohanty P (2002) Molecular mechanisms of quenching of reactive oxygen species by proline under stress in plants. Curr Sci 82:525–531

Mayak S, Tirosh T, Glick BR (2004) Plant growth promoting bacteria that confer resistance to water stress in tomato and pepper. Plant Sci 166:525–530

Meyer JM, Geoffroy VA, Baida N (2002) Siderophore typing, a powerful tool for the identification of fluorescent and nonfluorescent pseudomonads. Appl Environ Microbiol 68(6):2745–2753

Molla AH, Shamsuddin ZH, Halimi MS, Morziah M, Puteh AB (2001) Potential for enhancement of root growth and nodulation of soybean co-inoculated with *Azospirillum* and *Bradyrhizobium* in laboratory systems. Soil Biol Biochem 33:457–463

Nadeem SM, Zahir ZM, Naveed M, Arshad M (2007) Preliminary investigation on inducing salt tolerance in maize through inoculation with rhizobacteria containing ACC-deaminase activity. Can J Microbiol 53:1141–1149

Peck SC, Kende H (1995) Sequential induction of the ethylene biosynthetic enzymes by indole-3-acetic acid in etiolated peas. Plant Mol Biol 28:293–301

Potarzycki J, Grzebisz W (2010) Effect of zinc foliar application on grain yield of maize and it's yielding components. Plant Soil Environ 55(12):519–527

Qurashi AW, Sabri AN (2012) Bacterial exopolysaccharide and biofilm formation stimulate chick-pea growth and soil aggregation under salt stress. Braz J Microbiol 43(3):1183–1191

Ramesh A, Sharma SK, Sharma MP, Namrata Y, Joshi OP (2014) Inoculation of zinc solubilizing *Bacillus aryabhattai* strains for improved growth, mobilization and biofortification of zinc in soybean and wheat cultivated in vertisols of central India. Appl Soil Ecol 73:87–96

Richardson AE, Barea JM, McNeill AM, Prigent-Combaret C (2009) Acquisition of phosphorus and nitrogen in the rhizosphere and plant growth promotion by microorganisms. Plant Soil 321:305–339

Saharan BS, Nehra V (2011) Plant growth promoting rhizobacteria: a critical review. LSMR 21:1–30

Sandhya V, Ali Sk Z, Grover M, Reddy G, Venkateswarlu B (2009) Alleviation of drought stress effects in sunflower seedlings by exopolysaccharides producing *Pseudomonas putida* strain P45. Biol Fertil Soil 46:17–26

Sarode PD, Rane MR, Chaudhari BL, Chincholkar SB (2009) Siderophoregenic *Acinetobacter calcoaceticus* isolated from wheat rhizosphere with strong PGPR activity. Malayas J Microbiol 5:6–12

Shahzad SM, Khalid A, Arif MS, Riaz M, Ashraf M, Iqbal Z, Yasmeen T (2014) Co-inoculation integrated with P-enriched compost improved nodulation and growth of Chickpea (*Cicer arietinum* L.) under irrigated and rainfed farming systems. Biol Fertil Soils 50:1–12

Sharma A, Johri BN (2003) Growth promoting influence of siderophore-producing *Pseudomonas* strains GRP3A and PRS9 in maize (*Zea mays* L.) under iron limiting conditions. Microbiol Res 158:243–248

Sharma P, Khanna V, Kumari P (2013) Efficacy of aminocyclopropane-1-carboxylic acid (ACC)-deaminase-producing rhizobacteria in ameliorating water stress in chickpea under axenic conditions. Afr J Microbiol Res 7(50):5749–5757

Sharma P, Khanna V, Kumari S (2015) Potential of ACC-deaminase producing plant growth promoting rhizobacteria on water stress mitigation in lentil (*Lens culinaris L. Medikus*) under axenic conditions. Int J Adv Res 3(12):59–67

Shrivastava UP, Kumar A (2013) Characterization and optimization of 1-Aminocyclopropane-1-Carboxylate deaminase (ACCD) activity in different rhizospheric PGPR along with *Microbacterium* sp. strain ECI-12A. J Appl Sci Biotechnol 1(1):11–15

Singh RP, Shelke GM, Kumar A, Jha PN (2015) Biochemistry and genetics of ACC deaminase: a weapon to "stress ethylene" produced in plants. Front Microbial. http://dx.doi.org/10.3389/fmicb.2015.00937

Spaepen S, Vanderleyden J (2010) Auxin and plant microbe interactions. Cold Spring Harb Perspect Biol. doi:10.1101/cshperspect.a001438

Sundaramoorthy S, Balabaskar P (2012) Consortial effect of endophytic and plant growth promoting rhizobacteria for the management of early blight of tomato incited by *Alternaria solani*. J Plant Pathol Microbiol 3:145. doi:10.4172/2157-7471.1000145

Timmusk S (2003) Mechanism of action of the plant growth promoting Bacterium *Paenibacillus polymyxa*. Comprehansive Summaries of Uppsala Dissertations from the Faculty of Science and Technology 908. *Acta Univ Upsaliensis Uppsala*, Sweden, pp 40

Vaid SK, Gangwar BK, Sharma A, Srivastava PC, Singh MV (2013) Effect of zinc solubilizing bioinoculants on zinc nutrition of wheat (*Triticum aestivum* L.). Int J Adv Res 1(9):805–820

Velu G, Ortiz-Monasterio I, Cakmak I, Hao Y, Singh RP (2013) Biofortification strategies to increase grain zinc and iron concentrations in wheat. J Cereal Sci. http://dx.doi.org/10.1016/j.jcs.2013.09.001

Venkateswarlu B, Shanker AK (2009) Climate change and agriculture: adaptation and mitigation strategies. Indian J Agron 54:226–230

Verbruggen N, Hermans C (2008) Proline accumulation in plants: a review. Amino Acids 35:753–759

Verma JP, Yadav J, Tiwari KN, Kumar A (2013) Effect of indigenous *Mesorhizobium spp.* and plant growth promoting rhizobacteria on yeilds and nutrients uptake of Chickpea (*Cicer areitinum L.*) under sustainable agriculture. Ecol Eng 51:282–86

Vu B, Chen M, Crawford RJ, Ivanova EP (2009) Bacterial extracellular polysaccharides involved in biofilm formation. Molecules 14:2535

Waller F, Achatz B, Baltruschat H, Fodor J, Becker K, Fischer M, Heier T, Huckelhoven R, Neumann C, Von Wettstein D, Franken P, Kogel KH (2005) The endophytic fungus *Piriformis indica* reprograms barley to salt-stress tolerance, disease resistance, and higher yield. Proc Natl Acad Sci 102:13386–13391

Wang KL, Li H, Ecker JR (2002) Ethylene biosynthesis and signaling networks. Plant Cell Suppl 14:131–151

WHO (2012) The world health report. World Health Organization, Geneva

Yang SF, Hoffman NE (1984) Ethylene biosynthesis and its regulation in higher plants. Annu Rev Plant Physiol Plant Mol Biol 35:155–189

Yang J, Kloepper JW, Ryu C (2004) Rhizosphere bacteria help plants tolerate abiotic stress. Trends Plant Sci 14(1)

Zafar-ul-Hye M, Farooq HM, Zahir ZA, Hussain M, Hussain A (2014) Application of ACC-deaminase containing rhizobacteria with fertilizer improves maize production under drought and salinity stress. Int J Agric Biol 16:591–596

Part III

Plant-Microbe Interaction and Plant Productivity

Growth Promotion Features of the Maize Microbiome: From an Agriculture Perspective

16

Ubiana de Cássia Silva, Christiane Abreu de Oliveira, Ubiraci Gomes de Paula Lana, Eliane Aparecida Gomes, and Vera Lúcia dos Santos

Abstract

Microorganisms associated to maize can present a diversity in its composition according to maize genotype and soil properties, such as pH, texture, water availability, nutritional status, weather conditions, and agricultural practices. These microorganisms can stimulate plant growth by nutrients acquisition in poor soils through nitrogen fixation, phosphate solubilizing, phytate mineralization besides of the phytohormone production that help in the survival stress and can stimulate growth of plant parts several. Some molecules produced by microrganisms inhibits the action of phytopathogenic agents or can induce the plant resistance. Thus, the maize microbiome investigation can contribute for prospecting of microorganisms with potential for use as plant inoculant focused on the development of cheaper, environmentally-sound and sustainable agricultural techniques.

U. de Cássia Silva (✉) • V.L. dos Santos
Laboratory of Applied Microbiology, Institute of Biological Science,
Federal University of Minas Gerais,
Block F4, 259. 6627 Avenida Presidente Antônio Carlos, Pampulha, Belo Horizonte
31270-901, MG, Brazil
e-mail: ubiana.microb.ufmg@gmail.com

C.A. de Oliveira • U.G. de Paula Lana • E.A. Gomes
Laboratory of Applied Microbiology, Institute of Biological Science,
Federal University of Minas Gerais,
Block F4, 259. 6627 Avenida Presidente Antônio Carlos, Pampulha, Belo Horizonte
31270-901, MG, Brazil

National Research Center Mayze and Sorghum, Brazilian Agricultural Research Company,
424. Sete lagoas MG, Belo Horizonte 35701-2 970, MG, Brazil

D.K. Choudhary et al. (eds.), *Plant-Microbe Interaction: An Approach to Sustainable Agriculture*, DOI 10.1007/978-981-10-2854-0_16

16.1 Introduction

Cereals have been the principal component of animal and human diet for thousands of years and have played a major role in the establishment of human civilization. Moreover, they can be used for numerous applications in general for fuels such as ethanol production industrially. From cereals, the maize (*Zea mays* L.) is the most produced worldwide, and it has an average of 800 million tons produced annually of which the United States is the largest producer, followed by China and Brazil. Maize has a large genetic variability, which allows its cultivation in tropical, subtropical, and temperate climates. However, production systems are highly dependent on chemical inputs, especially nitrogen and phosphate fertilizers, pesticides to control pests, diseases, and weeds, which raise production cost. In addition, intensive use of agricultural inputs can cause environmental impacts. Therefore, there is a large demand for technological innovations that reduce the production costs, the inputs, and energy consumption of nonrenewable sources.

Microorganisms are attractive and viable alternatives for the reduction of fertilizer and pesticide use, easing the burden farming imposes on the environment, and reduction of production costs. Microbial inoculants that promote plant growth (PGPM) may be developed based on microbiota residing inside plants without harming their host (endophytes) or only the surface of the plant organs (epiphytic) or those found in the rizosphere of plants (Wu et al. 2005; Montanez et al. 2012; Mendes et al. 2013). The study of these microorganisms involves the investigation of the microbial community profile associated to plants and evaluation of cultured strains *in vitro* in order to identify genes related to vegetal growth promotion and ability of tissues colonization of the plant besides genes that help in the vegetal survival to biotic and abiotic stress (Table 16.1).

The microorganisms that penetrate and colonize plant tissues have evolved an elaborate system to bypass the natural defense system of plants and persist in it, named "competent endophytes" (Rosenblueth and Martinez-Romero 2006; Hardoim et al. 2008). The system involves inactivation of reactive oxygen species (ROS) and plant secondary metabolites by anti-oxidative enzymes such as catalase and superoxide dismutase. The colonization of endosphere requires indeed mechanisms to increase the nutrient acquisition including siderophore production and membrane transporters (Barret et al. 2011; Loaces et al. 2011).

Microorganisms also show essential functions for effective bacterial colonization and survival of the rhizosphere ("rhizosphere competence"). Although considered a nutrient-rich environment, this region also exerts a selective pressure on microorganisms by releasing plant-derived toxic compounds such as indoles, terpenoids, benzoxazinones, flavonoids, and isoflavonoids that induce a stress response in certain bacteria (Miche et al. 2006; Bais et al. 2006). The strategies employed by bacteria to cope with a toxic compound are to extrude it out of cell by efflux pumps, production of oxidative enzymes of aromatic compounds present in exudates, and alteration of composition of fatty acids and phospholipids to compensate increased fluidity of membrane due to interaction with phenolic compounds. The presence of these adaptation mechanisms can explain the selection of specific microbial populations by diverse plant species or cultivars, which can show exudates of variable

Table 16.1 Gene list associated with plant growth-promoting characteristics, microbial colonization of the plants that helps in survival to stresses

Gene	Features
Promotion of plant growth	
pqq, gene glucose dehydrogenase, *pst*A, B, C	P solubilization
*ipd*C	AIA production
*nif*H	Nitrogen fixation
Histidine acid phosphatases (HAP)	Phytate mineralization
Purple acid phosphatases (PAP)	
β-Propeller phytases (BPP)	
Pvd (pyoverdine gene), *fpv*A, *mbt*H, *ocr*A, B, *fhu*	Siderophore production
*Acd*S, *rim*M, *dcy*D	Activity of the ACC-deaminase
*cys*C, J, I, N	H_2S production
Colonization ability	
als, *bud*A, C, *pox*B	Synthesis of acetoin and butanediol
Chitinase homolog gene	Chitin production
Operom lsr: LsrA, lsrB, lsrC, lsrD, lsrE, lsrF, lsrG	Transport, internalization, phosphorylation, and autoinducer processing in quorum sensing
luxS	Quorum sensing control
gacA, rsmA, rpoS	Regulation of LasRI and RhlRI in the quorum sensing
Secretion systems: type II, VI. Sec and twin arginine	Secretion systems can help both in promoting plant growth and colonization
Surviving to stress abiotics and biotics	
*phz*F	Fenazine, fungicidal action
*dna*J, K, *gro*E	Heat shock proteins
*csp*A, C, D, E	Cold shock proteins
*sox*B, *opu*, *pro*X, glycine betaine homologous gene	Glycine betaine production
Catalase homologous gene	Catalase. Protection against oxidative stress
*sod*B, C, superoxide dismutase homologous gene	Superoxide dismutase. Protection against oxidative stress
*tre*Y, Z: trehalose synthase homolog gene	Helps in stress by salinity and osmotic stress
Polyhydroxybutyrate (PHB) metabolism genes	PHB is a compound intracellularly stored granules. It helps in tolerating high temperatures, exposure to UV irradiation, and desiccation

chemical composition. The root exudation varies also with plant age (Aira et al. 2010; Ramachandran et al. 2011).

Beyond the microorganism competence, other factors such as geo-location, climatic conditions, host plant genotype, growth stage and physiological status, and type of plant tissue determine microbial community colonization (Hoffman and Arnold 2008; Sun et al. 2012; Giauque and Hawkes 2013). In general, the continuation of

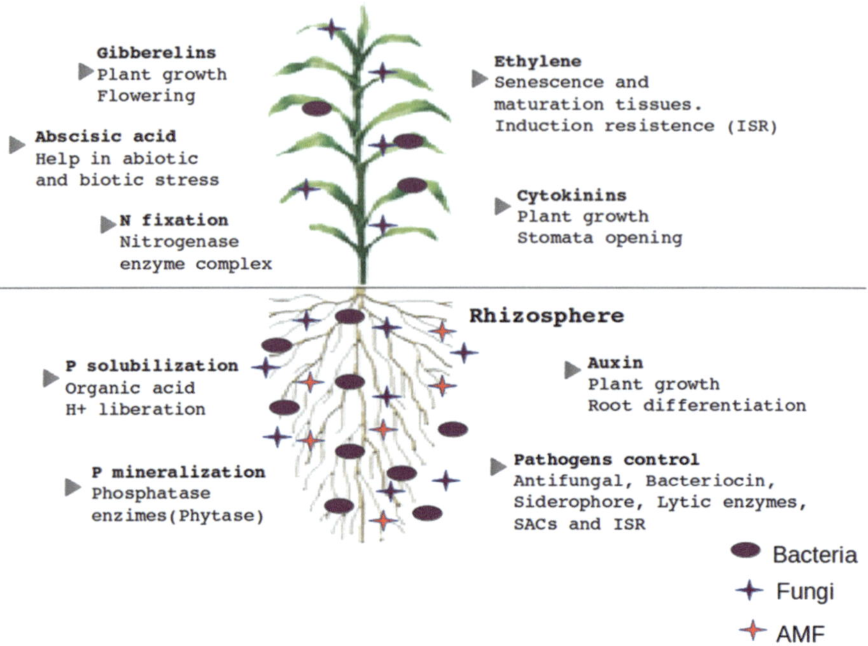

Fig. 16.1 Overview of growth-promoting factors triggered by microorganisms associated with maize

plant host–endophyte symbiotic interactions changes from antagonistic to mutualistic (Saikkonen et al. 1998); the mutualistic microbial processes involve nitrogen fixation, solubilization of phosphorus and potassium, the production of siderophores and phytohormones, surface-active compounds (SACs) and bioactive molecule production, and induction of systemic resistance (ISR) (Fig. 16.1) (Bulgarelli et al. 2013).

Several studies have prospected specific functional microbial groups associated with maize and adapted to the several ecosystems for development of PGPM inoculants with emphasis on the traits that improve the fitness of plants (Hameeda et al. 2008; Montanez et al. 2012; Szilagyi-Zecchin et al. 2014). However, the inconsistencies between results obtained in vitro compared with field trial tests have led to the unsuccessful commercialization of microbial inoculants.

Thus, to obtain effective inoculant is necessary to evaluate the ideal characteristics of candidate microorganisms, the environmental parameters that interfere with the success of colonization and sustaining bacterial life within host plants in field experiments. Soil health is another factor that affects the inoculation efficiency, due to several characteristics such as type and structure of soil, soil moisture and pH, nutrient level and toxic metal concentrations, microbial diversity, and soil disturbances caused by management practices. Furthermore, another important function of soil microbiota is the aggregation of soil particles, which can be promoted by microbial inoculants.

This chapter presents an overview of the importance of the microbiome to the plant growth promotion, focusing on the functional and taxonomic diversity of the microbiota associated with maize and the desirable characteristics of microorganism's candidates to the use in PGP formulations.

16.2 Maize Microbiome Diversity

Microbial community which colonizes maize has been intensively investigated by a number of strategies, such as cultivation of microorganisms, techniques that evaluate the community profile as terminal restriction fragment length polymorphism (T-RFLP) and others based in cloning and sequencing in small scale or high throughput as the next-generation DNA sequencing technologies. For rhizospheric community of maize, regardless of the strategy used, most of the work describes the phylum *Proteobacteria* as dominant, particularly classes α-, β-, and γ-*Proteobacteria* (Chauhan et al. 2011; Peiffer et al. 2013; Turner et al. 2013; Johnston-Monje et al. 2016). The genera *Burkholderia, Pantoea, Enterobacter, Pseudomonas, Massilia, Sphingobium, Sphingomonas, Agrobacterium, Rhizobium, Bradyrhizobium*, and *Ochrobactrum* are most commonly found in maize rhizosphere, many of them could also be found as endophytes (Johnston-Monje et al. 2016). *Proteobacteria* are also dominant in the maize rhizosphere of different regions in the world, as shown in the study that assessed soils of Canada and Brazil and of the states of Florida and Illinois, USA (Roesch et al. 2008). In this work, the β-*Proteobacteria* subphylum was dominant in all soils, except from Brazil, which is predominated by γ-*Proteobacteria*. Then, the phyla *Actinobacteria, Firmicutes, Bacteroidetes, Planctomycetes, Verrucomicrobia*, and *Acidobacteria* were frequent in all these regions (Roesch et al. 2008; Turner et al. 2013). Bacterial community rhizosphere of other plants has also displayed predominance of *Proteobacteria* phylum, such as potato, beet, and *Arabidopsis* (Weinert et al. 2011; Lundberg et al. 2012). Peiffer et al. (2013) characterized the bacterial diversity of the rhizosphere of maize lineages during flowering grown in the field in five different environments in the United States and noted that some orders such as *Burkholderiales, Oceanospirillales*, and *Sphingobacteriales* of the *Proteobacteria* phylum were consistently enriched in maize rhizosphere. According to authors, sampling area and its geographical origin were the main variation source in microbiota composition followed by root proximity (rhizospheric and non-rhizospheric soil). In the rhizosphere, strains belonging to *Proteobacteria* phylum predominated because in general show rapid growth capacity in response to sources of labile carbon released by plant. In contrast, non-rhizospheric soil is predominantly enriched by slow-growth microorganisms that has more stable populations, such as Acidobacteria, Chloroflexi, Planctomycetes, and Verrucomicrobia, which are also described as oligotrophic (Fierer et al. 2007).

Some studies suggest that maize plants can select specific bacterial communities depending on soil properties (Castellanos et al. 2009), genotypes (Aira et al. 2010), management techniques, such as fertilizers (Aira et al. 2010), and growth stage of the plant (Gomes et al. 2001). For example, Bouffaud et al. (2012) showed that the genotype influenced the microbiota composition of the maize rhizosphere. These authors evaluated the community rhizobacteria of five main genetic groups of maize by of microarray and 16S rRNA analysis that revealed a clear effect of genotype in the selection of rhizobacteria community. It was observed that main differences were related to the group of Betaproteobacteria, especially *Burkholderia*. However, other works show that the composition of the bacterial community is not dependent of the cultivar type or genotype (Schmalenberger and Tebbe 2002) or soil type (Johnston-Monje et al. 2016). This apparent disparity in results may be due to interactions between plants and soil types and according to methodology employed. The use of techniques that have different resolution and detection limits can also influence the results. Chelius and Triplet (2001), for example, observed that dominant bacteria group belongs to the *Actinobacteria* phylum when cultivation techniques were used, whereas the sequencing of clone libraries showed the α-*Proteobacteria* as predominant group.

The plant development stage has also been described as important factor to determine the microbial community. Gomes et al. (2001) evaluated the rizosphere bacterial community of two maize genotypes by cultivation-based techniques and TGGE. Differences were observed in the community composition of young roots when compared to mature plant, especially, in the α- and β-*Proteobacteria* population. Similarly, Li et al. (2014) using the pyrosequencing described alteration of the microbiome rhizosphere of maize with the plant growth stage. The genera *Massilia*, *Flavobacterium*, *Arenimonas*, and *Ohtaekwangia* were abundant in the early stages, while the population of *Burkholderia*, *Ralstonia*, *Dyella*, *Chitinophaga*, *Sphingobium*, *Bradyrhizobium*, and *Variovorax* was dominant in the later stages.

In addition, other characteristics can modulate the structural and functional diversity of the plant microbiome, as abiotic factors including soil properties such as pH, texture, water availability, nutritional status, weather conditions, and agricultural practices (Berg and Smalla 2009). The effect of different fertilizers in the bacterial and mycorrhizal fungi community of maize was assessed by T-RFLP, cloning, and sequencing, respectively (Toljander et al. 2008). Changes in microbial community were mainly correlated with pH changes induced by fertilization type, but other factors also contributed to the observed changes, including carbon and phosphate of the soil.

Furthermore, although the effects of the genotype and fertilization are important separately, the interaction between them can be determinant for the microbial community, as the exudation from plants is influenced by these factors; therefore, the microbial community structure of the rhizosphere will be modified. Aira et al. (2010) evaluated the effect of different strategies of maize fertilization and detected change in the composition of the exudates from the roots leading to an increase in biomass and modification in the bacterial community structure.

In addition to rhizosphere, a wide variety of endophytic microorganisms also colonizes maize. This microorganism group is characterized for living within the plant tissues at least part of the life cycle; they can colonize the apoplast, including intercellular spaces and cell walls of roots, stems, and leaves. They are generally nonpathogenic for the plant, but may include latent pathogens that depending on the environmental conditions can cause disease. The high number of bacteria into plant tissues are originated from soil suggesting that the roots as the main entry point of the endophyte in the host plants (Miliute et al. 2015).

In general, species of endophytic bacteria present in maize belong to subphylum α-, β-, and γ-*Proteobacteria*. Among these groups, γ-*Proteobacteria* is dominant and has greater diversity. But *Bacteroidetes*, *Actinomycetes*, and *Firmicutes* are also commonly observed in maize endophytic community. The genera *Rhizobium*, *Burkholderia*, *Enterobacter*, *Erwinia*, *Pseudomonas*, *Flavobacterium*, and *Bacillus* are the most described in studies of maize endophytic microbiome (Kobayashi and Palumbo 2000; Seghers et al. 2004; Rai et al. 2007; Pereira et al. 2011; Bulgarelli et al. 2013; Philippot et al. 2013).

Most studies of maize endophytic microorganisms were conducted in temperate climate (McInroy and Kloepper 1995; Fisher et al. 1992; Seghers et al. 2004). However, studies conducted by our research group in tropical conditions showed that the microbiota colonizing maize plant organs varied, being observed was the predominance of genera *Microbacterium* (16 isolates), *Pseudomonas* (three isolates), *Staphylococcus* (nine isolates), *Curtobacterium* and *Lactococcus lactis* (seven isolates each), *Pantoea* (four isolates), and *Psychrobacter* (three isolates) on the leaf. On the roots, the predominant genera were *Bacillus* (11 isolates), *Leuconostoc* (four isolates), *Enterobacter* (six isolates), *Pseudomonas* (seven isolates), and *Serratia* (five isolates) (Vieira 2015). Rai et al. (2007) also evaluated maize endophytic bacteria in tropical soils. They observed that the bacteria density in maize ranges from 1.36×10^5 colony-forming units per gram of fresh tissue (UFC/g) in the first week of seedling emergence to 6.12×10^5 UFC/g at the end of growing season, after 10 weeks. The peak of bacteria density was 12×10^5 UFC/g at 28 days after emergence. The predominant species were *Bacillus pumilus*, *B. subtilis*, *Pseudomonas aeruginosa*, and *P. fluorescens*. The genera *Pseudomonas* and *Bacillus* also predominated in the study of McInroy and Kloepper (1995), which recovered 232 endophytic bacteria of maize distributed in more than 40 distinct genera.

Although most studies report that the soil is an important source of endophytic microorganisms, they also may be transmitted through vectors like insects and vertically through propagules such as seeds, rhizomes, and cuttings (Hardoim et al. 2011). These sources are more relevant for the microbiota associated with stems and leaves of plants, both as epiphytic or endophytic lifestyle. Seeds can also be an important source of these microorganisms. Johnston-Monje et al. (2016) found that the most abundant bacteria in the rhizosphere originated from endophyte or epiphyte transmitted by seeds, colonizing approximately 55 % of the bacterial population when maize plants were grown in non-sterile conditions and up to 90 % of the bacterial population when maize was grown in sterile sand (Johnston-Monje et al. 2016).

16.3 The Microbial Role in the Maize Growth Promotion

16.3.1 Acquisition of Nutrients

Nitrogen (N) and phosphorus (P) are nutrients required to plant growth and to ensure the productivity of crops. Farming systems are highly dependent on fertilizers, especially nitrogen and phosphorus. These fertilizers are the largest part of grain production costs as maize and may negatively impact the environment (Novais and Smyth 1999).

Microorganism prokaryotes have the capacity of N fixing of the air and convert it to chemical species readily utilized by plants, such as ammonium and nitrate (Nunes et al. 2003). N-fixing bacteria are called diazotrophs and are found free-living in the rhizosphere, endophytically associated with some species or establishing mutualistic symbiosis with the plants as observed in the nodulating bacteria of Leguminosae (Didonet et al. 2000).

N biological fixation is a complex process that requires joint expression of various genes, including the *nif* genes that are responsible for the synthesis of the nitrogenase complex, and its homologous *nif*H I are the most widely used for phylogenetic studies of diazotrophic symbionts (Zhao et al 2010; Gaby and Buckley 2012). Several groups of endophytic N-fixing bacteria in plants have been reported, including the genera *Acetobacter*, *Azoarcus*, *Gluconacetobacter*, *Herbaspirillum*, *Methylobacterium*, and *Burkholderia* (Donato et al. 2005; Balachandar et al. 2006; Govindarajan et al. 2007).

Maize plants can be simultaneously colonized by a wide variety of diazotrophic bacteria (Table 16.2) (Chelius and Triplett 2001; Lodewyckx et al. 2002), and the genera *Azospirillum*, *Herbaspirillum*, and *Klebsiella* are the most frequently observed (Baldani et al. 1986; Chelius and Triplett 2001; Alves et al. 2015). The genus *Burkholderia* has also been appointed as widely spread in association with maize cultivated in tropical conditions, including the species *B. tropica* and *B. unamae* (Govindarajan et al. 2006; Perin et al. 2006). More recently, *Bacillus pumilus* was efficient in biological nitrogen fixation in maize in greenhouse conditions (Kuan et al. 2016). The N-fixing bacteria are located most often in roots, followed by stems and leaves of maize (Mendonça et al. 2006), and high densities are observed during the plant growth cycle that match with the N-fixing peak (Siqueira and Franco 1988). Additionally, several studies have indicated that the bacterial N fixation efficiency in maize is strongly influenced by the plant genotype (Montanez et al. 2009; Araújo et al. 2014).

Arbuscular mycorrhizal fungi (AMF) play an essential role in the absorption of nutrients in the most land plants. Although the contribution of the symbiosis of AMF with plants has been recognized for phosphate nutrition, its role in nitrogen nutrition is still controversial (Bucking and Kafle 2015). Several works have demonstrated an increased N uptake by roots infected with AMF (Saia et al. 2014; Correa et al. 2015; Mensah et al. 2015). However, in some cases, the N fixation has been attributed the bacteria associated with mycorrhizal fungi (Minerdi et al. 2001).

P is the second most nutrient limiting to plant development, participating as a structural component of nucleic acids, phospholipids, and adenosine triphosphate (ATP); it is a key element of metabolic and biochemical pathways that can also affect the grain yield in cereals (Khan et al. 2009). In the soil, P is distributed in the inorganic and organic forms. Although the organic P corresponds between 30 and 80 % of total P, found mainly in the form of phytate (Richardson and Simpson 2011), it is not readily available for uptake by plant root system (Mudge et al. 2003; Tarafdar and Gharu 2006). Insoluble mineral complexes are also important sources of P in the soil (Rodríguez et al. 2006), but the levels of reactivity of the P linked to iron and aluminum in acid soils ($pH < 5$) and linked to calcium in alkaline soils

Table 16.2 Species or genus related to factors promoting plant growth or biocontrol in maize

PGP[a] features or biocontrol	Species or genus	Reference
Nitrogen fixation	*Azospirillum*	Hungria et al. (2010)
	Herbaspirillum	Alves et al. (2015)
	Klebsiella, Gluconacetobacter	Riggs et al. (2001)
	Burkholderia	Perin et al. (2006)
	Bacillus pumilus	Kuan et al. (2016)
	Pseudomonas, Bacillus	Pal et al. (2001)
Phosphate solubilization	*Pantoea, Pseudomonas*	Kaur and Reddy (2015)
	Enterobacter	Chabot et al. (1996)
	Serratia, Bacillus	Hameeda et al. (2008)
	Burkholderia	Gomes et al. (2014)
	Aspergillus and *Penicillium*	Coutinho et al. (2012)
	Pseudomonas, Bacillus	Pal et al. (2001)
Phosphate mineralization	*Talaromyces rotundus*	
	Aspergillus terreus	Oliveira et al. (2009)
	Burkholderia cepacia	
	Glomus mosseae, Glomus deserticola	Vazquez et al. (2000)
Auxin	*Pseudomonas*	Picard and Bosco (2005)
	Bacillus, Burkholderia, Micrococcus	Pal et al. (2001)
		Naveed et al. (2015)
	Trichoderma harzianum	Akladious and Abbas (2012)
	Glomus intraradices	Ludwig-Müller et al. (1997)
	Pseudomonas, Bacillus	Pal et al. (2001)
Cytokinin	*Bacillus, Burkholderia, Micrococcus*	Raza and Faisal (2013)
Gibberellins	*Azospirillum brasilense*	Lucangeli and Bottini (1997)
	Azospirillum lipoferum	Cohen et al. (2009)
	Trichoderma harzianum	Akladious and Abbas (2012)
Abscisic acid	*Azospirillum lipoferum*	Cohen et al. (2009)
Antifungal antibiotics	*Pseudomonas, Bacillus*	Pal et al. (2001)
	Paenibacillus polymyxa	Mousa et al. (2015)
	Acremonium zeae	Wicklow et al. (2005)
Bacteriocin	*Luteibacter, Microbacterium, Arthrobacter, Cellulomonas*, and *Burkholderia*	Johnston-Monje and Raizada (2011)

(continued)

Table 16.2 (continued)

PGP[a] features or biocontrol	Species or genus	Reference
Siderophore	*Pseudomonas*	Pal et al. (2001)
	Bacillus	Szilagyi-Zecchin et al. (2014)
	Azospirillum brasilense	Tortora et al. (2011)
	Luteibacter, *Microbacterium*, *Arthrobacter*, *Cellulomonas*, and *Burkholderia*	Monje and Raizada (2011)
Lytic enzymes	*Luteibacter*, *Microbacterium*, *Arthrobacter*, *Cellulomonas*, and *Burkholderia*	Monje and Raizada (2011)
Surface-active compounds (SACs)	*Bacillus mojavensis*	Snook et al. (2009)
Induction of systemic resistance (ISR)	*Paenibacillus polymyxa*	Mei et al. (2014)
	Pseudomonas putida	Planchamp et al. (2014)
	Pseudomonas aurantiaca	Fang et al. (2013)
	Bacillus subtilis, *B. amyloliquefaciens*	Gong et al. (2015)
	Azospirillum brasilense	Santos et al. (2014)
	Trichoderma virens	Lamdan et al. (2015)

[a]Plant growth promotion

(pH > 7) are low (McLaughlin et al. 2011). This, only a small proportion of P, is readily available for uptake by plants (Tinker and Nye 2000).

Several microorganisms solubilizing inorganic P and mineralizing phytate have been identified and characterized (Table 16.2). *Acinetobacter*, *Bacillus*, *Burkholderia*, *Bradyrhizobium*, *Enterobacter*, *Mesorhizobium*, *Paenibacillus*, *Pantoea*, *Pseudomonas*, *Rhizobium*, *Serratia marcescens*, *Penicillium*, *Aspergillus*, and *Micromonospora* stand out within the group of bacteria, fungi, and actinomycetes (Marra 2012; Silva et al. 2014). Many of these have already been evaluated as biofertilizers (Goldstein et al. 2003; Oliveira et al. 2009; Jorguera et al. 2011; Mander et al. 2012).

The mechanisms associated with the increased P availability in the soil by microorganisms are varied and are related mainly to the release of organic acids (Whitelaw 2000; Goldstein et al. 2003), release of cations H^+ and consequent reduction of the soil pH (Villegas and Fortin (2002)), production of exopolysaccharides (Yi et al. 2008) and siderophores (Hamdali et al. 2008), and action of phosphatase enzymes (Richardson et al. 2009; Ogbo 2010), mainly as phytases (Greiner 2006). Some genes involved in the P solubilization and mineralization, including *pqq* and *bpp*, have been identified and isolated in different species of microorganisms (Table 16.1) (Rodríguez et al. 2006; Jorguera et al. 2011; Kim et al. 2003).

Several studies have reported increased growth and absorption of nutrients in maize plant inoculated with P-solubilizing microorganisms under greenhouse or filled conditions (Hameeda et al. 2008; Kumar et al. 2007). Some species have been cited such as *Pantoea cypripedii*, *Pseudomonas plecoglossicida* (Kaur and Reddy

2015), *Pseudomonas tolaasii* (Viruel et al. 2014), *Serratia marcescens*, *Bacillus coagulans*, and *Enterobacter asburiae* (Hameeda et al. 2008). These bacteria have significantly positive effects on grain yield, biomass and P content of the plants.

Microorganisms isolated from the maize rhizosphere have also been identified and showed effective solubilization and mineralization of sources of insoluble inorganic P (Table 16.2). Oliveira et al. (2009) evaluated bacteria and fungi of the maize rhizosphere to P solubilization and mineralization. The species of *Burkholderia* and *Bacillus* were more efficient to P solubilization releasing up to 67 % of total P in the medium. For phytate, the most effective were fungal species *Talaromyces rotundus* and *Aspergillus terreus* and *Burkholderia cepacia.* The genera *Aspergillus* and *Penicillium* have also been associated with P solubilization in other works (Coutinho et al. 2012; Gomes et al. 2014).

The mycorrhizal fungi (AMF) also play an important role in increasing the P availability for plants. These microorganisms contribute to higher P uptake due to the larger volume of soil explored by higher branching and extension their hyphae (Berbara et al. 2006). Additionally, AMF may contribute to the mineralization of organic P by phosphatase enzyme production (Yao et al. 2001; Cardoso and Kuyper 2006). Some studies suggest that the AMF interacting with plants help in the tolerance of crops to certain tensions present in many agricultural soils, especially in the nutrient starvation such as phosphorus and water deficiency (Abbott and Robson 1991, Williams and Sylvia in 1992; Chu et al. 2013). For maize, species of AMF of the *Glomus* genus have been used as inoculant (Chu et al. 2013; Dhawi et al. 2015). The positive effect of this interaction is gain in plant growth by increase of the biomass-infected plants in many crops such as maize (Hu et al. 2009; Souza et al. 2015; Cozzolino et al. 2013).

16.3.2 Phytohormone Production

Microorganisms are able to produce phytohormones that promote plant development and growth, including auxins, cytokinins, gibberellins, abscisic acid, and ethylene (Zahir et al. 2003). These molecules can also stimulate the reproduction and colonization of beneficial microorganisms, besides indirectly contributing to plant development by regulating the immune response of the plant against pathogens and herbivorous insects (Pieterse et al. 2012). For maize, bacteria and fungi, rhizosphere, and endophyte have been described by their capacity to produce phytohormones (Table 16.2) and help in its development directly or indirectly by increasing the protection of plants against pathogens (see Sect. 16.3.3).

Auxin is a group of aromatic ring compounds with a carboxyl group, and its main member is indole-3-acetic acid (IAA) (Korasick et al. 2013). IAA can act in the stimulation of stem growth, cell division, initiation and differentiation of the roots, differentiation of the vascular tissues in phloem and xylem, and promotion of the flowering and help in the formation of fruit. In addition, auxin can contribute to delay leaf senescence and fruit maturation (Davies 2010).

Cytokinins are adenine-type compounds, being cytokinin zeatin the most commonly found in plant tissues (Davies 2010). Cytokinins are present in all parts of the plants, but root region, apical part of shoots, and seeds have higher concentrations of this hormone, which is explained by its function of triggering cell divisions in these plant parts (Santner and Estelle 2010). Furthermore, cytokinin is also involved in the germination, initial formation of branches, growth of lateral buds, leaf expansion, opening of stomata, chloroplast development, and leaf senescence and still plays an important role in the formation of nodules during nitrogen fixation (Murray et al. 2007; Davies 2010).

Gibberellins are diterpene compounds (Bomke and Tudzynski 2009), and the gibberellic acid (GA) is the most known and active. However, there are over 130 different molecules belonging to gibberellin group (Dodd et al. 2010). Gibberellins are involved primarily in cell division and elongation within the apical meristem. It also stimulates seed germination, pollen tube growth, and plant flowering. Various functions triggered by gibberellins are a result of their interaction with DELLA proteins (repressors of transcription GA-dependent processes). Gibberellin operates in destabilization or degradation of DELLA, for example, DELLA degradation by GA suppresses the defense response dependent on jasmonic acid and stimulates dependent response of salicylic acid (Pieterse et al. 2012). As auxin and cytokinin molecules, gibberellins act in combination with other hormones and are affected mainly by auxin and ethylene (Tsavkelova et al. 2006).

Abscisic acid (ABA) is a phytohormone synthesized via isopentenyl diphosphate and carotenoids from glyceraldehyde-3-phosphate (Davies 2010) and is involved in plant responses to biotic and abiotic stresses, such as the inhibition of seed germination and flowering in response to stress by drought, salinity, and toxic metals (Smyth 2011). High ABA concentration, for example, stimulates the gene transcription involved in the protection against dehydration and osmotic stress. In this case, ABA leads to protein expression that helps in osmotic stabilization, enzymes for detoxification of reactive oxygen species, and aquaporins, which facilitate the water and ion movement across membranes, also regulating water loss rate by stomatal opening (Davies 2010).

Ethylene is a gaseous hormone (C_2H_4) synthesized from methionine. It is produced from conversion of S-adenosyl methionine to 1-aminocyclopropane-1-carboxylate (ACC) by activity of the ACC synthase enzyme (Giovanelli et al. 1980). Ethylene is synthesized by most tissues in response to abiotic and biotic stress being mainly produced in tissues undergoing senescence and maturation and in response to pathogen attack. It can act synergistically with jasmonic acid. Furthermore, ethylene may contribute to plant growth by stimulating rooting, opening of flowers, and release of dormancy and may act in the detoxification of reactive oxygen species in stressed cells (Davies 2010; Pieterse et al. 2012)

Rhizobacteria of the *Pseudomonas* genus, isolated from maize, presented capacity to produce AIA (Picard and Bosco 2005). In this work, the authors also found that in hybrid plants, roots have increased the AIA-positive *Pseudomonas* sp. colonization, showing a superiority of hybrid plants in comparison to their parental lines for the recruitment of beneficial bacteria. Besides *Pseudomonas*, rhizobacteria of the genera *Bacillus*, *Burkholderia*, and *Micrococcus* also presented the phytohormone production, such as AIA (Pal et al 2001; Naveed et al. 2015) and cytokinin (Raza and Faisal 2013), and contributed to the maize growth during its colonization. The *Azospirillum lipoferum* bacterium has also stimulated the development of maize plants; in this case, the production of the gibberellic acid and abscisic acid hormones stimulates the growth and tolerance of the plants during colonization under drought period (Cohen et al. 2009). In another study, endophytic bacteria *Azospirillum brasilense* inoculated in maize promoted its growth also due to production of gibberellic acid (Lucangeli and Bottini 1997). Other genera of endophytic bacteria from maize have shown potential to AIA production, such as *Microbacterium*, *Pseudomonas*, *Bacillus*, *Enterobacter*, *Curtobacterium*, *Serratia*, and *Pantoea* (Vieira 2015).

For fungi group, *Trichoderma harzianum* species showed capacity to produce phytohormones such as gibberellins and auxin and contributed to maize plant growth (Akladious and Abbas 2012). Furthermore, during maize colonization by *Glomus intraradices* (mycorrhizal fungus), there was also an observed increase in auxin production (IBA, indole-3-butyric acid), as well as increased activity of IBA synthetase enzyme (Ludwig-Müller et al. 1997). Given the beneficial effects of auxin on the root system structure, for example, to root growth, it can be important to AMF interaction establishment (Sukumar et al. 2013).

16.3.3 Biocontrol

Biocontrol of phytopathogens by plant-associated microbiota can be based on several mechanisms which include antibiosis, competition for nutrients and niches, as well as induction of host defense genes to avoid pathogen attack or reduce pathogen growth (Beneduzi et al. 2012). The act of antagonism against pathogen growth is viewed as the most powerful and best-characterized mechanism that explains the capacity of the microbiota to pathogen control. *Bacillus* and *Paenibacillus* species actually devote larger part of the genome (4–8 %) to antibiotic synthesis and, therefore, display potential to produce a vast array of structurally diverse antimicrobial compounds (Chen et al. 2009; Aleti et al. 2015). These genera and others plant-associated are considered a source of great biotechnological potential of bioactive metabolites that not yet fully known. Strains of *Pseudomonas* sp. EM85, *Bacillus* sp. MR-11(2), and *Bacillus* sp. MRF from maize rhizosphere antagonized the fungi pathogens as *Fusarium moniliforme*, *F. graminearum*, and *Macrophomina phaseolina* (Pal et al. 2001). *Pseudomonas* sp. produced antifungal compounds, siderophore, cianidric acid and fluorescent pigments; while *Bacillus* sp. MR-11(2) produced siderophore, antibiotics, and antifungal volatiles and *Bacillus* sp. MRF exhibited the production of antifungal, antibiotics and siderophores. In this study, the combined application of two bacilli reduced 56.04 % of the *Macrophomina*-induced charcoal rots of maize. Positive effects with purified antifungal, antibiotics and/or fluorescent pigment of *Pseudomonas* sp. EM85 and purified antifungal, antibiotics of bacilli along with the successful colonization of all the isolates might be involved in the biological suppression of the maize root diseases.

Paenibacillus sp. produces diverse antifungal compounds including polymyxins, fusaricidins, colistins, volatile compounds, and lytic enzymes (Raza et al. 2015; Naghmouchi et al. 2012). *Paenibacillus polymyxa* strain and *Citrobacter* sp. isolated from diverse maize genotypes suppressed the growth of *Fusarium graminearum* and other 20 fungi (Mousa et al. 2015). These microorganisms reduced deoxynivalenol mycotoxin concentrations produced by *F. graminearum* during storage to levels significantly below acceptable safety limit. *P. polymyxa* fungicidal action mechanism involved the fusaricidin production and induction of systemic host resistance (Mousa et al. 2015; Mei et al. 2014).

Pyrrocidines A and B are polyketide amino acid-derived antibiotics produced by endophytic fungus *Acremonium zeae* of *Zea mays* (Wicklow et al. 2005). These biomolecules displayed significant activity against kernel rotting and fungi mycotoxin produced by *Aspergillus flavus* and *Fusarium verticillioides* (Wicklow and Poling 2009). In equivalent assays performed with conidia or hyphal cells as inoculum, pyrrocidine A was active against major stalk and ear rot pathogens of maize, including *F. graminearum*, *Nigrospora oryzae*, *Stenocarpella* (*Diplodia*) *maydis*, and *Rhizoctonia zeae*, besides activity against *Clavibacter michiganensis* subsp. *nebraskensis*, causal agent of Goss's bacterial wilt of maize. Pyrrocidine A displayed also significant activity against seed rot saprophytes *A. flavus* and *Eupenicillium ochrosalmoneum*, as well as seed-infecting colonists of the phylloplane *Alternaria alternata*, *Cladosporium cladosporioides*, and *Curvularia lunata*, which produces a damaging leaf spot disease (Wicklow and Poling 2009). This antibiotic also exhibited potent activity against *Bacillus mojavensis* and *Pseudomonas fluorescens*, maize endophytes applied as biocontrol agents, but was ineffective against the wilt-producing bacterium *Pantoea stewartii*.

Bacteriocins are proteins or ribosomally bacteria-synthetized peptides with bactericidal or bacteriostatic effect against closely related species and strains unrelated (Klaenhammer 1993; Cotter et al. 2013). Most bacteriocins kill target cells by formation of pores or channels in the inner membrane that results in the leakage of cytoplasmic compounds, destruction of electrochemical gradient, ion loss, and cell death (Riley and Wertz 2002). Others interfere with DNA, RNA, and protein metabolism (Riley 1998) or contain DNase, 16S rRNase, and tRNase activities (Riley and Wertz 2002). Others can degrade the peptidoglycan precursor, leading to an inability to synthesize peptidoglycan and bacterial death (Cascales et al. 2007), or prevent spore outgrowth (Mazzotta et al. 1997). The higher number of described active biomolecules was associated with *B. thuringiensis*, such as thuricin (Favret and Yousten 1989), tochicin (Paik et al. 1997), entomocin 9 (Cherif et al. 2003), and bacthuricin F4 (Kamoun et al. 2005). Furthermore, the genus *Pseudomonas* produces bacteriocins that are structurally and mechanistically diverse, including polypeptides of middle size, such as colicin-like S pyocins produced by *P. aeruginosa* (Parret et al. 2003); large phage taillike multiprotein complexes, such as syringacin M and R- and F-type pyocins produced by *P. syringae* and *P. aeruginosa*, respectively (Nakayama et al. 2000; Michel-Briand and Baysse 2002); lectin-like bacteriocins, such as putidacin A (Parret et al. 2003); and colicin M-like colicins, such as PaeM produced from *P. aeruginosa*, PsyM from *P. syringae*, and PflM isolated from *P. fluorescens* (Barreteau et al. 2012).

Surface-active compounds (SACs) act also as biocide agents. Several of these biomolecules play essential roles for the survival of producer microorganisms in natural and artificial environments facilitating nutrient transport; they are important for gliding and swarming motility as well as de-adhesion from surfaces or microbe–host interactions (Compant et al. 2010; Chrzanowski et al. 2012). In addition, SACs have been reported to be involved in the stimulation of immunity in plants and animals. Due to a broad range of physicochemical properties of the SACs, their low toxicity, high biodegradability, and antimicrobial properties are promising molecules to be used against pathogens in agriculture (Sachdev and Cameotra 2013). The *Bacillus*-related lipopeptides, cyclopeptides (iturins) (Gong et al. 2015), or macrolactones (fengycins and surfactins) (Gong et al. 2015; Snook et al. 2009) are characterized by the presence of L- and D-amino acids and variable hydrophobic tails. They are among the most documented lipopeptides by their activity against plant pathogens in maize. As example, Leu(7)-surfactin, a cyclic heptapeptide linked to a β-hydroxy fatty acid, was identified as the inhibitory substance of the *Bacillus mojavensis* culture extracts, an endophytic bacterium patented for control of fungal diseases in maize. The bacteria antagonize the pathogenic and mycotoxic fungus *Fusarium verticillioides* (Snook et al. 2009).

The ability to produce siderophore and capture the iron allows a great competitive advantage against pathogens (Radzki et al. 2013). Siderophores are compounds with low molecular weight (200–2000 Da) produced by microorganisms and plants, especially under Fe-limiting conditions. There are three main kinds of siderophore: hydroxamate, catecholate and carboxylate. These molecules show high specificity and affinity for binding Fe^{3+} (Schwyn and Neilands 1987; Krewulak and Vogel 2008). Thus, siderophores can display role of biocontrols, biosensors, and chelation agents as well as helps in the plant growth in weathering soil (Dimkpa et al. 2008; Tortora et al. 2011). The excretion of siderophores by bacteria might stimulate plant growth by direct effect (improving nutrition) or indirectly by inhibiting of phytopathogens establishment through Fe sequestration from environment, limiting mineral available to the pathogen growth. Unlike microbial pathogens, plants are not affected by bacterial-mediated Fe depletion and some plants can also capture and utilize Fe^{3+} from bacterial siderophores complexes (Dimkpa et al. 2008). Bacteria of the genera *Pseudomonas*, *Azotobacter*, *Bacillus*, *Enterobacter*, *Serratia*, *Azospirillum* and *Rhizobium* are related to siderophore production (Loper and Buyer 1991; Neilands 1995). In maize, endophytic strains belonging to genus Bacillus showed siderophore production and were efficient against the growth of *Fusarium verticillioides*, *Colletotrichum graminicola*, *Bipolaris maydis*, and *Cercospora zeae-maydis* (Szilagyi-Zecchin et al. 2014). Siderophores produced by *A. brasilense* showed also antifungal activity in vitro against Colletotrichum acutatum, the causal agent of anthracnose (Tortora et al. 2011).

Moreover, microorganisms can parasitize disease-causing fungi by the production of hydrolytic enzymes. These can hydrolyze a wide variety of polymeric compounds, including chitin, proteins, cellulose, hemicellulose, and DNA. Production of extracellular cell wall-degrading enzymes such as glucanases and chitinases has been associated with biocontrol abilities of the producing bacteria (Fridlender et al. 1993; Valois et al. 1996). Ordentlich and coworkers (1988) have reported the potential of *Serratia marcescens* to control fungus *Sclerotium rolfsii* by degrading the cell walls of the fungus through production of chitinase enzymes.

Several studies have attributed the increased production of plant biomass to antagonist activity of microbial inoculants or the combination of this activity with other PGP features. Johnston-Monje and Raizada (2011) reported that maize endophytes of the genera Luteibacter, Microbacterium, Arthrobacter, Cellulomonas, and Burkholderia showed antagonist activity against *Escherichia coli* DH5α, *B. subtilis* spp., and yeast. These strains were also active against bacteria (Pseudomonas aeruginosa and *Bacillus subtilis*) and fungi (*Fusarium moniliforme, Aspergillus flavus* and *Helminthosporium sativum*) in *in vitro* assays, besides showing siderophore production, phosphorus solubilization and bacteriocin production. The high antioxidant enzyme activity including superoxide dismutase, peroxidase, catalase and ascorbate peroxidase were also enhanced in these bacteria inoculated to maize.

Besides the control of plant pathogens by production of bioactive molecules, many microorganisms can suppress plant diseases through induction of plant resistance against pathogens and herbivores. Various elicitor compounds produced by microorganisms can stimulate plant resistance systems, for example, surfactin, fengycin, rhamnolipids (Ongena et al. 2007; Sanchez et al. 2012), acyl-homoserine lactones, N-alkylated benzylamines (Van Loon et al. 1998), exopolysaccharides, volatile organic compounds (Ryu et al. 2004), phenylacetic acid (Akram et al. 2016), antibiotics such as 2,4-diacetylphloroglucinol (Weller et al. 2012) and pyocyanin (De Vleesschauwer et al. 2006), flagellum, phytohormones, and siderophores such as salicylic acid that has been reported as inducers of plants resistance (van Loon et al. 1998).

From the induction of plant defense genes, the response can be triggered by defense systems dependent on jasmonic acid (JA) or salicylic acid (SA) (Vlot et al. 2009). Briefly, the systemic acquired resistance (SAR) is a way that usually depends on the SA. The plant defense capability is acquired after the first infection, taking effect mainly against biotrophic pathogens. The NPR (non-expressor of PR genes) regulatory protein is activated after recognition of pathogen attack by receptors' extracellular surface that recognizes molecular patterns associated with pathogens (PAMPs) and acts as a transcriptional cofactor of genes related to defense plant (Moore et al. 2011), stimulating the production of reactive oxygen species (ROS), the SA accumulation, and increased expression of proteins related to direct attack of pathogens such as chitinase, proteases, and antimicrobial compounds (Vlot et al. 2009).

In the induced systemic resistance (ISR), JA can act synergistically with ethylene (ET) in response to necrotrophic and herbivore pathogens. The route of JA-dependent resistance, in general, has two branches: (1) triggered in response to herbivorous attack controlled by the MYC transcription factors (Lorenzo et al. 2004), there is induction of the expression of genes that affect digestive ability of the insects, defense protein genes with large antimicrobial spectrum, and the production of volatiles that attract carnivorous arthropods or parasites and herbivorous insects; (2) during response to necrotrophic pathogens regulated by JA/ET, there is induction of the expression of chitinase enzyme genes and glucanase that degrade the cell walls of phytopathogenic fungi and peroxidases to produce reactive oxygen species that has antimicrobial action (Berrocal-wolf et al. 2002; Sobrinho et al. 2005).

Among the microorganisms with potential already known to stimulate the resistance of plants, *Pseudomonas* and *Bacillus* within the bacteria group are most frequently reported. Aside from these, other species of bacteria such as *Serratia marcescens* (Press et al. 1997), *Rhizobium etli* (Reitz et al. 2002), *Streptomyces* sp. (Salla et al. 2016), and *Paenibacillus lentimorbus* (Kumar et al. 2016) also showed ability to stimulate plant resistance. For fungi, *Trichoderma* sp. (Saxena et al. 2015), *Piriformospora indica* (Wang et al. 2015) and *Penicillium simplicissimum* (Elsharkawy and Mousa 2015) have been related as stimulators of the resistance of plants.

In maize, some nonpathogenic bacteria have shown potential to stimulate resistance system to diseases. *Pseudomonas putida*, for example, induced resistance against maize anthracnose caused by *Colletotrichum graminicola* (Planchamp et al. 2014). This work shows a strong reduction of the pathogenic fungus growth and leaf necrosis on inoculated plants with *P. putida*. There was also a correlation between the induction of resistance during inoculation depending on the phospholipid metabolism and phytohormone production, indicating that these molecules stimulated the expression of ABA, ET, auxin, JA, and cytokines. All these phytohormones are involved with plant defense system stimulation. In another study, *Pseudomonas aurantiaca* showed ability to induce resistance of maize plants infected with the fungus *Bipolaris maydis*. Both extracts as supernatant of free cells of *P. aurantiaca* were effective for reduction of leaf necrosis caused by pathogenic fungus, leading to reduction of about 30 % disease severity (Fang et al. 2013).

Bacillus subtilis and *Azospirillum brasilense* were also related to capacity of inducing the maize defense system (Santos et al. 2014). *B. amyloliquefaciens* and *B. subtilis* stimulate the expression of resistance genes (PR-1 and PR-4) in maize plants infected with *Fusarium moniliforme* through the production of iturin A, fengycin, and bacillomycin and prevent the emergence of root lesions (Gong et al. 2015). *Azospirillum brasilense* induce also the maize resistance against the herbivorous insect (*Diabrotica speciosa*) attack (Santos et al. 2014).

For the fungi, *Trichoderma* genus is the most reported with the ability to induce the defense of maize plants. Many studies have demonstrated the effect of *Trichoderma* sp. in the induction of maize resistance against pathogens such as *Fusarium* sp. (Luongo et al. 2005), *Colletotrichum graminicola* (Djonovic et al. 2007), *Fusarium verticillioides* and *fumonisins* (Nayaka et al. 2010), and *Curvularia lunata* (Fan et al. 2015). In the working of Mukherjee et al. (2012), there was a detected activity of the polyketide synthase and phenylalanine ammonia lyase enzymes of *Trichoderma virens* interacting for inducing maize plant resistance. In an investigation of the *Trichoderma virens* secretome during interaction with maize plants, there was a detected presence of small cysteine-rich proteins that possibly act as effectors to reduce the level of plant stress, which can be part of a slight induction of systemic resistance against plant pathogen attack (Lamdan et al. 2015).

16.4 Microorganisms as Bioinoculants: Challenges to Microbial Diversity Use in the Sustainable Agriculture

Some microorganisms have shown the potential as bioinoculant. These microbial formulations are cheaper alternative according to requirement of sustainable environmental practice (Singh and Ratna 2016). Different microorganisms belonging to various taxa of bacteria, fungi, and possibly protozoa can colonize the rhizosphere or plant tissues and promote plant growth (Malusa et al. 2016; Szilagyi-Zecchin et al. 2016).

Among the main groups of microorganisms investigated for use as biofertilizer, it has been especially multifunctional strains. These microorganisms are both rhizosphere as endophytes that have capacity to produce many beneficial factors for the plant development (Montanez et al. 2012). Besides, protection to abiotic stresses such as drought in maize was observed after inoculation with microorganisms in experiments performed by Zoppellari et al. (2014).

Research related to the application of microbial biofertilizer has been carried out, and there are some bacteria-based commercial products (Hungria et al. 2010). However, only particular types of N-fixing bacteria have been used most extensively in agriculture as inoculants, such as those based in *Rhizobium* spp. to soybean in Brazil (Hungria et al. 2015). For grasses, as maize, the use of microorganisms of the rhizosphere have not been promising, probably due to the weak interaction with the plant comparing to endophytic microorganisms that have a much close symbiotic relationship with the plant, and it can be an advantage for its use as inoculant in relation to the rhizosphere (Sharma and Nowak 1998; Souza et al. 2015).

Maize is a C4 plant of annual cycle and therefore has a high nutrient demand. The maize association with bacteria of the *Azospirillum* genus has shown increasing up to 40 % of grain yield, this is equivalent to addition 80 kg/ha of N under field conditions (Marriel et al. 2008; Hungria et al. 2010). However, the inoculant based on the *Azospirillum* has shown variable effectiveness according to environmental conditions, plant genotype and especially bacteria species evaluated (Dobbelaere et al. 2001; Hungria et al. 2010).

Some phosphate fertilizer products for maize have been in the market in countries such as Canada, Australia, Egypt, and India, obtained from the rock phosphate mixture, solubilizing microorganisms, and a carbon source derived from sugarcane waste or cassava (Khalil et al. 2002; Faye et al. 2013). The combined use of rock phosphates and P-solubilizing microorganisms has been considered a promising strategy in environmental and economic terms (Singh and Ratna 2016). Results demonstrated productivity gains and plant mass of maize (Patil et al. 2016). Hameeda et al. (2008) observed increase approximately 30 % of the productivity gains related to control without microorganism inoculation and accumulation of 66

% of P in maize plants. Mineralizing microorganisms of phytate (MMP) has also been assessed for their use as biofertilizer in several crops (Goldstein et al. 2003; Jorguera et al. 2011; Mander et al. 2012) including maize (Oliveira et al. 2009). Under controlled conditions, in the maize inoculated with MMP and fertilized with rocks, an increase of the biomass of root up to 76 % and of the P content in the dry mass of maize was found (Oliveira et al. 2013). Regarding the inoculant-based AMF in maize, there are many difficulties for cultivation and application of AMF on a large scale. Part of this difficulty is because the interaction with AMF and plants is species specific, which affects the adaptation of the fungus according to the host plant (Bagyaraj et al. 2015; Zoppellari et al 2014). However, some studies also report the combined effect of biofertilizers containing microorganism solubilizing P, AMF, and diazotrophs in maize (Wu et al. 2005; Mohamed et al. 2014; Manzoor et al. 2016).

The use of microorganisms with ability to solubilize potassium (K) has also been investigated (Basak and Biswas 2009; Lopes-Assad et al. 2010; Meena et al. 2014; Zhang and Kong 2014; Silva et al. 2015), showing potential of the microorganisms to release K in the soil–plant system, promoting plant development (Alves et al. 2010; Verma et al. 2013; Prajapati et al. 2012; Zhang and Kong 2014). In maize, Singh et al. 2010 observed a higher biomass and K content accumulation in plants inoculated with *Bacillus mucilaginosus* in soil added with mica rock, as K source.

The microorganism selection process for inoculant formulation involves the isolation and identification of species responsible for promoting growth for a specific plant type and then evaluation of the survival rate, adaptation and multiplication of the microorganisms in the rhizosphere (rhizosphere competence), and infection and colonization of the host plant (endophyte competence) (Sathya et al. 2016), in laboratory and field tests. Thus, there are many challenges to achieve efficient inoculants on a large scale; any microorganisms found effective in in vitro studies can fail in promoting plant growth in the field conditions. Moreover, some strain characteristics are important to the successful growth promotion: ability to survive in the inoculant formulation, capacity to maintain its properties during storage, and tolerance to stress factors such as acidity, desiccation, high temperatures, chemical pesticides, and competition with other microorganisms. The high concentration of viable microbial cells and contaminant absence are essential factors for the quality of the inoculant (Leggett et al. 2007).

The adequate choice of the vehicle used in the formulation is another key factor to cell viability and inoculant quality (Silva et al. 2012). It can be used as vehicle in soil and inert material as turf or waste of industrials and of plant. Biodegradable polymers such as sodium alginate have also been identified as ecologically safe vehicles (Sahu and Brahmaprakash (2016)). These polymers promote encapsulation of cells and protection from environmental stress; the cells will be released after degradation in the environment.

The use of inoculants has found barriers with regard to reducing or replacement of industrial fertilizers by farmers, but its use as complement to fertilizer is becoming more acceptable and already has results from productivity gains (Oliveira et al. 2013). Considering the low cost of inoculant and its environmental role, the use of inoculants based in PGPM in the maize becomes a viable and promising practice within of the context of sustainable agriculture (Singh and Ratna 2016). However, many advances in understanding of the microbes, plant, and environment interactions are still required. Among the challenges, the methods to manage the introduction of the microorganism, its adaptation and colonization in several hosts, growth in regions with different soil and climatic conditions, and the determination of its effectiveness and agronomic validation stand out.

References

Abbott LK, Robson AD (1991) Factors influencing the occurrence of vesicular-arbuscular mycorrhizas. Agric Ecosyst Environ 35:121–150

Aira M, Gómez-Brandón M, Lazcano C, Baath E, Domínguez J (2010) Plant genotype strongly modifies the structure and growth of maize rhizosphere microbial communities. Soil Biol Biochem 42:2276–2281

Akladious SA, Abbas SM (2012) Application of *Trichoderma harziunum* T22 as a biofertilizer supporting maize growth. Afr J Biotechnol 11:8672–8683

Akram W, Anjum T, Ali B (2016) Phenylacetic acid is ISR determinant produced by *Bacillus fortis* IAGS162, Which involves extensive re-modulation in metabolomics of tomato to protect against *Fusarium* Wilt. Front Plant Sci 7. Doi: 10.3389/fpls.2016.00498

Aleti G, Sessitsch A, Brader G (2015) Genome mining: prediction of lipopeptides and polyketides from *Bacillus* and related Firmicutes. Comput Struct Biotechnol J 24:192–203

Alves L, Oliveira VL, Silva Filho GN (2010) Utilization of rocks and ectomycorrhizal fungi to promote growth of eucalypt. Braz J Microbiol 41:676–684

Alves GC, Videira SS, Urquiaga S, Reis VM (2015) Differential plant growth promotion and nitrogen fixation in two genotypes of maize by several *Herbaspirillum inoculants*. Plant Soil 387:307–321

Araújo EO, Martins MR, Mercante FM, Vitorino ACT, Urquiaga SS (2014) *Herbaspirillum seropedicae* inoculation and nitrogen fertilization on nitrogen use efficiency of different maize genotypes. Afr J Agric Res 9:3025–3031

Bagyaraj DJ, Sharma MP, Maiti D (2015) Phosphorus nutrition of crops through arbuscular mycorrhizal fungi. Curr Sci 108:1288–1293

Bais HP, Weir TL, Perry LG, Gilroy S, Vivanco JM (2006) The role of root exudates in rhizosphere interactions with plants and other organisms. Annu Rev Plant Biol 57:233–266

Balachandar D et al (2006) Flavonoids and growth hormones influence endophytic colonization and in plant nitrogen fixation by a diazotrophic *Serratia sp.* in rice. World J Microbiol Biotechnol 22:707–712

Baldani VL et al (1986) Establishment of inoculated *Azospirillum* spp in the rhizosphere and in roots of field grown wheat and sorgum. Plant Soil 90:35–46

Barret M, Morrissey JP, O'Gara F (2011) Functional genomics analysis of plant growth-promoting rhizobacterial traits involved in rhizosphere competence. Biol Fertil Soils 47:729–743

Barreteau H, Tiouajni M, Graille M et al (2012) Functional and structural characterization of PaeM, a Colicin M-like bacteriocin produced by *Pseudomonas aeruginosa*. J Biol Cement 287:37395–37405

Basak BB, Biswas DR (2009) Influence of potassium solubilizing microorganism (Bacillus mucilaginosus) and waste mica on potassium uptake dynamics by sudan grass (Sorghum vulgare Pers.) grown under two Alfisols. Plant Soil 317:235–255

Beneduzi A, Ambrosini A, Passaglia LMP (2012) Plant growth-promoting rhizobacteria (PGPR): their potential as antagonists and biocontrol agents. Genet Mol Biol 35:1044–1051

Berbara RLL, Souza FA, Fonseca HMAC (2006) Fungos micorrízicos arbusculares: Muito além da nutrição. In: Fernandes MS (ed) Nutrição mineral de plantas. Sociedade Brasileira de Ciência do Solo, Viçosa, MG, pp 54–79

Berg G, Smalla K (2009) Plant species and soil type cooperatively shape the structure and function of microbial communities in the rhizosphere. FEMS Microbiol Ecol 68:1–13

Berrocal-wolf M, Molina A, Solano R (2002) Constitutive expression of ETHYLENE-RESPONSE-FACTOR1 in Arabidopsis confers resistance to several necrotrophic fungi. Plant J 29:23–32

Bomke C, Tudzynski B (2009) Diversity, regulation, and evolution of the gibberellin biosynthetic pathway in fungi compared to plants and bacteria. Phytochemistry 70:1876–1893

Bouffaud ML, Kyselkova M, Gouesnard B, Grundmann G, Muller D, Moenne-Loccoz Y (2012) Is diversification history of maize influencing selection of soil bacteria by roots? Mol Ecol 21:195–206

Bucking H, Kafle A (2015) Role of arbuscular mycorrhizal fungi in the nitrogen uptake of plants: current knowledge and research gaps. Agronomy 5:587–612

Bulgarelli D, Schlaeppi K, Spaepen S, van Themaat EVL, Schulze-Lefert P (2013) Structure and functions of the bacterial microbiota of plants. Annu Rev Plant Biol 64:807–838

Cardoso IM, Kuyper TW (2006) Mycorrhizas and tropical soil fertility. Agric Ecosyst Environ 116:72–84

Cascales E, Buchanan SK, Duché D et al (2007) Colicin biology. Microbiol Mol Biol Rev 71:158–229

Castellanos T, Dohrmann AB, Imfeld G, Baumgarte S, Tebbe CC (2009) Search of environmental descriptors to explain the variability of the bacterial diversity from maize rhizospheres across a regional scale. Eur J Soil Biol 45:383–393

Chabot R, AntouN H, Kloepper JW, Beauchamp CJ (1996) Root colonization of maize and lettuce by bioluminescent *Rhizobium leguminosarum* biovar. phaseoli. Appl Environ Microbiol 62:2767–2772

Chauhan PS, Chaudhry V, Mishra S, Nautiyal CS (2011) Uncultured bacterial diversity in tropical maize (*Zea mays* L.) rhizosphere. J Basic Microbiol 51:15–32

Chelius MK, Triplett EW (2001) The diversity of archaea and bacteria in association with the roots of Zea mays L. Microb Ecol 41:252–263

Chen XH, Koumoutsi A, Scholz R et al (2009) Genome analysis of *Bacillus amyloliquefaciens* FZB42 reveals its potential for biocontrol of plant pathogens. J Biotechnol 140:27–37

Cherif A, Chehimi S, Limem F et al (2003) Detection and characterization of the novel bacteriocin entomocin 9, and safety evaluation of its producer, *Bacillus thuringiensis* ssp. entomocidus HD9. J Appl Microbiol 95:990–1000

Chrzanowski L, Ławniczak L, Czaczyk K (2012) Why do microorganisms produce rhamnolipids? World J Microbiol Biotechnol 28:401–419

Chu Q, Wang X, Yang Y, Chen F, Zhang F, Feng G (2013) Mycorrhizal responsiveness of maize (*Zea mays* L.) genotypes as related to releasing date and available P content in soil. Mycorrhiza 23:497–505

Cohen AC, Travaglia CN, Bottini R, Piccoli PN (2009) Participation of abscisic acid and gibberellins produced by endophytic *Azospirillum* in the alleviation of drought effects in maize. Botany 87:455–462

Compant S, Clément C, Sessitsch A et al (2010) Plant growth-promoting bacteria in the rhizo and endosphere of plants: their role, colonization, mechanisms involved and prospects for utilization. Soil Biol Biochem 42:669–678

Correa A, Cruz C, Ferrol N (2015) Nitrogen and carbon/nitrogen dynamics in arbuscular mycorrhiza: the great unknown. Mycorrhiza 25:499–515

Cotter PD, Ross RP, Hill C (2013) Bacteriocins—a viable alternative to antibiotics? Nat Rev Microbiol 11:95–105

Coutinho FP, Felix WP, Yano-Melo AM (2012) Solubilization of phosphates in vitro by *Aspergillus* spp. and *Penicillium* spp. Ecol Eng 42:85–89

Cozzolino V, Dimeo V, Piccolo A (2013) Impact of arbuscular mycorrhizal fungi applications on maize production and soil phosphorus availability. J Geochem Explor 129:40–44

Davies PJ (2010) Regulatory factors in hormone action: level, location and signal transduction. In: Plant hormones. Springer, Dordrecht, pp 16–35

De Vleesschauwer D, Maizeelis P, Hofte M (2006) Redox-active pyocyanin secreted by *Pseudomonas aeruginosa* 7NSK2 triggers systemic resistance to *Magnaporthe grisea* but enhances *Rhizoctonia solani* susceptibility in rice. Mol Plant-Microbe Interact 19:1406–1419

Dhawi F, Datta R, Ramakrishna W (2015) Mycorrhiza and PGPB modulate maize biomass, nutrient uptake and metabolic pathways in maize grown in mining-impacted soil. Plant Physiol Biochem 97:390–399

Didonet DA, Lima OS, Candaten MH, Rodrigues O (2000) Realocação de nitrogênio e de biomassa para os grãos, em trigo submetido à inoculação de *Azospirillum*. Pesq Agropec Bras 35:401–411

Dimkpa C, Svatos A, Merten D, Buchel G, Kothe E (2008) Hydroxamate siderophores produced by Streptomyces acidiscabies E13 bind nickel and promote growth in cowpea (Vigna unguiculata L.) under nickel stress. Can J Microbiol 54:163–172

Djonovic S, Vargas WA, Kolomiets MV, Horndeski M, Wiestand A, Kenerley CM (2007) A proteinaceous elicitor Sm1 from the beneficial fungus *Trichoderma virens* is required for induced systemic resistance in maize. Plant Physiol 145:875–889

Dobbelaere S, Croonenborghs A, Thys A, Ptacek D, Vanderleyden J, Dutto BS (2001) Responses of agronomically important crops to inoculation with Azospirillum. Funct Plant Biol 28:871–879

Dodd IC, Zinovkina NY, Safronova VI, Belimov AA (2010) Rhizobacterial mediation of plant hormone status. Ann Appl Biol 157:361–379

Donato VMTS, Andrade AG, Takaki GMC, Mariano RLR, Maciel GA (2005) Plantas de cana-de-açúcar cultivadas *in vitro* com antibióticos. Ciência e Agrotecnologia 29:134–141

Elsharkawy MM, Mousa KM (2015) Induction of systemic resistance against Papaya ring spot virus (PRSV) and its vector Myzus persicae by Penicillium simplicissimum GP17–2 and silica (Sio 2) nanopowder. Int J Pest Manag 61:353–358

Fan L, Fu K, Yu C, Li Y, Li Y, Chen J (2015) Thc6 protein, isolated from *Trichoderma harzianum*, can induce maize defense response against *Curvularia lunata*. J Basic Microbiol 55:591–600

Fang R, Lin J, Yao S, Wang Y, Wang J, Zhou C, Wang H, Xiao M (2013) Promotion of plant growth, biological control and induced systemic resistance in maize by *Pseudomonas aurantiaca* JD37. Ann Microbiol 63:1177–1185

Favret ME, Yousten AA (1989) Thuricin: the bacteriocin produced by *Bacillus thuringiensis*. J Invertebr Pathol 53:206–216

Faye A, Dalpe Y, Ndung'u-Magiroi K, Jefwa J, Ndoye ID, Lesueur D (2013) Evaluation of commercial arbuscular mycorrhizal inoculants on maize in Kenya. Can J Plant Sci 93:1201–1208

Fierer N, Bradford MA, Jackson RB (2007) Toward an ecological classification of soil bacteria. Ecology 88:1354–1364

Fisher PJ, Petrini O, Scott HML (1992) The distribution of some fungal and bacterial endophytes in maize (*Zea mays* L.). New Phytol 122:299–305

Fridlender M, Inbar J, Chet I (1993) Biological control of soilborne plant pathogens by a ß-1, 3 glucanase producing *Pseudomonas cepacia*. Soil Biol Biochem 25:1211–1221

Gaby JC, Buckley DHA (2012) Comprehensive evaluation of PCR primers to amplify the *nifH* gene of nitrogenase. PLos One 7:e42149

Giauque H, Hawkes CV (2013) Climate affects symbiotic fungal endophyte diversity and performance. Am J Bot 100:1435–1444

Giovanelli J, Mudd SH, Datko AH (1980) Sulfur amino acids in plants. In: Miflin BJ (ed) Amino acids and derivatives in the biochemistry of plants: a comprehensive treatise. Academic Press, New York, pp 453–505

Goldstein A, Lester T, Brown J (2003) Research on the metabolic engineering of the direct oxidation pathway for extraction of phosphate from ore has generated preliminary evidence for PQQ biosynthesis in *Escherichia coli* as well as a possible role for the highly conserved region of quinoprotein dehydrogenases. Biochem Biophys Acta 1647:266–271

Gomes NCM, Heuer H, Schonfeld J, Costa R, Mendonca-Hagler L, Smalla K (2001) Bacterial diversity of the rhizosphere of maize (*Zea mays*) grown in tropical soil studied by temperature gradient gel electrophoresis. Plant Soil 232:167–180

Gomes EA, Silva UC, Marriel IE, Oliveira CA, Lana UGP (2014) Rock phosphate solubilizing microorganisms isolated from maize rhizosphere soil. Revista brasileira de Milho e Sorgo 13:69–81

Gong AD, Li HP, Yuan QS, Song XS, Yao W, He WJ, Liao YC (2015) Antagonistic mechanism of iturin A and plipastatin A from *Bacillus amyloliquefaciens* S76-3 from wheat spikes against Fusarium graminearum. PLoS One 10, e0116871

Govindarajan M, Balandreau J, Muthukumarasamy R, Revathi G, Lakshminarasimhan C (2006) Improved yield of micropropagated sugarcane following inoculation by endophytic *Burkholderia vietnamiensis*. Plant Soil 280:239–252

Govindarajan M, Kwon SW, Weon HY (2007) Isolation, molecular characterization and growth-promoting activities of endophytic sugarcane diazotroph *Klebsiella* sp. GR9. World J Microbiol Biotechnol 23:997–1006

Greiner R (2006) Phytate-degrading enzymes: regulation of synthesis in microorganisms and plants. In: Turner BL, Richardson AE, Mullaney EJ (eds) Inositol phosphates: linking agriculture and environment. CAB International, London, pp 78–96

Hamdali H, Bouizgarne B, Hafidi M, Lebrihi A, Virolle MJ, Ouhdouch Y (2008) Screening for rock phosphate solubilizing Actinomycetes from Moroccan phosphate mines. Appl Soil Ecol 38:12–19

Hameeda B, Harini G, Rupela OP, Wani SP, Reddy G (2008) Growth promotion of maize by phosphate-solubilizing bacteria isolated from composts and macrofauna. Microbiol Res 163:234–242

Hardoim PR, Van Overbeek LS, Elsas JD (2008) Properties of bacterial endophytes and their proposed role in plant growth. Trends Microbiol 16:463–471

Hardoim PR, Andreote FD, Reinhold-Hurek B, Sessitsch A, Van Overbeek LS, Van Elsas JD (2011) Rice root-associated bacteria: insights into community structures across 10 cultivars. FEMS Microbiol Ecol 77:154–164

Hoffman MT, Arnold AE (2008) Geographic locality and host identity shape fungal endophyte communities in cupressaceous trees. Mycol Res 112:331–344

Hu J, Lin X, Wang J, Dai J, Cui X, Chen R, Zhang J (2009) Arbuscular mycorrhizal fungus enhances crop yield and P-uptake of maize (*Zea mays L.*): a field case study on a sandy loam soil as affected by long-term P-deficiency fertilization. Soil Biol Biochem 41:2460–2465

Hungria M, Campo RJ, Souza EM, Pedrosa FO (2010) Inoculation with selected strains of *Azospirillum brasilense* and *A. lipoferum* improves yields of maize and wheat in Brazil. Plant Soil 331:413–425

Hungria M, Nogueira MA, Araujo RS (2015) Alternative methods of soybean inoculation to overcome adverse conditions at sowing. Afr J Agric Res 10:2329–2338

Johnston-Monje D, Raizada MN (2011) Conservation and diversity of seed associated endophytes in Zea across boundaries of evolution, ethnography and ecology. PLoS One 6:20396–20396

Johnston-Monje D, Lundberg D S, Lazarovits G, Reis V M, Raizada M N (2016) Bacterial populations in juvenile maize rhizospheres originate from both seed and soil. Plant and Soil 405:1–19

Jorguera MA, Crowley DE, Marschner P, Greiner R, Fernandez MT, Romero D, Menezes-Blackburn D, Mora ML (2011) Identification of β-propeller phytase-encoding genes in culturable *Paenibacillus* and *Bacillus* spp. from the rhizosphere of pasture plants on volcanic soils. FEMS Microbiol Ecol 75:163–172

Kamoun F, Mejdoub H, Aouissaoui H et al (2005) Purification, amino acid sequence and characterization of Bacthuricin F4, a new bacteriocin produced by *Bacillus thuringiensis*. J Appl Microbiol 98:881–888

Kaur G, Reddy MS (2015) Effects of phosphate-solubilizing bacteria, rock phosphate and chemical fertilizers on maize-wheat cropping cycle and economics. Pedosphere 25:428–437

Khalil MY, Naguib NY, El-Sherbeny SE (2002) Response of *Tagetes erecta L.* to compost and foliar application of some microelements. Arab Univ J Agric Sci, Ain Shams Univ, Cairo 10:939–964

Khan AA, Jilani G, Akhtar MS, Naqvi SMS, Rasheed M (2009) Phosphorus solubilizing bacteria: occurrence, mechanisms and their role in crop production. J Agric Biol Sci 1:48–58

Kim CH, Han SH, Kim KY, Cho BH, Kim YH, Koo BS (2003) Cloning and expression of Pyrroloquinoline Quinone (*PQQ*) genes from a phosphate-solubilizing bacterium *Enterobacter intermedium*. Curr Microbiol 47:457–461

Klaenhammer TR (1993) Genetics of bacteriocins produced by lactic acid bacteria. FEMS Microbiol Rev 12:39–85

Kobayashi DY, Palumbo JD (2000) Bacterial endophytes and their effects on plants and uses in agriculture. In: Microbial endophytes. Marcel Dekker Inc., New York, pp 199–233

Korasick DA, Enders TA, Strader LC (2013) Auxin biosynthesis and storage forms. J Exp Bot 64: 2541–2555

Krewulak HD, Vogel HJ (2008) Structural biology of bacterial iron uptake. Biochim Biophys Acta 1778:1781–1804

Kuan KB, Othman R, Abdul-Rahim K, Shamsuddin ZH (2016) Plant growth-promoting rhizobacteria inoculation to enhance vegetative growth, nitrogen fixation and nitrogen remobilisation of Maize under greenhouse conditions. PLoS One 11:e0152478

Kumar B, Trivedi P, Pandey A (2007) *Pseudomonas corrugate*: a suitable bacterial inoculant for maize grown under rainfed conditions of Himalayan region. Soil Biol Biochem 39:3093–3100

Kumar S, Chauhan PS, Agrawal L, Raj R, Srivastava A, Gupta S, Mishra S K, Yadav S, Singh PC, Raj SK, Nautiyal CS (2016) *Paenibacillus lentimorbus* inoculation enhances tobacco growth and extenuates the virulence of *Cucumber mosaic* virus. PLoS One 11:e0149980

Lamdan NL, Shalaby S, Ziv T, Kenerley CM, Horwitz BA (2015) Secretome of Trichoderma interacting with maize roots: role in induced systemic resistance. Mol Cell Proteomics 14:1054–63

Leggett M, Cross J, Hnatowich G, Holloway G (2007) Challenges in commercializing a phosphate-solubilizing microorganism: Penicillium bilaiae, a case history. In: Velázquez E, Rodrıguez-Barrueco C (eds) First international meeting on microbial phosphate solubilization, pp 215–222

Li X, Rui J, Mao Y, Yannarell A, Mackie R (2014) Dynamics of the bacterial community structure in the rhizosphere of a maize cultivar. Soil Biol Biochem 68:392–401

Loaces I, Ferrando L, Scavino AF (2011) Dynamics, diversity and function of endophytic siderophore-producing bacteria in rice. Microb Ecol 61:606–618

Lodewyckx C et al (2002) Endophytic bacteria and their potential applications. Crit Rev Plant Sci 21:583–606

Loper JE, Buyer JS (1991) Siderophores in microbial interactions of plant surfaces. Mol Plant-Microbe Interact 4:5–13

Lopes-Assad M, Avansini SH, Rosa MM, Carvalho JRP, Antonini SRC (2010) The solubilization of potassium-bearing rock powder by Aspergillus niger in small-scale batch fermentations. Can J Microbiol 56:598–605

Lorenzo O, Chico JM, Sanchez-Serrano JJ, Solano R (2004) Jasmonate-insensitive1 encodes a MYC transcription factor essential to discriminate between different jasmonate-regulated defense responses in Arabidopsis. Plant Cell 16:1938–1950

Lucangeli C, Bottini R (1997) Effects of Azospirillum spp. on endogenous gibberellin content and growth of maize (Zea mays L.) treated with uniconazole. Symbiosis 23:63–72

Ludwig-Müller J, Kaldorf M, Sutter EG, Epstein E (1997) Indole-3-butyric acid (IBA) is enhanced in young maize (*Zea mays* L.) roots colonized with the arbuscular mycorrhizal fungus *Glomus intraradices*. Plant Sci 125:153–162

Lundberg DS, Lebeis SL, Paredes SH et al (2012) Defining the core Arabidopsis thaliana root microbiome. Nature 488:86

Luongo L, Galli M, Corazza L, Meekes E, De Haas L, Van der Plas CL, Kohl J (2005) Potential of fungal antagonists for biocontrol of Fusarium spp. in wheat and maize through competition in crop debris. Biocontrol Sci Tech 15:229–242

Malusa A, Pinzari E, Canfora F, Singh L, Singh PD, Prabha BH (2016) Efficacy of biofertilizers: challenges to improve crop production. In: Microbial inoculants in sustainable agricultural productivity, vol 2. 978-81-322-2644-4

Mander C, Wakelin S, Young S, Condron I, O'Callaghan M (2012) Incidence and diversity of phosphate-solubilizing bacteria are linked to phosphorus status in grassland soils. Soil Biol Biochem 44:93–101

Manzoor M, Kaleem Abbasi M, Sultan T (2016) Isolation of phosphate solubilizing bacteria from maize rhizosphere and their potential for rock phosphate solubilization–mineralization and plant growth promotion. Geomicrobiol J. doi:10.1080/01490451.2016.1146373

Marra LM (2012) Solubilização de fosfato por bactérias e sua contribuição no crescimento de leguminosas e gramíneas. 142f. Tese (Doutorado em Ciência do solo). Universidade Federal de Lavras, Lavras – MG

Marriel IE, Coelho AM, Guimaraes LJM, Guimaraes PE, Pacheco CA, Magalhaes PC, Oliveira AC (2008) Bactérias diazotróficas selecionadas contribuem com economia de fertilizantes nitrogenados em milho In: Simpósio sobre inovação e criatividade científica na Embrapa, 1., 2008, Brasília, DF. Resumos. Embrapa, Brasília, DF

Mazzotta AS, Crandall AD, Montville TJ (1997) Nisin resistance in *Clostridium botulinum* spores and vegetative cells. Appl Environ Microbiol 63:2654–2659

McInroy JA, Kloepper JW (1995) Survey of indigenous bacterial endophytes from cotton and sweet maize. Plant Soil 173:337–342

McLaughlin MJ, McBeath TM, Smernik R, Stacey SP, Ajiboye B, Guppy C (2011) The chemical nature of P accumulation in agricultural soils—implications for fertiliser management and design: an Australian perspective. Plant Soil 349:69–87

Meena VS, Maurya BR, Verma JP (2014) Does a rhizospheric microorganism enhance K + availability in agricultural soils? Microbiol Res 169:337–347

Mei L, Liang Y, Zhang L, Wang Y, Guo Y (2014) Induced systemic resistance and growth promotion in tomato by an indole-3-acetic acid-producing strain of *Paenibacillus polymyxa*. Ann Appl Biol 165:270–279

Mendes R, Garbeva P, Raaijmakers JM (2013) The rhizosphere microbiome: significance of plant-beneficial, plant-pathogenic and human-pathogenic microorganisms. FEMS Microbiol Rev

Mendonça MM, Urquiaga SS, Reis VM (2006) Variabilidade genotípica de milho para acumulação de nitrogênio e contribuição da fixação biológica de nitrogênio. Pesq Agrop Brasileira 41:1681–1685

Mensah JA, Koch AM, Antunes PM, Hart MM, Kiers ET, Bücking H (2015) High functional diversity within arbuscular mycorrhizal fungal species is associated with differences in phosphate and nitrogen uptake and fungal phosphate metabolism. Mycorrhiza 25:533–546

Miche L, Battistoni FJ, Gemmer S, Belghazi M, Reinhold-Hurek B (2006) Upregulation of jasmonate-inducible defense proteins and differential colonization of roots of *Oryza sativa* cultivars with the endophyte *Azoarcus* sp. Mol Plant-Microbe Interact 19:502–511

Michel-Briand Y, Baysse C (2002) The pyocins of *Pseudomonas aeruginosa*. Biochimie 84:499–510

Miliute I, Buzaite O, Baniulis D, Stanys V (2015) Bacterial endophytes in agricultural crops and their role in stress tolerance: a review. Zemdirbyste-Agriculture 102:465–478

Minerdi D, Fani R, Gallo R, Boarino A, Bonfante P (2001) Nitrogen fixation genes in an endosymbiotic Burkholderia strain. Appl Environ Microbiol 67(2):725–732

Mohamed AA, Eweda WEE, Heggo AM, Hassan EA (2014) Effect of dual inoculation with arbuscular mycorrhizal fungi and sulphur-oxidising bacteria on onion (Allium cepa L.) and maize (Zea mays L.) grown in sandy soil under green house conditions. Ann Agric Sci 59:109–118

Montanez A, Abreu C, Gill PR, Hardarson G, Sicardi M (2009) Biological nitrogen fixation in maize (*Zea mays* L.) by ^{15}N isotope dilution and identification of associated culturable diazotrophs. Biol Fertil Soils 45:253–263

Montanez A, Rodriguez Blanco A, Barlocco C, Beracochea M, Sicardi M (2012) Characterization of cultivable putative endophytic plant growth promoting bacteria associated with maize cultivars (Zea mays L.) and their inoculation effects in vitro. Appl Soil Ecol 58:21–28

Moore JW, Loake GJ, Spoel SH (2011) Transcription dynamics in plant immunity. Plant Cell 23:2809–2820

Mousa WK, Shearer CR, Limay-Rios V, Zhou T, Raizada MN (2015) Bacterial endophytes from wild maize suppress *Fusarium graminearum* in modern maize and inhibit mycotoxin accumulation. Front Plant Sci 6:805

Mudge SR, Frank WS, Richardson AE (2003) Root-specific and phosphate-regulated expression of phytase under the control of a phosphate transporter promoter enables Arabidopsis to grow on phytate as a sole P source. Plant Sci 165:871–878

Mukherjee PK, Buensanteai N, Moran-Diez ME, Druzhinina IS, Kenerley CM (2012) Functional analysis of non-ribosomal peptide synthetases (NRPSs) in Trichoderma virens reveals a polyketide synthase (PKS)/NRPS hybrid enzyme involved in the induced systemic resistance response in maize. Microbiology 158(1):155–165

Murray JD et al (2007) A cytokinin perception mutant colonized by Rhizobium in the absence of nodule organogenesis. Science 315:101–104

Naghmouchi K, Hammami R, Fliss I, Teather R, Baah J, Drider D (2012) Colistin A and colistin B among inhibitory substances of *Paenibacillus polymyxa* JB05-01-1. Arch Microbiol 194:363–370

Nakayama K, Takashima K, Ishihara H et al (2000) The R-type pyocin of *Pseudomonas aeruginosa* is related to P2 phage, and the F-type is related to lambda phage. Mol Microbiol 38:213–231

Naveed M, Qureshi MA, Zahir ZA, Hussain MB, Sessitsch A, Mitter B (2015) L-Tryptophan-dependent biosynthesis of indole-3-acetic acid (IAA) improves plant growth and colonization of maize by *Burkholderia phytofirmans* PsJN. Ann Microbiol 65:1381–1389

Nayaka C, Niranjana SR, Uday ACS, Reddy MS, Prakash HS, Mortensen CN (2010) Seed biopriming with novel strain of Trichoderma harzianum for the control of toxigenic Fusarium verticillioides and fumonisins in maize. Arch Phytopathol Plant Protect 43:264–282

Neilands JB (1995) Siderophores: structure and function of microbial iron transport compounds. J Biol Chem 270:26723–26726

Novais RF, Smyth TJ (1999) Fósforo em solo e planta em condições tropicais. Universidade Federal de Viçosa, Viçosa. 399p

Nunes FS, Raimondi AC, Niedwieski AC (2003) Fixação de nitrogênio: estrutura, função e modelagem bioinorgânica das nitrogenases. Quim Nova 26:872–879

Ogbo FC (2010) Conversion of cassava wastes for biofertilizer production using phosphate solubilizing fungi. Bioresour Technol 101:4120–4124

Oliveira CA, Alves VMC, Marriel IE, Gomes EA, Scotti MRSM, Carneiro NM, Guimarães CT, Schaffert RE, Sá NMA (2009) Phosphate solubilizing microorganisms isolated from rhizosphere of maize cultivated in an oxisol of the brazilian cerrado biome. Soil Biol Biochem 25:1–6

Oliveira CA, Marriel IE, Gomes EA, Mattos BB, Santos FC, Oliveira MC, Alves VMC (2013) Metodologia de aplicação de microrganismos solubilizadores de fósforo em sementes visando melhor aproveitamento deste nutriente pelas plantas. Sete Lagoas: Embrapa Milho e Sorgo, 25 p. (Embrapa Milho e Sorgo. Boletim de Pesquisa e Desenvolvimento, 88)

Ongena M, Jourdan E, Adam A, Paquot M, Brans A, Joris B, Arpigny JL, Thonart P (2007) Surfactin and fengycin lipopeptides of Bacillus subtilis as elicitors of induced systemic resistance in plants. Environ Microbiol 9:1084–1090

Ordentlich A, Elad Y, Chet I (1988) The role of chitinase of Serratia marcescens in the biocontrol of Sclerotium rolfsii. Phytopathology 78:84–88

Paik HD, Bae SS, Park SH, Pan JG (1997) Identification and partial characterization of tochicin, a bacteriocin produced by *Bacillus thuringiensis* subsp tochigiensis. J Ind Microbiol Biotechnol 19:294–298

Pal KK, Tilak KV, Saxena AK, Dey R, Singh CS (2001) Suppression of maize root diseases caused by Macrophomina phaseolina, Fusarium moniliforme and Fusarium graminearum by plant growth promoting rhizobacteria. Microbiol Res 156:209–23

Parret AHA, Schoofs G, Proost P, De Mot R (2003) Plant lectin-like bacteriocin from a rhizosphere-colonizing *Pseudomonas* isolate. J Bacteriol 185:897–908

Patil et al (2016) Microbial inoculants in sustainable agricultural productivity, vol 1. Springer, New Delhi, pp 319–343

Peiffer J et al (2013) Diversity and heritability of the maize rhizosphere microbiome under field conditions. Proc Natl Acad Sci U S A 110:6548–6553

Pereira F, Ibañez M, Rosenblueth M, Etcheverry E, Martínez-Romero (2011) Analysis of the bacterial diversity associated with the roots of maize (*Zea mays* L.) through culture-dependent and culture-independent methods. International Scholarly Research Network Ecology – ISRN, 10 p

Perin L et al (2006) Diazotrophic Burkholderia species associated with field-grown maize and sugarcane. Appl Environ Microbiolo 72:3103–3110

Philippot L, Raaijmakers JM, Lemanceau P, van der Putten WH (2013) Going back to the roots: the microbial ecology of the rhizosphere. Nat Rev Microbiol 11:789–799

Picard C, Bosco M (2005) Maize heterosis affects the structure and dynamics of indigenous rhizospheric auxins-producing *Pseudomonas* populations. FEMS Microbiol Ecol 53:349–357. doi:10.1016/j.femsec.2005.01.007

Pieterse CM, Van der Does D, Zamioudis C, Leon-Reyes A, Van Wees SC (2012) Hormonal modulation of plant immunity. Ann Rev Cell Dev Biol 28:489–521

Planchamp C, Glauser G, Mauch-Mani B (2014) Root inoculation with *Pseudomonas putida* KT2440 induces transcriptional and metabolic changes and systemic resistance in maize plants. Front Plant Sci 5:719

Prajapati KB, Sharma MC, Modi HA (2012) Isolation of two potassium solubilizing fungi from ceramic industry soils. Life Sci Leaflets 5:71–75

Press CM, Wilson M, Tuzun S, Kloepper JW (1997) SA produced by S. marcescens 90–166 is not the primary determinant of ISR in cucumber/ tobacco. Mol Plant-Microbe Interact 10:761–768

Radzki W, Manero FG, Algar E, Garcia JL, Garcia VA, Solano BR (2013) Bacterial siderophores efficiently provide iron to iron-starved tomato plants in hydroponics culture. Antonie Van Leeuwenhoek 104:321–330

Rai R, Dash PK, Prasanna BM, Singh A (2007) Endophytic bacterial flora in the stem tissue of a tropical maize (Zea mays L.) genotype: isolation, identification and enumeration. World J Microbiol Biotechnol 23:853–858

Ramachandran VK, East AK, Karunakaran R, Downie JA, Poole PS (2011) Adaptation of *Rhizobium leguminosarum* to pea, alfalfa and sugar beet rhizosphere investigated by comparative transcriptomics. Genome Biol 12:106–106

Raza FA, Faisal M (2013) Growth promotion of maize by desiccation tolerant Micrococcus luteus-chp37 isolated from Cholistan desert, Pakistan. AJCS 7:1693–1698

Raza W, Yuan J, Ling N, Huang Q, Shen Q (2015) Production of volatile organic compounds by an antagonistic strain *Paenibacillus polymyxa* WR-2 in the presence of root exudates and organic fertilizer and their antifungal activity against *Fusarium oxysporum* f. sp. *Niveum*.

Reitz M, Oger P, Meyer A, Niehaus K, Farrand SK, Hallmann J, Sikora RA (2002) Importance of the O-antigen, core-region and lipid A of rhizobial LPS for the induction of SR in potato to Globodera pallida. Nematology 4:73–79

Richardson AE, Simpson RJ (2011) Soil microorganisms mediating phosphorus availability update on microbial phosphorus. Plant Physiol 156:989–996

Richardson AE, Barea J, Mcneill AM, Prigent-Combaret C (2009) Acquisition of phosphorus and nitrogen in the rhizosphere and plant growth promotion by microorganisms. Plant Soil 321:305–339

Riggs PJ, Chelius MK, Iniguez AL, Kaeppler SM, Triplett EW (2001) Enhanced maize productivity by inoculation with diazotrophic bacteria. Funct Plant Biol 28:829–836

Riley MA (1998) Molecular mechanisms of bacteriocin evolution. Annu Rev Genet 32:255–278

Riley MA, Wertz JE (2002) Bacteriocins: evolution, ecology, and application. Ann Rev Microbiol 56:117–137

Rodríguez H, Fraga R, Gonzalez T, Bashan Y (2006) Genetics of phosphate solubilization and its potential applications for improving plant growth-promoting bacteria. Plant Soil 287:15–21

Roesch LFW, Camargo FA, Bento FM, Triplett EW (2008) Biodiversity of diazotrophic bacteria within the soil, root and stem of field-grown maize. Plant Soil 302:91–104
Rosenblueth M, Martinez-Romero E (2006) Bacterial endophytes and their interactions with hosts. Mol Plant-Microbe Interact 19:827–837
Ryu CM, Farag MA, Hu CH, Reddy MS, Kloepper JW, Paré PW (2004) Bacterial volatiles induce systemic resistance in Arabidopsis. Plant Physiol 134:1017–1026
Sachdev DP, Cameotra SS (2013) Biosurfactants in agriculture. Appl Microbiol Biotechnol 97:1005–16
Sahu PK, Brahmaprakash GP (2016) Formulations of biofertilizers–approaches and advances. In Microbial inoculants in sustainable agricultural productivity. Springer, New York, pp 179–198
Saia S, Benitéz E, Garcia-Garrido JM, Settanni L, Amato G, Giambalvo D (2014) The effect of arbuscular mycorrhizal fungi on total plant nitrogen uptake and nitrogen recovery from soil organic material. J Agric Sci 152:370–378
Saikkonen K, Faeth SH, Helander M, Sullivan TJ (1998) Fungal endophytes: a continuum of interactions with host plants. Annu Rev Ecol Syst 29:319–343
Salla TD, Astarita LV, Santarém ER (2016) Defense responses in plants of Eucalyptus elicited by Streptomyces and challenged with Botrytis cinerea. Planta 243:1055–1070
Sanchez L, Courteaux B, Hubert J, Kauffmann S, Renault JH, Clément C, Dorey S (2012) Rhamnolipids elicit defense responses and induce disease resistance against biotrophic, hemibiotrophic, and necrotrophic pathogens that require different signaling pathways in Arabidopsis and highlight a central role for salicylic acid. Plant Physiol 160:1630–1641
Santos F, Peñaflor MFG, Paré PW, Sanches PA, Kamiya AC, Tonelli M, Bento JMS (2014) A novel interaction between plant-beneficial rhizobacteria and roots: colonization induces corn resistance against the root herbivore *Diabrotica speciosa*. PLoS One 9:e113280
Santner A, Estelle M (2010) The ubiquitin-proteasome system regulates plant hormone signaling. Plant J 61:1029–1040
Sathya A, Gopalakrishnan R, Singh S, Dhananjaya P (2016) Soil microbes: the invisible managers of soil fertility. In: Singh BH, Ratna P (eds) Microbial inoculants in sustainable agricultural productivity, vol 2. Springer, New Delhi, pp 1–16
Saxena A, Raghuwanshi R, Singh HB (2015) Elevation of defense network in Chilli against Colletotrichum capsici by Phyllospheric Trichoderma strain. J Plant Growth Regul 35:1–13
Schmalenberger A, Tebbe CC (2002) Bacterial community composition in the rhizosphere of a transgenic, herbicide-resistant maize (Zea mays) and comparison to its non-transgenic cultivar Bosphore. FEMS Microbiol Ecol 40:29–37
Schwyn B, Neilands JB (1987) Universal chemical assay for the detection and determination of siderophores. Anal Biochem 160:47–56
Seghers D, Wittebolle L, Top EM, Verstraete W, Siciliano SD (2004) Impact of agricultural practices on the Zea mays L. endophytic community. Appl Environ Microbiol 70:1475–1482
Sharma VK, Nowak J (1998) Enhancement of verticillium wilt resistance in tomato transplants by in vitro co-culture of seedlings with a plant growth promoting rhizobacterium (Pseudomonas sp. strain PsJN). Can J Microbiol 44:528–536
Silva MF, Souza AC, Oliveira PJ (2012) Survival of endophytic bacteria in polymer-based inoculants and efficiency of their application to sugarcane. Plant Soil 356:231–243
Silva UC, Mendes GO, Silva NMR, Duarte JL, Silva IR, Tótola MR, Costa MD (2014) Fluoride-tolerant mutants of Aspergillus niger show enhanced phosphate solubilization capacity. PLoS One 9, e110246
Silva UC, Marriel IE, Oliveira CA, Gomes EA, Rezende AV, Lana UGP (2015) Biossolubilização de Potássio In Vitro a Partir da Rocha Fonolito por Microrganismos do Solo. Série Documentos, Embrapa Milho e Sorgo, 177–7
Singh BH, Ratna P (2016) Microbial inoculants in sustainable agricultural productivity: vol. 2: Functional applications. Springer, New Delhi, 308 p
Singh G, Biswas DR, Marwah TS (2010) Mobilization of potassium from waste mica by plant growth promoting rhizobacteria and its assimilation by maize (*Zea mays)* and wheat (T*riticum aestivum* L.). J Plant Nutr 33:1236–1251

Siqueira JO, Franco AA (1988) Biotecnologia do solo: fundamentos e perspectivas. MEC Ministério da Educação, ABEAS/ESAL, FAEPE, Brasília/Lavras, 236 p

Smyth E (2011) Selection and analysis of bacteria on the basis of their ability to promote plant development and growth. PhD thesis, University College Dublin

Snook ME, Mitchell T, Hinton DM, Bacon CW (2009) Isolation and characterization of Leu(7)-surfactin from the endophytic bacterium Bacillus mojavensis RRC 101, a biocontrol agent for *Fusarium verticillioides*. J Agric Food Chem 57:4287–4292

Sobrinho CA, Ferreira PTO, Cavalcanti LS, Indutores Abióticos (2005) In: Cavalcanti LS, Di Piero RM, Cia P, Pascholati SF, Resende MLV, Romeiro RS, Piracicaba SP (eds) Indução de Resistência em plantas a patógenos e insetos. FEALQ, pp 51–80

Souza R, Ambrosini A, Passaglia LMP (2015) Passaglia plant growth-promoting bacteria as inoculants in agricultural soils. Genet Mol Biol 38(4):401–419

Sukumar P, Legue V, Vayssieres A, Martin F, Tuskan GA, Kalluri UC (2013) Involvement of auxin pathways in modulating root architecture during beneficial plant-microorganism interactions. Plant Cell Environ 36:909–919

Sun X, Ding Q, Hyde KD, Guo LD (2012) Community structure and preference of endophytic fungi of three woody plants in a mixed forest. Fungal Ecol 5:624–632

Sylvia DM, Williams SE (1992) Vesicular-arbuscular mycorrhizae and environmenlal stress. In: Beihlenfalvay GJ, Linderman RG (eds) Mycorrhizae in susíainable agriculture. American Society of Agronomy, Special Publication 54, American Society of Agronomy, Madison, pp 101–124

Szilagyi-Zecchin VJ, Ikeda AC, Hungria M, Adamoski D, Kava-Cordeiro V, Glienke C, Galli-Terasawa LV (2014) Identification and characterization of endophytic bacteria from maize (*Zea mays* L.) roots with biotechnological potential in agriculture. AMB Express 4:2–9

Szilagyi-Zecchin VJ et al. (2016) LIVRO microbial inoculants in sustainable agricultural productivity, vol 1. Springer, New Delhi, pp 1–16

Tarafdar JC, Gharu A (2006) Mobilization of organic and poorly soluble phosphates by Chaetomium globosum. Appl Soil Ecol 32:273–283

Tinker P, Nye P (2000) Solute movement in the rhizosphere. Oxford University Press, New York

Toljander JF, Santos-González JC, Tehler A, Finlay RD (2008) Community analysis of arbuscular mycorrhizal fungi and bacteria in the maize mycorrhizosphere in a long-term fertilization trial. FEMS Microbiol Ecol 65:323–338

Tortora ML, Díaz-Ricci JC, Pedraza RO (2011) *Azospirillum brasilense* siderophores with antifungal activity against *Colletotrichum acutatum*. Arch Microbiol 193:275–286

Tsavkelova EA, Klimova SY, Cherdyntseva TA, Netrusov AI (2006) Microbial producers of plant growth stimulators and their practical use: a review. Appl Biochem Microbiol 42:117–126

Turner TR, James EK, Poole PS (2013) The plant microbiome. Genome Biol 14:209

Valois D, Fayad K, Barasubiye T et al (1996) Glucanolytic Actinomycetes antagonistic to *Phytophthora fragariae* var. *rubi*, the causal agent of raspberry root rot. Appl Environ Microbiol 62:1630–1635

Van Loon LC, Bakker PAHM, Pieterse CMJ (1998) Systemic resistance induced by rhizosphere bacteria. Annu Rev Phytopathol 36:453–483

Vázquez MM, César S, Azcón R, Barea JM (2000) Interactions between arbuscular mycorrhizal fungi and other microbial inoculants (*Azospirillum, Pseudomonas, Trichoderma*) and their effects on microbial population and enzyme activities in the rhizosphere of maize plants. Appl Soil Ecol 15:261–272

Verma JP, Yadav J, Tiwari KN, Kumar A (2013) Effect of indigenous Mesorhizobium spp. and plant growth promoting rhizobacteria on yields and nutrients uptake of chickpea (Cicer arietinum L.) under sustainable agriculture. Ecol Eng 51:282–286

Vieira JAC (2015) Bactérias endofíticas de milho e seu potencial como promotoras de crescimento vegetal e agentes de controle biológico. Dissertação de mestrado -Universidade Federal de Minas Gerais

Villegas J, Fortin JA (2002) Phosphorus solubilization and pH changes as a result of the interactions between soil bacteria and arbuscular mycorrhizal fungi on a medium containing NO3-as nitrogen source. Can J Bot 80:571–576

Viruel E, Erazzú LE, Martínez-Calsina L, Merrero AF, Lucca ME, Siñeriz F (2014) Inoculation of maize with phosphate solubilizing bacteria: effect on plant growth and yield. J Soil Sci Plant Nutr 14:819–831

Vlot AC, Dempsey DMA, Klessig DF (2009) Salicylic acid, a multifaceted hormone to combat disease. Annu Rev Phytopathol 47:177–206

Wang H, Zheng J, Ren X, Yu T, Varma A, Lou B, Zheng X (2015) Effects of *Piriformospora indica* on the growth, fruit quality and interaction with Tomato yellow leaf curl virus in tomato cultivars susceptible and resistant to TYCLV. Plant Growth Regul 76:303–313

Weinert N, Piceno Y, Ding GC et al (2011) PhyloChip hybridization uncovered an enormous bacterial diversity in the rhizosphere of different potato cultivars: many common and few cultivar-dependent taxa. FEMS Microbiol Ecol 75:497–506

Weller DM, Mavrodi DV, van Pelt JA, Pieterse CM, van Loon LC, Bakker PA (2012) Induced systemic resistance in *Arabidopsis thaliana* against *Pseudomonas syringae* pv. tomato by 2, 4-diacetylphloroglucinol-producing *Pseudomonas fluorescens*. Phytopathology 102:403–412

Whitelaw MA (2000) Growth promotion of plants inoculated with phosphate-solubilizing fungi. Adv Agron 69:99–151

Wicklow DT, Poling SM (2009) Antimicrobial activity of pyrrocidines from *Acremonium zeae* against endophytes and pathogens of maize. Phytopathology 99:109–115

Wicklow DT, Roth S, Deyrup ST, Gloer JB (2005) A protective endophyte of maize: *Acremonium zeae* antibiotics inhibitory to *Aspergillus flavus* and *Fusarium verticillioides*. Mycol Res 109:610–618

Wu SC, Cao ZH, Li ZG, Cheung KC, Wong MH (2005) Effects of biofertilizer containing N-fixer, P and K solubilizers and AM fungi on maize growth: a greenhouse trial. Geoderma 125:155–166

Yao Q, Li XL, Feng G, Christie P (2001) Mobilization of sparingly soluble inorganic phosphates by the external mycelium of an arbuscular mycorrhizal fungus. Plant Soil 230: 279–285

Yi Y, Huang W, Ge H (2008) Exopolysaccharide: a novel important factor in the microbial dissolution of tricalcium phosphate. World J Microbiol Biotechnol 24:1059–1065

Zahir ZA, Arshad M, Frankenberger WT (2003) Plant growth promoteng rhizobacteria: applications and perspectives in agriculture. Adv Agron 81:97–168

Zhang C, Kong F (2014) Isolation and identification of potassium-solubilizing bacteria from tobacco rhizospheric soil and their effect on tobacco plants. Appl Soil Ecol 82:18–25

Zhao L, Deng Z, Yang W, Cao Y, Wang E, Wei G (2010) Diverse rhizobia associated with *Sophora alopecuroides* grown in different regions of Loess Plateau in China. Syst Appl Microbiol 33:468–477

Zoppellari F, Malusà E, Chitarra W, Lovisolo C, Spanna F, Bardi L (2014) Improvement of drought tolerance in maize (*Zea maiz* L.) by selected rhizospheric microorganisms. Ital J Agrometeorol 18:5–18

Biofertilizers: A Timely Approach for Sustainable Agriculture

17

Supriya Tomer, Deep Chandra Suyal, and Reeta Goel

Abstract

Chemical fertilizers are extensively being used all around the globe in order to get the high yield of the crops. However, their indiscriminate use has imposed detrimental impact on soil animal health as well. A better alternative of these chemicals might be to exploit the microbial capabilities to be served as biofertilizer. Crop growth and yield are closely related to the soil microbiota, especially those in close proximity to plant roots, generally termed as "rhizosphere." These microbes are known to play a number of vital roles in soil fertility, crop productivity, and production in agriculture and are the best supplement of chemical fertilizers.

17.1 Introduction

Agriculture is the largest private enterprise in India and will continue to be the lifeline of Indian economy in the future. Indian agriculture sector has only 0.2 % growth rate. It comprises 13.7 % of total GDP in 2015 and half of the total work force (BANR/NRC 2015; Roychowdhury et al. 2014). It is estimated that overall food demand will rise in the proportion of world population. Global population is continuously increasing at the rate of above 1.8 % annually, and it will reach to the point from today's calculated 7.4 billion to an anticipated demographical data of 9.6 billion by 2050 (United Nations 2013). After green revolution, the chemical-based fertilizers and pesticides have enormously boosted the agricultural production. However, their indiscriminate use, besides imposing a detrimental effect on

S. Tomer • D.C. Suyal • R. Goel (✉)
Department of Microbiology, College of Basic Science and Humanities, G.B.P.U.A&T, Pantnagar Pin Code-263145, India
e-mail: rg55@rediffmail.com

D.K. Choudhary et al. (eds.), *Plant-Microbe Interaction: An Approach to Sustainable Agriculture*, DOI 10.1007/978-981-10-2854-0_17

atmosphere has developed the resistance in insects against common pesticides. Therefore, establishing an environmental friendly coexisting mechanism on our planet Earth is of utmost importance.

Most of the fertile soils in favorable environments are being diverted toward alternate uses due to increasing urbanization which creates pressure to expand agriculture. The agrochemicals cannot increase crop yield beyond a threshold level. Moreover, due to lack of knowledge, farmers use more agrochemicals than the recommended level. This excessive application of agrochemicals leads to negative impact on consumer health and loss of soil fertility as they increase salt content of the soil (Swapna 2013). This poor cropping has resulted in new challenges for agricultural productivity. External environments such as biotic (plant pathogens) and abiotic (excess or deficient water and nutrients, high or low temperature, and soil salinity) are dominant factors affecting crop production. Therefore, to increase actual yield of various crops, there is a need to modify abiotic and biotic factors in such a way that they will become capable to fulfill our agricultural demands, where the conventional agriculture practices disappoint. Recently, toxic effects of agrochemicals on human life and environment have shifted the focus on eco-friendly alternatives. Promotion of biofertilizer-based organic farming is need of the hour as demand for safe and residue-free food is gradually rising with high pace.

17.2 Biofertilizers: An Alternative of Chemical Fertilizers

Biofertilizers are the microbial inoculants that colonize the rhizosphere and improve plant growth by enhancing nutrient accessibility to plants. Microorganisms residing in rhizosphere immensely facilitate trace element's uptake. They may act as biocontrol agent, by means of antagonistic activity against phytopathogenic microorganisms, interfering in the bacterial quorum sensing systems, etc. However, biofertilizers perform more than one mechanism for accomplishing plant growth enhancement (Rani et al. 2013; Suyal et al. 2014a; Kumar et al. 2014). These abilities are of great agriculture importance as far as crop yield and soil fertility improvement is concerned, thus decreasing the ill effects of chemical-based fertilizers in our environment. For instance, excessive use of chemical N fertilizers causes soil acidification and, thus, groundwater and atmospheric pollution. Nonetheless, synthesis of chemical fertilizers is highly energy-consuming processes. Chemical-based fertilizers impose long-lasting effects on the atmosphere in terms of carbon footprint, eutrophication, and soil fertility decline. Vast research program on beneficial microbes has resulted in the formulation of biofertilizers, which are capable to satisfy the needs of sustainable agricultural plans (Table 17.1). Sustainable agricultural practices using biofertilizers and biopesticides consisting potential microbes elevate plant health by multiple means in comparison of their synthetic counterparts. Such agricultural practice uses special farming techniques in order to fully utilize environmental resources besides ensuring that no harm was done to it.

Table 17.1 List of selected bioinoculants available in the literature

S. No.	Bioinoculant	PGPR trait	Tested crop	References
1.	*Acinetobacter rhizosphaerae*	P-solubilization, production of ACC-deaminase, IAA, ammonia, siderophore	Pea	Gulati et al. (2009)
2.	*Bacillus subtilis*	P-solubilization, biocontrol	Lentil	Pandey (2009)
3.	*Azospirillum brasilense* Az39	Phytostimulation	Maize	Cassan et al. (2009)
4.	*P. fluorescens*, *Chryseobacterium balustinum*	Biocontrol (*Magnaporthe grisea*), salinity	Rice	Lucas et al. (2009)
5.	*B. japonicum* E109	Phytostimulation	Soybean	Cassan et al. (2009)
6.	*Rahnella* sp.	P-solubilization, production of ACC-deaminase, IAA, ammonia, siderophore	Pea	Vyas et al. (2010)
7.	*Paenibacillus rhizosphaerae*	Phytostimulation	Soybean	Bidondo et al. (2011)
8.	*Arthrobacter* sp. *and B. subtilis*	Stress controller (salinity)	Wheat	Upadhyay et al. (2012)
9.	*Providencia* sp.	Enhancement 18.6 % protein content	Wheat	Rana et al. (2012)
10.	*R. tropici* CIAT899	Enhanced (N and P)	Bean	Tajini et al. (2012)
11.	*Chryseobacterium* sp.	N_2 fixation, P-solubilization	Horse gram	Singh et al. (2012)
12.	*Pseudomonas putida* 710A and *Comamonas aquatica* 710B	P-solubilization, heavy metal bioremediation	Mung bean	Rani et al. (2013)
13.	*Glomus fasciculatum*, *Rhizobium japonicum*, and *Trichoderma harzianum*	Enhanced (N and P) biocontrol	Green gram	Rajeshkannan et al. (2008)
14.	*Pseudomonas jessenii* strain MP1	N_2 fixation	Chickpea, black gram, green gram, pigeon pea, finger millet	Kumar et al. (2014)
15.	*Pseudomonas migulae* S10724	N_2 fixation	Green gram	Suyal et al. (2014a, b)

17.3 The Rhizosphere Microbiome

The soil portion which is in immediate vicinity of root is termed as "rhizosphere," while soil-inhabiting bacteria, able to colonize the rhizosphere, are termed as "rhizobacteria." Besides giving the physical support and facilitating nutrient and water absorption, plants secrete a versatile range of compounds. These compounds may attract a wide range of metabolically diverse soil-inhabiting microbial communities.

Root exudates are organic compounds secreted by plant roots. Various chemical compounds which are found in root exudation alter the soil's physiochemical qualities. By this way root exudates regulate microbial community structure and function. Furthermore, root exudates secreted by different plant species have versatile array of chemicals. Thus, amounts and composition of exudates may constitute the variation in microbial community dynamics. Root exudates provide a highly nutritious carbon diet to the rhizobacteria. Predominant molecules found in exudates are organic acids, sugars, amino acids, nucleobases, and vitamins (Keiluweit et al. 2015) (Table 17.2). However, root exudates composition is dependent upon both plant and microorganism's species and their physiological condition (Wang et al. 2015). Additionally, root exudates help in developing symbiotic plant microbe interactions. Root exudates also hamper the growth of the competing plant (Wang et al. 2015). Some amount of exudates are utilized by adjacent microbes as nutrient source in their metabolic processes, while few microorganism-derived compounds are further absorbed by plants for their developmental benefits (Keiluweit et al. 2015).

Table 17.2 Various compounds in root exudates of different plant species

Amino acids	α-Alanine, β-alanine, asparagines, aspartates, cysteine, cystine, glutamate, glycine, isoleucine, leucine, lysine methionine, serine, threonine, proline, valine tryptophan, arginine, ornithine, histidine, phenylalanine, homoserine, α-aminoadipic acid, γ-aminobutyric acid
Organic acids	Citric acid, oxalic acid, malic acid, fumaric acid, succinic acid, acetic acid, butyric acid, tetronic acid, aldonic acid, erythronic acid, valeric acid, glycolic acid, formic acid, piscidic acid, pyruvic acid, lactic acid, aconitic acid, malonic acid, glutamic acid
Sugars	Glucose, galactose, fructose, rhamnose, ribose, xylose, maltose, arabinose, raffinose, oligosaccharides
Vitamins	Thiamine, riboflavin, niacin, pantothenate, biotin
Nucleosides (purines)	Adenine, guanine, cytidine, uridine
Enzymes	Amylase, protease, invertase, acid phosphatase, alkaline phosphatase
Inorganic ions	H^+, OH^-, HCO_3^-
Gaseous molecules	CO_2, H_2

Adapted from Dakora and Phillips (2002)

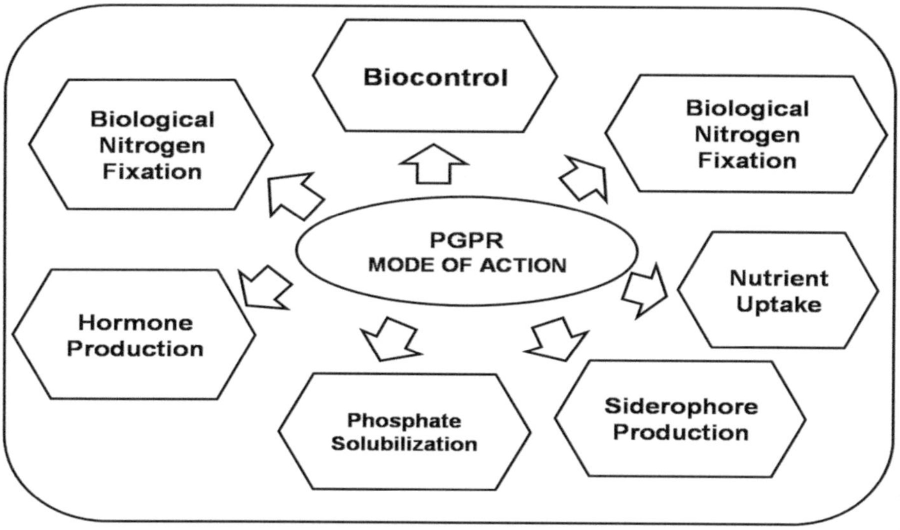

Fig. 17.1 Mode of action of PGPR

Plant growth-promoting rhizobacteria (PGPR) are the group of soil bacteria residing either around or at the plant root surface and benefitting plant through different means.

PGPR have direct or indirect role in plant growth and development through synthesis of different chemical regulators in the rhizosphere's close proximity. PGPR directly help plant development by helping them in macro and micro mineral nutrient uptake as well as by modulating plant hormone levels. Indirectly PGPR boost plant health through declining the detrimental effects of different phytopathogens on plant development (Fig. 17.1). The rhizobacterial potential to utilize organic acids as carbon resources is correlated with rhizosphere competence (Lagos et al. 2015). *Pseudomonas* spp. has the potential to metabolize malate and succinate by greater efficiency than glucose and fructose, during the course of rhizosphere competence. Flagellar mobility, lipopolysaccharide (LPS) structure, chemotaxis, the outer membrane protein OprF, and, to some extent, pili are all crucial for competitive root colonization (Lugtenberg and Kamilova 2009). Agglutinin is a glycoprotein complex of roots thought to facilitate short-term adherence of *Pseudomonas* sp. (Zhang et al. 2014).

17.4 Endophytic Bacteria as Potent Biofertilizer

Endophytic bacteria enhance plant development in nonleguminous crops and enhance their nutritional level through N_2 fixation, phosphate solubilization, and siderophore production (Szilagyi-Zecchin et al. 2014). Besides biofertilization, endophytic bacteria also increase plant growth and yield through producing phytostimulators, like phytohormones, the cofactor pyrroloquinoline quinone (PQQ), and the volatile acetoin. Endophytic bacteria can help plant to combat stress through the production of stress modulators like the enzyme ACC-deaminase, which helps in plant growth and development by reducing plant ethylene levels as well as indirectly through biological control or biotization. Some fungi are also associated in endophytic association, viz., *Trichoderma stromaticum*, *T. evansi*, *T. amazonicum*, *T. martiale*, *T. theobromicola*, *T. taxi*, etc. Few reports reveal that *Trichoderma* spp. induce transcriptomic changes in plants when associated with them as endophytes. Few species of Trichoderma aid plants to escape diseases and abiotic stresses (Bae et al. 2009). Few fungal endophytes prefer to inhabit surface of glandular trichomes and form structures termed as appressoria (Bailey et al. 2009).

17.5 Groups of Biofertilizers

17.5.1 Nitrogen Fixers

Nitrogen plays a vital role to sustain life on Earth. It is a major component of nucleic acids, proteins, and macromolecules. Nitrogen contributes to 4 % and 3 % dry weight of plants and human body, respectively (Cheng 2008). Nitrogen facilitates photosynthesis in plants as it is essential for chlorophyll synthesis. Even though nitrogen is one of the most abundant elements (nitrogen gas (N_2) contributes to 78 % of the Earth's atmosphere), plants can only utilize reduced forms of this element.

To sustain life processes, nitrogen gets converted from one form to another. During this transformation nitrogen moves in between the atmosphere, land, and living system, and this is called nitrogen cycle. Thus, nitrogen cycle results in the conversion of nitrogen to distinct chemical forms (Fig. 17.2). This conversion may happen *via* biological as well as physical means. Main steps in the nitrogen cycle are fixation, nitrification, denitrification, and ammonification.

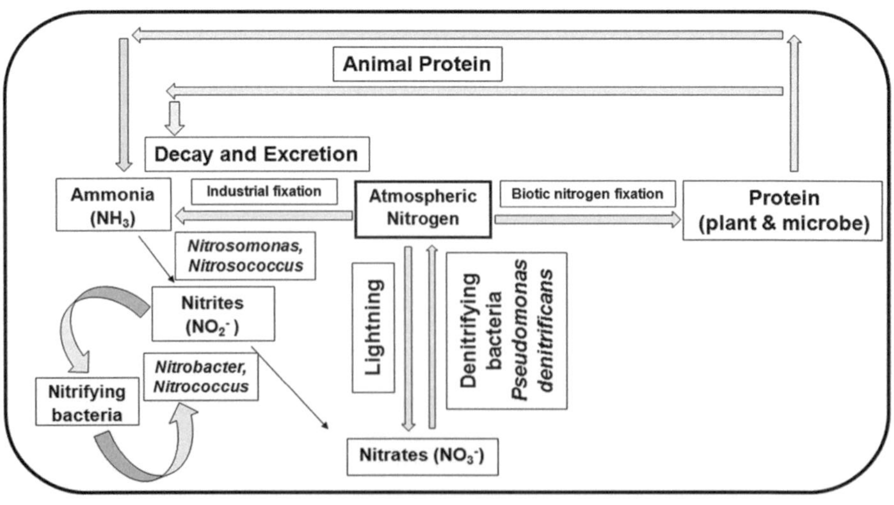

Fig. 17.2 Schematic representation of nitrogen cycle

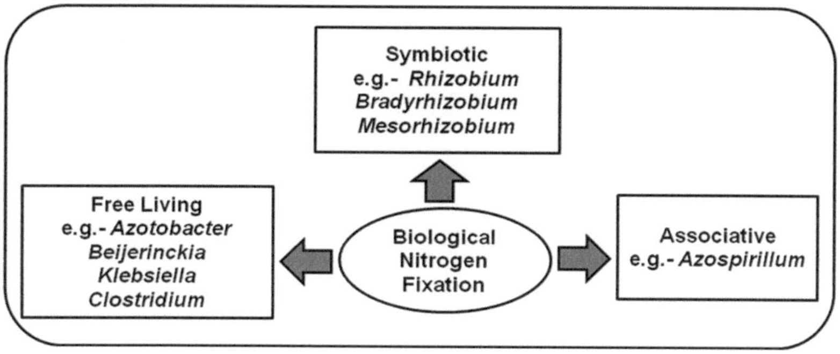

Fig. 17.3 Nitrogen-fixing microorganisms

There are many routes to fix atmospheric nitrogen. Lightning fix nearly 1 % of the total N_2 per year (~3 × 10^{14} g/year). In order to fix large amount of nitrogen for commercial utility, Haber–Bosch process is used especially for the production of nitrogenous fertilizer (~49 % of the total N_2 fixed/year). In 1908, Fritz Haber invented the process of industrial N_2 fixation. Later Carl Bosch increased the efficiency by using high pressure (200 atm) and temperature (450 °C) to convert atmospheric N_2 to NH_3 in the presence of Fe as catalysts (Smil 2001). Every year early 50 % of the nitrogen is biologically fixed. Among microbes, diazotrophs play a crucial role in this process. They utilize various metabolic pathways in the presence of nitrogenase that is a chief metalloenzyme which facilitates the conversion of N_2 to NH_3.

Diazotrophs are widely distributed in nature and are emerging as an economically beneficial alternative against chemical fertilizers. They can be classified as:

(a) Symbiotic nitrogen fixers are from rhizobiaceae family. They form symbiotic association with leguminous plants (e.g., *Rhizobium*, *Mesorhizobium*, *Bradyrhizobium*) while *Frankia* with nonleguminous trees. Symbiotic relation is established by unique interaction between host and bacteria that results in nodulation. Rhizobium resides intracellularly within root nodules.
(b) Nonsymbiotic (free-living/associative or endophytic) example of nitrogen-fixing nonsymbiotic forms are *Cyanobacteria* (*Anabaena*, *Nostoc*), *Azospirillum*, *Azotobacter*, *Beijerinckia*, *Klebsiella*, *Clostridium*, etc. However, they provide a little fraction of the total nitrogen fixed to the host plant (bacterially associated) (Fig. 17.3).

17.5.1.1 Nitrogenases

Biological conversion of N_2 to NH_3 is by nitrogenases which act as catalyst. Nitrogenases are metalloenzymes with complex and conserved structural features. They comprise of two parts: the small dimeric component called iron (Fe) protein and a heterotetrameric component, molybdenum–iron (Mo–Fe) protein. Fe protein actively donates electron to Mo–Fe protein-containing catalytic site. Other families of nitrogenase share common features except different central metal atoms (Mo, V, Fe). All nitrogenase contains a small component called Fe protein, i.e., dinitrogenase reductase. Fe protein comprises of [4Fe-4S] cluster that actively delivers electron to the Mo–Fe protein (dinitrogenase) in the presence of Mg ion. Larger component protein contains two metal clusters. One is active metal cluster and another is P-cluster, i.e., [8Fe-7S] that is an intermediate during electron transfer. In general nitrogen fixation reaction is represented as:

$$N_2 + 8e^- + 8H^+ + 16MgATP \rightarrow 2NH_3 + H_2 + 16MgADP + 16P_i$$

A set of operons encode the nitrogenase in nitrogen-fixing bacteria that is comprised of structural genes (*nif*HDK), regulatory genes (*nif*LA), and rest which are supplementary genes. Free-living diazotrophs, *K. pneumoniae*, are studied as a model organism to examine nitrogenase regulation, its biosynthesis, and how it assembles (Desnoues et al. 2003). *K. pneumonia nif* cluster consists of 20 genes located in 24 kbp DNA region (Fischer 1994). Group of structural genes, i.e., *nif*-HDK encodes for three entities of Mo nitrogenase. In maximum nitrogen-fixing prokaryotic microorganism, *nif*HDK genes form one unit to transcribe *nif*H gene. To ensure the activation of apo–Fe protein, i.e., NifH, products of *nif*M *nif*H, *nif*S, and *nif*U are essential, while activation of apo–MoFe protein needs minimum six genes *nif*H, *nif*Q *nif*E, *nif*B, and *nif*N to facilitate FeMoCo biosynthesis. There is significant similarity between *nif*DK and *nif*EN. Now it is well established that product of *nif*EN forms a scaffold for the synthesis of FeMoCo which is then transferred to *nif*DK complex. The gene product of *nif*B, termed NifB-co, is an iron- and sulfur-containing precursor of FeMoCo. Gene product of *nif*Q is suspected to be engaged in forming molybdenum sulfur precursor and has specific metal-binding sequences (Cys-X4-Cys-*X*2-Cys-X5-Cys) (Einsle et al. 2002). The gene product of *nif*V is homocitrate synthase and is essential for FeMoCo biosynthesis. The *nif*W is involved in the early assembly of MoFe protein but gene product of *nif*W prevents MoFe protein from oxygen. The function of *nif*Y gene product is same as that of γ protein. The *nif*F and *nif*J gene products are required to synthesize specific components of electron transfer chain. In this electron transfer chain, electrons are transferred from pyruvate to flavodoxin and received by Fe protein of nitrogenase. *nif*M gene product is important for both stabilization and maturation of *nif*H gene product though its major role is still unknown. Further, many organisms have *nif*S and *nif*U. Thus, the products of at least 12 *nif* genes are necessary to initiate synthesis of active and stable molybdenum nitrogenase (Rubio and Ludden 2008). The *nifH* gene is a useful tool to characterize the diazotrophic communities (Suyal et al. 2014b).

The bioavailable form of nitrogen is limited and crop growth depends upon bioavailable nitrogen. Because of all these facts, nitrogen fertilizer-manufacturing industries have flourished all over the world (Reich et al. 2014). Nowadays nearly 60 % synthetic nitrogen fertilizer is being consumed solely by cereals, while irrigated paddy cultivation uses approximately 10 % of such fertilizers. Half of the total fertilizer applied to field is used by plants, and the rest half results in nitrate contamination in ground and soil water, therefore leading to serious health problems and imparting threat to the concept of sustainable development. Furthermore, manufacturing N fertilizer is highly energy-consuming process. It consumes six times more energy than that is required to produce either potassium or phosphate fertilizers. Nevertheless, the efficiency of added nitrogen fertilizer is very low. The primary causes of low nitrogen fertilizer efficiency are denitrification, leaching losses, and NH_3 volatilization. Denitrification and NH_3 volatilization produce greenhouse gases like N_2O and NH_3 and thus cause atmospheric pollution and groundwater toxicity. Moreover, the long-term use of nitrogen fertilizer depletes the soil organic matter (Liao et al. 2015). One way to overcome the harmful effects of synthetic nitrogen fertilizer is to use plant growth-promoting diazotrophic bacteria as bioinoculants (Suyal et al. 2014a; Kumar et al. 2014). The need of the hour is to promote sustainable agricultural practices by making use of PGPR (plant growth-promoting rhizobacteria) especially in economically important crops.

17.5.2 Phosphate Solubilizers

P is present in abundance in both organic and inorganic form in soil; still plant available form of P is usually low. Insoluble forms of P are found relatively in higher amount. But, plants only absorb P that is available in two forms, one in monobasic form ($H_2PO_4^-$) and other in form of diabasic (HPO_4^{-2}) ions (Bagyaraj et al. 2015). However, as per the reports, of the total P existing, only 0.1 % is present in soluble form and is free for plant assimilation.

The unavailable phosphorus exists either as an inorganic mineral such as apatite or as one of the many organic forms such as soil phytate, phosphotriesters, and phosphomonesters (Bagyaraj et al. 2015). Farmers apply phosphatic fertilizers in agricultural fields to combat the P deficiency in soils. Plants have restricted efficiency toward utilization of applied phosphatic fertilizers. Remaining unused phosphatic fertilizers get quickly transformed into inaccessible P complexes. Regular long-term application of phosphate fertilizers is detrimental to environment, and sometimes it is unaffordable to the farmers of developing nations. Microorganisms having the potential to convert plant unavailable insoluble P into plant available soluble P are called P-solubilizing microorganisms (PSMs). PSMs are environmentally safe and economically feasible alternative of chemical-based phosphatic fertilizers. Most significant phosphate-solubilizing bacterial genera are reported *Azotobacter*, *Beijerinckia*, *Bacillus*, *Burkholderia*, *Erwinia*, *Enterobacter*, *Flavobacterium*, *Pseudomonas*, *Microbacterium*, *Serratia*, *Rhizobium*, etc.

Solubilization of inorganic phosphorus involves the synthesis of organic acids which are of low-molecular weight by different soil-inhabiting bacteria (Singh et al. 2012; Rani et al. 2013). Conversion of organic phosphorus into inorganic phosphorus is called phosphorus mineralization. Phosphorus mineralization takes place via several phosphatases synthesized by soil microorganisms (Bagyaraj et al. 2015). However, a sole bacterial strain is found to be able to mineralize organic phosphorus and phosphate solubilization as well (Tao et al. 2008). Inoculation of phosphate solubilizer either alone (Singh et al. 2012) or in combination with some other potential PGPR has been well reported (Zaidi and Khan 2005; Vikram and Hamzehzarghani 2008). Besides increasing accessibility of plant available P to the plants, PSB also made other elemental nutrients available through synthesizing plant growth promontory substances (Ahemad and Kibret 2014).

17.5.3 Siderophore Producers

Iron is required nearly by all life forms except certain lactobacilli to carry out respiration and DNA synthesis. Regardless of being the fourth most plentiful element on Earth (approximately 5 % by weight), the bioavailability of iron is limited. In non-acidic, aqueous, and oxygenated environment, ferric ion is the widespread state of iron. Iron is accumulated to form common mineral phases. These common mineral phases are iron oxides and oxyhydroxides. Hence, it is not easy for organisms to readily utilize it. Microbes obtain Fe^{3+} by secreting low-molecular mass iron scavengers known as siderophores. To solubilize iron, these mineral compounds bind in the form of soluble ferric ion. Fe^{3+} complex uptake by microbial cell is through active transport system and, thus, makes iron unavailable for phytopathogens; however, plants can assimilate iron from bacterial siderophores by special mechanisms. Thus, phytopathogens are deprived of essential macronutrient while plants gain iron through siderophore producing PGPR.

Iron uptake plays a crucial role in microbial competition particularly in the regions of intense competition like rhizosphere. Siderophores are classified into two types, viz., extracellular and intracellular. These are mostly soluble in water. The intracellular siderophore ferricrocin is responsible for iron storage and prevents cellular oxidative stress (Wallner et al. 2009). Rhizobacteria differs regarding cross-utilizing ability of siderophore. Some can utilize siderophores of the like genera, i.e., homologous siderophores, whereas others are proficient in using siderophores produced by rhizobacteria of unlike genera, i.e., heterologous siderophores. In rhizobacterial cell membrane, iron in ferric ion-siderophore complex is converted to ferrous ion that is released in cell (Khan et al. 2009). This release of iron from such complexes takes place through gating mechanism l. In the course of reduction, siderophore's iron complex is either destroyed or recycled (Rajkumar et al. 2010). Thus, when there is iron-limiting situation, siderophores help in iron solubilization from complex compounds (Indiragandhi et al. 2008). Siderophores make stable complexes with several heavy metals which could have been otherwise dangerous to environment, viz., aluminum, cadmium, copper, gallium, lead, zinc, as well as radionuclides (Rajkumar et al. 2010).

Siderophores produced by bacteria alleviate plant stress due to high concentrations of heavy metals in soil. Assimilation of iron by plants via bacterial siderophores is done by various mechanisms such as through ligand exchange, chelation, and release of iron and direct acquisition of iron siderophore complexes (Colo et al. 2014).

17.5.4 Phytohormone Producers

Phytohormones are organic substances produced in specific plant organs. They are either transferred to different sites or active in the same tissue where they are synthesized. They are signal molecules which control overall plant growth. Expression of plant intrinsic genes is regulated by them.

Phytohormones are available for the plants by two sources. They are either endogenously synthesized by plant tissues itself or exogenously synthesized by associated PGPR. Many species of soil-inhabiting bacteria and fungi are known for phytohormone production, viz., *Galactomyces geotrichum*, *Pseudomonas*, *Azospirillum*, *Bacillus*, and *Arthrobacter* (Waqas et al. 2014; Lei and Ya-qing 2015).

17.5.4.1 Indole-3-Acetic Acid/Indoleacetic Acid (IAA)

IAA is synthesized as secondary metabolites. Endogenous pool of plant indoleacetic acid is modified through rhizobacterial indoleacetic acid. This alters plant growth (colo et al. 2014). IAA plays prominent role in rhizobacteria–plant interactions. It is also a signaling molecule which is involved in the plant's defense mechanisms. It influences all aspects of plant development and growth in overall cell cycle of plant by regulating cell elongation, differentiation, cell division, apical dominance, root initiation (lateral and adventitious), flowering, fruit ripening, and senescence. It stimulates tuber and seed germination, promotes xylem development rate and root development rate, regulates vegetative growth of plants, mediates tropistic responses (to gravity, florescence, and light), and affects formation of pigment and photosynthesis. Auxin's synthesis, transport, and signaling pathways are complex. Microbial auxins enhance length and surface area of root. Thus, plant will be able to utilize more soil nutrients due to larger root surface. Auxins secreted by PGPR act as crucial hormone for interaction of plant and microbe. It plays roles in pathogenesis as well as in phytostimulation (Spaepen and Vanderleyden 2011).

Main precursor for IAA biosynthesis is the amino acid tryptophan. Tryptophan synthesis initiates from chorismate. It takes place through five stepped reactions. The trp operon encodes the enzymes which catalyze these reactions. Chorismate is biosynthesized by PEP (phosphoenolpyruvate) and erythrose 4-phosphate through shikimate pathway. This shikimate pathway is a common pathway for the biosynthesis of aromatic amino acids (phenylalanine, tyrosine, tryptophane) and several other secondary metabolites.

17.5.4.2 Gibberellins/Gibberellic Acid/Gibberellin-A3/GA_3 (GA)

Gibberellic acid has molecular formula $C_{19}H_{22}O_6$. These are hormones found in fungi and plants. Gibberellins are the derivatives of gibberellic acids commonly known as GAs. *Gibberella fujikuroi* is the most common fungi which synthesize about 20 different types of gibberellins. GA_3 is the most abundant gibberellic acid among all. The number in the molecular formula is according to the approximate order of their discovery. GA_1 is the most common gibberellic acid responsible for stem elongation. Gibberellins promote stem growth and root growth, instigate mitotic division in some plant leaves, and trigger seed germination. These hormones are also used in greenhouse and laboratory to initiate sprouting in dormant seeds. The biosynthetic pathway of GA synthesis consists of three classes of enzymes. The first class comprises terpene cyclases which synthesize ent-kaurene. The second class consists of cytochrome P450 monooxygenase, which catalyzes formation of GA_{12} from ent-kaurene, and the third class is comprised of dioxygenases. The final step is catalyzed by these dioxygenases. Recently, Ullah et al. 2014 gave affirmation that unlike forms of GA are produced by entomo-pathogenic bacterium named *Photorhabdus temperata*.

17.5.4.3 Cytokinins

This phytohormone stimulates cell division, enlargement of cell, and expansion of tissue in specific plant parts. Cytokinins are responsible mainly for cellular growth and cellular differentiation. They influence apical dominance and leaf senescence and axillary bud growth. It doesn't influence parenchyma cells only by itself. Auxi-to-cytokinin ratio influences plant growth. When auxin and cytokinin are present equally, the cells of parenchyma form an undifferentiated callus but, when cultured only with auxin cell, grow large but do not divide. When cytokinins are introduced, the cells differentiate along with increment in size; it is observed that more cytokinin stimulates shoot bud growth, while more auxins stimulate formation of root. The most common producers are *Azotobacter chroococcum*, *Azospirillum* sp., *Azotobacter beijerinckii*, *Paenibacillus polymyxa*, *Pseudomonas putida*, and *P. fluorescens* (Kudoyarova et al. 2014; Arkhipova et al. 2007).

17.5.4.4 ACC-Deaminase/1-Aminocyclopropane-1-Carboxylic Acid

Ethylene is crucial stress hormone that works as growth regulator (Magnucka and Pietr 2015). Ethylene is synthesized by almost all plants through biotic and abiotic processes in soils. It is essential for normal growth, development, and induction of various physiological changes of plants (Magnucka and Pietr 2015).

Drought, salinity, flooding, extreme high temperature, heavy metal contamination, and pathogenicity like stress conditions induce the ethylene overproduction. Significant increase in ethylene hampers the root growth and defoliation and thus retards overall plant growth. Ethylene biosynthesis is initiated by methionine. Initial step is synthesis of S-adenosyl-methionine which is then transformed into ACC (1-aminocyclopropane-1-carboxylic acid) (Fig. 17.4).

In plants, ACC is an intermediate precursor for ethylene biosynthesis. ACC-deaminase is the enzyme which catalyzes ACC cleavage into α-ketobutyrate and NH_3, consequently decreasing ethylene levels in plants. ACC-deaminase which is in the form of homo-trimer protein needs a cofactor which is pyridoxal-phosphate. The *acdS* is transcribed into ACC-deaminase. *acdS* has already been cloned from *Pseudomonas*, *Rhizobium*, and *Enterobacter cloacae* (Magnucka and Pietr 2015). PGPR consisting of enzyme ACC-deaminase are able to enhance plant growth (Gontia-Mishra et al. 2014). PGPR synthesizing ACC-deaminase minimizes drought stress and induces tolerance against high salt concentration in plants (Zahir et al. 2008, 2009). Its activity has been observed in several rhizobacteria like *Acinetobacter*, *Azospirillum*, *Agrobacterium*, *Achromobacter*, *Ralstonia*, *Enterobacter*, *Bacillus*,

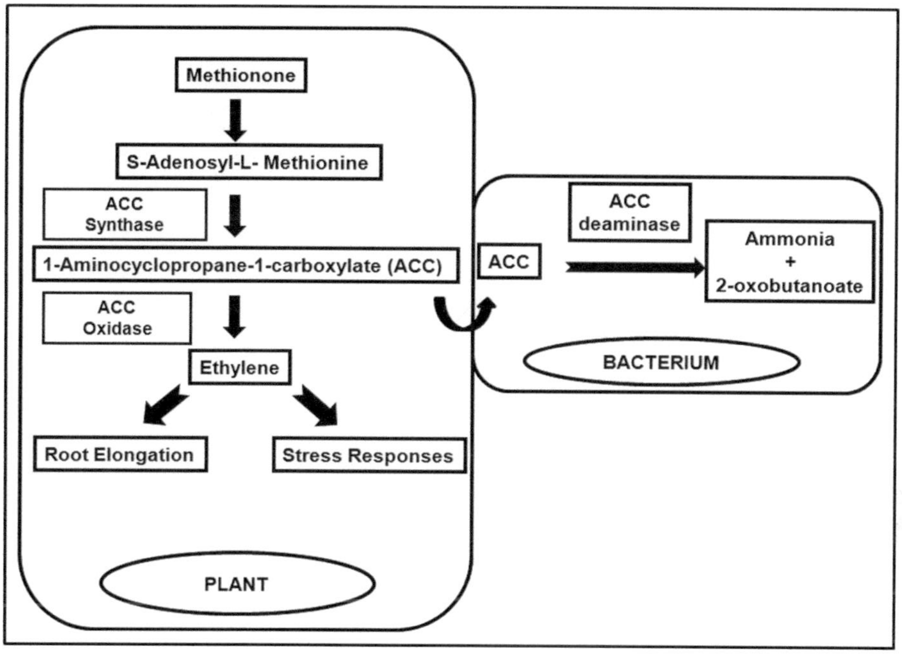

Fig. 17.4 Mode of action of ACC-deaminase (Kang et al. 2010)

Rhizobium, *Burkholderia*, *Serratia*, *Pseudomonas*, etc. (Kang et al. 2010). Recently, ACC-deaminase synthesizing *Pseudomonas stutzeri* A1501 is reported to facilitate rice cultivation in salt and heavy metal presence (Han et al. 2015). Rhizobacteria capable of synthesizing ACC-deaminase also provide stress resistance from radiation, poly-aromatic hydrocarbons, high salt concentration, insect predation, wounding, and high light intensity (Glick 2012). ACC-deaminase synthesizing rhizobacterial inoculation induces the plant shoot growth and root elongation and increases nodulations in rhizobium. It also promotes mycorrhizal colonization and thus nutrient uptake in several agricultural crops (Glick 2012).

17.6 Indirect Mechanism of Plant Growth Promotion

The traits involved in boosting plant health never work independently; rather, they act additively and synergistically. Biological control is environmentally friendly way to control diseases through application of microorganisms. One of the mechanisms adopted by rhizobacteria to improve plant health is an antagonistic action against plant pathogens. Biocontrol activity in PGPR is possible through conflict for food and space. Rhizobacterial synthesis of antifungal metabolites and induction of induced systemic resistance by them also suppress plant pathogens (Lugtenberg and Kamilova 2009). Chief antifungal metabolites effective in suppressing phytopathogens are HCN, 2,4-diacetylphloroglucinol, viscosinamide, pyoluteorin, tensin, pyrrolnitrin, phenazines, and peptaibols (Bhattacharyya and Jha 2012). A vast group of peptides having good antibiotic ability are reported and termed as "peptaibols." Soil-inhabiting fungi like *Trichoderma* sp., *Gliocladium* sp., and *Emericelopsis* sp. are reported to be able to synthesize peptaibols (Daniel and Filho 2007; Brito et al. 2014). Peptaibols having antimicrobial potential isolated from *Trichoderma pseudokoningii* regulate apoptosis (programmed cell death) in phytopathogens (Shi et al. 2012). Some other fungal metabolites, having biocontrol potential, are:

1. **Daucanes**—Daucane sesquiterpenes or carotanes, found in plant families Umbelliferae and Compositae, are rare fungal metabolites. *T. virens* one strain produces a novel metabolite consisting antifungal properties against various yeast and dermatophytes. Daucanes have a significant inhibitory effect on *Candida albicans*.
2. **6-Pentyl pyrone (pyrones)—**These compounds are obtained from *Trichoderma* spp. It has a specific coconut aroma. Pyrones suppress fungal phytopathogens and henceforth promotes plant health**.**
3. **Terpenoids/steroids—**Terpenoid or steroids are versatile group of compounds. They are produced from pentacarbon iso-pentenyl units. The *T. virens* synthesizes viridin. It is found to be an efficient fungistatic and anticarcinogenic compound. Trichodermin is trichothecene-type terpenoid toxin; chemically it is fungitoxic and phytotoxic produced by *Trichoderma brevicompactum* (Mukherjee et al. 2012).
4. **Polyketides**—These metabolites have anticarcinogenic properties. They possess antimicrobial activities and also suppress immune system. They are synthesized by filamentous fungi and by many other organisms. They facilitate communication between organisms and also promote struggle for substrate (Khosla 2009). Polyketides may have possible role in mycoparasitism. This phenomenon has been noticed while *T. atroviride* encounters with *Rhizoctonia solani.* Two polyketide genes were expressed in *Trichoderma* during the course of phytopathogen confrontation (Mukherjee et al. 2012).
5. **Gliotoxin and gliovirin**—These are fungistatic compounds. Their discovery was based on its antagonistic properties. Gliotoxin is beneficial to *Aspergillus fumigatus* in soil habitat due to its antagonistic properties (Giles et al. 2011). Gliotoxin producing microbes may be an excellent biopesticide to control plant disease caused by the soilborne phytopathogen (Mukherjee et al. 2012).

Induced systemic resistance is a well-known phenomenon in which plants defend themselves against phytopathogen infection through several mechanisms. These mechanisms might be local, constitutive, or inducible. Induced resistance mechanism is plant defense response which is biologically induced either by exposing plant with certain weak pathogenic strain or by exposing it to natural/synthetic chemicals. Bacterial components such as cell wall-derived oligouronides, glycoproteins, lipopolysaccharides (LPS), 2,4-diacetyl phloroglucinol, cyclic lipopeptides, homoserine lactones, volatile chemicals like 2,3-butanediol and acetoin, and siderophores and flagella induce defense responses in host plant (Lugtenberg and Kamilova

2009). These defense responses involve cell wall thickening by lignification, accumulation of callose, phytoalexins, and synthesis of various proteins (e.g., chitinases, peroxidases, glucanases, and other pathogenesis-related proteins).

17.7 Tripartite Relationship Between N_2 Fixers, P-Solubilizers, and Arbuscular Mycorrhizal Fungi

Little is known regarding the effects of inoculation of plant with phosphate solubilizers as well as diazotrophs simultaneously in the occurrence of arbuscular mycorrhizal fungi (AMF). PSMs increase plant growth by providing plant utilizable phosphates, while on the other hand, diazotrophs increase nitrogen content in the soil which is used by plants to enhance their growth. When there is deficiency of both phosphorous and nitrogen, AMF helps in the assimilation of phosphates, hence supporting plant growth. In plants high level of phosphorous supports the diazotrophs and increases nitrogenase activity. By this nitrogen fixation is enhanced which facilitates root growth and mycorrhizal development. This is the beauty of intergeneric interaction.

17.8 Biofertilizers for Mountain Ecosystems

Diazotrophy at mountain ecosystems is considered good plant growth-promoting property for cold climate agriculture and is reported by distinct research teams (Breitbarth et al. 2007; Suyal et al. 2014a, 2015). These ecosystems contain a series of different climatic zones within small distances and elevations, displaying the diverse microhabitats (Regato and Salman 2008). They are delicate and susceptible to any change and thus grow well only in specific environmental conditions. Moreover, changes caused by infrastructure development, excessive tourism, overutilization of natural resources, land use pattern, habitat loss, and long-term changes in the Earth's climate are exerting an additional pressure on these ecosystems. They harbor a variety of psychrophilic and psychrotolerant bacterial communities (Suyal et al. 2015). Kumar et al. 2014 and Shukla et al. 2015 reported the effect of psychrotolerant diazotrophic bacteria isolated from Western Himalayan region on plant growth promotion of various hill cultivated crops. Few psychrotolerant species of genus *Pseudomonas* have been found capable to fix nitrogen at low temperature 4–10 °C but with a significant reduction in the nitrogenase activity (Eckford et al. 2002; Kumar et al. 2014). Further, occurrence of *nif*H gene, an indication of diazotrophy, is also reported in psychrotolerant species of *Paenibacillus* (Rodríguez-Diaz et al. 2005), *Arthrobacter*, and *Rhodococcus* (Suyal et al. 2014b). Low temperature adversely affects the rate of N_2 fixation (Soni et al. 2015). Temperature below 9 °C is generally considered as the limiting temperature for nitrogen fixation (Simon et al. 2014). In recent past, seven diazotrophs which are able to survive in low temperature have been isolated from kidney bean's rhizosphere from Kumaun, Himalaya. Among them proteome of *Pseudomonas migulae* strain S10724 has already been listed (Suyal et al. 2014a).

17.9 Indirect Mechanism of Crop Improvement

Increased level of salt in soil possesses major threat to crops and other important plants. The bacteria *Pseudomonas* is of utmost importance in these areas as it is able to use many substrates, produces a diverse range of compounds, as well as easily conquers rhizosphere, thus helping plant to withstand unfavorable situations. PGPR can act as prominent substitute to overcome plant stress caused by salinity, as well as it induces the host resistance mechanism and direct antagonistic interaction with pathogens.

17.10 Application of Genetic Engineering in Developing New Strain

Biofertilizer performance mainly depends on its potential to colonize a certain rhizospheric habitat. Biofertilizer colonization study gives information about its success. Plate count enumeration method and most probable number (MPN) method are helpful in understanding those rhizospheric bacterial communities which act as biofertilizers. Not even 1 % of the total bacterial communities inhabiting in the environment are cultivated by known standard techniques. Strain-specific DNA probes help to estimate the diverse microbial population in the rhizosphere. With the advancement of techniques in molecular biology, it is now possible for microbiologists to decipher the non-culturable microbial communities (Soni et al. 2010; Suyal et al. 2015). Thus, molecular biology techniques are extensively useful to characterize microbial communities in different habitats.

Cloning and sequencing techniques are commonly used techniques which are helpful in determining composition of microbial community. Besides them, hybridization and probing techniques can also determine the same with the advantage that they are less time consuming; however, it is mandatory to have adequate information of microbial community in order to select desired target sequence. Alternative methods such as ribosomal intergenic spacer analysis (RISA) and amplified ribosomal DNA restriction analysis (ARDRA) are helpful to analyze community structure and colonization ability of biofertilizers. They are often used to analyze bacterial diversity from varied environments. ARDRA can mark genotypic transformation occurring in a community with respect to time. However, RISA aids in comparative analysis of microbial communities exposed to unlike habitats or treatment without any kind of biasness enforced through culture-dependent methods. In short, it includes PCR amplification of intergenic spacer region (ISR). These techniques involving molecular biology have higher quantitative effectiveness and are further used to characterize bioinoculants in situ.

In case of PSM, organic acids help to solubilize mineral phosphate. Solubilization of phosphorus is carried out by reducing pH or by chelation of cations accompanying phosphorus. Knowledge of the genes governing the production of organic acids would make it possible to transfer the ability of P-solubilization to those bacteria which are able to colonize a specific rhizosphere. It is now understood that rhizosphere competence is an important factor to determine the fate of victory or defeat of microbial inoculum. Rhizosphere has versatile carbon sources that can be utilized by the diversified microbial communities residing in soil for the production of different organic acids. Available phosphate is readily utilized by plant before its precipitation to unavailable form. In gram-negative bacteria, gluconic acid has been proved to aid phosphate solubilization (Oteino et al. 2015). Oxidative metabolism of glucose to gluconic acid occurs in the presence of glucose dehydrogenase (GDH). GDH in turn needs pyrroloquinoline quinone (PQQ) as a cofactor. Therefore, genes associated in the transfer and synthesis of PQQ are cloned from bacteria of one type and passed to other type (Bruto et al. 2015). If the PQQ-synthesizing genes are transferred to *Rhizobium* strain that possess apo-GDH and that is rhizosphere competent too, then, the resulting strains will possess the ability to solubilize phosphate activity as well as will fix atmospheric nitrogen. Enhanced expression of GDH genes along with PQQ biosynthesis in *Pseudomonas* spp. also produces gluconic acid via oxidative metabolism of glucose. Similarly, Ambrose et al. (2015) have successfully characterized salicylate hydroxylase gene from the fungal endophyte *Epichloë festucae*.

17.11 Conclusion

The integrated approach is vital to improve crop productivity and to maintain the soil fertility. As PGPR have multidimensional approach toward sustainable agricultural system, it is important to determine the most favorable plant–microbial interaction. Biofertilizers not only exhibit plant growth promotion but are also effective in bioremediation by detoxifying detrimental pollutants such as pesticides and heavy metal pollutants. Nevertheless, they are potential biopesticides, as they can control a wide variety of phytopathogens. Few reports reveal that in case of controlled soil conditions, significant enhancement in crop production was achieved through biofertilizer applications. But soil is an unpredictable natural ecosystem. Biofertilizer efficacy in crop productivity may vary between controlled condition of laboratory and farm, and therefore, desired results are not always attained. Besides it, effectiveness of biofertilizers also depends upon climatic changes among different geographical locations. However, their performance can be optimized through acclimatization according to prevailing natural soil environment. In present scenario, where genetically modified food crops have a big question mark, biofertilizers may be a boon to humanity. This is a technology which is easy access even to the farmers of developing nations including India. PGPR-mediated organic farming would pave the way to prosperous, healthy, and sustainable nation. Thus, this trend of least possible input of chemicals in sustainable agricultural systems may help to achieve the goal of holistic well-being of planet Earth.

Acknowledgment The work mentioned in this chapter from author group was supported by the National Bureau of Agriculturally Important Microorganisms, India (NBAIM/ICAR), grant to R. G.

References

Ahemad M, Kibret M (2014) Mechanisms and applications of plant growth promoting rhizobacteria: current perspective. J King Saud Univ Sci 26:1–20

Ambrose KV, Tian Z, Wang Y et al (2015) Functional characterization of salicylate hydroxylase from the fungal endophyte Epichloë festucae. Sci Rep 5:10939. doi:10.1038/srep10939

Arkhipova TN, Prinsen E, Veselov SU et al (2007) Cytokinin producing bacteria enhance plant growth in drying soil. Plant Soil 292:305–315

Bae H, Sicher RC, Kim MS et al (2009) The beneficial endophyte *Trichoderma hamatum* isolate DS 219b promotes growth and delays the onset of the drought response in *Theobroma cacao*. J Exp Bot 60:3279–3295

Bagyaraj DJ, Sharma MP, Maiti D (2015) Phosphorus nutrition of crops through arbuscular mycorrhizal fungi. Curr Sci India 108(7):1288–1293

Bailey BA, Stream MD, Wood D (2009) *Trichoderma* species form endophytic associations within *Theobroma cacao* trichomes. Mycol Res 113:1365–1376

Bhattacharyya PN, Jha DK (2012) Plant growth-promoting rhizobacteria (PGPR): emergence in agriculture. World J Microbiol Biotechnol 28:1327–1350

Bidondo LF, Silvani V, Colombo R (2011) Pre-symbiotic and symbiotic interactions between *Glomus intraradices* and two *Paenibacillus species* isolated from AM propagules. In vitro and in vivo assays with soybean (AG043RG) as plant host. Soil Biol Biochem 43(9):1866–1872

Board on Agriculture and Natural Resource, National Research Council (2015) A framework for assessing effects of the food system. The National Academies Press. doi:10.1017226/18846

Breitbarth E, Oschlies A, LaRoche J et al (2007) Physiological constraints on the global distribution of Trichodesmium-effect of temperature on diazotrophy. Biogeosciences 4(1):53–61

Brito JP, Ramada MH, de Magalhaes MT et al (2014) Peptaibols from *Trichoderma asperellum* TR356 strain isolated from Brazilian soil. Springerplus 3:600. doi:10.1186/2193-1801-3-600

Bruto M, Prigent-Combaret C, Muller D et al (2015) Analysis of genes contributing to plant-beneficial functions in plant growth-promoting rhizobacteria and related proteobacteria. Sci Rep 4:6261. doi:10.1038/srep06261

Cassan F, Perrig D, Sgroy V et al (2009) *Azospirillum brasilense* Az39 and *Bradyrhizobium japonicum* E109, inoculated singly or in combination, promote seed germination and early seedling growth in corn (Zea mays L.) and soybean (Glycine max L.). Eur J Soil Biol 45(1):28–35

Cheng Q (2008) Perspectives in biological nitrogen fixation research. J Integr Plant Biol 50(7):786–798. doi:10.1111/j.1744-7909.2008.00700

Colo J, HajnaL-Jafari TI, Duric S et al (2014) Plant growth promotion rhizobacteria in onion production. Pol J Microbiol 63(1):83–88

Dakora FD, Phillips DA (2002) Root exudates as mediators of mineral acquisition in low-nutrient environments. Plant Soil 245:35–47

Daniel JFS, Filho ER (2007) Peptaibols of *Trichoderma*. Nat Prod Rep 24:1128–1141

Desnoues N, Lin M, Guo X et al (2003) Nitrogen fixation genetics and regulation in a *Pseudomonas stutzeri* strain associated with rice. Microbiology 149(8):2251–2262

Eckford R, Cook FD, Saul D et al (2002) Free-living heterotrophic nitrogen-fixing bacteria isolated from fuel-contaminated Antarctic soils. Appl Environ Microbiol 68(10):5181–5185

Einsle O, Tezcan FA, Andrade SL et al (2002) Nitrogenase MoFe-protein at 1.16 A resolution: a central ligand in the FeMo-cofactor. Science 297(5587):1696–1700

Fischer HM (1994) Genetic regulation of nitrogen fixation in rhizobia. Microbiol Rev 58(3):352–386

Giles SS, SoukupAA LC et al (2011) Cryptic *Aspergillus nidulans* antimicrobials. Appl Environ Microbiol 77:3669–3675

Glick BR (2012) Plant growth-promoting bacteria: mechanisms and applications. Hindawi Publishing Corporation, Scientifica. http://dx.doi.org/10.6064/2012/963401

Gontia-Mishra I, Sasidharan S, Tiwari S (2014) Recent developments in use of 1-aminocyclopropane-1- carboxylate (ACC) deaminase for conferring tolerance to biotic and abiotic stress. Biotechnol Lett. doi:10.1007/s10529-014-1458-9

Gulati A, Vyas P, Rahi P et al (2009) Plant growth-promoting and rhizosphere-competent *Acinetobacter rhizosphaerae* strain BIHB 723 from the cold deserts of the Himalayas. Curr Microbiol 58(4):371–377

Han Y, Wang R, Yang Z et al (2015) 1-Aminocyclopropane-1-Carboxylate Deaminase from Pseudomonas stutzeri A1501 facilitates the growth of rice in the presence of salt or heavy metals. J Microbiol Biotechnol 25(7):1119–1128

Indiragandhi P, Anandham R, Madhaiyan M et al (2008) Characterization of plant growth-promoting traits of bacteria isolated from larval guts of diamondback moth *Plutella xylostella* (Lepidoptera: Plutellidae). Curr Microbiol 56:327–333

Kang BG, Kim WT, Yun HS et al (2010) Chang use of plant growth-promoting rhizobacteria to control stress responses of plant roots. Plant Biotechnol Rep 4:179–183

Keiluweit M, Bougoure JJ, Nico PS et al (2015) Mineral protection of soil carbon counteracted by root exudates. Nat Clim Chang 5:588–595. doi:10.1038/nclimate2580

Khan MS, Zaidi A, Wani PA et al (2009) Role of plant growth promoting rhizobacteria in the remediation of metal contaminated soils. Environ Chem Lett 7:1–19

Khosla C (2009) Structures and mechanisms of polyketide synthases. J Org Chem 74:6416–6420

Kudoyarova GR, Melentiev AI, Martynenko EV et al (2014) Cytokinin producing bacteria stimulate amino acid deposition by wheat roots. Plant Physiol Biochem 83:285–291

Kumar S, Suyal DC, Dhauni N, Bhoriyal M, Goel R (2014) Relative plant growth promoting potential of Himalayan psychrotolerant *Pseudomonas jesenii* Strain MP1 against Native *Cicer arietinum* L., *Vigna mungo* (L.) Hepper; *Vigna radiata* (L.) Wilczek., *Cajanus cajan* (L.) Millsp. and *Eleusine coracana* (L.) Gaertn. Afr J Microbiol 8(50):3931–3943

Lagos ML, Maruyama F, Nannipieri P et al (2015) Current overview on the study of bacteria in the rhizosphere by modern molecular techniques: a mini–review. J Soil Sci Plant Nutr 15(2):504–523

Lei Z, Ya-qing Z (2015) Effects of phosphate solubilization and phytohormone production of *Trichoderma asperellum* Q1 on promoting cucumber growth under salt stress. J Integr Agric 14(8):1588–1597

Liao Y, WuWL MFQ et al (2015) Increase in soil organic carbon by agricultural intensification in northern China. Biogeosciences 12:1403–1413

Lucas JA, Solano BR, Montes F et al (2009) Use of two PGPR strains in the integrated management of blast disease in rice (*Oryza sativa*) in Southern Spain. Field Crop Res 114(3):404–410

Lugtenberg B, Kamilova F (2009) Plant-growth-promoting rhizobacteria. Annu Rev Microbiol 63:541–556

Magnucka EG, Pietr SJ (2015) Various effects of fluorescent bacteria of the genus Pseudomonas containing ACC deaminase on wheat seedling growth. Microbiol Res 181:112–119

Mukherjee PK, Buensanteai N, Moran-Diez ME et al (2012) Functional analysis of non-ribosomal peptide synthetases (NRPSs) in *Trichoderma virens* reveals a polyketide synthase (PKS)/NRPS hybrid enzyme involved in induced systemic resistance response in maize. Microbiology 158:155–165

Oteino N, Lally RD, Kiwanuka S et al (2015) Plant growth promotion induced by phosphate solubilizing endophytic *Pseudomonas* isolates. Front Microbiol 6:745. doi:10.3389/fmicb.2015.00745

Pandey A (2009) *Bacillus subtilis* NRRL B-30408 inoculation enhances the symbiotic efficiency of *Lens esculenta* Moench at a Himalayan location. J Plant Nutr Soil 172(1):134–139

Rajkumar M, Ae N, Prasad MN et al (2010) Potential of siderophore-producing bacteria for improving heavy metal phytoextraction. Trends Biotechnol 28(3):142–149

Rana A, Joshi M, Prasanna R (2012) Biofortification of wheat through inoculation of plant growth promoting rhizobacteria and cyanobacteria. Eur J Soil Biol 50:118–126

Rajeshkannan V, Thangapandian V, Muthukumar T et al (2008) Influence of bioinoculants on growth, nutrient uptake and yield of green gram [Vigna radiata (L.) Wilczek]. J Sustain Agr 31(3):85–109

Rani A, Souche Y, Goel R (2013) Comparative in situ remediation potential of *Pseudomonas putida* 710A and *Commamonas aquatica* 710B using plant (*Vigna radiata* (L.) wilczek) assay. Ann Microbiol 63(3):923–928

Regato SR (2008) Mediterranean mountains in a changing world: guidelines for developing action plans. https://cmsdata.iucn.org/downloads/mediterranean_mountains.pdf. Accessed 25 Apr 2016

Reich PB, Hobbie SE, Lee TD (2014) Plant growth enhancement by elevated CO2 eliminated by joint water and nitrogen limitation. Nat Geosci 7:920–924

Rodríguez-Diaz M, Lebbe L, Rodelas B et al (2005) *Paenibacillus wynnii* sp. nov., a novel species harbouring the *nifH* gene, isolated from Alexander Island, Antarctica. Int J Syst Evol Microbiol 55:2093–2099

Roychowdhury D, PauL M, Banerjee SK (2014) A review on the effects of biofertilizers and biopesticides on rice and tea cultivation and productivity. Int J Eng Sci Technol 2(8):96–106

Rubio LM, Ludden PW (2008) Biosynthesis of the iron-molybdenum cofactor of nitrogenase. Annu Rev Microbiol 62:93–111

Shi M, Chen L, Wang XW et al (2012) Antimicrobial peptaibols from Trichoderma pseudokoningii induce programmed cell death in plant fungal pathogens. Microbiology 158:166–175

Shukla A, Dhauni N, Suyal DC, Kumar S, Goel R (2015) Comparative plant growth promoting potential of psychrotolerant diazotrophs, *Pseudomonas* sp. JJS2 and Enterobacter sp. AAB8 against native *Cajanus cajan* (L.) and *Eleusine coracana* (L.). Afr J Microbiol 9(20):1371–1375

Simon Z, Mtei K, Gessesse A et al (2014) Isolation and characterization of nitrogen fixing rhizobia from cultivated and uncultivated soils of Northern Tanzania. Am J Plant Sci 5:4050–4067

Singh AV, Chandra R, Goel R et al (2012) Phosphate solubilization by *Chryseobacterium* sp. and their combined effect with N and P fertilizers on plant growth promotion. Arch Agron Soil Sci 59(5):641–651

Smil V (2001) Enriching the earth: Fritz Haber, Carl Bosch, and the transformation of world food production. The MIT Press, Cambridge, MA

Soni R, Shaluja B, Goel R (2010) Bacterial community analysis using temporal temperature gradient gel electrophoresis (TTGE) of 16S rDNA PCR products of soil metagenome. Ekologija 56(3–4):94–98

Soni R, Suyal DC, Agrawal K, Yadav A, Souche Y, Goel R (2015) Differential proteomic analysis of Himalayan psychrotolerant diazotroph *Pseudomonas palleroniana* N26 Strain under low temperature diazotrophic conditions. CryoLetters 36(2):74–82

Spaepen S, Vanderleyden J (2011) Auxin and plant-microbe interactions. Cold Spring Harb Perspect Biol 3(4). doi:10.1101/cshperspect.a001438

Suyal DC, Shukla A, Goel R (2014a) Growth promotory potential of the psychrophilic diazotroph *Pseudmonas migulae* S10724 against Native *Vigna radiata* (L.) Wilczek. 3 Biotechnol 4:665–668

Suyal DC, Yadav A, Shouche Y, Goel R (2014b) Diversified diazotrophs associated with the rhizosphere of Western Indian Himalayan native red kidney beans (*Phaseolus vulgaris* L.). 3 Biotechnol. doi 10.1007/s13205-014-0238-5

Suyal DC, Yadav A, Shouche Y, Goel R (2015) Bacterial diversity and community structure of Western Indian Himalayan red kidney bean (*Phaseolus vulgaris* L.) rhizosphere as revealed by 16S rRNA gene sequences. Biologia 70(3):305–313

Swapna AL (2013) Development of biofertilizers and its future perspective. J Pharm 4:327–332

Szilagyi-Zecchin VJ, Ikeda AC, Hungria M et al (2014) Identification and characterization of endophytic bacteria from corn (Zea mays L.) roots with biotechnological potential in agriculture. AMB Exp 4(1):26. doi:10.1186/s13568-014-0026-y

Tajini F, Trabelsi M, Drevon JJ (2012) Combined inoculation with *Glomus intraradices* and *Rhizobium tropici* CIAT899 increases phosphorus use efficiency for symbiotic nitrogen fixation in common bean (*Phaseolus vulgaris* L.). Saudi J Biol Sci 19(2):157–163

Ullah I, Khan AR, Jung BK et al (2014) Gibberellins synthesized by the entomopathogenic bacterium, *Photorhabdus temperata* M1021 as one of the factors of rice plant growth promotion. J Plant Interact 9(1):775–782

United Nations (2013) World population prospects: the 2012 revision, highlights and advance tables. Working paper no. ESA/P/WP. 228. http://www.un.org/en/development/desa/population. Accessed 22 Apr 2016

Upadhyay SK, Singh JS, Saxena AK et al (2012) Impact of PGPR inoculation on growth and antioxidant status of wheat under saline conditions. Plant Biol 14(4):605–611

Vikram A, Hamzehzarghani H (2008) Effect of phosphate solubilizing bacteria on nodulation and growth parameters of greengram (*Vigna radiata* L. Wilczek). Res J Microbiol 3(2):62–72

Vyas P, Joshi R, Sharma KC et al (2010) Cold-adapted and rhizosphere-competent strain of *Rahnella sp.* with broad-spectrum plant growthpromotion potential. J Microbiol Biotechnol 20(12):1724–1734

Wallner A, Blatzer M, Schrettl M et al (2009) Ferricrocin, a siderophore involved in intra- and transcellular iron distribution in *Aspergillus fumigates*. Appl Environ Microbiol 75(12):4194–4196

Wang Y, Tang S, Jin H (2015) Effect of glucose, root exudates and N forms in mycorrhizal symbiosis using *Rhizophagus intraradices*. J Soil Sci Plant Nutr 15(3):726–736

Waqas M, Khan AL, Kang S-M, Kim Y-H, Lee I-J (2014) Phytohormone producing fungal endophytes and hardwood-derived biochar interact to ameliorate heavy metal stress in soybeans. Biol Fertil Soils 50(7):1155–1167

Zahir ZA, Munir A, Asghar HN et al (2008) Effectiveness of rhizobacteria containing ACC-deaminase for growth promotion of pea (*Pisum sativum*) under drought conditions. J Microbiol Biotechnol 18:958–963

Zahir ZA, Ghani U, Naveed M et al (2009) Comparative effectiveness of *Pseudomonas* and *Serratia* sp. containing ACC-deaminase for improving growth and yield of wheat (*Triticum aestivum* L.) under salt-stressed conditions. Arch Microbiol 191:415–424

Zaidi A, Khan MS (2005) Interactive effect of rhizospheric microorganisms on growth, yield and nutrient uptake of wheat. J Plant Nutr 28:2079–2092

Zhang J, Meng L, Cao Y et al (2014) The role of wheat germ agglutinin in the attachment of *Pseudomonas* sp. WS32 to wheat root. J Microbiol 52(12):1020–1024

Role of Beneficial Fungi in Sustainable Agricultural Systems

18

Mehrnaz Hatami and Fereshteh Ahangarani

Abstract

Sustainable agriculture is a farming technique on the basis of knowledge of ecosystem services, the study of relationships/interactions between organisms and their physical environment. In sustainable agricultural systems, the inhabitant soil microflora is more crucial for ecosystem processes including nutrient availability and pest/disease suppression.

The rising demand for environmental friendly, organic, and sustainable agricultural practices are driving the application of fertilizers based on beneficial biological products. The use of beneficial fungi in agriculture sector is potentially useful for improved plant health and growth, water uptake, nutrient availability, stress tolerance, and biocontrol. Fungi also play a fundamental role in multifarious physiological processes including mineral and water uptake, photosynthesis, stomatal movement, and biosynthesis of compounds termed biostimulants, auxins, lignan, and ethylene to enhance the ability of plants to establish and cope environmental stresses such as drought, salinity, heat, cold, and heavy metals. This chapter describes the mechanisms underlying beneficial impacts of fungi on growth promotion of the host plant.

M. Hatami (✉)
Department of Medicinal Plants, Faculty of Agriculture and Natural Resources, Arak University, 38156-8-8349 Arak, Iran
e-mail: hatamimehrnaz@yahoo.com; hatamimehrnaz@gmail.com; m-hatami@araku.ac.ir

F. Ahangarani
Master of Science of Weed Identification and Management, Arak, Iran
e-mail: F_ahangarani@yahoo.com

D.K. Choudhary et al. (eds.), *Plant-Microbe Interaction: An Approach to Sustainable Agriculture*, DOI 10.1007/978-981-10-2854-0_18

18.1 Introduction

Many of the world's ecosystems are in different moods of decline affirmed by erosion, low fertility and productivity, poor water quality, etc. caused by forest clearing, high-input agricultural production, and stay using land resources for multiple purposes that are not sustainable. However, sustainable agriculture prepares high yields without causing weakness or damages the natural systems and resources that productivity depends on. This kind of agriculture utilizes a special cultivation technique wherein the environmental resources may be entirely used and at the same time guarantee that no injury was executed to it. Therefore, the method of performance is environmental friendly and assures safe and healthy agricultural crops and products. In low-input organic and sustainable agroecosystems, the natural function of microorganisms in supporting soil fertility and/or control of plant pathogens may be more important than in conventional agriculture practices where their importance has been marginalized using high inputs of chemical fertilizers and other agrochemicals.

The status of soil microbial populations is beneficial to vital processes that compel fertility, productivity, and stability of agricultural systems. Several investigations are directed to enhancing knowledge of the diversity, structure, dynamics, and significance of soil microbial communities and their advantageous and cooperative actions in soil fertility and crop productivity.

There is a growing body of evidence that exhibits the potential of different microorganisms to improve plant growth and productivity in agricultural systems. Understanding the potential of soil-beneficial microorganisms needs comprehension of the action of microorganisms in growth enhancement, especially in terms of nutrient supplement and disease suppression, the underlying mechanisms, and the challenges in implementation and commercialization of plant growth-promoting (PGP) microorganisms (Johansson et al. 2004; Pereg and McMillan 2015).

Soil microorganisms possess a close, symbiotic (reciprocal) relationship with plants. They are the most plentiful of all the biota in soil and responsible for promoting nutrition and organic matter cycling, soil fertilization and restoration, and plant health and growth as well as ecosystem primary production. Two examples include rhizobia and mycorrhizae. Mycorrhizae are known as very host-specific fungi that create symbiotic associations with roots of host plant. The beneficial fungi play an important role in improving plant growth and increasing plant yield and also involved in biotic and abiotic stress tolerance, hazardous materials remediation, sustainable crop production, and food safety (Borde et al. 2009). Some of the fungi do not permit the deleterious fungi to colonize the root surface area and are beneficial as biocontrol agents (Ha 2010). In this chapter, we explain the mechanisms behind the positive role of symbionts to host partners.

18.2 General Mechanisms Involved in Plant Growth Promotion Elicited by Microorganisms

Plant growth and performance is remarkably affected by the interactions between plant roots and the surrounding relative soil involving the microbial community inside the soil. The rhizosphere supports microorganisms that may have both positive and negative or inconsequential impact on plant growth and productivity. Although most rhizospheric microorganisms seem to be desirable, harmful microbes such as pathogens and microorganisms generate toxins that prevent root growth or those that eliminate essential substances from the soil. In contrary, the major mechanisms for promoting plant growth include improvement of nutrient availability (biofertilization), suppression of parasitic and nonparasitic pathogens (biocontrol), and production of plant hormones/and or plant growth-promoting substances (phytostimulation) (Martinez-Viveros et al. 2010; Bhattacharyya and Jha 2012).

Many factors are affecting the population of the indigenous rhizospheric microbes including agricultural practices (e.g., soil cultivation, stubble maintenance, burning, season, and so on), plant species, variety/cultivar and genotype, and soil type (Berg and Smalla 2009; Reeve et al. 2010). Plant exudates may cause alterations to soil characteristics including carbon availability and pH, influencing the diversity and activity of microbial communities (Haichar et al. 2008). It is acknowledged that the addition of microorganisms to cropping systems and agricultural soils (bioaugmentation) exhibit an important action on soil microbial processes. The application of agrochemicals such as chemical fertilizers and pesticides/and or fungicides caused concerns about their potential risks to living organisms and pathogen resistance, imposing continuous expansion of novel agents (Fernando et al. 2006). Rhizospheric microorganisms that prevent plant pathogens could be applied as biocontrol agents and may be considered as efficient and alternative to chemical pesticides. Some of the mechanisms for suppression of plant pathogen are direct inhibition of pathogen growth via production of antibiotics, hydrogen cyanide (HCN), and toxins and activation of hydrolytic enzymes (e.g., lipases, proteases, and chitinases) that degrade toxicity agents or pathogen cell-wall components (Whipps 2001; Compant et al. 2005).

18.2.1 Mechanisms of Biofertilization

"Biofertilizers" are beneficial microbes that improve nutrient uptake and availability to inoculated plants, contributing to plant nutrition through increasing nutrient uptake and/or through accelerating primary nutrient availability in the rhizosphere. Also, they could be applied for improving crop yield when used complementary to, or as alternative for, synthetic fertilizers.

Nitrogen (N) is an important plant macronutrient that is frequently limited in agricultural soils because of high losses through leaching and emission. Biological nitrogen fixation can be performed by nonsymbiotic bacteria including *Azospirillum*, *Gluconacetobacter*, *Burkholderia*, and *Pseudomonas* species (Dobbelaere et al.

2003) and may be employed in biofertilization of nonleguminous plants including wheat (Egamberdiyeva and Hoflich 2002), rice (Mirza et al. 2006; Muthukumarasamy et al. 2007), maize (Estrada et al. 2005), and sugarcane (Suman et al. 2005). Also, the *Azotobacter* Azo-8 strain was introduced as an efficient bio-inoculant for wheat plant grown under water scarcity conditions along with urea and manure (Singh et al. 2013).

Although agricultural soils usually have considerable total phosphorus, available phosphorus is frequently exhausted from the rhizosphere (Richardson et al. 2009). Soil microorganisms play an essential role in the phosphorus cycle and, therefore, in facilitating phosphorus availability to plants, improving the capacity of plants to obtain phosphorus from the soil through solubilizing and mineralizing inorganic phosphorus, or via accelerating the mobility of organic phosphorus by microbial turnover and/or extending the root system of crop species (Richardson and Simpson 2011). A great number of soil microorganisms with the ability of solubilizing inorganic phosphorus have been isolated, such as *Actinomycetes*, *Pseudomonas*, *Rhizobium*, and *Bacillus* spp. (Richardson et al. 2009; Richardson and Simpson 2011; Bhattacharyya and Jha 2012). Moreover, some fungal from the *Penicillium* genus excrete compounds (organic acids) that expedite the conversion of immobilized phosphorus into soluble forms available for root uptake and plants (Wakelin et al. 2004).

The response of root growth and the flexibility of root system architecture along with the expansion of the rhizosphere, via either root growth or root hair development, are obviously significant for impressive exploration of soil environment and interruption of nutrients. Root hair may form up to 70 % of root volume and can take up to 80 % of phosphorus in non-mycorrhizal inoculated plants (Fohse et al. 1991). Mycorrhizal fungi generally colonize the root cortex of plant and enlarge exteriorly, joining the roots with soil environment and enhancing efficacy of phosphorus absorption by mycorrhizal inoculated plants (Barea et al. 2008).

Mycorrhizal symbiosis can increase plant growth by improvement of plant establishment, protection against different types of stress, and enhanced soil structure and nutrient uptake, especially as fundamental macronutrients (e.g., P, Mg, Ca, K) and micronutrients (e.g., Zn, Cu,) depend on soil pH (Clark and Zeto 2000; Richardson et al. 2009).

It has been reported that enhanced absorptive surface area of the inoculated plant-root systems caused increased area for interactions with other soil microorganisms through formation of hyphae of these symbiotic fungi which also act as a significant route for the translocation of energy-rich plant assimilates to the soil (Johansson et al. 2004). Generally, the effect of plant assimilates on microbial populations has been described in relation to the rhizosphere (Hiltner 1904). The rhizosphere (the narrow region of soil surrounding living plant roots) is characterized by improved microbial activity stimulated by root exudates (Grayston et al. 1997). However, since plant roots in natural habitats are prevalently mycorrhizal, the concept of rhizosphere has been broadened to comprise the fungal component of the symbiosis, causing the term "mycorrhizosphere" (Rambelli 1973; Johansson et al. 2004) (Fig. 18.1).

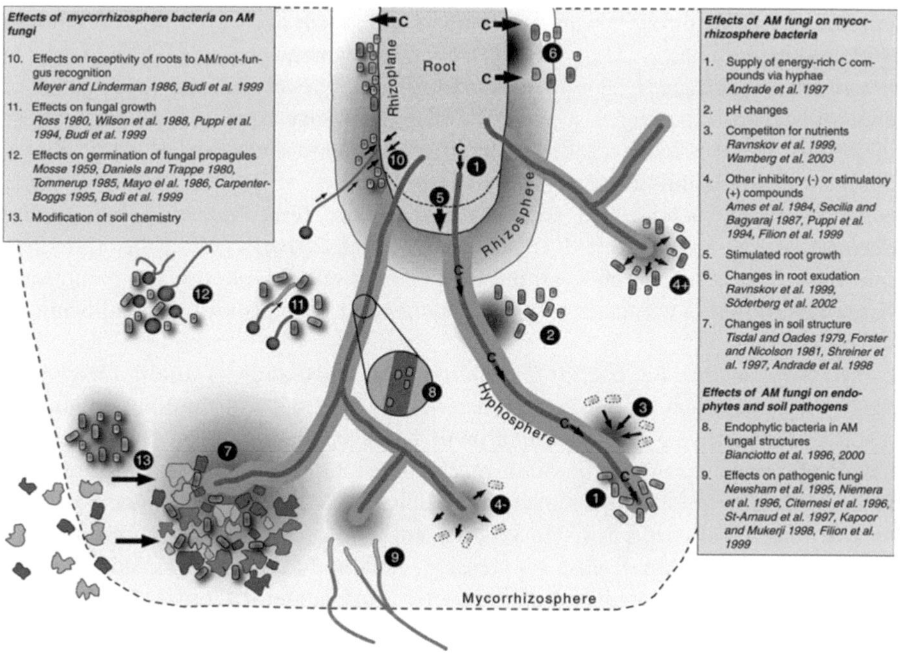

Fig. 18.1 Schematic model of possible interactions between various components of the mycorrhizosphere (Johansson et al. 2004)

18.2.2 Mechanisms of Disease Suppression

Generally, protection of plant growth is consecutively challenged through emerging, reemerging, and indigenous plant pathogens (Miller et al. 2009). As mentioned above, there are a number of mechanisms for plant-pathogen suppression including direct inhibition of pathogen growth by production of antibiotics and other toxins (Whipps 2001; Compant et al. 2005).

Antibiotics are a common section of the self-protective arsenals of bacteria, including *Pseudomonas* species (e.g., *Pseudomonas fluorescens* strains) (Haas and Defago 2005) and *Bacillus* species (e.g., *Bacillus subtilis*) (Kim et al. 2003), as well as fungal species including *Trichoderma*, *Gliocladium*, *Ampelomyces*, and *Chaetomium* (Kaewchai et al. 2009); therefore, these living organisms have great potential for soil conditioning process.

Multifunctional microorganisms including *Trichoderma harzianum* Rifai appear to improve plant growth through solubilizing phosphate and essential micronutrients required for plants, including iron (Fe) and manganese (Mn), and suppress plant pathogens (Altomare et al. 1999). Hydrogen cyanide production inhibits microbial growth and may suppress pathogens including root-knot, black rot, and bacterial canker in tobacco and tomato plants (Lanteigne et al. 2012; Voisard et al. 1989; Siddiqui et al. 2006). However, it has been reported that HCN might be injurious to plants through reducing energy metabolism and inhibiting root growth (Siddiqui et al. 2006). Many various bacterial genera produce HCN, such as *Rhizobium*, *Alcaligenes*, *Bacillus*, *Aeromonas*, and *Pseudomonas* spp. (Ahmad et al. 2008). Also, pathogen suppression may take place competitively by indirect inhibition. There is evidence that a number of bacteria and fungi produce siderophores as iron-chelating sources particularly in times of iron deficiency (Sharma and Johri 2003), such as *Azospirillum*, *Pseudomonas*, *Bradyrhizobium*, *Rhizobium*, *Serratia*, and *Streptomyces* (Martinez-Viveros et al. 2010). Their ability to reduce iron from their surrounding environment makes it unavailable form to pathogenic fungi, resulting in a competitive benefit (Loper and Henkels 1999; O'Sullivan and O'Gara 1992).

Carrillo-Castaneda et al. (2002) reported that inoculation of alfalfa (*Medicago sativa*) with siderophore-producing bacteria grown under iron-limiting conditions resulted in a positive effect on plant growth; however, the possible role for a combination of several growth-promoting mechanisms and not siderophore generation alone cannot be ignored. Moreover, activation of the plant's own defense system, termed induced systemic resistance (ISR), may be considered as another mechanism that is involved in disease suppression. Release of a blend of volatile organic compounds by plant growth-promoting bacteria and fungi may initiate ISR, causing enhanced expression of defense-related genes in the inoculated plants (Naznin et al. 2014).

18.2.3 Mechanisms of Phytostimulation

One of the most important mechanisms involved in plant growth enhancement through some rhizospheric living microorganisms is the production of plant hormones, or phytostimulation; plant growth-promoting microbes stimulate plant growth by producing growth hormones, including auxins, gibberellins, and cytokinins in the adjacency of the roots, or by regulating the levels of ethylene produced by plants. Root characteristics, size, and depth affect the capacity of plants to effectively entrap nutrients from soil environment and vice versa; root elongation and morphology can alter in response to soil nutrient availability (Wijesinghe et al. 2001). Plants bearing both deep and shallow root systems are able to acquire mineralized nitrogen available in top of soils and leached nitrogen in the subsoils (Ho et al. 2005). Therefore, application of phytostimulation for improving plant-root growth could play an important role in facilitating nutrient uptake, particularly if employed in combination with biofertilization.

The main phytohormone, auxin (IAA), promotes root growth and root architecture attributes such as root hair cluster positioning, lateral root extension, and root vascular tissue development (Aloni et al. 2006). Many rhizobacteria, such as beneficial, pathogenic, associative, and free living, are potentially proficient to produce IAA (Tsavkelova et al. 2006). These include *Azospirillum*, *Azotobacter*, *Aeromonas*, *Bacillus*, *Burkholderia*, *Enterobacter*, *Rhizobium*, and *Pseudomonas* (Spaepen et al. 2006; Martinez-Viveros et al. 2010). Cytokinins promote plant cell division and regulate root growth and development through reducing primary root elongation and lateral root formation and stimulating root hair formation (Werner et al. 2003; Riefler et al. 2006). These substances are produced by some plant growth-promoting rhizobacteria, including *Arthrobacter*, *Azospirillum*, *Pseudomonas*, and *Paenibacillus* species, but their participation in plant growth promotion is not well known (Richardson et al. 2009).

Likewise, gibberellins induce the development of stem tissue, root growth, and lateral root branching and are produced by different species of plant growth-promoting rhizobacteria, including *Azotobacter*, *Azospirillum*, *Bacillus*, *Herbaspirillum*, *Rhizobium*, and *Gluconobacter* (MacMillan 2002; Bottini et al. 2004).

Another important hormone involved in plant growth and development is ethylene, although it may have distinctive impacts on plant growth depending on its doses in plant roots (Pierik et al. 2006). Ethylene synthesis is necessary for the initiation of systemic resistance during interaction with associative microorganisms, and higher doses are mediated in plant defense against a range of stresses and upon pathogen infection (Broekaert et al. 2006). Certain growth-promoting bacteria, including *Azospirillum brasilense*, may produce lower levels of ethylene, which may enhance root hair formation (Ribaudo et al. 2006). In plants, ethylene has previously been found to be produced from the precursor 1-aminocyclopropane-1-carboxylate (ACC), which is released by roots within the rhizosphere during stress and reabsorbed through the roots, which is in turn converted to ethylene. However, accumulation of ethylene in the roots causes decreased root growth, aggravating plant stress (Babalola 2010). Rhizospheric plant growth-promoting fungi and bacteria that can break down ethylene precursor ACC diminish the adsorption of ethylene through the root tissues and permit the plant to reestablish a healthy root and cope with environmental biotic and abiotic stresses (Glick 2005). Plant growth enhancement by ACC (a source of nitrogen) degrading microorganisms seems to be specifically significant under stress conditions including drought, cold, salinity, or heavy metal stress (Mayak et al. 2004; Grichko and Glick 2001). Root zone microorganisms are able to degrade ACC including *Azospirillum*, *Achromobacter*, *Enterobacter*, *Bacillus*, *Pseudomonas*, and *Rhizobium* strains (Martinez-Viveros et al. 2010).

18.3 The Arbuscular Mycorrhizal Symbiosis

The oldest (>460 million years BP) and the most popular kind of mycorrhizal association is *arbuscular mycorrhizal* (AM) symbiosis. Many terrestrial plants (~250,000 species of plants worldwide) are capable of forming the symbiosis (Smith and Read 1997).

According to Schussler et al. (2001), almost 160 fungal taxa of the order *Glomales* (*Glomeromycota*) have been explained in terms of their spore morphology, although new molecular technique exhibit that the real number of AM taxa may be much higher (Daniell et al. 2001; Vandenkoornhuyse et al. 2002).

It has been reported that in time of AM symbiosis formation, the fungus permeates the cell walls of root cortical and constitutes haustoria-like structures that interact with the host cytoplasm (Smith and Read 1997). These fungal structures and highly branched arbuscules in particular prepare an improved surface area for metabolic conversions between the fungus and the plant. Moreover, production of certain structures, vesicles, by some of the AM believed to act as storage tissues (Smith and Read 1997). According to Johnson et al. (2002), plants in natural terrestrial ecosystems colonized with AM may instate about 10–20 % of the fixed carbon through photosynthetic process in their fungal confederate. Obviously, this exhibits an important input of energy to the soil environment, and this source of carbon may be vital to living microbes associated with the mycorrhizosphere.

It has been acknowledged that AM fungi also interact directly with the soil through producing extra radical hyphae that may spread out several centimeters within the soil (Rhodes and Gerdemann 1975). The extra radical hyphae structures constitute a greater total surface area than that of roots solely, which enhances the potential for nutrient and water uptake (Auge 2001; Rhodes and Gerdemann 1975). Besides, the extra radical hyphae structures seem to be significant to the plants for phosphorus acquisition efficiency and other mineral nutrients uptake (Read and Perez-Moreno 2003). Also, Hodge et al. (2001) suggested that the extra radical mycelium of AM fungi may improve mobilization of organic nitrogen forms from plant residue. It was previously believed that mycorrhizal symbiosis may also mitigate adverse effects of plant pathogens (Newsham et al. 1995; Niemira et al. 1996; St-Arnaud et al. 1997; Azcon-Aguilar and Barea 1996) and negative role of metals (Khan et al. 2000). Moreover, the extra radical hyphae may interface with other soil microbes either directly through physically and/or metabolically interfacing with other soil-living organisms in the mycorrhizosphere or indirectly through altering host plant physiology (e.g., root morphogenesis and patterns of exudation within the mycorrhizosphere) (Johansson et al. 2004).

18.3.1 Effects of AM Fungi on Fungal Pathogens

Arbuscular mycorrhizal fungi may interact with other root-associated microbes, like pathogenic fungi. The potential mechanisms of interaction are similar as those mentioned above. The differential impacts of a crude extract from the growth medium of the AM fungus *Glomus intraradices* were studied on spore formation of two pathogenic fungi and on the growth of two bacterial species (Filion et al. 1999). Conidial germination of *Fusarium oxysporum* (a plant-root pathogen) was inhibited, while conidial germination of *Trichoderma harzianum* (a mycoparasitic fungus) and the growth of *Pseudomonas chlororaphis* were promoted, and *Clavibacter michiganensis* growth was uninfluenced. The assayed impacts were correlated with extract dose, and no significant effect of pH on germination or growth was found. They concluded that the unspecified substances released by the AM fungus to the growth medium were the major factor describing the differential growth of the employed microbes.

Citernesi et al. (1996) screened bacteria separated from 17-year-old *Glomus mosseae* pot culture. They reported that many of the bacterial strains within the various zones of the mycorrhizosphere were vigorously antagonistic against in vitro growth of *Fusarium* and *Phytophthora*. Their findings also suggest the probability of integrated application of AM fungi and their associated bacteria in biocontrol of soilborne pathogenic fungi. Many researchers have mentioned that the ability of AM-inoculated plants to better stand up to an attack from root pathogens may be described to an improved nutritional status in the host plant because of the attendance of the AM fungus. However, there are contradictory reports on this theory. In a field experiment, Newsham et al. (1995) transplanted *Glomus* sp.-treated and *Glomus* sp.-non-treated seedlings of *Vulpia ciliata* into a natural ecosystem and found that inoculation of AM did not influence phosphorous content in the plants. However, the AM protected the plants from the adverse impacts of *Fusarium oxysporum* attack on root and shoot growth. Obviously, the AM inhibited pathogen development in the root tissues. The results also showed that root-infecting mycofloras of AM plants had fewer naturally occurring infections of *F. oxysporum* and *Embellisia chlamydospora* compared to AM plants following transplantation (Newsham et al. 1995). They suggested that the main advantage granted by AM fungi to *V. ciliata* seedlings is the protection from deleterious fungi, rather than enhanced phosphorous acquire. In a study, Niemira et al. (1996) employed a peat-based medium containing *Glomus intraradices* to test whether it could inhibit *Fusarium sambucinum* (a common tuber dry rot) in minitubers of potato plants. Results revealed less (20–90 %) tuber dry rot for minitubers grown in this medium. Furthermore, St-Arnaud et al. (1997) reported that the presence of *Tagetes patula* plants inoculated with AM fungus *G. intraradices* may suppress root pathogen development in soil and by means of that decrease severity of disease in cocultured non-mycorrhizal carnation (*Dianthus caryophyllus*). In other study, Caron (1989) found significant decrease in Fusarium populations in the soil surrounding mycorrhizal tomato (*Lycopersicon esculentum*) roots and subsequently proposed a possible role for AM fungi in biocontrol of the soilborne pathogens.

18.3.2 Soil Fungal Communities Confer Agroecosystem Stability

Beneficial soil microorganisms such as AM fungi are key component in natural agroecosystem through providing crucial ecosystem services including nutrient uptake, organic matter recycling, and antagonism versus plant pests/disease (Borie et al. 2010; Pozo et al. 2009; Ramos-Zapata et al. 2012). *Arbuscular mycorrhizal* fungi, saprophytes, use up the destroyed organic materials in soil and are definitely innocuous and often beneficial for mobilization of mineral nutrients (Hodge et al. 2001; Lopez-Roez and Pozo 2013).

The ecological importance of *Perisporiopsis lateritia*, *Phanerochaete velutina*, and *Pleurotus* sp. can be described by their function in dead vegetation recycling process through converting hard wood to usable forms (Chaverri and Gazis 2010; Wells et al. 1998; Cohen et al. 2002). Some fungi (e.g., *Navisporus floccosus*), characterized in digesting the secondary compounds such as lignin, tannin, and cellulose in soils; preparing nutrients availability for offspring of the present plant generation

(Parihar et al. 2012). It has been reported that soil AM fungi may be considered as an essential component of an ecosystem to help the carbon and nitrogen recycling in soil environment (Phillips et al. 2012).

Generally, mycorrhizal fungi include higher value of carbon (10/1 C/N) and less amount of nitrogen (N ¼ 10 %) in their cells than those of bacteria (Hoorman 2011). Mycorrhizal fungi contribute in recycling of both nitrogen and phosphorus to improve availability of mineral nutrients for the plants. Their properties, small size and high surface area, are more efficient in mineral acquisition from the soils when compared to the plant-root hairs (Hoorman 2011). Evidence suggests, however, that plant's roots cultivated with particular species of fungi caused significant nutrient acquisition and higher nutritional levels of crop plants (Yaseen et al. 2011; Albrechtova et al. 2012). In addition to symbiont, free-living soil fungi including *Trichoderma* are established to be responsible for improved plant growth and development, higher biomass production, and lateral root branching via the mechanism mediated by synthesis of auxins (Contreras-Cornejo et al. 2009). Specific fungi, including *Piriformospora indica* and *Trichoderma*, are beneficial in plant-soil systems and act as biocontrol agents (Harman and Mastouri 2010; Serfling et al. 2007) to support agricultural crops from severe injury caused by pathogen attack (Ha 2010). According to Chalot and Brun (1998), ectomycorrhizal fungi can effectively degrade the undesirable phenolic constituents in the forest soils.

The ecto- and endo-AM fungi may prepare definite advantages to host plants through expanding surface area for sufficient water and nutrient uptake, improving stomatal regulation to preserve proper water potential, and increasing twofold the minimal stomatal conductance (Aroca et al. 2008; Arnold and Engelbrecht 2007) for better gaseous exchange which subsequently result in amplified photosynthetic quantum yield (Wu and Xia 2006; Xian-Can et al. 2010) (Table 18.1).

18.3.3 Relevance of Mycorrhizosphere Interactions to Sustainable Agriculture

Soil-beneficial fungi are specifically useful for the plant partner in agriculture and take part in several services including water levels, nutrient improvement, stress tolerance, pest and disease protection, and weed control. Sustainable agroecosystem relies on beneficial fungi due to its contribution in decomposition of soil organic matter, nutrient uptake, organic matter and nutrient recycling, antagonism against plant pathogens/pests, and crop management (Ansari et al. 2013). Generally, two main groups of soil endophytic fungi have been previously recognized, exposing dissimilarity in evolutionary interaction: (1) the clavicipitaceous endophytes (C-endophytes), which associate with grasses and systemically infect their hosts, and (2) the non-clavicipitaceous endophytes (NC-endophytes), which can be reproduced from asymptomatic parts of a broad range of plant (nonvascular) hosts, belonging to angiosperms (Singh et al. 2011).

Mycorrhizal fungi colonizing the plant-root systems (rhizosphere) extend within the rhizosphere and are efficiently involved in enhancing soil fertility and crop

Table 18.1 Soil-beneficial fungi effects on different physiological and catabolic processes in various host plant species

Fungal species/strain	Plant type	Fungi-mediated response/activities	Beneficial effects on plant species	References
AM fungi	Dead vegetation in soil	Degrade of dead organic	Nutrient mobilization	Hodge et al. (2001)
Phanerochaete velutina	Wood	Decomposing wood	Phosphorus translocation	Wells et al. (1998)
Pleurotus sp.	Wood	Wood decay	Nutrient mobilization	Cohen et al. (2002)
Perisporiopsis lateritia	Leaves of *Hevea* sp.	Leaves decay	Nutrient mobilization	Chaverri and Gazis (2010)
Navisporus floccosus	Wood	Wood decay	Nutrient mobilization	Phillips et al. (2012)
M fungi	*Pinus taeda*	Decomposing organic matter	Carbon and nitrogen cycling	Hoorman (2011)
AM fungi	*Vigna unguiculata*	Mineral uptake	Improved nutritional status	Yaseen et al. (2011)
M fungi	*Allium cepa*	Plant growth	Improved nutritional status	Albrechtova et al. (2012)
Trichoderma sp.	Arabidopsis sp.	Auxins dependent mechanism	Higher biomass production and increased lateral roots formation	Contreras-Cornejo et al. (2009)
Trichoderma sp.	Agriculturally important crops	Biocontrol	Crop management	Chalot and Brun (1998), Harman and Mastouri (2010), and Serfling et al. (2007)
Ectomycorrhizal fungi	Higher plant species	Phenolic compounds degradation	Plant protection	Ha (2010)
Ectomycorrhizal fungi and AM fungi	Agricultural crops	Stomatal physiology and water relation	Improved water potential status and increased photosynthesis rate	Arnold and Engelbrecht (2007) and Wu and Xia (2006)

productivity in natural and agroecosystems (Bonfante and Genre 2010). It has been reported that mutualistic symbioses (intimate interspecies interactions) contribute to plant's life cycle through supplying micro- and macronutrients, enhanced growth, and improved thermotolerance and resistance from different environmental biotic and abiotic stresses such as drought, salinity, herbivore, and pathogen infection (Lingua et al. 2012; Singh et al. 2011; Rodriguez et al. 2008). A strong growth-promoting activity was found during the symbiosis of *Piriformospora indica*, a plant-root-colonizing basidiomycete fungus, with a broad spectrum of plant species

(Verma et al. 1998). The *P. indica* produces thin-walled, white color, and hyaline hyphae throughout the life cycle which exhibits multinucleated character. Following to anastomosis between various types of hyphae, the formation of certain chlamydospores (thick-walled big resting spore) occurs either separately or in clusters at their tip. The released chlamydospores then trigger germination in the soil and subsequently infect other host through creating intra- and intercellular hyphal network (Das et al. 2012). Moreover, *P. indica* imitate the potentiality of specific AM fungi in different morphological, functional, and growth promotional points of view (Das et al. 2013) with supplemental profit that it can grown axenically. Tsimilli-Michael and Strasser (2013) confirmed that the *P. indica* may be a novel candidate symbiont for supplying immense growth-promoting activity with a broad spectrum of plants species. This symbiotic interaction caused in higher biomass production of the shoots and floral parts of the plant which can be used for biologically active compounds in pharmaceutical industries (Kumar et al. 2011; Oelmuller et al. 2009). *P. indica* (the growth-promoting endophyte) in many cases acts as a biofertilizer, bioregulator, and bioprotector both in mono- and dicotyledonous plant species (Das et al. 2012). The mutually beneficial relations between *P. indica* and roots are valuable being its wide spectrum of uses in farming systems (Franken 2012). It is acknowledged that specific biochemical and/or genetic processes are involved in biosynthesis of ethylene and signaling to maintain an interaction between the symbionts and host plants (Khatabi et al. 2012).

18.3.4 Sustainable Nutrient Supply

It has been well known that AM may improve phosphorous level, enhance nitrogen uptake, or increase disease resistance in their host partners. Other soil microorganisms, such as nitrogen fixing and or phosphate solubilizing bacteria, can synergistically interact with AM fungi and stimulate plant growth through a range of mechanisms (Puppi et al. 1994). The symbiotic association between fungi and host plants becomes even more important in low sustainable input and organic agricultural systems. Under these situations, AM mycelium may act an influential role in nutrient mobilization from plants litter (Johansson et al. 2004).

Hodge et al. (2001) suggested that the inoculation with AM *Glomus hoi* improved decomposition of plant litter in soil and caused increased nitrogen acquisition from the litter as well. Hyphal growth of the fungal symbiont was also improved in the presence of the complex organic material in soil (Fig. 18.2). Bacteria associated with the AM may assist the nutrient cycling in soil. Several examples of this kind of association are available from bacterial-AM fungal-legume tripartite symbiosis, where diazotrophic bacteria prepare fixed nitrogen for both the plant and the fungus. Interestingly, legume nodulation by nitrogen-fixing bacteria and AM establishment often take place synchronously and synergistically.

The presence of nitrogen fixation genes in endosymbiotic bacteria (*Burkholderia*) in AM hyphae has been previously showed by Minerdi et al. (2001) who suggest that there may be a potential for enhanced nitrogen source to mycorrhizal infected

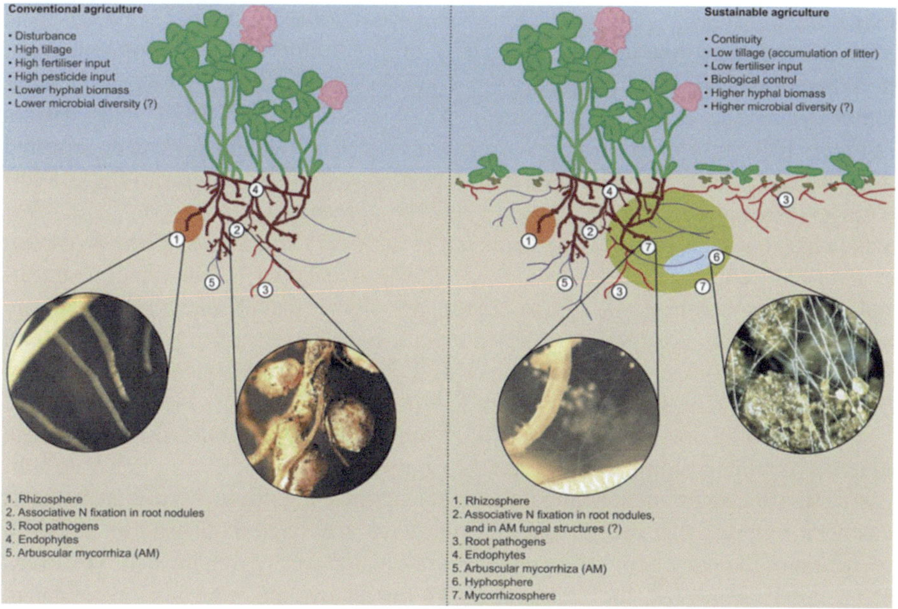

Fig. 18.2 Schematic model of the mycorrhizosphere concept compared to the rhizosphere concept: characteristics of conventionally managed agricultural soils (*left*) in contrast to sustainably managed agricultural soils (*right*) (Johansson et al. 2004)

plants through atmospheric nitrogen fixation. More study is also required on the possible interactions of mycorrhizal fungi with decomposition processes. There is now a growing consciousness of functional differences among various AM fungi, and as our ability to perceive of their functional capacities raises we may be able to select appropriate species for maximize of nutrient recycling (Johansson et al. 2004).

18.3.5 Biocontrol

Microbial inoculants may be utilized as alternative and efficient tools for suppression of disease and pests in agricultural cropping systems, allowing the decreased application of chemical pesticides that could otherwise display threats to human health and nontargeted living organisms. The biological control agents may influence AM fungi, or be influenced themselves by AM fungi, in the same manner to the interactions as mentioned above. Biocontrol organisms against particular pathogenic fungi may have adverse impacts on nontarget soil AM fungi. Studies have shown that the mechanisms of antagonistic interactions causing biocontrol may involve competition for soil nutrients and colonization sites as well as production of fungistatic compounds. However, few researches have explicitly investigated interactions involving AM fungi (Johansson et al. 2004). Some beneficial impacts of rhizobacteria on AM fungal colonization of roots could be because of antagonistic effects on competing pathogens (Azcon-Aguilar and Barea 1996), as well as direct synergistic effects on mycorrhizal colonization itself (Budi et al. 1999). Different plant-root-colonizing or seed-borne *Pseudomonas* spp. have been reported to be influential microbial control agents in plant-pathogen systems under in vitro (Chin-A-Woeng et al. 2003; Leeman et al. 1996), greenhouse (Knudsen et al. 1999), and field (Johnsson et al. 1998; Knudsen et al. 1997) conditions. Chin-A-Woeng et al. (1998) reported that *Pseudomonas chlororaphis* PCL1391, an effective bacterial strain for colonizer of tomato roots, revealed efficient antagonistic activity against *Fusarium oxysporum*. The aforesaid bacterial strain produced a broad spectrum of antifungal compounds, such as hydrogen cyanide, phenazine-1 carboxamide, proteases, and chitinases (Chin-A-Woeng et al. 1998).

By knocking out the phenazine biosynthetic operon, it was shown that the mutants exhibited significantly lower biocontrol activity, indicating that this substance was an important antifungal factor for suppressing disease in tomato roots. It has been reported that the presence of the biocontrol bacteria caused in 70–80 % reduction of the density of the hyphal network inside part of tomato roots (Bolwerk et al. 2003). However, the effects on AM fungal hyphae were not investigated. Besides producing antifungal substances, the capacity of bacteria to colonize root surfaces and thereby closely interact with pathogens may further promote pathogenic suppression (Lagopodi et al. 2002).

Despite the rising number of studies over the last years, the underlying mechanisms are poorly understood. Some fundamental mechanisms have been previously proposed: enhancement of plant nutrition and competition for photosynthates (Azcon-Aguilar and Barea 1996); however, AM caused suppression of root pathogens and promotion of saprotrophs and plant growth (Kapoor and Mukerji 1998). The other mechanisms that tend to be inconsistent among studies include changes in morphological and anatomical features of root system induced by the AM fungus and triggering plant defense mechanisms by AM fungi (Gianinazzi-Pearson et al. 1994). The combination of AM fungi with growth-promoting rhizobacteria may prefer the inoculum production (Singh 1992).

Many studies have shown that some AM fungi present biocontrol characteristics (Niemira et al. 1996; Caron 1989; Newsham et al. 1995) against plant-root pathogens. Whether AM fungi may be applied as biological control agents virtually or potentially act as vectors for associated bacteria with biocontrol characteristics remains to be further explored.

18.4 Conclusions

Interactions between symbiotic microbial and host plant are prominent to keep the continued existence of both microorganism as well as the host under environmental restrictions. These interactions are important for soil-plant-water relations, mineral uptake, stomatal regulation, gas exchange, and photosynthetic process. Moreover, symbiotic root-associated fungi are critically efficient in enhancing plant growth and conferring plant tolerance to different types of stresses. In view of the fact that process of plant growth and development, which conclude crop yield, cannot be correctly described without possessing idea of microbial interactions. Thus, it is required to study plants from a symbiotic systems attitude to understand the contributions of all organisms in a symbiotic relationship for better plant health, growth, and survival.

References

Ahmad F, Ahmad I, Khan MS (2008) Screening of free-living rhizospheric bacteria for their multiple plant growth promoting activities. Microbiol Res 163:232–238

Albrechtova J, Latr A, Nedorost L, Pokluda R, Posta K, Vosatka M (2012) Dual inoculation with mycorrhizal and saprotrophic fungi applicable in sustainable cultivation improves the yield and nutritive value of onion. Sci World J 2012:1–8

Aloni R, Aloni E, Langhans M, Ullrich CI (2006) Role of cytokinin and auxin in shaping root architecture: regulating vascular differentiation, lateral root initiation, root apical dominance and root gravitropism. Ann Bot 97:883–893

Altomare C, Norvell W, Bjorkman T, Harman GE (1999) Solubilization of phosphates and micronutrients by the plant-growth promoting and biocontrol fungus Trichoderma harzianum Rifai 1295–22. Appl Environ Microbiol 65:2926–2933

Ansari MW, Trivedi DK, Sahoo RK, Gill SS, Tuteja N (2013) A critical review on fungi mediated plant responses with special emphasis to Piriformospora indica on improved production and protection of crops. Plant Physiol Biochem 70:403–410

Arnold AE, Engelbrecht BMJ (2007) Fungal endophytes nearly double minimum leaf conductance in seedlings of a neotropical tree species. J Trop Ecol 23:369–372

Aroca R, Vernieri P, Ruiz-Lozano JM (2008) Mycorrhizal and non-mycorrhizal Lactuca sativa plants exhibit contrasting responses to exogenous ABA during drought stress and recovery. J Exp Bot 59:2029–2041

Auge RM (2001) Water relations, drought and vesicular-arbuscular mycorrhizal symbiosis. Mycorrhiza 11:3–42

Azcon-Aguilar C, Barea JM (1996) Arbuscular mycorrhizas and biological control of soil-borne plant pathogens – an overview of the mechanisms involved. Mycorrhiza 6:457–464

Babalola O (2010) Ethylene quantification in three rhizobacterial isolates from Striga hermonthica-infested maize and sorghum. Egypt J Biol 12:1–5

Barea J, Ferrol N, Azcon-Aguilar C, Azcon R (2008) Mycorrhizal symbiosis. In: White P, Hammond J (eds) The ecophysiology of plant-phosphorus interactions. Springer, Dordrecht, pp 143–163

Berg G, Smalla K (2009) Plant species and soil type cooperatively shape the structure and function of microbial communities in the rhizosphere. FEMS Microbiol Ecol 6:1–13

Bhattacharyya P, Jha D (2012) Plant growth-promoting rhizobacteria (PGPR): emergence in agriculture. World J Microbiol Biotechnol 28:1327–1350

Bolwerk A, Lagopodi A, Wijfjes A, Lamers G, Chin-AWoeng TC, Lugtenberg B, Bloemberg G (2003) Interactions in the tomato rhizosphere of two Pseudomonas biocontrol strains with the phytopathogenic fungus Fusarium oxysporum f. sp. radicis-lycopersici. Mol Plant-Microbe Interact 16:983–993

Bonfante P, Genre A (2010) Mechanisms underlying beneficial plant fungus Piriformospora indica on wheat under greenhouse and field conditions. Phytopathology 97:523–531

Borde M, Dudhane M, Jite PK (2009) Role of bioinoculant (AM fungi) increasing in growth, flavor content and yield in Allium sativum L. under field condition. Not Bot Horti Agrobot Cluj 37:124–128

Borie F, Rubio R, Morales A, Cornejo P (2010) Arbuscular mycorrhizae in agricultural and forest ecosystems in Chile. J Soil Sci Plant Nutr 10:204–223

Bottini R, Cassan F, Piccoli P (2004) Gibberellin production by bacteria and its involvement in plant growth promotion and yield increase. Appl Microbiol Biotechnol 65:497–503

Broekaert W, Delaure S, De Bolle M, Cammue BPA (2006) The role of ethylene in host-pathogen interactions. Annu Rev Phytopathol 44:393–416

Budi SW, van Tuinen D, Martinotti G, Gianinazzi S (1999) Isolation from the Sorghum bicolor mycorrhizosphere of a bacterium compatible with arbuscular mycorrhiza development and antagonistic towards soilborne fungal pathogens. Appl Environ Microbiol 65:5148–5150

Caron M (1989) Potential use of mycorrhizae in control of soilborne diseases. Can J Plant Pathol 11:177–179

Carrillo-Castaneda G, Juarez Munos J, Peralta-Videab JR, Gomezb E, Tiemannb KJ, Duarte-Gardeac M, Gardea-Torresdeyb JL (2002) Alfalfa growth promotion by bacteria grown under iron limiting conditions. Adv Environ Res 6:391–399

Chalot M, Brun A (1998) Physiology of organic nitrogen acquisition by ectomycorrhizal fungi and ectomycorrhizas. FEMS Microbiol Rev 22:21–44

Chaverri P, Gazis RO (2010) Perisporiopsis lateritia, a new species on decaying leaves of Hevea spp. from the Amazon basin in Peru. Mycotaxon 113:163–169

Chin-A-Woeng TFC, Bloemberg GV, van der Bij AJ, van der Drift K, Schripsema J, Kroon B, Scheffer RJ, Keel C, Bakker P, Tichy HV, de Bruijn FJ, Thomas-Oates JE, Lugtenberg BJJ (1998) Biocontrol by phenazine-1-carboxamide- producing Pseudomonas chlororaphis PCL1391 of tomato root rot caused by Fusarium oxysporum f. sp. radicis-lycopersici. Mol Plant-Microbe Interact 11:1069–1077

Chin-A-Woeng TFC, Bloemberg GV, Lugtenberg BJJ (2003) Phenazines and their role in biocontrol by Pseudomonas bacteria. New Phytol 157:503–523

Citernesi AS, Fortuna P, Filippi C, Bagnoli G, Giovannetti M (1996) The occurrence of antagonistic bacteria in Glomus mosseae pot cultures. Agronomie 16:671–677

Clark R, Zeto S (2000) Mineral acquisition by arbuscular mycorrhizal plants. J Plant Nutr 23:867–902

Cohen R, Persky L, Hadar Y (2002) Biotechnological applications and potential of wood-degrading mushrooms of the genus Pleurotus. Appl Microbiol Biotechnol 58:582–594

Compant S, Duffy B, Nowak J, Clement C, Barka EA (2005) Use of plant growth promoting bacteria for biocontrol of plant diseases: principles, mechanisms of action and future prospects. Appl Environ Microbiol 71(9):4951–4959

Contreras-Cornejo H-A, Macias-Rodriguez L, Cortes-Penagos C, Lopez-Bucio J (2009) Trichoderma virens, a plant beneficial fungus, enhances biomass production and promotes lateral root growth through an auxin-dependent mechanism in Arabidopsis. Am Soc Plant Biol 149:1579–1592

Daniell TJ, Husband R, Fitter AH, Young JPW (2001) Molecular diversity of arbuscular mycorrhizal fungi colonizing arable crops. FEMS Microbiol Ecol 36:203–209

Das A, Kamal S, Shakil NK, Sherameti I, Oelmüller R, Dua M, Tuteja N, Johri AK, Varma A (2012) The root endophyte fungus Piriformospora indica leads to early flowering, higher biomass and altered secondary metabolites of the medicinal plant, Coleus forskohlii. Plant Signal Behav 7:103e112

Das A, Prasad R, Srivastava RB, Deshmukh S, Rai MK, Varma A (2013) Cocultivation of Piriformospora indica with medicinal plants: case studies. Soil Biol 33:149–171

Dobbelaere S, Vanderleyden J, Okon Y (2003) Plant growth promoting effects of diazotrophs in the rhizosphere. Crit Rev Plant Sci 22:107–149

Egamberdiyeva D, Hoflich G (2002) Root colonization and growth promotion of winter wheat and pea by Cellulomonas spp. at different temperatures. Plant Growth Regul 38:219–224

Estrada P, Mavingui P, Cournoyer B, Fontaine F, Balandreau J, Caballero-Mellado J (2005) A N2-fixing endophytic Burkholderia sp. associated with maize plants cultivated in Mexico. Int J Syst Evol Microbiol 55:1233–1237

Fernando W, Nakkeeran S, Zhang Y (2006) Biosynthesis of antibiotics by PGPR and its relation in biocontrol of plant disease. In: Siddiqui Z (ed) PGPR: biocontrol and biofertilization. Springer, Netherlands, pp 67–109

Filion M, St-Arnaud M, Fortin JA (1999) Direct interaction between the arbuscular mycorrhizal fungus Glomus intraradices and different rhizosphere microorganisms. New Phytol 141:525–533

Fohse D, Claassen N, Jungk A (1991) Phosphorus efficiency of plants II. Significance of root radius, root hairs and cation-anion balance for phosphorus influx in seven plant species. Plant Soil 132:261–272

Franken P (2012) The plant strengthening root endophyte Piriformospora indica: potential application and the biology behind. Appl Microbiol Biotechnol 96:1455–1464

Gianinazzi-Pearson V, Gollotte A, Dumas-Gaudot E, Franken P, Gianinazzi S (1994) Gene expression and molecular modifications associated with plant responses to infection by arbuscular mycorrhizal fungi. In: Daniels M, Downic JA, Osbourn AE (eds) Advances in molecular genetics of plant–microbe interactions. Kluwer, Dordrecht, pp 179–186

Glick B (2005) Modulation of plant ethylene levels by the bacterial enzyme ACC deaminase. FEMS Microbiol Lett 252:1–7

Grayston SJ, Vaughan D, Jones D (1997) Rhizosphere carbon flow in trees, in comparison with annual plants: the importance of root exudation and its impact on microbial activity and nutrient availability. Appl Soil Ecol 5:29–56

Grichko V, Glick B (2001) Amelioration of flooding stress by ACC deaminase containing plant growth-promoting bacteria. Plant Physiol Biotechnol 39:11–17

Ha TN (2010) Using Trichoderma species for biological control of plant pathogens in Vietnam. J ISSAAS 16:17–21

Haas D, Defago G (2005) Biological control of soil-borne pathogens by fluorescent pseudomonads. Nat Rev Microbiol 3:307–319

Haichar F, Marol C, Berge O, Rangel-Castro JI, Prosser JI, Balesdent J, Heulin T, Achouak W (2008) Plant host habitat and root exudates shape soil bacterial community structure. ISME J 2:1221–1230

Harman G, Mastouri F (2010) The role of Trichoderma in crop management systems. Phytopathology 100:165

Hiltner L (1904) € Uber neuere Erfahrungen und Probleme auf dem Gebiet der Bodenbakteriologie unter besonderer Ber€uksichtigung der Gr€und€ungung und Brache (On recent insights and problems in the area of soil bacteriology under special consideration of the use of green manure and fallowing). Arb Dtsch Landwirt Ges 98:59–78

Ho M, Rosas J, Brown KM, Lynch JP (2005) Root architectural tradeoffs for water and phosphorus acquisition. Funct Plant Biol 32:737–748

Hodge A, Campbell CD, Fitter AH (2001) An arbuscular mycorrhizal fungus accelerates decomposition and acquires nitrogen directly from organic matter. Nature 413:297–299

Hoorman JJ 2011 The role of soil fungus (report no. SAG-14-11). Ohio State University, Columbus, Ohio, USA. http://ohioline.osu.edu/sag-fact/

Johansson JF, Paul LR, Finlay RD (2004) Microbial interactions in the mycorrhizosphere and their significance for sustainable agriculture. FEMS Microbiol Ecol 48:1–13

Johnson D, Leake JR, Ostle N, Ineson P, Read DJ (2002) In situ (CO2)–C-13 pulse-labelling of upland grassland demonstrates a rapid pathway of carbon flux from arbuscular mycorrhizal mycelia to the soil. New Phytol 153:327–334

Johnsson L, Hokeberg M, Gerhardson B (1998) Performance of the Pseudomonas chlororaphis biocontrol agent MA 342 against cereal seed-borne diseases in field experiments. Eur J Plant Pathol 104:701–711

Kaewchai S, Soytong K, Hyde KD (2009) Mycofungicides and fungal biofertilizers. Fungal Divers 38:25–50

Kapoor R, Mukerji KG (1998) Microbial interactions in mycorrhizosphere of Anethum graveolens L. Phytomorphology 48:383–389

Khan AG, Kuek C, Chaudhry TM, Khoo CS, Hayes WJ (2000) Role of plants, mycorrhizae and phytochelators in heavy metal contaminated land remediation. Chemosphere 41:197–207

Khatabi B, Molitor A, Lindermayr C, Pfiffi S, Durner J, Wettstein DV, Kogel KH, Schafer P (2012) Ethylene supports colonization of plant roots by the mutualistic fungus Piriformospora indica. PLoS One 0035502, http://www.plosone.org/article/info:doi/10.1371/journal.pone

Kim H, Park J, Choi SW, Choi KH, Lee G, Ban S, Lee C, Kim CS (2003) Isolation and characterization of Bacillus strains for biological control. J Microbiol 41(3):196–201

Knudsen IMB, Hockenhull J, Jensen DF, Gerhardson B, Hokeberg M, Tahvonen R, Teperi E, Sundheim L, Henriksen B (1997) Selection of biological control agents for controlling soil and seed-borne diseases in the field. Eur J Plant Pathol 103:775–784

Knudsen IMB, Debosz K, Hockenhull J, Jensen DF, Elmholt S (1999) Suppressiveness of organically and conventionally managed soils towards brown foot rot of barley. Appl Soil Ecol 12:61–72

Kumar M, Yadav V, Singh A, Tuteja N, Johri AK (2011) Piriformospora indica enhances plant growth by transferring phosphate. Plant Signal Behav 6:723–725

Lagopodi AL, Ram AFJ, Lamers GEM, Punt PJ, Van den Hondel C, Lugtenberg BJJ, Bloemberg GV (2002) Novel aspects of tomato root colonization and infection by Fusarium oxysporum f. sp. radicis-lycopersici revealed by confocal laser scanning microscopic analysis using the green fluorescent protein as a marker. Mol Plant-Microbe Interact 15:172–179

Lanteigne C, Gadkar V, Wallon T, Novinscak A, Filion M (2012) Production of DAPG and HCN by Pseudomonas sp. LBUM300 contributes to the biological control of bacterial canker of tomato. Phytopathology 102(10):967–973

Leeman M, DenOuden FM, VanPelt JA, Cornelissen C, MatamalaGarros A, Bakker P, Schippers B (1996) Suppression of Fusarium wilt of radish by co-inoculation of fluorescent Pseudomonas spp. and root-colonizing fungi. Eur J Plant Pathol 102:21–31

Lingua G, Bona E, Todeschini V, Cattaneo C, Marsano F, Berta G, Cavaletto M (2012) Effects of heavy metals and arbuscular mycorrhiza on the leaf proteome of a selected poplar clone: a time course analysis. PLoS One 7:4–25

Loper J, Henkels M (1999) Utilization of heterologous siderophores enhances levels of iron available to Pseudomonas putida in the rhizosphere. Appl Environ Microbiol 65:5357–5363

Lopez-Roez JA, Pozo MJ (2013) Chemical signalling in the arbuscular mycorrhizal symbiosis. In: Aroca R (ed) Progress in symbiotic endophytes. Springer, Dordrecht, pp 215–232

MacMillan J (2002) Occurrence of gibberellins in vascular plants, fungi and bacteria. J Plant Growth Regul 20:387–442

Martinez-Viveros O, Jorquera M, Crowley DE, Gajardo G, Mora ML (2010) Mechanisms and practical considerations involved in plant growth promotion by rhizobacteria. J Soil Sci Plant Nutr 10(3):293–319

Mayak S, Tirosh T, Glick BR (2004) Plant growth-promoting that confer resistance to water stress in tomatoes and peppers. Plant Sci 166:525–530

Miller S, Beed F, Harmon CL (2009) Plant disease diagnostic capabilities and networks. Annu Rev Phytopathol 47:15–38

Minerdi D, Fani R, Gallo R, Boarino A, Bonfante P (2001) Nitrogen fixation genes in an endosymbiotic Burkholderia strain. Appl Environ Microbiol 67:725–732

Mirza M, Mehnaz S, Normand P, Prigent-Combaret C, Moenne-Loccoz Y, Bally R, Malik KA (2006) Molecular characterization and PCR detection of a nitrogen-fixing Pseudomonas strain promoting rice growth. Biol Fertil Soils 43:163–170

Muthukumarasamy R, Kang U, Park KD, Jeon W, Park CY, Cho Y, Kwon S, Song J, Roh D, Revathi G (2007) Enumeration, isolation and identification of diazotrophs from Korean wetland rice varieties grown with long-term application of N and compost and their short term inoculation effect on rice plants. J Appl Microbiol 102:981–991

Naznin H, Kiyohara D, Kimura M, Miyazawa M, Shimiz M, Hyakumachi M (2014) Systemic resistance induced by volatile organic compounds emitted by plant-growth promoting fungi in Arabidopsis thaliana. PLoS One 9(1)

Newsham KK, Fitter AH, Watkinson AR (1995) Arbuscular mycorrhiza protect an annual grass from root pathogenic fungi in the field. J Ecol 83:991–1000

Niemira BA, Hammerschmidt R, Safir GR (1996) Postharvest suppression of potato dry rot (Fusarium sambucinum) in prenuclear minitubers by arbuscular mycorrhizal fungal inoculum. Am Potato J 73:509–515

Oelmuller R, Sherameti I, Tripathi S, Varma A (2009) Piriformospora indica, a cultivable root endophyte with multiple biotechnological applications. Symbiosis 49:1–17

O'Sullivan D, O'Gara F (1992) Traits of fluorescent Pseudomonas spp. involved in suppression of plant root pathogens. Microbiol Rev 56:662–676

Parihar J, Tiwari CK, Ayachi A, Verma RK (2012) Biodegradation of cellulose by wood decaying fungi. J Appl Sci Environ Sanit 7:209–214

Pereg L, McMillan M (2015) Scoping the potential uses of beneficial microorganisms for increasing productivity in cotton cropping systems. Soil Biol Biochem 80:349–358

Phillips RP, Meier IC, Bernhardt ES, Grandy S, Wickings K, Finzi AC (2012) Roots and fungi accelerate carbon and nitrogen cycling in forests exposed to elevated CO2. Ecol Lett 15:1042–1049

Pierik R, Tholen D, Poorter H, Visser E, Voesenek LA (2006) The Janus factor of ethylene: growth inhibition and stimulation. Trends Plant Sci 11:176–183

Pozo MJ, Verhage A, García-Andrade J, García JM, Azcón-Aguilar C (2009) Priming plant defences against pathogens by arbuscular mycorrhizal fungi. In: Azcón-Aguilar C, Barea JM, Gianinazzi S, Gianinazzi-Pearson V (eds) Mycorrhizas: functional processes and ecological impact. Springer, Heidelberg, pp 137–149

Puppi G, Azcon R, H€oflich G (1994) Management of positive interactions of arbuscular mycorrhizal fungi with essential groups of soil microorganisms. In: Gianinazzi S, Schuepp H (eds) Impact of arbuscular mycorrhizas on sustainable agriculture and natural ecosystems. Birkhauser Verlag, Basel, pp 201–215

Rambelli A (1973) The rhizosphere of mycorrhizae. In: Marks GL, Koslowski TT (eds) Ectomycorrhizae. Academic, New York, pp 299–343

Ramos-Zapata JA, Marrufo-Zapata D, Guadarrama P, Carrillo-Sánchez L, Hernández Cuevas L, Caamal-Maldonado A (2012) Impact of weed control on arbuscular mycorrhizal fungi in a tropical agroecosystem: a long term experiment. Mycorrhiza 22:653–661

Read J, Perez-Moreno J (2003) Mycorrhizas and nutrient cycling in ecosystems-a journey towards relevance? New Phytol 157:475–492

Reeve J, Schadt C, Carpenter-Boggs L, Kang S, Zhou J, Reganold JP (2010) Effects of soil type and farm management on soil ecological functional genes and microbial activities. ISME J 4:1099–1107

Rhodes LH, Gerdemann JW (1975) Phosphate uptake zones of mycorrhizal and non-mycorrhizal onions. New Phytol 75:555–561

Ribaudo C, Krumpholz E, Cassan F, Bottini R, Cantore M, Cura JA (2006) Azospirillum sp. promotes root hair development in tomato plants through a mechanism that involves ethylene. J Plant Growth Regul 25:175–185

Richardson A, Simpson R (2011) Soil microorganisms mediating phosphorus availability. Plant Physiol 156:989–996

Richardson A, Barea J, McNeill A, Prigent-Combaret C (2009) Acquisition of phosphorus and nitrogen in the rhizosphere and plant growth promotion by microorganisms. Plant Soil 321:305–339

Riefler M, Novak O, Strnad M, Schmulling T (2006) Arabidopsis cytokinin receptor mutants reveal functions in shoot growth, leaf senescence, seed size, germination, root development and cytokinin metabolism. Plant Cell 18(1):40–54

Rodriguez RJ, Henson J, Volkenburgh EV, Hoy M, Wright L, Beckwith F, Kim YO, Redman RS (2008) Stress tolerance in plants via habitat-adapted symbiosis. ISME J 2:404–416

Schussler A, Schwarzott D, Walker C (2001) A new fungal phylum, the Glomeromycota: phylogeny and evolution. Mycol Res 105:1413–1421

Serfling A, Wirsel SG, Lind V, Deising HB (2007) Performance of the biocontrol fungus Piriformospora indica on wheat under greenhouse and field conditions. Phytopathology 97:523–531

Sharma A, Johri B (2003) Growth promoting influence of siderophore-producing Pseudomonas strains GRP3A and PRS9 in maize (Zea mays L.) under iron limiting conditions. Microbiol Res 158:243–248

Siddiqui I, Shaukat S, Sheikh IH, Khan A (2006) Role of cyanide production by Pseudomonas fluorescens CHAO in the suppression of root-knot nematode, Meloidogyne javanica in tomato. World J Microbiol Biotechnol 22:641–650

Singh CS (1992) Mass inoculum production of vesicular-arbuscular (VA) mycorrhizae. 2. Impact of N2-fixing and P-solubilizing bacterial inoculation on VA-mycorrhiza. Zentralblatt F€ur Mikrobiologie 147:503–508

Singh LP, Gill SS, Tuteja N (2011) Unraveling the role of fungal symbionts in plant abiotic stress tolerance. Plant Signal Behav 6:175–191

Singh NK, Chaudhary FK, Patel DB (2013) Effectiveness of Azotobacter bioinoculant for wheat grown under dryland conditions. J Environ Biol 34:927–932

Smith SE, Read DJ (1997) Mycorrhizal symbiosis. Academic, San Diego, 605 pp

Spaepen S, Vanderleyden J, Remans R (2006) Indole-3-acetic acid in microbial and microorganism-plant signalling. FEMS Microbiol Rev 31:425–448

St-Arnaud M, Hamel C, Vimard B, Caron M, Fortin JA (1997) Inhibition of Fusarium oxysporum f. sp. dianthi in the non-VAM species Dianthus caryophyllus by co-culture with Tagetes patula companion plants colonized by Glomus intraradices. Can J Bot 75:998–1005

Suman A, Gaur A, Shrivastava A, Yadav RL (2005) Improving sugarcane growth and nutrient uptake by inoculating Gluconacetobacter diazotrophicus. Plant Growth Regul 47:155–162

Tsavkelova E, Klimova S, Cherdyntseva A, Netrusov I (2006) Microbial producers of plant growth stimulators and their practical use: a review. Appl Biochem Microbiol 42:117–126

Tsimilli-Michael M, Strasser RJ (2013) Biophysical phenomics: evaluation of the impact of mycorrhization with Piriformospora indica. Soil Biol 33:173–190

Vandenkoornhuyse P, Husband R, Daniell TJ, Watson IJ, Duck JM, Fitter AH, Young JPW (2002) Arbuscular mycorrhizal community composition associated with two plant species in a grassland ecosystem. Mol Ecol 11:1555–1564

Verma S, Varma A, Rexer KH, Hassel A, Kost G, Sarbhoy A, Bisen P, Bütehorn B, Franken P (1998) Piriformospora indica, gen. et sp. nov., a new root-colonizing fungus. Mycologia 90:896–903

Voisard C, Keel C, Haas D, Defago G (1989) Cyanide production by Pseudomonas fluorescens helps suppress black root rot of tobacco under gnotobiotic conditions. EMBO J 8(2):351–358

Wakelin S, Warren R, Harvey P, Ryder M (2004) Phosphate solubilization by Penicillium spp. closely associated with wheat roots. Biol Fertil Soils 40:36–43

Wells JM, Boddy L, Donnelly DP (1998) Wood decay and phosphorus translocation by the cord forming basidiomycete Phanerochaete velutina: the significance of local nutrient supply. New Phytol 138:607–617

Werner T, Motyka V, Laucou V, Smets R, Van Onckelen H, Schmulling T (2003) Cytokinin-deficient transgenic Arabidopsis plants show multiple developmental alterations indicating opposite functions of cytokinins in the regulation of shoot and root meristem activity. Plant Cell 15:2532–2550

Whipps J (2001) Microbial interactions and biocontrol in the rhizosphere. J Exp Bot 52:487–511

Wijesinghe D, John E, Beurskens S, Hutchings M (2001) Root system size and precision in nutrient foraging: responses to spatial patterns of nutrient supply in six herbaceous species. J Ecol 89:972–983

Wu QS, Xia RX (2006) Arbuscular mycorrhizal fungi influence growth, osmotic adjustment and photosynthesis of citrus under well-watered and water stress conditions. J Plant Physiol 163:417–425

Xian-Can Z, Feng-Bin S, Hong-Wen X (2010) Arbuscular mycorrhizae improves low temperature stress in maize via alterations in host water status and photosynthesis. Plant Soil 331:129–137

Yaseen T, Burni T, Hussain F (2011) Effect of Arbuscular mycorrhizal inoculation on nutrient uptake, growth and productivity of cowpea (Vigna unguiculata) varieties. Afr J Biotechnol 10:8593–8598

Significance of Arbuscular Mycorrhizal Fungi and Rhizosphere Microflora in Plant Growth and Nutrition

19

Hindumathi Amballa and Narasimha Reddy Bhumi

Abstract

Arbuscular mycorrhizal fungi are common components of soil microorganisms inhabiting the rhizosphere. The rhizosphere is a dynamic microhabitat where microorganisms, plant roots and soil constituents interact with constituting root-soil interface. The rhizosphere of mycorrhizal plants – the 'mycorrhizosphere' – harbours a wide range of microbial activities responsible for several ecosystem processes. Arbuscular mycorrhizal fungi interact with microorganisms colonizing the rhizosphere. The microbial interactions in the mycorrhizosphere are the primary determinants of plant health and soil quality. This chapter summarizes various microbial interactions between mycorrhizal fungi and other soil microbial communities. This chapter discusses (1) microbial communities in the soil, (2) arbuscular mycorrhizal fungal interaction with plants, (3) interaction with rhizosphere microorganisms, (4) interaction with soilborne pathogens, (5) potential benefits of arbuscular mycorrhizal fungi in plant growth and disease control and (6) effect of soil microorganisms on mycorrhizal symbiosis. The main conclusion is that the microbial population interactions with arbuscular mycorrhizal fungi in the rhizosphere majorly influence plant health, crop productivity and soil fertility. Arbuscular mycorrhizal fungi in corporation with other rhizosphere microbial organisms can contribute to improve plant growth and nutrition.

19.1 List of Abbreviations

A	*Acaulospora*
AM	Arbuscular mycorrhiza

H. Amballa • N.R. Bhumi (✉)
Mycology and Plant Pathology Laboratory, Department of Botany, Osmania University, Hyderabad, Telangana 500007, India
e-mail: reddybn1@yahoo.com

D.K. Choudhary et al. (eds.), *Plant-Microbe Interaction: An Approach to Sustainable Agriculture*, DOI 10.1007/978-981-10-2854-0_19

AMF	Arbuscular mycorrhizal fungi
AMB	Bacteria associated with arbuscular mycorrhizal fungi
AV	Auxiliary cells
BLO	Bacteria-like organisms
Br	Bromine
C	Carbon
C-source	Carbon source
Ca	Calcium
CBE	Chlorazol black E
cfu	Colony-forming units
Cl	Chlorine
CO_2	Carbon dioxide
Cu	Copper
DNA	Deoxyribonucleic acid
ERH	Extraradical hyphae
F	*Funneliformis*
Fe	Iron
G	*Glomus*
Gig.	*Gigaspora*
GM	Genetically modified
rDNA	Ribosomal deoxyribonucleic acid
H	Soil hyphae
µm	Micrometre
Mg	Magnesium
Mn	Manganese
K	Potassium
MHB	Mycorrhizal-helper bacteria
N	Nitrogen (Elemental Nitrogen)
N_2	Nitrogen (Molecular Nitrogen)
NH_4^+	Ammonium ion
Ni	Nickel
NO_3^-	Nitrate ion
O_2	Oxygen
P	*Pseudomonas*
Pb	Lead
PGPR	Plant growth-promoting rhizobacteria
pH	Hydrogen ion concentration
PR	Pathogenesis-related proteins
PSB	Phosphate-solubilizing bacteria
S	Sulphur
SAR	Systemic acquired resistance
S	*Scutellospora*
Spp.	Species
T	*Trichoderma*
VAM	Vesicular arbuscular mycorrhizal fungi

WT	Wild type
Zn	Zinc
^{15}N	Isotope of nitrogen with atomic mass 15
^{31}P	Isotope of phosphorus with atomic mass 31
^{32}P	Isotope of phosphorus with atomic mass 32

19.2 Introduction

Soil is a dynamic medium and supports different microbial communities such as bacteria, fungi, actinomycetes, algae, viroids, viruses, protozoans, nematodes, etc., which play a vital role in maintaining soil fertility, cycling of nutrient elements in the biosphere, humus formation, biological conversions, geochemical cycling, ecosystem sustenance, etc., besides supporting plant life and plant productivity. Plant-microbe interactions are an integral part of our terrestrial ecosystem that contribute to sustainable agriculture. The different interactions of rhizosphere microorganisms with each other and with plants influence plant health and soil quality.

The plant roots grow into the soil creating plant-root interface named 'rhizosphere', the term first coined by Hiltner (1904). The rhizosphere probably represents a unique habitat on the Earth. In particular, the major factor that determines the availability of nutrients to the plants in the rhizosphere is the microbial activity that has significant influence on plant growth as well as soil health and productivity. It is very important to understand the basic principles of rhizosphere microbial ecology, viz. the function and diversity of the microorganisms that reside there, before soil microbial technologies can be applied.

19.3 The Rhizosphere

The rhizosphere is best defined as the volume of soil in close proximity to roots characterized by high microbial populations of active microorganisms than the soil away from the plant roots (Hiltner 1904). The rhizosphere is under continuous influence of living roots and the microbial activities in such microsite makes rhizosphere the most dynamic habitat for soil microorganisms on Earth. It differs from the bulk soil in its physicochemical characteristics such as low pH, low water potential, low partial pressure of O_2 and higher concentrations of soluble carbohydrates due to root exudates.

The rhizosphere is known to be a hot spot of microbial activities as the plant roots influence microbial communities by depositing photosynthates (amino acids, low molecular weight organic acids or phytosiderophores) into the rhizosphere (rhizodeposition). The root activities can also modify soil physicochemical properties. Rhizodeposition is influenced by plant and soil biotic and abiotic factors. Some plant biotic and abiotic factors that influence rhizosphere comprise mycorrhiza, root architecture, nutrient deficiency, photosynthesis, temperature, light intensity and physical disturbance, while soil biotic and abiotic factors include pathogens, biocontrol agents, root herbivores, metal toxicity, soil pH, soil texture and water availability (Jones et al. 2004).

Root exudates constitute a major part of rhizodeposition, mainly composed of flavonoids, phenolic compounds, carbohydrate monomers, organic acids and plant hormones (Lynch and Whipps 1990). Rhizodeposition also corresponds to 15–30 % of total carbon produced by plants during photosynthesis transferred towards microorganisms of the rhizosphere.

The organic materials released as root exudates act as signal molecules or growth substrates to the heterotrophic microbial communities (Werner 1998) and regulate different kinds of associations between the plant and soil microorganisms affecting microbial composition and diversity. Rhizosphere functioning is known to significantly influence plant fitness and soil quality because microbial activities in such habitat can help the host plant to adapt to stress conditions like water and mineral deficit and also soilborne plant pathogens (Lynch 1990; Bowen and Rovira 1999).

The overall influence of plant roots on soil microorganisms is termed 'rhizosphere effect'. The microbiological activity is greater in rhizosphere than in the soil away from the plant roots. The intensity of such activity depends on the distance to which exudations from the root system can diffuse. Hence, the rhizosphere microflora differ both qualitatively and quantitatively from that of bulk soil, i.e. beyond the influence of roots (Parkinson 1967). The number of microorganisms (cfu) per gram soil is greater by two- to threefolds in the rhizosphere than in the non-rhizosphere soil (Mehrag and Killham 1995). The rich nutrient supply and close contact to the living roots enable rhizosphere microorganisms to have a direct influence on plant growth and phytopathogens. The rhizosphere has been described as both a 'playground' and a 'battlefield' for beneficial microorganisms and soilborne pathogens (Raaijmakers et al. 2008).

Though it may be difficult to physically separate rhizosphere and bulk soils, they differ in inherent biological, chemical and physical characters (Barea et al. 2005). The rhizosphere is characterized by altered microbial diversity and activity.

The major soil ecological environment for plant-microbe interactions is the plant rhizosphere which involves colonization of different microorganisms in and around growing roots which result either in associative, symbiotic, neutralistic or parasitic interactions depending upon plant nutrient status, soil environment, plant defence mechanism and the type of microorganism proliferating in the rhizosphere. In response to the adhesion of microorganisms very close to the epidermis, plants secrete signal molecules for protection against invasion of the heterogeneous microbes into the root zone. Plant signal molecules such as flavonoids and flavones produced in the rhizosphere in response to microbial adhesion remain attached to plant cell walls to act as antimicrobial agents (phytoalexins). The microorganisms inhabiting the rhizosphere produce a variety of compounds that stimulate plant growth or can be antagonistic to plant pathogens. These interactions may be beneficial or detrimental. The beneficial interactions are caused by symbiotic and non-symbiotic bacteria and by a highly specialized type of fungi, the mycorrhizae. The pathogenic or detrimental interactions involve microbes such as viroids, viruses, bacteria and fungi.

19.4 Microbial Communities in the Rhizosphere

Plants live in association with a rich diversity of microorganisms during their entire development. Of the various microbial communities colonizing rhizosphere, mycorrhizal fungi, nitrogen-fixing bacteria, soilborne pathogens, free-living fungi and bacteria and antagonistic/plant growth-stimulating fungi and bacteria are commonly known to share the microhabitat.

Of the different interactions of rhizosphere microorganisms with each other and also with the plants, the beneficial plant microbial interactions in the rhizosphere are the primary determinants of plant health and soil fertility (Jeffries et al. 2003).

The most abundant and important members of the soil microbial community that develop mutually beneficial relationship with plant roots and contribute majorly to plant growth by nutrient acquisition and pathogen control are the mycorrhizal fungi. Mycorrhizal symbiosis plays a key role in the productivity and diversity of natural plant ecosystem. Mycorrhizal fungi are the relevant members of the rhizosphere, mutually symbiotic population known to carry out many critical ecosystem functions such as improvement of plant establishment, enhancement of plant nutrient uptake, plant protection against various abiotic and biotic stresses and improvement of soil structure (Smith and Read 1997).

The term 'mycorrhiza' that literally means 'fungus root' was coined by Frank (1885) to describe symbiotic relationship of plant roots with certain soilborne fungi. Mycorrhizal fungi differ from other plant-fungus associations because of their ability to create an interface for nutrient exchange which occurs within living cells of the plant (Brundrett 2002, 2004). There are different types of mycorrhizal interactions which have been broadly classified into ectomycorrhiza, endomycorrhiza and ecto-endomycorrhiza based on the presence of various extraradical or intraradical hyphal structures. Based on the type of fungus involved and the resulting structures produced by the root-fungus combination, seven different types of mycorrhizal associations have been identified. They are ectomycorrhiza, ecto-endomycorrhiza, monotropoid mycorrhiza, ericoid mycorrhiza, arbutoid mycorrhiza, orchid mycorrhiza and vesicular arbuscular mycorrhiza (VAM) (Fig 19.1). All these mycorrhizal types differ from each other by the characteristic host plant that they associate with, fungal species involved and morphology within the roots (Brundrett 2002).

Of the seven mycorrhizal types that have been identified, the most common type of endomycorrhizas are the vesicular arbuscular mycorrhizas that are most commonly found in agriculturally important crop plants and are one of the most important symbiotic associations on earth linking the root and the soil system (Koide and Mosse 2004). VAM fungal association is the oldest and probably the most abundant plant-microbe association on earth (Simon et al. 1993; Smith and Read 1997).

19.5 Arbuscular Mycorrhizal Fungi

Vesicular arbuscular mycorrhiza (VAM) is the most common type of mycorrhiza found associated with some bryophytes, pteridophytes, gymnosperms (excluding Pinaceae which have sheathing mycorrhizas) and in virtually all families of

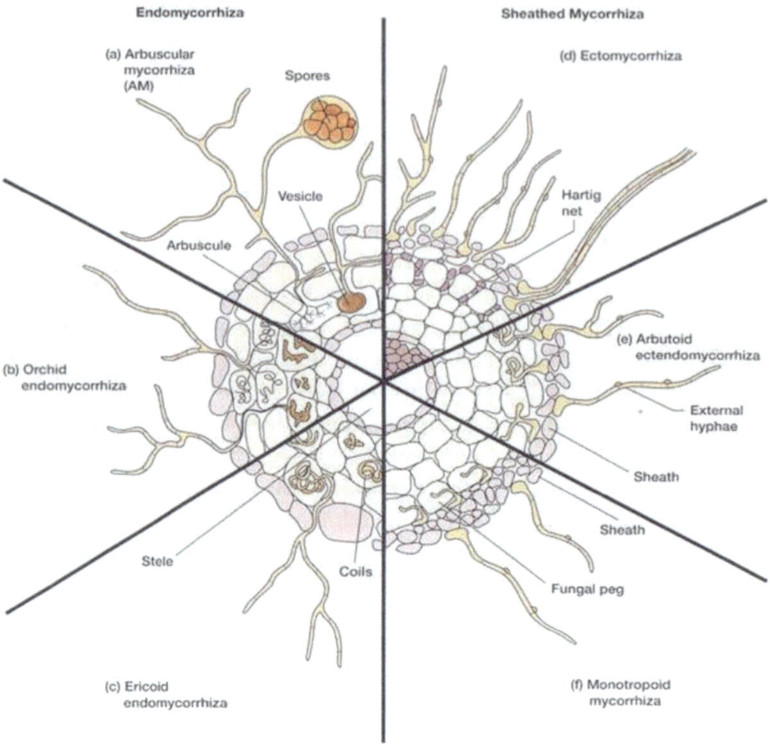

Fig. 19.1 Root cross section: illustrating different types of mycorrhizal relationships that exist within plants (Courtesy: Saved from davidmoore.org.uk)

angiosperms. They generally occur in Gramineae, Palmae, Rosaceae and Leguminosae which include many crop plants. They have a very ancient origin, dating back to early land plants (Simon et al. 1993), and they could even have been a major factor in the colonization of land.

Several aspects of plant physiology such as mineral nutrient composition, hormonal balance, C allocation patterns, etc. are responsible to be modified by the mycorrhizal establishment (Harley and Smith 1983; Smith et al. 1994). Bidirectional movement of nutrients characterizes this symbiosis where photosynthetically derived carbohydrates flow to the fungi and inorganic nutrient get transported to the plant through mycelial network, thereby providing a critical linkage between the plant root and rhizosphere. They are of ecological significance for higher plants as they govern their growth and survival. Hence, they are employed to augment the production of forest and crop ecosystem.

AM fungi are widely distributed being abundant in phosphorus- and mineral-deficient soils. AM fungi are characterized by the formation of arbuscules and vesicles inside the cortex cells. Their presence led to the former common name vesicular arbuscular mycorrhizal (VAM) fungi, but the term arbuscular mycorrhiza (AM) is now preferred (Friberg 2001) because not all fungi produce vesicles. But there is some disagreement about the two terms as arbuscules are not always present in the mycorrhizal roots. Vesicles are not formed by the genera belonging to the order Gigasporales but are found in the other genera of the Glomeromycota (Isaac 1992).

The arbuscular mycorrhizal (AM) fungi are the most common obligate symbiotic fungi, belonging to phylum Glomeromycota (Schüßler et al. 2001). This association is geographically ubiquitous, occurring in arctic, temperate and tropical regions over a broad ecological range from aquatic to desert environment (Gerdemann 1975). From fossil records of Ordovician age, the evolution of symbiotic fungi was thought to have existed at least 470 million years ago.

19.6 Taxonomy

Traditionally the taxonomy of AM fungi was largely based on the morphological and anatomical characteristics of their spores and sporocarps, spore germination and the method of spore formation on the hypha (Morton 1988). The spores are relatively large (40–800 μm) with layered walls and lipids in their cytoplasm. Spores are important for identification of AM fungi. AM fungal species isolated from the rhizosphere soils of safflower were identified based on their morphological characters are represented in Figs. 19.2 and 19.3. Among them *Acaulospora alpina, A. myriocarpa, G. australe, G. diaphanum, G. heterosporum, G. manihotis, G. microaggregatum, G. multicaule* and *Gigaspora rami sporopora* are the first reports from Telangana state (Hindumathi and Reddy 2016a).

Now, several modern methods like serology, isozyme variation by electrophoresis (Hepper et al. 1988), fatty acid variation (Bentivenga and Morton 1994) and molecular techniques such as DNA-based methods (Helgason et al. 1999; Schüßler et al. 2001) have aided in a clearer phylogenetic analysis that was possible using morphological and microscopic identification.

19.7 Classification

In earlier systems of classification, the AM fungi were placed in the order Glomales within the division Zygomycota. They have nonseptate hyphae, a similar characteristic to that found in hyphae of most Zygomycota. However, AM fungi are distinguished from the zygomycotan lineages due to some spore characteristics, e.g. mutualistic symbiotic nutrient habit and lack of formation of characteristic zygospores. The rDNA analysis exposed a clear separation of AM fungi from other fungal groups, and the AM fungi have been elevated and now placed in a separate new phylum Glomeromycota (Schüßler et al. 2001). Three glomeromycotean classes (Archaeosporomycetes, Glomeromycetes and Paraglomeromycetes), five orders (Archaeosporales, Diversisporales, Gigasporales, Glomerales, Paraglomerales), 14 families, 29 genera and approximately 230 species have been recognized (Table 19.1). The classification of arbuscular mycorrhizal fungi (Oehl et al. 2011) up to genus level is presented in Table 19.1.

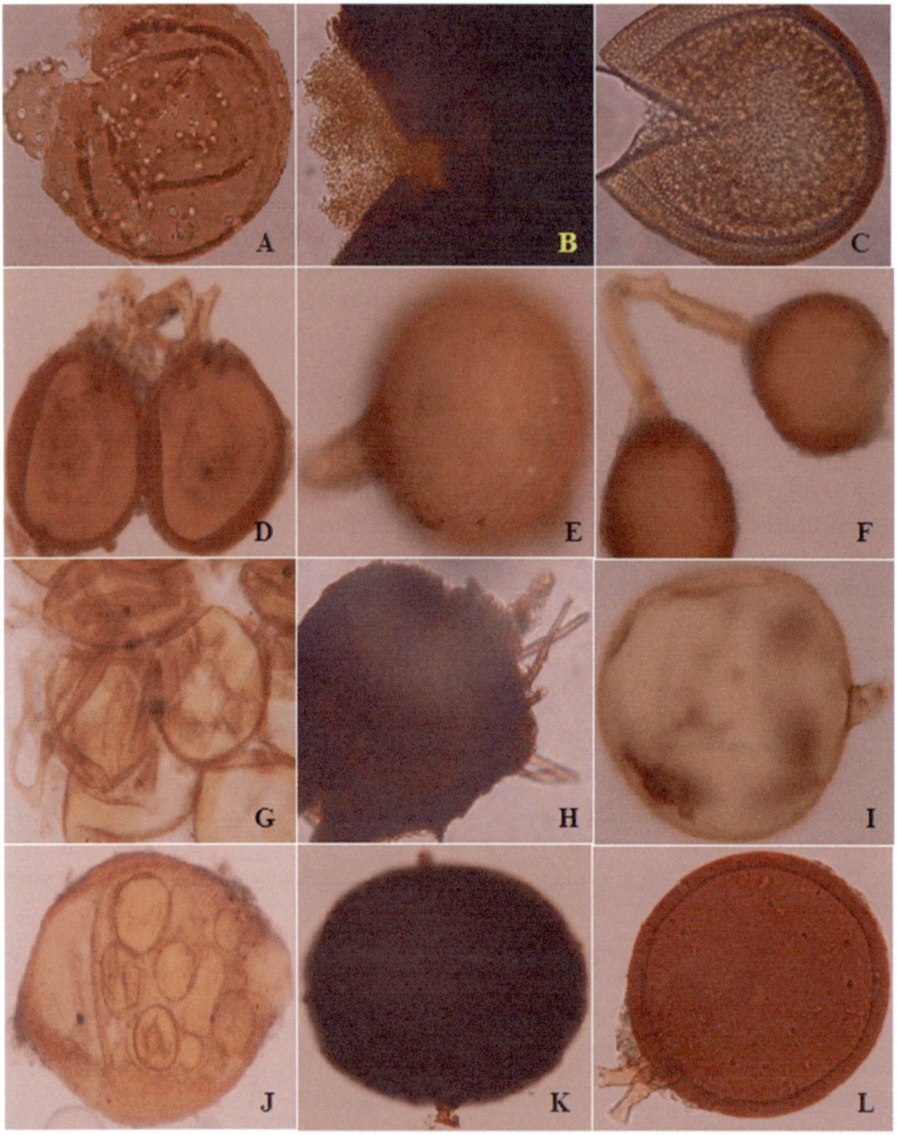

Fig. 19.2 (**A**) *Acaulospora alpina*, (**B**) *A. myriocarpa*, (**C**) *A. scrobiculata*, (**D**) *G. ambisporum*, (**E**) *G. australe*, (**F**) *G. diaphanum*, (**G**) *G. fasciculatum*, (**H**) *G. heterosporum*, (**I**) *G. manihotis*, (**J**) *G. microaggregatum*, (**K**) *G. multicaule*, (**L**) *G. multisubstensum*

The distribution and occurrence of AM fungi differ both qualitatively as well as quantitatively with changes in seasonal variation, climatic and edaphic factors and type of soil vegetation. Physicochemical factors and microbiological components of

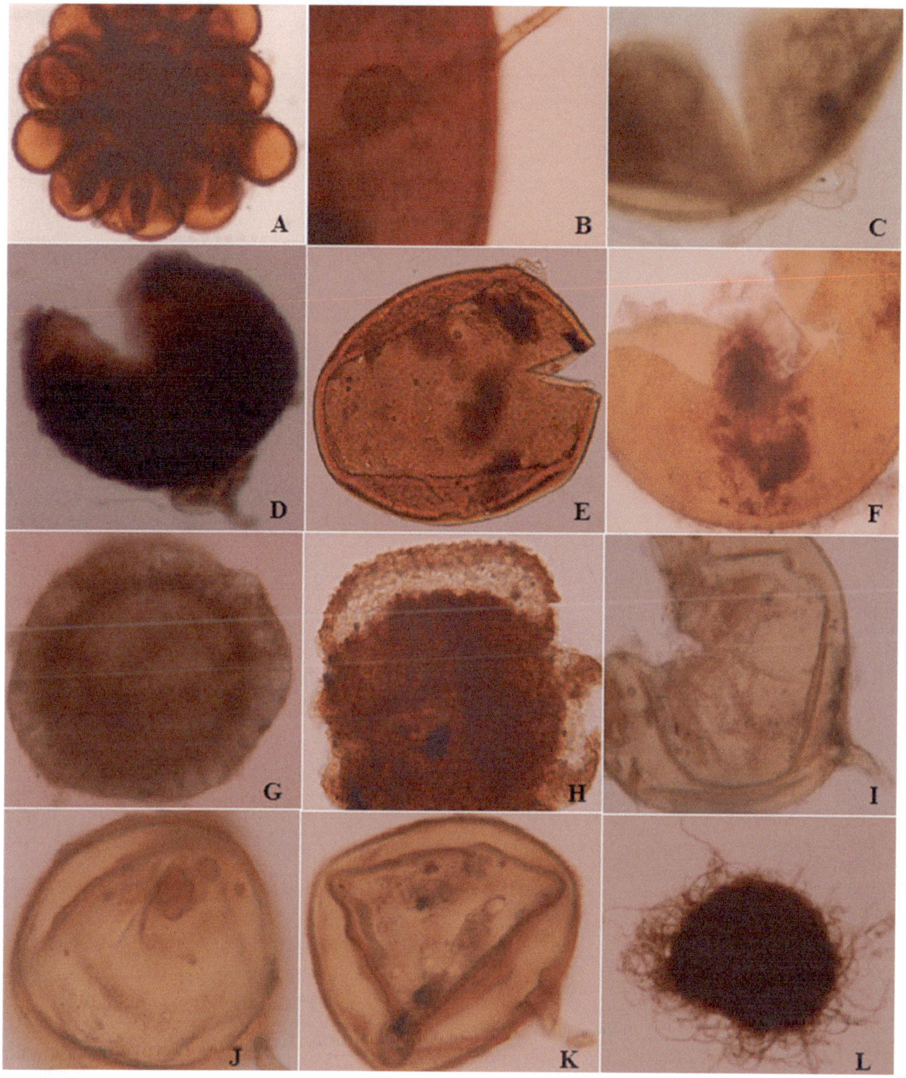

Fig. 19.3 (**A**) *G. rubiforme*, (**B**) *Gigaspora gigantea*, (**C**) *Gig. rami sporopora*, (**D**) *Scutellospora nigra*, (**E**) *S. pellucida*, (**F**) *Scutellospora* sp., (**G**) *Entrophospora schenckii*, (**H**) *Entrophospora* sp., (**I**) *Funneliformis caledonius*, (**J**) *F. geosporum*, (**K**) *F. mosseae*, (**L**) Unidentified genus/sp1

the soil are reported to play significant role in the distribution, density, composition and activity of AM fungi (Nicolson 1959; Bagyaraj et al. 1979b; Mukerji et al. 1982; Manjunath and Bagyaraj 1984; Hindumathi 1999; Sreevani and Reddy 2005; Reddy et al. 2006a, b, 2007; Hindumathi and Reddy 2011a, b, 2012b, 2015, 2016a; Satya Vani 2012; Satya Vani et al. 2014a).

Table 19.1 Classification of Glomeromycota up to genus level
Phylum: Glomeromycota

Class	Order	Family	Genus
Glomeromycetes	Glomerales Morton and Benny	Glomeraceae Pirozynski and Dalpe	*Glomus* Tulasne and Tulasne
			Funneliformis
			Septoglomus
			Simiglomus
		Entrophosporaceae	*Claroideoglomus*
			Albahyphae
			Viscospora
			Entrophospora Ames and Schneider
	Diversisporales	Diversisporaceae	*Diversispora* Walker & Schüßler
			Redeckera
			Otospora
			Tricospora
		Sacculosporaceae	*Sacculospora*
		Pacisporaceae	*Pacispora* Oehl & Sieverding
		Acaulosporaceae Morton & Benny	*Kuklospora*
			Acaulospora (Gerdemann and Trappe) Berch
	Gigasporales	Scutellosporaceae	*Orbispora*
			Scutellospora Walker and Sanders
		Dentiscutataceae	*Fuscutata*
			Dentiscutata
			Quatunica
		Racocetraceae	*Cetraspora*
			Racocetra
		Gigasporaceae Morton & Benny	*Gigaspora* (Gerd. & Trappe) Walker & Sanders
Archaeosporomycetes	Archaeosporales	Ambisporaceae	*Ambispora* Walker, Vestberg & Schussler
		Archaeosporaceae	*Archaeospora* Morton and Redecker
			Intraspora
		Geosiphonaceae	*Geosiphon*
Paraglomeromycetes	Paraglomerales	Paraglomeraceae	*Paraglomus* Morton and Redecker

Phylum: Glomeromycota

19.8 Interaction of Arbuscular Mycorrhizal Fungi with Plants

Arbuscular mycorrhizal fungal associations are established in the presence of a variety of microorganisms, and some of these microbes interact in rather specific ways to influence this association and its effect on plant growth. About 80 % of plant families of land plants and majority of agricultural crops are estimated to be colonized by AM fungi. The remaining plant species are either non-mycorrhizal or non-hosts of AM fungi. Plant species belonging to the Cruciferae and Chenopodiaceae are not known to form AM fungal symbiosis (Smith and Read 1997). Giovannetti and Sbrana (1998) suggested that this is due to the lack of any recognition event leading to the establishment of a functional symbiosis. The inability of these plants to support mycorrhizas may also be related to accumulation of chemicals like alkaloids, cyanogenic glucosinolates and antifungal compounds in the root cortical tissues or in root exudates (Brundrett 2002). The degree of host specificity could be under the genetic control of the host, the AM fungus or more likely a complex interaction of both symbionts with the soil environment (Sylvia et al. 2003).

19.9 Arbuscular Mycorrhizal Fungal Root Colonization

The spores of most species do not require host factors for germination and initiation of the hyphal growth, but continuous hyphal growth, differentiation into infection structures and penetration into the host are reported to be affected by plant signals (Bécard and Piché 1989). Three major parameters such as specificity, infectivity and effectivity determine root colonization. The process and rate of colonization determine the effectiveness of an AM fungus or a mycorrhizal association.

As the infection spreads within the root cortical cells of the host, extraradical hyphae grow out into the soil, play an important role in nutrient acquisition and, furthermore, form a source of secondary colonization (Harley and Smith 1983).

The other important structures involved in the colonization of roots are spores and extraradical auxiliary bodies (Fig. 19.4) produced in the soil and unique structures such as hyphae, arbuscules and vesicles produced inside the roots (Fig. 19.5). Arbuscules are dichotomously branched intracellular structures and are considered as the major sites of carbon needed for energy and nutrient exchange between the fungus and host plant. Vesicles are storage organs and store phosphorus as phospholipids and sometimes help in vegetative reproduction. The establishment, development, survival and performance of AM fungi are affected by soil fertility, cropping patterns, environmental factors and host-plant genotype.

The extraradical hyphae (ERH) also known as soil hyphae or external hyphae associated with the root radiate out into the soil. These hyphae are distinguished as thin highly branched 'absorptive' hyphae responsible for nutrient acquisition (Friese and Allen 1991) and thick 'runner' or 'distributive' hyphae (infective hyphae) running towards and along the root surface to establish new entry points. Maximum root colonization and sporulation is most prevalent in soils of low fertility. External

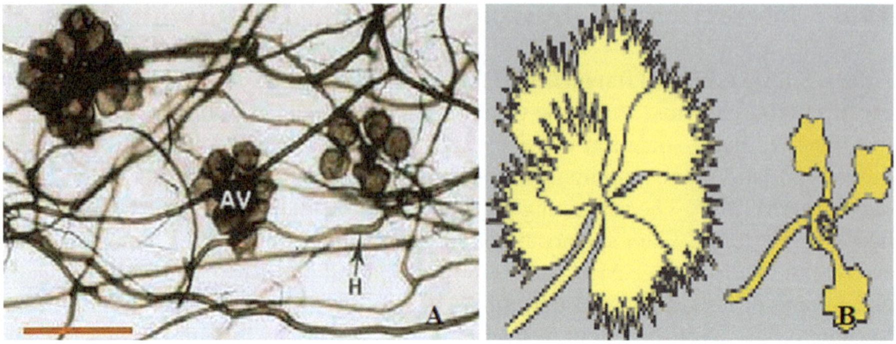

Fig. 19.4 (**A**) Darkly pigmented soil hyphae (H) of a *Scutellospora* species with auxiliary cells (AV), (**B**) auxiliary bodies on soil hyphae (Source Brundrett et al. 1996)

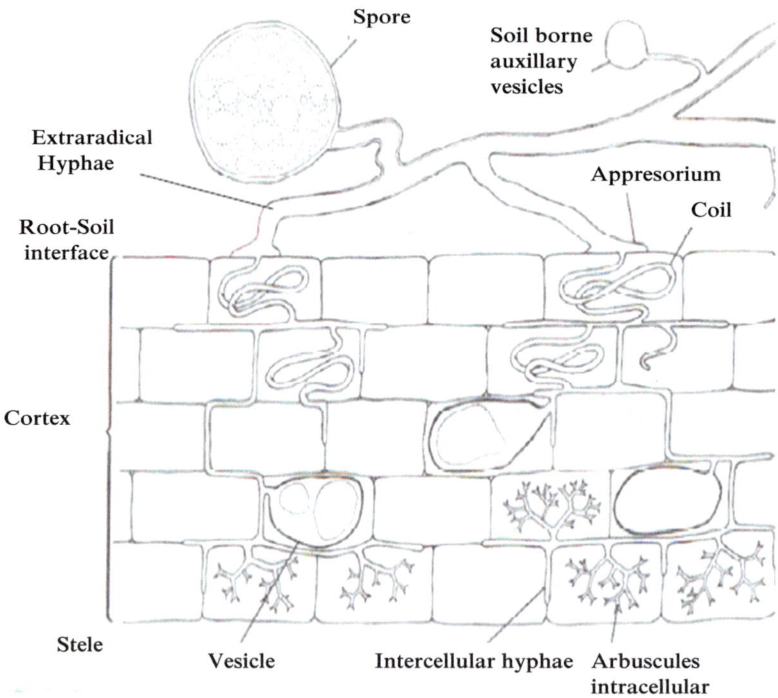

Fig. 19.5 Diagrammatic representation of the characteristic structures of arbuscular mycorrhizal fungi as identified in the cortical cell of a plant host when viewed under a microscope (Source Isaac 1992)

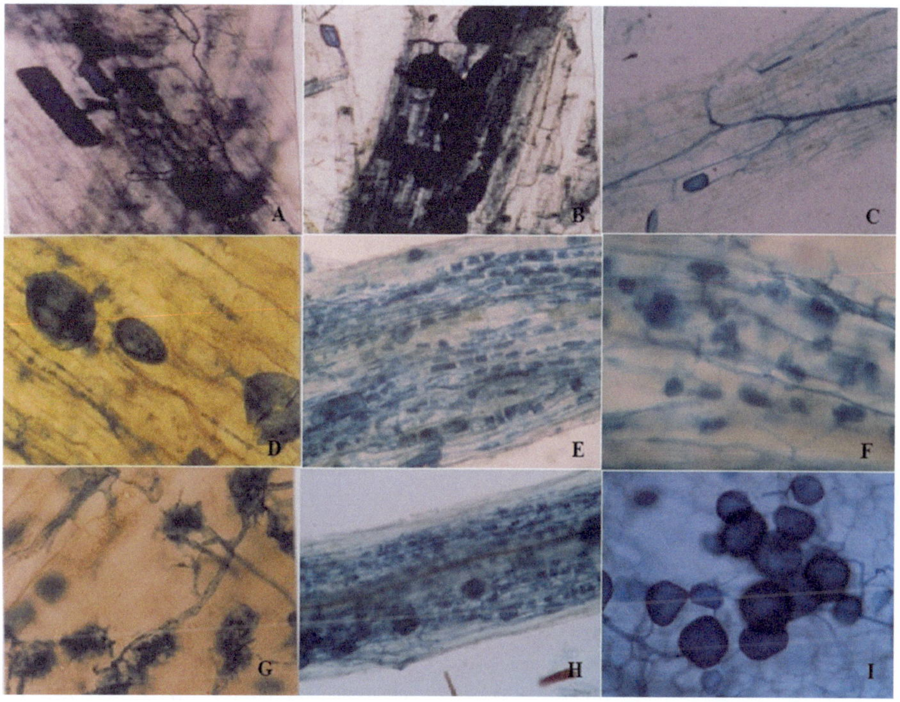

Fig. 19.6 (**A–D**, **H**) Colonization of root cortical cells of sorghum by arbuscular mycorrhizal fungi showing mycelium and different shaped vesicles in agricultural field conditions; (**E–H**) mycelium and arbuscules; (**I**) sporocarp in root tissue (Source Hindumathi and Reddy 2011c)

input of mineral N or P to the soil decreased mycorrhizal development in several legumes (Abbott and Robson 1977) and nonlegumes (Krishna and Bagyaraj 1982).

Plant growth hormones like auxins and other compounds such as flavonoids, phenolics and carbon dioxide are known to play an important role in spore germination, development, proliferation and stimulation of hyphal growth and mycorrhizal colonization. The association/colonization of AM fungi is usually not detected by the naked eye as there are no external or morphological root changes. Root colonization structures (Fig. 19.6) are visible only when they are cleared, stained and examined under the microscope following the most commonly and frequently adopted method (Phillips and Hayman 1970). Detection and quantification of AM fungal colonization in the roots is very essential for mycorrhizal research. A range of light microscopy-based, biochemical and molecular techniques were also used for identification and/or quantification of AM fungi in roots. Nonvital staining with various stains such as trypan blue, cotton blue, aniline blue, ink, vinegar and chlorazol black E (CBE) are some of the methods used to visualize AM fungi in roots (Vierheilig et al. 2005).

19.10 Mycorrhizal Dependency

In natural ecosystems, plants have varying degrees of dependence on mycorrhizal associations, based on the availability of nutrients in the soil in which they naturally occur. Mycorrhizal dependency is a measure of the benefit provided by mycorrhizae and depends on relative contribution of root and mycorrhizal-mediated nutrient uptake to plants. AM fungi are not host specific because many species have been shown to colonize a wide range of hosts, and the same plant root may be colonized by a mixture of AM fungal species (Helgason et al. 1999; Klironomos 2000). Vandenkoornhuyse et al. (2002) demonstrated that distinct AM fungal communities are associated with different plant hosts, and Van der Heijden et al. (1998) reported plants might select AM fungal species. The magnitude of response to root colonization varied with different cultivars of sorghum (Hindumathi and Reddy 2011c).

19.11 Mycorrhizosphere

The mycorrhizal symbiotic status of the plant changes the chemical composition of root exudates and modifies the root functions and microbial communities. This zone of soil influenced by combined activities of the root and mycorrhizal fungus is termed as 'mycorrhizosphere' (Linderman 1992, 2000; Barea 1997, 2000; Gryndler 2000). The fungal soil mycelium serves as carbon source to the rhizosphere microbial communities. These changes therefore affect the microbial communities in the rhizosphere of mycorrhizal plants both qualitatively and quantitatively producing the 'mycorrhizosphere effect' (Linderman 1988). As the mycorrhizal fungus uses the root exudates and modifies root functions, microbial communities in the mycorrhizosphere differ from those in the rhizosphere soil (Garbaye 1991; Barea et al. 2002c).

Mycorrhizosphere has two components: one is the layer surrounding the mycorrhizal roots (colonized by AM fungi) referred to as the 'mycorrhizosphere', and the other is the layer of the soil surrounding the hyphae of AM fungi referred to as 'hyphosphere' (Marschner 1995), 'hyphorhizosphere' (Klynchnikov and Kozherin 1990) or 'mycosphere' (Gilbert and Linderman 1971). Microbial interactions in the rhizosphere of mycorrhizal plants improve plant fitness and soil quality, the critical issues for a sustainable agricultural development and ecosystem functioning (Barea et al. 2002b).

19.12 Potential Benefits of AM Fungi

The beneficial effects of AM fungal inoculation on plant growth and yield promotion have generally been attributed to improved nutrition uptake, mobilization of nutrients (Abbott and Robson 1977), production of enzymes (Tarafdar and Claassen 1988) and plant protection against pathogen infection (Dehne 1982). AM fungal partner has been shown to acquire mineral nutrients from the soil, especially immobile elements such as P, Zn and Cu, and also more mobile ions such as S, Ca, K, Fe, Ni, Pb, Mg, Mn, Cl, Br and N (Tinker 1984). In the soils where such elements may be deficient or otherwise less available, mycorrhizal fungi increase efficiency of mineral uptake, resulting in enhanced plant growth.

The fungal hyphae in ERH phase extending into the soil serve as extensions of the root systems, which are both physiologically and geometrically more effective in the absorption of nutrients, especially phosphorus (P), than the roots themselves, and translocate to the root. AM fungal hyphae are not only structurally efficient in extraction of nutrients from exchange sites in soil, they also produce exogenous enzymes such as phosphatases, phytases and nitrate reductase, which are important in uptake and metabolism of nutrients (Ho and Trappe 1980).

The absorbed 'P' is probably converted into polyphosphate granules in the external hyphae (Callow et al. 1978) and passed to the arbuscules for transfer to the host. This flow of phosphates is known to occur in the presence of acid phosphatases (Gianinazzi et al. 1979) during arbuscule life span or senescence.

Exogenous enzymes, like phosphatases produced by AM fungal extraradical hyphae, hydrolyse unavailable sources of P and release P from organic P complexes and facilitate absorption of P especially under humid tropical conditions (Koide and Kabir 2000; Carlile et al. 2001).

AM fungal extraradical hyphae obtained nitrogen in different forms such as amino acids, peptides, ions (No_3^- or NH_4^+) and recalcitrant organic nitrogen forms (Hawkins et al. 2000; Giri and Mukerji 2004).

AM fungal influence on plant N nutrition is not as high as 'P', but they give their host access to different forms of N, thereby increasing plant N uptake (Hodge et al. 2001). Hodge et al. (2001) demonstrated that the ability of AM fungi to decompose organic matter and acquire N from organic source. They also found that AM fungi increased N diffusion rate into its host. Hence, mycorrhizal plants have additional access to N sources compared to non-mycorrhizal plants. AM fungal extraradical hyphae were reported to absorb inorganic N and transfer it to intraradical hyphae as amino acids (arginine). They have also evidenced intraradical hyphae decompose the amino acids to access the C and then transfer the remaining N as ammonium to the host plant.

ERH or external fungal mycelial network extend beyond the nutrient depletion zone for enhanced nutrient acquisition and water uptake and provide to the plant. In addition, fungal hyphae are better adapted to explore patch nutrients through rapid proliferation and competitive ability with soil microbe (Smith and Read 1997).

AM fungi also play an important role in the water economy of plants. Their association improves hydraulic conductivity of roots which contributes towards better uptake of water by the plants. It has been suggested that mycorrhizal fungi help the plants in better absorption of water by the roots by exploiting in wider zones of soil and result in better performance (Safir et al. 1971; Kehri and Chandra 1990). It has been demonstrated that in extremely dry conditions, mycorrhizal plants showed a better survival over non-mycorrhizal ones. The most established benefit from AM fungus to the host plant is due to the widespread mycelial network that penetrates deeper and wider in the soil in search of water and nutrients, thereby widening the zone of activity.

AM fungal effects extend to production of phytohormones such as gibberellins, cytokinins, ethylene and other growth hormones suggesting that mycorrhizae play a key role by influencing regulatory systems in plants.

Other major benefits include improved tolerance to drought and salinity (Augé 2004; Augé et al. 2015), high soil temperatures, adverse soil pH, heavy metal toxicity alleviation (Lingua et al. 2008; Meier et al. 2015), toxicities related to mine spoils or landfills, toxicities due to minor element imbalance such as Mn, increased uptake of macronutrients (N, K and Mg) other than P as well as uptake of micronutrients and overcoming transplantation shock compared to non-mycorrhizal plants.

It was demonstrated that AM fungal inoculation increased mineral nutrient uptake with consequent increase in plant growth and seed yield over control plants (Bagyaraj et al. 1979a; Hindumathi and Reddy 2012a; Satya Vani 2012; Satya Vani et al. 2015).

Mycorrhizal symbiosis plays an important role in the tropical agriculture because the soils are phosphorus deficient and P fixing. The soil phosphate (P) availability is the most limiting factor in legume growth and biological N_2 fixation, and AM fungal symbiosis with legume can overcome this limitation. Nodules require relatively large amounts of P indicating high demand for P by the nodules. It has been suggested that P level influences not only mycorrhizal infection frequency but also process of nodulation in legume species since legumes are poor competitors for soil phosphates. It was demonstrated that mycorrhizal nodulated plants exhibited higher levels of nitrogenase and nitrate reductase activity compared to non-mycorrhizal plants (Carling et al. 1978).

Inoculation of AM fungi in legume-*Rhizobium* symbiosis in the presence of the pathogen resulted in better nodulation, biological N_2 fixation, enhanced plant growth and nutrition and biological control of root rot pathogens and increased soil nitrogen content (Bagyaraj et al. 1979a; Hindumathi and Reddy 2012a; Hindumathi et al. 2016b). Experimental evidence showed that dual inoculation of AM fungi with *Rhizobium* strain in legume tripartite (AM + *Rhizobium* + legume) symbiosis enhanced nutrient uptake (Krishna and Bagyaraj 1982; Morton et al. 1990; Hindumathi and Reddy 2012a; Hindumathi et al. 2016b) compared to single inoculations. Legumes cultivated in soil with low P were most responsive to combined inoculation of AM fungi and *Rhizobium* as increased 'P' availability stimulates biological N_2 fixation and growth of the host legume.

Furthermore, AM fungi improve soil structure through the secretion of proteinaceous substance called glomalin (Steinberg and Rillig 2003). It can have a direct effect on the ecosystem, as they improve the soil aggregation by forming structure of macroaggregates through physical binding of soil particles and organic material (Rillig and Mummey 2006; Leifheit et al. 2014, 2015; Rillig et al. 2015). Such aggregates enhance carbon and nutrient storage and create conducive environment for survival and growth of soil microorganisms. They also influence soil porosity, which promotes aeration and water movement, essential for better root growth, root development and microbial activity, thereby driving the structure of plant communities and productivity.

Some mycorrhizal fungi produce metabolites that can alter the plant's ability to produce roots from cuttings or alter root regeneration and root morphology resulting in greatly increased absorptive surface area and feeder root longevity (Linderman and Call 1977).

One of the major changes in mycorrhizal plants is reduced membrane permeability primarily due to increased P nutrition affecting quality and quantity of root exudates which, in turn, has the potential to induce significant changes in the rhizosphere microflora.

19.13 AM Fungi in Disease Tolerance

AM fungal colonization in plant roots has been found to increase plant tolerance to root/soilborne plant pathogens, thereby acting as a biocontrol agent (Chhabra et al. 1992; Azcón-Aguilar and Barea 1996). Several mechanisms or combination of mechanisms could account for the observed bioprotection of plants by AM fungi.

Smith (1988) proposed that the interaction of AM fungi with soil root pathogens enhanced uptake of P and other nutrients. Through this action, the fungus increases the plant tolerance to pathogens, through mechanisms such as alteration of root exudates, increased root growth and function and competition for space of infection sites. Chhabra et al. (1992) reported that increased nutritional status of plants with AM fungi might increase tolerance to root pathogens. But no effect was observed on the development of leaf diseases caused by *Helminthosporium maydis* and *Acremonium kiliense* in maize. AM fungi were found to increase *Zea mays* tolerance to leaf rust showing less than 5 % over pathogen-inoculated plants with 80 % leaf rust.

In addition, microbial changes in the mycorrhizosphere and anatomical changes in the root induced by AM formation may bring about stimulation of specific functional groups in the microbiota that are antagonistic to pathogens (Azcón-Aguilar et al. 2002; Sylvia et al. 1998; Azcón-Aguilar and Barea 1996; Linderman 1994). Several studies on biocontrol potential of AM fungi used as inoculant on root/plant pathogens proved to increase plant tolerance to the pathogen.

Systemic acquired resistance (SAR) plays an important role in the ability of plants to defend themselves against pathogens. SAR occurs in most plants in response to colonization of AMF. A number of biochemical and physiological changes have been associated with AM colonization including the production of antifungal or oxidative enzymes, cell death and deposition of lignin.

The AM fungus-plant combinations also proved to be useful in conferring localized or induced systemic protection against pathogens to plants. It was indicated that this mechanism is signalled by modulations such as lignifications, induction of cell wall appositions containing callose and accumulation of pathogenesis-related (PR) proteins or phenolic compounds (Pozo et al. 2002).

AM fungi not only have synergistic interaction with beneficial soil microorganisms but also exihibit antagonistic interaction with root pathogenic microflora and microfauna, thereby promoting plant growth (Dehne 1982).

19.14 Interaction of Arbuscular Mycorrhizal Fungi with Rhizosphere Microorganisms

19.14.1 Interaction of AM Fungi with Bacteria

A variety of microorganisms interact with mycorrhizal fungi which include phosphate solubilizers, free-living and symbiotic nitrogen fixers, antibiotic, plant growth hormone, siderophore and chitinase producers, saprophytes, plant pathogens, predators and parasites. These soil bacteria possess the ability to produce antibiotics or siderophores which are Fe chelators that may act as inhibitors against pathogens or stimulate plant growth.

AM fungal hyphae, in addition to having enhanced nutrient absorption capability of their host plant, provide area for the interaction of plants with other soil microorganisms that have effect on root development and performance. These interactions can be found at all stages of AM fungal life cycle, from spore formation and germination through root colonization to external hyphae (Bianciotto and Bonfante 2002; Bianciotto et al. 1996, 2003; Toljander et al. 2006). The nature of these interactions may be inhibitory or stimulatory and competitive or mutualistic to each other for the plant.

Mycorrhizal establishment changes the microbial population in the rhizosphere both quantitatively and qualitatively (Azcón-Aguilar and Barea 1992; Linderman 1992; Barea 1997; Cordier et al. 1999). Mycorrhizal formation can directly or indirectly affect microbial communities in the rhizosphere through induced changes on root exudates and transport of carbon compounds to the mycorrhizosphere (mycorrhizosphere effect).

Two main groups of microorganisms, the saprophytes and symbionts, interact with mycorrhizal fungi in the rhizosphere environment, both of them comprising detrimental, neutral and beneficial bacteria and fungi. Detrimental microbes include the major plant pathogens, as well as minor parasitic and nonparasitic deleterious rhizosphere organisms (Weller and Thomashow 1994; Nehl et al. 1996). Beneficial microbes include nitrogen fixers, phosphate solubilizers, growth promoters and biocontrol agents which are known to play a major role in soil-plant systems (Barea 1997). Important among them are rhizobacteria (Kloepper 1994, 1996) that are known to show a specific ability for root colonization, some of them able to improve plant development, therefore they are termed as *plant growth-promoting rhizobacteria* (PGPR).

19.14.2 Interaction of AM Fungi with Plant Growth-Promoting Rhizobacteria

Interaction of AM fungi with PGPR carries out many important ecosystem processes and contributes to the productivity of agricultural system (Adesemoye and Kloepper 2009), and also they are known to involve in the biological control of plant pathogens, nutrient cycling and/or seedling establishment and soil quality (Kloepper et al. 1991; Barea 2000; Jeffries and Barea 2001).

PGPR belonging to genera *Paenibacillus*, *Burkholderia*, *Pseudomonas* and *Bacillus* spp. exert direct or indirect effects on plant growth. The direct effects are through the release of phytohormones, nitrogen fixation and mineralization of organic phosphates into available forms for plants, while the indirect effect on plant growth is by decreasing or preventing deleterious effects of pathogens mainly through synthesis of antibiotics or production of siderophores.

The AM fungi interact with different types of soil bacteria that can influence their development and symbiotic establishment. The interaction between AM fungi and bacteria can be positive (Gryndler et al. 1996), negative (Gryndler et al. 1996) or neutral (Edwards et al. 1998). The positive interactions include enhanced mycorrhizal development and function. Synergistic positive interactions have been reported between AM fungi and PGPR such as N_2 fixers, fluorescent pseudomonads and sporulating bacilli (Hameeda et al. 2007). Negative interactions include reduced spore germination and hyphal length in the extrametrical stage, decreased root colonization and a reduction in metabolic activity of the internal hyphae. Studies of Walley and Germida (1997) on dual inoculation of *Pseudomonas* strains with AM fungi evidenced varying effects, i.e. *Pseudomonas* strains hindered AM fungal germination. Hence, this indicates that not all PGPR are mycorrhizal-helper bacteria (MHB) or vice versa.

Pseudomonas strains produce non-volatile diffusible compounds such as methane, acetaldehyde, acetoin and diacetyl that may or may not reduce mycorrhizal volume (Aspray et al. 2006; Gryndler 2000; Linderman 1992). It was demonstrated that incorporation of fungus *Trichoderma harzianum* with *Pseudomonas fluorescens*, *Azospirillum* sp. and AM fungal species *G. mosseae* and *G. deserticola* did not affect the establishment of AM fungal spp. in maize (Vázquez et al. 2000). However, an increase in phosphatase, esterase, trehalase and chitinase enzymatic activity was observed. These soil enzymes are mainly used as indicator to detect microbial functioning in the rhizosphere as influenced by AM fungi. Phosphatases which are produced both by bacteria and AM fungi catalyse organic bound P into inorganic P. Esterases indicate catabolic activity in the soil which is directly correlated with microbial activity (Vázquez et al. 2000). Trehalase hydrolyses trehalose, a common sugar found in plant symbiosis, while chitinase degrades chitin, a major compound of fungal cell walls that plays a major role in plant defence mechanisms (Pozo et al. 2002; Vázquez et al. 2000).

Ravnskov et al. (1999) observed that *G. intraradices* showed negative effect on the growth and survival of *Pseudomonas putida* under controlled conditions which may likely be due to competition for nutrients. The bacterial population composition in the mycorrhizosphere of AM plants can affect the interaction between plant and AM fungi (Andrade et al. 1997), or alternatively the AM fungi can influence a shift in specific groups of bacteria in the rhizosphere of mycorrhizal plants towards more facultative anaerobic bacteria and fewer fluorescent pseudomonads. Klyuchnikov and Kozherin (1990) demonstrated proliferation of fluorescent pseudomonads in the hyphosphere. Vosatka and Gryndler (1999) reported the most common bacteria in the mycorrhizosphere were *Pseudomonas*, while *Arthrobacter* and *Bacillus* were common in the hyphosphere. It was suggested that AM fungi can regulate the microflora for its own benefit which, in turn, can benefit the host plant.

Marschner et al. (1997) studied the effect of mycorrhizal colonization by *G. deserticola* and *G. intraradices* on the changes in root exudation pattern and rhizosphere microflora using split root system. The results evidenced the latter species colonizing the root on one-half of the split root system significantly altered the root exudation pattern compared to the former. It was found that root colonization by *G. intraradices* in one-half of the root system reduced the population of *Pseudomonas fluorescens* on both sides of the split root system. Further, it was observed that *G. deserticola* could also reduce population density of *P. fluorescens* on the side where fungus colonized the root system, suggesting that colonization by AM fungi could significantly influence the organisms colonizing the rhizosphere.

19.14.3 Interaction of AM Fungi with Phosphate-Solubilizing Bacteria

Phosphate-solubilizing bacteria (PSB) have great potential to improve plant growth under P-deficient conditions when used in combination with AM fungi (Gryndler 2000). They are known to solubilize sparingly soluble organic and inorganic P sources and mobilize phosphate ions to the plants. However, the released P does not reach root surface due to inadequate diffusion (Barea et al. 2005; Azcón-Aguilar and Barea 1992). It was demonstrated that AM fungi could improve uptake of solubilized P. Hence, combined interaction should improve P nutrition and transfer to the plants (Barea et al. 2002a, b, c). Dual inoculation of PSB such as *Bacillus circulans* together with AM fungi resulted in increased uptake of 'P' and enhanced plant yield.

Raj et al. (1981) studied the effect of *G. fasciculatum* and non-phytohormone-producing PSB strain *Bacillus circulans* on phosphate solubilization, growth of finger millet and 'P' uptake from isotope labelled ^{32}P-tricalcium phosphate and super phosphate. Their results clearly indicated that though AM fungi did not solubilize unavailable form of 'P', it enhanced 'P' uptake, which was attributed to better exploration of soil. The synergistic interaction effect between AM fungi and PSB was further confirmed by Karthikeyan et al. (1995) on neem and by Singh (1995) on *Pennisetum* grass.

By using split-dish in vitro carrot mycelial system, it was demonstrated that AM fungi are capable of hydrolyzing organic phosphorus sources and are able to translocate 'P' to plant roots (St. Arnaud et al. 1996). These findings indicated that mycorrhizal plants have access to organic 'P' sources and successfully compete with soil microorganisms for 'P'. Van der Heijden (2010) established that AM fungi have the capacity to increase available soil 'P' and reduce losses of 'P'.

Free-living microorganisms solubilize phosphate ions from sparingly soluble organic and inorganic P compounds (Whitelaw 2000), increase soil phosphate pools and make available for extraradical AM fungal mycelium to absorb. Barea et al. (1997) and Kim et al. (1998) demonstrated in their experiments the symbiotic microbial interaction involving phosphate-solubilizing bacteria (PSB) and AM fungi.

By using a soil microcosm system integrated with ^{32}P dilution, the interactive effects of AM fungi and PSB were studied on plant use of soil in the form of either endogenous or added rock P. The results revealed that the PSB (*Enterobacter* sp. and *Bacillus subtilis*) promoted mycorrhizal establishment of *G. intraradices*, and their dual inoculation increased biomass and N and P accumulation in plant tissues of onion (Toro et al. 1997). They also found that mycorrhizal formation increased the density of PSB population.

The interactive effect of PSB, AM fungi and *Rhizobium* with regard to agronomic efficiency of rock phosphate for legume crops (*Medicago sativa*) was evaluated using isotopic ^{32}P and ^{15}N dilution technique under controlled conditions and further validated under field conditions (Barea et al. 2002a, b). They have observed that the tested microbial interaction improved plant growth and N and P acquisition under normal cultivated conditions. Similar results were obtained by using *Medicago arborea*, a woody legume of interest for revegetation and biological reactivation of desertified semiarid Mediterranean ecosystem (Valdenegro et al. 2001).

Multi-microbial interactions between AM fungi, PSB and *Azospirillum* when inoculated combinedly have reported to show synergistic effect (Muthukumar et al. 2001). They have confirmed by inoculating *G. intraradices*, *G. geosporum*, *Azospirillum brasilense* and PSB individually or in various combinations on neem tree seedlings under nursery conditions. Mycorrhizal colonization, leaf area and number, plant height, biomass, nutrient content (N, P, K) and seedling quality showed significant increase because of combined interaction of microbial inoculants.

Dual inoculation of mycorrhizal fungi and mycorrhizal-helper bacteria (MHB) showed significant increase in biomass and N and P accumulation in plant tissues compared to controls. The dual inoculated plants also showed lower specific activity ($^{32}P/^{31}P$) than their controls suggesting that these mycorrhizal interaction contributed to biogeochemical ‘P’ cycling, thereby promoting plant nutrition.

The synergistic interaction of mycorrhizal fungi and N_2-fixing bacteria and phosphate-solubilizing microorganism has been demonstrated to improve the bioavailability of major plant nutrients N and P. These interaction effects are a promising approach for low-input agricultural technologies (Bethlenfalvay and Linderman 1992; Jeffries and Barea 2001).

19.14.4 Interaction of AM Fungi with *Rhizobium*

The interaction of AM fungi with *Rhizobium* has received much attention due to high ‘P’ demand for N_2 fixation. Studies have shown that co-inoculation of legumes with AM fungi and *Rhizobium* increased plant growth compared to plants inoculated with *Rhizobium* alone (Hindumathi and Reddy 2012a; Hindumathi et al. 2016b). This was attributed to the fact that under N- and P-limiting conditions, AM fungi improves P uptake, thereby enhancing the plant nitrogenase activity, which in turn promotes root and mycorrhizal development (Sylvia et al. 1998; Fitter and Garbaye 1994). Several reported results on synergistic interaction between AM fungi and *Rhizobium* showed that AM fungi have been found to improve nodulation and N_2 fixation, with consequential benefit to plant growth and soil quality (Azcón-Aguilar and Barea 1992; Barea 2000; Hindumathi and Reddy 2012a; Hindumathi et al. 2016b). Thus the symbiotic effect of *Rhizobium* is said to be dependent on the beneficial nutrient effect of AM fungi. Apart from enhanced P uptake, other nutrients such as Zn, Cu and Ca by AMF can also influence the symbiotic effectiveness of *Rhizobium* as well as other microbial processes that occur at root nodule level (Barea et al. 2002a, b; Azcón-Aguilar and Barea 1992).

Rhizobium is a well-known inoculant for legumes; they have also been used as inoculants for nonleguminous plants (Chabot et al. 1996). Galal et al. (2003) studied effect of P and N fertilization on the growth and yield of wheat on inoculation with AM fungi and *Rhizobium* using radiolabelled ^{15}N technique. They found increase in growth of wheat when both AM fungi and *Rhizobium* co-inoculated at high levels of N and P. This dual inoculation also showed increase in the uptake of N and P, while plants inoculated with AM fungi alone increased grain yield of wheat indicating the ability of both organisms to stimulate plant growth and accumulate P and N.

Harrison (1997) reported that certain nod factor stimulated mycorrhizal colonization in soybean. Dual inoculation by the two symbionts showed greater advantage in field experimental studies. These studies have additional advantages in the tropics because of the grain legume programmes introduced to increase protein content of the diet. The reason is that the tropical soils are deficient in 'P'.

Genetically modified (GM) *Rhizobium* developed to improve the nodulation competitiveness of the wild type (WT) strain (Sanjuan and Olivares 1991) was inoculated with *Glomus mosseae* on *Medicago sativa* (Tobar et al. 1996). The results indicated that GM *Rhizobium* strain did not interfere with any process related to mycorrhizal formation (spore germination, mycelial growth) and 'entry point' formation on developing root system of host plant. Indeed, the GM *Rhizobium* increased the number of colonization units and the nutrient acquisition ability in mycorrhizal plants compared with WT *Rhizobium* strain. This symbiotic interaction establishment also induced changes in the root morphology; particularly the degree of branches increased and the number of lateral roots was higher in mycorrhizal plants inoculated with the GM *Rhizobium* strain (Barea et al. 1996).

Rhizobium strains have also been shown to colonize the rhizosphere of nonlegume hosts and established interactions with mycorrhizal fungi (Galleguillos et al. 2000). Several experiments have demonstrated a positive effect on the interaction between mycorrhizal fungi and nodule rhizobacteria under drought stress conditions.

A synergistic effect was observed between *Glomus fasciculatum* and *Azotobacter chroococcum* in tomato plants. The latter helped to enhance fungal colonization and spore production, while the former increased the bacterial population in the rhizosphere (Bagyaraj 1984). Similar interactions have also been observed between *Azotobacter paspali* and AM fungi in *Paspalum* (Barea et al. 1973) and *A. chroococcum* and *G. fasciculatum* in tall fescue (Ho and Trappe 1979). Biro et al. (2000) reported an increase in nodulation of alfalfa plants with combined inoculation of *G. fasciculatum, Azospirillum* and *Rhizobium* under sterile and normal soil conditions.

Synergistic interaction was reported between AM fungi and *Azospirillum* species (Saxena and Tilak 1997) and *Acetobacter diazotrophicus* (Paula et al. 1992). An interaction study between the *Beijerinckia mobilis*, phosphate-solubilizing fungi *Aspergillus niger* and *G. fasciculatum* was reported to show symbiotic beneficial effect on the growth of onions with all three organisms (Manjunath et al. 1981). It was attributed to the production of hormones. Mosse et al. (1981) suggested that hormones produced by these three bacteria could exert symbiotic effect on plant growth or mycorrhizal effect.

In the studies conducted between free-living N_2-fixing bacteria and AM fungi in the rhizosphere, a positive interaction was observed with consequent improvement in plant growth.

The actinomycetes *Frankia* is known to produce N_2-fixing nodules on roots of nonlegumes like *Alnus*, *Casuarina*, *Ceanothus*, *Myrica*, etc., fix atmospheric nitrogen and made available to the host plant. Dual inoculation of AM fungi with *Frankia* increased total dry weight of shoots and roots, number of nodules, weight of nodular tissues, as well as levels of N and P in *Casuarina* (Vasantha Krishna et al. 1994).

The importance of this type of symbiotic fungal association for plant mineral nutrition and more generally plant health makes it one of the potentially more useful biotechnological means of assuring plant production with a minimum input of chemicals such as fertilizers or pesticides.

19.15 Effect of Rhizosphere Microorganism on Mycorrhizal Symbiosis

Rhizosphere microorganisms are known to either interfere with or benefit mycorrhizal development and symbiotic establishment. AM fungi interacting with different types of rhizosphere bacteria can influence their development and symbiotic establishment. These interactions can be found at all stages of the AM fungal life cycle, from spore formation and germination through root colonization to external hyphae (Bianciotto and Bonfante 2002; Bianciotto et al. 1996, 2003; Roesti et al. 2005; Toljander et al. 2006). The nature of interaction between AM fungi and bacteria can be positive, negative or neutral. Negative effects are reduced spore germination and hyphal length in the extramatrical stage, decreased root colonization and a decline in the metabolic activity of the internal mycelium. Positive beneficial interactive effects include enhanced mycorrhizal formation and function. One example among the beneficial effects is that exerted by mycorrhizal-helper bacteria (MHB) known to stimulate mycelial growth of mycorrhizal fungi and/or enhance mycorrhizal formation (Garbaye 1994; Barea 1997; Gryndler et al. 2000).

Rhizosphere microorganism can produce compounds that can influence increased rates of root exudation, which in turn stimulated mycorrhizal fungal mycelium in the rhizosphere or facilitated root penetration by the fungus. Plant hormone production by rhizosphere microorganisms is known to affect mycorrhizal establishment (Azcón-Aguilar and Barea 1992, 1995; Barea 1997, 2000).

Rhizobacteria are reported to affect the pre-symbiotic stages (Giovannetti 2000) of mycorrhizal development, like spore germination and mycorrhizal growth rate of AM fungi *G. mosseae* (Azcón-Aguilar and Barea 1992), which resulted in higher influence on plant root.

Dual inoculation with different species of AM fungi and *Pseudomonas putida* on subterranean clover and maize showed increase in plant growth and AM fungal colonization (Gryndler and Vosatka 1996). Azcón (1987) observed increased growth of emerging mycelium from *G. mosseae* spores in the presence of PGPR. The nodule-forming bacteria, *Frankia*, *Rhizobium* and *Bradyrhizobium*, generally form symbiotic interactions with AM fungi. It is evidenced that AM fungal symbiosis reduced phosphate stress for the plant, which is essential for N_2-fixing and nitrogenase activity of the bacteria, resulting in enhanced fixation and improved N status of the plant. This in turn promotes plant growth and mycorrhizal development. Thus, this type of interaction between AM fungi and bacteria depends on the soil environment, bacterial spp., AM fungal spp. and plant spp.

MHB are organisms that specifically promote mycorrhizal formation especially ectomycorrhiza by producing growth metabolites that encourages easy proliferation of fungal hyphae, thereby increasing its chances to colonize plant roots with a large surface area for absorption. When PGPR are found to stimulate mycorrhizal formation, they can be regarded as MHB (Fitter and Garbaye 1994).

Spanish workers reported that cell-free extracts of *Rhizobium* enhanced colonization of host by AM fungi and was attributed to the presence of extracellular polysaccharide production by *Rhizobium*, which might have increased the number of entry points of AM fungi per unit length of root (Azcón-Aguilar and Barea 1992).

Okon (1994) reported *Azospirillum* bacteria benefit plant development and yield under appropriate conditions and suggested that these bacteria mainly act by influencing the morphology, geometry and physiology of the root system. Volpin and Kaputnik (1994) demonstrated in their interaction study between AM fungi and *Azospirillum* that *Azospirillum* could enhance mycorrhizal formation and response, while AM fungi may improve *Azospirillum* establishment in the rhizosphere.

PGPR bacteria are extensively studied for their role in improvement of crop production in agriculture. Biological active substances such as amino acids, plant growth hormones, vitamins, volatile substances (CO_2) and other organic compounds produced by rhizosphere microorganisms can stimulate the growth rates of AM fungi. Positive interactive effect between AM fungi and PGPR was evidenced by Chanway et al. (1991).

19.16 AMF-Associated Bacteria

Mosse (1962) first showed that bacteria colonize the spores of AM fungi (AMB). Different studies have shown that the spore-associated bacteria can influence AM fungal spore germination, their growth (Walley and Germida 1996; Bianciotto and Bonfante 2002; Hildebrandt et al. 2002; Xavier and Germida 2003) and the formation of the mycorrhizosphere (Budi et al. 1999).

AM fungal structures such as external hyphae (Toljander et al. 2006) and spore or spore walls have been found to be associated with some bacteria (AMB) (Xavier and Germida 2003; Roesti et al. 2005). Spores of *Glomus fasciculatum* were found associated with *Azotobacter* sp. (Ho and Trappe 1979). Mosse (1962) reported certain bacteria on the surface of AM fungal spore aid infection and colonization of alfalfa roots by *G. mosseae*. *Pseudomonas* sp. associated with AM fungal spores was found to help mycorrhizal fungi in infecting the roots suggesting that it could be either due to production of enzymes or growth-promoting substances. There are also reports that bacteria associated with AM (AMB) fungal spores have the ability to influence spore germination and hyphal growth (Mosse 1962; Walley and Germida 1996; Xavier and Germida 2003). The AMB can degrade biopolymers such as protein, chitin and cellulose (Roesti et al. 2005), inhibit the growth of different plant pathogens (Budi et al. 1999) and improve the soil structure (Andrade et al. 1995).

Hildebrandt et al. (2002) reported that AMB have the potential to stimulate the growth of AMF up to the formation of fertile spores in the absence of a host. These reports indicate that AMB might be one important factor involved in AMF development, plant growth and plant protection.

Budi et al. (1999) found that some AMB have antagonistic potential against several soilborne plant pathogens. The antagonistic potential of spore-associated bacteria against pathogens is to be studied in order to obtain information on the plant health-promoting effect of the mycorrhizae.

Xavier and Germida (2003) reported the ability of *Bacillus pabuli* to enhance AMF root colonization and also improve plant growth (Artursson et al. 2006). Budi et al. (1999) reported that *Paenibacillus* sp. isolated from surface-sterilized *G. mosseae* spores significantly stimulated mycorrhizal colonization in *Sorghum bicolor*. Thus, AMB from spores can have potential both as mycorrhizal-helper bacteria (MHB) and PGPR. The multifunctional traits could confer an advantage to the AMB in colonizing the spore surface and spore walls and ensure their survival in specific microhabitats in competition for nutrients and space with other soil microbes.

Interest in research has been increasing on spore-associated bacteria because these have shown the potential to support AMF to complete spore production in vitro in the absence of a host (Hildebrandt et al. 2002).

19.17 Bacteria-Like Organisms (BLO)

Bacteria have also been reported to live inside the spores of certain AM fungal isolates (Bianciotto et al. 1996, 2003). The AMF also harbour bacteria-like organisms (BLO) in their cytoplasm. They are referred to as BLO because they are actually of true bacterial origin and have endobacterial properties, i.e. they complete their life cycle within fungal cells (Bianciotto et al. 1996). The BLOs are gram negative and rod-shaped and present in several AM fungal species such as *Acaulospora laevis, Gigaspora margarita* and *Glomus versiforme*. They are usually found in the cytoplasm of intracellular hyphae, arbuscules and resting spores.

19.18 Interaction of AM Fungi with Soilborne Pathogens

The role of AM fungi in biological control has been studied in a number of plant pathogen-host species combinations. Several reports suggested that mycorrhizal establishment has been shown to reduce damage caused by soilborne plant pathogens (Azcón-Aguilar and Barea 1996; Schenck 1987). Few reports have also indicated that there is either no effect or an increase in the severity of disease due to mycorrhizal colonization. Primarily, the ability of AM fungi to enhance plant vigour due to increased nutrient uptake enables it to resist pathogen infection. Different AM fungal species have been studied and found to be effective in reducing plant disease caused by the pathogens such as species of *Cylindrocladium*, *Fusarium*, *Macrophomina*, *Phytophthora*, *Pythium*, *Rhizobium*, *Sclerotium*, and *Verticillium* on different host species. AM fungi-mediated biocontrol potential of *Macrophomina* root rot on cowpea (Bagyaraj 1984; Caron et al. 1986), *Pythium aphanidermatum* on tomato (Reddy et al. 2006c) and *Verticillium* wilt in brinjal, chilli (Satya Vani 2012), tomato (Satya Vani et al. 2014b), sorghum (Hindumathi 1999), soybean (Hindumathi and Reddy 2012a) and mung bean (Hindumathi et al. 2016b) proved to increase plant tolerance to the pathogen, thereby acting as biocontrol agents.

Several mechanisms or a combination of mechanisms is reported to involve in bioprotection of plants by AM fungi against soilborne pathogens. One of the proposed mechanisms is based on the microbial population changes produced in the mycorrhizosphere. Azcón-Aguilar and Barea (1992) and Linderman (1994, 2000) evidenced changes in the microbial population shift in the mycorrhizosphere and suggested that the resulting microbial equilibrium could influence the general health of the plants. Earlier studies reported that mycorrhizal formation induced changes in the microbial population may lead to the stimulation of certain organisms of the resident microbiota that can be antagonistic to the root/plant pathogens. Caron et al. (1986) reported a reduction in the population density of *Fusarium oxysporum* f. sp. *lycopersici* on tomato colonization by *G. intraradices*.

Meyer and Linderman (1986) observed lower number of sporangia and zoospores of *Phytophthora cinnamomi* by adding extracts of rhizosphere soil from mycorrhizal plants. Secilia and Bagyaraj (1987) found that there were more pathogen-antagonistic actinomycetes in the rhizosphere of plants inoculated with mycorrhizal fungi than in that of non-mycorrhizal controls. Further studies have ascertained these findings and demonstrated that such an effect is dependent on the mycorrhizal fungi involved, as well as the substrate and the host plant (Azcón-Aguilar and Barea 1996; Linderman 2000).

Various synergistic effects of AM fungi and bacteria can also be exploited for pathogen control and nutrient acquisition in low-input agricultural systems (Artursson et al. 2006). Rhizosphere microbes antagonistic to soilborne pathogens are being used as biological control agents. Therefore, the prophylactic ability of mycorrhizal fungi has been exploited in association with these antagonists (Linderman 1994, 2000; Barea et al. 1998; Budi et al. 1999). Several studies have demonstrated that microbial antagonism of fungal pathogens, either fungi or PGPR, exerts no microbial effect against mycorrhizal fungi (Barea et al. 1998; Vázquez et al. 2000).

19.19 Interaction of AM Fungi with Fungi

Saprophytic fungi are common in the rhizosphere of plants and live on dead organic material utilizing a wide range of complex organic molecules such as lignins, proteins, glycoproteins, cellulose and other polysaccharides. AM fungi have the ability to utilize stored C-related products in the hyphae of saprophytic fungi in the absence of plant photosynthates (Suresh and Bagyaraj 2002).

Saprophytic fungi can be classified into ecological functional groups such as phosphate solubilizers, antagonists or symbiotic organisms. As antagonists, they may affect the germination of AM fungal spores and development of mycorrhizal colonization by their competition for space and nutrients (Gryndler 2000).

Fusarium and *Trichoderma koningii* were tested for their effect on the growth and mycelial formation of *Glomus mosseae* in maize and lettuce (Mc Allister et al. 1994). They evidenced that mycorrhizal root colonization by *G. mosseae* was decreased in maize when inoculated before or at the same time with *T. koningii*, while *Fusarium* had no effect on colonization of maize. However, *T. koningii* showed no effect on mycorrhizal colonization in lettuce.

Gliocladium virens used as a biocontrol agent was tested for its effect on the pathogen *Pythium ultimatum* and colonization of AM fungi *Glomus etunicatum* in cucumber plants. *Gliocladium virens* showed no deleterious effect on AM fungi ascertained by the colonization of cucumber roots, while it showed biocontrol activity on *P. ultimatum*. This indicates synergistic interaction between AM fungi and *Gliocladium virens*.

Several reports suggest that the organisms such as *Trichoderma viride* (Reddy et al. 2016) are potential plant growth promoters and *Trichoderma harzianum*, *Aspergillus niger*, *Penicillium* variable, white-rot fungi and other filamentous fungi are capable of solubilizing P along with exhibiting biocontrol activity. This potential was known to be exerted by the production of siderophores, organic acids, lytic enzymes, glucose oxide and melanin-degrading enzymes.

The synergistic effects of phosphate-solubilizing fungi (*Penicillium bilaji*) and mycorrhizal fungi to effectively increase the absorption of P by the plant root system of wheat and bean plants were confirmed by Kucey (1987).

19.20 Conclusion

From the present information, it can be stated that interactions of arbuscular mycorrhizal fungi and rhizosphere microflora of plant roots play an important role in enhancing plant growth. AM fungi are promising for their potential use in sustainable agriculture.

Acknowledgments Dr. A. Hindumathi is grateful to DST, New Delhi for providing fellowship under Women Scientist Scheme-A (WOS-A) with grant No. *SR/WOS-A/LS-498/2011*.

References

Abbott LK, Robson AD (1977) Growth stimulation of subterranean clover with vesicular–arbuscular mycorrhizas. Aust J Agric Res 29:639–649

Adesemoye AO, Kloepper JW (2009) Plant-microbes interactions in enhanced fertilizer use efficiency. Appl Microbiol Biotechnol 85:1–12. doi:10.1007/s00253-009-2196

Andrade G, Azcón R, Bethlenfalvay GJ (1995) A rhizobacterium modifies plant and soil responses to the mycorrhizal fungus *Glomus mosseae*. Appl Soil Ecol 2:195–202

Andrade G, Mihara KL, Linderman RG, Bethlenfalvay GJ (1997) Bacteria from rhizosphere and hyphorhizosphere soils of different arbuscular mycorrhizal fungi. Plant Soil 192:71–79

Artursson V, Finlay RD, Jansson JK (2006) Interactions between arbuscular mycorrhizal fungi and bacteria and their potential for stimulating plant growth. Environ Microbiol 8:1–10

Aspray TJ, Frey-Klett P, Jones JE, Whipps JM, Garbaye J, Bending GD (2006) Mycorrhization helper bacteria: a case of specificity for altering ectomycorrhiza architecture but not ectomycorrhiza formation. Mycorrhiza 16:533–541. doi:10.1007/s00572-006-0068-3

Augé RM (2004) Arbuscular mycorrhizae and soil/plant water relations. Can J Soil Sci 84:373–381

Augé RM, Toler HD, Saxton AM (2015) Arbuscular mycorrhizal symbiosis alters stomatal conductance of host plants more under drought than under amply watered conditions: a meta-analysis. Mycorrhiza 25:13–24. doi:10.1007/s00572-014-0585-4

Azcón R (1987) Germination and hyphal growth of *Glomus mosseae in vitro*: effects of rhizosphere bacteria and cell-free culture media. Soil Biol Biochem 19:417–419

Azcón-Aguilar C, Barea JM (1992) Interactions between mycorrhizal fungi and other rhizosphere microorganisms. In: Allen MJ (ed) Mycorrhizal functioning. An integrative plant-fungal process. Routledge/Chapman & Hall Inc, New York, pp 163–198

Azcón-Aguilar C, Barea JM (1995) Saprophytic growth of arbuscular-mycorrhizal fungi. In: Hock B, Varma A (eds) Mycorrhiza structure function, molecular biology and biotechnology. Springer-Verlag, Heidelberg, pp 391–407

Azcón-Aguilar C, Barea JM (1996) Arbuscular mycorrhizas and biological control of soil-borne plant pathogens. An overview of the mechanisms involved. Mycorrhiza 6:457–464

Azcón-Aguilar C, Jaizme-Vega MC, Calvet C (2002) The contribution of arbuscular mycorrhizal fungi for bioremediation. In: Gianinazzi S, Schuepp H, Barea JM, Haselwandter K (eds) Mycorrhizal technology in agriculture. From genes to bioproducts. Birkhauser Verlag, Berlin, pp 187–197. ISBN 10: 0-89054-245-71

Bagyaraj DJ (1984) Biological interactions with VA mycorrhizal fungi. In: Powell CL, Bagyaraj DJ (eds) VA mycorrhiza. CRC, Boca Raton, pp 131–153

Bagyaraj DJ, Manjunath A, Patil RB (1979a) Interaction between a vesicular-arbuscular mycorrhiza and *Rhizobium* and their effects on soybean in the field. New Phytol 82:141–145

Bagyaraj DJ, Manjunath A, Patil RB (1979b) Interaction of vesicular-arbuscular mycorrhiza with root knot nematode in tomato. Plant Soil 51:397

Barea JM (1997) Mycorrhiza/bacteria interactions on plant growth promotion. In: Ogoshi A, Kobayashi L, Homma Y, Kodama F, Kondon N, Akino S (eds) Plant growth-promoting rhizobacteria, present status and future prospects. OECD, Paris, pp 150–158

Barea JM (2000) Rhizosphere and mycorrhiza of field crops. In: Toutant JP, Balazs E, Galante E, Lynch JM, Schepers JS, Werner D, Werry PA (eds) Biological resource management: Connecting science and policy (OECD). Springer/INRA, Berlin/Paris, pp 110–125

Barea JM, Brown ME, Mosse B (1973) Association between VA mycorrhiza and *Azotobacter*. Rothamsted Exp Stat Annu Rep 1:82

Barea JM, Azcón-Aguilar C, Azcón R (1996) Interactions between mycorrhizal fungi and rhizosphere microorganisms within the context of sustainable soil-plant system. In: Ganga AC, Brown VK (eds) Multitrophic interactions in terrestrial systems. Blackwell Science, Oxford, pp 65–77

Barea JM, Azcón-Aguilar C, Azcón R (1997) Interactions between mycorrhizal fungi and rhizosphere microorganisms within the context of sustainable soil-plant systems. In: Gange AC, Brown VK (eds) Multitrophic interactions in terrestrial systems. Blackwell Science, Oxford, pp 65–77

Barea JM, Andrade G, Bianciotto V, Dowling D, Lohrke S, Bonfante P, O'Gara F, Azcón-Aguilar C (1998) Impact on arbuscular mycorrhiza formation of *Pseudomonas* strains used as inoculants for the biocontrol of soil-borne plant fungal pathogens. Appl Environ Microbiol 64:2304–2307

Barea JM, Toro M, Orozco MO, Campos E, Azcón R (2002a) The application of isotopic (^{32}P and ^{15}N) dilution techniques to evaluate the interactive effect of phosphate-solubilizing rhizobacteria, mycorrhizal fungi and *Rhizobium* to improve the agronomic efficiency of rock phosphate for legume crops. Nutr Cycl Agroecosyst 63:35–42

Barea JM, Azcón R, Azcón-Aguilar C (2002b) Mycorrhizosphere interactions to improve plant fitness and soil quality. Anton Leeuw Int J G 81:343–351

Barea JM, Gryndler M, Lemanceau P, Schüepp H, Azcón R (2002c) The rhizosphere of mycorrhizal plants. In: Gianinazzi S, Schüepp H, Barea JM, Haselwandter K (eds) Mycorrhiza technology in agriculture: from genes to bioproducts. Birkhäuser Verlag, Basel, pp 1–18

Barea JM, Pozo MJ, Azcon R, Azcon-Aguilar C (2005) Microbial co-operation in the rhizosphere. J Exp Bot 56:1761–1778

Bécard G, Piché Y (1989) Fungal growth stimulation by CO_2 and root exudates in vesicular-arbuscular mycorrhizal symbiosis. Appl Environ Microbiol 55:2320–2325

Bentivenga SP, Morton JB (1994) Systemics of glomalean endomycorrhizal fungi: current views and future directions. In: Pfleger FL, Linderman RG (eds) Mycorrhizae and plant health. APS press, Minnesota, pp 283–308

Bethlenfalvay GJ, Linderman RG (1992) Mycorrhizae in sustainable agriculture. ASA Special publication no. 54, Madison

Bianciotto V, Bonfante P (2002) Arbuscular mycorrhizal fungi: a specialised niche for rhizospheric and endocellular bacteria. Anton Leeuw Int J G 81:365–371

Bianciotto V, Bandi C, Minerdi D, Sironi M, Tichy HV, Bonfante P (1996) An obligately endosymbiotic mycorrhizal fungus itself harbors obligately intracellular bacteria. Appl Environ Microbiol 62:3005–3010

Bianciotto V, Lumini E, Bonfante P, Vandamme P (2003) '*Candidatus* Glomeribacter gigasporarum' *gen. nov. sp nov.* an endosymbiont of arbuscular mycorrhizal fungi. Int J Syst Evol Microbiol 53:121–124

Biro B, Koves-Pechy K, Voros I, Takacs T, Eggenberg P, Strasser RJ (2000) Interrelations between *Azospirillum* and *Rhizobium* nitrogen-fixers and arbuscular mycorrhizal fungi in the rhizosphere of alfalfa at sterile, AMF-free or normal soil conditions. Appl Soil Ecol 15:159–168

Bowen GD, Rovira AD (1999) The rhizosphere and its management to improve plant growth. Adv Agron 66:1–102

Brundrett MC (2002) Coevolution of roots and mycorrhizas of land plants. New Phytol 154:275–304

Brundrett M (2004) Diversity and classification of mycorrhizal associations. Biol Rev 79:473–495

Brundrett M, Bougher N, Dell B, Grove T, Malajczuk N (1996) Working with mycorrhizas in forestry and agriculture. ACIAR, Canberra

Budi SW, Van Tuinen D, Martinotti G, Gianinazzi S (1999) Isolation from *Sorghum bicolor* mycorrhizosphere of a bacterium compatible with arbuscular mycorrhiza development and antagonistic towards soilborne fungal pathogens. Appl Environ Microbiol 65:5148–5150

Callow JA, Capaccio LCM, Parish G, Tinker PB (1978) Detection and estimation of polyphosphate in vesicular-arbuscular mycorrhiza. New Phytol 80:125–134

Carlile MJ, Watkinson SC, Gooday GW (2001) The fungi, 2nd edn. Academic Press, San Diego

Carling DE, Richie WG, Brown MF, Tinker PB (1978) Effects of vesicular-arbuscular mycorrhizal fungus on nitrate reductase and nitrogenase activities in nodulating and non-nodulating soybeans. Phytopathol 68:1590–1596
Caron M, Richard C, Fortin JA (1986) Effect of preinfestation of the soil by a vesicular-arbuscular mycorrhizal fungus, *Glomus intraradices*, on Fusarium crown and root rot of tomatoes. Phytoprotection 67:15–19
Chabot R, Antoun H, Cescas MC (1996) Growth promotion of maize and lettuce by phosphate solubilizing *Rhizobium leguminosarum, Biovar phaseoli*. Plant Soil 184:311–321
Chanway CP, Turkington R, Holl FB (1991) Ecological implications of specificity between plants and rhizosphere microorganisms. Adv Ecol Res 21:121–169
Chhabra ML, Bhatnagar MK, Sharma MP (1992) Influence of vesicular arbuscular (VA) mycorrhizal fungus on important diseases of maize. Indian Phytopathol 45:235–236
Cordier C, Lemoine MC, Lemanceau P, Gianinazzi-Pearson V, Gianinazzi S (1999) The beneficial rhizosphere: a necessary strategy for microplant production. Acta Hortic 530:259–265
Dehne HW (1982) Interaction between vesicular-arbuscular mycorrhizal fungi and plant pathogens. Phytopathol 72:1115–1119
Edwards SG, Young JPW, Fitter AH (1998) Interactions between *Pseudomonas fluorescens* biocontrol agents and *Glomus mosseae*, an arbuscular mycorrhizal fungus, within the rhizosphere. FEMS Microbiol Lett 116:297–303
Fitter AH, Garbaye J (1994) Interactions between mycorrhizal fungi and other soil organisms. Plant Soil 159:123–132
Frank B (1885) Über die auf Wurzelsymbiose beruhende Ernährung gewisser Bäume durch unterirdische Pilze. Ber der Deut Bot Ges 3:128–145
Friberg S (2001) Distribution and diversity of arbuscular mycorrhizal fungi in traditional agriculture on the Niger inland delta, Mali, West Africa. CBM's Skriftserie 3:53–80
Friese CF, Allen MF (1991) The spread of VA mycorrhizal fungal hyphae in the soil: inoculum types and external hyphal architecture. Mycologia 83:409–418
Galal YGM, El-Ghandour IA, Osman ME, Abdel Raouf AMN (2003) The effect of inoculation by mycorrhizae and rhizobium on the growth and yield of wheat in relation to nitrogen and phosphorus fertilization as assessed by 15N techniques. Symbiosis 34:171–183
Galleguillos C, Aguirre C, Barea JM, Azcón R (2000) Growth promoting effect of two *Sinorhizobium meliloti* strains (a wild type and its genetically modified derivative) on a nonlegume plant species in specific interaction with two arbuscular mycorrhizal fungi. Plant Sci 159:57–63
Garbaye J (1991) Biological interactions in the mycorrhizosphere. Experientia 47:370–375
Garbaye J (1994) Helper bacteria: a new dimension to the mycorrhizal symbiosis. New Phytol 128:197–210
Gerdemann JW (1975) Vesicular arbuscular mycorrhizae. In: Torrey JG, Clarkson DT (eds) The development and function of roots. Academic Press, New York, pp 575–595
Gianinazzi S, Gianinazzi-Pearson V, Dexheimer J (1979) Enzymatic studies on the metabolism of vesicular-arbuscular mycorrhiza. III. Ultrastructural localization of acid and alkaline phosphatase in onion roots infected by *Glomus mosseae* (Nicol. & Gerd.). New Phytol 82:127–132
Gilbert RG, Linderman RG (1971) Increased activity of soil microorganisms near sclerotia of *Sclerotium rolfsii* in soil. Can J Microbiol 17:557–562
Giovannetti M (2000) Spore germination and pre-symbiotic mycelial growth. In: Kapulnik Y, Douds DD Jr (eds) Arbuscular mycorrhizas: physiology and function. Kluwer Academic Publishers, Dordrecht, pp 3–18
Giovannetti M, Sbrana C (1998) Meeting a nonhost: the behaviour of AM fungi. Mycorrhiza 8:123–130
Giri B, Mukerji KG (2004) Mycorrhizal inoculant alleviates salt stress in *Sesbania aegyptiaca* and *Sesbania grandiflora* under field conditions: Evidence for reduced sodium and improved magnesium uptake. Mycorrhiza 14:307–312

Gryndler M (2000) Interactions of arbuscular mycorrhizal fungi with other soil organisms. In: Kapulnik Y, Douds DD Jr (eds) Arbuscular mycorrhizas: physiology and function. Kluwer Academic Publishers, Dordrecht, pp 239–262

Gryndler M, Vosatka M (1996) The response of *Glomus fistulosum*-maize mycorrhiza to treatments with culture fractions from *Pseudomonas putida*. Mycorrhiza 6:207–211

Gryndler M, Hrselova H, Chvatalova I (1996) Effect of free-soil-inhabiting or root associated microfungi on the development of arbuscular mycorrhizae and on proliferation of intraradical mycorrhizae hyphae. Folia Microbiol 41:193–196

Gryndler M, Hrselová H, Stríteská D (2000) Effect of soil bacteria on growth of hyphae of the arbuscular mycorrhizal (AM) fungus *Glomus claroideum*. Folia Microbiol 45:545–551

Hameeda B, Srijana M, Rupela OP, Reddy G (2007) Effect of bacteria isolated from composts and macrofauna on sorghum growth and mycorrhizal colonization. World J Microbiol Biotechnol 23:883–887

Harley JL, Smith SE (1983) Mycorrhizal symbiosis. Academic Press, London

Harrison MJ (1997) The arbuscular mycorrhizal symbiosis: an underground association. Trends Plant Sci Rev 2:54–60

Hawkins HJ, Johansen A, George E (2000) Uptake and transport of organic and inorganic nitrogen by arbuscular mycorrhizal fungi. Plant Soil 226:275–285

Helgason T, Fitter AH, Young JPW (1999) Molecular diversity of arbuscular mycorrhizal fungi colonising *Hyacinthoides non-scripta* (bluebell) in a semi natural woodland. Mol Ecol 8:659–666

Hepper CM, Sen R, Azcon-Aguilar C, Grace C (1988) Variation in certain isozymes amongst different geographical isolates of the vesicular-arbuscular mycorrhizal fungi *Glomus clarum*, *Glomus monosporum* and *Glomus mosseae*. Soil Biol Biochem 20:51–59

Hildebrandt U, Janetta K, Bothe H (2002) Towards growth of arbuscular mycorrhizal fungi independent of a plant host. Appl Environ Microbiol 68:1919–1924

Hiltner L (1904) Uber neuere erFahrungen und probleme auf dem gebiet der bodenbakteruiligie und unter besonderer berucksichtiguang der grundungung und brache. Arb Deutsch Landwirt ges 98:59–78

Hindumathi A (1999) Role of arbuscular mycorrhizae in plant growth and biocontrol of charcoal rot in sorghum. PhD thesis, Department of Botany, Osmania University, Hyderabad, India

Hindumathi A, Reddy BN (2011a) Occurrence and distribution of arbuscular mycorrhizal fungi and microbial flora in the rhizosphere soils of mungbean [*Vigna radiata* (L.)] and soybean [*Glycine max* (L.) Merr.] from Adilabad, Nizamabad and Karimnagar districts of Andhra Pradesh state, India. Adv Biosci Biotech 2:275–286

Hindumathi A, Reddy BN (2011b) Influence of arbuscular mycorrhizal fungi on plant growth and nutrition of Sorghum. In: Proccedings of II Asian PGPR Congress at Beijing, China, August 21–24

Hindumathi A, Reddy BN (2011c) Dependency of Sorghum on arbuscular mycorrhizal colonization for growth and development. Indian J Mycol Plant Pathol 41:537–542

Hindumathi A, Reddy BN (2012a) Synergistic effect of arbuscular mycorrhizal fungi and *Rhizobium* on the growth and charcoal rot of soybean [*Glycine max* (L.) Merr.]. World J Sci Technol 2:63–70

Hindumathi A, Reddy BN (2012b) Systematics and occurrence of arbuscular mycorrhizal fungi. Lap Lambert Academic Publishing. GmbH & Co. K.G. Dudweiler Landstr, Saarbrücken, p 168

Hindumathi A, Reddy BN (2015) Species diversity and population density of arbuscular mycorrhizal fungi associated with *Carthamus tinctorius* L. Rhizosphere Soils of Telangana, India. Mycorrhiza News 7:5–17

Hindumathi A, Reddy BN (2016a) Dynamics of arbuscular mycorrhizal fungi in the rhizosphere soils of safflower from certain areas of Telangana. Indian Phytopath 69:67–73

Hindumathi A, Reddy BN, Sabitha Rani A, Reddy AN (2016b) Associative effect of arbuscular mycorrhizal fungi and *Rhizobium* on plant growth and biological control of charcoal rot in green gram [*Vigna radiata* L. (Wilczek)]. In: Bhima B, Anjana Devi T, Taylor and Francis

Group (eds) Microbial biotechnology: technological challenges and developmental trends. Apple Academic Press, Milton, pp 155–170

Ho I, Trappe JM (1979) Interaction of a VA-mycorrhizal fungus and a free-living nitrogen fixing bacterium on growth of tall fescue. In: Abstracts of the 4th North American Conference on Mycorrhizae. Fort Collins, Colorado

Ho I, Trappe JM (1980) Nitrate reductase activity of nonmycorrhizal Douglas fir rootlets and of some associated mycorrhizal fungi. Plant Soil 54:395–398

Hodge A, Campbell CD, Fitter AH (2001) An arbuscular mycorrhizal fungi accelerates decomposition and acquires nitrogen directly from organic matter. Nature 413:297–299

Isaac S (1992) Fungal – plant interactions. Chapman & Hall, London

Jeffries P, Barea JM (2001) Arbuscular mycorrhiza – a key component of sustainable plant-soil ecosystems. In: Hock B (ed) The mycota, vol IX, Fungal associations. Springer-Verlag, Berlin, pp 95–113

Jeffries P, Gianinazzi S, Perotto S, Turnau K, Barea JM (2003) The contribution of arbuscular mycorrhizal fungi in sustainable maintenance of plant health and soil fertility. Biol Fertil Soils 37:1–16

Jones DL, Hodge A, Kuzyakov Y (2004) Plant and mycorrhizal regulation of rhizodeposition. New Phytol 163:459–480. doi:10.1111/j.1469-8137.2004.01130.x

Karthikeyan B, Pandiyarajan P, Santhana Krishnan P (1995) Effect of dual inoculation of phosphobacteria and VA mycorrhizal fungi on the growth of neem. In: Adholeya A, Singh S (eds) Mycorrhizal biofertilizers for the future, Proceedings of the 3rd national conference on mycorrhiza, New Delhi

Kehri HK, Chandra S (1990) Mycorrhizal association in crops under sewage farming. J Ind Bot Soc 69:267–270

Kim KY, Jordan D, McDonald GA (1998) Effect of phosphate solubilizing bacteria and vesicular-arbuscular mycorrhizae on tomato growth and soil microbial activity. Biol Fertil Soils 26:79–87

Klironomos JN (2000) Host-specificity and functional diversity among arbuscular mycorrhizal fungi. In *Microbial Biosystems: New Frontiers*. In: Bell CR, Brylinski M, Johnson-Green P (eds) Proceedings of the eighth international symposium on microbial ecology. Atlantic Canada Society for Microbial Ecology, Halifax, pp. 845–851

Kloepper JW (1994) Plant growth-promoting rhizobacteria (other systems). In: Okon Y (ed) *Azospirillum*/plant associations. CRC Press, Boca Raton, pp 111–118

Kloepper JW (1996) Host specificity in microbe-microbe interactions. Bioscience 46:406–409

Kloepper JW, Zablotowick RM, Tipping EM, Lifshitz R (1991) Plant growth promotion mediated by bacterial rhizosphere colonizers. In: Keister DL, Cregan PB (eds) The rhizosphere and plant growth. Kluwer Academic Publishers, Dordrecht, pp 315–326

Klyuchnikov AA, Kozherin PA (1990) Dynamics of *Pseudomonas flourescens* and *Azospirillum brasilense* population during the formation of vesicular arbuscular mycorrhiza. Microbiol 59:449–452

Koide RT, Kabir Z (2000) Extraradical hyphae of the mycorrhizal fungus *Glomus intraradices* can hydrolyse organic phosphate. New Phytol 148:511–517

Koide RT, Mosse B (2004) A history of research on arbuscular mycorrhiza. Mycorrhiza 14:145–163. doi:10.1007/s00572-004-0307-4

Krishna KR, Bagyaraj DJ (1982) Influence of VA mycorrhiza on growth and nutrition of *Arachis hypogea*. Legume Res 5:18–22

Kucey RMB (1987) Increased phosphorus uptake by wheat and field beans inoculated with a phosphorus-solubilising *Penicillium bilaj*i strain and vesicular-arbuscular mycorrhizal fungi. Appl Environ Microbiol 53:2699–2703

Leifheit EF, Veresoglou SD, Lehmann A, Morris EK, Rillig MC (2014) Multiple factors influence the role of arbuscular mycorrhizal fungi in soil aggregation- a meta-analysis. Plant Soil 374:523–537

Leifheit EF, Verbruggen E, Rillig MC (2015) Arbuscular mycorrhizal fungi reduce decomposition of woody plant litter while increasing soil aggregation. Soil Biol Biochem 81:323–328

Linderman RG (1988) Mycorrhizal interactions with the rhizosphere microflora: the mycorrhizosphere effect. Phytopathol 78:366–371

Linderman RG (1992) Vesicular-arbuscular mycorrhizae and soil microbial interactions. In: Bethlenfalvay GJ, Linderman RG (eds) Mycorrhizae in sustainable agriculture. ASA Spec. Publ., Madison, pp 45–70

Linderman RG (1994) Role of VAM fungi in biocontrol. In: Pfleger FL, Linderman RG (eds) Mycorrhizae and plant health. APS Press, St Paul, pp 1–26

Linderman RG (2000) Effects of mycorrhizas on plant tolerance to diseases. In: Kapulnik Y, Douds DD Jr (eds) Arbuscular mycorrhizas: physiology and function. Kluwer Academic Publishers, Dordrecht, pp 345–365

Linderman RG, Call CA (1977) Enhanced rooting of woody plant cuttings by mycorrhizal fungi. J Am Soc Hortic Sci 102:629–632

Lingua G, Franchin C, Todeschini V, Castiglione S, Biondi S, Burlando B (2008) Arbuscular mycorrhizal fungi differentially affect the response to high zinc concentrations of two registered poplar clones. Environ Pollut 153:137–147

Lynch JM (1990) The rhizosphere. John Wiley, New York

Lynch JM, Whipps JM (1990) Substrate flow in rhizosphere. Plant Soil 129:1–10

Manjunath A, Bagyaraj DJ (1984) Response of pigeonpea and cowpea to phosphate and dual inoculation in vesicular-arbuscular mycorrhiza and *Rhizobium*. Trop Agric 61:48–52

Manjunath A, Mohan R, Bagyaraj DJ (1981) Interaction studies between *Beijerinckia mobilis*, *Aspergillus niger* and *Glomus fasciculatus* and their effects on growth of onion. New Phytol 87:723–727

Marschner H (1995) Mineral nutrition of higher plants. Academic Press, London

Marschner P, Crowley DE, Higash RM (1997) Root exudation and physiological status of root colonizing fluorescent pseudomonad, in mycorrhizal and non-mycorrhizal pepper (*Capsicum annum*). Plant Soil 189:11–20

McAllister CB, Garcia-Romera I, Godeas A, Ocampo JA (1994) Interaction between Trichoderma koningii, Fusarium solani and Glomus mosseae: effect on plant growth, arbuscular mycorrhizas and the saprophytic population. Soil Biol Biochem 26:1363–1367

Mehrag AA, Killham K (1995) Loss of exudates from the roots of perennial ryegrass inoculated with a range of microorganisms. Plant Soil 170:345–349

Meier S, Cornejo P, Cartes P, Borie F, Medina J, Azcón R (2015) Interactive effect between Cu-adapted arbuscular mycorrhizal fungi and biotreated agrowaste residue to improve the nutritional status of *Oenothera picensis* growing in Cu-polluted soils. J Plant Nutr Soil Sci 178:126–135

Meyer JR, Linderman RG (1986) Selective influence on populations of rhizosphere or rhizoplane bacteria and actinomycetes by mycorrhizas formed by *Glomus fasciculatum*. Soil Biol Biochem 18:191–196

Morton JB (1988) Taxonomy of VA mycorrhizal fungi: classification, nomenclature and identification. Mycotaxon 32:267–324

Morton JB, Yarger JE, Wright SF (1990) Soil solution phosphorus requirement for nodulation and nitrogen fixation in mycorrhizal and nonmycorrhizal red clover (*Trifolium pratense* L.). Soil Biol Biochem 22:128–129

Mosse B (1962) The establishment of vesicular mycorrhizal under aseptic conditions. J Gen Microbiol 27:509–520

Mosse B, Stribley DP, Tacon L (1981) Ecology of mycorrhizae and mycorrhizal fungi. Adv Microbial Ecol 5:137–210

Mukerji KG, Bhattacharjee M, Mohan M (1982) Ecology of the Indian Endogonaceae. Angew Botanik 56:121–131

Muthukumar T, Udaiyan K, Rajeshkannan V (2001) Response of neem (*Azadirachta indica* A. Juss) to indigenous arbuscular mycorrhizal fungi, phosphate - solubilizing and asymbiotic nitrogen-fixing bacteria under tropical nursery conditions. Biol Fertil Soils 34:417–426

Nehl DB, Allen SJ, Brown JF (1996) Deleterious rhizosphere bacteria: an integrating perspective. Appl Soil Ecol 5:1–20

Nicolson TH (1959) Mycorrhiza in the Gramineae. I. Vesicular-arbuscular endophytes, with special reference to the external phase. Trans Br Mycol Soc 42:421–438

Oehl F, Sieverding E, Palenzuela F, Ineichen K, Silva GA (2011) Advances in Glomeromycota taxonomy and classification. IMA Fungus 2:191–199. doi:10.5598/imafungus.2011.02.02.10

Okon Y (ed) (1994) *Azospirillum* plant associations. CRC Press, Boca Raton

Parkinson D (1967) Soil microorganisms and plant roots. In: Burges A, Raw F (eds) Soil biology. Academic Press, London, pp 449–478

Paula MA, Drguiaga S, Sequera JO, Dobereiner J (1992) Vesicular arbuscular mycorrhizal fungi and diazotrophic bacteria on nutrition and growth of sweet potato. Biol Fertil Soils 14:61–66

Phillips JM, Hayman DS (1970) Improved procedures for cleaning roots and staining parasitic and vesicular-arbuscular mycorrhizal fungi for rapid assessment of infection. Trans Br Mycol Soc 55:158–161

Pozo MJ, Slezack-Deschaumes S, Dumas-Gaudot E (2002) Plant defense responses induced by arbuscular mycorrhizal fungi. In: Gianinazzi S, Scüepp H, Barea JM, Haselwandter K (eds) Mycorrhizal technology in agriculture: from genes to bioproducts. Birkhäuser Verlag, Basel, pp 103–111

Raaijmakers JM, Paulitz TC, Steinberg C, Alabouvette C, Moënne-Loccoz Y (2008) The rhizosphere: a playground and battlefield for soilborne pathogens and beneficial microorganisms. Plant Soil 321:341–361

Raj J, Bagyaraj DJ, Manjunath A (1981) Influence of soil inoculation with vesicular arbuscular mycorrhizas and a phosphate dissolving bacterium on plant growth and ^{32}P uptake. Soil Biol Biochem 13:105–108

Ravnskov S, Nybroe O, Jakobsen I (1999) Influence of an arbuscular mycorrhizal fungus on *Pseudomonas fluorescens* DF57 in rhizosphere and hyphosphere soil. New Phytol 142:113–122

Reddy BN, Hindumathi A, Raghavender CR (2006a) Influence of Physico-chemical factors on arbuscular mycorrhizal population associated with sorghum. Indian J Bot Res 2:75–82

Reddy BN, Sreevani A, Raghavender CR (2006b) Association of AM fungi in three solanaceous vegetable crops. Indian J Mycol Plant Pathol 36:52–56

Reddy BN, Raghavender CR, Sreevani A (2006c) Approach for enhancing mycorrhiza-mediated disease resistance of tomato damping-off. Indian Phytopathol 59:299–304

Reddy BN, Hindumathi A, Raghavender CR (2007) Occurrence and systematics of arbuscular mycorrhizal fungi associated with sorghum. J Phytol Res 20:11–22

Reddy BN, Saritha K, Hindumathi A (2016) Potential use of *Trichoderma* species as promising plant growth stimulator in Tomato (*Lycopersicum esculantum* L.). In: Bhima B, Anjana Devi T, Taylor and Francis Group (eds) Microbial biotechnology: technological challenges and developmental trends. Apple Academic Press, Milton, pp 185–198

Rillig MC, Mummey DL (2006) Mycorrhizas and soil structure. New Phytol 171:41–53

Rillig MC, Aguilar-Trigueros CA, Bergmann J, Verbruggen E, Veresoglou SD, Lehmann A (2015) Plant root and mycorrhizal fungal traits for understanding soil aggregation. New Phytol 205:1385–1388

Roesti D, Ineichen K, Braissant O, Redecker D, Wiemken A, Aragno M (2005) Bacteria associated with spores of arbuscular mycorrhizal fungi *Glomus geosporum* and *Glomus constrictum*. Appl Environ Microbiol 71:6673–6679

Safir GR, Boyer JS, Gerdemann JW (1971) Mycorrhizal enhancement of water transport in soybean. Science 172:581–583

Sanjuan J, Olivares J (1991) Multicopy plasmids carrying the *Klebsiella pneumoniae* nifA gene enhance *Rhizobium meliloti* nodulation competitiveness on alfalfa. Mol Plant Micr Interact 4:365–369

Satya Vani M (2012) AM fungi as bio-fertilizer and bio-control agent of Verticillium wilt of some solanaceous crops. PhD thesis, Department of Botany, Osmania University, Hyderabad, India

Satya Vani M, Hindumathi A, Reddy BN (2014a) Influence of arbuscular mycorrhizal fungi on plant growth promotion and biological control of Verticillium wilt of Tomato (*Lycopersicum esculantum*). Inter J Pharm and Bio Sci 5:1000–1009

Satya Vani M, Hindumathi A, Reddy BN (2014b) Arbuscular myorrhizal fungi associated with rhizosphere soil of Brinjal cultivated in Andhra Pradesh, India. Int J Curr Microbiol App Sci 3:519–529

Satya Vani M, Hindumathi A, Reddy BN (2015) Application of arbuscular mycorrhizal fungi to improve plant growth in *Solanum melongena* L. Ann Biol Res 6:21–28 http://scholarsresearchlibrary.com/archive.html

Saxena AK, Tilak KVBR (1997) Interaction of soil microorganisms with vesicular arbuscular mycorrhiza. In: Tiwari JP, Saxena G, Tewari I, Mittal N, Chamola BP (eds) New approaches in microbial ecology. Aditya Books Pvt. Ltd, New Delhi

Schenck NC (1987) VA mycorrhizal fungi and the control of fungal root disease. In: Chet I (ed) Innovative approaches to plant disease control. Wiley, New York, pp 179–191

Schüßler A, Schwarzott D, Walker C (2001) A new phylum, the *Glomeromycota*: phylogeny and evolution. Mycol Res 105:1413–1421

Secilia J, Bagyaraj DJ (1987) Bacteria and actinomycetes associated with pot cultures of vesicular-arbuscular mycorrhizas. Can J Microbiol 33:1069–1073

Simon L, Bousquet J, Levesque RC, Lalonde M (1993) Origin and diversification of endomycorrhizal fungi and coincidence with vascular land plants. Nature 363:67–69

Singh CS (1995) Impact of nitrogen fixing and phosphate-solubilizing bacteria on mycorrhizal root colonization and spore production. In: Mycorrhizae: biofertilizers for the future, Proceedings of the 3rd national conference on Mycorrhiza, New Delhi

Smith GS (1988) The role of phosphorus nutrition in interactions of vesicular–arbuscular mycorrhizal fungi with soilborne nematodes and fungi. Phytopathol 78:371–374

Smith DE, Read DJ (1997) Mycorrhizal symbiosis. Academic Press, London

Smith SE, Gianinazzi-Pearson V, Koide R, Cairney JWG (1994) Nutrient transport in mycorrhizas: structure, physiology and consequences for efficiency of the symbiosis. In: Robson AD, Abbott LK, Malajczuk N (eds) Management of mycorrhizas in agriculture, horticulture and forestry. Kluwer Academic Publishers, Dordrecht, pp 103–113

Sreevani A, Reddy BN (2005) Arbuscular mycorrhizal fungi with tomato (*Lycopersicum esculentum* Mill.) as influenced by soil physico chemical properties. Philipp J Sci 133:115–129

St-Arnaud M, Hamel C, Vimard B, Caron M, Fortin JA (1996) Enhanced hyphal growth and spore production of the arbuscular mycorrhizal fungus *Glomus intraradices* in an *in vitro* system in absence of host roots. Mycol Res 100:328–332

Steinberg PD, Rillig MC (2003) Differential decomposition of arbuscular mycorrhiza fungal hyphae and glomalin. Soil Biol Biochem 35:191–194

Suresh CK, Bagyaraj DJ (2002) Mycorrhiza-microbe interface: effect on rhizosphere. In: Sharma AK, Johri BN (eds) Arbuscular mycorrhizae. Scientific Publishers, Enfield, pp 7–28

Sylvia DM, Fuhrmann JJ, Hartel PT, Zuberer E (1998) Principles and applications of soil microbiology. Prentice Hall, London 550 p

Sylvia DM, Alagely AK, Kane ME, Philman NL (2003) Compatible host/mycorrhizal fungus combinations for micropropagated sea oats. I. Field sampling and green house evaluations. Mycorrhiza 13:177–183

Tarafdar JC, Claassen N (1988) Organic phosphorus compounds as a phosphorus source for higher plants through the activity of phosphatases produced by plant roots and microorganisms. Biol Fertil Soils 5:308–312

Tinker PB (1984) The role of microorgansims in mediating and facilitating the uptake of plant nutrients from soil. Plant Soil 76:77–91

Tobar RM, Azcón-Aguilar C, Sanjuán J, Barea JM (1996) Impact of a genetically modified *Rhizobium* strain with improved nodulation competitiveness on the early stages of arbuscular mycorrhiza formation. Appl Soil Ecol 4:15–21

Toljander JF, Artursson V, Paul LR, Jansson JK, Finlay RD (2006) Attachment of different soil bacteria to arbuscular mycorrhizal fungal extraradical hyphae is determined by hyphal vitality and fungal species. FEMS lett 254:34–40

Toro M, Azcón R, Barea JM (1997) Improvement of arbuscular mycorrhizal development by inoculation with phosphate solubilizing rhizobacteria to improve rock phosphate bioavailability (^{32}P) and nutrient cycling. Appl Environ Microbiol 63:4408–4412

Valdenegro M, Barea JM, Azcón R (2001) Influence of arbuscular mycorrhizal fungi, *Rhizobium meliloti* strains and PGPR inoculation on the growth of *Medicago arborea* used as model legume for re-vegetation and biological reactivation in a semi-arid Mediterranean area. Plant Growth Regul 34:233–240

Van der Heijden MGA (2010) Mycorrhizal fungi reduce nutrient loss from model grassland ecosystems. Ecology 91:1163–1171

Van der Heijden MGA, Klironomos JN, Ursic M, Moutoglis P, Streitwolf-Engel R, Boller T, Wiemken A, Sanders IR (1998) Mycorrhizal fungal diversity determines plant biodiversity, ecosystem variability and productivity. Nature 396:69–72

Vandenkoornhuyse P, Husband R, Daniell TJ, Watson IJ, Duck M, Fitter AH, Young JPW (2002) Arbuscular mycorrhizal community composition associated with two plant species in a grassland ecosystem. Mol Ecol 11:1555–1564

Vasantha Krishna M, Bagyaraj DJ, Nirmalnath PJ (1994) Response of *Casuarina equisetifolia* to inoculation with *Glomus fasciculatum* and/or Frankia. For Ecol Manage 68:399–402

Vázquez MM, Cesar S, Azcón R, Barea JM (2000) Interactions between arbuscular mycorrhizal fungi and other microbial inoculants (*Azospirillum, Pseudomonas, Trichoderma*) and their effects on microbial population and enzyme activities in the rhizosphere of maize plants. Appl Soil Ecol 15:261–272

Vierheilig H, Scheweiger P, Brundrett M (2005) An overview of methods for the detection and observation of arbuscular mycorrhizal fungi in roots. Physiol Plant 125:393–404

Volpin H, Kapulnik Y (1994) Interaction of *Azospirillum* with beneficial soil microorganisms. In: Okon Y (ed) *Azospirillum*/plant associations. CRC Press, Boca Raton, pp 111–118

Vosátka M, Gryndler M (1999) Treatment with culture fractions from *Pseudomonas putida* modifies the development of *Glomus fistulosum* mycorrhiza and the response of potato and maize plants to inoculation. Appl Soil Ecol 11:245–251

Walley FL, Germida JJ (1996) Failure to decontaminate *Glomus clarum* NT4 spores is due to spore wall-associated bacteria. Mycorrhiza 6:43–49

Walley FL, Germida JJ (1997) Response of spring wheat (*Triticum aestivum*) to interactions between Pseudomonas species and *Glomus clarum* NT4. Biol Fertil Soils 24:365–371

Weller DM, Thomashow LS (1994) Current challanges in introducing beneficial microorganisms into the rhizosphere. In: O'Gara F, Dowling DN, Boesten B (eds) Molecular ecology of rhizosphere microorganisms biotechnology and the release of GMOs. VCH, Weinheim, pp 1–18

Werner D (1998) Organic signals between plants and microorganisms. In: Pinton R, Varanini Z, Nannipieri P (eds) The rhizosphere: biochemistry and organic substances at the soil-plant interfaces. Marcel Dekker Inc., New York

Whitelaw MA (2000) Growth promotion of plants inoculated with phosphate-solubilizing fungi. Adv Agron 69:99–151

Xavier LJC, Germida JJ (2003) Bacteria associated with *Glomus clarum* spores influence mycorrhizal activity. Soil Biol Biochem 35:471–478

Prospect of Phyllosphere Microbiota: A Case Study on Bioenergy Crop *Jatropha curcas*

20

Santosh Ranjan Mohanty, Garima Dubey, Usha Ahirwar, Ashok Kumar Patra, and Bharati Kollah

Abstract

Phyllosphere is the most abundant environment that supports microbial life. This ecosystem is a stress environment characterized by fluctuation in moisture, nutrients, radiation and plant's own immune system. However, plants support microbial community on the phyllosphere as a strategy for its survival and growth. This chapter addresses general characterization of phyllospheric environment, microbial association process, microbial population structure, quorum sensing and cross talk between plant and microbes. This chapter provides information on the microbial diversity of the phyllosphere of bioenergy crop *Jatropha curcas*. Major bacterial groups prevalent on the *J. curcas* phyllosphere and plant growth-promoting activities are addressed.

20.1 Introduction

Phyllosphere is one of the large habitats for microbial population accounting for ~6.4×10^8 km^2 on the earth. Leaf surface supports extensive bacterial populations which can be as high as 10^7 cm^{-2}. It is estimated that phyllosphere bacterial population could be of 10^{26} cells in the tropical plants (Morris et al. 2002). The phyllosphere is a stress environment because it is controlled by external factors like temperature, moisture and solar radiation. It is also a low-nutrient environment. The phyllosphere provides environmental niche for different microorganisms. Bacteria are the most predominant groups of the phyllosphere. Plant modifies phyllospheric bacterial community by changing leaf exudates and moisture like regulating leaf moisture

S.R. Mohanty (✉) • G. Dubey • U. Ahirwar • A.K. Patra • B. Kollah
ICAR-Indian Institute of Soil Science, Nabi bagh, Berasia Road, Nabibagh, Bhopal 462038, India
e-mail: mohantywisc@gmail.com

D.K. Choudhary et al. (eds.), *Plant-Microbe Interaction: An Approach to Sustainable Agriculture*, DOI 10.1007/978-981-10-2854-0_20

through stomata opening and closing. Some microorganisms invade into leaf tissue as a strategy to survive under water stress condition while others manage to survive under low moisture condition (Hardoim et al. 2008). The phyllospheric microorganisms play crucial role in plant's growth. These bacteria fix N and C protect plant from pathogens and produce phytohormones (Bulgarelli et al. 2013). Thus, the phyllospheric organisms provide ecosystem services like C sequestration, N fixation, and bioremediation, enhance crop yield and improve soil health (Bulgarelli et al. 2013). These microbial groups hold key to plant's solvability under extreme condition and sustainability in future climate change.

Jatropha curcas is also known as the biofuel/biodiesel crop. The *J. curcas* belongs to Euphorbiaceae family and is characterized as a drought-resistant perennial plant. It is propagated in tropical and subtropical countries for augmenting renewable energy. In addition, it has several beneficial properties that are significant for agriculture, ecology and environment. Plantation of *J. curcas* is generally recommended to manage degraded wasteland. *J. curcas*-derived biodiesel is a biodegradable and non-toxic fuel compared to petroleum-based diesel. *J. curcas* grows well in low to high rainfall areas. It is cultivated as a commercial crop or as a hedge plant to protect agricultural fields from grazing animals. It can improve socio-economic status of poor farmers in Third World countries as a resource for biodiesel. It is interesting that this plant doesn't need much input like other plants. It can grow under various stress condition like low nutrient and water input. It is hypothesized that microbiome of this plant may hold the key for its sustainability and growth under limited environmental conditions.

Recently many studies elucidate the diversity of microorganisms prevalent in the rhizosphere of *J. curcas*. It harbours significant numbers of arbuscular mycorrhizae like *Acaulospora* sp., *Gigaspora* sp., *Glomus* sp., *Sclerocystis* sp. and *Scutellospora* sp. The major bacterial groups are *Pseudomonas* sp., *Enterobacter* sp. and several gram-negative and gram-positive bacteria. In addition, many plant growth-promoting bacteria including *Azotobacter*, *Rhizobium*, *Pleomorphomonas diazotrophica*, *Bacillus megaterium* and *Bacillus thuringiensis* have been isolated from its rhizosphere. *J. curcas* can fix 5100–6100 kg C ha^{-1} *as* biomass. Incorporation of *J. curcas* biomass into the soil results into significant increase in soil macro- and micronutrients. Keeping in view of the extent of microbial diversity in the rhizosphere of *J. curcas*, the basis of such microbial diversity is unclear. Further studies are required to link microbial diversity with the plant and environment. It is hypothesized that phyllospheric microbes might have significant role in framing the rhizosphere microbial community. To understand the plant-microbial interaction, this paper aims to address the diversity of phyllospheric microbes in different terrestrial plants and microbial species associated with the *J. curcas*.

20.2 Phyllosphere Environment and Microbial Diversity

The phyllosphere is a much intricate environment than the rhizosphere. It is a nutrient-poor environment for microbial activity compared to the belowground rhizosphere. The microbial colonization on leaves is not homogenous because leaf veins, hairs and stomata affect surface uniformity. Microbial communities of phyllosphere are under constant variation of temperature, moisture and radiation over the day and night. These external factors also affect the phyllosphere microbiome by altering plant metabolism. Precipitation and wind also contribute to the temporal changes in the phyllospheric microbes.

There is limited information on the chemical characteristics of leaf surface that would explain for the high microbial activity. The unevenness of cuticle and complicated structures of veins and trichomes are the adverse structures for microbial growth. In addition, rain, dew and leaf exudates, pollutants and removal of nutrients after rain inhibit microbial proliferation. On the leaf surface, microorganisms assimilate carbon and nitrogen mainly from the leaf exudates. These organic compounds are generally glucose, sucrose and fructose. In addition to organic acids, alcohols and amino acids are also released from leaf. The concentration of the nutrients on the leaves is very low and occurs in the range of 1–20 ug/g leaf. The analysis of protein and genomic data revealed that phyllospheric microorganisms assimilate plant-derived NH_4-N, amino acids and simple carbohydrates as primary N and C sources. Microbial stress response protein porins, the component of ATP-binding cassette transporters and TonB-dependent receptors, remain at high level among the phyllospheric microorganisms. This suggests that phyllosphere is a nutrient-poor environment for the growth of microorganisms.

The methylotroph species are generally found on the phyllosphere of many plants. These methylotroph species actively assimilate and metabolize methanol from plant pectin. Several phyllospheric microbial species have rhodopsins. These light-sensing proteins and proton pumps have different absorption spectra than the host plants. This indicates that energy metabolism of the phyllospheric microorganisms is not dependent on the plant. Phyllospheric microbes are capable of coping with UV radiation. These species possess pigments which help them to withstand UV radiations. A group of bacteria isolated from peanut produced pink or orange pigments when exposed to UV as protectant mechanism. Limitation of water and nutrients is compensated by microorganisms with the help of various mechanisms. Some epiphytic *Pseudomonas* sp. produces surfactants that increase the water retention ability of leaf surfaces. This increases solubilization and diffusion of nutrients for the microbial metabolism. Some bacteria increase the diffusion potential of the leaf cuticle by producing toxins. These toxins affect the ion transport potential of cell plasma membranes and improve water and nutrient availability for the phyllospheric microorganisms (Quigley and Gross 1994; Hutchison et al. 1995; Schreiber et al. 2005). Epiphytes also produce extracellular polysaccharides. These compounds protect the bacteria from water stress and help in binding to the leaf surface (Morris et al. 1997; Gal et al. 2003).

Using molecular techniques, phyllospheric microbial diversity has been studied. Microbial species richness on the phyllosphere is high in warmer and humid climates than the temperate regions. The alpha and gamma classes of *Proteobacteria* are the dominant bacterial phyla in phyllosphere. The *Bacteroidetes* and *Actinobacteria* represent the most common species of these phyla. The phyllosphere of several Mediterranean plants is dominated by lactic acid bacteria. In summer, the most dominant species associated with these plants are *Firmicutes*. The growth of *Firmicutes* on the phyllosphere increases plant's tolerance to the hot and dry weather. At higher taxa level, phyllospheric microbiomes of different plants are similar, but at the species level, strains vary significantly. This suggests that bacterial diversity of the phyllosphere is linked with the micro-environment.

The environmental parameters like UV radiation, relative humidity and temperature influence the association of *E. coli* with plants (Seo and Matthews 2014). The indigenous phyllosphere microorganisms are also influenced by these environmental parameters. Changes in the indigenous microflora contribute to the promotion or prevention of *E. coli*. However, the role of the indigenous microorganisms on the long-term persistence of the pathogens is unclear. However, the interaction between pathogens and indigenous microbiota is difficult to understand because the diversity of phyllosphere microbiota varies with geographical locations and environment.

20.3 Ecological Niche of Phyllospheric Bacteria

In the phyllosphere, bacteria colonize typically as aggregates or clusters. In a study, it was found that up to 50 % of *Pseudomonas syringae* bacteria on bean leaves were present in clusters of 10^3 cells or more. It is assumed that before colonization, the incoming bacteria first reach at the leaf as single cells on different parts of the leaf. Only a few sites on the leaf are suitable for bacterial multiplication. The growth of cells in the favourable sites of leaf results into microbial colonies. Secondly, the bacterial species colonize on the leaf surface vary in their ability to produce offspring. This suggests that the leaf surface consists of sites with different conduciveness for bacterial cluster formation.

Moisture is one of the major factors that shape the bacterial clustering on leaf surfaces. Water stays for longer period in the veins and trichomes of leaf than other parts (Esser et al. 2015). Prolonged presence of water at these sites increases the nutrient availability. Most leaf nutrients available on the leaf surface are the photosynthetic compounds diffused from the leaf cuticle. Water droplets on a leaf surface also act as the effective sink for the diffused nutrients. The rate of diffusion of nutrients from leaf to water droplet depends on the volume of water and the activity of bacteria in consuming the nutrients. It also depends on the hydrophobicity and thickness of the cuticle (van der Wal et al. 2013). These factors regulate nutrient availability for bacterial community and act as the major driving factors for the spatial and temporal variation in bacterial population on the leaf surface.

20.4 Microbial Communities on the Phyllosphere

The phyllospheric microbial communities represent bacteria, filamentous fungi, yeasts, algae, protozoa and nematodes. Filamentous fungi are considered transient inhabitants of leaf surfaces because they are present predominantly as spores. However, the rapidly sporulating fungal species and yeasts colonize easily on the leaf surface. Phyllospheric bacterial populations differ sharply among and within the same plant species. Bacterial population vary with the growth phase of the host plant as the plant's growth is associated with colonization of microorganisms. Variation in bacterial population in phyllosphere is caused by the extensive fluctuations in the physical and nutritional status of the phyllosphere. Plant species vary with different carrying capacity of the leaf microbiota. For example, the broader leaves of cucumber and beans carry high number of bacteria than grasses or waxy broad-leaf plants. The physicochemical environments of phyllosphere substantially cause variation in the bacterial flora.

The phyllosphere-dominating microorganisms are unique, but their community can be reproduced with the same plant system. However, the biogeography of these phyllospheric bacteria is less known. In a study, the bacterial communities on the leaves of *Magnolia grandiflora* were analysed by sequencing the 16S ribosomal RNA (rRNA) gene. Bacterial assemblages were dominated by members of the *Alphaproteobacteria*, *Bacteroidetes* and *Acidobacteria*. Patterns in community composition are measured by both relative abundance and Jaccard metrics. Distance based on the analyses indicated that trees positioned closely had more similar bacterial communities than the distantly placed. Indirect gradient analyses indicated that environmental parameters like canopy cover, slope, elevation and aspect of the ground beneath trees significantly influence bacterial community (Finkel et al. 2012).

20.5 Phyllospheric Microbes and Plant Growth-Promoting Activities

Generally it is considered that the community composition of phyllospheric microbes to some extent is random. It is the plant who selects microbes by providing favourable environment in the rhizosphere or phyllosphere for the specific microbial communities. Phyllosphere microbial communities influence plant growth. These microbial communities also contribute to the ecosystem function. However, the host plant is mainly responsible to modulate the plant-microbial interaction. Environmental factors also affect biosynthesis of many photosynthates within the plants. This change in metabolite concentration in plant affects the association of the rhizospheric microbes and alters plant development. Occurrence of certain microbial groups on the leaves suppresses feeding by insect larvae. Some signal molecules produced by phyllospheric microorganisms enhances plant growth under abiotic stress. For instance, the phytomicrobiome of *Arabidopsis* senses drought to maintain growth.

Many phylloplane-inhabiting microbes produce phytohormones. Among different phytohormones, auxin is the most commonly found molecule in the phylloplane. Like phyllospheric microorganisms belowground, PGPRs also produce auxin. This phytohormone plays an important role in the development of root system and overall plant growth. Indole acetic acid (IAA), another potential phytohormone, is also produced by the phyllospheric microorganisms. Many of these phytohormones stimulate root growth that eventually enhances plant's root contact surface with soil and increases nutrient uptake. Due to this ability, microbial inoculants are recommended as a substitute or supplement for chemical fertilizers.

Yeasts are also widely distributed in the nature and coexist with other microorganisms. In a study, 12 yeast strains were isolated from leaf samples of a carnivorous plant *Drosera indica* L. This plant is currently endangered because of restricted habitats and use in herbal industries. The 16S rRNA gene sequence revealed that these yeasts belong to the phylum *Ascomycota* and *Basidiomycota*. The isolated yeasts produced indole-3-acetic acid (IAA). The IAA produced by wild yeasts modifies auxin-inducible gene expression in *Arabidopsis*. Phyllospheric yeasts can promote plant growth and may be considered for inclusion into biofertilizer for sustainable agriculture.

20.6 Phyllospheric Microbes and Plant Protection

Generally plants are exposed to the attack of herbivorous insects and pathogens. Herbivorous insects induce production of phytohormone jasmonic acid in plant, while many phyllospheric bacteria induce salicylic acid production in the host plant. The proportion of the two phytochemicals decides whether a plant would be susceptible or resistant against the pathogens. In an experiment, the bittercress plant (*Cardamine cordifolia*, Brassicaceae) was applied with jasmonic acid or salicylic acid prior to damage. Changes in abundance of phyllosphere bacteria were monitored to examine if chewing of herbivores correlates with the bacterial abundance on leaves. Study revealed that jasmonic acid treatment reduced herbivory, while salicylic acid treatment increased herbivory. Phyllospheric bacterial abundance was higher in herbivore-damaged plants than the undamaged plants. It is hypothesized that the abundance and the complex diversity of phyllospheric microorganisms have significant role in the plant's defence mechanism.

The phyllosphere acts as a media that supports the survival or proliferation of diverse microorganisms that are epiphytes, saprophytes and pathogens. Some phyllospheric microorganisms complete their life cycle along with the plant's growth. On the contrary, pathogens enter the leaf and multiply in the interior leaf tissue. Natural surface openings, such as stomata, are important entry ports for microorganisms. Stomata are the key organ for water transpiration and gaseous exchange. This activity is important for plant's growth. Recent studies show that stomata can limit pathogen entry as part of the plant innate defence process. Some plant pathogens have developed counter defence system. For example, the plant pathogen *Pseudomonas syringae* produces coronatine which suppress plant's stomata-based defence system.

20.7 Quorum Sensing in Phyllosphere

The microbial community dynamics of phyllosphere is complex. Cross talk or signal exchange occurs among the various microbial groups present on the phyllosphere. These signals regulate activities and community dynamics of various phyllospheric microbial groups. These signals either help the plants to initiate immune responses to the harmful pathogens or facilitate the entry of beneficial microbes (Hartmann et al. 2015). Some *Bacillus* sp. secretes antibiotic in the presence of plant root exudates. This process keeps off pathogens in the rhizosphere. Phyllospheric bacteria also interfere with signalling between plants and microbial strains. Lipo-chito-oligosaccharides produced by many microorganisms are cleaved by certain bacteria which produce chitinases. In this way, these bacteria interfere with plant-microbial interaction. Plant signalling compounds are carbohydrates, proteins, organic acids or the secondary metabolites like flavonoids, phenol, phytohormones etc. The PGPR-related signalling compounds are phytohormones, acyl homoserine lactones, phenols and peptides.

Like other ecological niches, bacteria in the phyllosphere communicate by quorum sensing. One of the best studied quorum-sensing molecules is *N*-acyl homoserine lactone. These molecules trigger immune responses and change the phytohormone profile of plants. Plants also detect signal molecules from pathogens and respond by activating their own defence systems. Aboveground microorganisms communicate with the belowground microorganisms and shape plant's microbiomes. It has been observed that change in aboveground microbial communities due to environmental factor or even herbivore activity alters microbial community composition of the below ground.

20.8 Phyllosphere Microbial Diversity of *J. curcas*

Bioenergy crop *J. curcas* is a renewable energy plant. In a study the diversity of bacteria prevalent in phylloplane and rhizosphere of *J. curcas* compared. The diversity of bacterial 16S rRNA gene was estimated by molecular technique known as terminal restriction fragment length polymorphism (T-RFLP). The terminal restriction fragments (ribotypes) obtained from both rhizosphere and phylloplane were affiliated to *Firmicutes*, *Actinobacteria*, *Bacillus*, *Chloroflexi*, *Acidobacteria*, *Verrucomicrobia* and *Methylobacteria*. Fluorescence intensity of TRFs was high in the phylloplane than the rhizospheric soil. The ribotypes TRF56, TRF65, TRF95 and TRF423 were the main variables in soil. The ribotypes TRF466, TRF475 and TRF483 were major TRFs in the phylloplane of *J. curcas*. Diversity indices were high in soil than phylloplane. Study indicated that both belowground and aboveground plant parts harbour selective bacterial groups with different level of diversity and abundance. In a study, it was observed that *Jatropha* plantation increased the members of *Proteobacteria* and *Bacteroidetes* compared to unplanted soil (Agarwal et al. 2015). Several diazotrophic bacterial species, like *Azospirillum*, *Herbaspirillum*, *Burkholderia* and *Gluconacetobacter*, are present in the rhizosphere of *J. curcas* (Zehr 2011). These PGPRs occur on the rhizoplane or as endophytes. Some of these PGPRs fix nitrogen and promotes plant growth (Liu et al. 2011). These strains have the ability to enhance *J. curcas* through the production of IAA, solubilize inorganic P and produce ACC-deaminase and siderophore (Jha and Saraf 2012).

Phylloplane of *J. curcas* possesses many gram-positive and gram-positive bacteria. Plants like *Mangifera indica* have more of gram-positive *Bacillus* sp. and *Corynebacterium* sp. than gram-negative bacteria (Jager et al. 2001). Gram-positive *Bacillus* sp. has been isolated from leaves of groundnut, and these strains stimulate plant growth when inoculated as seed coat (Kishore et al. 2005). Methylotrophs are found from the phylloplane of *J. curcas*. Many species of *Methylobacteria* colonize plants as epiphytes and endophytes (Kwak et al. 2014; Dourado et al. 2015). *Methylobacterium* has been isolated from bamboo phylloplane (Madhaiyan and Poonguzhali 2014). These methylotrophs use methanol as C source released by the plant during pectin demethylation process (Galbally and Kirstine 2002). These bacteria can also multiply using other simpler photosynthathes (Iguchi et al. 2015).

The role of phyllospheric microbes and their interaction with the growth of *J. curcas* is not clearly known. In an experiment, the dominant phylloplane bacteria of *J. curcas* were isolated, and their plant growth-promoting activities were evaluated. The 16S rRNA gene sequences of these bacteria were similar to *Alphaproteobacteria*, *Betaproteobacteria*, *Gammaproteobacteria*, *Firmicutes* and *Actinobacteria*. Most of the isolates were motile and gram variable. Many novel species closely related to *Ralstonia*, *Methylobacter* and *Actinomycetes* detected. The isolates exhibited PGPR activities like ACC-deaminase, phosphatase, K solubilization and indole acetic acid (IAA) production activity. These isolates were further tested on maize plants to check their plant growth-promoting activities. The isolates significantly increased the shoot and root length of the maize seedlings. Linear regression model of the PGPR activities significantly correlated with growth parameters. Among the plant growth-promoting attributes, ACC-deaminase and IAA production were the major growth factors for improving the maize growth.

In *J. curcas* phylloplane, the most abundant species were *Firmicutes*, *Alphaproteobacteria*, *Gammaproteobacteria* and *Actinobacteria*. Species belonging to *Betaproteobacteria* were the least abundant. The 16S rRNA gene sequences of the *Alphaproteobacteria* were similar to *Brevibacterium* sp., *Methylobacterium extorquens* and *Agrobacterium tumefaciens*. Phylloplane of many terrestrial plants contains *Alphaproteobacteria*. Some of these plants are *Magnolia grandiflora*, *Prunus* species and bamboo. *Firmicutes* are predominant on the phylloplane of plum and dessert plants. *Actinobacteria* are a group of important plant-associated spore-forming bacteria, known for their role in the biocontrol of pathogens, plant growth promotion and interaction with plants. *Actinobacteria* were mostly related to *Nocardia*, *Micrococcus*, *Brevibacterium* and *Agromyces*. This group has been found on the phylloplane of apricot (Jo et al. 2015) and many salt-tolerant plants (del Rocío Mora-Ruiz et al. 2015). *Gammaproteobacteria* has been found on the phylloplane of *Prunus* species (Jo et al. 2015) and dessert tree (Belkin and Qvit-Raz 2010). The isolates stimulated the growth of the maize seedling through various plant growth-promoting attributes. Such relation between PGPR activities of phylloplane microorganisms and plant growth has been found in agroforestry plants (George et al. 2002). Probably, the phosphates and indole acetic acid (IAA) production potential of the phylloplane bacteria stimulated plant growth. IAA stimulates cells present on root tip and shoot tip. Further studies are essential to explore the phyllospheric microbes of *Jatropha curcas* to develop microbial inoculants for agriculturally important crops.

Acknowledgment This work is part of the project entitled "Metagenomic characterization and spatiotemporal changes in the prevalence of microbes involved in nutrient cycling in the rhizoplane of bioenergy crop *Jatropha curcas.*" The authors acknowledge the Department of Biotechnology, Government of India, for financial support to SRM. The authors declare no conflict of interest of any type.

References

Agarwal P, Singh S, Mastan SG et al (2015) Soil microbial diversity shift as affected by conversion of shallow and rocky wastelands to Jatropha curcas L. plantation. Int J Environ Stud 72:1–19
Belkin S, Qvit-Raz N (2010) Life on a leaf: bacterial epiphytes of a salt-excreting desert tree. In: Symbioses and stress. Springer, Dordrecht, pp 393–406
Bulgarelli D, Schlaeppi K, Spaepen S et al (2013) Structure and functions of the bacterial microbiota of plants. Annu Rev Plant Biol 64:807–838
del Rocío Mora-Ruiz M, Font-Verdera F, Díaz-Gil C et al (2015) Moderate halophilic bacteria colonizing the phylloplane of halophytes of the subfamily Salicornioideae (Amaranthaceae). Syst Appl Microbiol 38:406–416
Dourado MN, Aparecida Camargo Neves A, Santos DS, Araújo WL (2015) Biotechnological and agronomic potential of endophytic pink-pigmented Methylotrophic Methylobacterium spp. Biomed Res Int 2015:909016
Esser DS, Leveau JH, Meyer KM, Wiegand K (2015) Spatial scales of interactions among bacteria and between bacteria and the leaf surface. FEMS Microbiol Ecol 91:fiu034
Finkel OM, Burch AY, Elad T et al (2012) Distance-decay relationships partially determine diversity patterns of phyllosphere bacteria on Tamrix trees across the Sonoran Desert. Appl Environ Microbiol 78:6187–6193
Gal M, Preston GM, Massey RC et al (2003) Genes encoding a cellulosic polymer contribute toward the ecological success of Pseudomonas fluorescens SBW25 on plant surfaces. Mol Ecol 12:3109–3121
Galbally IE, Kirstine W (2002) The production of methanol by flowering plants and the global cycle of methanol. J Atmos Chem 43:195–229
George TS, Gregory PJ, Wood M et al (2002) Phosphatase activity and organic acids in the rhizosphere of potential agroforestry species and maize. Soil Biol Biochem 34:1487–1494
Hardoim PR, van Overbeek LS, van Elsas JD (2008) Properties of bacterial endophytes and their proposed role in plant growth. Trends Microbiol 16:463–471
Hartmann A, Rothballer M, Hense BA, Schröder P (2015) Bacterial quorum sensing compounds are important modulators of microbe-plant interactions. Plant Microb Import Plant Hum Health 41:131
Hutchison ML, Tester MA, Gross DC (1995) Role of biosurfactant and ion channel-forming activities of syringomycin in transmembrane ion flux: a model for the mechanism of action in the plant-pathogen interaction. Mol Plant-Microbe Interact 8:610–620
Iguchi H, Yurimoto H, Sakai Y (2015) Interactions of methylotrophs with plants and other heterotrophic bacteria. Microorganisms 3:137–151
Jager ES, Wehner FC, Korsten L (2001) Microbial ecology of the mango phylloplane. Microb Ecol 42:201–207
Jha CK, Saraf M (2012) Evaluation of multispecies plant-growth-promoting Consortia for the growth promotion of Jatropha curcas L. J Plant Growth Regul 31:588–598
Jo Y, Cho JK, Choi H et al (2015) Bacterial communities in the phylloplane of Prunus species. J Basic Microbiol 55:504–508

Kishore GK, Pande S, Podile AR (2005) Phylloplane bacteria increase seedling emergence, growth and yield of field grown groundnut (Arachis hypogaea L.). Lett Appl Microbiol 40:260–268

Kwak M-J, Jeong H, Madhaiyan M et al (2014) Genome information of Methylobacterium oryzae, a plant-probiotic methylotroph in the phyllosphere. PLoS One 9:e106704

Liu Y, Wang H, Sun X et al (2011) Study on mechanisms of colonization of nitrogen-fixing PGPB, Klebsiella pneumoniae NG14 on the root surface of rice and the formation of biofilm. Curr Microbiol 62:1113–1122

Madhaiyan M, Poonguzhali S (2014) Methylobacterium pseudosasicola sp. nov. and Methylobacterium phyllostachyos sp. nov., isolated from bamboo leaf surfaces. Int J Syst Evol Microbiol 64:2376–2384

Morris CE, Monier J, Jacques M (1997) Methods for observing microbial biofilms directly on leaf surfaces and recovering them for isolation of culturable microorganisms. Appl Environ Microbiol 63:1570–1576

Morris CE, Barnes MB, McLean RJC et al (2002) Biofilms on leaf surfaces: implications for the biology, ecology and management of populations of epiphytic bacteria. Phyllosphere microbiology. American Phytopathological Society, St. Paul, pp 139–155

Quigley NB, Gross DC (1994) Syringomycin production among strains of Pseudomonas syringae pv. syringae: conservation of the syrB and syrD genes and activation of phytotoxin production by plant signal molecules. Mol Plant Microb Interact 7:78

Schreiber L, Krimm U, Knoll D et al (2005) Plant–microbe interactions: identification of epiphytic bacteria and their ability to alter leaf surface permeability. New Phytol 166:589–594

Seo S, Matthews KR (2014) Exposure of Escherichia coli O157: H7 to soil, manure, or water influences its survival on plants and initiation of plant defense response. Food Microbiol 38:87–92

van der Wal A, Tecon R, Kreft J-U et al (2013) Explaining bacterial dispersion on leaf surfaces with an individual-based model (PHYLLOSIM). PLoS One 8:e75633

Zehr JP (2011) Nitrogen fixation by marine cyanobacteria. Trends Microbiol 19:162–173

Sinker Root System in Trees with Emphasis on Soil Profile

21

S. Devi, R. Angrish, S. Madaan, O.P. Toky, and S.S. Arya

Abstract

Anchorage, water and nutrient uptake and transport are well-known functions of tree roots. However, recent studies ascribe more complex physiological and ecological role to tree roots. This is more particularly so in many tree species where roots have a characteristic dimorphic spread having (1) the surface roots that have a subterranean horizontal spread a few metres around the trunk and (2) sinker roots that go vertically downwards to 10 m and beyond. Increasing evidence is accumulating that the surface and sinker roots form a very dynamic water facilitating system in the soil. This is discussed under three main heads. Firstly, the sinker roots have access to groundwater moisture and even the capillary fringe of the deep water tables making the transpiration and the vital shoot processes sustainable, even when the upper soil profiles are dry. Such roots also cause biodrainage of the water table preventing it from rising to surface layers and making the soil waterlogged. Secondly, the sinker and surface roots form an integrated conduit in the soil that causes upward hydraulic redistribution of the deep soil water to soil surface. Interestingly, this water may also be used by shallow-rooted herbaceous vegetation for its sustainability during episodes of drought. Thirdly, a downward hydraulic redistribution from the surface roots in moist topsoil to the deep soil through the sinker roots may recharge the deep dry soil profiles for future use. The sinker root system, therefore, enables hydraulic redistribution sustaining dry season transpiration and photosynthetic rates of the

S. Devi • R. Angrish • S. Madaan
Department of Botany and Plant Physiology, CCS HAU, Hisar, Haryana, India

O.P. Toky
Department of Forestry, CCS HAU, Hisar, Haryana, India

S.S. Arya (✉)
Department of Botany, Maharshi Dayanand University, Rohtak, Haryana, India
e-mail: aryasunder.hau@gmail.com

D.K. Choudhary et al. (eds.), *Plant-Microbe Interaction: An Approach to Sustainable Agriculture*, DOI 10.1007/978-981-10-2854-0_21

parent tree and surrounding shallow-rooted vegetation, prolonging the life span of fine roots and maintaining root–soil contact in dry soils and storing rainwater down into deeper soil layers for dry season utilization.

21.1 Introduction

Roots of a tree are its dynamic hidden half. These are essential conduits for water and mineral transport. They are equally important in anchorage of the shoot. Tree roots also store carbohydrates and are also involved in signalling to the shoot system in response to the soil–environment cues (Pallardy 2010). Tree roots have been found to be wonderfully opportunistic in their search for water and nutrients and follow moisture gradients up to 20 m horizontally or even vertically downwards (Knight 1999). Study of the structure of tree root spread in the soil has been limited as destructive soil excavation and safety issues are involved in such studies. However, the increasing use of non-destructive techniques like geoelectric measurements is becoming increasingly popular for detecting root placement in soil (Zanetti et al. 2011). Likewise the increasing use of sensitive thermal probes involving heat ratio method is being suitably used to measure up- and downstream movement of water (Hultine et al. 2003, 2004) through the root. In this article we specifically discuss some lesser-known ecophysiological functions attributed to the sinker roots that explore profiles of the soil some 5 m or more vertically downwards. Sinker roots have been known to be responsible for biodrainage or the vertical drainage of the groundwater through evapotranspiration so as to stabilize the water table. Further the sinker root-based phenomenon of hydraulic redistribution of water has been described in which water moves from the wetter soil profiles to the drier ones in both upward and downward directions. The ecological implications of these processes have been discussed along with.

21.2 Tree Root Spread

21.2.1 Surface and Sinker Roots

Trees have a diverse root spread. If water and nutrition requirements are available in the topsoils, roots remain confined to 1–2 m depth and a few metres around the trunk below the surface soil. In other cases, the tap root may extend deeper and deeper vertically downwards into the soil and explores the soil for water and nutrients. While the subterranean root system around the tree trunk is designated as surface root system, the singularly vertically downward roots are designated as sinker roots. Trees may have a dimorphic root system comprising of both the surface and sinker root system.

It may be noted that much length of a tree root is inactive on its surface so far as water uptake from the soil is concerned. This is due to secondary growth-related

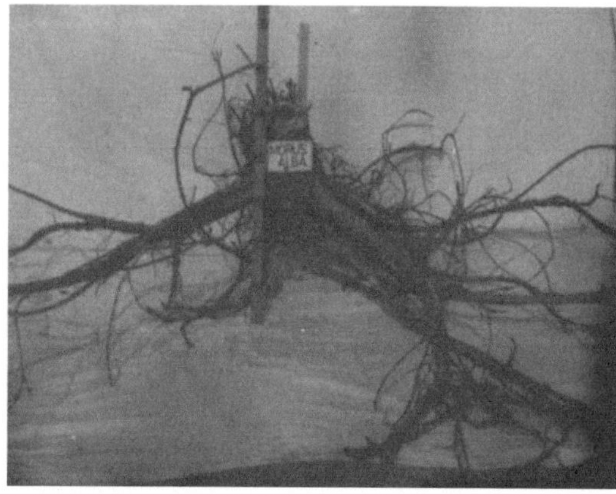

Fig. 21.1 Root system of *Morus alba*. Note the horizontal surface root system and lack of sinker roots

bark insulating greater lengths of the tree root system. These remain unaffected even if the soil becomes dry. It is only at the ends of the surface and sinker roots that non-lignified primary roots with root hairs make dynamic contact with the soil particles. These fine roots have to be in optimal soil moisture to remain functional. Mycorrhizal associations that assist in nutrient uptake are also found on the fine roots. In the perennial tree system, fine root viability must be maintained in soil profiles that may go dry.

21.2.2 Extent of Depth of Sinker Roots

An intensive study of 6-year-old tree species in semiarid north–west India was studied by one of the authors (Toky and Bisht 1992). It was seen that species like *Melia azedarach*, *Morus alba* (Fig. 21.1) and *Populus deltoides* showed a more or less horizontal surface root system confined to the top 80 cm of the soil profile and extending up to a radius of 120 cm.

On the other hand, species like *Prosopis cineraria*, *Acacia nilotica* and *Eucalyptus tereticornis* (Fig. 21.2) had a distinct sinker root system that penetrated more deeply to 250 cm in addition to a horizontal surface root system.

In a monograph on *Prosopis juliflora–Prosopis* complex, Pasiecznik et al. (2001) highlighted that the root system in these arid land species is dimorphic. It, characteristically, has a deep sinker root system and a superficial root system, both having different functions during different seasons. The sinker root system is made up of one, two or three (rarely more) main tap roots, which may divide at lower depths. They have the function of anchoring the tree but are primarily for sourcing groundwater reserves, whether a water table or other subterranean supply. They can become very thick and tens of metres long until a permanent water source is found. These authors quoted reports to show that in certain cases, *P. pallida* tap roots reach water tables at 20–25 m depth. Such plants are also designated as phreatophytes (Hultine et al. 2003). Phreatophytes are deep-rooted trees and shrubs that obtain a dependable water supply from the 'phreatic surface', i.e. from the

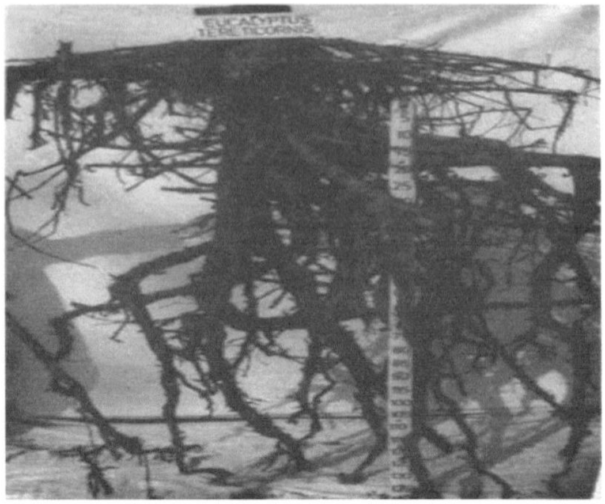

Fig. 21.2 Root system of *Eucalyptus tereticornis*. Note the presence of a surface root system and a deep sinker root system

saturated water table, and thus maintain water status that is largely independent of soil water derived from incident precipitation. These plants develop a zone of maximum root development in the capillary fringe above the water table, rather than in the oxygen-poor saturated zone within the water table. Jackson et al. (1996) surveyed literature and reported root depths of 253 plant species including trees. Maximum depth striking sinker roots were reported to be 68 m for *Boscia albitrunca* in the central Kalahari Desert. They reported 194 species had roots at least 2 m deep, 50 species had roots at a depth of 5 m or more, and 22 species had roots as deep as 10 m or more. Tropical grassland/savanna had maximum root depth of 15.0 ± 5.4 m. They computed that trees had an average root depth of 7.0 ± 1.2 m and concluded that deep root habits are quite common in woody species across most of the terrestrial biomes worldwide.

21.3 Sinker Roots and Biodrainage

21.3.1 Soil Water Use by Trees and Biodrainage

Biodrainage may be defined as the vertical drainage of water table through evapotranspiration of strategically planted vegetation, particularly deep-rooted trees. There is nothing new in the concept of consumptive use of water by trees. Also the fantastic volumes of water the trees can transpire are a matter of record. For example, an overstory *Eperua purpurea* tree in Amazonian rainforest was estimated to transpire 1180 kg day^{-1} of water (see Wullschleger et al. 1998). Equally noteworthy is the fact that at the ecological level, interaction of deep-rooted tree flora with groundwater table is recognized since long. Thus, Wilde et al. (1953) noted that tree species influence groundwater table by acting as biological pumps. However, large-scale scientific use of trees in water table control seems to be of more recent origin. The concept of biological drainage or biodrainage appears to have originated from the waterlogged agricultural areas where the conventional surface and subsurface drainage techniques were in vogue. Frequent use of the term 'biodrainage' in scientific literature is only post 2000. Trees as plantations along the canal banks, in fields as agroforestry components and as commercial block plantations, have had always been a common sight. Their contribution as simple biological pumps cannot be disputed. It is the strategic component of their plantation that revolves around the concept of biodrainage (Anonymous 2003; Angrish et al. 2006; Toky and Angrish 2014).

21.3.2 Conventional Drainage and Biodrainage

Conventionally the control of the problems of waterlogging and soil salinity has been obtained through civil engineering techniques like surface drainage and horizontal subsurface drainage. In surface drainage, excavation of open trenches is done to immediately drain away surface water and to prevent ponded conditions, flooding and consequent damage to the crops. In the horizontal subsurface drainage, removal of soil water below the crop root zone is done through a network open tile drains or underground perforated pipes. However, these techniques, particularly horizontal subsurface drainage, are costly to install and maintain (Tanji 1991; Ritzema et al. 2008).

Biodrainage or the use of trees as a drainage system in problem areas is a green concept that is catching the fancy of technoscientific community in agriculture and even urban development. Its merits are economy in cost and environment friendliness. The limitations are requirement of land for tree plantations, slow lowering of water table, limited evacuation of salts from the system and vulnerability of trees to highly saline conditions. In planning of a biodrainage system, the concept of recharge and discharge zones should be clearly understood. Recharge areas are locations from where water seeps into the water table, e.g. leaky canals or distributaries, elevated areas receiving rainfall with runoff water. However, the most significant recharge areas are the agricultural fields where liberal canal irrigation is applied. The areas where biodrainage plantations are raised to offset the recharge water are known as discharge areas. On an average, about 10 % of land in a waterlogged agricultural landscape is to be marked as discharge area (Heuperman et al. 2002).

21.3.3 Impact of Biodrainage on Depression of Water Table

Biodrainage certainly depresses the water table immediately underneath the plantations, but in agroforestry set-up, the objective is to take the water table to a safer depth well below the crop root zone in the vast cultivated area that surrounds the plantation. Pumping from a well in water table aquifer (unconfined aquifer) is known to develop a cone of water table depression with lowest point near the cavity of the well. Further if two wells are operating simultaneously at suitable distance, two 'interfering' cones of depression shall be formed. It was clearly demonstrated

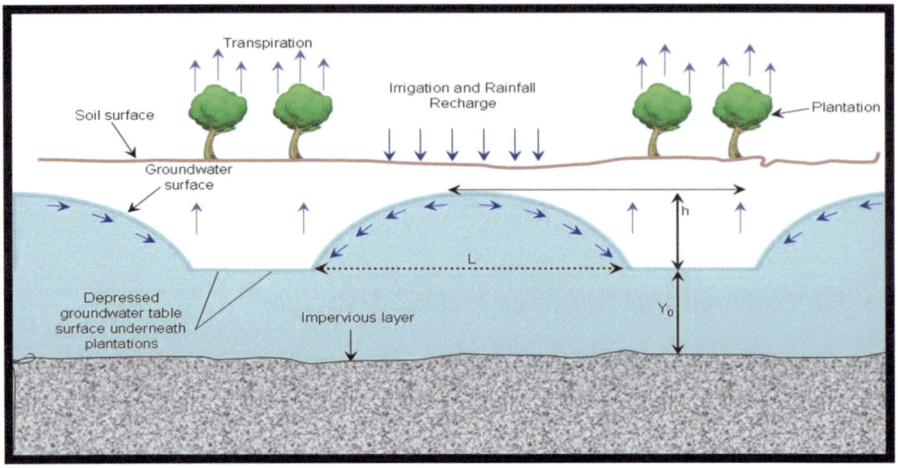

Fig. 21.3 Water table contours due to biodrainage by strip plantations and the associated flux of water (*arrows*). See text for symbols details (Use authorized by FAO, Heuperman et al. 2002)

by Ram et al. (2007) that the drawdown effect of two adjacent *Eucalyptus tereticornis* block plantations was similar to the combined interacting cones of depression of two pumping wells. Another interesting analogy was made by Heuperman et al. (2002), who showed that in case of parallel strip plantations of trees, the water table contours would be similar to the contours found between parallel open drain ditches. The relationship between water table depression, rate of recharge, hydraulic conductivity of soil, depth of barrier layer and distance between plantations can be computed by applying the equation developed by Donnan (see Heuperman et al. 2002) as follows (Fig. 21.3):

$$L = \sqrt{\frac{8KY_o h}{R} + \frac{4Kh^2}{R}}$$

Here L = distance between parallel strip plantations (m)
R = rate of recharge (m/day)
Y_o = water table height above impervious layer under the tree plantations (m)e
K = soil hydraulic conductivity (m/day)
h = head difference (m)

Taking rate of recharge (R) equal to 0.5 mmd^{-1}, head difference (h) of 10.0 m, depth of impervious layer underneath biodrainage plantations (Y_0) as 10 m and hydraulic conductivity value of (1) 10 m d^{-1} (2) 100 m d^{-1} and 1000 m d^{-1}, the distance between plantations for the three values of h is worked out to be 150, 500 and 1500 m, respectively. Thus, plantations shall provide effective biodrainage to greater distances in soils with greater permeability as compared to impermeable soils.

21.3.4 Practical Examples of Biodrainage Systems

First example of biodrainage by sinker roots pertains to the Australian continent. Here a pristine tree system with sinker roots was in a cut-off state of existence with underlying brackish water aquifers for the past thousands of years. This was because the annual rainfall was intercepted and evapotranspired by the native vegetation. Recent introduction of intensive agriculture in the past 100 years necessitated the clearing of this tree vegetation and its replacement with shallow-rooted annual crop plants. The annual consumptive water use of this vegetation was less than the rainfall, and as result water percolated to the underlying saline groundwater table causing its gradual rise. The twin menace of salinity and waterlogging appeared. Now suitable development of agroforestry systems incorporating tree flora with deep-rooted sinker roots has been planted to recede the salinity and water table down away from the root zone of commercially important annual crops. The Australian system is the most exhaustively studied disturbed agroecosystem that unambiguously demonstrates the necessity of harmony between water use by vegetation vis-à-vis its root depth and groundwater table (Heuperman et al. 2002; Crosbie et al. 2008)

In the second case, introduction of canal irrigation and intensive agriculture caused gradual seepage of the liberally used irrigation water which caused rise of saline water table. Productive lands became waterlogged and saline. For example, in western zone of Haryana, average water table depth was static at about 28 m from the ground surface from the 1930s to the early 1950s. After the commissioning of Bhakra canal system in 1956, a sharp increase in saline water table has brought the water table up to a peaked average of only 6 m from the ground surface towards 2002. As a matter of fact during the past two decades, nearly 50 % of the area of south-west Haryana has been critically waterlogged with water table hitting <3 m of the ground surface at one stage or the other. The phenomenon is worldwide but where biodrainage systems involving Eucalyptus trees which have robust sinker roots are being planted. This adoption of such biodrainage systems in Haryana and elsewhere in India has resulted in widespread drop of water table (Anonymous 2003; Angrish et al. 2006; Toky and Angrish 2014).

21.4 Hydraulic Redistribution

21.4.1 Definition of Hydraulic Redistribution

In literature different workers have used different terms and here the terms used have been defined after Neumann and Cardon (2012). When root systems span soil layers of different moisture content, water is moved in soil by roots in the direction of the difference in water potential involving the phenomenon of hydraulic redistribution. This phenomenon has been increasingly demonstrated in woody perennials with elaborate root systems with the help of sap flow movements and soil moisture measurements. Hydraulic redistribution is of two types. Firstly, when the surface

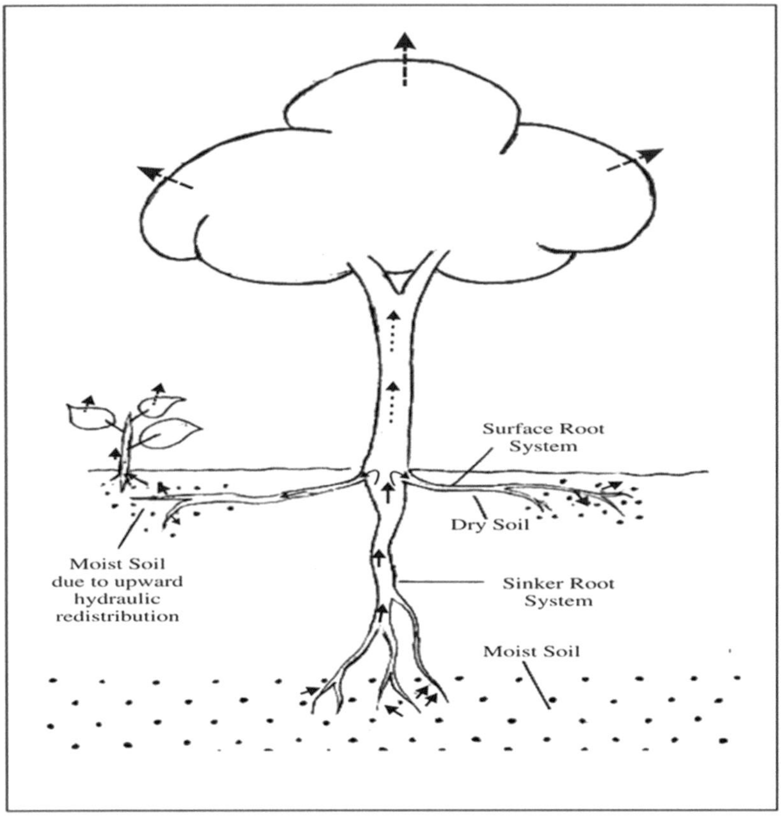

Fig. 21.4 Diagrammatic representation of upward hydraulic redistribution in trees (*dark dots* denote optimal soil moisture, *broken line arrows* indicate routine water movement due to transpiration pull, and line arrows denote water movement due to upward hydraulic redistribution). Note that the surface root system is in dry soil and the terminal ends of the sinker root system are in moist soil profile of the groundwater. Due to upward hydraulic redistribution (*line arrows*), water accumulation occurs in upper soil profile near the ends of the surface root system. Other shallow-rooted vegetation also utilizes this water

root system is in dry topsoil and the sinker roots are in wet soil near to the water table, a movement of water from the soil surrounding the sinker roots to the tips of the surface root occurs and is denoted as upward hydraulic redistribution or hydraulic lift (Fig. 21.4).

Conversely, when the surface roots are in wet topsoil and the ends of the sinker roots are in dry deep soil, a net movement of water occurs from wet topsoil to the dry deep soil at the end of the sinker roots due to downward hydraulic redistribution or hydraulic descent (Fig. 21.5).

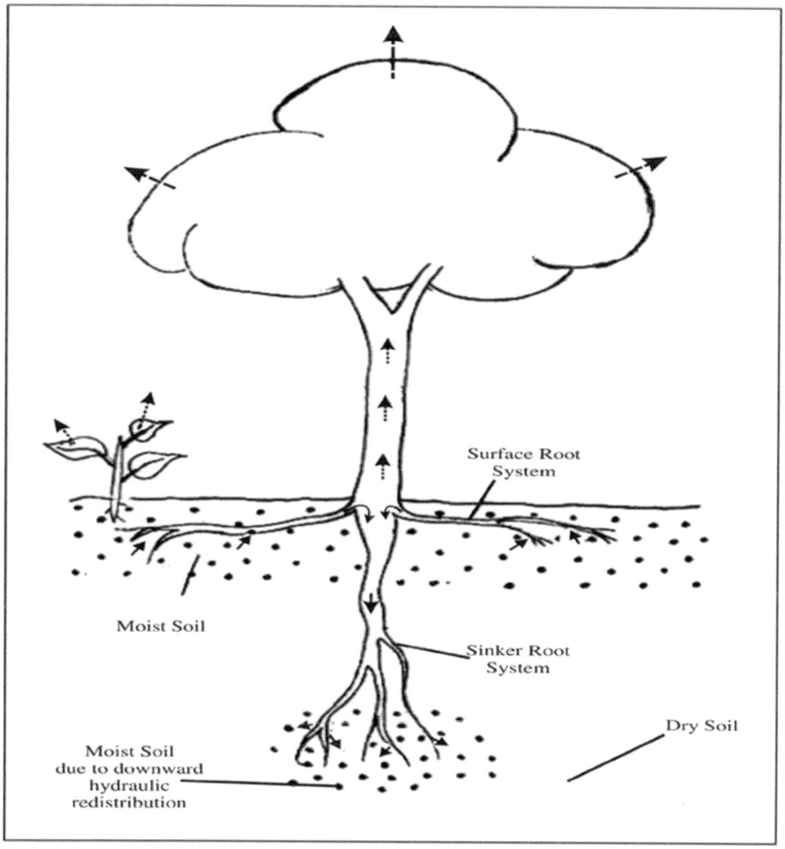

Fig. 21.5 Diagrammatic representation of downward hydraulic redistribution in trees (*dark dots* denote optimal soil moisture, *broken line arrows* indicate routine water movement due to transpiration pull, and *line arrows* denote water movement due to upward hydraulic redistribution). Note that the surface root system is in proper moist soil and the terminal ends of the sinker root system are in the dry soil profile. Downward hydraulic redistribution is due to which water accumulation occurs in the deeper soil profiles

21.4.2 Case Studies in Hydraulic Redistribution

Australian workers (Burgess et al. 1998; Stephen et al. 2001) were the first to provide convincing evidence on hydraulic redistribution using heat ratio method for sap flow measurements on a river red gum (*Eucalyptus camaldulensis*) and a coastal moort (*Eucalyptus platypus*). It was demonstrated that when root systems span soil layers of different moisture content, water is redistributed by roots in the direction of the difference in water potential. Tree sinker roots were shown to transfer significant quantities of water downwards to dry soil layers when surface soil layers become wet following rain. The benefits of this modification to the external environment include reduced waterlogging in surface soils and increased moisture content in dry subsoils.

Hultine et al. (2003) measured sap flow in sinker-tap roots, lateral roots and stems within a single individual in each of three co-occurring tree species *Fraxinus velutina*, *Juglans major* and *Celtis reticulata* in a Chihuahuan Desert to assess the seasonality and magnitude of hydraulic redistribution. The species showed hydraulic redistribution, but the patterns were not essentially similar. Nocturnal downward hydraulic redistribution in surface roots of *Fraxinus* was 0–120 g h^{-1} and 0–18 g h^{-1} in *Juglans*. No such downward hydraulic redistribution was recorded in *Celtis*. The workers concluded that species differences in nocturnal root function may have significant impacts on ecosystem hydrological fluxes and should be considered when scaling fluxes to catchment, landscape and regional levels.

Priyadarshini et al. (2015) demonstrated very interesting tree–grass coexistence involving hydraulic redistribution in semiarid savanna vegetation in South Africa. Experiment involved labelling deep soil (2.5-m depth) with a deuterium tracer. Trees and grasses used water from the topsoil after rainfall. All tree species shifted to groundwater or subsoil water use when there was no water in the topsoil indicating partitioning of water use. Grasses always used water from the topsoil. The seasonal changes in water source used by trees and grasses indicated possible shifts in tree–grass interactions during different periods of the year. The tracer experiment confirmed upward hydraulic redistribution in all the three tree species and water transfer to grasses via the topsoil. However, this occurred only in the dry season. An important facilitative mechanism maintaining tree–grass coexistence in savannas involving upward hydraulic redistribution was described for the first time.

Neumann and Cardon (2012) used data on hydraulic redistribution from 29 published papers focused on 16 different ecosystems and concluded that the movement of water from moist to dry soil through plant roots, both as upward and downward hydraulic redistribution, occurs worldwide within a range of different ecosystems and plant species. They computed average magnitude of hydraulic redistribution and reported it to vary by nearly two orders of magnitude across ecosystems, from 0.04 to 1.3 mm H_2O d^{-1} in the empirical literature and from 0.1 to 3.23 mm H_2O d^{-1} in the modelling literature. The authors considered these upward and downward hydraulic redistribution rates to be ecologically and hydrologically significant in many ecosystems, enhancing transpiration and photosynthetic carbon gain and conducting precipitation to deep soil layers.

21.4.3 Biological Significance of Hydraulic Redistribution

Some additional significance of the hydraulic redistribution is enlisted pointwise as follows:

During dry seasons, moisture content of the topsoil layers is rapidly depleted. Bulk of the root biomass of the shallow-rooted vegetation lies here and is prone to water stress. Growth during dry season may stop, and the very survival of the shallow-rooted plants may be at stake. By moving deep soil water to topsoils through upward hydraulic redistribution where bulk of root spread of the vegetation, particularly the shallow-rooted one, exists, the vital transpiration and photosynthesis of the vegetation are sustained (Hawkins et al. 2009). Excessive dryness of the topsoils may also restrict or abolish the active microbial populations. In the event of upward hydraulic redistribution, the nutrient availability through microbial, particularly mycorrhizae, is improved in the moist soil zones (Aanderud and Richards 2009; Lehto and Zwiazek 2011).

Active root–soil contact through fine roots with root hairs is of utmost importance. During dry spells in soil, this fine root system is the first casualty. Upward and downward hydraulic redistribution can prevent fine root damage in upper or lower soil profiles, respectively, as the case may (Bauerle et al. 2008).

More recently, it has been shown by Yu and Foster (2016) that deep-rooted CAM plants in CAM-grass associations could perform upward hydraulic redistribution at a higher rate than trees in tree–grass associations in a relatively wet environment, as explained by a significant increase in grass transpiration rate in the shallow soil layer, balancing a lower transpiration rate by CAM plants. By comparison, trees in tree-CAM associations may perform downward hydraulic redistribution at a higher rate than those in tree–grass associations in a dry environment.

21.5 Concluding Remarks

This article, incorporating some work done in the authors' laboratory, provides information regarding some lesser-known but important function of the tree root system involving sinker roots. These are (1) biodrainage that stabilizes the groundwater tables and (2) hydraulic redistribution that causes upward or downward movement of soil water through the tree roots. Both the processes have ecophysiological implications and need more investigations.

References

Aanderud ZT, Richards JH (2009) Hydraulic redistribution may stimulate decomposition. Biogeochemistry 95:323–333

Angrish R, Toky OP, Datta KS (2006) Biological water management: biodrainage. Curr Sci 90:897

Anonymous (2003) Biodrainage: status in India and other countries. Indian National committee on Irrigation and Drainage, New Delhi, pp 1–4

Bauerle TL, Richards JH, Smart DR, Eissenstat DM (2008) Importance of internal hydraulic redistribution for prolonging the lifespan of roots in dry soil. Plant Cell Environ 31:177–186

Burgess SSO, Adams MA, Turner NC, Ong CK (1998) The redistribution of soil water by tree root systems. Oecologia 115:306–311

Crosbie RS, Wilson B, Hughes JD, McCulloch C, King WM (2008) A comparison of the water use of tree belts and pasture in recharge and discharge zones in a saline catchment in the central west of NSW, Australia. Agric Water Manag 95:211–223

Hawkins H-J, Hettasch H, West AG, Cramer MD (2009) Hydraulic redistribution by Protea "Sylvia" (Proteaceae) facilitates soil water replenishment and water acquisition by an understorey grass and shrub. Funct Plant Biol 36:752–760

Heuperman AF, Kapoor AS, Denecke HW (2002) Biodrainage – principles, experiences and applications. Knowledge synthesis report No. 6. International Programme for Technology and

Research in Irrigation and Drainage (IPTRID), IPTRID Secretariat, Food and Agriculture Organization of the United Nations, Rome
Hultine KR, Williams DG, Burgess SOS, Keefer TO (2003) Contrasting patterns of hydraulic redistribution in three desert phreatophytes. Oecologia 135:167–175
Hultine KR, Scott RL, Cable WL, Goodrich DC, Williams DG (2004) Hydraulic redistribution by a dominant, warm-desert phreatophyte: seasonal patterns and response to precipitation pulses. Funct Ecol 18:530–538
Jackson RB, Ehleringer JR, Mooney HA, Sala OE, Schulze ED (1996) Maximum rooting depth of vegetation types at the global scale. Oecologia 108:583–595
Knight JH (1999) Roots distributions and water uptake patterns in *Eucalypts* and other species. In: Landsberg J (ed) The ways trees use water. Rural Industries Research and Development Corporation (RIRDC) Australia, Publication No. 99/37, pp 55–75
Lehto T, Zwiazek JJ (2011) Ectomycorrhizas and water relations of trees: a review. Mycorrhiza 21:71–90
Neumann RB, Cardon ZG (2012) The magnitude of hydraulic redistribution by plant roots: a review and synthesis of empirical and modeling studies. New Phytol 194:337–352
Pallardy SG (2010) Physiology of woody plants, 3rd edn. Academic, New York, 464 pp
Pasiecznik NM, Felker P, Harris, PJC, Harsh LN, Cruz G, Tewari JC, Cadoret K, Maldonado LJ (2001) The *Prosopis juliflora – Prosopis pallida* complex: a monograph. HDRA, Coventry, UK, p 172. ISBN: 0 905343 30 1
Priyadarshini KVR, Prins HHT, de Bie S, Heitkönig IMA, Woodborne S, Gort G, Kirkman K, Ludwig F, De Kroon H (2015) Seasonality of hydraulic redistribution by trees to grasses and changes in their water-source use that change tree–grass interactions. Ecohydrology 9:218–228
Ram J, Garg VK, Toky OP, Minhas PS, Tomar OS, Dagar JC, Kamra SK (2007) Biodrainage potential of E*ucalyptus tereticornis* for reclamation of shallow water table areas in north-west India. Agrofor Syst 69:147–165
Ritzema HP, Satyanarayana TV, Raman S, Boonstra J (2008) Subsurface drainage to combat waterlogging and salinity in irrigated lands in India: lessons learned in farmer's fields. Agric Water Manag 95:179–189
Stephen SO, Burgess SSO, Adams MA, Turner NC, White DA, Ong CK (2001) Tree roots: conduits for deep recharge of soil water. Oecologia 126:158–165
Tanji KK (1991) Agricultural salinity assessment and management. American Society of Civil Engineers, New York
Toky OP, Angrish R (2014) Mitigation of water logging and salinity through biodrainage: potential and practice. World Agric (UK) 4:72–77
Toky OP, Bisht RP (1992) Observations on the rooting patterns of some agroforestry trees in arid region of north-western India. Agrofor Syst 18:245–263
Wilde SA, Steinbrenner RS, Pierce RS, Dozen RC, Pronin DT (1953) Influence of forest cover on the state of groundwater table. Proc Soil Sci Soc Am 17:65–67
Wullschleger SD, Meinzer FC, Vertessy RA (1998) A review of whole plant water use studies in trees. Tree Physiol 18:499–512
Yu K, Foster A (2016) Modeled hydraulic redistribution in tree-grass, CAM-grass, and tree-CAM associations: the implications of crassulacean acid metabolism (CAM). Oecologia 180:1113–1125
Zanetti C, Weller A, Vennetier M, Menaux P (2011) Detection of buried tree root samples by using geoelectrical measurements: a laboratory experiment. Plant Soil 339:273–283

Plant Growth-Promoting Rhizobacteria Play a Role as Phytostimulators for Sustainable Agriculture

22

Sapna Gupta, Ruchi Seth, and Anima Sharma

Abstract

During the past few decades, increasing use of chemical fertilizers has caused many negative effects in agriculture: development of infectious agent resistance, adverse impact on nontarget species, and reduction in crop yield resulting from the harmful effects of chemical fertilizers on soil quality parameters. Thus, the search for an eco-friendly approach has been emphasized during the past several years. Plant growth-promoting rhizobacteria (PGPR) perform varied functions as (1) biofertilizers, (2) phytostimulators, (3) rhizoremediators, and (4) biopesticides. Plants do not seem to be axenic in natural conditions, and typically are influenced directly by completely different microorganisms such as rhizobacteria, of which several have the ability to provide phytohormones. This chapter sums up data relating to the synthesis, metabolism, regulation, physiological role, and agronomic impact of plant products made by plant growth-promoting rhizobacteria. We have included information regarding the auxins, cytokinins, gibberellins, and ethylene.

22.1 Introduction

Soil is a mixture of organic matter, minerals, gases, liquids, and innumerable microorganisms and macroorganisms that can support plant life. Soil operates as an engineering medium, a locale for soil organisms, a reprocessing system for organic dissipation of nutrients, a means to modify atmospheric composition, a manager of water supply, and a medium for plant growth. Over the years crop demands have increased as the world's population has increased. Thus, dependency on

S. Gupta • R. Seth (✉) • A. Sharma
JECRC University, Jaipur, Rajasthan, India
e-mail: gsapna309@gmail.com; ruchi.seth@jecrcu.edu.in; Ruchikool@gmail.com

D.K. Choudhary et al. (eds.), *Plant-Microbe Interaction: An Approach to Sustainable Agriculture*, DOI 10.1007/978-981-10-2854-0_22

agrochemicals such as chemical pesticides and fertilizers has increased many fold. The increased use of chemical fertilizers and pesticides to support crop yield and to manage plant diseases has resulted in serious issues such as the accumulation of chemicals in plant products and their entry into the food chain. Moreover, their continuous use leads to the generation of resistant strains of pests and pathogens (Agrios 1988). Further, chemical fertilizers contaminate water resources and adversely affect the populations of naturally occurring beneficial organisms, which has resulted in the depletion of soil fertility (Kuhajek et al. 2003). Chemically based products generally persist in the environment: they are cyanogenic and nonbiodegradable in nature and also exert harmful effects on animals, human health, and our environment. These factors have generated alarms for limiting the usage of agrochemicals in regard to food quality and safety.

Soil fertility is directly or indirectly related to the microorganisms residing in the soil because they are a vital part of the soil ecosystem with a major role in plant growth. Microorganisms improve the value of the soil and are involved in different biotic activities of the soil bionetwork to enhance it for nutrient turnover and sustainable crop production (Ahemad et al. 2009; Chandler et al. 2008). Microorganisms support plant growth through collecting the nutrients in soils, fabricating many plant growth regulators, defending plants from phytopathogens by dominating or inhibiting them, enhancing soil structure, and bioremediating impure soils by sequestering toxic heavy metals and degrading xenobiotic compounds (such as pesticides) (Chandler et al. 2008; Podile and Kishore 2006; Barea et al. 2005; Kloepper et al. 1991; Kloepper and Okon 1994).

22.2 Rhizosphere and Rhizobacteria

The term "rhizosphere" was derived by the German expert and plant life scientist Lorenz Hiltner in 1904 to explain the plant–root interface: this word is fabricated from the Greek word "rhiza," which means root (Hiltner 1904; Hartmann et al. 2008). According to wide-ranging opinion, the area around a plant root that is colonized by a distinctive population of microorganisms and roots of plants releasing chemical compounds would be termed the rhizosphere. The rhizosphere includes three major zones (endorhizosphere, rhizoplane, and ectorhizosphere), which are defined on the basis of their relative closeness to plant tissue. The endorhizosphere includes parts of the cortex and endodermis within which microbes and cations occupy the "free space" between cells (apoplastic space). The medial zone directly together with the foundation cuticle and mucilage is the rhizoplane. The outermost zone is the ectorhizosphere, which extends from the rhizoplane out into the bulk soil (Barea et al. 2005; Kloepper et al. 1991; Kloepper and Okon 1994). High levels of water content and nutrients within the rhizosphere attract larger numbers of microorganisms than those further away within the soil. The composition and pattern of root exudates have an effect on microorganism action and population numbers that, in turn, have an effect on the different soil organisms which share this environment. The rhizosphere zone is about 1 mm wide and is supplemented with sugars, amino acids, secondary metabolites, DNA, and polysaccharides.

As plant roots grow through the soil, they release water-soluble compounds such as amino acids, sugars, organic acids, vitamins, enzymes, inorganic ions, and gaseous compounds that offer food for the microorganisms. All these activities make the rhizosphere the most active surroundings within the soil because the roots are underground and rhizosphere activity has been for the most part been unseen because of the occurrence of the advanced interactions of microbes among the roots. The exclusive biological, chemical, and physical properties of soils that are related to roots, compared to the soils far away from the root and root surface, are accountable for improved microorganism populations at the site of enhanced numbers and microorganism activity within the rhizosphere (Zaidi et al. 2009).

Bacteria present in the rhizospheric soil, called rhizobacteria, have an important function in plant growth and development. Rhizobacteria are divided into those that form a dependent relationship with the plants and those which do not. Those that do not form a dependent relationship are referred to as free-living (nonsymbiotic), closely connected with the root surface, or existing within the roots as endophytic bacteria (Kloepper et al. 1989). Rhizobacteria exert a beneficial effect through increasing soil fertility and crop improvement by their various direct and indirect mechanisms under various environmental conditions. The microorganisms lodging around or in the plant roots (rhizobacteria) are very adjustable in reworking, mobilizing, and solubilizing the nutrients as compared to different microbes present in bulk soils. Therefore, the rhizobacteria are the dominant deriving forces in utilization of soil nutrients and, as a consequence, they have a significant role in soil fertility and plant growth (Vaishnav et al. 2014).

22.3 Plant Growth-Promoting Rhizobacteria (PGPR)

Soil bacterial species growing in the plant rhizosphere that grow in, on, or around plant tissues and stimulate plant growth by numerous mechanisms are jointly referred to as PGPR (plant growth-promoting rhizobacteria). The term PGPR was coined by Kloepper and Schroth to explain the helpful microorganism population that colonizes the roots of plants within the rhizosphere, once attached to the root surface, to support plant growth and plant growth promotion activities (Chandler et al. 2008). PGPR can be classified into completely different subgroups on the basis of their application: (1) biofertilizers (increasing the availability of nutrients to plants), (2) phytostimulators (plant growth promotion, usually through

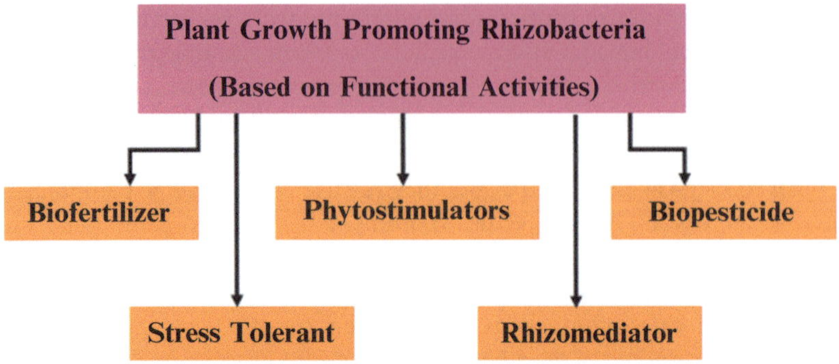

Fig. 22.1 Classification of plant growth-promoting rhizobacteria (PGPR)

phytohormones such as indole-3-acetic acid), (3) rhizoremediators (degrading organic pollutants), and (4) biopesticides (controlling diseases, chiefly by the production of antibiotics and antifungal metabolites) (Fig. 22.1) (Antoun and Prévost 2005). All PGPR perform either directly or indirectly to facilitate or support plant growth under nutritionary, biotic (biocontrol, PGPB), or abiotic stress conditions in three different ways by synthesizing explicit compounds for the plants (Dobbelaere et al. 2003; Zahir et al. 2004), facilitating the uptake of bound nutrients from the soil (Lucas et al. 2004a, b; Çakmakçi et al. 2006), and reducing or preventing plant diseases (Guo et al. 2004; Jetiyanon and Kloepper 2002; Raj et al. 2003) (Fig. 22.2). These mechanisms can work autonomously or simultaneously with one another. The indirect plant growth promotion inspired by biocontrol PGPB includes a variety of mechanisms such as rhizosphere competition, rhizospheric engineering, quorum sensing, production of volatile organic compounds, enzyme production, induced systematic resistance (ISR), reduction or prevention of deleterious effects of phytopathogens on plant growth by biosynthesis of stress-related phytohormones such as jasmonic acid (JA) or ethylene, and biosynthesis of antimicrobial molecules (Jain et al. 2014). In direct plant growth promotion mechanisms, PGPR assists the uptake of nutrients from the environment by nitrogen fixation, diminishes toxic compounds, provides phytohormones such as auxins, gibberellins (GAs), cytokinins (CK), and nitric oxide (NO), phosphate solubilization, and iron sequestration by siderophore production (Choudhary et al. 2015).

To provide tolerance to host plants under different environmental conditions, various bacteria belonging to different genera have been reported in the last decade, including *Achromobacter*, *Bacillus*, *Pseudomonas*, *Methylobacterium*, *Pantoea*, *Paenibacillus*, *Variovorax*, *Azospirillum*, *Microbacterium*, *Burkholderia*, *Rhizobium*, and *Enterobacter*, similar to the diagram (Choudhary 2012). PGPR enjoy a close in-depth organization with the plant and so they are vitally important candidates to be developed as tools for improving plant stimulatory factors and plant health, growth, and development.

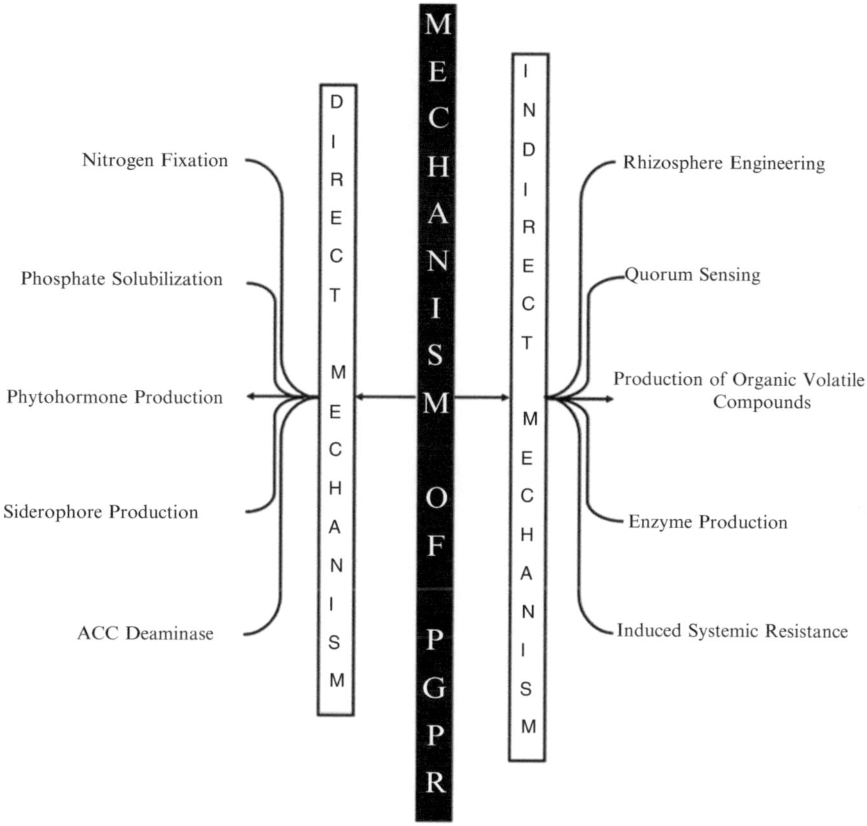

Fig. 22.2 Mechanisms of PGPR

22.4 PGPR: Phytohormone Producers

Phytohormones are naturally available to the plants from two sources: endogenous production by the plant tissues, and exogenous production by associated microorganisms. These phytohormones are iinvolved in several restrictive functions that comprise biological process and enlargement, cell division and expansion, cell elongation, stem elongation, root growth, activation of bud growth, branch maturity, promotion or delay of leaf senescence, and chlorophyll production (Kumar and Lonsane 1989; Arshad and Frankenberger 1991; Costacurta and Vanderleyden 1995; Patten and Glick 1996).

Phytohormone production by PGPR is one of the foremost imperative mechanisms that promote plant growth (Spaepen et al. 2007). Phytohormones are natural signal molecules acting as chemical messengers. They have an associated necessary role as growth and development regulators in extraordinarily low concentrations; as a result they influence biochemical, physiological, and morphological processes in plants, and their synthesis is finely regulated (Fuentes-Ramírez and

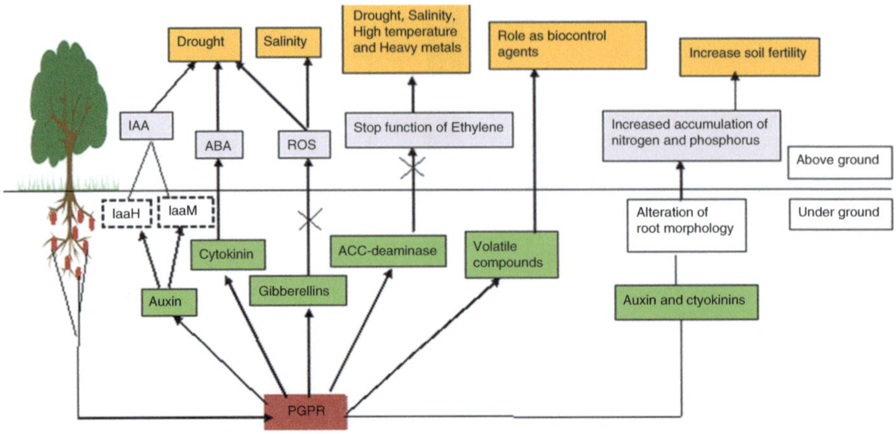

Fig. 22.3 Graphic representation of bacterially produced phytohormones in plant growth regulation

Caballero-Mellado 2006). The phytohormone manufacturing ability is widely distributed among microbes related to soil and plants. Various fungal and bacterial species, such as cyanobacteria, can produce phytohormones (Tsavkelova et al. 2006). Rhizospheric, epiphytic, and symbiotic bacteria known as PGPR are able to secrete hormones. However, free-living microorganisms are reported to provide phytohormones (Spaepen et al. 2007). Growth regulators (indole-acetic acid), cytokinins, gibberellins, and ethylene are the most important plant hormones. These plant hormones are also synthesized by bacteria that are directly associated with plant growth (Pirlak and Kose 2009; Bloemberg and Lugtenberg 2001; Bottini et al. 2004). The level of phytohormones in plant-associated bacteria is a critical factor in phytostimulation on the plant side (Spaepen et al. 2007) (Fig. 22.3).

22.4.1 Auxins

An auxin is a crucial cluster of chemical compounds distinguished by their capability to encourage cell elongation in the subapical region of the stem and to reproduce the physiological impact. Auxins have a crucial role in the regulation of numerous plant growth processes: (a) gravitropism and tropism, (b) plant tissue differentiation, (c) apical dominance, (d) lateral and adventitious root initiation, (e) stimulation of cell division, and (f) stem and root elongation (Teale et al. 2006). Indole-3-acetic acid (IAA) is a naturally occurring auxin molecule that is the most abundant chemical compound.

A number of molecules are categorized as auxins; however, IAA is among the foremost established and vigorous in biological systems. Different molecules are sensitive to indole-3-butyric acid (IBA) and phenylacetic acid (PAA), and in addition to the precursor indole-3-acetonitrile (IAN), are considered active auxins. A variety of inactive molecules, together with IAA halogenate compounds such

4-chloroindole-3-acetic acid and conjugated forms with sugars, alcohols, amino acids, and glycoproteins, are known in plants and bacteria (Glick et al. 1999; Korasick et al. 2013).

Various very important plant–microbial communications center on the fabrication of auxins; among them, IAA is the chief plant growth regulator (auxin). The ability to synthesize IAA has been detected in several bacterial species such as rhizobacteria, as in pathogenic, symbiotic, and free-living bacteria (Tsavkelova et al. 2006; Costacurta and Vanderleyden 1995). At present, auxin-synthesizing rhizobacteria are the foremost well-studied phytohormone producers (Spaepen et al. 2007; Tsavkelova et al. 2006).

The rhizobacteria synthesize IAA by two major pathways: the tryptophan (Trp)-independent and Trp-dependent pathways. In plants, two major pathways have been postulated for Trp-dependent IAA biosynthesis: (1) the indole-3-acetamide (IAM) pathway and (2) the indole-3-pyruvic acid (IPyA) pathway (Spaepen et al. 2007). The indole acetamide pathway is principally utilized by phytopathogenic bacteria for the production of IAA, which is responsible for tumor induction in plants. Utilization of the IAM pathway by beneficial bacteria for IAA biosynthesis is not clear. In contrast, the PGPB make use of the IPyA pathway for IAA biosynthesis (Patten and Glick 2002).

Azospirillum is one of the simplest IAA producers among the PGPR species studied (Dobbelaere et al. 1999). From different rhizospheric soils, other IAA-producing bacteria belonging to the genera *Aeromonas*, *Azotobacter*, *Bacillus*, *Burkholderia*, *Enterobacter*, *Pseudomonas*, and *Rhizobium* have been isolated (Halda-Alija 2003; Ahmad et al. 2008; Swain et al. 2007; Shoebitz et al. 2009; Hariprasad and Niranjana 2009; Ghosh et al. 2008). IAA-manufacturing PGPR has been inoculated to stimulate seed germination, to extend the root biomass, to accelerate root growth, and to modify the design of the root system. An IAA-producing *Mycobacterium* sp. strain has been reported to extend the germination of orchid seeds (*Dendrobium moschatum*) (Tsavkelova et al. 2006). Besides stimulating root growth, IAA-producing bacteria can also be used to stimulate tuber growth. In one study, an IAA-producing *Bacillus subtilis* strain had a positive effect on the edible tubercle *Dioscorea rotundata* L. and increased the length and fresh weight of root/shoot and the root:stem ratio and numbers of sprouts as compared with noninoculated plants. There are several techniques for detection of IAA and related indole compounds (Fig. 22.4).

22.4.2 Detection Techniques of IAA

Indole is generated by indole pyruvic acid via subtractive deamination of tryptophan. Through the deamination reaction, during which tryptophanase catalyzes the amino alkane ($-NH_2$) group, the tryptophan molecule is removed, and therefore the final products of the reaction are indole, pyruvic acid, ammonium (NH_4^+), and energy. Pyridoxal phosphate is needed as a coenzyme. To determine the assembly of IAA from the culture, the most necessary requirement is that the medium

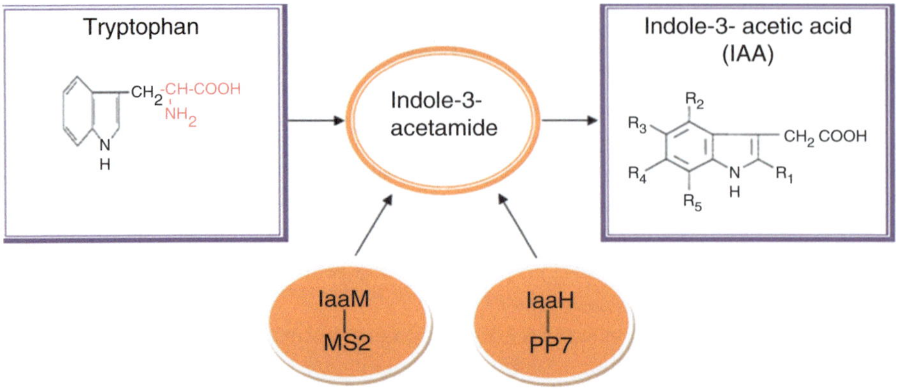

Fig. 22.4 Formation of indole-acetic acid via tryptophan

contains an adequate amount of tryptophan and a pinch of sodium chloride as per Difco to culture an organism before the indole test (Mac Faddin 1976). For an associated alternate approach to IAA production, casein peptone, sodium chloride, and tryptone are also used as a medium (Mac Faddin 1976). At the same time, deciding alternative characteristics such as motility and therefore the ability to produce hydrogen sulfide as a by-product of metabolism of the bacteria, the sulfide indole motility (SIM) medium could be a multi-test agar used to test for indole production (Mac Faddin 1976). Another multi-test agar is motility-indole ornithine (MIO) medium. In addition to testing for indole production, it is used to test for motility and ornithine decarboxylase, causing an rise in pH in the tube. The positive results of this test are indicated by the purple-gray color throughout the tube (Hiroya et al. 2004).

We can also detect the assembly of IAA production by many alternative strategies, such as detection IAA with biochemical assay, detection of IAA by capillary electrophoresis, qualitative detection of IAA by thin-layer chromatography (TLC) and paper chromatography, quantitative determination by high pressure liquid chromatography (HPLC), detection by the HPTLC method, detection of auxin (IAA) by chromatography/mass spectroscopy, and detection by Fourier transform infrared (FTIR) analysis.

22.4.3 Cytokinin

Cytokinins are units of phytohormones that control cell division, the cell cycle, and differentiation and stimulate developmental processes in plants (Srivastava 2002). By structure, cytokinins are divided into two subgroups: the adenine-type and the phenylurea-type cytokinin group. The adenine-type cytokinin group is pictured by natural and artificial compounds such as kinetin (K), zeatin (Z), or 6-benzylaminopurine (6-BAP), and therefore the phenylurea-type cytokinin group is represented by the synthetic molecules diphenylurea and thidiazuron (TDZ). Chemically, adenine-type cytokinins are mostly purines and derived from adenine and modified by substitutions on the N6, which also contains their several ribotides, ribosides, and glycosides. The stimulatory or repressing functions of cytokinins are related to several physiological and biological processes, including senescence delay by chlorophyll accumulation and organ formation in a wide range of tissues, root and shoot development, leaf expansion, control of apical dominance in the shoot, and chloroplast development (Sakakibara 2006; Werner et al. 2001). By definition, once these compounds are combined with an optimal auxin concentration they induce cell division in plants. Miller et al. (1955) discovered the primary artificial cytokinin molecule that was named kinetin (K). In 1963, Letham knew 50 molecules referred to as zeatin (Z), and their metabolites have been classified as CKs. The biological activity for all CK-like compounds is not uniform and normally depends on many structural aspects such as a purine ring within the molecule, substitution of N6 with a simple ribosyl chain isopurine-derived unit, and substitution on positions two and nine of the ring for H, CH_3–S, or an unsaturated side chain (optimally five carbons). The natural and artificial adenine-type cytokinin molecules with confirmed biological activity on plant tissues are zeatin, isopentenyl adenine, kinetin, and 6-benzylaminopurine, and all have a double alkyl bridge at position N6.

By altering the size and activity of meristems, cytokinin influences cell division activity in embryonic as well as mature plants (Werner et al. 2001). Yang et al. (2002) demonstrated that the pace of reproductive structure cell division is intimately connected with the cytokinin level in the endosperm. They also reported that exogenous kinetin significantly increased the number of endosperm cells and grain weight. These various rhizobacterial strains (*Halomonas desiderata*, *Proteus mirabilis*, *P. vulgaris*, *Klebsiella pneumoniae*, *Bacillus megaterium*, *B. cereus*, *B. subtilis*, and *Escherichia coli*) were detected as phytohormones producing cytokinins (Arkhipova et al. 2005; Karadeniz et al. 2006; Ali et al. 2009). Different cytokinins are detected not only in the culture medium but also in the biomass of microorganisms (in free state or bound to certain tRNAs) within the type of either adenine derivatives, isoprenylated at N6 position, or their ribosides, such as 6-benzyladenine, N6-isopentenyl adenosine, and zeatin riboside (Serdyuk et al. 2003). Krall et al. repoted that *trans*-zeatine has also been found in the culture of *Agrobacterium tumefaciens* (Krall et al. 2002). Ryu et al. (2003) reported that cytokinin from a bacterial origin improves growth in *Arabidopsis*. Inoculation of a plant with bacteria-producing cytokinin has been shown to stimulate shoot growth and reduce root:shoot ratio in drought-stricken plants (Arkhipova et al. 2007). *Rhizobium* was reported to enhance cytokinin production in plants by regulation of expression of the communication pathway and to trigger cortical cells to divide in plants (Oldroyd 2007). A variety of these effects has been observed in wheat when cultured with *Paenibacillus polymyxa* strains B1 and B2 (Lindberg and Granhall 1984; Lindberg and Granhall 1986). One of every strain (B2) was, therefore, chosen for further investigation.

22.4.4 Detection of Cytokinin

Reverse section column under acidic condition is used for separation of cytokinin-like compounds. Acetic acid or formic acid and their ammonium salts are added to the solvent methanol/acetonitrile for better separation (Ge et al. 2005). UV detection is appropriate for detection of cytokinin because it exhibits strong UV absorbance between 200 and 300 nm. These differing kinds of ionization techniques were used for mass analysis of cytokinin together with reversed-phase (RP)-HPLC including thermospray, electrospray, atmospheric pressure chemical ionization, and fast atom bombardment (Novák et al. 2003).

22.4.5 Gibberellins (GAs)

Gibberellins (GAs) are a vast range of tetracyclic diterpene acids that elicit numerous metabolic functions needed throughout plant growth, at the site of seed germination, stem elongation, sex expression, flowering, fruiting, and senescence (Hedden 1997; Hedden and Kamiya 1997; Davies 1995). GAs are known and isolated from higher plants and from bacterial and fungal species. To date, 136 gibberelins from higher plants (128 species), 28 GAs from fungi (7 species), and only 4 GAs (GA1, GA3, GA4, GA20) from bacteria (7 species) are known (MacMillan 2002). Plant growth promotion and development by PGPR species that produce GAs has been previously reported (Atzhorn et al. 1998; Bastian et al. 1998; Gutierrez-Manero et al. 2001).

Gibberellins are present in two main forms, the free form and the conjugated form. Free gibberellins are subdivided into two subgroups. One subgroup possesses the entire complement of carbon atoms, or is referred to as C20-GAs; within the alternative subgroup, the C20 is lost so it is called the C19-GAs. Except GA12-aldehyde, all gibberellins are carboxylated at C7, and possess one (G4), two (GA1), three (GA8), or four (GA32) hydroxyl functions. The biological activity of the molecule is determined by the position of the hydroxylation (OH). Hydroxylation of C3 and C13 in their β- and α-positions, respectively, results in activation of the molecule, whereas the hydroxylation of C2 in the β-position has a strong negative effect on activity (Pearce et al. 1994). In addition to the opposite types of GAs, such as free forms, conjugated forms are known in plants: these include organic glycoside ethers (GA-G), in which a sugar molecule is connected to the structure of the GA by a hydroxyl group, and glycoside esters (GA-GE), in which a sugar residue is bound to the hormone through a carboxyl group on C7 (Sembder et al. 1980). The organic chemistry and physiological aspects of the GA conjugates have been discussed by Rood and Pharis (1987), who recommend that the most notable feature of those compounds is the lack of biological activity and therefore the potential reversibility to the active forms by hydrolytic enzyme activity. GA production by PGPR promotes the expansion and yield of many crop plants by deconjugation of gibberellin glucosyl in the root zone, causing 3β-hydroxylation of inactive 3-deoxy GAs to active forms such as GA1, GA3, and GA4 bacterial enzymes (Cassan et al. 2001a, b; Piccoli et al. 1996).

Bottini et al. (1989) were the first to validate the ability of *Azospirillum* sp. to produce gibberellins in a chemically outlined medium. Using gas chromatography–mass spectroscopy (GC-MS) analysis, they reported the production of GA1 and GA3 in a nitrogen-free medium culture of *Azospirillum lipoferum* Op33. Similar results were reported in *Azospirillum brasilense* Cd and in *A. lipoferum* AZm5 and *A. brasilense* VS9 (Janzen et al. 1992; Esquivel-Cote et al. 2010). In addition, the assembly of inactive precursors GA19 and GA9 in a chemically defined medium of *A. lipoferum* Op33 was reported (Piccoli et al. 1996). Kang et al. (2014) isolated *Leifsonia soli* sp. SE134 and detected different GAs by chromatographic analysis. Application of *L. soli* culture filtrate was found to have considerably increased biomass, hypocotyl, and root lengths of cucumber seeds as compared to noninoculated. Similarly, Pandya and Desai (2014) isolated and identified *Pseudomonas monteilii*. The culture filtrate of this bacterium was bioassayed on wheat and chana bean crops and was found to significantly promote growth in both plants. In the same manner, *Acinetobacter calcoaceticus*-inoculated cucumber plants exhibited higher GAs (GA1, GA4, GA9, GA20) as compared to noninoculated plants. The PGPR activated the GAs biosynthesis pathway, thereby promoting cucumber plant growth (Kang et al. 2012).

22.4.6 Detection of Gibberellin

The estimation of gibberellic acid should be done in a specific medium, wherever an organism can increase its biomass that results in the production of gibberellin or gibberellic acid-like substances. Gibberellic acid from *Fusarium* species was determined by the acid–base volumetric analysis technique, in which gibberellic acid was titrated with 0.1 or 0.25 N NaOH solution using phenolphthalein as an indicator and measured in 10^{-3} gram equivalents (milliequivalent) of gibberellic acid. The gibberellin phytohormone was also detected by several other techniques such as spectrophotometric assay: qualitative estimation by TLC, HPLC, and paper chromatography.

22.4.7 Ethylene

Another necessary hormone in plant growth and development is ethylene (Et). A protected Et was not thought to be a phytohormone because of its gaseous state under physiological conditions. Currently, totally different studies have shown that its synthesis and action are vital for certain physiological processes. A large number of publications have been found related to the synthesis of Et in higher plants, but only a few studies have been published on the microbial biosynthesis of ethylene (Arshad and Frankenberger 1993). It is a simple and symmetrical molecule composed of two carbon atoms (joined by a double bond) and four hydrogen atoms. It is water soluble, and at very low concentrations in plant tissues (about 0.1 ppm) will exert physiological effects. In higher plants, all tissues have the potential to manufacture this hormone; however, in general its concentration is related to the developmental state and growth phase of the plant, with a higher concentration in those tissues involved in vigorous cell division, those which are under a stressful environment, or those in a senescence stage (Burg and Burg 1968). It is produced in principally all plants and mediates in an exceedingly wide selection of various responses and developmental processes (Arshad and Frankenberger 2002; Belimov et al. 2002). The presence of ethylene also in some instances has a stimulatory result, whereas in others it represses this result, depending upon its concentration in the tissues of the roots, the physiological nature being processed, and the stage of plant growth. The endogenous level of ethylene, which causes a change in a plant tissue, may modify growth and development (Arshad and Frankenberger 2002).

Ethylene influences varied physiological processes together with liberation of dormancy, shoot and root growth differentiation, adventitious root development, initiation of flowering, and amplified female function in dioecious plants, flowering, fruit ripening, leaf senescence, and leaf and fruit abscission (Abeles et al. 1992; Johnson and Ecker 1998). Different studies showed that Et was involved in premature shedding of leaves, the geotropism of etiolated pea seedlings on exposure to a revealing gas, premature flowering of pineapples treated with smoke, and maturation of oranges exposed to gas from fuel combustion (Arshad and Frankenberger 2002; Abeles et al. 1992). Aside from the positive effects of ethylene, its overproduction may lead to abnormal growth of roots and induce defoliation and cellular processes that also result in inhibition of stem and root growth as early senescence, all of which lead to reduced crop performance because of adverse affects on plant growth and development (Ovakim et al. 2000). Ethylene production in plant roots is accelerated in response to both biotic and abiotic stresses (Arshad and Frankenberger 2002; Abeles et al. 1992). Senescence of plant leaves is among the most important symptoms of accelerated ethylene levels (Arshad and Frankenberger 2002). There is a dire need to regulate ethylene production in the rhizosphere for normal growth and development of the plants.

Plants respond to different stresses by synthesizing 1-aminocyclopropane-1-carboxylate (ACC), which is the precursor for ethylene (Chen et al. 2002; Glick et al. 2007). Some of the ACC is secreted into the rhizosphere and is readsorbed by the roots, where it is converted again into ethylene. This accumulation of ethylene results in a downward-sloping spiral effect, because poor root growth leads to a diminished capability to obtain water and nutrients, which in turn leads to the promotion of stress. Thus, PGPR, with the ability to degrade ACC in the rhizosphere, can help to break this downward cycle and reinstate the healthy root system that is required to manage environmental stress. The primary mechanism that is used by rhizobacteria that degrade ethylene is the destruction of ethylene via the enzyme ACC-deaminase. This enzyme can diminish or prevent some of the harmful effects of high levels of ethylene (Glick et al. 1998). ACC-deaminase acts on ACC, an on-the-spot ethylene precursor in higher plants, degrading this chemical to alpha-ketobutyrate and ammonium (Glick et al. 1998; Grichko and Glick 2001; Mayak et al. 2004). Rhizosphere bacteria with ACC-deaminase activity belonging to the

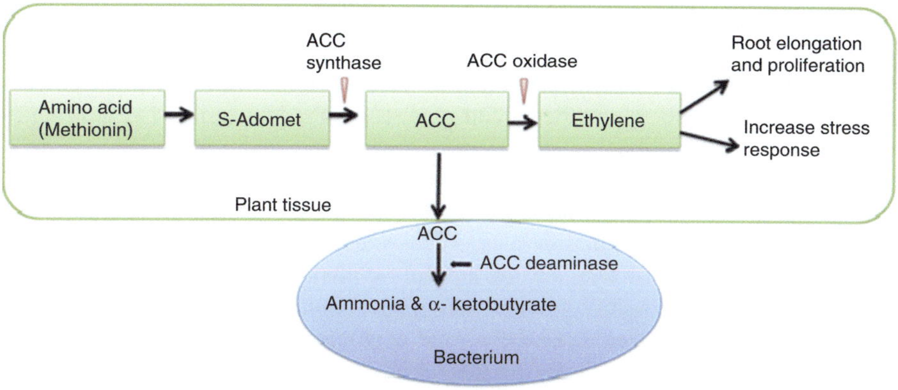

Fig. 22.5 Mechanism of bacteria that reduce ethylene levels in the plant root using bacterial 1-aminocyclopropane-1-carboxylate (ACC) deaminase

genera *Achromobacter*, *Azospirillum*, *Bacillus*, *Enterobacter*, *Pseudomonas*, and *Rhizobium* have been isolated from different soils (Ovakim et al. 2000; Govindasamy et al. 2008; Ghosh et al. 2003; Duan et al. 2009). Various studies have demonstrated that plants treated with PGPR bacteria that produce ACC-deaminase have increased their resistance to environmental stress. Grinchko and Glick (2001) inoculated tomato seeds with the ACC-deaminase expressing the bacteria *Enterobacter cloacae* and *Pseudomonas putida* and registered a rise in plant resistance on 55 days of aging to 9 consecutive days of flooding. Ghosh et al. (2003) found ACC-deaminase activity in three *Bacillus* species (*Bacillus circulans* DUC1, *Bacillus firmus* DUC2, and *Bacillus globisporus* DUC3), which stimulated root elongation of *Brassica campestri* plants. Mayak et al. (2004) evaluated tomato plants inoculated with the bacterium *Achromobacter piechaudii* under water and saline stress conditions. The authors reported a major increase in fresh and dry weight of inoculated plants. In soils with a high copper content, Reed and Glick (2005) reported an increase in dry matter content of the root and the air part in rape seeds inoculated with the ACC-deaminase-producing bacterium *Pseudomonas asplenii* (Fig. 22.5).

22.4.8 Ethylene/ACC Detection Assay

Ethylene can be detected by gas chromatography (GC) or GC–mass spectroscopy (GC-MS). We can indirectly check the amount of ACC by HPLC and spectrophotometric analysis. Several studies also detected ACC-deaminase (ACC-D) enzyme activity in ACC-D-producing bacteria. Bacterial cells were induced by ACC for a time period and then labeled by toluene, a supernatant used for the quantification of ACC-deaminase activity by observing the amount of α-ketobutyrate produced by ACC through the ACC-deaminase enzyme (Penrose and Glick 2003; Kumari et al. 2016).

22.5 Conclusion

Phytohormone production by bacteria has been a groundwork topic for several decades in either infective or beneficial plant-associated bacteria. Phytohormone-producing PGPR contribute to eco-friendly, sustainable, and organic farming, providing high yield and quality in sustainable agriculture and therefore alleviating food deficiencies. These phytohormones act as effector metabolites in plant–microbe interactions and phytostimulation in agro-ecosystems. Bacteria manufacturing phytohormones within the rhizosphere are helpful for plant growth and development by triggering nutrient accessibility, encouraging root colonization, and imparting protection from phytopathogens. However, the ecological significance of bacterially produced phytohormones still needs exploration. As each plant and bacteria secretes these hormones in the rhizosphere, it is difficult to determine the contribution of each one.

Acknowledgments Some of the research in the present review has been supported in part by JECRC University, Jaipur, and Rajasthan. The author acknowledges Mr. Anukool Vaishnav for his careful review of the content and for improvement of the text. The author also thanks Mr. Mohit Agrawal and Mr. Gaurav Kaushik for valuable suggestions and guidance.

References

Abeles FB, Morgan PW, Saltveit ME (1992) Ethylene in plant biology. Academic, San Diego

Agrios GN (1988) Plant pathology, 3rd edn. Academic, San Diego

Ahemad M, Khan MS, Zaidi A, Wani PA (2009) Remediation of herbicides contaminated soil using microbes. In: Khan MS, Zaidi A, Musarrat J (eds) Microbes in sustainable agriculture. Nova Science, New York, p 358

Ahmad F, Ahmad I, Khan MS (2008) Screening of free-living rhizospheric bacteria for their multiple plant growth promoting activities. Microbiol Res 163:173–181

Ali B, Sabri AN, Ljung K, Hasnain S (2009) Auxin production by plant associated bacteria: impact on endogenous IAA content and growth of *Triticum aestivum* L. Lett Appl Microbiol 48:542–547

Antoun H, Prévost D (2005) Ecology of plant growth promoting rhizobacteria. In: Siddiqui ZA (ed) PGPR: biocontrol and biofertilization. Springer, Dordrecht, pp 1–38

Arkhipova TN, Veselov SU, Melentiev AI, Martynenko EV, Kudoyarova GR (2005) Ability of bacterium *Bacillus subtilis* to produce cytokinins and to influence the growth and endogenous hormone content of lettuce plants. Plant Soil 272:201–209

Arkhipova TN, Prinsen EA, Veselov SU, Martinenko EV, Melentiev AI, Kudoyarova GR (2007) Cytokinin producing bacteria enhance plant growth in drying soil. Plant Soil 292:305–315

Arshad M, Frankenberger WTJ (1991) Microbial production of plant hormones. In: Keister DL, Cregan PB (eds) The rhizosphere and plant growth. Kluwer Academic, Dordrecht, pp 327–334

Arshad M, Frankenberger R (1993) Microbial production of plant growth regulators. In: Meeting BF (ed) Soil microbial ecology. Dekker, New York, pp 307–347

Arshad M, Frankenberger WT (2002) Ethylene: agricultural sources and applications. Kluwer Academic, New York, pp 342

Atzhorn R, Crozier A, Wheeler CT, Sandberg G (1998) Production of gibberellins and indole-3-acetic acid by *Rhizobium phaseoli* in relation to nodulation of *Phaseolus vulgaris* roots. Planta 175:532–538

Barea JM, Pozo MJ, Azcon R, Aguilar CA (2005) Microbial co-operation in the rhizosphere. J Exp Bot 56:1761–1778

Bastian F, Cohen A, Piccoli P, Luna V, Baraldi R, Bottini R (1998) Production of indole-3-acetic acid and gibberellins A1and A3 by *Acetobacter diazotrophicus* and *Herbaspirillum seropedicae* in chemically defined media. Plant Growth Regul 24:7–11

Belimov AA, Safranova VI, Mimura T (2002) Response of spring rape (*Brassica napus*) to inoculation with PGPR containing ACC-deaminase depends on nutrient status of plant. Can J Microbiol 48:189–199

Bloemberg GV, Lugtenberg BJ (2001) Molecular basis of plant growth promotion and biocontrol by rhizobacteria. Curr Opin Plant Biol 4:343–350

Bottini R, Fulchieri M, Pearce D, Pharis R (1989) Identification of gibberellins A1, A3, and Iso-A3 in cultures of *A. lipoferum*. J Plant Physiol 90:45–47

Bottini R, Cassán F, Piccoli P (2004) Gibberellin production by bacteria and its involvement in plant growth promotion and yield increase. Appl Microbiol Biotechnol 65:497–503

Burg S, Burg P (1968) Ethylene formation in pea seedlings: its relation to the inhibition of bud growth caused by indole-acetic acid. Plant Physiol 43:1069–1074

Çakmakçi R, Dönmez F, Şahin F (2006) Growth promotion of plants by plant growth promoting rhizobacteria under greenhouse and two different field soil conditions. Soil Biol Biochem 38:1482–1487

Cassan F, Bottini R, Schneider G, Piccoli P (2001a) *Azospirillum brasilense* and *Azospirillum lipoferum* hydrolyze conjugates of GA20 and metabolize the resultant aglycones to GA1 in seedlings of rice dwarf mutants. Plant Physiol 125:2053–2058

Cassan F, Lucangeli C, Bottini R, Piccoli P (2001b) *Azospirillum* spp. metabolize [17,17-2H2] gibberellin A20 to [17,17-2H2] gibberellin A1 in vivo in dry rice mutant seedlings. Plant Cell Physiol 42:763–767

Chandler D, Davidson G, Grant WP, Greaves J, Tatchell GM (2008) Microbial biopesticides for integrated crop management: an assessment of environmental and regulatory sustainability. Trends Food Sci Technol 19(5):275–283

Chen H, Quails RG, Miller GC (2002) Adaptive responses of *Lepidium latifolium* to soil flooding: biomass allocation, adventitious rooting, aerenchyma formation and ethylene production. Environ Exp Bot 48:119–128

Choudhary DK (2012) Microbial rescue to plant under habitat-imposed abiotic and biotic stresses. Appl Microbiol Biotechnol 96:1137–1155

Choudhary DK, Kasotia A, Jain S, Vaishnav A, Kumari S, Sharma KP, Varma A (2015) Bacterial-mediated tolerance and resistance to plants under abiotic and biotic stresses. Plant Growth Regul 35:276–300

Costacurta A, Vanderleyden J (1995) Synthesis of phytohormones by plant-associated bacteria. Crit Rev Microbiol 21:1–18

Davies P (1995) Plant hormones, physiology, biochemistry and molecular biology. Kluwer Academic, Dordrecht, pp 833

Dobbelaere S, Croonenborghs A, Thys A, Vande Broek A, Vanderleyden J (1999) Phytostimulatory effect of *Azospirillum brasilense* wild type and mutant strains altered in IAA production on wheat. Plant Soil 212:153–162

Dobbelaere S, Vanderleyden J, Okon Y (2003) Plant growth promoting effects of diazotrophs in the rhizosphere. Crit Rev Plant Sci 22:107–149

Duan J, Miiller KM, Charles TC, Vesely S, Glick BR (2009) 1-Aminocyclopropane-l-carboxylate (ACC) deaminase genes in rhizobia from southern Saskatchewan. Microb Ecol 57:423–436

Esquivel-Cote R, Ramírez-Gama R, Tsuzuki-Reyes G, Orozco-Segovia A, Huante P (2010) *Azospirillum lipoferum* strain AZm5 containing 1-aminocyclopropane-1-carboxylic acid deaminase improves early growth of tomato seedlings under nitrogen deficiency. Plant Soil 337:65–75

Fuentes-Ramírez LE, Caballero-Mellado J (2006) Bacterial biofertilizers. In: Siddiqui ZA (ed) PGPR: biocontrol and biofertilization. Springer, Dordrecht, pp 143–172

Ge L, Yong JWH, Goh NK, Chia LS, Tan SN, Ong ES (2005) Identification of kinetin and kinetin riboside in coconut (*Cocos nucifera* L.) water using a combined approach of liquid chromatography–tandem mass spectrometry, high performance liquid chromatography and capillary electrophoresis. J Chromatogr 829:26–34

Ghosh S, Penterman JN, Little RD, Chavez R, Click BR (2003) Three newly isolated plant growth-promoting bacilli facilitate the seedling growth of canola, *Brassica campestris*. Plant Physiol Biochem 41:277–281

Ghosh S, Sengupta C, Maiti TK, Basu PS (2008) Production of 3-indolylacetic acid in root nodules and culture by a *Rhizobium* species isolated from root nodules of the leguminous pulse *Phaseolus mungo*. Folia Microbiol 53:351–355

Glick BR, Penrose DM, Li J (1998) A model for lowering plant ethylene concentration by plant growth promoting rhizobacteria. J Theor Biol 190:63–68

Glick B, Patten C, Holguin G, Penrose D (1999) Biochemical and genetic mechanisms used by plant growth promoting bacteria. Imperial College Press, London, 267pp

Glick BR, Todorovic B, Czarny J, Cheng Z, Duan J, McConkey B (2007) Promotion of plant growth by bacterial ACC deaminase. Crit Rev Plant Sci 26:227–242

Govindasamy V, Senthilkumar M, Gaikwad K, Annapurna K (2008) Isolation and characterization of ACC deaminase gene from two plant growth-promoting rhizobacteria. Curr Microbiol 57:312–317

Grichko VP, Glick BR (2001) Amelioration of flooding stress by ACC deaminase-containing plant growth-promoting bacteria. Plant Physiol Biochem 39:11–17

Guo J-H, Qi H-Y, Guo Y-H, Ge H-L, Gong L-Y, Zhang L-X (2004) Biocontrol of tomato wilt by plant growth promoting rhizobacteria. Biol Control 29:66–72

Gutierrez-Manero FJ, Ramos-Solano B, Probanza A, Mehouachi J, Tadeo FR, Talon M (2001) The plant growth promoting rhizobacteria *Bacillus pumilis* and *Bacillus licheniformis* produce high amounts of physiologically active gibberellins. Physiol Plant 111:206–211

Halda-Alija L (2003) Identification of indole-3-acetic acid producing freshwater wetland rhizosphere bacteria associated with *Juncus effusus* L. Can J Microbiol 49:781–787

Hariprasad P, Niranjana SR (2009) Isolation and characterization of phosphate solubilizing rhizobacteria to improve plant health of tomato. Plant Soil 316:13–24

Hartmann A, Rothballer M, Schmid M, Lorenz H (2008) A pioneer in rhizosphere microbial ecology and soil bacteriology research. Plant Soil 312:7–14

Hedden P (1997) The oxidases of gibberellin biosynthesis: their function and mechanism. Physiol Plant 101:709–719

Hedden P, Kamiya Y (1997) Gibberellin biosynthesis: enzymes, genes and their regulation. Annu Rev Plant Physiol Plant Mol Biol 48:431–460

Hiltner L (1904) Ueber neuere erfahrungen und probleme auf dem gebiete der bodenbakteriologie und unter besonderer berucksichtigung der grundungung und brache. Arb Deut Landw Gesell 98:59–78

Hiroya K, Itoh S, Sakamoto T (2004) Development of an efficient procedure for indole ring synthesis from 2-ethynylaniline derivatives catalyzed by Cu (II) salts and its application to natural product synthesis. J Org Chem 69:1126–1136

Jain S, Vaishnav A, Kasotia A, Kumari S, Choudhary DK (2014) Plant growth-promoting bacteria elicited induced systemic resistance and tolerance in plants. Emerg Technol Manage Crop Stress Tolerance 109

Janzen R, Rood S, Dormaar J, McGill W (1992) *Azospirillum brasilense* produces gibberellins in pure culture on chemically defined medium and in co-culture on straw. Soil Biol Biochem 24:1061–1064

Jetiyanon K, Kloepper JW (2002) Mixtures of plant growth promoting rhizobacteria for induction of systemic resistance against multiple plant diseases. Biol Control 24:285–291

Johnson B, Ecker JR (1998) The ethylene gas signal: transduction pathway. Annu Rev Genet 32:227–254

Kang SM, Khan AL, Hamayun M, Shinwari ZK, Kim YH, Joo GJ, Lee IJ (2012) *Acinetobacter calcoaceticus* ameliorated plant growth and influenced gibberellins and functional biochemicals. Pak J Bot 44(1):365–372

Kang SM, Khan AL, Hyun YY, Kim JG, Kamran M, Lee IJ (2014) Gibberellin production by newly isolated strain *Leifsonia soli* SE134 and its potential to promote plant growth. J Microbiol Biotechnol 24:106–112

Karadeniz A, Topcuoglu SF, Inan S (2006) Auxin, gibberelin, cytokinin and abscisic acid production in some bacteria. World J Microbiol Biotechnol 22:1061–1064

Kloepper JW, Okon Y (1994) Plant growth-promoting rhizobacteria (other systems). In: Okon Y (ed) *Azospirillum*/plant associations. CRC Press, Boca Raton, pp 111–118

Kloepper JW, Lifshitz R, Zablotwicz RM (1989) Free-living bacterial inocula for enhancing crop productivity. Trends Biotechnol 7:39–44

Kloepper JW, Zablotowick RM, Tipping EM, Lifshitz R (1991) Plant growth promotion mediated by bacterial rhizosphere colonizers. In: Keister DL, Cregan PB (eds) The rhizosphere and plant growth. Kluwer, Dordrecht, pp 315–326

Korasick DA, Enders TA, Strader LC (2013) Auxin biosynthesis and storage forms. J Exp Bot 64:2541–2555

Krall L, Raschke M, Zenk MH, Baron C (2002) The Tzs protein from *Agrobacterium tumefaciens* C58 produces zeatin riboside 50-phosphate from 4-hydroxy-3-methyl-2-(*E*)-butenyl diphosphate and AMP. FEBS Lett 527:315–318

Kuhajek JM, Jeffers SN, Slattery M, Wedge DE (2003) A rapid microbioassay for discovery of novel fungicides for *Phytophthora* sp. Phytopathology 93:46–53

Kumar PKR, Lonsane BK (1989) Microbial production of gibberellins: state of the art. Adv Appl Microbiol 34:29–139

Kumari S, Vaishnav A, Jain S, Varma A, Choudhary DK (2016) Induced drought tolerance through wild and mutant bacterial strain *Pseudomonas simiae* in mung bean (*Vigna radiata* L.). World J Microbiol Biotechnol 32:1–10

Letham D (1963) Zeatin: a factor inducing cell division from *Zea mays*. Life Sci 8:569–573

Lindberg T, Granhall U (1984) Isolation and characterization of nitrogen-fixing bacteria from the rhizosphere of temperate cereals and forage grasses. Appl Environ Microbiol 48:683–689

Lindberg T, Granhall U (1986) Acetylene reduction in gnotobiotic cultures with rhizosphere bacteria and wheat. Plant Soil 92:171–180

Lucas GJA, Probanza A, Ramos B, Colon Flores JJ, Gutierrez Mañero FJ (2004a) Effect of plant growth promoting rhizobacteria (PGPRs) on biological nitrogen fixation, nodulation and growth of *Lupinus albus* L. cv. Multolupa. Eng Life Sci 7:1–77

Lucas GJA, Probanza A, Ramos B, Palomino MR, Gutierrez Mañero FJ (2004b) Effect of inoculation of *Bacillus licheniformis* on tomato and pepper. Agronomie 24:169–176

MacFaddin JF (1976) Biochemical tests for identification of medical bacteria. Williams & Wilkins, Baltimore

MacMillan J (2002) Occurrence of gibberellins in vascular plants, fungi, and bacteria. J Plant Growth Regul 20:387–442

Mayak S, Tirosh T, Click BR (2004) Plant growth-promoting bacteria that confer resistance to water stress in tomatoes and peppers. Plant Sci 166:525–530

Miller C, Skoog F, Von Saltza M, Strong F (1955) Kinetin, a cell division factor from deoxyribonucleic acid. J Am Chem Soc 77:1392

Novák O, Tarkowski P, Tarkowská D, Doležal K, Lenobel R, Strnad M (2003) Quantitative analysis of cytokinins in plants by liquid chromatography-single-quadrupole mass spectrometry. Anal Chim Acta 480:207–218

Oldroyd GED (2007) A hormone-signaling pathway is crucial to the ability of certain plants to form nodules when stimulated by nitrogen fixing bacteria. Science 315:52–53

Ovakim Li J, Charles TC, Glick BR (2000) An ACC deaminase minus mutant of *Enterobacter cloacae* UW4 no longer promotes root elongation. Curr Microbiol 41:101–105

Pandya ND, Desai PV (2014) Screening and characterization of GA3 producing *Pseudomonas monteilii* and its impact on plant growth promotion. Int J Curr Microbiol Appl Sci 3:110–115
Patten CL, Glick BR (1996) Bacterial biosynthesis of indole-3-acetic acid. Can J Microbiol 42:207–220
Patten CL, Glick BR (2002) Role of *Pseudomonas putida* indole-acetic acid in development of the host plant root system. Appl Environ Microbiol 68:3795–3801
Pearce D, Koshioka M, Pharis R (1994) Chromatography of gibberellins. J Chromatogr 658:91–122
Penrose DM, Glick BR (2003) Methods for isolating and characterizing ACC deaminase containing plant growth-promoting rhizobacteria. Physiol Plant 18:10–15
Piccoli P, Masciarelli O, Bottini R (1996) Metabolism of 17,[2H2]-gibberellins A4, A9, and A20 by *Azospirillum lipoferum* in chemically-defined culture medium. Symbiosis 21:167–178
Pirlak L, Kose M (2009) Effects of plant growth promoting rhizobacteria on yield and some fruit properties of strawberry. J Plant Nutr 32:1173–1184
Podile AR, Kishore GK (2006) Plant growth-promoting rhizobacteria. In: Gnanamanickam SS (ed) Plant-associated bacteria. Springer, New York, pp 195–230
Raj SN, Deepak SA, Basavaraju P, Shetty HS, Reddy MS, Kloepper J (2003) Comparative performance of formulations of plant growth promoting rhizobacteria in growth promotion and suppression of downy mildew in pearl millet. Crop Prot 22:579–588
Reed MLE, Click BR (2005) Growth of canola (*Brassica napus*) in the presence of plant growth-promoting bacteria and either copper or polycyclic aromatic hydrocarbons. Can J Microbiol 51:1061–1969
Rood S, Pharis R (1987) Evidence for reversible conjugation of gibberellins in higher plants. In: Schreiber H, Schutte H, Semder G (eds) Conjugated plant hormones. Structure, metabolism and function. Proceedings of the international symposium on conjugated plant hormones: structure, metabolism and function, Gera, Germany. VEB Deustcher Verlag der Wissenschaften, Berlin, pp 183–190
Ryu CM, Farag MA, Hu CH, Reddy MS, Wei HX, Pare PW, Kloepper JW (2003) Bacterial volatiles promote growth in *Arabidopsis*. Proc Natl Acad Sci USA 100:4927–4932
Sakakibara H (2006) Cytokinins: activity, biosynthesis, and translocation. Annu Rev Plant Biol 57:431–449
Sembdner G, Gross D, Liebisch H, Schneider G (1980) Biosynthesis and metabolism of plant hormones. In: MacMillan J (ed) Encyclopedia of plant physiology, New series. Springer, Berlin, pp 281–444
Serdyuk OP, Smolygna LD, Ianova EP, Adanin M (2003) Phototrophic purple bacterium *Chromatium minutissimum* does not synthesize cytokinins under optimal growth conditions. Dokl Biochem Biophys 392:700–702
Shoebitz M, Ribaudo CM, Pardo MA, Cantore ML, Ciampi L, Curá JA (2009) Plant growth promoting properties of a strain of *Enterobacter ludwigii* isolated from *Lolium perenne* rhizosphere. Soil Biol Biochem 41:1768–1774
Spaepen S, Vanderleyden J, Remans R (2007) Indole-3-acetic acid in microbial and microorganism–plant signaling. FEMS Microbiol Rev 31:425–448
Srivastava LM (2002) Plant growth and development: hormones and environment. Academic, San Diego
Swain MR, Naskar SK, Ray RC (2007) Indole-3-acetic acid production and effect on sprouting of yam (*Dioscorea rotundata* L.) minisetts by *Bacillus subtilis* isolated from culturable cow dung microflora. Pol J Microbiol 56:103–110
Teale WW, Paponov I, Palme K (2006) Auxin in action: signalling, transport and the control of plant growth and development. Nat Rev Mol Cell Biol 7:847–859
Tsavkelova EA, Klimova SY, Cherdyntseva TA, Netrusov AI (2006) Microbial producers of plant growth stimulators and their practical use: a review. Appl Biochem Microbiol 42:117–126

Vaishnav A, Jain S, Kasotia A, Kumari S, Gaur RK, Choudhary DK (2014) Molecular mechanism of benign microbe-elicited alleviation of biotic and abiotic stresses for plants. In: Approaches to plant stress and their management. Springer, New Delhi, pp 281–295

Werner T, Motyka V, Strnad M, Schmulling T (2001) Regulation of plant growth by cytokinin. Proc Natl Acad Sci USA 98:10487–10492

Yang J, Zhang J, Huang Z, Wang Q, Zhu L, Liu L (2002) Correlation of cytokinin levels in the endosperms and roots with cell number and cell division activity during endosperm development in rice. Ann Bot 90:369–377

Zahir AZ, Arshad M, Frankenberger WT (2004) Plant growth promoting rhizobacteria: application and perspectives in agriculture. Adv Agron 81:97–168

Zaidi A, Khan MS, Ahemad M, Oves M (2009) Plant growth promotion by phosphate solubilizing bacteria. Acta Microbiol Immunol Hung 56:263–284

23 Diversity, Quorum Sensing, and Plant Growth Promotion by Endophytic Diazotrophs Associated with Sugarcane with Special Reference to *Gluconacetobacter diazotrophicus*

Iqbal Ahmad, Mohd. Musheer Altaf, Jyoti Sharma, and Abdullah Safar Al-thubiani

Abstract

Endophytic bacteria are widely distributed among plants and colonize both intracellular and intercellular spaces and do not harm the host plant. However, the distributions of endophytic diazotrophs are limited. Endophytic diazotrophs like *Gluconacetobacter diazotrophicus* are mainly associated with sugarcane and some other plants and responsible for significant contribution of biological nitrogen fixation with sugarcane. In this article, we described the diversity and role of quorum sensing. We also discussed the contributions of different bacterial traits that are necessary for successful colonization of the plant interior part. Further mechanisms of plant growth promotion are elaborated. Molecular characterization and identification of endophytic diazotrophs will further help in better understanding of plant colonization and plant growth promotion.

I. Ahmad • M.M. Altaf (✉)
Department of Agricultural Microbiology, Aligarh Muslim University, Aligarh 202002, India
e-mail: mohdmusheer@rediffmail.com

J. Sharma
Department of Science and Technology, Technology Bhavan, New Delhi 110016, India

A.S. Al-thubiani
Department of Biology, School of Applied Science, Umm Al-Qura University, Makkah 21955, Saudi Arabia

D.K. Choudhary et al. (eds.), *Plant-Microbe Interaction: An Approach to Sustainable Agriculture*, DOI 10.1007/978-981-10-2854-0_23

23.1 Introduction

The Green Revolution had increased the agricultural productivity to a great extent by the increased application of high-yielding crop varieties, heavy farm equipments, synthetic fertilizers, pesticides, improved irrigation, better soil administration, and massive conversion of forest to agricultural lands (Tilman et al. 2002; Gomiero et al. 2011). But there is a growing apprehension that intensive practices employed for increasing agricultural output promote ecosystem deterioration and loss of yield. Adverse environmental effects include deforestation, soil degradation, large-scale greenhouse gas emissions, accumulation of pesticides and chemical fertilizers, pollution of groundwater, and decreased water table due to excessive irrigation (Tilman et al. 2002; Foley et al. 2011). The International Fertilizer Industry Association (IFA) agriculture committee projected that the global fertilizer consumption is expected to grow, and it will reach 199.4 million metric tons (Mt) of nutrients in 2019 (Heffer and Prud'homme 2015). The projected increase will be at the rate of 1.3, 2.1, and 2.4 % for nitrogen, phosphorus, and potassium, respectively. In reality, an intensive agricultural practice is considered to be the main source of loss of global biodiversity. Traditional agricultural practices like organic farming, which considerably decrease the input of chemical fertilizers, pesticides, energy, and mechanic stress, help us in extenuating the harmful effects of intensive agricultural practices and simultaneously boost the sustainable agriculture production (Gomiero et al. 2011).

World total population is presently around 7 billion, and this is anticipated to grow to approximately 8 billion people until the year 2025 and 9 billion by 2050. Considering the increase in worldwide population with the increase in environmental damage due to ever-increasing industrialization, it is clear that, in the coming next 50 years, it will be a daunting task to feed the existing population, a problem that will increase with time. Therefore, to provide food for the ever-growing population, there is an urgent need for tremendous increase in agricultural productivity in a sustainable and environmentally friendly manner. To produce more food, humankind will need a range of diverse schemes and approaches which should consist of feasible and environmentally favorable biological solutions (Glick 2014). The effective use of PGPR in agriculture in an integrated manner is an interesting technology to tackle these problems.

Microbes provide help to plants either directly by increasing crop nourishment or indirectly by minimizing the damage created by pathogens or environmental stress. Plants live in a close relationship with microbes that fulfill important functions in agricultural ecosystems. Microorganisms may live as free-living organisms in soils or may be associated with the surface of the roots or phyllosphere and may establish symbiotic relations with plants (Smith and Goodman 1999). Endophytic bacteria are a class of endosymbiotic microbes that live in inner plant tissues of apparently healthy host plants (Schulz and Boyle 2006). Unlike phytopathogens, generally such bacteria do not create any substantial disease symptoms, and occurrence of endophytes is not associated with morphological changes of plant tissues such as caused by root-nodule symbionts.

Endophytes inhabit plant apoplast, as well as the intercellular regions of the cell walls and xylem vessels present in roots, stems, and leaves; in addition to this, these bacteria also reside in tissues, flowers, fruits, and seeds (Compant et al. 2011; Pereira et al. 2012). Populations of endophytes are uneven in different plant parts and have been shown to vary from 10^2 to 10^9 of bacterial cells per gram of plant tissue (Jacobs et al. 1985; Chi et al. 2005). This density is governed by several factors, like the plant under study, the respective part under investigation, the developmental period of the plant, the genotype, and the cooperation with other microorganisms (Costa et al. 2012). Normally, the plant roots and other belowground tissues contain higher densities of endophytes as compared to aboveground plant parts.

The cooperation that takes place among sugarcane and other grasses with nitrogen-fixing endophytic bacteria has increased their importance for their utilization in agriculture, because of their positive response on root development and enhanced biomass and productivity. It is well known that close association between host plant and endophytes takes places through various compounds secreted by the microbes and the host plant (Reinhold-Hurek and Hurek 2011; Brader et al. 2014). The endophytes enhance nutrient availability and uptake, augment stress tolerance, and offer disease resistance (Ryan et al. 2008).

Sugarcane (*Saccharum officinarum* L.) is a high-rising, long-standing grass that belongs to the family Poaceae and subfamily Panicoideae, which is generally grown in tropical and warm-temperate regions between 35°N and 35°S. It belongs to the C4 plant category and has high photosynthetic efficiency, increased rate of biomass conversion from solar energy, and high efficiency of water use (Ward et al. 1999; Reis et al. 2007). Endophytic bacterial isolates have been obtained from sugarcane (*Saccharum* spp.) that promotes plant growth. The most distinguished sugarcane endophytic diazotrophs are *Gluconacetobacter diazotrophicus* (alpha subclass of Proteobacteria), *Herbaspirillum seropedicae*, *Herbaspirullum rubrisubalbicans*, and *Burkholderia* sp. (beta subclass of the Proteobacteria) (Reis et al. 2000).

The communication that takes place between the target plant and nitrogen-fixing endophytes depicts a different organization regarding advantageous plant–microbe associations, which showed distinctive properties that need description. In this article, we have attempted to review the scientific literature available on endophytic bacteria and their identification and impact on plant growth and health.

23.2 Diversity of Endophytic Diazotrophs Associated with Sugarcane

It has been proposed that the planet Earth consists of approximately 300,000 species of plants, the bulk of which consists of endophytes (Smith et al. 2008). Actually, endophytes (bacteria and fungi) have been reported to be present in each and every plant species that has been investigated. Partida-Martínez and Heil (2011) reported that an endophyte-free plant is a rare exception in nature. Timmusk et al. (2011) observed that a plant devoid of endophytes was unable to cope with environmental stress conditions and susceptible to pathogen attack. In case of sugarcane, the majority of the study on endophytic microbes has been directed on diazotrophs. The major representative includes *Gluconacetobacter diazotrophicus*, *Herbaspirillum* spp. (Baldani et al. 1986; Cavalcante and Döbereiner 1988; James 2000; Boddey et al. 2003), and *Azospirillum amazonense* (Reis Júnior et al. 2000). However, the Indian sugarcane was found to contain a low population of diazotrophic bacteria among the entire populations of microbes (Suman et al. 2001).

Gluconacetobacter genus was suggested by Yamada et al. (1997) as one of the four genera for acetic acid bacteria. *Gluconacetobacter diazotrophicus* is known to be a plant growth-promoting bacteria (PGPB) which colonizes sugarcane, pineapple, wetland rice, sweet potato, corn, sorghum, coffee, wheat, and tomato plants (Cavalcante and Döbereiner 1988; Cocking et al. 2006; Saravanan et al. 2008; Luna et al. 2010). In addition to this, a significant feature of *G. diazotrophicus* is the tolerance to high sucrose level (30 %), demonstrating high osmotolerance, which is constant with its continued existence in sugarcane stems naturally (Cavalcante and Döbereiner 1988). However, on the contrary, this organism is susceptible to NaCl (Tejera et al. 2003; Boniolo et al. 2009).

G. diazotrophicus, associated with sugarcane, is known as "obligate endophytes," because it is not possible to isolate it from non-rhizospheric soils and can only be isolated from plants, fungi, insects, etc. The isolation of this bacterium can only be possible from roots, stems, and leaves of sugarcane (Gillis et al. 1989) and coffee. It normally inhabits the sugarcane tissues that lack dissolved carbon compounds, like root and stem xylem vessels. It was also found to inhabit intercellular apoplastic stem areas that contain the sucrose niche and phloem sieve tubes involved in sucrose translocation. The cells of this bacterium were reported to inhabit inside plant stems as microcolonies and haphazardly distributed on the plant surface in an apolar direction forming a monolayer wrapped around roots and leaves. *G. diazotrophicus* was also found compiling the lateral root junctions and inhabiting the damaged epidermal cells, where it does not penetrate beyond the epidermis of root. *G. diazotrophicus* is not capable to exist in the soil in the absence of their host plants and can only grow in low-pO2 environment, which is essential for the expression and normal performance of the nitrogenase system (James and Olivares 1997). Ahmad et al. (2004) isolated a total of eleven isolates of *Acetobacter diazotrophicus* (*Gluconacetobacter diazotrophicus*) from roots, stems, and leaves of the four locally grown varieties of sugarcane (Co-1148, UP 39, Satha-676, and Satha-91269) of Aligarh. These isolates along with three Brazilian strains of *Acetobacter diazotrophicus* (PR2-ATCC49039, Pal5-ATCC 49037, Ppe-4 ATCC49038) were considered for their biochemical characteristics and resistance traits. These isolates use sucrose, glucose, and ethanol (1 %), whereas all the isolates were found negative for maltose and ethanol (10 %). Fructose was consumed inconsistently. Similarly all the isolates were found positive for catalase and H_2S production and negative for oxidase, nitrate reduction, denitrification, gelatin liquefaction, and indole test. Antibiotic resistance was expressed by 12 isolates only. Metal resistance at MIC 100 $\mu g\ ml^{-1}$ was found highest against lead followed by cadmium, mercury, nickel, and copper. Majority of the indigenous isolates demonstrate resistance to both antibiotics and heavy metals. However, three Brazilian isolates of *Acetobacter diazotrophicus* were found sensitive to all five (Cu, Cd, Ni, Pb, and Hg) metals. Further, tolerance to salt (NaCl) was higher (1.0–1.5 %) within indigenous isolates compared to Brazilian isolates, which showed NaCI tolerance up to 0.5 % in agar medium.

Azospirillum species is a native soil bacteria commonly found as root-associated diazotrophs. They are attached to the roots by fibrillar material and are occasionally reported from the superficial film of the root cortex (Bashan and Levanony 1990). Majority of the *Azospirillum* species have been isolated from the surface-sterilized roots, which signifies that a portion of these cells get shielded from different sterilizing agents and are located inside the root tissues (Dobereiner and Day 1976; Hallmann et al. 1997). These bacterial species were obtained from different cash crops like sugarcane, palm trees, forage grasses, tuber plants, cereals, and sweet potato (Mohanta et al. 2010). Farrar et al. (2014) reported several different endophytic bacteria from sugarcane like *Gluconacetobacter diazotrophicus* (syn. *Acetobacter diazotrophicus*), *Burkholderia*, *Pantoea*, *Pseudomonas*, *Microbacterium*, *Citrobacter*, *Enterobacter*, *Klebsiella*, *Erwinia*, *Brevibacillus*, *Staphylococcus*, *Curtobacterium*, *Pseudomonas* sp., *Bacillus*, *Paenibacillus*, *Brevibacillus*, and *Burkholderia australis*. Velázquez et al. (2008) reported the genetic diversity of 29 endophytic bacteria from healthy grown sugarcane plant from Cuba and investigated using two primers, random amplified polymorphic DNA fingerprinting (TP-RAPD) and 16S rRNA gene sequencing, demonstrating that these isolates are associated to different phylogenetic groups being strongly connected to species of genera *Bacillus* and *Staphylococcus* from *Firmicutes*; *Microbacterium*, *Micrococcus*, and *Kokuria* from *Actinobacteria*; *Rhizobium* and *Gluconacetobacter* from α-Proteobacteria; *Comamonas* and *Xanthomonas* from β-Proteobacteria; and *Acinetobacter* and *Pantoea* from γ-Proteobacteria.

23.3 Cell-to-Cell Communication Among Endophytes

Gluconacetobacter diazotrophicus is a well-known endophyte obtained from the inner tissues of many crop plants. The plant growth-promoting capability of this *Alphaproteobacterium* has been associated not only to its ability to carry out biological nitrogen fixation but also through the production of siderophores, antimicrobial compounds, and solubilization of phosphate and other minerals by the production of gluconic acid (Saravanan et al. 2008). Colonization and persistence of an endophyte involve intricate regulatory pathways. Among them, quorum sensing systems (QS) are signaling methods connected with the regulation of numerous genes associated with microbial communications, host establishment, and survival under stress conditions. Quorum sensing is connected with the capability of a bacterium to react to autoinducers, hormone-like compounds which are capable of altering gene expression at a critical threshold population (Reading and Sperandio 2006).

The genes related to quorum sensing in *G. diazotrophicus* consist of one luxI autoinducer synthase gene and two luxR-type transcriptional regulator genes, which are associated with the expression of three N-acyl homoserine lactones (AHLs)

(Eskin et al. 2014). Analysis of *G. diazotrophicus* AHLs identified 8 different signaling molecules: C6-homoserine lactone (HSL), C8-HSL, C10-HSL, C12-HSL, C14-HSL, 3-oxo-C10-HSL, 3-oxo-C12- HSL, and 3-oxo-C14-HSL (Nieto-Penalver et al. 2012).

Research should also continue to focus on the importance of quorum sensing. It has been revealed that *G. diazotrophicus* contains 3 different AHLs; their precise roles have yet to be identified. Recently recognized molecular methods used for studying *G. diazotrophicus* such as mutational studies via Tn5 transposon mutagenesis could assist in identification of quorum sensing genes.

23.4 Methods for Detection and Characterization of Endophytes

The endophytic habitat provides defense against the adverse environmental conditions for bacteria that are capable of colonizing and residing *in planta*. These bacterial isolates usually take possession of the areas/voids present between cells, and they have been reported from almost all the sections of plant together with seeds (Posada and Vega 2005). Endophytic microbes have been described and obtained from both monocotyledonous and dicotyledonous plants. Classical research related to the diversity of endophytes (*Gluconacetobacter diazotrophicus*, *Herbaspirillum seropedicae*, and *H. rubrisubalbicans*) had paid attention on classification of isolates acquired from inner spaces of tissues after sterilization of plant outer areas using sodium hypochlorite or by culturing serial dilutions prepared from plant tissue macerates in nitrogen-free semisolid media which acts as semi-selective media for the species/genera (Miche and Balandreau 2001; Silva-Froufe et al. 2009). Fundamentally, the most probable number (MPN) procedure, using a McCrady table, has been used to enumerate the quantity of bacteria (Paula et al. 1991; Eskin et al. 2014). But, the MPN technique is not recognized to be very precise and should be directed for additional testing to authenticate isolates at a species level. The weakness of MPN method can be mitigated by using enzyme-linked immunosorbent assay (ELISA) (Silva-Froufe et al. 2009). Further, by the use of different microscopies like optical and transmission electron microscopy (TEM), the presence of *G. diazotrophicus* endophytic colonization can be confirmed (Luna et al. 2010). In addition to these methods used for localization of *G. diazotrophicus*, one of the most widespread techniques used is green fluorescent protein (GFP) labeling (Gaiero et al. 2013). Eskin et al. (2014) described the application of *gusA* and *gfp* reporter genes from strains containing pHRGFPGUS (*gfp::gusA*) and pHRGFPTC (*gfp*) plasmids, respectively. *G. diazotrophicus* UAP5541/pRGS561 constitutively expressing GUS and UAP5541/pRGS562 with a *nif*H::*gus*A transcriptional fusion is two supplementary strains that have been used in different investigations in which both intercellular and intracellular localization have been determined. Rouws and colleagues (2010) in their study related to G. *diazotrophicus* strain Pal5 carrying *gfp*::*gus*A plasmid pHRGFPGUS and *gfp* plasmid pHRGFPTC proved the validity of these techniques for colonization and localization of endophytes.

Different techniques are used to study plants for the occurrence of *G. diazotrophicus*. One of the key techniques for identification of *G. diazotrophicus* is through polymerase chain reaction (PCR). While a simple PCR is sufficient in identifying the bacterium at high colony numbers, a nested PCR in which a second round of PCR is used to amplify the product from the first round of PCR is instrumental in detecting the bacterium when found at very low colony numbers (Tian et al. 2009). While PCR is proficient to authenticate the presence of the bacterium, it is not capable of determining the number of bacterium present within a sample. Bacterial populations colonizing the stems, roots, and tubers of different plants were studied by 16S rRNA gene-associated methods like terminal restriction fragment length polymorphism analysis, denaturing gradient gel electrophoresis, as well as 16S rRNA gene cloning and sequencing. Characterization related to endophytes involves not merely the separation from sterilized tissues but also visualization by various types of microscopy within plant tissues (Sagarika et al. 2010). Endophytes normally present can be seen using different types of microscopy such as fluorescence in situ hybridization (FISH) together with confocal laser scanning microscopy (CLSM) using specific probes, scanning electron microscopy (SEM), transmission electron microscopy (TEM), and phase contrast microscopy (Amann et al. 1990; Loy et al. 2007). Nautiyal (2000) has reassessed new development which includes the use of various markers for studying root colonization. Microscopic researches related to gfp tagged endophytic inoculants disclose extremely diverse colonization arrangements.

23.5 Colonization of Plant by Endophytic Diazotrophs

Generally, the communities of endophytes were observed to be present in lower concentration as compared to root-associated bacteria or phytopathogens (Hallmann et al. 1997; Rosenblueth and Martínez-Romero 2006). Endophytes similar to their rhizospheric counterparts are hardened by biotic and abiotic components (Fuentes-Ramírez et al. 1999; Hallmann et al. 1997; Seghers et al. 2004). However, the endophytes received better protection from biotic and abiotic stress compared to root-associated bacteria (Hallmann et al. 1997). These bacteria also perform a crucial task in inhabitation.

Colonization related to bacterial species either in roots or on plant surfaces is a complicated procedure facilitated by the exchange of numerous bacterial character and genes. Colonization process is a step-by-step procedure, which involves (1) movement toward root surface, (2) adhesion, (3) dispersal along the root, and (4) persistence and survival of the bacterial communities. Among endophytes, besides the above-said steps, the colonization process involved one extra step, which involves the access inside the root and development of small colonies within and between tissues, which can be tracked by labeling the target bacteria with molecular indicator such as (gfp) or β-glucosidase (gus) and finally visualizing with electron or confocal laser scanning microscopy (Reinhold-Hurek and Hurek 2011). The mechanism implicated with endophytic colonization process is not fully inferred.

Experimental data based on genomic and other techniques revealed a similarity in the colonization process among pathogenic and plant growth-enhancing bacteria (Hardoim et al. 2008). The endophytic bacteria get entry into the plants using two modes: vertical transmission and horizontal transmission. In vertical transmission, endophytic bacteria might be transferred by seeds (vegetatively), and during horizontal transmission, endophytes may be selected by soil and most likely need a stage of rhizoplane colonization, for which they require properties assigning "rhizosphere competence."

Apart from sugarcane, *G. diazotrophicus* has been isolated from several different crops like coffee, pineapple, and wetland rice. The large part of these hosts consists of comparatively higher levels of sucrose which appeared to be a requirement for colonization by this bacterium (Riggs et al. 2001). *G. diazotrophicus*, an obligate endophyte, is incompetent of surviving in soil without a plant host for more than 2 days, with the exception of being capable of surviving within the spores of the vesicular arbuscular mycorrhizal fungus *Glomus clarum* and within the root hairs of a host plant's rhizosphere (James et al. 2001; Luna et al. 2010).

The bacterium is capable to achieve entry into its host plant through the roots, stems, or leaves. With respect to the roots, *G. diazotrophicus* enters through the root tips and cells of the root cap and meristem, at areas of lateral root emergence, and by root hairs (Rouws et al. 2010). In case of stems of host plants, especially sugarcane, the bacterium gets entry at cracks generated by the disjointing of plantlets into individuals. Finally, through the leaves, the endophyte gets entry via injured stomata. An added pathway used by *G. diazotrophicus* is accomplished through an insect vector, the pink sugarcane mealybug (*Saccharicoccus sacchari*), a plant sap-sucking insect (Franke-Whittle et al. 2005; Ortega-Rodes et al. 2011). Once inside the host plant, *G. diazotrophicus* was observed to mainly occupy intercellular apoplastic spaces, the xylem, and the xylem parenchyma (Boddey et al. 2001). However, new research showed that β -glucuronidase (GUS)-labeled *G. diazotrophicus* bacterium is also proficient in intracellular colonization with in membrane-bound vesicles in its host plant (Cocking et al. 2006). Recognized *G. diazotrophicus* colonies can grow up to 10^8 CFU per gram of tissue, as found within sugarcane. The above-mentioned methods of passage for *G. diazotrophicus* were found to be assisted by hydrolytic enzymes (Adriano-Anaya et al. 2005). Adriano-Anaya et al. (2005) established the secretion of endoglucanase, endopolymethylgalacturonase, and endoxygluconase within both PAL5 and UAP5541 strains of *G. diazotrophicus*, which utilizes sucrose as carbon source. These enzymes perform a key function in the entry of endophytic bacteria to host and its mobility inside host plant tissue.

The colonization of root surface by endophytes involves the arrival of bacteria to the rhizoplane as a result of chemotactic response, to surpass the additional microbes in order to get entry into root surface, express genes in a synchronized manner for intrusion in the plant, prevent host plant immune responses, and protect a place inside the plant tissue (Bais et al. 2006; Rosenblueth and Martínez-Romero 2006; Compant et al. 2010). Moreover, microbe–microbe communication and microbe–plant signaling are implicated at every point involved in the process of root colonization. The root endophytic bacterial populations can vary considerably compared to the rhizospheric communities, signifying identification and collection of helpful microbes by roots (Compant et al. 2005). Lipopolysaccharides, flagella, pili, and twitching motility have been reported to influence endophytic colonization and bacterial motion inside host plants (Böhm et al. 2007). Additionally, the discharge of cell wall-degrading enzymes (CWDEs) is reported to be implicated in bacterial infiltration (Lodewyckx et al. 2002) and dispersal inside the plant. To inhabit the inner plant tissues, it has been suggested that bacterial endophytes contain genomic differences compared to root-colonizing bacteria, despite the fact that so far no ultimate group of genes has been recognized that is accountable for the endophytic way of life. However, a directory of genes which play a potential role in endophytic behavior was recently recognized by Ali et al. (2014) by analyzing the complete genomes of nine proteobacterial endophytes. At this junction, barely a few genes have been experimentally revealed to be implicated in colonization by endophytes.

Several defense reactions have been implicated throughout plant–endophyte associations. Reinforcement of cell walls, organization of adjoining substance within the cortex or xylem, as well as gum secretion inside vessels have been reported (Miché et al. 2006). Although several defense reactions have been reported for plant response to plant pathogens, merely a few defense responses have been explained in plant reply to endophytes. These differences may be illustrated through the discharge of several substances, which may be of extremely small quantity for endophytes (James et al. 2002). On the other hand, it has been observed that plants may demonstrate resistance response regulating colonization by endophytes (Iniguez et al. 2005). Dicotyledonous plants species are acknowledged for using salicylic acid (SA) and ethylene in communication, which manage colonization of some endophytes, as confirmed under laboratory conditions (Iniguez et al. 2005). However, among monocotyledonous plants like rice, accumulation of jasmonic acid (JA) but not ethylene was observed to obstruct the colonization of the diazotroph *Azoarcus* sp., signifying that plant protection reactions linking the JA signaling pathways may also be involved in managing endophytic colonization within the root system (Miché et al. 2006). However, in a well-matched endophytic organization, JA-related plant reactions were negligible and do not limit endophytic establishment (Miché et al. 2006).

23.6 Plant Growth-Promoting Effect of Endophytes

Plant growth-enhancing bacterial endophytes inhabit the inner part of plant and are capable of establishing a unique type of association, in which both the participants get benefitted from their relationship (Hallmann et al. 1997; Reiter and Sessitsch 2006). Bacterial endophytes increased plant growth using different mechanisms like production of phytohormones, siderophores, solubilization of phosphate, nitrogen fixation, and accessibility of key nutrients to their host plants (Lodewyckx et al. 2002; Lee et al. 2004; Puente et al. 2009), as shown in Fig. 23.1. Endophytes can also enhance plant growth as a result of the bacterium secreting the enzyme 1-aminocyclopropane-1-carboxylate (ACC) deaminase which dissects ACC to

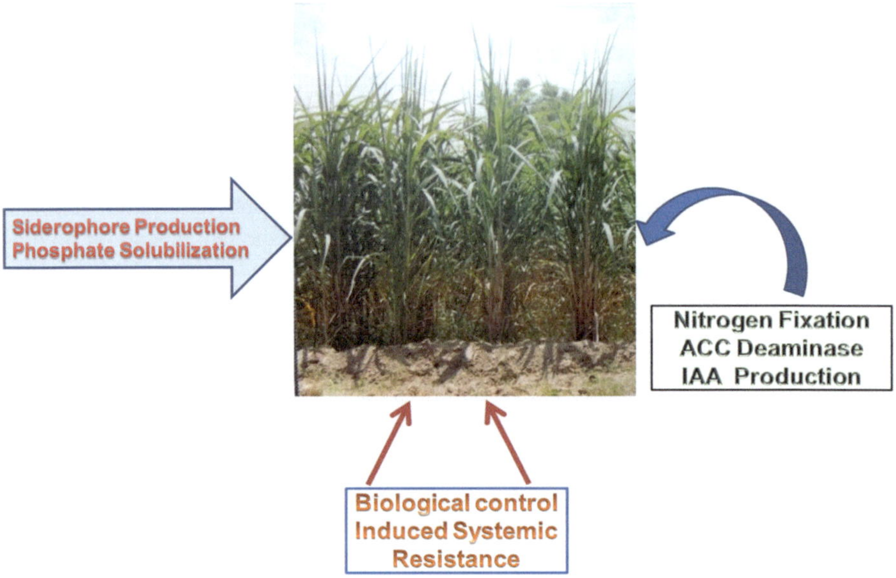

Fig. 23.1 Mechanism of sugarcane growth promotion by endophytic microbes

α-ketobutyrate and ammonia and thus diminishes ethylene levels in host plants (Glick 2014). Moreover, these plant growth-promoting endophytes are also playing a significant role in forest restoration and phytoremediation of polluted soils (Ryan et al. 2008).

Earlier researchers reported that endophytes promote plant growth by changing plant physiology (osmotic pressure regulation), modifying stomatal reaction, altering root dimensions and make, changing nitrogen buildup and metabolism, and enhancing uptake of necessary minerals (Compant et al. 2005). Endophytic microbes also work as a biocontrol agent. The mechanism used by the endophytes for working as biocontrol agents involves the secretion of antibiotics and lytic enzymes like hydrolases and chitinases (Chernin and Chet 2002; Ezra et al. 2004). Further, the bacterial endophytes were also found to activate induced systemic resistance (ISR)-based plant growth promotion (Ait Barka et al. 2000, 2002). Several workers reported the beneficial effect of application of *Gluconacetobacter* on enhancement of sugarcane growth under field conditions (Fig. 23.2).

Chauhan et al. (2010) isolated 11 species of *Gluconacetobacter* from different varieties of sugarcane, and under field trial, it was observed that the endophytes significantly increased plant height, chlorophyll content, cane girth, number of millable canes, and total nitrogen, ensuing the enhancement in cane yield by 42 % compared to control plant. Murumkar et al. (2016) under their field experiments observed that the use of *Gluconacetobacter diazotrophicus* + phosphate-solubilizing bacteria along with 75 % recommended N and 75 % recommended P_2O_5 considerably enhanced the growth, two eye bud-set yield, and quality of sugarcane. Hari (1995) and Srinivasan and Naidu (1987) observed that application of N-fixing bacteria to sugarcane has

Fig. 23.2 Plant growth-promoting effect of *Gluconacetobacter diazotrophicus* on sugarcane grown under field conditions

augmented the cane yield by 5–15 %, which helps in saving of 25 kg fertilizer N ha^{-1}, and also enhanced the juice quality factors, like sucrose and clarity. Schultz et al. (2014) also reported the similar results on sugarcane yield by inoculating endophytic diazotrophs. Oliveira et al. (2003, 2006) demonstrated the increased contribution of biological nitrogen fixation (BNF) in micropropagated sugarcane by inoculating with endophytic diazotrophic bacterial isolates. In addition to providing beneficial plant growth-promoting effect on sugarcane, the endophytic bacteria also conferred drought tolerance to sugarcane (Vargas et al. 2014).

23.7 Conclusion and Future Direction

Considerable research work has been carried out on endophytes including *G. diazotrophicus*. It is known that the endophytic bacteria *G. diazotrophicus* possess 3 different AHLs, but their precise function is yet to be discovered. Recently discovered molecular methods, employed in research for endophytes, can aid in discovering the role of quorum sensing genes and their role in biological nitrogen fixation. Although plenty of scientific data is available on endophytes, still there is a lot more to discover on how a PGPR changes onto plant endophytes. The sufficient understanding of plant–endophyte interaction will certainly play a significant part not only in the enhancement of plant growth and health but also in sustainable agriculture and in obtaining the biotechnological efficiency for various tasks.

References

Adriano-Anaya M, Salvador-Figueroa M, Ocampo JA et al (2005) Plant cell-wall degrading hydrolytic enzymes of *Gluconacetobacter diazotrophicus*. Symbiosis 40:151–156

Ahmad I, Sharma J, Ahmad F (2004) Isolation and characterization of resistance traits of indigenous strains of *Acetobacter diazotrophicus* associated with sugarcane. Sugar Tech 6:41–46

Ait Barka E, Belarbi A, Hachet C et al (2000) Enhancement of *in vitro* growth and resistance to gray mould of *Vitis vinifera* cocultured with plant growth-promoting rhizobacteria. FEMS Microbiol Lett 186:91–95

Ait Barka E, Gognies S, Nowak J et al (2002) Inhibitory effect of endophyte bacteria on *Botrytis cinerea* and its influence to promote the grapevine growth. Biol Control 24:135–142

Ali S, Duan J, Charles TC (2014) A bioinformatics approach to the determination of genes involved in endophytic behavior in *Burkholderia* spp. J Theor Biol 343:193–198

Amann RI, Binder BJ, Olson RJ et al (1990) Combination of 16S rRNA-targeted oligonucleotide probes with flow cytometry for analyzing mixed microbial populations. Appl Environ Microbiol 56:1919–1925

Bais HP, Weir TL, Perry LG et al (2006) The role of root exudates in rhizosphere interactions with plants and other organisms. Annu Rev Plant Biol 57:233–266

Baldani JI, Baldani VLD, Seldin L et al (1986) Characterization of *Herbaspirillum seropedicae* gen. nov., sp. nov., a root-associated nitrogen-fixing bacterium. Int J Syst Bacteriol 36:86–93

Bashan Y, Levanony H (1990) Current status of *Azospirillum* inoculation technology: *Azospirillum* as a challenge for agriculture. Can J Microbiol 36:591–608

Boddey RM, Polidoro JC, Resende AS et al (2001) Use of the 15N natural abundance technique for the quantification of the contribution of N2 fixation to sugar cane and other grasses. Aus J Plant Physiol 28:889–895

Boddey RM, Urquiaga S, Alves BJR et al (2003) Endophytic nitrogen fixation in sugarcane: present knowledge and future applications. Plant Soil 252:139–149

Böhm M, Hurek T, Reinhold-Hurek B (2007) Twitching motility is essential for endophytic rice colonization by the N_2-fixing endophyte *Azoarcus* sp. strain BH72. Mol Plant-Microbe Interact 20:526–533

Boniolo FS, Rodrigues RC, Delatorre EO et al (2009) Glycine betaine enhances growth of nitrogen-fixing bacteria *Gluconacetobacter diazotrophicus* PAL5 under saline stress conditions. Curr Microbiol 59:593–595

Brader G, Compant S, Mitter B et al (2014) A: metabolic potential of endophytic bacteria. Curr Opin Biotechnol 27:30–37

Cavalcante VA, Döbereiner J (1988) A new acid-tolerant nitrogen-fixing bacterium associated with sugarcane. Plant Soil 108:23–31

Chauhan H, Sharma A, Saini SK (2010) Response of sugarcane to endophytic bacterial inoculation. Ind J Sugar Tech 25:1–4

Chernin L, Chet I (2002) Microbial enzymes in biocontrol of plant pathogens and pests. In: Burns RG, Dick RP (eds) Enzymes in the environment: activity, ecology, and applications. Marcel Dekker, New York, pp 171–225

Chi F, Shen SH, Cheng HP et al (2005) Ascending migration of endophytic rhizobia, from roots to leaves, inside rice plants and assessment of benefits to rice growth physiology. Appl Environ Microbiol 71(11):7271–7278

Cocking EC, Stone PJ, Davey MR (2006) Intracellular colonization of roots of Arabidopsis and crop plants by *Gluconacetobacter diazotrophicus*. In Vitro Cell Dev Biol 42:74–82

Compant S, Duffy B, Nowak J (2005) Use of plant growth-promoting bacteria for biocontrol of plant diseases: principles, mechanisms of action, and future prospects. Appl Environ Microbiol 71:4951–4959

Compant S, Clément C, Sessitsch A (2010) Plant growth-promoting bacteria in the rhizo- and endosphere of plants: their role, colonization, mechanisms involved and prospects for utilization. Soil Biol Biochem 42:669–678

Compant S, Mitter B, Colli-Mull JG (2011) Endophytes of grapevine flowers, berries, and seeds: identification of cultivable bacteria, comparison with other plant parts, and visualization of niches of colonization. Microb Ecol 62:188–197

Costa LEO, Queiroz MV, Borges AC et al (2012) Isolation and characterization of endophytic bacteria isolated from the leaves of the common bean (*P. vulharis*). Br J Microbiol 43:1562–1575

Dobereiner J, Day JM (1976) Associative symbiosis and free-living systems. In: Newton WE, Nyman CJ (eds) Proceedings of the 1st international symposium on nitrogen fixation. Washington State University Press, Pullman, pp 518–538

Eskin N, Vessey K, Tian L (2014) Research progress and perspectives of nitrogen fixing bacterium, *Gluconacetobacter diazotrophicus*, in monocot plants. Int J Agron 2014(4):1–13

Ezra D, Castillo UF, Strobel GA et al (2004) Coronamycins, peptide antibiotics produced by a verticillate *Streptomyces* sp. (MSU-2110) endophytic on *Monstera* sp. Microbiology 150:785–793

Farrar K, Bryant D, Cope-Selby N (2014) Understanding and engineering beneficial plant microbe interactions: plant growth promotion in energy crops. Plant Biotechnol J 12:1193–1206

Foley JA, Ramankutty N, Brauman KA et al (2011) Solutions for a cultivated planet. Nature 478:337–342

Franke-Whittle IH, O'Shea MG, Leonard GJ et al (2005) Design, development, and use of molecular primers and probes for the detection of Gluconacetobacter species in the pink sugarcane mealy bug. Microb Ecol 50:128–139

Fuentes-Ramírez LE, Caballero-Mellado J, Sepúlveda J et al (1999) Colonization of sugarcane by *Acetobacter diazotrophicus* is inhibited by high N-fertilization. FEMS Microbio Ecol 29:117–128

Gaiero JR, McCall CA, Thompson KA et al (2013) Inside the root microbiome: bacterial root endophytes and plant growth promotion. Am J Bot 100(9):1738–1750

Gillis M, Kersters K, Hoste B et al (1989) *Acetobacter diazotrophicus* sp. nov., a nitrogen-fixing acetic acid bacterium associated with sugarcane. Int J Syst Bacteriol 39:361–364

Glick BR (2014) Bacteria with ACC deaminase can promote plant growth and help to feed the world. Microbiol Res 169:30–39

Gomiero T, Pimentel D, Paoletti MG (2011) Environmental impact of different agricultural management practices: conventional vs. organic agriculture. Crit Rev Plant Sci 30:95–124

Hallmann J, Quadt-Hallmann A, Mahaffee WF et al (1997) Bacterial endophytes in agricultural crops. Can J Microbiol 43:895–914

Hardoim PR, van Overbeek LS, Elsas JD (2008) Properties of bacterial endophytes and their proposed role in plant growth. Trends Microbiol 16:463–471

Hari K (1995) Biofertilizers in sugarcane. Lead paper presented in 10th sugarcane research and development workers' meeting for South Karnataka, Shimoga, Karnataka, India

Heffer P, Prud'homme M (2015) Fertilizer outlook 2015–2019. In: 83rd IFA annual conference, Istanbul (Turkey), 25–27 May 2015. International Fertilizer Industry Association (IFA), Paris, France, p 4

Iniguez AL, Dong Y, Carter HD (2005) Regulation of enteric endophytic bacterial colonization by plant defenses. Mol Plant-Microbe Interact 18:169–178

Jacobs MJ, Bugbee WM, Gabrielson DA (1985) Enumeration, location, and characterization of endophytic bacteria within sugar-beet roots. Can J Bot 63(7):1262–1265

James EK (2000) Nitrogen fixation in endophytic and associative symbiosis. Field Crops Res 65:197–209

James EK, Olivares FB (1997) Infection and colonization of sugarcane and other graminaceous plants by endophytic diazotrophs. Crit Rev Plant Sci 17:77–119

James EK, Olivares FL, De Oliveira ALM et al (2001) Further observations on the interaction between sugar cane and *Gluconacetobacter diazotrophicus* under laboratory and greenhouse conditions. J Exp Bot 52:747–760

James EK, Gyaneshwar P, Manthan N (2002) Infection and colonization of rice seedlings by the plant growth-promoting bacterium *Herbaspirillum seropedicae* Z67. Mol Plant-Microbe Interact 15:894–906

Lee S, Flores-Encarnacion M, Contreras-Zentella M et al (2004) Indole-3-acetic acid biosynthesis is deficient in Gluconacetobacter diazotrophicus strains with mutations in cytochrome C biogenesis genes. J Bacteriol 186:5384–5391
Lodewyckx C, Vangronsveld J, Porteous F et al (2002) Endophytic bacteria and their potential applications. Crit Rev Plant Sci 21:583–606
Loy A, Maixner F, Wagner M et al (2007) probe Base-an online resource for rRNA-targeted oligonucleotide probes: new features. Nucleic Acids Res 35:800–804
Luna MF, Galar ML, Aprea J et al (2010) Colonization of sorghum and wheat by seed inoculation with *Gluconacetobacter diazotrophicus*. Biotechnol Lett 32:1071–1076
Miche L, Balandreau J (2001) Effects of rice seed surface sterilization with hypochlorite on inoculated *Burkholderia vietnamiensis*. Appl Environ Microbiol 67:3046–3052
Miché L, Battistoni F, Gemmer S et al (2006) Upregulation of jasmonate-inducible defense proteins and differential colonization of roots of Oryza sativa cultivars with the endophyte *Azoarcus* sp. Mol Plant-Microbe Interact 19:502–511
Mohanta S, Sharma GD, Deb B (2010) Diversity of endophytic diazotrophs in non-leguminous crops – a review. Assam Univ J Sci Technol 6:109–122
Murumkar DR, Nalawade SV, Indi DV (2016) Response of sugarcane seed plot to microbial inoculation by *Gluconacetobacter diazotrophicus* and phosphate-solubilizing bacteria. Sugar Tech. doi:10.1007/s12355-016-0432-3
Nautiyal CS (2000) Plant beneficial rhizosphere competent bacteria. Proc Natl Acad Sci India 70:107–123
Nieto-Penalver CG, Bertini EV, de Figueroa LIC (2012) Identification of N-acyl homoserine lactones produced by Gluconacetobacter diazotrophicus PAL5 cultured in complex and synthetic media. Arch Microbiol 194:615–622
Oliveira ALM, Canuto EL, Reis VM et al (2003) Response of micropropagated sugarcane varieties to inoculation with endophytic diazotrophic bacteria. Braz J Microbiol 34:59–61
Oliveira ALM, Canuto EDL, Urquiaga S et al (2006) Yield of micropropagated sugarcane varieties in different soil types following inoculation with diazotrophic bacteria. Plant Soil 284:23–32
Ortega-Rodes P, Ortega E, Kleiner D et al (2011) Low recovery frequency of *Gluconacetobacter diazotrophicus* from plants and associated mealybugs in Cuban sugarcane fields. Symbiosis 54:131–138
Partida-Martínez LP, Heil M (2011) The microbe-free plant: fact or artifact? Front Plant Sci 2:100
Paula MA, Reis VM, Dobereiner J (1991) Interactions of *Glomus clarum* with *Acetobacter diazotrophicus* in infection of sweet potato (*Ipomoea batatas*), sugarcane (*Saccharum* spp.), and d sweet sorghum (*Sorghum vulgare*). Biol Fertil Soils 11:111–115
Pereira GVM, Magalhães KT, Lorenzetii ER et al (2012) A multiphasic approach for the identification of endophytic bacterial in strawberry fruit and their potential for plant growth promotion. Microb Ecol 63:405–417
Posada F, Vega FE (2005) Establishment of the fungal entomopathogen *Beauveria bassiana* (Ascomycota: Hypocreales) as an endophyte in cocoa seedlings (*Theobroma cacao*). Mycologia 97:1195–1200
Puente ME, Li CY, Bashan Y (2009) Endophytic bacteria in cacti seeds can improve the development of cactus seedlings. Environ Exp Bot 66:402–408
Reading NC, Sperandio V (2006) Quorum sensing: the many languages of bacteria. FEMS Microbiol Lett 254:1–11
Reinhold-Hurek B, Hurek T (2011) Living inside plants: bacterial endophytes. Curr Opin Plant Biol 14:435–443
Reis Júnior FB, Silva LG, Reis VM et al (2000) Occurrence of diazothrophic bacteria in different sugar cane genotypes. [Ocorrência de bactérias diazotróficas em diferentes genótipos de cana-de-açúcar]. Pesq Agropec Bras 35:985–994
Reis VM, Baldani JI, Baldani VLD et al (2000) Biological dinitrogen fixation in gramineae and palm trees. Crit Rev Plant Sci 10:227–247
Reis V, Lee S, Kennedy C (2007) Biological nitrogen fixation in sugarcane. In: Emerich C, Newton WE (eds) Associative and endophytic nitrogen-fixing bacteria and cyanobacterial associations. Springer, Dordrecht, pp 213–232

Reiter B, Sessitsch A (2006) Bacterial endophytes of the wildflower *Crocus albiflorus* analyzed by characterization of isolates and by a cultivation-independent approach. Can J Microbiol 52:140–149

Riggs PJ, Chelius MK, Iniguez AL et al (2001) Enhanced maize productivity by inoculation with diazotrophic bacteria. Aus J Plant Physiol 28:829–836

Rosenblueth M, Martínez-Romero E (2006) Bacterial endophytes and their interactions with hosts. Mol Plant-Microbe Int 19:827–837

Rouws LFM, Meneses CHSG, Guedes HV et al (2010) Monitoring the colonization of sugarcane and rice plants by the endophytic diazotrophic bacterium Gluconacetobacter diazotrophicus marked with gfp and gusA reporter genes. Lett Appl Microbiol 51:325–330

Ryan RP, Germaine K, Franks A et al (2008) Bacterial endophytes: recent developments and applications. FEMS Microbiol Lett 278:1–9

Sagarika M, Sharma GD, Deb B (2010) Diversity of endophytic diazotrophs in non leguminous crops – a review. Assam Univ J Sci Technol 6:109–122

Saravanan VS, Madhaiyan M, Osborne J et al (2008) Ecological occurrence of Gluconacetobacter diazotrophicus and nitrogen-fixing Acetobacteraceae members: their possible role in plant growth promotion. Microb Ecol 55:130–140

Schultz N, Silva JA, Sousa JS (2014) Inoculation of sugarcane with diazotrophic bacteria. R Bras Ci Solo 38:407414

Schulz B, Boyle C (2006) What are endophytes? In: Schulz BJE, Boyle CJC, Sieber TN (eds) Microbial root endophytes. Springer, Berlin, pp 1–13

Seghers D, Wittebolle L, Top EM et al (2004) Impact of agricultural practices on the *Zea mays* L. endophytic community. Appl Environ Microbiol 70:1475–1482

Silva-Froufe LG, Boddey RM, Reis VM (2009) Quantification of natural populations of *Gluconacetobacter diazotrophicus* and *Herbaspirillum* spp. in sugarcane (*Saccharum* spp.) using different polyclonal antibodies. Braz J Microbiol 40:866–878

Smith KP, Goodman RM (1999) Host variation for interactions with beneficial plant-associated microbes. Annu Rev Phytopathol 37:473–491

Smith SA, Tank DC, Boulanger LA et al (2008) Bioactive endophytes warrant intensified exploration and conservation. PLoS One 3, e3052

Srinivasan TR, Naidu KM (1987) Response of sugarcane varieties to biofertilizers under different soil conditions. Sugarcane 3:5–11

Suman A, Shasany AK, Singh M et al (2001) Molecular assessment of diversity among endophytic diazotrophs isolated from subtropical Indian sugarcane. World J Microbiol Biotechnol 17:39–45

Tejera NA, Ortega E, González-López J et al (2003) Effect of some abiotic factors on the biological activity of *Gluconacetobacter diazotrophicus*. J Appl Microbiol 95:528–535

Tian G, Pauls P, Dong Z et al (2009) Colonization of the nitrogen-fixing bacterium Gluconacetobacter diazotrophicus in a large number of Canadian corn plants. Can J Plant Sci 89:1009–1016

Tilman D, Cassman KG, Matson PA et al (2002) Agricultural sustainability and intensive production practices. Nature 418:671–677

Timmusk S, Paalme V, Pavlicek T et al (2011) Bacterial distribution in the rhizosphere of wild barley under contrasting microclimates. PLoS ONE 6(3), e17968

Vargas L, Santa Brígida AB, Mota Filho JP et al (2014) Drought tolerance conferred to sugarcane by association with *Gluconacetobacter diazotrophicus*: a transcriptomic view of hormone pathways. PLoS ONE 9(12), e114744

Velázquez E, Rojas M, Lorite MJ (2008) Genetic diversity of endophytic bacteria which could be found in the apoplastic sap of the medullary parenchyma of the stem of healthy sugarcane plants. J Basic Microbiol 48:118–124

Ward JK, Tissue DT, Thomas RB et al (1999) Comparative responses of model C3 and C4 plants to drought in low and elevated CO_2. Glob Chang Biol 5:857–867

Yamada Y, Hoshino KI, Ishikawa T (1997) The phylogeny of acetic acid bacteria based on the partial sequences of 16S ribosomal RNA: the elevation of the subgenus gluconacetobacter to the generic level. Biosci Biotechnol Biochem 61:1244–1251

MIX
Papier aus verantwortungsvollen Quellen
Paper from responsible sources
FSC® C105338

If you have any concerns about our products,
you can contact us on
ProductSafety@springernature.com

In case Publisher is established outside the EU,
the EU authorized representative is:
Springer Nature Customer Service Center GmbH
Europaplatz 3, 69115 Heidelberg, Germany

Printed by Libri Plureos GmbH
in Hamburg, Germany